Discrete Multivariate Analysis

Discrete Multivariate Analysis: Theory and Practice

Yvonne M. M. Bishop, Stephen E. Fienberg, and Paul W. Holland
with the collaboration of
Richard J. Light and Frederick Mosteller

The MIT Press   Cambridge, Massachusetts, and London, England

CREDITS

Figures 2.7-2, 2.7-3, 2.7-4, and 11.2-2 originally appeared in the *Journal of the American Statistical Association*, vol. 65, pp. 694–701, 1970, and are reprinted by permission of the publisher.

Figures 12.3-1, 12.4-1, 12.4-2, 12.4-3, 12.4-4, 12.5-1, 12.5-2, and 12.5-3 originally appeared in the *Journal of the American Statistical Association*, vol. 68, pp. 683–691, 1973, and are reprinted by permission from the publisher.

Figure 3.8-1 originally appeared in the *New England Journal of Medicine*, vol. 285, pp. 641–647, 1971, and is reprinted by permission from the publisher.

Tables 5.2-1, 5.2-12, 5.2-13, and Figure 5.2-1 originally appeared in *Biometrika*, vol. 7, pp. 452–504, 1910, and are reprinted by permission of the publisher.

Tables 6.5-1 and material in sections 6.3, 6.5, and 6.6 originally appeared in slightly different form in *Biometrika*, vol. 59, pp. 591–603, 1972, and are reprinted by permission of the publisher.

Tables 3.4-2 and 4.4-12 are from "Effects of collapsing multidimensional contingency tables" by Y. M. M. Bishop; *Biometrics* Volume 27, Number 3, pages 545–62, and are reprinted by permission of the Editor of *Biometrics*.

Tables 5.2-6 and 5.2-7 and example 5.2-4 are from "Incomplete two-dimensional contingency tables" by Y. M. M. Bishop and S. E. Fienberg; *Biometrics* Volume 22, Number 1, pages 119–127, and are reprinted by permission of the Editor of *Biometrics*.

Much of the material in sections 5.2 and 5.4 and all of the tables in section 5.4 are from "The analysis of incomplete multiway contingency tables" by S. E. Fienberg; *Biometrics* Volume 28, Number 1, pages 177–202, and are reprinted by permission of the Editor of *Biometrics*.

This book was set in Times New Roman, and printed and bound by The Alpine Press in the United States of America

Second printing, 1976
Third printing, first MIT Press paperback edition, 1977
Fourth printing, 1977
Fifth printing, 1978

Library of Congress Cataloging in Publication Data

Bishop, Yvonne M. M.
   Discrete multivariate analysis.

   Bibliography: p.
   1. Multivariate analysis.   I. Fienberg, Stephen E., joint author.   II. Holland, Paul W., joint author.
III. Title.
QA278.B57   519.5'3   74-5177
ISBN 0-262-02113-7   (hardcover)
ISBN 0-262-52040-0   (paperback)

CONTENTS

Preface

The analysis of discrete multivariate data, especially in the form of cross-classifications, has occupied a prominent place in the statistical literature since the days of Karl Pearson and Sir R. A. Fisher. Although Maurice Bartlett's pioneering paper on testing for absence of second-order interaction in $2 \times 2 \times 2$ tables was published in 1935, the widespread development and use of methods for the analysis of multidimensional cross-classified data had to await the general availability of high-speed computers. As a result, in the last ten years statistical journals, as well as those in the biological, social, and medical sciences, have devoted increasing space to papers dealing with the analysis of discrete multivariate data. Many statisticians have contributed to this progress, as a glance at the reference list will quickly reveal. We point, especially, to the sustained and outstanding contributions of Joseph Berkson, M. W. Birch, I. J. Good, Leo A. Goodman, James E. Grizzle, Marvin Kastenbaum, Gary G. Koch, Solomon Kullback, H. O. Lancaster, Nathan Mantel, and R. L. Plackett.

The one person most responsible for our interest in and continued work on the analysis of cross-classified data is Frederick Mosteller. It is not an overstatement to say that without his encouragement and support in all phases of our effort, this book would not exist. Our interest in the analysis of cross-classified data goes back to 1964 and the problems which arose during and after Mosteller's work on the National Halothane study. This work led directly to the doctoral dissertations of two of us (Bishop and Fienberg), as well as to a number of published papers. But Fred's contributions to this book are more than just encouragement; he has read and copiously commented on nearly every chapter, and while we take complete responsibility for the final manuscript, if it has any virtues they are likely to be due to him.

Richard Light enthusiastically participated in the planning of this book, and offered comments on several chapters. He prepared the earlier drafts of Chapter 11, Measures of Association and Agreement, and he made the major effort on the final version of this chapter.

We owe a great debt to many of our colleagues and students who have commented on parts of our manuscript, made valuable suggestions on aspects of our research, and generally stimulated our interest in the subject. Those to whom we are indebted include Raj Bahadur, Darrell Bock, Tar Chen, William Cochran, Joel Cohen, Arthur Dempster, O. Dudley Duncan, Hillel Einhorn, Robert Fay, John Gilbert, Anne Goldman, Shelby Haberman, David Hoaglin, Nathan Keyfitz, William Kruskal, Kinley Larntz, Siu-Kai Lee, Lincoln Moses, I. R. Savage, Thomas Schoener, Michael Sutherland, John Tukey, David Wallace, James Warram, Sanford Weisberg, Janet Wittes, and Jane Worcester.

For the production of the manuscript we are indebted to Holly Grano, Kathi Hirst, Carol Lambert, and Mary Jane Schleupner.

The National Science Foundation has provided substantial support for our research and writing under grant GS-32327X1 to Harvard University. We have also received extensive support from other research grants. These include: research grants CA-06516 from the National Cancer Institute and RR-05526 from the Division of Research Facilities and Resources, National Institutes of Health to the Children's Cancer Research Foundation; National Science Foundation research grants GP-16071, GS-1905, and a grant from the Statistics Branch, Office of Naval Research, to the Department of Statistics, University of Chicago, as well as a grant from the Alfred P. Sloan Foundation to the Department of Theoretical Biology, University of Chicago; National Science Foundation research grant GJ-1154X to the National Bureau of Economic Research, Inc., and a faculty research grant from the Social Science Research Council to Paul W. Holland.

Earlier versions of material in several chapters appeared in *The Annals of Statistics, Biometrics, Biometrika*, and *The Journal of the American Statistical Association.*

Brookline, Massachusetts	Y.M.M.B.
New Brighton, Minnesota	S.E.F.
Hingham, Massachusetts	P.W.H.

February 1974

1 Introduction

1.1 The Need

The scientist searching for structure in large systems of data finds inspiration in his own discipline, support from modern computing, and guidance from statistical models. Because large sets of data are likely to be complicated, and because so many approaches suggest themselves, a codification of techniques of analysis, regarded as attractive paths rather than as straitjackets, offers the scientist valuable directions to try. In statistical specialities such as regression and the analysis of variance, codifications are widely available and sometimes keyed to special disciplines. In discrete multivariate statistics, however, the excellent guides already available, for example Cox [1970], Fleiss [1973], Good [1965], Lancaster [1969], and Maxwell [1961], stop short of giving a systematic treatment of large contingency tables, and especially tables that have troublesome irregularities. This book offers such a treatment.

1.2 Why a Book?

The literature on discrete multivariate analysis, although extensive, unfortunately is widely scattered. This book brings that literature together in an organized way. Although we do report a few new results here, that is not our primary purpose. Our purpose is to organize the materials needed by both theoretical and practical workers so that key ideas stand out. By presenting parametric models, sampling schemes, basic theory, practical examples, and advice on computation, this book serves as a ready reference for various users.

To bring together both the theory and practice of discrete multivariate analysis, a good deal of space is required. We need to relate various techniques of analysis, many of which are quite close to one another in both concept and result, so that the practitioner can know when one method is essentially the same as another, and when it is not. We need to provide basic theory, both for understanding and to lay a basis for new variations in the analysis when conditions do not match the ones presented here.

When we deal with several variables simultaneously, the practical examples we analyze tend to be large—larger than those ordinarily treated in the standard texts and monographs. An exploratory analysis of a set of data often leads us to perform several separate parallel analyses. Sometimes one analysis suggests another. Furthermore, we are obliged to discuss computing to some extent because these large-scale analyses are likely to require iterative methods, which are best done by high-speed computers. The availability of high-speed computing facilities has encouraged investigators to gather and ready for analysis substantial

1

sets of data. Applications and examples play a central role in most of the chapters in this book, and they take considerable space because we illustrate calculations, present alternative analyses, and discuss the conclusions the practitioner might draw for various data sets.

These reasons all lead to a treatment of book length.

1.3 Different Users

The applied statistician or quantitative research worker looking for comprehensive analyses of discrete multivariate data will find here a variety of ways to attack both standard and nonstandard sets of data. As a result, he has available a systematic approach to the analysis of multiway contingency tables. Naturally, new difficulties or constraints raise new problems, but the availability of a flexible approach should strengthen the practitioner's hand, just as the ready availability of analysis of variance and regression methods has for other data. He will understand his computer output better and know what kinds of computer analyses to ask for.

By skillful use of one computer program for obtaining estimates, the researcher can solve a wide range of problems. By juxtaposing practical examples from a variety of fields, the researcher can gain insight into his own problem by recognizing similarities to and differences from problems that arise in other fields. We have therefore varied the subject matter of the illustrations as well as the size of the examples. We have found the methods described in this book useful for small as well as large data sets.

On many occasions we have helped other people analyze sets of discrete multivariate data. In such consulting work we have found some of the material in this book helpful in guiding the practitioner to suitable analyses. Of course, several of the examples included here are drawn directly from our consulting experiences.

Parts of several chapters have grown out of material used in different university courses or sets of lectures we have given. Some of these courses and lectures stressed the application of statistical methods and were aimed at biological, medical, or social scientists with less preparation than a one-year course in statistics. Others stressed the statistical theory at various graduate levels of mathematical and statistical sophistication.

For the student we have included some exercise work involving both the manipulation of formulas and the analysis of additional data sets. In the last few years, certain examples have been analyzed repeatedly in the statistical literature, gradually bringing us a better understanding of what various methods accomplish. By making more examples of varied character readily available, we hope this healthy tradition of reanalyzing old problems with new methods will receive a substantial boost.

Finally, of course, we expect the book to provide a reference source for the methods collected in it. Although we do not try to compete with the fine bibliography provided by Lancaster [1969], some of the papers we cite have appeared since the publication of that work.

1.4 Sketch of the Chapters

Although each chapter has its own introduction, we present here a brief description of the contents and purposes of each chapter. We have organized the chapters into three logical groups of unequal size. The first group introduces the log-linear model, presents the statistical theory underlying its use in the analysis of contingency-table data, and illustrates the application of the theory to a wide variety of substantive problems. The second group of chapters deals with approaches and methods not relying directly on the log-linear model. The final pair of chapters contains basic statistical results and theory used throughout the book.

Section 1 Log-Linear Models, Maximum Likelihood Estimation, and Their Application

Chapter 2

With one exception, the example on the relation of survival of mothers to prenatal care, this is a theoretical chapter. It is meant for the practitioner as well as the theoretician, although they may read it from different points of view. The chapter develops the notation and ideas for the log-linear model used so extensively in this book. It begins with two-by-two (2×2) tables of counts and works up to tables of four or more dimensions. The emphasis is, first, on describing structure rather than sampling, second, on the relation of the log-linear model approach to familiar techniques for testing independence in two-way contingency tables, and third, on the generalization of these ideas to several dimensions. Fourth, the chapter shows the variety of models possible in these higher dimensions.

Chapter 3 (Preparation for most readers: Chapter 2)

Although from its title this chapter sounds like a theoretical one, its main emphasis is on important practical devices for computing estimates required for the analysis of multidimensional tables of counts. These devices include both how to recognize when simple direct estimates for cells are and are not available, and how to carry out iterative fitting when they are not. The chapter explains how to count degrees of freedom, occasionally a trickier problem than many of us are used to. All aspects of the analysis—models, estimates, calculation, degrees of freedom, and interpretation—are illustrated with concrete examples, drawn from history, sociology, political science, public health, and medicine, and given in enough detail that the practical reader should now have the main thrust of the method.

Chapter 4 (Preparation: Chapter 3)

Chapter 4 provides tools and approaches to the selection of models for fitting multidimensional tables. While it includes some theory, this is the key chapter for the reader who must actually deal with applications and choose models to describe data.

First, it summarizes the important large sample results on chi square goodness-of-fit statistics. (These matters are more fully treated in Chapter 14.) Second, it explains how the partitioning of chi square quantities leads to tests of special hypotheses and to the possibility of more refined inferences. Third, although there is no "best" way to select a model, the chapter describes several different approaches to selecting log-linear models for multidimensional tables, using large-sample results and the partitioning method.

Chapter 5 (Preparation: Chapter 3)

Cell counts can be zero either because of sampling variability even when observations can occur in a cell, or because constraints make the cell automatically zero (for example, the losing football team does not score more than the winning team). In addition, for some problems, certain cells are not expected to fit smoothly into a simple model for the table of counts, so the counts for these cells, although available, are set aside for special treatment. One danger is that the researcher may not recognize that he is afflicted with an incomplete table.

This profusely illustrated chapter offers standard ways of handling such incomplete tables and is oriented to problems of estimation, model selection, counting of degrees of freedom, and applications.

Chapter 6 (Preparation: Chapter 5)

This chapter deals with a special application: If, as sometimes happens, we have several samplings or censuses, we may wish to estimate a total count. For example, we may have several lists of voluntary organizations from the telephone book, newspaper articles, and other sources. Although each list may be incomplete, from the several lists we want to estimate the total number of voluntary organizations (including those on *none* of these lists). This chapter offers ways to solve such multiple-census problems by treating the data sets as incomplete multidimensional tables. The method is one generalization of the capture-recapture method of estimation used in wildlife and other sampling operations.

Chapter 7
(Preparation: Chapter 3)

Since Markov chains depend on a relation between the results at one stage and those at later stages, there are formal similarities with contingency tables. Consequently, analysis of Markov chains using the log-linear model is an attractive possibility, treated here along with other methods.

This is a practical chapter, containing illustrations which come from market research and studies of political attitudes, language patterns (including bird songs), number-guessing behavior, and interpersonal relations.

Chapter 8
(Preparation: Chapter 3 and some of Chapter 5)

Although square tables are discussed elsewhere, this chapter focuses on the special questions of symmetry and marginal homogeneity. These problems arise in panel studies when the same criteria are used at each point in time and in psychological studies, as when both members of a pair are classified simultaneously according to the same criteria. The chapter gives methods for assessing symmetry and homogeneity, illustrated with practical examples. It also treats multidimensional extensions of the notions of symmetry and marginal homogeneity, relating them to the basic approach of this book using log-linear models.

Chapter 9

For practitioners, this chapter gives numerous suggestions and examples about how to get started in building models for data. The attitude is that of exploratory data analysis rather than confirmatory analysis. When models are fitted, the problem still remains of how to view the fit. What if it is too good? How bad is too bad? Rough approximations may be just what is desired. The beginner may wish to start reading the book with this chapter.

Section II   Related models and methods

Chapter 10

Although this book offers a systematic treatment of contingency tables through the approach of log-linear models and maximum likelihood estimation, the reader may want to know what alternative methods are available. This chapter offers introductions to some of the more widely used of these methods and points the reader to further literature.

Chapter 11

This chapter differs from the others in the book because it deals only with two-way tables, and also because the main thrust of measuring association is to summarize many parameters in one. The basic principles for the choice and use of measures of association depend on the purposes of the user. The chapter also treats a special case of association, referred to as "agreement." Because we view interaction and association as due primarily to many different parameters, this chapter presents a different outlook than does the rest of the book.

Chapter 12

In a contingency table of many dimensions, the number of cells is often high while the average number of observations per cell in many practical problems is small, and so many cells may have zero entries. We wish to estimate the probabilities associated with the cells. Extra information about these probabilities may be available from the general distribution of the counts or from the margins. Bayesian approaches offer methods of estimating these probabilities, but they usually leave to the user the problem of choosing the parameter of the prior distribution, a job he may be ill equipped to do. Theoretical investigations can help the reader choose by showing him some methods that will protect him from bad errors. This chapter reviews the literature and provides new results, together with some applications. The treatment deals with the case where the cells merely form a list, and the case where the cells form a complete two-way table.

Section III   Theoretical background

Chapter 13

In working with contingency tables, both the practitioner and the theorist face certain standard sampling distributions repeatedly: the binomial, Poisson, multinomial, hypergeometric, and negative binomial distributions, and some of their generalizations. This chapter offers a ready source of reference for information about these distributions.

Chapter 14

In discrete problems, exact calculations are notoriously difficult, especially when the sample sizes are large. This difficulty makes the results of this chapter especially important, since asymptotic (large-sample) methods are so widely used in this work.

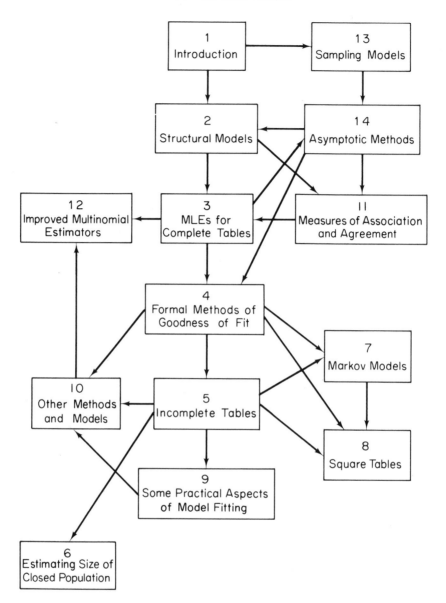

Figure 1.6-1.

The mathematics of asymptotic methods useful to the theoretician is scattered throughout the literature. For getting asymptotic distributions, mathematical orders of magnitude play an important role. This chapter provides a convenient short course in the devices and theorems about the O, o notation, for sequences of real numbers and about the analogous O_p, o_p notation, for random variables. The treatment includes vector random variables. The material is illustrated with examples and practice exercises, enabling the student to derive many of the theoretical results in this book. Moreover, the devices discussed are useful in every branch of theoretical statistics.

1.5   Computer Programs

As we have already noted, one general-purpose computer program can be used to carry out most of the calculations described in this book. Many researchers working with multiway table data have prepared such programs to carry out estimation using the method of iterative proportional fitting and to compute various goodness-of-fit statistics. These programs are now available at a large number of computer centers and research installations. We refer those who would like to use programs which are not available at their institutions to the Fortran listings in Bishop [1967, appendix I] and Haberman [1972, 1973b].

1.6   How to Proceed from Here

Readers of this book come to it with different backgrounds and different interests. We have ordered Chapters 2 through 9 so that each chapter builds only on preceding ones, with the exception that most of them use material from the two theoretical background chapters (Chapters 13 and 14). Thus a good basic sequence consists of Chapters 2 through 9. Nevertheless, readers may choose to work with chapters in different orders.

Graduate students in theoretical statistics may choose to begin with a review of the sampling distribution properties and large-sample theory in Chapters 13 and 14, then proceed to Chapters 2, 3, 4, 5, 9, 10, 11, and 12. Chapters 6, 7, and 8 can be handled either after Chapter 5 or at the end of the indicated sequence.

Quantitative biological or social scientists interested in analyzing their own data will most likely profit from a quick reading of Chapters 2 and 3, followed by Chapter 9 (Sections 9.1–9.5) and Chapters 4 and 5. Then they might turn to Chapter 7, 8, or 11, depending on their interests.

Other sequences of chapters come to mind. Figure 1.6-1 gives a schematic representation of alternative orderings for different readers and an indication of how the chapters are linked.

2 Structural Models for Counted Data

2.1 Introduction

As soon as a problem is clearly defined, its solution is often simple. In this chapter we show how complex qualitative data may be described by a mathematical model. Questions that the data were designed to answer may then be stated precisely in terms of the parameters of the model.

In multivariate qualitative data each individual is described by a number of attributes. All individuals with the same description are enumerated, and this count is entered into a cell of the resulting contingency table. Descriptive models with as many independent parameters as the table has cells are called "saturated." They are useful in reducing complexity only if the parameters can be readily interpreted as representing "structural" features of the data, because most of the questions of importance may be interpreted as being questions about the data structure.

The complexity of the data is reflected by the number of parameters in the model describing its structure. When the structure is simple, the model has few parameters. Whenever the model has fewer parameters than the number of data cells, we say that the model is "unsaturated." For some unsaturated models we can reduce the number of cells in the table without distorting the structure. Such reduction we refer to as "collapsing" and we give theorems defining those structures that are collapsible. Before proceeding to describe models for the simplest four-cell table, we enlarge on this concept of structure and on the development and uses of models.

2.1.1 Structure

If every individual in the population under study can be classified as falling into one and only one of t categories, we say that the categories are mutually exclusive and exhaustive. A randomly selected member of the population will fall into one of the t categories with probability p_i, where $\{p_i\}$ is the vector of cell probabilities

$$\{p_i\} = (p_1, p_2, \ldots, p_t) \tag{2.1-1}$$

and

$$\sum_{i=1}^{t} p_i = 1.$$

Here the cells are strung out into a line for purposes of indexing only; their arrangement and ordering does not reflect anything about the characteristics of individuals falling into a particular cell. The p_i reflect the relative frequency of each category in the population.

When the cells are defined in terms of the categories of two or more variables, a structure relating to the nature of the data is imposed. The natural structure for two variables is often a rectangular array with columns corresponding to the categories of one variable and rows to categories of the second variable; three variables create layers of two-way tables, and so on. As soon as this structure is imposed, the position of the cells tells us something about the characteristics of individuals falling into them: For instance, individuals in a specific cell have one characteristic in common with individuals in all cells of the same row, and another characteristic in common with all individuals in cells in the same column. A good mathematical model should reflect this structure.

As soon as we consider more than one randomly selected individual we must consider the sampling plan. If the second and all subsequent individuals are sampled "with replacement," that is, the first is replaced in the population before the second is randomly drawn, and so on, then the vector of probabilities (2.1-1) is unchanged for each individual. Alternatively, the vector of probabilities is unchanged if the population is infinitely large. In either of these circumstances, if we take a simple random sample of size N, we obtain a sample of counts $\{x_i\}$ such that

$$\{x_i\} = (x_1, x_2, \ldots, x_t), \tag{2.1-2}$$

where

$$\sum_i x_i = N.$$

The corresponding expected counts are $\{m_i\}$, such that

$$\{m_i\} = (m_1, m_2, \ldots, m_t), \tag{2.1-3}$$

where

$$E(x_i) = m_i \quad \text{for } i = 1, \ldots, t,$$

$$m_i = N p_i.$$

In Chapter 3 we deal with estimating the $\{m_i\}$ from the $\{x_i\}$ under a variety of sampling schemes and for different models. In Chapter 13 we consider different sampling distributions and the relationships between the $\{m_i\}$ and the $\{p_i\}$. In this chapter we are not concerned with the effects of sampling, but only with the underlying data structure. Thus we are interested in models that specify relationships among the cell probabilities $\{p_i\}$ or among the expected counts $\{m_i\}$. Some sampling schemes impose restrictions on the $\{m_i\}$, and so we also discuss situations where these constraints occur without considering how they arise. The constraints occur in situations where we are in effect taking several samples, each drawn from one segment of the population. We then have probabilities for each segment summing to 1, but we cannot relate probabilities in different segments to the population frequency in different segments unless we know the relative size of the segments.

2.1.2   Models

The smallest rectangular table is based on four cells, and the saturated model describing it has four independent parameters. In Section 2.2 we give a four-term model for this table that is linear in the logarithmic scale, and we give an interpretation of each of the four terms. In Section 2.3 we extend this four-term model to larger two-dimensional tables by enlarging the number of parameters encompassed by each term of the model.

Log-linear models are not new; they are implicit in the conventional χ^2 test for independence in two-way contingency tables. The notation of Birch [1963] is convenient for such models, as the number of terms depends on the dimension and the interdependencies between dimensions, rather than on the number of cells. Each term encompasses as many parameters as are needed for the total number of independent parameters in the saturated model to equal the number of cells in the table. When the model is unsaturated, the reduction is generally achieved by removing one or more terms completely, because the terms rather than the parameters correspond to effects of interest. In Section 2.4 we show that an s-dimensional table of any size is described by a model with 2^s terms. Thus the models reflect the structure imposed on the data, and the terms are closely related to hypotheses of interest.

2.1.3 *Uses of structural models*

The interpretation of the terms of saturated models that fully specify an array leads to interpretation of models with fewer terms. The investigator faced with data of an unknown structure may wish to determine whether they are fitted well by a particular unsaturated model, that is he may wish to test a particular hypothesis. Alternatively, he may wish to obtain good estimates for some or all of the cells and may obtain such estimates by fitting an unsaturated model. Using unsaturated models to obtain stable cell estimates is akin to fitting an approximate response curve to quantitative data; the investigator gains knowledge of important underlying trends by reducing the number of parameters to less than that required for perfect fit. Thus comprehension is increased by focusing on the most important structural features.

If the data can be described by models with few terms, it may be possible to condense the data without either obscuring important structural features or introducing artifactitious effects. Such condensation is particularly pertinent when the data are sparse relative to the magnitude of the array. In addition to focusing on parameter and model interpretation, we look in each section of this chapter at the problem of determining when such condensation is possible without violating important features of the underlying structure.

In this chapter we do not discuss fitting models; we discuss procedures that yield maximum likelihood estimates in Chapter 3 and assessment of goodness of fit in Chapter 4. The concern here is with such questions as:

1. What do we mean by "independence" and "interaction"?
2. Why is it necessary to look at more than two dimensions at a time?
3. How many variables should be retained in a model and which can safely be removed?

2.2 Two Dimensions—The Fourfold Table

The simplest contingency table is based on four cells, and the categories depend on two variables. The four cells are arranged in a 2×2 table whose two rows correspond to the categorical variable A and whose two columns correspond to the second categorical variable B. We consider first the different constraints that we may use to specify the cell probabilities, then the effect of rearranging the cells. This leads to formulation of a model, the log-linear model, that we can readily interpret in terms of the constraints and apply to any arrangement of the four cells.

We discuss features of the log-linear model for the 2×2 table in detail. Important features that also apply to larger tables are:

1. Only one parameter of the model is changed when it is used to describe expected cell counts m instead of probabilities p;
2. the model is suitable for a variety of sampling schemes;
3. the ready interpretability of the terms of the model is not shared by models that are linear in the arithmetic scale.

In Section 2.7 we give a geometric interpretation of the 2×2 table and show how the parameters of the log-linear model are related to the structural features of a three-dimensional probability simplex.

2.2.1 *Possible constraints for one arrangement*

Double subscripts refer to the position of the cells in our arrangement. The first subscript gives the category number of variable A, the second of variable B, and the two-dimensional array is displayed as a grid with two rows and two columns:

$$
\begin{array}{c}
 & & B & \\
 & 1 & & 2 \\
A \quad
\begin{array}{c} 1 \\[2em] 2 \end{array}
&
\begin{array}{|c|c|}
\hline
p_{11} & p_{12} \\
\hline
p_{21} & p_{22} \\
\hline
\end{array}
\end{array}
\qquad (2.2\text{-}1)
$$

We consider first a simple random sample such that the cell probabilities sum to 1, that is, we have the linear constraint

$$
\sum_{i=1}^{2} \sum_{j=1}^{2} p_{ij} = 1. \qquad (2.2\text{-}2)
$$

By displaying the cells as in expression (2.2-1), we introduce a structure to the corresponding probabilities, and it is natural for us to examine the row and column marginal totals:

$$
p_{i+} = \sum_{k=1}^{2} p_{ik} \qquad i = 1, 2 \qquad (2.2\text{-}3)
$$

$$
p_{+j} = \sum_{k=1}^{2} p_{kj} \qquad j = 1, 2. \qquad (2.2\text{-}4)
$$

These totals give the probabilities of an individual falling in categories i and j of variables A and B, respectively. (Throughout this book, when we sum over a subscript we replace that subscript by a " $+$ ".) At this point, we can expand our tabular display (2.2-1) to include the marginal totals and the basic constraint (2.2-2):

$$B$$

		1	2	Totals
	1	p_{11}	p_{12}	p_{1+}
A	2	p_{21}	p_{22}	p_{2+}
Totals		p_{+1}	p_{+2}	1

(2.2-3)

The marginal probabilities p_{i+} and p_{+j} are the unconditional probabilities of belonging to category i of variable A and category j of variable B, respectively. Each set of marginal probabilities must sum to 1. As we have only two categories, once we know one row total, p_{1+}, we also know the other row total, p_{2+}, because $p_{2+} = 1 - p_{1+}$, and similarly for column totals. Thus if we know the values of p_{1+} and p_{+1}, the two linear constraints on the marginal probabilities lead to a complete definition of all the marginal probabilities. We need only one further constraint involving the internal cells to specify completely the structural relationships of the table.

We refer to the internal cells as "elementary" cells. The probability p_{ij} is the probability of an individual being in category i of variable A and category j of variable B. Most questions of interest related to the fourfold table are concerned with differences between such internal probabilities and the marginal probabilities. A variety of functions of the probabilities are commonly used, and others can readily be devised, that will produce the further constraint needed for complete specification of the table. Commonly used are:

(i)   the difference in column proportions

$$\frac{p_{11}}{p_{+1}} - \frac{p_{12}}{p_{+2}};$$

(ii)   the difference in row proportions

$$\frac{p_{11}}{p_{1+}} - \frac{p_{21}}{p_{2+}};$$

(iii)   the cross-product ratio

$$\alpha = \frac{p_{11}p_{22}}{p_{12}p_{21}}.$$

A natural choice if we wish to continue to use linear constraints is:

(iv)   the diagonal sum

$$S_d = p_{11} + p_{22}.$$

Finally, we can choose:

(v)   the ratio of an elementary cell probability to the product of its row and column probabilities

$$\frac{p_{11}}{p_{1+}p_{+1}}.$$

Other measures appear in Chapter 11. Specifying the value of any one of the five statistics in this list is equivalent to specifying the remaining four, given p_{1+} and p_{+1}. Such specification completely determines the values of the cell probabilities $\{p_{ij}\}$. The third function, α, has desirable properties not possessed by the others. We consider its properties in detail because they lead us to the formulation of our model for the fourfold table.

Properties of the cross-product ratio

Since the rows of the table correspond to one variable, A, and the columns to a second variable, B, it is natural for us to be interested in the relationship between these underlying categorical variables. We first consider the behavior of the statistics (i)–(v) under independence. If the state of A is independent of the state of B, then

$$p_{ij} = p_{i+}p_{+j} \qquad i = 1, 2; \quad j = 1, 2; \tag{2.2-6}$$

but this relationship is not satisfied for all i and j if A and B are dependent.

As any of the functions, when combined with the marginal totals, completely specify the table, they also measure dependence between the underlying variables. For instance, the independence relationship (2.2-6) is equivalent to stating that the proportional difference (i) or (ii) is 0, or that the measure (v) has the value 1. The measure (iv) becomes a less attractive function of the marginal probabilities, namely,

$$S_d = 1 - p_{+1} - p_{1+} + 2p_{1+}p_{+1}.$$

When we focus on the product relationship (2.2-6), it is reasonable for us to choose the cross-product ratio instead of the linear functions. The cross-product ratio α, like measure (v), attains the value 1 when the condition of independence holds, and it has two properties not possessed by measure (v), or any of the other measures:

1. α is invariant under the interchange of rows and columns;
2. α is invariant under row and column multiplications. That is, suppose we multiply the probabilities in row 1 by $r_1 > 0$, those in row 2 by $r_2 > 0$, those in column 1 by $c_1 > 0$, and those in column 2 by $c_2 > 0$. Then we get

$$\alpha = \frac{p_{11}p_{22}}{p_{12}p_{21}} = \frac{(r_1 c_1 p_{11})(r_2 c_2 p_{22})}{(r_1 c_2 p_{12})(r_2 c_1 p_{21})}. \tag{2.2-7}$$

This result holds regardless of whether we normalize so that the new cell entries sum to 1. An important implication is that we obtain the same value of α when we use either the cell probabilities or the expected counts in each cell.

Interpretation of cross-product ratio

The cross-product ratio α is also known as the "odds ratio." For the first level of variable A, the odds on being in the first level of variable B are p_{11}/p_{12}, and for the second level of variable A they are p_{21}/p_{22}. The cross-product ratio is the ratio of these odds,

$$\alpha = \frac{p_{11}/p_{12}}{p_{21}/p_{22}} = \frac{p_{11}p_{22}}{p_{12}p_{21}}.$$

This definition is also invariant under interchange of the variables. It should not be confused with another measure used by epidemiologists, the relative risk r, defined as the ratio of the row proportion $p_{11}/(p_{11} + p_{12})$ to the corresponding row proportion $p_{21}/(p_{21} + p_{22})$. Thus we have

$$r = \frac{p_{11}(p_{21} + p_{22})}{p_{21}(p_{11} + p_{12})} = \frac{p_{11}p_{2+}}{p_{21}p_{1+}}. \tag{2.2-8}$$

We can define r similarly in terms of column proportions, but then we obtain a different value. The relative risk does not have the invariance properties possessed by the relative odds, although its interpretation when dealing with the risk of contracting disease in two population groups makes it a useful parameter.

The logarithm of the relative odds is also a linear contrast of the log-probabilities of the four elementary cells, namely

$$\log \alpha = \log p_{11} - \log p_{12} - \log p_{21} + \log p_{22}, \tag{2.2-9}$$

and when $\log \alpha = 0$ we have independence between variables A and B.

The cross-product ratio and bivariate distributions

As soon as we consider the cross-product ratio as a measure of departure from independence, the question of its relationship to the correlation coefficient arises. Mosteller [1968] takes bivariate normals with different selected values of ρ and shows that the value of α differs according to the breaking point chosen. Thus α is not easily related to ρ for bivariate normals, but Plackett [1965] shows that it is possible to construct a class of distributions where the value of α is unchanged by the choice of breaking point.

2.2.2   *Effect of rearranging the data*

Suppose that the two underlying variables A and B for the 2×2 table actually have the same categories and simply represent measurements on one variable at two points in time. We can then refer to them as A_1 and A_2. There are three different arrays that may be of interest:

1.   the basic table

$$
\begin{array}{cc}
 & A_2 \\
\begin{array}{c|c|c|}
 & 1 & 2 \\
\hline
1 & p_{11} & p_{12} \\
\hline
2 & p_{21} & p_{22} \\
\hline
\end{array}
\end{array}
\tag{2.2-10}
$$

with A_1 labeling the rows 1 and 2.

2. the table measuring changes from the first measurement to the second. This table preserves the margins for the first variable:

		Same	Different
A_1	1	p_{11}	p_{12}
	2	p_{22}	p_{21}

(2.2-11)

3. the table measuring changes going back from the second measurement to the first. This table preserves the margins for the second variable:

A_2

	1	2
Same	p_{11}	p_{22}
Different	p_{21}	p_{12}

(2.2-12)

For each of these 2×2 tables we have a cross-product ratio. Taking tables (2.2-10)–(2.2-12) in order, these ratios are

$$\alpha_3 = \frac{p_{11}p_{22}}{p_{12}p_{21}}, \tag{2.2-13}$$

$$\alpha_2 = \frac{p_{11}p_{21}}{p_{12}p_{22}}, \tag{2.2-14}$$

$$\alpha_1 = \frac{p_{11}p_{12}}{p_{22}p_{21}}. \tag{2.2-15}$$

The reason for this ordering of the subscripts will become apparent shortly. For the moment we note that these three expressions suggest a class of structural models based on α_3, α_2, and α_1, rather than on one of the cross products together with the margins of one of the tables.

Taking logarithms of the $\{\alpha_i\}$, we get three linear contrasts

$$\log \alpha_3 = \log p_{11} - \log p_{12} - \log p_{21} + \log p_{22}, \tag{2.2-16}$$

$$\log \alpha_2 = \log p_{11} - \log p_{12} + \log p_{21} - \log p_{22}, \tag{2.2-17}$$

$$\log \alpha_1 = \log p_{11} + \log p_{12} - \log p_{21} - \log p_{22}. \tag{2.2-18}$$

If we specify values for these three contrasts and recall that

$$\sum p_{ij} = 1, \tag{2.2-19}$$

we have completely defined the four cell probabilities. This formulation suggests that we look for a model that is linear in the log scale.

2.2.3 The log-linear model

A simple way to construct a linear model in the natural logarithms of the cell probabilities is by analogy with analysis of variance (ANOVA) models. We write

$$\log p_{ij} = u + u_{1(i)} + u_{2(j)} + u_{12(ij)} \qquad i = 1, 2; j = 1, 2, \qquad (2.2\text{-}20)$$

where u is the grand mean of the logarithms of the probabilities:

$$u = \tfrac{1}{4}(\log p_{11} + \log p_{12} + \log p_{21} + \log p_{22}), \qquad (2.2\text{-}21)$$

$u + u_{1(i)}$ is the mean of the logarithms of the probabilities at level i of first variable:

$$u + u_{1(i)} = \tfrac{1}{2}(\log p_{i1} + \log p_{i2}) \qquad i = 1, 2, \qquad (2.2\text{-}22)$$

and similarly for the jth level of the second variable:

$$u + u_{2(j)} = \tfrac{1}{2}(\log p_{1j} + \log p_{2j}) \qquad j = 1, 2. \qquad (2.2\text{-}23)$$

Since $u_{1(i)}$ and $u_{2(j)}$ represent deviations from the grand mean u,

$$u_{1(1)} + u_{1(2)} = u_{2(1)} + u_{2(2)} = 0. \qquad (2.2\text{-}24)$$

Similarly, $u_{12(ij)}$ represents a deviation from $u + u_{1(i)} + u_{2(j)}$, so that

$$u_{12(11)} = -u_{12(12)} = -u_{12(21)} = u_{12(22)}. \qquad (2.2\text{-}25)$$

We note that the additive properties (2.2-24) and (2.2-25) imply that each u-term has one absolute value for dichotomous variables. Thus we introduce no ambiguity by writing, for instance, $u_1 = 0$ without specifying the second subscript.

If we define $l_{ij} = \log p_{ij}$, then by analogy with ANOVA models we can write the grand mean as

$$u = \frac{l_{++}}{4} = \sum_{i,j} \frac{l_{ij}}{4}. \qquad (2.2\text{-}26)$$

Similarly, the main effects are

$$u_{1(i)} = \frac{l_{i+}}{2} - \frac{l_{++}}{4}, \qquad (2.2\text{-}27)$$

$$u_{2(j)} = \frac{l_{+j}}{2} - \frac{l_{++}}{4}, \qquad (2.2\text{-}28)$$

and the interaction term becomes

$$u_{12(ij)} = l_{ij} - \frac{l_{i+}}{2} - \frac{l_{+j}}{2} + \frac{l_{++}}{4}. \qquad (2.2\text{-}29)$$

We note that the main effects are functions of the marginal sums of the logarithms but do not correspond to the marginal sums p_{i+} and p_{+j} in the original scale.

We now consider properties of the log-linear model.

Relationship of u-terms to cross-product ratios

From equations (2.2-18) and (2.2-27) we have

$$u_{1(1)} = \tfrac{1}{4}(\log p_{11} + \log p_{12} - \log p_{21} - \log p_{22})$$

$$= \tfrac{1}{4} \log \alpha_1. \qquad (2.2\text{-}30)$$

Similarly, from expressions (2.2-17) and (2.2-28) we have

$$u_{2(1)} = \tfrac{1}{4}(\log p_{11} - \log p_{12} + \log p_{21} - \log p_{22})$$
$$= \tfrac{1}{4}\log \alpha_2, \tag{2.2-31}$$

and from (2.2-16) and (2.2-29) we have

$$u_{12(11)} = \tfrac{1}{4}(\log p_{11} - \log p_{12} - \log p_{21} + \log p_{22})$$
$$= \tfrac{1}{4}\log \alpha_3. \tag{2.2-32}$$

Thus the main effects in the log-linear u-term model are directly related to the two cross-product ratios described above, u_1 to α_1 and u_2 to α_2. The choice of subscripts for the α_i now becomes apparent. We note that for $u_{1(1)}$ the terms in p appear with positive sign whenever variable 1 is at level 1, and similarly for $u_{2(1)}$ and variable 2 at level 1. For $u_{12(11)}$, the positive sign appears whenever both variables are on the same level. Thus the u-terms can be regarded as measures of departure from independence for the three different data arrangements.

Effect of imposing constraints

To assess the effect on the u-terms of imposing constraints on the $\{p_{ij}\}$, we need to revert to the arithmetic scale.

We can rewrite the model (2.2-20) for cell (1, 1) as

$$\log p_{11} = u + \tfrac{1}{4}\log \alpha_1 + \tfrac{1}{4}\log \alpha_2 + \tfrac{1}{4}\log \alpha_3, \tag{2.2-33}$$

and hence

$$p_{11} = \lambda \alpha_1' \alpha_2' \alpha_3', \tag{2.2-34}$$

where $\log \lambda = u$ and $\alpha_i' = (\alpha_i)^{1/4}$ for $i = 1, 2, 3$. Then the basic table can be rewritten as

		A_2		
		1	2	Totals
A_1	1	$\lambda \alpha_1' \alpha_2' \alpha_3'$	$\dfrac{\lambda \alpha_1'}{\alpha_2' \alpha_3'}$	$\lambda \alpha_1'\left(\alpha_2' \alpha_3' + \dfrac{1}{\alpha_2' \alpha_3'}\right)$
	2	$\dfrac{\lambda \alpha_2'}{\alpha_1' \alpha_3'}$	$\dfrac{\lambda \alpha_3'}{\alpha_1' \alpha_2'}$	$\dfrac{\lambda}{\alpha_1'}\left(\dfrac{\alpha_2'}{\alpha_3'} + \dfrac{\alpha_3'}{\alpha_2'}\right)$
Totals		$\lambda \alpha_2'\left(\alpha_1' \alpha_3' + \dfrac{1}{\alpha_1' \alpha_3'}\right)$	$\dfrac{\lambda}{\alpha_2'}\left(\dfrac{\alpha_1'}{\alpha_3'} + \dfrac{\alpha_3'}{\alpha_1'}\right)$	1

Setting $p_{1+} = p_{2+} = 1/2$ implies that the $\{\alpha_i\}$ must satisfy the relationship

$$\alpha_1^{1/2} - \alpha_2^{1/2} - \alpha_3^{1/2} + (\alpha_1 \alpha_2 \alpha_3)^{1/2} = 0. \tag{2.2-35}$$

If we set $\alpha_1' = 1$, which is equivalent to setting $u_1 = 0$, the condition (2.2-35) becomes

$$\left(\alpha_2' - \dfrac{1}{\alpha_2'}\right)\left(\alpha_3' - \dfrac{1}{\alpha_3'}\right) = 0, \tag{2.2-36}$$

which is satisfied by either $\alpha_2' = 1$ or $\alpha_3' = 1$, or both. Equivalently, we must have $u_2 = 0$ or $u_{12} = 0$, or both.

This result also holds in larger tables; constant marginal probabilities do not imply that $u_1 = 0$ unless we also have $u_2 = 0$ or $u_{12} = 0$, or both. Consequently, when we move from simple random sampling to sampling different segments of the population independently, we cannot specify that a margin is fixed by placing constraints on a single u-term.

Model describes probabilities or expected counts

So far we have dealt entirely with a table of probabilities that sum to 1. If we consider instead a table of expected counts $\{m_{ij}\}$ that sum to a grand total $N = \sum_{i,j} m_{ij}$, we have $m_{ij} = N p_{ij}$, and hence

$$\log m_{ij} = \log N + \log p_{ij}$$
$$= u' + (u_{1(i)} + u_{2(j)} + u_{12(ij)}), \qquad (2.2\text{-}37)$$

where $u' = u + \log N$. Thus for the single sample we can describe the structure of the expected counts instead of the structure of the probabilities by changing the value of u from the mean of the logarithms of the $\{p_{ij}\}$ to the mean of the logarithms of the $\{m_{ij}\}$, and henceforth we denote the constant by u in both cases. In other words, the equations (2.2-26)–(2.2-29) are applicable if we define $l_{ij} = \log m_{ij}$ instead of $l_{ij} = \log p_{ij}$.

It follows that α can be defined similarly as the cross-product ratio of expected counts instead of probabilities.

Model applicable in varied sampling situations

So far we have considered taking a single sample of size N, with p_{ij} the probability of an individual falling into the cell (i, j). This is the simple random sampling scheme. A fourfold table can also be generated by other sampling schemes. Suppose that we take a sample of N_1 individuals from the first category of variable A and N_2 from the second category, and then count how many fall into the different categories of variable B. Our table of expected counts becomes

		B		
		1	2	Totals
A	1	m_{11}	m_{12}	N_1
	2	m_{21}	m_{22}	N_2
Totals		m_{+1}	m_{+2}	N

$$(2.2\text{-}38)$$

and we have

$$m_{11} + m_{12} = N_1,$$
$$m_{21} + m_{22} = N_2,$$
$$N_1 + N_2 = N.$$

Corresponding to this table, there is a table of probabilities $P_{j(i)}$ the probability of being in category j for sample i. Thus

$$N_1 P_{j(1)} = m_{1j},$$
$$N_2 P_{j(2)} = m_{2j},$$

(2.2-39)

for $j = 1, 2$. We write these probabilities with capital letters, as they are no longer the probabilities giving the frequency of occurrence of the four types of individuals in the population. Instead of the four probabilities summing to 1, we have

$$P_{1(1)} + P_{2(1)} = 1,$$
$$P_{1(2)} + P_{2(2)} = 1.$$

(2.2-40)

We have taken two independent samples from different segments of the population and cannot get back to the population p_{ij} unless we know the relative magnitude of the two segments of the population.

Our log-linear model is still applicable to the table of expected counts (2.2-38), but the restriction (2.2-35) derived for equal row margins applies, so the relative magnitudes of the u-terms are constrained. In other sampling plans the restrictions on the probabilities differ in other ways. For simplicity, in the rest of this chapter we discuss log-linear models in terms of expected counts, not probabilities.

Before comparing the log-linear model with other models, we give an example of sampling that gives a 2×2 table with a fixed margin.

Example 2.2-1   Sensitivity, specificity, and predictive value

The problem of evaluating a new laboratory procedure designed to detect the presence of disease affords an example not only of sampling so that a 2×2 table has a fixed margin, but also of rearranging four cells for three different purposes.

1. *Natural arrangement for laboratory data*

To determine how effectively the laboratory procedure identifies positives and negatives, the investigator evaluates N_1 persons known to have the disease and N_2 persons known to be free of the disease. The results are designed to estimate the expected counts in array (2.2-41). In this array we no longer enclose every elementary cell in a box, but the arrangement of cells is the same as in array (2.2-10).

True State	Laboratory Procedure		Totals	
	Disease	No Disease		
Disease	m_{11}	m_{12}	N_1	(2.2-41)
No Disease	m_{21}	m_{22}	N_2	

A perfect laboratory procedure correctly identifies as diseased all those persons who are truly diseased and none of those who are not diseased; this situation corresponds to $m_{21} = m_{12} = 0$. Thus $\alpha_3 = m_{11}m_{22}/m_{21}m_{12}$ tells us whether the laboratory procedure is of any value. Unless α_3 is large, the laboratory procedure is abandoned.

2. *Measuring sensitivity and specificity*

When the evaluation of the laboratory procedure is described, laboratory results indicating disease are considered positive, the others negative. The term "sensitivity" is used for the proportion of positive results that agree with the true state, and the term "specificity" for the proportion of negative results that agree with the true state. These are the proportions described by the rearranged array:

True State	Laboratory Procedure Correct	Incorrect	Totals	
Disease	m_{11}	m_{12}	N_1	(2.2-42)
No Disease	m_{22}	m_{21}	N_2	

Now each row yields one of the proportions of interest:

$$\text{sensitivity} = P_{1(1)} = \frac{m_{11}}{N_1} = 1 - P_{2(1)} = 1 - \frac{m_{12}}{N_1},$$

$$\text{specificity} = P_{2(2)} = \frac{m_{22}}{N_2} = 1 - P_{1(2)} = 1 - \frac{m_{21}}{N_2}.$$

(2.2-43)

The relative magnitude of the sensitivity and specificity is measured by

$$\alpha_1 = \frac{m_{11}m_{21}}{m_{22}m_{12}}.$$

Such laboratory procedures are often used on large populations to find diseased persons. When a choice is to be made between two competitive procedures for screening a population, the prevalence and nature of the disease determines which characteristic, sensitivity or specificity, should be maximized.

3. *Assessing predictive value*
The third arrangement of the array does not preserve the fixed margins N_1 and N_2:

Agrees with True State	Laboratory Procedure Disease	No Disease	
Yes	m_{11}	m_{22}	(2.2-44)
No	m_{21}	m_{12}	

Unless the sample sizes N_1 and N_2 are proportional to the prevalence of the disease in the population where the laboratory procedure is to be used as a screening device, $\alpha_2 = m_{11}m_{12}/m_{21}m_{22}$ does not measure the relative odds on a correct prediction according to the outcome of the laboratory procedure.

To assess whether the cost of screening a population is worthwhile in terms of the number of cases detected, the health official needs to know the positive predictive value $PV+$ and the negative predictive value $PV-$. To compute predictive values we need to know the proportion D of diseased persons in the population to be screened. Then we multiply the first row of the original table (2.2-41) by D/N_1 and the second row by $(1 - D)/N_2$ to obtain

True State	Laboratory Procedure Disease	No Disease	
Disease	$DP_{1(1)}$	$DP_{2(1)}$	(2.2-45)
No Disease	$(1 - D)P_{1(2)}$	$(1 - D)P_{2(2)}$	

The cross-product ratio α_3 is the same in array (2.2-45) as in array (2.2-41). Similarly, if we rearrange array (2.2-45) to correspond with array (2.2-42), we obtain the same values for sensitivity and specificity. When we rearrange array (2.2-45) to correspond with array (2.2-44), a difference occurs. We obtain

| Agrees with True State | Laboratory Procedure | |
	Disease	No Disease
Yes	$DP_{1(1)}$	$(1 - D)P_{2(2)}$
No	$(1 - D)P_{1(2)}$	$DP_{2(1)}$

(2.2-46)

The cross product in array (2.2-46) differs from that in array (2.2-44) by the factor $D^2/(1 - D)^2$ and measures the relative odds in the population of having the disease according to the results of the laboratory procedure. For the positive laboratory results we have

$$PV+ = \frac{DP_{1(1)}}{(1 - D)P_{1(2)} + DP_{1(1)}}$$

$$= \frac{1}{1 + \dfrac{1 - D}{D} \dfrac{N_1}{N_2} \dfrac{m_{21}}{m_{11}}}, \tag{2.2-47}$$

and for the negative laboratory results

$$PV- = \frac{1}{1 + \dfrac{D}{1 - D} \dfrac{N_2}{N_1} \dfrac{m_{12}}{m_{22}}}. \tag{2.2-48}$$

When the two predictive values are equal we have independence in array (2.2-46).

Thus we have shown that rearranging tables has practical applications. It is helpful in assessing the relationships between predictive values, and between sensitivity and specificity for particular disease prevalences, as discussed by Vechio [1966]. (See exercises 1 and 2 in Section 2.6 for further details.)* ■■

2.2.4 Differences between log-linear and other models

Models other than the log-linear have been proposed for describing tables of counts. We now discuss two contenders and show that the logit model can be regarded as a different formulation of the log-linear model, but models that are linear in the arithmetic scale have different advantages and disadvantages.

Logit models

Suppose that the row totals m_{1+} and m_{2+} are fixed and that we are interested in the relative proportions in the rows. We have, as before, $P_{1(i)} = m_{i1}/m_{i+}$ for $i = 1, 2$.

Then the logit for the *i*th row is defined as

$$L_i = \log \frac{P_{1(i)}}{1 - P_{1(i)}} = \log \frac{m_{i1}}{m_{i2}}. \tag{2.2-49}$$

From the saturated model

$$\log(m_{ij}) = u + u_{1(i)} + u_{2(j)} + u_{12(ij)}, \tag{2.2-50}$$

* The symbol ■■ marks the end of an example.

we find that

$$L_i = u_{2(1)} - u_{2(2)} + u_{12(i1)} - u_{12(i2)}$$

$$= 2u_{2(1)} + 2u_{12(i1)},$$

and letting $w = 2u_{2(1)}$ and $w_{1(i)} = 2u_{12(i1)}$, we get

$$L_i = w + w_{1(i)}, \tag{2.2-51}$$

with $w_{1(1)} + w_{1(2)} = 0$. Thus we have transformed the log-linear model for the expected cell counts into a linear model for the logits.

We can now compare the linear logit model with the linear model for the one-way analysis of variance, because we can think of the row variable A as being an independent variable and the column variable B as being a dependent variable. As $w_{1(i)}$ measures the structural relationship between A and B (i.e., because $u_{12(ij)}$ measures this relationship), we can speak of the effect of A on B.

We discuss other aspects of logits in Section 2.3.5, where we show that the logit model is appropriate primarily for stratified samples. It is unduly restrictive for a simple random sample, as it requires that one margin be fixed. In Chapter 10, Section 10.4, we discuss uses that have been made of the logistic model for mixtures of quantitative and qualitative variables.

Additive models

It is natural to explore the possibility of using a linear model in the cell probabilities instead of their logarithms. Suppose we let

$$p_{ij} = \mu + \beta_i + \gamma_j + \varepsilon_{ij} \qquad i = 1, 2; \quad j = 1, 2, \tag{2.2-52}$$

with

$$\beta_+ = \gamma_+ = \varepsilon_{i+} = \varepsilon_{+j} = 0.$$

Since the $\{p_{ij}\}$ must sum to 1, $\mu = \frac{1}{4}$. By examining the marginal totals, we also have

$$\beta_i = \tfrac{1}{2}(p_{i+} - \tfrac{1}{2}) \qquad i = 1, 2,$$

$$\gamma_j = \tfrac{1}{2}(p_{+j} - \tfrac{1}{2}) \qquad j = 1, 2. \tag{2.2-53}$$

Thus, unlike the u-terms, the β_i and γ_j are directly interpretable in terms of the marginal totals p_{i+} and p_{+j}. This advantage brings with it the range restrictions

$$-\tfrac{1}{4} \leq \beta_i \leq \tfrac{1}{4},$$

$$-\tfrac{1}{4} \leq \gamma_j \leq \tfrac{1}{4}, \tag{2.2-54}$$

$$-\tfrac{1}{4} \leq \varepsilon_{ij} \leq \tfrac{1}{4}.$$

The major problem comes in the interpretation of ε_{11}, which we can write as

$$\varepsilon_{11} = \tfrac{1}{4}(p_{11} + p_{22} - p_{12} - p_{21})$$

$$= \tfrac{1}{4}(4p_{11} - 2p_{1+} - 2p_{+1} + 1). \tag{2.2-55}$$

Setting $\varepsilon_{11} = 0$ does not imply independence of the underlying variables unless $p_{1+} = \frac{1}{2}$ or $p_{+1} = \frac{1}{2}$, nor does setting $p_{ij} = p_{i+}p_{+j}$ imply that ε_{11} takes on any specific value.

We have found that the cross-product ratio α_3 is a simple function of u_{12} and has useful invariance properties. We cannot express α_3 as a simple function of the $\{\varepsilon_{ij}\}$, nor can we find a simple alternative function of the $\{\varepsilon_{ij}\}$ that has the invariance properties. We conclude that the difficulty of relating the additive model to the concept of independence makes it less attractive than the log-linear model.

2.3   Two Dimensions—The Rectangular Table

The log-linear model used to describe the structure of the 2×2 table is unaltered in appearance when applied to larger two-way tables. The number and interpretation of parameters differ for larger tables. The applicability of the model to expected cell counts or to probabilities and its suitability for a variety of sampling schemes persist, as does the relationship to logit models. The rearrangement of cells demonstrated for the 2×2 table is not as useful for interpreting parameters in larger tables, except when the arrays, instead of being rectangular, take some other shape such as triangular.

2.3.1   *The log-linear model*

Suppose the cells from a single sample of size N form a rectangular array with I rows and J columns, corresponding to the I categories of variable 1 and J categories of variable 2. We have already seen that the log-linear model describes either probabilities or expected counts. As we wish to consider later a variety of sampling schemes, we define the model in terms of expected counts. A given sampling scheme places constraints on the expected counts, but for every scheme we have

$$\sum_{i,j} m_{ij} = N, \tag{2.3-1}$$

and define $l_{ij} = \log m_{ij}$ for $i = 1, \ldots, I; j = 1, \ldots, J$.

The log-linear model is unchanged from the form used for the 2×2 table:

$$l_{ij} = u + u_{1(i)} + u_{2(j)} + u_{12(ij)}. \tag{2.3-2}$$

The number of parameters contained in each u-term is a function of I and J, but the constraints are unaltered:

$$\sum_i u_{1(i)} = \sum_j u_{2(j)} = \sum_i u_{12(ij)} = \sum_j u_{12(ij)} = 0. \tag{2.3-3}$$

By analogy with analysis of variance, we define

$$\text{overall mean, } u = \frac{l_{++}}{IJ}, \tag{2.3-4}$$

$$\text{main effect of variable 1, } u_{1(i)} = \frac{l_{i+}}{J} - \frac{l_{++}}{IJ}, \tag{2.3-5}$$

$$\text{main effect of variable 2, } u_{2(j)} = \frac{l_{+j}}{I} - \frac{l_{++}}{IJ}, \tag{2.3-6}$$

two-factor effect between variables,

$$u_{12(ij)} = l_{ij} - \left(\frac{l_{+j}}{I} + \frac{l_{i+}}{J} \right) + \frac{l_{++}}{IJ}. \tag{2.3-7}$$

Degrees of freedom

The constraints (2.3-3) reduce the number of independent parameters represented by each u-term. Thus the value of $u_{1(i)}$ differs for each of the I categories of variable 1, but the constraints reduce the number of independent parameters to $I - 1$. Similarly, u_{12} is an $I \times J$ array of parameters that sum to zero across each row and column, and so has $(I - 1)(J - 1)$ independent values. To verify that the total number of independent parameters equals the total number of elementary cells, we add the contributions from each u-term. The numbers of parameters for each term are listed under "degrees of freedom" because this is how we view parameters when we fit models to data in later chapters.

u-Term	Degrees of Freedom	
u	1	
u_1	$I - 1$	(2.3-8)
u_2	$J - 1$	
u_{12}	$IJ - I - J + 1$	
Total	IJ	

The sum IJ of all parameters in the saturated model matches the number of elementary cells (i, j).

Constructing a table

We can substantiate the claim that a log-linear model describes the structure of a table by using a model to build a table. Table 2.3-1 helps us illustrate the process

Table 2.3-1　Construction of Artificial 2×3 Table

Cell	e^u	e^{u_1}	e^{u_2}	$e^{u_{12}}$	$e^{u_1 + u_2 + u_{12}}$	m_{ij}
1, 1	60*	2*	6*	4*	48	2,880
2, 1	60	1/2	6	1/4	3/4	45
1, 2	60	2	1/2*	5*	5	300
2, 2	60	1/2	1/2	1/5	1/20	3
1, 3	60	2	1/3	1/20	1/30	2
2, 3	60	1/2	1/3	20	10/3	200
Total						3,430

*Selected values.

for a 2×3 table. As there are six cells, we select values for six parameters (indicated by an asterisk in the table), and the constraints enable us to derive the other parameters from these six. For the main effect u_1 we put $e^{u_{1(1)}} = 2$; then its reciprocal is the value for $e^{u_{1(2)}}$, and the u_1-term is completely defined. The second main effect u_2 has two degrees of freedom, so we select $e^{u_{2(1)}} = 6$, $e^{u_{2(2)}} = 1/2$, and derive $e^{u_{2(3)}}$ as the reciprocal of their product. Similarly for u_{12} we select $e^{u_{12(11)}} = 4$, derive $e^{u_{12(21)}} = 1/4$, and then select $e^{u_{12(12)}} = 5$ and derive the remaining parameters.

Multiplying e^{u_1}, e^{u_2}, and $e^{u_{12}}$ gives $e^{u_1 + u_2 + u_{12}}$. For a simple example, we use integers for the $\{m_{ij}\}$, and we can get them by selecting $e^u = 60$. The $\{m_{ij}\}$ are given in the last column. If we wish to construct values for p_{ij} instead of m_{ij}, we sum the values for $e^{u_1 + u_2 + u_{12}}$ and define e^u as the reciprocal of this sum to ensure that the $\{p_{ij}\}$ sum to 1.

Extension of this construction process to larger tables follows the same procedure. Such dummy tables of known structure are used for checks of computing procedures, empirical investigations of the effects of collapsing tables, and theoretical investigations utilizing Monte Carlo methods.

2.3.2   Cross-product ratios and linear contrasts

When the number of categories per variable exceeds 2, we find that we can still express each u-term as a function of cross-product ratios, or equivalently as a linear constrast. We consider first increasing the number of categories for one variable only.

The $2 \times J$ table

With two rows and J columns, the log odds for any column j are

$$\log(m_{1j}/m_{2j}) = l_{1j} - l_{2j} = 2(u_{1(1)} + u_{12(1j)}). \tag{2.3-9}$$

If we let $\alpha_{r \cdot s}$ denote the cross-product ratio for columns r and s, we have

$$\log \alpha_{r \cdot s} = \log(m_{1r}/m_{2r}) - \log(m_{1s}/m_{2s}). \tag{2.3-10}$$

For the first two columns,

$$\log(\alpha_{1 \cdot 2}) = 2(u_{12(11)} - u_{12(12)}), \tag{2.3-11}$$

and for the first and jth column,

$$\log(\alpha_{1 \cdot j}) = 2(u_{12(11)} - u_{12(1j)}). \tag{2.3-12}$$

Taking the logarithm of the product of all $J - 1$ such α-terms yields

$$\log(\alpha_{1 \cdot 2} \, \alpha_{1 \cdot 3} \dots \alpha_{1 \cdot J}) = 2(J - 1)u_{12(11)} - 2(u_{12(12)} + u_{12(13)} + \dots + u_{12(1J)})$$
$$= 2Ju_{12(11)}. \tag{2.3-13}$$

Thus each parameter of u_{12} is a function of $J - 1$ cross products. It is a matter of rearrangement to obtain the linear contrast

$$u_{12(11)} = \frac{1}{2J} \sum_{j=2}^{J} \log(\alpha_{1 \cdot j})$$

$$= \frac{J-1}{2J}(l_{11} - l_{21}) + \frac{1}{2J} \sum_{j=2}^{J} (l_{2j} - l_{1j}). \tag{2.3-14}$$

Alternatively, we can write

$$e^{u_{12(11)}} = \prod_{j=2}^{J} \left(\frac{m_{11} m_{2j}}{m_{21} m_{1j}} \right)^{1/2J}. \tag{2.3-15}$$

The contrast (2.3-14) could have been derived more directly from expression (2.3-7). Using this more direct approach for the other terms, we have, from expression (2.3-5),

$$u_{1(1)} = \frac{\sum_j l_{1j}}{J} - \frac{\sum_{i,j} l_{ij}}{2J} = \sum_j \frac{(l_{1j} - l_{2j})}{2J}, \qquad (2.3\text{-}16)$$

or the corresponding relationship

$$e^{u_{1(1)}} = \prod_j \left(\frac{m_{1j}}{m_{2j}}\right)^{1/2J}. \qquad (2.3\text{-}17)$$

The other main-effect term is similarly defined, from (2.3-6), as

$$u_{2(1)} = \frac{(J-1)}{2J}(l_{11} + l_{21}) - \sum_{j=2}^{J} \frac{(l_{1j} + l_{2j})}{2J}, \qquad (2.3\text{-}18)$$

or can be written in product form

$$e^{u_{2(1)}} = \prod_{j=2}^{J} \left(\frac{m_{11}m_{21}}{m_{1j}m_{2j}}\right)^{1/2J}, \qquad (2.3\text{-}19)$$

with each term a rearrangement of the four cells used to define $\alpha_{1 \cdot j}$.

In any rectangular table we can use the definitions (2.3-4)–(2.3-7) to express parameters of subscripted u-terms as linear contrasts of the $\{l_{ij}\}$ or to give the corresponding multiplicative form.

2.3.3   *Effect of combining categories*

As soon as a table has more than two categories per variable, the possibility of amalgamating categories arises. We need to consider when we can combine categories without changing the structure of the table.

The 2 × 3 table

Consider taking a 2 × 2 table and subdividing the second category of the second variable to give a 2 × 3 table. The new table has cell entries m'_{ij}, where

$$m_{i2} = m'_{i2} + m'_{i3}, \qquad (2.3\text{-}20)$$

$$m_{i1} = m'_{i1} \qquad (2.3\text{-}21)$$

for $i = 1, 2$, and is described by the new model

$$\log m'_{ij} = u' + u'_{1(i)} + u'_{2(j)} + u'_{12(ij)}. \qquad (2.3\text{-}22)$$

Although cell (1, 1) is unchanged, the parameters in the new model will in general differ from those in the original model. For instance, in the original table from (2.3-15) we obtain

$$4u_{12(11)} = \log\left(\frac{m_{11}m_{22}}{m_{21}m_{12}}\right), \qquad (2.3\text{-}23)$$

and in the expanded table

$$6u'_{12(11)} = \log\left[\left(\frac{m'_{11}}{m'_{21}}\right)^2 \left(\frac{m'_{22}m'_{23}}{m'_{12}m'_{13}}\right)\right]. \qquad (2.3\text{-}24)$$

Comparing expressions (2.3-23) and (2.3-24) shows us that in general $u_{12(11)} \neq u'_{12(11)}$.

Putting all three odds ratios in the new table equal to one another, i.e., setting

$$\frac{m'_{11}}{m'_{21}} = \frac{m'_{12}}{m'_{22}} = \frac{m'_{13}}{m'_{23}},$$

gives $u'_{12(11)} = 0$, and we find that for the original table

$$\frac{m_{12}}{m_{22}} = \frac{m'_{12} + m'_{13}}{m'_{22} + m_{23}} = \frac{m_{11}}{m_{21}}.$$

Thus we also have $u_{12(11)} = 0$ and have shown that independence in the expanded table implies independence in the condensed table. The converse does not hold: a smaller table fitting the independence model can be derived from a larger table with more complex structure, as we show in the following example.

Example 2.3-1   Condensing categories
Consider three tables:

Table A				Table B			Table C		
4	2	6		4	8		4	1	7
6	3	9		6	12		6	6	6

In table A the row odds m_{1j}/m_{2j} are constant for all j, so the table fits the independence model. Pooling the last two columns gives table B, with the same constant row odds. Thus both tables are fitted by the independence model. A new partitioning of the second column of table B gives table C, and the row odds are no longer constant. In other words, table C does not fit the independence model, but in the condensed table B we have independence. ■■

The independent rectangular table
For any rectangular array, we can express the independence of rows and columns in two equivalent forms:

$$m_{ij} = \frac{m_{i+}m_{+j}}{N}, \tag{2.3-25}$$

$$l_{ij} = u + u_{1(i)} + u_{2(j)}. \tag{2.3-26}$$

The independence model (2.3-26) is derived from the model (2.3-2) by putting $u_{12(ij)} = 0$ for all i and j, or, more briefly, by putting $u_{12} = 0$. Independent tables have special properties.

From model (2.3-26), the difference between the logarithms of any two cells in the same column is

$$l_{ij} - l_{rj} = u_{1(i)} - u_{1(r)}. \tag{2.3-27}$$

Changing from the log scale to the original scale, this yields

$$\frac{m_{ij}}{m_{rj}} = e^{u_{1(i)} - u_{1(r)}}$$

$$= \frac{m_{i+}}{m_{r+}} \qquad (2.3\text{-}28)$$

for all j. Thus the ratio of internal cells in a given column is the same as the ratio of corresponding row sums. By symmetry, this is also true for row elements and column sums. For tables described by the independence model (2.3-26), we have the following conclusions:

1. independence is not lost if we combine some of the categories of either variable;
2. the parameters of u_1 can be determined from the vector of row sums $\{m_{i+}\}$, and similarly for u_2 from the column sums.

We have thus established that two-way tables with independent structures are *collapsible*. We can combine one or more categories of either variable, and the same model describes the structure of the reduced table. If we combine all the categories of one variable, say variable 2, then we have a string of I cells, and from these cells we can compute the parameters of u_1. The values of u_1 that we obtain from the reduced table are identical to those obtained, using expression (2.3-20) for instance, from the full table. Conversely, if the structure is not independent, combining categories gives a reduced table with parameter values that differ from those of the parent table; the parameters of u_{12} may even become zero, thus giving a reduced table of different structural form from the parent table.

Identifying collapsible structures in more than two dimensions is a useful tool for handling large tables. When the structures are not collapsible, the analyst who inspects only two-way tables of sums can be led to false conclusions about the interaction patterns between the variables.

2.3.4 *Different sampling schemes*
When we constructed a table by specifying values of the parameters we found that the size N was dependent on the value u selected for the overall mean. This constraint is imposed by simple random sampling. When we have stratified sampling, the size of the sample selected from each stratum is fixed. We can still use model (2.3-2) to describe the structure of the table of expected counts, but further constraints are imposed on the u-terms.

Consider I strata with N_i observations in the ith stratum, where $\sum_i N_i = N$. If the J categories for each stratum are the same, putting

$$\log m_{ij} = l_j^{(i)} = v^{(i)} + v_{2(j)}^{(i)} \qquad (2.3\text{-}29)$$

with $\sum_j v_{2(j)}^{(i)} = 0$ defines each stratum in terms of the log mean of the stratum and deviations from it.

We now compare the expressions obtained from model (2.3-29) for average values of the $l_j^{(i)}$ with values obtained from the u-term model (2.3-2) applied to the whole table. Taking the mean of all cells gives

$$u = \sum_i \frac{v^{(i)}}{I}. \tag{2.3-30}$$

Using this relation, we obtain from the mean of the jth column

$$u_{2(j)} = \sum_i \frac{v^{(i)}_{2(j)}}{I}, \tag{2.3-31}$$

from the mean of the ith row

$$u_{1(i)} = v^{(i)} - u, \tag{2.3-32}$$

and finally, from the expressions for single cells and relations (2.3-30)–(2.3-32),

$$u_{12(ij)} = v^{(i)}_{2(j)} - u_{2(j)}. \tag{2.3–33}$$

Thus we can relate the two-dimensional log-linear model (2.3-2) to the one-dimensional analogue (2.3-29). The only constraint on the two-dimensional model introduced by stratification is that the magnitudes of the N_i restrict the magnitudes of the $u + u_{1(i)}$.

2.3.5   The logit model

The logit formulation is equivalent to the log-linear model for describing the structure of the $2 \times J$ table with one fixed margin $\{m_{+j}\}$. Thus we can write $m_{+j} = N_j$. By definition, we have

$$\text{logit}(j) = \log\left(\frac{m_{1j}}{m_{2j}}\right) = l_{1j} - l_{2j} \tag{2.3-34}$$

and can write $\text{logit}(j) = 2(u_{1(1)} + u_{12(1j)})$, or equivalently,

$$\text{logit}(j) = w + w_{2(j)}, \tag{2.3-35}$$

where $w = 2u_{1(1)}$ and $w_{2(j)} = 2u_{12(1j)}$.

When the sampling determines the J marginal totals N_j, we have only J degrees of freedom among the $2J$ cells, and this is adequately reflected by the two-term model (2.3-35). For a single sample with only N fixed, model (2.3-35) does not describe the structure of the full array because it gives no information about the margins $\{m_{+j}\}$.

Although we most often use the logit model for stratified sample schemes where each stratum has only two categories, we can also use it for the multiple-category case by defining

$$\text{logit}(ij) = \log \frac{m_{ij}}{\sum_{r \neq i} m_{rj}}$$

for $i = 1, \ldots, (I - 1)$. We can thus regard the logit model as a special case of the more general log-linear model, suitable only for stratified sampling schemes. In Section 10.4 we discuss useful extensions of the logit model for handling mixtures of discrete and continuous variables.

2.3.6   *Irregular arrays*

When rows and columns correspond to two variables it is sometimes impossible to observe particular category combinations. If this logical impossibility occurs in the (i, j) cell, we say that we have a *structural zero* and $p_{ij} = m_{ij} = 0$. We discuss tables with structural zeros in Chapter 5 and show that we can use log-linear models to describe the structure of the cells where $m_{ij} \neq 0$. We refer to the array of nonzero cells as an *incomplete* table.

In two dimensions, incomplete tables can often be arranged in a triangular array. Triangular arrays arise commonly from

1. measurements on paired individuals where no other distinction is made between members of each pair, e.g., measurements on two eyes of an individual that are not distinguished as right eye and left eye but only sorted by the measurement itself as better eye or worse eye;
2. paired measurements on a single individual, where the value of one measurement sets bounds on the possible values of the second. Commonly, we first assign each individual to a particular category of an ordered set, and any subsequent assignment must place the individual in a higher category; e.g., individuals graded by severity of disease on admission to hospital are not discharged alive until the disease grade has improved.

We can also create triangular tables by folding square tables along a diagonal if this helps in our analysis. We discuss the usefulness of this device in Chapter 8. Further discussion of incomplete tables is deferred to Chapters 5 and 8.

2.4   Models for Three-Dimensional Arrays

As the number of variables measured on each individual increases, the resulting multidimensional contingency tables become more unwieldy. The investigator is apt to give up on multiway tables and look instead at large numbers of two-way tables derived by adding over the categories of all except two variables. In this section we show for three-dimensional tables the dangers inherent in examining only such tables of sums and advocate instead construction of models that describe the full array. We discuss first the $2 \times 2 \times 2$ table and then proceed to general $I \times J \times K$ rectangular tables, with the main focus on

1. interpreting each parameter of the saturated model;
2. interpreting unsaturated models as descriptions of hypotheses;
3. determining when the size of the table may be reduced without distorting the structural relationships between variables of interest.

We extend the notation describing the cells of a two-dimensional array to encompass multiway tables simply by adding more subscripts. The number of subscripts normally matches the number of variables, but exceptions occur. It may sometimes be convenient to split or combine the categories of a single variable, while in other instances it may be useful to fold a two-dimensional array so that it forms an irregular three-dimensional shape. We therefore define subscripts to match the dimension of a particular arrangement of the cells. In three dimensions, the probability of a count falling in cell (i, j, k) is p_{ijk} and the expected count is m_{ijk}, where $i = 1, \ldots, I; j = 1, \ldots, J; k = 1, \ldots, K$.

In complete three-dimensional arrays we have a rectangular parallelepiped of size $I \times J \times K$ containing IJK cells. The subscript "+" denotes addition across all cells with a common value for one or more subscripts. Thus $m_{+jk} = \sum_{i=1}^{I} m_{ijk}$, the sum of all cells with a common value for j and k.

2.4.1   The 2 × 2 × 2 table

When we take measurements on three dichotomous variables, we have eight possible combinations of outcome. We can arrange the eight cells in a $2 \times 2 \times 2$ cube with each dimension corresponding to one variable. For display in two dimensions, we split the cube into two 2×2 arrays. If we make the split by dividing the categories of the third variables we have:

		First Category of Variable 3 Variable 2		Second Category of Variable 3 Variable 2	
		1	2	1	2
Variable 1	1	m_{111}	m_{121}	m_{112}	m_{122}
	2	m_{211}	m_{221}	m_{212}	m_{222}

We can now describe each 2×2 array by a separate log-linear model: for array k, we have

$$l_{ijk} = v^{(k)} + v^{(k)}_{1(i)} + v^{(k)}_{2(j)} + v^{(k)}_{12(ij)} \qquad k = 1, 2, \qquad (2.4\text{-}1)$$

with all subscripted v-terms summing to zero across each subscript variable as usual. We combine these models into a single linear expression by taking means across the tables. Remembering that in this array $K = 2$, we have the following mean effects:

overall mean,
$$u = \frac{1}{K} \sum_k v^{(k)},$$

main effect of variable 1,
$$u_{1(i)} = \frac{1}{K} \sum_k v^{(k)}_{1(i)},$$

$$(2.4\text{-}2)$$

main effect of variable 2,
$$u_{2(j)} = \frac{1}{K} \sum_k v^{(k)}_{2(j)},$$

interaction between variables 1 and 2,
$$u_{12(ij)} = \frac{1}{K} \sum_k v^{(k)}_{12(ij)}.$$

Thus u_{12}, sometimes called the "partial association," is the average interaction between variables 1 and 2.

The deviations from these means depend on the third variable and are defined as follows:

main effect of variable 3,

$$u_{3(k)} = v^{(k)} - u,$$

interactions with variable 3,

$$u_{13(ik)} = v_{1(i)}^{(k)} - u_{1(i)}, \qquad (2.4\text{-}3)$$

$$u_{23(jk)} = v_{2(j)}^{(k)} - u_{2(j)},$$

three-factor effect,

$$u_{123(ijk)} = v_{12(ij)}^{(k)} - u_{12(ij)}.$$

We can now write the single linear model for the whole $2 \times 2 \times 2$ cube,

$$l_{ijk} = u + u_{1(i)} + u_{2(j)} + u_{3(k)} + u_{12(ij)} + u_{13(ik)} + u_{23(jk)} + u_{123(ijk)}. \qquad (2.4\text{-}4)$$

The subscripted u-terms of equations (2.4-2) are derived by taking means of u-terms that sum to zero within tables; this derivation preserves the property that subscripted terms sum to zero. The terms with subscript k are all deviations and so also have this property. For instance,

$$\sum_i u_{123(ijk)} = \sum_j u_{123(ijk)} = \sum_k u_{123(ijk)} = 0. \qquad (2.4\text{-}5)$$

Consequently, we have one absolute value for the parameters of each u-term in the $2 \times 2 \times 2$ table. Thus each of the eight u-terms in the saturated model contributes one degree of freedom, and the total number of degrees of freedom matches the total number of cells as required.

We now consider interpretations of the parameters of model (2.4-4) which can be extended to $I \times J \times K$ tables of any size.

Interpretation of parameters and the hierarchy principle

a. *Two-factor effects*

By splitting the cube according to the categories of the third variable, we introduced an apparent difference in the definitions of the two-factor effects that disappears when we consider different partitions. We defined the two-factor effect u_{12} as representing the interaction between variables 1 and 2 averaged over tables, and it was defined solely in terms of v_{12}. By partitioning the cube differently, we get the same interpretation as an average for the other two-factor terms u_{13} and u_{23}, instead of defining them as deviations as in (2.4-3). Conversely, the symmetry permits us to define u_{12} as a deviation from the overall effect of a single variable, similar to the definitions given for the other two-factor terms in (2.4-3). We can also define two-factor effects as products of cross-product ratios. If we define

$$\alpha^{(k)} = \frac{m_{11k}m_{22k}}{m_{12k}m_{21k}}, \qquad (2.4\text{-}6)$$

we can use expression (2.4-6) to get

$$u_{12} = \frac{1}{8}\log(\alpha^{(1)}\alpha^{(2)}). \qquad (2.4\text{-}7)$$

b. *Three-factor effect*

The three-dimensional model has one new feature that the two-dimensional model did not have, namely, the three-factor effect u_{123}. We derived u_{123} as the difference between the average value of u_{12} across tables and the particular value exhibited by table k. The symmetry of this definition becomes apparent when we write u_{123} in terms of cross-product ratios:

$$u_{123(111)} = \frac{1}{8} \log\left(\frac{\alpha^{(1)}}{\alpha^{(2)}}\right) = \frac{1}{8} \log\left(\frac{m_{111}m_{221}m_{122}m_{212}}{m_{121}m_{211}m_{112}m_{222}}\right). \qquad (2.4\text{-}8)$$

All the cells whose subscripts sum to an odd number appear in the numerator and those whose subscripts sum to an even number in the denominator. This formulation is independent of the direction of partitioning; α-terms corresponding to u_{13} or u_{23} give the same ratio (2.4-8).

The three-factor effect is sometimes called the "second-order interaction." It measures the difference in the magnitude of the two-factor effect between tables for any of the three partitions of the cube into two 2×2 tables. Such an interpretation of the meaning of the three-factor effect has the natural converse that if any two-factor effect is constant between tables, then the three-factor effect is zero. In particular, setting any two-factor effect equal to zero implies that the three-factor effect is zero. This leads us to a definition of the hierarchy principle, but before we can state this principle in general terms we need a more formal definition of the relationships between u-terms.

c. *Alternate interpretation of two-factor effects*

When dealing with two-dimensional tables we showed how to interpret all the subscripted u-terms in the log-linear model as functions of cross-product ratios. In particular, for a 2×2 table we showed how these cross-product ratios arise naturally from rearrangements of the table. For $2 \times 2 \times 2$ tables we can also show that all subscripted u-terms can be written as functions of ratios of cross-product ratios, and there are rearrangements of tables such that each subscripted u-term takes the form (2.4-8) and corresponds to the three-factor term for the rearrangement.

d. *Relationships between terms*

Consider two u-terms, one with r subscripts and the other with s subscripts, where $r > s$. We say that the terms are *relatives* if the r subscripts contain among them all the s subscripts. Thus u_{123} is a higher-order relative of all the other u-terms in the three-dimensional model, and u_{12} is a higher-order relative of both u_1 and u_2.

e. *The hierarchy principle*

The family of hierarchical models is defined as the family such that if any u-term is set equal to zero, all its higher-order relatives must also be set equal to zero. Conversely, if any u-term is not zero, its lower-order relatives must be present in the log-linear model. Thus if $u_{12} = 0$, we must have $u_{123} = 0$; also, if u_{13} is present in the model, then u_1 and u_3 must be present also.

f. *Linear contrasts*

We showed for the 2×2 table that every u-term can be written as a linear contrast of the logarithms of the four cells. By rearranging the cells of the table, all the subscripted terms can also be written as cross-product ratios. For the $2 \times 2 \times 2$ table we can similarly write every u-term as a linear contrast of the eight cells, or as a function of cross-product ratios.

Remembering that every u-term has one absolute value, in table 2.4-1 we assume that each term is positive in cell $(1, 1, 1)$. Thus the first row has all plus signs. We can fill in the columns corresponding to each u-term so that each term sums to zero over each of its subscripts. These columns give us the linear contrasts, for instance,

$$u_{12(11)} = \tfrac{1}{8}(l_{111} + l_{221} - l_{121} - l_{211} + l_{112} + l_{222} - l_{122} - l_{212}). \qquad (2.4\text{-}9)$$

This contrast has positive sign for all terms with subscripts i and j adding to an even number and negative for the remaining terms.

Table 2.4-1 Sign of u-terms of Fully Saturated Model for Three Dichotomous Variables

Cell	u	u_1	u_2	u_3	u_{12}	u_{13}	u_{23}	u_{123}
1, 1, 1	+	+	+	+	+	+	+	+
2, 1, 1	+	−	+	+	−	−	+	−
1, 2, 1	+	+	−	+	−	+	−	−
2, 2, 1	+	−	−	+	+	−	−	+
1, 1, 2	+	+	+	−	+	−	−	−
2, 1, 2	+	−	+	−	−	+	−	+
1, 2, 2	+	+	−	−	−	−	+	+
2, 2, 2	+	−	−	−	+	+	+	−

2.4.2 The $I \times J \times K$ model

We derived model (2.4-4) for the cube of side 2 by averaging across 2×2 tables. We can use a similar procedure for any three-dimensional array, namely, averaging across K tables of size $I \times J$. Expressions (2.4-1)–(2.4-3) are unaltered, and model (2.4-4) is the appropriate saturated model for any rectangular array in three dimensions.

When the sample is such that $\sum_{i,j,k} m_{ijk} = N$, the only further constraints are that each u-term sums to zero over each variable listed among its subscripts. We then have the following degrees of freedom associated with each level of u-terms:

Number in Level	Level	Degrees of Freedom	
1	overall mean	1	
3	one-factor terms	$(I - 1) + (J - 1) + (K - 1)$	
3	two-factor terms	$(I - 1)(J - 1) + (I - 1)(K - 1)$ $+ (J - 1)(K - 1)$	(2.4-10)
1	three-factor term	$(I - 1)(J - 1)(K - 1)$	
Total		IJK	

Thus the number of independent parameters corresponds to the number of elementary cells.

The interpretation of the individual u-terms is the same for any $I \times J \times K$ table as for the $2 \times 2 \times 2$ table except that when a u-term has more than one degree of freedom it is possible for some of its parameters to be zero while others are large. When this occurs, we need to consider whether some categories can be combined without violating the structure by introducing spurious interaction effects or masking effects that are present. Sometimes we are interested in combining only two categories of a multicategory variable, sometimes in combining more than two. In the extreme case we collapse the variable by combining all its categories. In theorem 2.4-1 below we consider when we can collapse a variable without violating the structure of the three-dimensional array. If the structure is such that we can safely collapse a variable, then we can also combine any subset of its categories without violating the structure.

Before turning to the problem of collapsing, we discuss briefly the effect of sampling design in fixing parts of the structure. We also consider the interpretation of the set of hierarchical models. In any particular data set, our decisions regarding the desirability of collapsing must be governed by what is meaningful in terms of the sampling design and the interpretation—as well as by the model structure.

Constraints imposed by sampling design

When we take a simple random sample and arrange the counts in a three-dimensional array, only the total sample size N is fixed. Stratified samples that yield a three-dimensional array of counts usually fall into two main types, each with a different structure of fixed subtotals:

(i) if one of the three variables, say variable 1, determines the strata, then the components of the one-dimensional array $\{m_{i++}\}$ are fixed, and we have $m_{i++} = N_i$ for all i, and $\sum_i N_i = N$;

(ii) if two of the variables, say variables 1 and 2, are required to describe the strata, then only the third variable, sometimes called the response variable, is measured on each individual. This scheme fixes the two-dimensional array $\{m_{ij+}\}$, and we have $m_{ij+} = N_{ij}$, for all i, j, and $\sum_{i,j} N_{ij} = N$. We show in Chapter 3 that only the hierarchical models that include u_{12} are appropriate for this design.

These are the two types of design that arise most frequently, but more complex designs can occur. For instance, it is possible to have more than one two-dimensional array of sums fixed by the sampling plan, and it is only appropriate to use a subset of the hierarchical models to describe them. We note that the saturated model describes all such arrays, but the constraints on the $\{n_{ijk}\}$ impose further constraints on the u-terms.

For two-dimensional tables we showed that the parameters of the logit model where one margin was fixed correspond to the relevant u-terms in the log-linear model. We can also use logit models for either of the designs (i) and (ii), and again the logit parameters correspond to the relevant u-terms in the log-linear model (see exercise 4 in Section 2.6).

Interpretation of models

In the following discussion of the interpretation of three-dimensional models, we assume that the sampling scheme is such that all the models are meaningful. As each subscripted u-term can be interpreted as measuring the deviations from lower-order terms, removal of higher-order terms has the effect of simplifying the structure. Starting with the most complex unsaturated model, we discuss each of the models that conform to the hierarchy principle defined in Section 2.4.1.

a.   *Three-factor effect absent*

As u_{123} measures the difference between two-factor effects attributable to the third variable, putting $u_{123} = 0$ enables us to describe a table with constant two-factor effects. Thus the model

$$l_{ijk} = u + u_{1(i)} + u_{2(j)} + u_{3(k)} + u_{12(ij)} + u_{13(ik)} + u_{23(jk)} \qquad (2.4\text{-}11)$$

states that there is "partial association" between each pair of variables, to use the terminology of Birch [1963].

In other chapters we fit such unsaturated models to observed data and compute measures of goodness of fit. These measures indicate whether the model is an adequate description of the structure of the population yielding the observed data. To test the null hypothesis that there is no three-factor effect or "second-order interaction" we fit model (2.4-11). The measure of goodness of fit is our test statistic, with $(I - 1)(J - 1)(K - 1)$ degrees of freedom.

b.   *Three-factor and one two-factor effect absent*

There are three versions of the model with the three-factor effect and one two-factor effect missing. Selecting u_{12} as the absent two-factor effect gives

$$l_{ijk} = u + u_{1(i)} + u_{2(j)} + u_{3(k)} + u_{23(jk)} + u_{13(ik)}. \qquad (2.4\text{-}12)$$

This model states that variables 1 and 2 are independent for every level of variable 3. but each is associated with variable 3. In other words, variables 1 and 2 are *conditionally independent*, given the level of variable 3.

c.   *Three-factor and two two-factor effects absent*

There are also three versions of the model with the three-factor effect and two two-factor effects missing. Selecting

$$u_{123} = u_{12} = u_{13} = 0$$

gives

$$l_{ijk} = u + u_{1(i)} + u_{2(j)} + u_{3(k)} + u_{23(jk)}. \qquad (2.4\text{-}13)$$

Variable 1 is now *completely independent* of the other two variables; variables 2 and 3 are associated.

Later theorems show we can collapse any variable that is independent of all other variables without affecting any of the remaining parameters of the subscripted u-terms. For the simple model (2.4-13) we can prove this directly by writing

$$\log(m_{+jk}) = w + u_{2(j)} + u_{3(k)} + u_{23(jk)}, \qquad (2.4\text{-}14)$$

where

$$w = u + \log\left(\sum_i e^{u_{1(i)}}\right) \qquad (2.4\text{-}15)$$

is a constant independent of any variable category. Equation (2.4-14) shows that the table of sums $\{m_{+jk}\}$ has the same subscripted u-terms as the overall model (2.4-13).

d. Three-factor and all two-factor effects absent

The model that represents *complete independence* of all three variables is

$$l_{ijk} = u + u_{1(i)} + u_{2(j)} + u_{3(k)}. \tag{2.4-16}$$

In this model none of the two-dimensional faces $\{m_{ij+}\}$, $\{m_{i+k}\}$, or $\{m_{ij+}\}$ exhibit any interaction. Summing over two variables gives

$$m_{i++} = \exp(u + u_{1(i)}) \sum_{j,k} \exp(u_{2(j)} + u_{3(k)})$$

$$= \exp(u + u_{1(i)}) \sum_{j} \exp(u_{2(j)}) \sum_{k} \exp(u_{3(k)}), \tag{2.4-17}$$

and similarly for each of the other one-dimensional sums. Summing over all three variables, we have for the total number of counts:

$$N = \sum_{i} m_{i++} = \exp(u) \sum_{i} \exp(u_{1(i)}) \sum_{j} \exp(u_{2(j)}) \sum_{k} \exp(u_{3(k)}). \tag{2.4-18}$$

Combining the three variants of (2.4-17) with (2.4-18), we find

$$m_{ijk} = \frac{m_{i++}m_{+j+}m_{++k}}{N^2}. \tag{2.4-19}$$

e. Noncomprehensive models

Proceeding further with deletion of u-terms gives models that are independent of one or more variables, which we call *noncomprehensive* models. Suppose we put $u_3 = 0$. Then the model becomes

$$l_{ijk} = u + u_{1(i)} + u_{2(j)}. \tag{2.4-20}$$

It is apparent that

$$m_{ijk} = \frac{m_{ij+}}{K}$$

and we have the same $I \times J$ array for all k. We can always sum over any variables not included in the model and describe the condensed structure by a *comprehensive* model that includes all the remaining variables in the resulting lower-dimensional array.

f. Nonhierarchical models

We defined the hierarchy principle in the Section 2.4.1. Most arrays can be described by the set of hierarchical models. Exceptions do occur, but generally the interpretation of nonhierarchical models is complex. In example 3.7-4 of Chapter 3 there is a $2 \times 2 \times 2$ model with $u_{12} = u_{23} = u_{13} = 0$, but $u_{123} \neq 0$. We show there that this model can be related to the concept of "synergism," where a response occurs when two factors are present together but not when either occurs alone.

In larger tables with nonhierarchical structure a possible strategy is to partition and look at smaller sections of data. If $u_{123} \neq 0$ but $u_{12} = 0$ when we partition according to the categories of variable 3, then we have some tables with interaction between variables 1 and 2 in one direction and others with interaction in the opposite direction. We can either consider the structure of each set of tables separately or rearrange the cells to form new compound variables.

Reduction to two dimensions

We derived the three-dimensional model by averaging across a set of two-dimensional tables, each relating to a category of the third variable. We found that all the u-terms involving variables 1 and 2 were averages of the corresponding terms in the two-way tables, as in expression (2.4-2), and all the u-terms involving variable 3 were deviations from these averages, as in expression (2.4-3). We now consider when we can obtain valid estimates of the two-factor terms u_{12}, u_{13}, and u_{23} from the two-way table of sums $\{m_{ij+}\}$, $\{m_{i+k}\}$, and $\{m_{+ik}\}$ formed by adding across such sets of two-way tables.

The rows of table 2.4-1 yield expressions for the sums of the l_{ijk}. For example, adding the first two rows gives

$$l_{+11} = 2(u + u_{2(1)} + u_{3(1)} + u_{23(11)}), \tag{2.4-21}$$

with no u-terms involving variable 1 remaining. These u-terms do not disappear, however, when we sum m_{ijk} over variable 1:

$$m_{+jk} = \exp(u + u_{2(j)} + u_{3(k)} + u_{23(jk)})$$

$$\times \sum_i \exp(u_{1(i)} + u_{12(ij)} + u_{13(ik)} + u_{123(ijk)})$$

$$= \exp(\text{terms independent of variable 1})$$

$$\times \sum_i \exp(\text{terms dependent on variable 1}). \tag{2.4-22}$$

Consequently, if we describe the table of sums by a saturated log-linear model

$$\log m_{+jk} = v + v_{2(j)} + v_{3(k)} + v_{23(jk)}, \tag{2.4-23}$$

we find that, in general,

$$v_{23(jk)} \neq u_{23(jk)}.$$

If $v_{23(jk)} = u_{23(jk)}$ for all j, k, then we say that in the three-dimensional table, variable 1 is *collapsible* with respect to the two-factor effect u_{23}. We now prove that variable 1 is only collapsible with respect to u_{23} if the $I \times J \times K$ table is described by an unsaturated model with $u_{123} = 0$ and either $u_{12} = 0$ or $u_{13} = 0$ or both.

THEOREM 2.4-1 *In a rectangular three-dimensional table a variable is collapsible with respect to the interaction between the other two variables if and only if it is at least conditionally independent of one of the other two variables given the third.*

Proof Without loss of generality we can consider the model with one interaction absent. We choose the model with $u_{12} = u_{123} = 0$, and so we have

$$l_{ijk} = u + u_{1(i)} + u_{2(j)} + u_{3(k)} + u_{13(ik)} + u_{23(jk)}. \tag{2.4-24}$$

We can write the logarithms of the marginal sums as

$$\log m_{ij+} = u + u_{1(i)} + u_{2(j)} + \lambda_{ij}, \tag{2.4-25}$$

$$\log m_{i+k} = u + u_{1(i)} + u_{3(k)} + u_{13(ik)} + \lambda_k, \tag{2.4-26}$$

$$\log m_{+jk} = u + u_{2(j)} + u_{3(k)} + u_{23(jk)} + \lambda'_k, \tag{2.4-27}$$

where

$$\lambda_{ij} = \log\left(\sum_k \exp(u_{3(k)} + u_{13(ik)} + u_{23(jk)})\right),$$

$$\lambda_k = \log\left(\sum_j \exp(u_{2(j)} + u_{23(jk)})\right), \tag{2.4-28}$$

$$\lambda'_k = \log\left(\sum_i \exp(u_{1(i)} + u_{13(ik)})\right).$$

The subscripts on the terms λ indicate which variables λ depends on. For m_{i+k} and m_{+jk} only one variable is involved, so we can also describe these sums by a saturated four-term model. For instance, if we consider collapsing over variable 2,

$$\log(m_{i+k}) = w + w_{1(i)} + w_{3(k)} + w_{13(ik)}. \tag{2.4-29}$$

We now compare (2.4-26) and (2.4-29). Summing over both i and k gives

$$w = u + \sum_k \frac{\lambda_k}{K}. \tag{2.4-30}$$

Summing over k only, we have

$$w_{1(i)} = u_{1(i)}, \tag{2.4-31}$$

and summing over i only gives

$$w_{3(k)} = u_{3(k)} + \lambda_k - \sum_k \frac{\lambda_k}{K}. \tag{2.4-32}$$

From (2.4-30) through (2.4-32), we have

$$w + w_{1(i)} + w_{3(k)} = u + u_{1(i)} + u_{3(k)} + \lambda_k, \tag{2.4-33}$$

and so from (2.4-26) and (2.4-29) we have

$$w_{13(ik)} = u_{13(ik)}. \tag{2.4-34}$$

The analogous result for m_{+jk} is

$$w_{23(jk)} = u_{23(jk)}. \tag{2.4-35}$$

This proves that summing over either of the unrelated variables 1 or 2 gives a table of sums that does not distort the two-factor effects present in the complete array.

Proof of Converse We need only observe from expression (2.4-25) that $\log m_{ij+}$ has a term λ_{ij} dependent on both variables 1 and 2. The table of sums $\{m_{ij+}\}$ thus exhibits a two-factor effect not present in the structural model for the elementary cells. We conclude that summing over a variable associated with both the other variables yields a table of sums that exhibits a different interaction than that shown by its component tables.* ∎

* The symbol ∎ marks the end of a proof.

Notes on Theorem 2.4-1

From expression (2.4-31) we observe that collapsing over variable 2 when $u_{12} = u_{123} = 0$ preserves the value of u_1 as well as the value of u_{13}. Contrastingly, from (2.4-30) and (2.4-32) we find that u and u_3 are changed by collapsing. Thus if a three-dimensional array is collapsed over a single variable (say variable 2), those u-terms involving a specific remaining variable (say u_1) are unchanged if and only if the variable collapsed over is independent of the specific variable (i.e., $u_{12} = 0$).

Clearly, if variable 2 is independent of variables 1 and 3 jointly, none of u_1, u_3, and u_{13} is changed by collapsing over variable 2.

Theorem 2.4-1 in a weaker form is analogous to a result in the theory of partial correlation. We recall a well-known formula from standard multivariate statistical analysis regarding partial correlation coefficients:

$$\rho_{12\cdot3} = \frac{\rho_{12} - \rho_{13}\rho_{23}}{\sqrt{(1 - \rho_{13}^2)(1 - \rho_{23}^2)}},$$

where $\rho_{12\cdot3}$ is the partial correlation between variables 1 and 2, controlling for variable 3, and ρ_{12}, ρ_{23}, and ρ_{13} are the simple correlations. If $\rho_{13} = 0$ or $\rho_{23} = 0$, then $\rho_{12\cdot3}$ is a scalar multiple of the ρ_{12}, and we can test the hypothesis $\rho_{12\cdot3} = 0$ by testing for $\rho_{12} = 0$.

For three-dimensional contingency tables, theorem 2.4-1 says that the term u_{12} in the three-dimensional log-linear model is the same as the u_{12}-term in the log-linear model for the two-dimensional table $\{m_{ij+}\}$, provided that u_{13} or $u_{23} = 0$ in the three-dimensional model.

Example 2.4-1 Collapsing a table

The data of table 2.4-2, analyzed by Bishop [1969], have been used for class exercises at the Harvard School of Public Health, but the original source is unfortunately lost. They relate to the survival of infants (variable 1) according to the amount of prenatal care received by the mothers (variable 2). The amount of care is classified as "more" or "less." The mothers attend one of two clinics, denoted here as A and B. Thus we have a three-dimensional array with the clinic as the third variable.

Table 2.4-2 Three-Dimensional Array Relating Survival of Infants to Amount of Prenatal Care Received in Two Clinics

Place where Care Received	Amount of Prenatal Care	Infants' Survival		Mortality Rate (%)
		Died	Survived	
Clinic A	Less	3	176	1.7
	More	4	293	1.4
Clinic B	Less	17	197	7.9
	More	2	23	8.0

Source: Bishop [1969].

The table for mothers who attended clinic A has the cross-product ratio

$$\frac{(3)(293)}{(4)(176)} = 1.2$$

and that for mothers who attended clinic B,

$$\frac{(17)(23)}{(2)(197)} = 1.0.$$

Both of these values are close to 1, and we conclude that u_{12} is very small. Thus the data could reasonably be considered to be a sample from a population where $u_{12} = 0$; in other words, survival is unrelated to amount of care.

Table 2.4-3 Two-Dimensional Array Relating Survival of Infants to Amount of Prenatal Care; Array Obtained by Pooling Data from Two Clinics

Amount of Prenatal Care	Infants' Survival		Mortality Rate (%)
	Died	Survived	
Less	20	373	5.1
More	6	316	1.9

If we collapse over the third variable (clinic), we obtain table 2.4-3, and we have the cross product

$$\frac{(20)(316)}{(6)(373)} = 2.8,$$

which does not reflect the magnitude of u_{12}. If we were to look only at this table we would erroneously conclude that survival was related to the amount of care received. Theorem 2.4-1 tells us that we cannot evaluate u_{12} from the collapsed table because both u_{13} and u_{23} are nonzero. ■■

2.5 Models for Four or More Dimensions

We proceeded from two dimensions to three dimensions by writing a model for each two-way table defined by the categories of the third variable. Similarly, when we have a four-dimensional array we can write a model in $w^{(l)}$-terms for each three-way array defined by the L categories of the fourth variable. The averages across three-way arrays give the corresponding u-terms for the overall four-dimensional model, and the deviations from these averages give new terms with subscripts that include variable 4, as shown in table 2.5-1. We can continue this process for any number of dimensions, say s. The expected counts in elementary cells of a four-dimensional array have four subscripts i, j, k, l, and in general in s dimensions they have s subscripts. In going from two to three dimensions we doubled the number of u-terms from 4 (i.e., u, u_1, u_2, and u_{12}) to 8. When we go to four dimensions another doubling brings the number of u-terms to 16. In general, for s dimensions the log-linear model has 2^s u-terms.

For the saturated s-dimensional model we have s single-factor u-terms, $s - 1$ of them from the first $s - 1$ dimensions and a term representing deviations from the overall mean due to the last variable to be added. All possible combinations of two variables at a time give $\binom{s}{2}$ two-factor terms. Proceeding thus we find that in general the number of r-factor terms is $\binom{s}{r}$ for $r = 0, \ldots, s$ if we define $\binom{s}{0}$ as 1.

When all variables have only two categories, we can readily interpret the parameters of the u-terms as functions of cross-product ratios involving four cells. In larger tables this formulation becomes more difficult, but the interpretation of unsaturated models is independent of the number of categories per variable. Thus it is sufficient for us to consider the interpretation of parameters in the 2^s table and discuss models that are often encountered in any dimension. We follow with a theorem defining in general terms when we can collapse multidimensional tables to fewer dimensions and still assess the magnitude of u-terms of interest from the condensed array.

Table 2.5-1   Relationship of Three-Dimensional w-terms to
Four-Dimensional u-terms in an $I \times J \times K \times L$ Array

		u-terms	
Order	w-terms for Each of L Three-Way Tables	Average of w-terms	Deviations from Average[a]
Mean	w	u	u_4
Single-factor	w_1, w_2, w_3	u_1, u_2, u_3	u_{14}, u_{24}, u_{34}
Two-factor	w_{12}, w_{23}, w_{13}	u_{12}, u_{23}, u_{13}	$u_{124}, u_{234}, u_{134}$
Three-factor	w_{123}	u_{123}	u_{1234}

[a] Note each deviation is one order higher than other terms on same line.

2.5.1   *Parameters in the 2^s table*

The additive constraints on all subscripted u-terms ensure that the number of independent parameters in the saturated model is the same as the total number of elementary cells. For the 2^s table each of the 2^s u-terms has one absolute value, and each of these can be expressed as a function of the log odds.

Suppose the array is split into 2^{s-2} two-dimensional tables relating variables 1 and 2. The cross-product ratio in each of these 2×2 tables is $\alpha^{(r)}$, where $r = 1, \ldots, 2^{s-2}$, and the tables are ordered so that the subscripts corresponding to the third variable change before the subscripts corresponding to the fourth, and so on. We can now define u_{12} and all higher-order relatives.

Two-factor terms

The two-factor term relating variables 1 and 2 is the average of the two-factor effects in each table:

$$u_{12} = \frac{1}{2^s} \log\left(\prod_r \alpha^{(r)} \right). \tag{2.5-1}$$

Three-factor terms

The definition of the three-factor term u_{123} varies according to the number of dimensions. Using superscript notation, the definition given in expression (2.4-8) for three dimensions becomes

$$u_{123} = \frac{1}{8} \log \frac{\alpha^{(1)}}{\alpha^{(2)}}, \tag{2.5-2}$$

the ratio of cross-product ratios. In four dimensions u_{123} is the average of two such three-dimensional terms:

$$u_{123} = \frac{1}{16}\left(\log\frac{\alpha^{(1)}}{\alpha^{(2)}} + \log\frac{\alpha^{(3)}}{\alpha^{(4)}}\right)$$

$$= \frac{1}{16}\left(\frac{\alpha^{(1)}\alpha^{(3)}}{\alpha^{(2)}\alpha^{(4)}}\right). \tag{2.5-3}$$

We can continue for further dimensions, and in general, for s dimensions

$$u_{123} = \frac{1}{2^s}\log\left(\frac{\displaystyle\prod_{i\,odd}\alpha^{(i)}}{\displaystyle\prod_{j\,even}\alpha^{(j)}}\right). \tag{2.5-4}$$

The α-terms derived from tables corresponding to the first category of variable 3 are in the numerator and those corresponding to the second category in the denominator.

Other three-factor terms u_{124}, u_{125}, etc., are similarly defined by selecting α-terms according to the category of the third variable.

Four-factor terms

In four dimensions the term u_{1234} measures differences of two three-factor terms from their average; thus, from (2.5-2) and (2.5-3),

$$u_{1234} = \frac{1}{8}\log\frac{\alpha^{(1)}}{\alpha^{(2)}} - \frac{1}{16}\log\frac{\alpha^{(1)}\alpha^{(3)}}{\alpha^{(2)}\alpha^{(4)}}$$

$$= \frac{1}{16}\log\frac{\alpha^{(1)}\alpha^{(4)}}{\alpha^{(2)}\alpha^{(3)}}. \tag{2.5-5}$$

This expression is a cross-product ratio of the $\{\alpha^{(r)}\}$, themselves cross-product ratios. In five dimensions two such terms are averaged to give

$$u_{1234} = \frac{1}{32}\log\frac{\alpha^{(1)}\alpha^{(4)}\alpha^{(5)}\alpha^{(8)}}{\alpha^{(2)}\alpha^{(3)}\alpha^{(6)}\alpha^{(7)}}. \tag{2.5-6}$$

In s dimensions the general form is

$$u_{1234} = \frac{1}{2^s}\log\left(\prod_r\frac{\alpha^{(4r-3)}\alpha^{(4r)}}{\alpha^{(4r-2)}\alpha^{(4r-1)}}\right), \tag{2.5-7}$$

where the product is taken from $r = 1$ to $r = 2^{s-4}$. In the numerator the $\{\alpha^{(r)}\}$ are derived from those tables where the categories of variables 3 and 4 are the same, while the remaining $\{\alpha^{(r)}\}$ are in the denominator.

Higher-order terms

Hierarchical models require that if a term is not set equal to zero, none of its lower-order relatives are set equal to zero. This requirement is reasonable when we consider expressing u-terms as functions of cross-product ratios. By continuing the process described previously, we can express any u-term involving variables 1 and 2 as a function of the $\{\alpha^{(r)}\}$. Similarly, by partitioning the array in different directions we can express all related terms as functions of a set of cross products.

Even within the set conforming to the hierarchical requirements we have a large choice of models available. We consider in more detail only a few of the many possibilities.

2.5.2   Uses and interpretation of models

We can divide the hierarchical models into two broad classes, those with all two-factor effects present and those with at least one two-factor effect absent.

All two-factor effects present

The hypotheses most frequently encountered relate to the independence of variables. Even conditional independence requires that at least one two-factor term is absent. Thus models with all two-factor effects present are more likely to be used not for hypothesis testing but for obtaining elementary cell estimates that are more stable than the observed cell counts. Successively higher-order terms can be regarded as deviations from the average value of related lower-order terms, and so models with only the higher-order terms removed are useful in describing the gross structure of an array. Such models describe general trends, and hence can be regarded as "smoothing" devices.

   In other chapters we show that these models are primarily used for

1.   obtaining cell estimates for every elementary cell in a sparse array. In Chapter 3 we show that fitting unsaturated models gives estimates for elementary cells that have a positive probability but a zero observed count.
2.   detecting outliers. In some circumstances the detection of sporadic cells that are unduly large may be of importance. For example, in some investigations it may be desirable to determine what combination of variable categories gives an excessive number of deaths. In Chapter 4 we describe how to detect cells that show large deviations from a hierarchical model applied to the whole array.

Hierarchical models with one two-factor effect absent

Restriction to the class of hierarchical models still permits many structures with one two-factor effect absent and all others present.

   If we put $u_{12} = 0$, in five dimensions the hierarchy principle requires that u_{123}, u_{124}, u_{125}, u_{1234}, u_{1245}, u_{1235}, and u_{12345} are also set equal to zero. Thus the total of 32 terms is reduced by one-fourth to 24. Those remaining fall into three groups: Eight terms involve neither variable 1 nor variable 2, eight involve variable 1 but not variable 2, and eight involve variable 2 but not variable 1. These three groups appear in the first three columns of table 2.5-2, and the fourth column gives u_{12} and its higher-order relatives. We define the sums of the four groups as A, B, C, and D. Thus we have

$$A = u + u_3 + u_4 + u_5 + u_{34} + u_{35} + u_{45} + u_{345}, \qquad (2.5\text{-}8)$$

and similarly for the other columns. When $u_{12} = 0$, the model describing the five-dimensional structure is

$$l_{ijklm} = A + B + C. \qquad (2.5\text{-}9)$$

Table 2.5-2   Grouping Terms of Fully Saturated Five-Dimensional Model

	Groups			
	A	B	C	D
	Includes Neither	Includes Variable	Includes Variable	Includes Both
Row	Variable 1 nor 2	1, but Not 2	2, but Not 1	Variables 1 and 2
1	u	u_1	u_2	u_{12}
2	u_3	u_{13}	u_{23}	u_{123}
3	u_4	u_{14}	u_{24}	u_{124}
4	u_5	u_{15}	u_{25}	u_{125}
5	u_{34}	u_{134}	u_{234}	u_{1234}
6	u_{35}	u_{135}	u_{235}	u_{1235}
7	u_{45}	u_{145}	u_{245}	u_{1245}
8	u_{345}	u_{1345}	u_{2345}	u_{12345}

The sums obtained by adding over variable 1, over variable 2, and over both variables are, respectively,

$$m_{+jklm} = \exp(A + C) \sum_i \exp(B),$$

$$m_{i+klm} = \exp(A + B) \sum_j \exp(C), \qquad (2.5\text{-}10)$$

$$m_{++klm} = \exp(A) \sum_i \exp(B) \sum_j \exp(C),$$

and hence we find

$$m_{ijklm} = \frac{m_{i+klm} m_{+jklm}}{m_{++klm}}. \qquad (2.5\text{-}11)$$

This is a frequently encountered model. In subsequent chapters we show that the relationship (2.5-11) between the expected elementary cell counts and the sums of counts enables us to fit this model readily. When the array is split into KLM two-way tables relating variables 1 and 2, the margins of each table are members of $\{m_{+jklm}\}$ and $\{m_{i+klm}\}$, and the total in each table is a member of $\{m_{++klm}\}$. Thus expression (2.5-11) describes independence in each table, and we say that variables 1 and 2 are "conditionally independent."

There are many other models with variables 1 and 2 conditionally independent. Starting at the bottom of column B or C and moving up, any number of terms can be set equal to zero to give a different hierarchical model. When terms in the same row are removed from columns B and C, the term in column A of this row can also be removed. These models all have fewer u-terms than model (2.5-9), but some are more complex in that the expected counts in the elementary cells cannot be derived from sets of sums as in expression (2.5-11). We give rules in Chapter 3 for determining when such relationships exist.

Even within the set of hierarchical models, the hypothesis $u_{12} = 0$ is thus consistent with a variety of structures. Consequently, it is never adequate to describe a hypothesis only in terms of the absence of one u-term. The model underlying the null hypothesis must be stated in full. The verbal interpretation of many high-dimensional models becomes more cumbersome than useful, and the simplest approach is to write out the log-linear model and examine which terms are included. One of the purposes of such inspection is to determine whether the size of the array can be reduced by summing over some of the variables without distorting the u-terms of interest.

2.5.3 *Collapsing arrays*

Theorem 2.4-1 deals with collapsing in three dimensions, and states that variable 3 is collapsible with respect to u_{12} if variable 3 is unrelated to either variable 1 or variable 2 or both. Thus in three dimensions at least one two-factor term must be absent for any collapsibility to exist. We now give a general theorem for collapsibility in s dimensions, which indicates for a given model which u-terms remain unchanged in the collapsed table. Following the statement of the theorem, we discuss its implications and consider some examples of its application. We first review the definition of collapsibility.

Definition of collapsibility

We say that the variables we sum over are *collapsible* with respect to specific u-terms when the parameters of the specified u-terms in the original array are identical to those of the same u-terms in the corresponding log-linear model for the reduced array.

THEOREM 2.5-1 *Suppose the variables in an s-dimensional array are divided into three mutually exclusive groups. One group is collapsible with respect to the u-terms involving a second group, but not with respect to the u-terms involving only the third group, if and only if the first two groups are independent of each other (i.e., the u-terms linking them are 0).*

Proof We regard the three groups in the statement of this theorem as being three compound variables. We then apply theorem 2.4-1 to these compound variables, and the result follows. ∎

Implications of collapsibility theorems

Independence of two variables implies that the model describing the overall structure has the two-factor term relating the variables and all its higher-order relatives set equal to zero. If a variable is collapsible with respect to specific u-terms, it may be removed by adding over all its categories, or condensed by combining some of its categories, without changing these u-terms.

This definition has two important implications.

1. If all two-factor effects are present, collapsing any variable changes all the u-terms;
2. if any variable is independent of all other variables, it may be removed by summing over its categories without changing any u-terms.

Thus the practice of examining all two-way marginal tables of a complex data base may be very misleading if any of the variables are interrelated. By contrast, the dimensionality of any array may be safely reduced by collapsing over all completely independent variables. The extent to which collapsing or condensing is permissible is determined by the absence of two-factor effects in the structural model, provided the model is hierarchical.

Illustration of collapsing in five dimensions

Suppose we have a five-variable array and wish to know whether we can safely sum over the fifth variable. If $u_{15} = 0$, the hierarchy principle implies that all higher-order relatives are also absent. Theorem 2.5-1 tells us that we can examine the four-dimensional array of sums and obtain valid estimates of all the u-terms involving variable 1, such as u_{12}, u_{123}, and u_{1234}, but we cannot obtain estimates of the terms that do not involve variable 1, such as u_2, u_{23}, and u_{234}.

Suppose now that we wish to know whether we can safely sum over the fourth and fifth variables. Referring to table 2.5-2, we find that $u_{14} = 0$ and $u_{15} = 0$ imply that all the entries in columns B and D from the third line downward are zero. Theorem 2.5-1 tells us that the terms remaining in columns B and D are unchanged by collapsing over variables 4 and 5, but the other terms in the first two rows are altered.

2.5.4   *Irregular tables*

In this chapter we have considered the log-linear model as a useful description of data structure and discussed its interpretation and properties. With the exception of a brief discussion of a simple triangular table in Section 2.3.6, we have dealt only with complete tables. Most of the interpretation and properties we discuss are also applicable for incomplete arrays, but difficulties can arise, and elaboration of these is deferred to Chapter 5.

Similarly, we defer examples of rearranging or partitioning a seemingly complete table to form an incomplete array to Chapters 3, 6, and 8. We also defer special interpretations, such as symmetry and marginal homogeneity, to other chapters.

2.6   Exercises

1. If we tried to assess the predictive value $PV+$ of a test from (2.2-44) without knowing the proportion of diseased persons in the population, we would obtain the pseudovalue

$$PPV+ = \frac{m_{11}}{m_{11} + m_{21}}.$$

   Show that this is only equal to the true predictive value when $D/(1 - D) = N_1/N_2$.

2. Remembering that $P_{1(1)}$ is sensitivity and $P_{2(2)}$ is specificity, show

   (i) for a disease prevalence D of 50% and sensitivity equal to specificity, we have $PV+ = P_{1(1)} = P_{2(2)}$;

   (ii) more generally, the positive predictive value is equal to the sensitivity when $D(1 - P_{1(1)}) = (1 - D)(1 - P_{2(2)})$;

   (iii) if the positive predictive value is equal to the sensitivity, then the negative predictive value is equal to the specificity.

3. Take a square table with four categories per variable, which is described by the log-linear model with $u_{12} = 0$. Fold the table along the diagonal to obtain a triangular table with expected counts m'_{ij}, where

$$m'_{ij} = m_{ij} + m_{ji} \qquad i \neq j,$$
$$m'_{ii} = m_{ii} \qquad\qquad i = 1, 2, 3, 4.$$

Consider the cross product α, defined as

$$\alpha = \frac{m'_{13}m'_{24}}{m'_{14}m'_{23}},$$

and show that a necessary and sufficient condition for $\alpha = 1$ is

$$\frac{m_{1+}}{m_{2+}} = \frac{m_{+1}}{m_{+2}}.$$

Hence show that for an independent two-dimensional structure to exhibit independence when folded, we must have homogeneous margins, i.e., $m_{i+} = m_{+i}$ for all i.

4. Suppose in a $2 \times 3 \times 3$ array the two-dimensional array $\{m_{+jk}\}$ is fixed.
   (i) Show that by defining

$$\text{logit}(j, k) = \log \frac{m_{1jk}}{m_{2jk}},$$

the no three-factor effect model can be written

$$\text{logit}(j, k) = w + w_{2(j)} + w_{3(k)}.$$

   (ii) How many degrees of freedom are associated with each w-term? Show that the sum for the three terms differs from JK by $(J - 1)(K - 1)$, and compare this difference with the three-dimensional equivalent.
   (Answer: The fully saturated three-dimensional model has IJK parameters. The no three-factor effect model differs by $(I - 1)(J - 1)(K - 1)$, which is equal to the logit difference when $I = 2$.)

5. In three dimensions, what structural models permit evaluating each of the three two-factor effects from the corresponding two-dimensional table of sums?
   (Answer: Models with three-factor effect and two two-factor effects absent.)

6. In a $2 \times 3 \times 4 \times 5$ array, write down the most highly parametrized model with variable 4 collapsible with respect to u_{123}.
   (Answer: Any model with two-factor effect involving variable 4 and higher-order relatives absent will do. We keep most parameters if we choose $u_{14} = 0$.)

7. In four dimensions, if we have a hierarchical model structure and we know $u_{14} = u_{23} = 0$, can we assess any of the other two-factor effects from the corresponding two-way tables of sums?
   (Answer: No. If we sum over variable 1 from $\{m_{+jkl}\}$ we can assess u_{24} and u_{34}, but we have w_{23} different from u_{23}. As $w_{23} \neq 0$ we cannot collapse any further.)

2.7 Appendix: The Geometry of a 2 × 2 Table

In this appendix we discuss structure in a 2×2 table in terms of the geometry of the tetrahedron. In particular, we derive the loci of (i) all points corresponding to tables whose rows and columns are independent, (ii) all points corresponding to tables with a given degree of association as measured by the cross-product ratio, (iii) all points corresponding to tables with a fixed set of marginal totals, and (iv) all points corresponding to tables exhibiting symmetry (or marginal homogeneity).

The geometric ideas discussed here allow us to visualize the properties of the various models discussed in Section 2.2 and are used explicitly in Chapters 11 and

12. The geometric model can also be used to provide a general proof of the convergence of iterative procedures used throughout this book (see Fienberg [1970a]).

2.7.1   *The tetrahedron of reference*

Suppose we have only two cells with probabilities p_1 and p_2, where $p_1 + p_2 = 1$. Any value of the $\{p_i\}$ may be represented in two dimensions by a straight line joining the point $(0, 1)$ on the y-axis to the point $(1, 0)$ on the x-axis, as we show in figure 2.7-1a.

When we add another cell, so that our probabilities are p_1, p_2, and p_3, with $p_1 + p_2 + p_3 = 1$, we can represent any set of the $\{p_i\}$ in three dimensions by the surface that joins the points $(1, 0, 0)$, $(0, 1, 0)$, and $(0, 0, 1)$. This surface is a 2-flat (see figure 2.7-1b) and becomes an isosceles triangle when we draw it in two dimensions (see figure 2.7-1c).

Analogously, four cells can be represented in three dimensions by a tetrahedron with vertices $(1, 0, 0, 0)$, $(0, 1, 0, 0)$, $(0, 0, 1, 0)$, and $(0, 0, 0, 1)$. These points, labeled $\mathbf{A}_1, \mathbf{A}_2, \mathbf{A}_3$, and $\mathbf{A}_4$ in figure 2.7-2, represent extreme 2×2 tables of probabilities with probability 1 in one cell and 0 in the others:

$$\mathbf{A}_1 = \begin{array}{|c|c|} \hline 1 & 0 \\ \hline 0 & 0 \\ \hline \end{array} \quad \mathbf{A}_2 = \begin{array}{|c|c|} \hline 0 & 1 \\ \hline 0 & 0 \\ \hline \end{array} \quad \mathbf{A}_3 = \begin{array}{|c|c|} \hline 0 & 0 \\ \hline 1 & 0 \\ \hline \end{array} \quad \mathbf{A}_4 = \begin{array}{|c|c|} \hline 0 & 0 \\ \hline 0 & 1 \\ \hline \end{array}$$

The general point $\mathbf{P} = (p_{11}, p_{12}, p_{21}, p_{22})$ within the tetrahedron corresponds to the general 2×2 table.

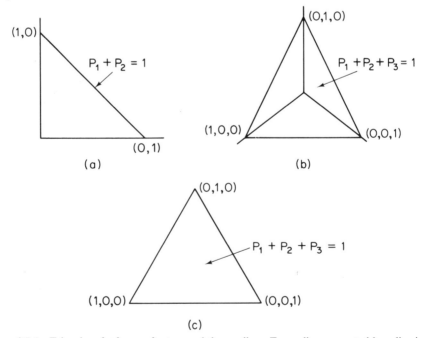

Figure 2.7-1   Triangles of reference for two and three cells. a. Two cells represented by a line in two dimensions. b. Three cells represented by a surface in three dimensions. c. Three cells represented by a surface in two dimensions.

2.7.2 *Surface of independence*

We can now define a surface in the tetrahedron that gives the locus of all tables exhibiting independence, i.e., those tables for which $\alpha_3 = 1$. (See (2.2-13).)

We take any point $\mathbf{T}$ on the line $\mathbf{A_1 A_2}$, defined by a distance t such that $1 \geq t \geq 0$. By taking weighted averages of $\mathbf{A_1}$ and $\mathbf{A_2}$, we obtain

$$\mathbf{T} = \begin{array}{|c|c|} \hline t & 1-t \\ \hline 0 & 0 \\ \hline \end{array} \qquad (2.7\text{-}1)$$

We can similarly choose a point $\mathbf{T'}$ on $\mathbf{A_3 A_4}$ so that

$$\mathbf{T'} = \begin{array}{|c|c|} \hline 0 & 0 \\ \hline t & 1-t \\ \hline \end{array} \qquad (2.7\text{-}2)$$

Any point $\mathbf{I}$ on the line $\mathbf{TT'}$ within the tetrahedron corresponds to a second number s such that $1 \geq s \geq 0$, and this point is derived as a weighted average of $\mathbf{T}$ and $\mathbf{T'}$, so we have

$$\mathbf{I} = \begin{array}{|c|c|c} \hline st & s(1-t) & s \\ \hline (1-s)t & (1-s)(1-t) & 1-s \\ \hline t & 1-t \end{array} \qquad (2.7\text{-}3)$$

The row and column marginal totals are independent, as required. By allowing s and t to take on all possible values between 0 and 1, we can find all points which correspond to tables whose rows and columns are independent. The lines $\mathbf{TT'}$ defined by different values of t ($0 \leq t \leq 1$) lie on this *surface of independence*. Alternatively, we could define the point $\mathbf{S}$ on $\mathbf{A_1 A_3}$ with coordinates $(s, 0, 1-s, 0)$ and $\mathbf{S'}$ on $\mathbf{A_2 A_4}$ with coordinates $(0, s, 0, 1-s)$. Any point $\mathbf{I'}$ on $\mathbf{SS'}$ would also have coordinates given by expression (2.7-3). Thus the lines $\mathbf{SS'}$ defined by different values of s ($0 \leq s \leq 1$) also lie on the surface of independence. This surface is completely determined by either family of lines (see figure 2.7-3). The surface of independence is a section of a hyperbolic paraboloid, and its saddle point is at the center of the tetrahedron, $\mathbf{C} = (\frac{1}{4}, \frac{1}{4}, \frac{1}{4}, \frac{1}{4})$. The hyperbolic paraboloid is a *doubly ruled* surface, since its surface contains two families of straight lines or "rulings." The tables corresponding to points on any one of the lines $\mathbf{TT'}$ have the same column margins (totals), while the tables corresponding to points on any one of the lines $\mathbf{SS'}$ have the same row margins.

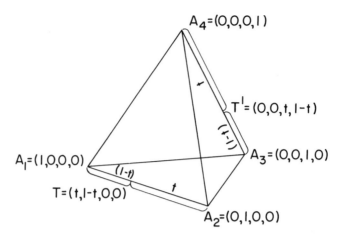

Figure 2.7-2 Tetrahedron of reference. Reproduced from Fienberg and Gilbert [1970].

2.7.3 The surface of constant association

To derive the surface with constant cross-product ratio α_3, we choose the point $\mathbf{T}$ as before on the line joining $\mathbf{A}_1$ to $\mathbf{A}_2$. On the line joining $\mathbf{A}_3$ to $\mathbf{A}_4$ we choose another point $\mathbf{T}^*$, where

$$\mathbf{T}^* = \begin{array}{|c|c|} \hline 0 & 0 \\ \hline t^* & 1 - t^* \\ \hline \end{array} \qquad (2.7\text{-}4)$$

and

$$\frac{t}{1-t} = \alpha_3 \frac{t^*}{1-t^*}. \qquad (2.7\text{-}5)$$

Any point $\mathbf{I}^*$ on the line $\mathbf{TT}^*$ is again a weighted average of $\mathbf{T}$ and $\mathbf{T}^*$, where the weights are s and $1 - s$. Thus we have

$$\mathbf{I}^* = \begin{array}{|c|c|c|} \hline st & s(1 - t) & s \\ \hline (1 - s)t^* & (1 - s)(1 - t^*) & 1 - s \\ \hline \end{array} \qquad (2.7\text{-}6)$$

Expression (2.7-6) gives the cross-product ratio α_3 for any s such that $1 \geqq s \geqq 0$. We note that the column totals are $t^* + s(t - t^*)$ and $1 - t^* - s(t - t^*)$ and so vary with s. Thus the lines $\mathbf{TT}^*$ which generate the surface of constant association are not the loci of points corresponding to tables with constant column totals as were the lines $\mathbf{TT}'$. The surface of constant α_3 intersects the surface of independence along the two edges $\mathbf{A}_1\mathbf{A}_2$ and $\mathbf{A}_3\mathbf{A}_4$ of the tetrahedron (see figure 2.7-4). We can similarly generate another surface of constant α_3 by choosing points $\mathbf{S}$ on $\mathbf{A}_1\mathbf{A}_3$ and $\mathbf{S}^*$ on $\mathbf{A}_2\mathbf{A}_4$. The surfaces of constant α_3 are sections of hyperboloids of one sheet and are again doubly ruled surfaces.

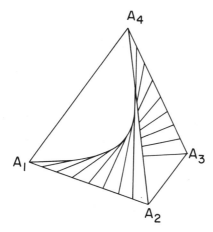

Figure 2.7-3 Surface of independence defined by family of lines TT'. Reproduced from Fienberg and Gilbert [1970].

We previously showed that α_3 was a function of u_{12}, and that by rearranging the cells of the fourfold table we could derive α_1 as a function of u_1 and α_2 as a function of u_2. It follows that by choosing points on different pairs of edges of the tetrahedron we can generate surfaces for constant values of α_1 and α_2 and so for constant values of u_1 and u_2. The lines $\mathbf{TT^*}$ and $\mathbf{SS^*}$ are the intersections of the surface of constant α_3 with surfaces of constant α_1 and α_2, respectively.

2.7.4 Line of constant margins

A table with fixed margins t and $1 - t$ for rows and s and $1 - s$ for columns can be written as the point $\mathbf{P}_a$, where

$$\mathbf{P}_a = \begin{array}{|c|c|c}
\hline
st + a & (1 - s)t - a & t \\
\hline
s(1 - t) - a & (1 - s)(1 - t) + a & 1 - t \\
\hline
s & 1 - s &
\end{array} \qquad (2.7\text{-}7)$$

for any a that gives nonnegative probabilities. To determine the locus of $\mathbf{P}_a$ we find the limiting values of a that give zero cell entries:

$$\begin{aligned}
a_1 &= -st, \\
a_2 &= (1 - s)t, \\
a_3 &= s(1 - t), \\
a_4 &= -(1 - s)(1 - t),
\end{aligned} \qquad (2.7\text{-}8)$$

and obtain the corresponding points $\mathbf{P}^{(1)}$, $\mathbf{P}^{(2)}$, $\mathbf{P}^{(3)}$, and $\mathbf{P}^{(4)}$, for each of which different conditions must be satisfied by s and t. The coordinates and restrictions on the $\mathbf{P}^{(i)}$ are

$$\mathbf{P}^{(1)} = (0, t, s, 1 - s - t) \qquad\qquad s + t \leqq 1,$$

$$\mathbf{P}^{(2)} = (t, 0, s - t, 1 - s) \qquad\qquad s \geqq t,$$

$$\mathbf{P}^{(3)} = (s, t - s, 0, 1 - t) \qquad\qquad t \geqq s, \qquad\qquad (2.7\text{-}9)$$

$$\mathbf{P}^{(4)} = (s + t - 1, 1 - s, 1 - t, 0) \qquad s + t \geqq 1.$$

If s and t are such that the conditions for $\mathbf{P}^{(1)}$ and $\mathbf{P}^{(2)}$ are satisfied, $\mathbf{P}^{(1)}$ is a point on the surface $\mathbf{A}_2\mathbf{A}_3\mathbf{A}_4$ of the tetrahedron, and $\mathbf{P}^{(2)}$ is a point on the surface $\mathbf{A}_1\mathbf{A}_3\mathbf{A}_4$. We choose a point $\mathbf{W}$ on the line $\mathbf{P}^{(1)}\mathbf{P}^{(2)}$ corresponding to a distance w, where $1 \geqq w \geqq 0$, i.e.,

$$\mathbf{W} = \begin{array}{|c|c|c}
\hline
wt & (1-w)t & t \\
\hline
s - wt & 1 - s - t + wt & 1 - t \\
\hline
\end{array} \qquad\qquad (2.7\text{-}10)$$

Then we may easily verify that $\mathbf{P}^{(3)}$ and $\mathbf{P}^{(4)}$ are also on the line defined by $\mathbf{W}$ for varying values of w. $\mathbf{W}$ lies on the line of constant margins, and we can confirm that it goes through the surface of independence by putting $w = s$.

Fienberg and Gilbert [1970] show that the line of constant margins is orthogonal to $\mathbf{A}_1\mathbf{A}_4$ and $\mathbf{A}_2\mathbf{A}_3$ and is parallel to the line connecting the mid-points of $\mathbf{A}_1\mathbf{A}_4$ and $\mathbf{A}_2\mathbf{A}_3$. Hence the line of constant margins is not perpendicular to the surface of independence unless the marginal totals all equal $1/2$.

When we have homogeneous margins we put $s = t$ and derive the locus of the point $\mathbf{W}^*$ (for varying w and t), where

$$\mathbf{W}^* = \begin{array}{|c|c|c}
\hline
wt & (1-w)t & t \\
\hline
(1-w)t & 1 - 2t + wt & 1 - t \\
\hline
t & 1 - t & \\
\end{array}$$

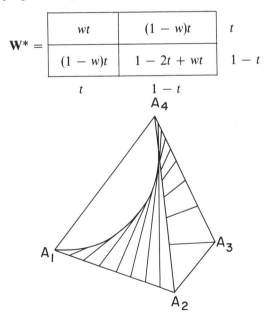

Figure 2.7-4   Surface of constant $\alpha(\alpha = 3)$ defined by family of lines TT*. Reproduced from Fienberg and Gilbert [1970].

All tables exhibiting marginal homogeneity correspond to points on the intersection of the tetrahedron with the plane through $A_1 A_4$ and the midpoint of $A_2 A_3$.

For further details on geometric interpretations, we refer the reader to Fienberg [1968] and Fienberg and Gilbert [1970].

3 Maximum Likelihood Estimates for Complete Tables

3.1 Introduction

Models that are linear in the logarithmic scale can describe the structure of multi-dimensional contingency tables. In Chapter 2 we showed the advantages of such models and described the uses of a special class of models, called hierarchical. In this chapter we discuss how to get maximum likelihood estimates of the cell counts expected for these hierarchical models.

Both the sampling scheme and the interrelationships between the variables determine the structure of tables. So in Section 3.2 we briefly review the most common forms of sampling that give rise to contingency tables; further features of these sampling distributions are discussed in Chapter 13. In Section 3.3 we derive, for each of these distributions, the sufficient statistics for estimating the parameters of hierarchical models.

The sufficient statistics for log-linear models have two desirable attributes: they are easy to obtain, and they readily yield the expected cell counts. The intermediate step of computing the model parameters, which in turn are used to obtain cell estimates, is unnecessary. For some models the cell estimates are explicit closed functions of the sufficient statistics, while for others we need an iterative procedure. In Section 3.4 we give simple rules for distinguishing between the two kinds of models and for obtaining explicit closed-form estimates, when this is possible. In Section 3.5 we describe an iterative procedure for obtaining estimates; this procedure is necessary when direct estimates do not exist and always yields the desired cell estimates. The computing algorithm has a long history, and examples of the use for which it was originally developed are given in Section 3.6.

Sometimes it is advantageous to rearrange the data before fitting a model. We give examples in Section 3.7 of rearranging, relabeling, and partitioning. Finally, in Section 3.8 we give rules and formulas for deriving degrees of freedom.

The ordering of sections follows the procedural steps taken when analyzing data. Chapters 4 and 9 contain a full discussion of the next step, assessing goodness of fit. In this chapter, to complete our examples we use either the Pearson chi square statistic X^2, defined as

$$X^2 = \sum_{\text{all cells}} \frac{(O - E)^2}{E},$$

where O stands for observed cell count and E for the maximum likelihood (ML) estimate in each cell, or minus twice the log-likelihood ratio

$$G^2 = -2 \sum O \log\left(\frac{E}{O}\right)$$
$$= 2 \sum O \log\left(\frac{O}{E}\right),$$

which is also distributed, under the null hypothesis, as a central χ^2 with appropriate degrees of freedom.

Although the scope of this chapter is confined to maximum likelihood estimation (MLE) and complete tables, we generalize results to cover tables of any dimension. In all sections the development is parallel, first the results for low-dimensional tables, then the generalization. Generalization requires notation that is not dependent on dimension. We now give reasons for choosing maximum likelihood estimation, our definition of completeness, and a description of both the individually subscripted and the generalized notation.

3.1.1   *Maximum likelihood estimation*

As he was for so many statistical concepts, R. A. Fisher was the first to study and establish optimum properties of estimates obtained by maximizing the likelihood function, using criteria such as consistency and efficiency (involving asymptotic variances) in large samples. Neyman [1949] pointed out that these large-sample criteria were also satisfied by other estimates. He defined a class of best asymptotically normal (BAN) estimates, all having the same asymptotic properties as the maximum likelihood estimates. More recently Rao [1961, 1962] proposed a concept of second-order efficiency for judging the performance of an estimate. He showed that for a multinomial distribution, maximum likelihood estimation is the only BAN method with an optimum second-order efficiency, under regularity conditions of the sort satisfied in log-linear models.

Maximum likelihood estimates (MLEs) are thus satisfactory on theoretical grounds. We show in this chapter that they also have some important practical advantages:

1. The MLEs for log-linear models are relatively easy to compute (having closed-form expressions in some cases);
2. the MLEs satisfy certain intuitive marginal constraints not intrinsic to other methods of estimation;
3. the method of maximum likelihood can be applied directly to multinomial data with several observed cell values of zero, and almost always produces nonzero estimates for such cells (an extremely valuable property in small samples).

These properties notwithstanding, alternative methods of estimation are sometimes useful and perhaps more appropriate. In Chapter 10, we discuss other estimates and show that most BAN estimates typically lead to the same conclusions we obtain by maximizing the likelihood.

3.1.2   *Completeness*

A complete table has a nonzero probability of a count occurring in every cell of a rectangular parallelepiped. For example, in an $I \times J$ two-way array with observed cell counts $\{x_{ij}\}$, the standard test for "association" in the body of the table begins by obtaining expected cell estimates $\{\hat{m}_{ij}\}$ under the model of independence between the two variables. For this model,

$$\hat{m}_{ij} = \frac{(\text{row total})(\text{column total})}{\text{grand total}}$$

$$= \frac{(x_{i+})(x_{+j})}{x_{++}}, \tag{3.1-1}$$

where

$$x_{i+} = \sum_{j} x_{ij}, \qquad x_{+j} = \sum_{i} x_{ij}, \qquad x_{++} = \sum_{i,j} x_{ij}.$$

The summations extend from $i = 1$ to I for all j, and similarly from $j = 1$ to J for all i. We show in this chapter that these are the maximum likelihood estimates under the model

$$\log m_{ij} = u + u_{1(i)} + u_{2(j)} \tag{3.1-2}$$

described in Chapter 2. If the cells $\{x_{i+}\}$ and $\{x_{+j}\}$ all contain counts, formula (3.1-1) provides a nonzero estimate for every cell of the array, even if some of the x_{ij} are zero. This is desirable if the table is complete.

An incomplete table has some cells that contain structural zeros; these are cells that must logically remain empty. For example, in a table of surgical procedures by sex, it would not make very good sense to obtain an entry for the "male hysterectomy" cell. In Chapter 5 we show that the model (3.1-2) can be applied to tables with structural zeros; we do not deal with them in this chapter, as they sometimes exhibit characteristics that need special attention. The rules that we derive for distinguishing between models that may be fitted directly and those that require iteration are only applicable to complete tables.

Complete tables may have some cells with no observed counts. The pattern of these random zeros determines which log-linear models can be fitted. We can illustrate this situation, using the letters a, b, c, and d to denote positive counts. Suppose first we have the two-way table of observed counts with two empty cells

$$\{x_{ij}\} = \begin{Bmatrix} a & 0 \\ 0 & d \end{Bmatrix},$$

but we believe that members of all four cells exist in the population sampled. We get estimates for the model (3.1-2) by using expression (3.1-1):

$$\{\hat{m}_{ij}\} = \begin{Bmatrix} \dfrac{a^2}{a+d} & \dfrac{ad}{a+d} \\[3mm] \dfrac{ad}{a+d} & \dfrac{d^2}{a+d} \end{Bmatrix}. \tag{3.1-3}$$

From these cell estimates, expressions (2.2-31)–(2.2-33) of Chapter 2 give estimates of the u-terms:

$$\hat{u} = \log \frac{ad}{a+d},$$

$$\hat{u}_{1(1)} = -\hat{u}_{1(2)} = \frac{1}{2}\log\left(\frac{a}{d}\right),$$

$$\hat{u}_{2(1)} = -\hat{u}_{2(2)} = \frac{1}{2}\log\left(\frac{a}{d}\right).$$

We can readily verify that the estimates (3.1-3) agree with these u-term estimates. For instance, for cell $(1, 1)$ the model (3.1-2) is

$$\log m_{11} = u + u_{1(1)} + u_{2(1)},$$

and on substituting our estimates $\hat{m}_{11}$, $\hat{u}$, $\hat{u}_{1(1)}$, and $\hat{u}_{2(1)}$, we have the consistent relationship

$$\log\left(\frac{a^2}{a+d}\right) = \log\left(\frac{ad}{a+d}\right) + \frac{1}{2}\log\left(\frac{a}{d}\right) + \frac{1}{2}\log\left(\frac{a}{d}\right).$$

In this example the presence of random zeros does not affect our ability to fit the model. When the two zeros are differently arranged, this ability may be lost. For example, if we have for the observed 2×2 table

$$\{x_{ij}\} = \begin{Bmatrix} 0 & 0 \\ c & d \end{Bmatrix},$$

then the sum $x_{1+} = 0$ and the table of estimates obtained by using expression (3.1-1) is identical to the observed table. In this table we cannot compute the u-terms under model (3.1-2). This does not mean that it is impossible to fit other models. If, for instance, we select the model that variable 1 has no effect, namely,

$$\log m_{ij} = u + u_{2(j)}, \tag{3.1-4}$$

by using the expression

$$\hat{m}_{ij} = \frac{x_{+j}}{I}, \tag{3.1-5}$$

we get the estimates

$$\{\hat{m}_{ij}\} = \begin{Bmatrix} \dfrac{c}{2} & \dfrac{d}{2} \\ \dfrac{c}{2} & \dfrac{d}{2} \end{Bmatrix}.$$

Thus some complete tables may have random zeros so arranged that we cannot fit the full spectrum of models to the entire array. We discuss this further in Section 3.8, but in general assume that we are not in the situation where the model estimates coincide with the observed counts.

3.1.3 *Notation*

In the preceding paragraphs we used a subscript notation for two dimensions that may readily be extended to three or four dimensions by increasing the number of subscripts. For a large number of dimensions this device becomes cumbersome, and we introduce instead single symbols to denote sets of subscripts. We define below the extended subscript notation in three dimensions, then illustrate the generality of the abbreviated notation.

Notation for three dimensions

a. *Cells*

For an $I \times J \times K$ table we have

$$x_{ijk} = \text{observed count in cell } (i, j, k),$$
$$m_{ijk} = \text{expected count in cell } (i, j, k),$$
$$\hat{m}_{ijk} = \text{MLE of } m_{ijk}.$$

These three symbols all refer to counts of the number of individuals classified as belonging to category i of the first variable or dimension, where $i = 1, \ldots, I$, category j of the second, where $j = 1, \ldots, J$, and category k of the third, where $k = 1, \ldots, K$. They are the counts associated with the *elementary cell* (i, j, k).

b. *Sums*

Sums of elementary cell counts have a " + " replacing one or more subscripts:

$$x_{i++} = \sum_{j,k} x_{ijk} = \sum_{j} x_{ij+}$$

and

$$N = x_{+++} = \text{total number of observations.}$$

Sums are naturally arranged in tables of *nonelementary* cells. These tables have fewer dimensions than the array of elementary cells. We call tables of sums *configurations*, and denote them by the letter C. For example, the two-dimensional array $\{x_{ij+}\}$, obtained by summing over the third variable, is the configuration C_{12}. Each of the IJ nonelementary cells contains the sum of K elementary cell counts. As the third variable has been removed by summing, the subscripts of C refer only to the remaining two variables.

c. *Models*

As in expression (3.1-2), when we write out a model we use the doubly subscripted u-term notation introduced in Chapter 2. When we wish to refer to all the parameters of a particular term, we drop the second set of subscripts. Thus $u_{12} = 0$ means $u_{12(ij)} = 0$ for $i = 1, \ldots, I, j = 1, \ldots, J$.

Notation for unlimited number of dimensions

To generalize results we denote the complete set of subscripts by a single symbol, usually θ. Thus x_θ is the observed count in an elementary cell.

We add a subscript to θ to denote a reduced dimensionality, so that x_{θ_1} is the observed sum in a cell of configuration C_{θ_1}. We use the second subscript, 1, solely to distinguish between different configurations; we do not attempt to distinguish between cells within a configuration, as this is apparent from the context.

Example of generalized notation

We can use the generalized notation in two dimensions by stating that θ_1 contains only the row variable A and θ_2 only the column variable B. We then denote N by x_{θ_3}, and expression (3.1-1) for the cell estimate obtained under the hypothesis of no two-way interaction between variables A and B becomes

$$\hat{m}_\theta = \frac{x_{\theta_1} x_{\theta_2}}{x_{\theta_3}}. \tag{3.1-6}$$

If we redefine the θ_i by exclusion, instead of by inclusion, we can interpret expression (3.1-6) in any dimension. We might, for instance, say that θ_1 contains all variables except variable B, θ_2 contains all variables except variable A, and θ_3 contains all variables common to θ_1 and θ_2, that is, $\theta_3 = \theta_1 \cap \theta_2$. The three-dimensional interpretation of expression (3.1-6) is then

$$\hat{m}_{ijk} = \frac{x_{i+k} x_{+jk}}{x_{++k}},$$

and in four dimensions the equivalent form is

$$\hat{m}_{ijkl} = \frac{x_{i+kl} x_{+jkl}}{x_{++kl}}.$$

When writing dimension-free models, we refer to both single u-terms u_θ and sums of u-terms U_θ. The sum U_θ contains u_θ and all its lower-order relatives, that is, all u-terms with sets of subscripts contained in θ. For example, if θ_1 includes only variables 1 and 2, then

$$u_{\theta_1} = u_{12}, \qquad U_{\theta_1} = u_1 + u_2 + u_{12}.$$

3.2 Sampling Distributions

To obtain the maximum likelihood estimates for a particular model we must consider the sampling distribution. Three commonly encountered sampling plans lead to different distributions. We describe them for the $I \times J \times K$ array and indicate how to generalize them, always using x for observed and m for expected cell counts.

3.2.1 Usual schemes

Independent Poisson sampling

With no restriction on the total sample size, each cell has an independent Poisson distribution. For the three-dimensional array, the probability density function (PDF) is

$$f(\{x_{ijk}\}) = \prod_{i,j,k} \frac{m_{ijk}^{x_{ijk}} e^{-m_{ijk}}}{x_{ijk}!}. \tag{3.2-1}$$

Such a distribution arises if observations are made over a period of time with no a priori knowledge of the total number of observations. By substituting θ for ijk throughout, we can generalize to any number of dimensions.

Simple multinomial sampling

Sometimes the total sample size N is fixed. This restriction imposed on a series of independent Poisson distributions gives the multinomial distribution (Cochran [1952], Fisher [1922]), with **PDF**

$$f(\{x_{ijk}\}) = \frac{N!}{\prod\limits_{i,j,k} x_{ijk}!} \prod_{i,j,k} \left(\frac{m_{ijk}}{N}\right)^{x_{ijk}}. \tag{3.2-2}$$

This function is often written with p_{ijk} representing the probability of a count falling in cell (i, j, k). We have written the equivalent expression m_{ijk}/N because it is less ambiguous when used with other sampling schemes. As N is the only number fixed, to generalize we again substitute θ for ijk throughout.

Product multinomial sampling

Although in observational studies only a single sample may be examined, in experimental situations it is more usual to have several groups, with the total number of individuals in each group determined by the sampling plan. When members of different groups are not matched, the elementary cell counts represent individuals and the group totals form a fixed configuration of sums.

If in three dimensions the group totals form the configuration C_{23}, the resulting distribution is the product of JK independent multinomials. We obtain this distribution by taking the multinomial distribution (3.2-2) and finding the conditional distribution for fixed C_{23}. The marginal distribution of C_{23} is multinomial with **PDF**

$$f(\{x_{+jk}\}) = \frac{N!}{\prod\limits_{j,k} x_{+jk}!} \prod_{j,k} \left(\frac{m_{+jk}}{N}\right)^{x_{+jk}}. \tag{3.2-3}$$

To obtain the conditional distribution we divide expression (3.2-2) by expression (3.2-3). We find the terms in N disappear, and rearranging gives

$$f(\{x_{ijk}\}|\{m_{+jk} = x_{+jk}\}) = \prod_{j,k} \left[\frac{x_{+jk}!}{\prod\limits_i x_{ijk}!} \prod_i \left(\frac{m_{ijk}}{x_{+jk}}\right)^{x_{ijk}}\right]. \tag{3.2-4}$$

The number of multinomials equals the number of cells in C_{23}.

When the total set of subscripts θ is divided into two subsets θ_1 and θ_2 and C_{θ_1} is fixed, the general expression is

$$f(\{x_\theta\}|\{m_{\theta_1} = x_{\theta_1}\}) = \prod_{\theta_1} \left[\frac{x_{\theta_1}!}{\prod\limits_{\theta_2} x_\theta!} \prod_{\theta_2} \left(\frac{m_\theta}{x_{\theta_1}}\right)^{x_\theta}\right], \tag{3.2-5}$$

where the subscript θ_1 implies that the product is taken over all categories of all variables in θ_1 and similarly for θ_2. In example 3.5-1 we describe data first analyzed by Bartlett (see table 3.5-1), where the sampling plan is of this type. We give a further illustration in example 3.7-4 (see table 3.7-10), where the number of cases and controls are fixed by design.

3.2.2   *Other sampling schemes*

Sampling schemes with more than one fixed configuration are less frequently encountered. When two configurations are fixed, we must consider whether they have any variables in common. When they have variables in common the resulting conditional distributions are more complicated than those given in (3.2-1)–(3.2-5) and they are closely related to the quasi-multinomial distribution of Rudolf [1967] discussed in Chapter 13. Such a distribution may arise if a study is designed with fixed numbers in certain groups initially and is continued until a specific proportion of all individuals achieve a particular value of another variable. For instance, groups of male and female patients may be assigned to different treatment groups and observations taken on them until half the total number of patients are discharged, at which point data collection ceases. Here the two-dimensional sex × treatment configuration is fixed and also the one-dimensional survival/discharge configuration.

In the next section we show that when $\sum_\theta m_\theta = N$, the kernel of the likelihood function is identical for the first three sampling schemes (see theorem 13.4-1 of Chapter 13 for proof). The likelihood differs when more than one configuration is fixed; Haberman [1974] obtained asymptotic results for this situation. The practical implications are that we may use the estimation procedures appropriate for the other sampling schemes, provided we stay within a restricted subset of models. The resulting estimates are close to the appropriate maximum likelihood estimates for large samples, and the usual measures of goodness of fit are asymptotically distributed as χ^2.

3.3 Sufficient Statistics

In this section we show that for the most common sampling schemes the sufficient statistics are trivially easy to obtain. They are the configurations of sums that correspond to the u-terms of the log-linear model. We then discuss the implications of Birch's [1963] results. These results enable us to derive estimates of the expected cell counts for all the elementary cells from the configurations of sums, without going through the intermediate step of estimating the u-terms themselves.

We deal first with Poisson and simple multinomial sampling in three dimensions and derive sufficient statistics for a selection of three-dimensional models. The selection is made according to the hierarchy principle, a necessary condition for the applicability of Birch's results. Generalization to any dimension and extension to product-multinomial sampling follows discussion of these results.

3.3.1 *Three dimensions and Poisson or simple multinomial sampling*

To derive the sufficient statistics we need to relate the log-linear model to the likelihood function. We derived the fully saturated model for three dimensions and discussed the interpretation of the u-terms in Chapter 2. The model for the expected cell count m_{ijk} is

$$\log m_{ijk} = u + u_{1(i)} + u_{2(j)} + u_{3(k)} + u_{12(ij)} + u_{13(ik)} + u_{23(jk)} + u_{123(ijk)}. \quad (3.3\text{-}1)$$

Each of the seven subscripted u-terms sums to zero over each lettered subscript $(i, j, \text{or } k)$, ensuring that the number of independent parameters equals the number of elementary cells. Thus the model fully defines the structure of any complete three-dimensional array.

The log-likelihood of the multinomial is readily obtained from the PDF (3.2-2) and may be written as

$$\log\left\{\frac{N!}{\displaystyle\prod_{i,j,k} x_{ijk}!}\right\} + \sum_{i,j,k} x_{ijk} \log m_{ijk} - N \log N. \tag{3.3-2}$$

The first and last terms are constant for any set of m, and we need only consider the remaining term, the kernel of this function.

Similarly, under Poisson sampling, from expression (3.2-1) we can distinguish three components of the log-likelihood:

$$\sum_{i,j,k} x_{ijk} \log m_{ijk} - \sum_{i,j,k} m_{ijk} - \sum_{i,j,k} \log(x_{ijk}!). \tag{3.3-3}$$

We may ignore the two negative components because the condition $\sum \hat{m}_{ijk} = N$ holds for all models of interest, and terms in x_{ijk} alone are constant for any model. We thus have the distribution proportional to the first term, which is the same kernel as for the single multinomial. This result is given in more general terms in Chapter 13, theorem 13.4-1.

Substituting for m_{ijk} from (3.3-1) into (3.3-2) or (3.3-3), we obtain for the kernel

$$\begin{aligned}
\sum_{i,j,k} x_{ijk} \log m_{ijk} = Nu &+ \sum_i x_{i++} u_{1(i)} + \sum_j x_{+j+} u_{2(j)} + \sum_k x_{++k} u_{3(k)} \\
&+ \sum_{i,j} x_{ij+} u_{12(ij)} + \sum_{i,k} x_{i+k} u_{13(ik)} \\
&+ \sum_{j,k} x_{+jk} u_{23(jk)} + \sum_{i,j,k} x_{ijk} u_{123(ijk)}.
\end{aligned} \tag{3.3-4}$$

The multinomial belongs to the class of exponential PDF's. It is a well-known result that for this class of functions the sufficient statistics are the x-terms adjacent to the unknown parameters, the u-terms. See, for instance, Hogg and Craig [1968] or Zacks [1971]. This formulation of the kernel of the log-likelihood is based on the saturated model and so is completely general. When we consider unsaturated models terms drop out and those that remain give the sufficient statistics. We describe below the procedure in detail for two hierarchical models. In practice the set of nonredundant or minimal sufficient statistics can be obtained by inspection of the log-linear model alone without writing out the expression comparable to (3.3-4).

No three-factor effect

If we put $u_{123(ijk)} = 0$ for all i, j, k, or more briefly $u_{123} = 0$, we are postulating a model with all two-factor effects present but no three-factor effect. The last term in expression (3.3-4) disappears and, as both the multinomial and Poisson distributions belong to the exponential family, the term N and the configurations with members x_{i++}, x_{+j+}, x_{++k}, x_{ij+}, x_{i+k}, and x_{+jk} are the sufficient statistics for this model. The last three configurations yield the others and form the complete *minimal set* (Haberman [1973a]). We say that the *minimal* sufficient statistics are C_{12}, C_{13}, and C_{23}.

In practice, the log-linear model is inspected to obtain the minimal set of sufficient statistics. Dropping the second set of subscripts for convenience, we write the model with families of terms grouped:

$$\begin{aligned}
\log m_{ijk} = u &+ (u_1 + u_2 + u_{12}) + (u_1 + u_3 + u_{13}) + (u_2 + u_3 + u_{23}) \\
&- (u_1 + u_2 + u_3).
\end{aligned}$$

In the generalized notation this becomes

$$\log m_{ijk} = u + U_{12} + U_{13} + U_{23} - (u_1 + u_2 + u_3), \qquad (3.3\text{-}5)$$

and expression (3.3-4) becomes

$$\sum_{i,j,k} x_{ijk} \log m_{ijk} = Nu + \left[\sum_{i,j} x_{ij+} U_{12} + \sum_{i,k} x_{i+k} U_{13} + \sum_{j,k} x_{+jk} U_{23} \right]$$
$$- \left[\sum_i x_{i++} u_1 + \sum_j x_{+j+} u_2 + \sum_k x_{++k} u_3 \right]. \qquad (3.3\text{-}6)$$

The terms in the first set of brackets give the minimal set of sufficient statistics because the terms in the second set of brackets are redundant. The configurations corresponding to U_{12}, U_{13}, and U_{23} are C_{12}, C_{13}, and C_{23}. We give below a generalized procedure for obtaining the families of u-terms corresponding to the minimal set of sufficient statistics. In practice, the minimal set can usually be determined from inspection of the model without formally going through the procedure of grouping the terms into families.

One two-factor effect absent

When we remove one more term by putting $u_{12} = u_{123} = 0$, the natural logarithm of the kernel of the likelihood function written in the form of expression (3.3-6) is

$$\sum_{i,j,k} x_{ijk} \log m_{ijk} = Nu + \left[\sum_{i,k} x_{i+k} U_{13} + \sum_{j,k} x_{+jk} U_{23} \right]$$
$$- \left[\sum_k x_{++k} u_3 \right]. \qquad (3.3\text{-}7)$$

Again, the x-terms in the first set of brackets give the minimal set of sufficient statistics because the terms in the second set of brackets are redundant: If we know either $\{x_{i+k}\}$ or $\{x_{+jk}\}$ we can readily sum to obtain $\{x_{++k}\}$. The sufficient configurations corresponding to U_{13} and U_{23} are C_{13} and C_{23}.

Correspondence of u-terms and configurations

In table 3.3-1 we list the four types of comprehensive, hierarchical models that can be fitted to three-dimensional data and give the corresponding sufficient configurations, to show the parallelism between the u-terms that are present in the model and the set of sufficient statistics. The models possible in three dimensions have the following interpretations.

Table 3.3-1   Complete Models in Three Dimensions

	Absent Terms	Sufficient Configurations	Degrees of Freedom
Model Type			
1	u_{123}	C_{12}, C_{13}, C_{23}	$(I-1)(J-1)(K-1)$
2	u_{12}, u_{123}	C_{13}, C_{23}	$(I-1)(J-1)K$
3	u_{12}, u_{13}, u_{123}	C_{23}, C_1	$(I-1)(JK-1)$
4	$u_{12}, u_{13}, u_{23}, u_{123}$	C_1, C_2, C_3	$IJK - (I+J+K) + 2$

1. *No three-factor effect*

For the no three-factor effect model we set $u_{123(ijk)} = 0$ for all i, j, k, or more briefly $u_{123} = 0$. This states that there is no three-factor effect or, in other words, each two-factor effect is unaffected by the level of the third variable.

2. *One two-factor effect absent*

We set $u_{12} = 0$, which states that there is no interaction between variables 1 and 2 at any level of variable 3, and hence, in hierarchical models, necessarily implies $u_{123} = 0$. In other words, it postulates conditional independence of variables 1 and 2, but, by not specifying the relationship, assumes that u_{13} and u_{23} are present. By interchanging variables, or otherwise, we may define three versions of this model. In each version, one of the three possible two-factor effects u_{12}, u_{13}, or u_{23} is set equal to zero, while the remaining two are retained.

3. *Two two-factor effects absent*

We set $u_{12} = u_{13} = 0$, and by implication $u_{123} = 0$. This model states that variable 1 is independent of the other variables, which are associated with each other. Again, three versions are possible.

4. *No two-factor effects*

We set $u_{12} = u_{13} = u_{23} = u_{123} = 0$, which states that there is no association. The three variables are completely independent.

5. *Noncomprehensive models*

If we continue to delete u-terms, we get incomplete models that do not include all variables. For instance, if $u_1 = 0$ then we are stating that the effect of this variable is simply to divide the configuration C_{23} into I identical subtables. It is important to recognize when such models are appropriate because they may help us to reduce the dimensionality of the data. However, they present no estimation problems, and so this chapter deals only with *comprehensive* models, namely, models that include at least the main effects for every variable.

6. *Nonhierarchical models*

We derived the set of models above by first setting the highest-order interaction equal to zero and then progressing to the lower-order terms. If we had set $u_{123} = 0$ and $u_1 = 0$ but retained all the two-factor effects, we would not have been able to write the model in the form of expression (3.3-5). This model violates the hierarchy principle, which we discussed informally in Chapter 2 and define formally now. The development of the remainder of this chapter is confined to hierarchical models.

3.3.2   *The hierarchy principle*

In Chapter 2 we showed how successively higher-order u-terms can be regarded as measuring deviations from lower-order terms. Thus in two dimensions we can write

$$u_{12(ij)} = \log m_{ij} - u - u_{1(i)} - u_{2(j)},$$

and in order to compute its value, we first compute u, $u_{1(i)}$, and $u_{2(j)}$. It is thus reasonable to limit our models to a hierarchical set; higher-order terms may only be included if the related lower-order terms are included. Thus the presence of u_{12}

implies that u_1 and u_2 are included in the model. Conversely, if a lower-order term is absent, then all its higher-order relatives must be absent. Thus if u_{12} is absent, so are u_{123}, u_{124}, and any other term that has both subscripts 1 and 2. This may be stated more formally as follows:

> Consider the general term u_{θ_1, θ_2}, where θ_1 and θ_2 correspond to any two mutually exclusive sets of subscripts. If u_{θ_1, θ_2} is present in the model, then u_{θ_1} and u_{θ_2} must be present. Conversely, if u_{θ_1} is set equal to zero, then u_{θ_1, θ_2} must be set equal to zero for all θ_2.

3.3.3   *Generalized method of deriving sufficient statistics*

The preceding discussion has defined the set of models that are comprehensive, unsaturated, and hierarchical. We deal only with this set until we reach Section 3.7. For each model, to obtain the minimal set of sufficient statistics we select the u-terms of order t, where t is the highest-order interaction exhibited by the model. If all possible terms of order t are included in the model, we stop our selection; otherwise, we continue by examining terms of order $t - 1$ and selecting those terms that are not lower-order relatives of the terms of order t. This process is continued for u-terms of every order, and at each step only those u-terms that are not included in higher-order terms are selected. Suppose the final set of s terms selected is $u_{\theta_1}, u_{\theta_2}, \ldots, u_{\theta_s}$. Then from the hierarchy principle, we may rewrite the model as

$$\log m_\theta = u + [U_{\theta_1} + U_{\theta_2} + \ldots + U_{\theta_s}] - [u_{\varphi_1} + u_{\varphi_2} + \ldots + u_{\varphi_r}]. \quad (3.3\text{-}8)$$

Here the r terms of the form u_{φ_t} are the redundant terms introduced by changing from the single term u_{θ_t} to the term U_{θ_t} that represents u_{θ_t} and all its lower-order relatives. Some of these terms may be duplicates. For instance, U_{12} and U_{13} have u_1 in common, and so do U_{14} and U_{15}. Thus for the model in five dimensions that includes all two-factor effects involving variable 1 but no other multiple-factor terms:

$$\log m_{ijklm} = u + u_{1(i)} + u_{2(j)} + u_{3(k)} + u_{4(l)} + u_{5(m)}$$
$$+ u_{12(ij)} + u_{13(ik)} + u_{14(il)} + u_{15(im)},$$

the highest-order terms are the two-factor terms, u_{12}, u_{13}, u_{14}, and u_{15}. If we replace u_{12} by U_{12} we also replace u_1 and u_2, and if we replace all the two-factor terms we include u_1 four times. So we must remove the three excess values, and our equation becomes

$$\log m_\theta = u + [U_{12} + U_{13} + U_{14} + U_{15}] - [u_1 + u_1 + u_1]. \quad (3.3\text{-}9)$$

This possible duplication of u_φ-terms does not prevent us from writing the kernel of the log-likelihood as

$$\sum_\theta x_\theta \log m_\theta = Nu + [\sum_{\theta_1} x_{\theta_1} U_{\theta_1} + \sum_{\theta_2} x_{\theta_2} U_{\theta_2} + \ldots + \sum_{\theta_s} x_{\theta_s} U_{\theta_s}]$$
$$- [\sum_{\varphi_1} x_{\varphi_1} u_{\varphi_1} + \sum_{\varphi_2} x_{\varphi_2} u_{\varphi_2} + \ldots + \sum_{\varphi_r} x_{\varphi_r} u_{\varphi_r}],$$

which can be more briefly expressed as

$$\sum_\theta x_\theta \log m_\theta = Nu + \sum_{t=1}^{s} \sum_{\theta_t} x_{\theta_t} U_{\theta_t} - \sum_{q=1}^{r} \sum_{\varphi_q} x_{\varphi_q} u_{\varphi_q}. \quad (3.3\text{-}10)$$

As the configurations with cells x_{φ_q} are contained in at least two of the configurations C_{θ_t}, they are redundant, and we are left with the configurations C_{θ_t}, of which there are s, as the set of minimal sufficient statistics. For the model (3.3-9), they are C_{12}, C_{13}, C_{14}, and C_{15}. This method of progressively selecting the highest-order u-terms gives us the corresponding minimal set of sufficient statistics for Poisson or simple multinomial sampling. It can be extended to product multinomial sampling provided the model is appropriate; for appropriate models, one of the U_{θ_t} or one of its lower-order relatives must correspond to the fixed configuration C_{θ_t}, as we show later.

3.3.4   *Birch's results*

Once a set of minimal sufficient statistics has been defined, we readily obtain maximum likelihood estimates by using two results of Birch. Before stating the results in general, let us consider the implications in three dimensions. For the model with $u_{123} = 0$, the sufficient configurations are C_{12}, C_{13}, and C_{23} with cell entries x_{ij+}, x_{i+k}, and x_{+jk}.

Birch showed that the maximum likelihood estimates of m_{ij+}, m_{i+k}, and m_{+jk} must correspond to these observed values, in other words, for all i, j, k,

$$\hat{m}_{ij+} = x_{ij+},$$

$$\hat{m}_{i+k} = x_{i+k},$$

$$\hat{m}_{+jk} = x_{+jk}. \qquad (3.3\text{-}11)$$

The second result of Birch states that there is a unique set of elementary cell estimates that satisfies the conditions of the model and these marginal constraints. Thus we know that there exists a set of estimates $\{\hat{m}_{ijk}\}$ that satisfies the equations (3.3-11) and also the hypothesis of no three-factor effect, defined as the equality of two-factor effects:

$$\frac{m_{ijk}m_{rsk}}{m_{rjk}m_{isk}} = \frac{m_{ijt}m_{rst}}{m_{rjt}m_{ist}} \qquad (3.3\text{-}12)$$

for $i \neq r, j \neq s, k \neq t$. Moreover, there is only one set of estimates satisfying relations (3.3-11) and (3.3-12), and this set maximizes the likelihood function. In Sections 3.4 and 3.5 we discuss methods of computing these cell estimates from the sufficient configurations. For the moment we state these results more formally before proceeding to consider product multinomial sampling.

RESULT 1   If x_{θ_i} is a member of the set of sufficient statistics, then the maximum likelihood estimate $\hat{m}_{\theta_i}$ of the sum m_{θ_i} is x_{θ_i}.

RESULT 2   There is a unique solution giving positive ML estimates $\hat{m}_\theta$ for every elementary cell in the original parallelepiped. This solution satisfies both the constraints of the model and the condition $\hat{m}_{\theta_i} = x_{\theta_i}$ for all θ_i such that x_{θ_i} is a member of the set of sufficient statistics.

To derive these results, Birch assumed that all the observed elementary cell counts are positive, but this is too restrictive, as we saw in Section 3.1. When the model is such that cell estimates can be derived directly, positive values for the sufficient statistics ensure that unique positive estimates are obtained for every elementary cell. When direct estimates are not available, it can sometimes occur that the sufficient statistics are all positive, but the constraints they impose can only be satisfied by zero estimates in some of the elementary cells that contained zero observations. Haberman [1974] has some results along this line.

Example (3.3-1)   Random zeros arranged so that the no-three-factor-effect model cannot be tested

Suppose that the observed table is

0	b
c	d

e	f
g	0

where b, c, d, e, f, and g are positive counts. Under the model $u_{123} = 0$, we obtain expected counts that have the observed marginal sums C_{12}, C_{13}, and C_{23}. So if we have a positive estimate Δ in the first cell, the other cells are determined by it, and we have

Δ	$b - \Delta$
$c - \Delta$	$d + \Delta$

$e - \Delta$	$f + \Delta$
$g + \Delta$	$-\Delta$

This gives us a negative value in the last cell. It is clearly impossible to find a value of Δ that yields a positive estimate for both cells with zero observed counts. In this example we cannot estimate u_{123} from the observed counts, and thus it is not surprising that we cannot test its magnitude. If we try to fit this model to these data, using methods described in Section 3.5-1 we find that the fitted values are equal to the observed values, in other words, $\Delta = 0$. We are alerted to our use of an invalid model when we try to compute the degrees of freedom by using expression (3.8-1) and obtain a negative value! ■■

3.3.5   Product multinomial sampling

The implications of Birch's results are that if C_{θ_1} is a sufficient statistic for a model under a Poisson or simple multinomial sampling scheme, then the cell estimates are such that $\hat{m}_{\theta_1} = x_{\theta_1}$. In other words, the configuration is fixed by the model. Conversely, in product multinomial sampling, we have a configuration C_{θ_2} which is fixed by the sampling design. For consistency, for this sampling scheme we consider only those models for which C_{θ_2} is fixed, which means that we must include the term u_{θ_2} in the model. If we confine ourselves to this subset of models, all the preceding results for Poisson and simple multinomial sampling also apply to product multinomial sampling because we can either fit the same model to each of the independent multinomials or, by formulating the model differently, fit one overall model. Instead of a formal proof we illustrate with a three-dimensional example—generalization requires only a change of subscripts.

Example (3.3-2)   Defining the subset of models appropriate for product multinomial sampling

Suppose in three dimensions the configuration C_1 is fixed. Then from expression

(3.2-4) we have a distribution that may be factored into I independent multinomials. If we fit the overall model with $u_{12} = u_{123} = 0$, we write

$$\log m_{ijk} = u + u_{1(i)} + u_{2(j)} + u_{3(k)} + u_{13(ik)} + u_{23(jk)}, \qquad (3.3\text{-}13)$$

and the minimal set of sufficient configurations consists of C_{13} and C_{23}. We know that the cell estimates for this model are

$$\hat{m}_{ijk} = \frac{x_{i+k} x_{+jk}}{x_{++k}}. \qquad (3.3\text{-}14)$$

If we consider the I independent multinomials, the corresponding model for the ith multinomial is

$$\log m_{jk}^{(i)} = w^{(i)} + w_{2(j)} + w_{3(k)}^{(i)} + w_{23(jk)} \qquad (3.3\text{-}15)$$

where the superscript (i) is appended to those terms that formerly included variable 1 and the cell m_{ijk} is represented as $m_{jk}^{(i)}$. The w-terms, like u-terms, sum to zero over their lettered subscripts. We can establish the equivalence of the two formulations by summing. Summing expressions (3.3-13) and (3.3-14) over j and k gives

$$u + u_{1(i)} = w^{(i)}. \qquad (3.3\text{-}16)$$

Summing expressions (3.3-13) and (3.3-14) over j and substituting from (3.3-16) gives

$$u_{3(k)} + u_{13(ik)} = w_{3(k)}^{(i)}, \qquad (3.3\text{-}17)$$

and similarly, summing over k and substituting from (3.3-16) gives

$$u_{2(j)} = w_{2(j)}, \qquad (3.3\text{-}18)$$

and so we have for the remaining terms

$$u_{23(jk)} = w_{23(jk)}. \qquad (3.3\text{-}19)$$

The kernel of the log-likelihood for the product of the I multinomials corresponding to model (3.3-15) may be written

$$\sum_i x_{ijk} \log m_{jk}^{(i)} = \sum_i \left[x_{i++} w^{(i)} + \sum_j x_{ij+} w_{2(j)} + \sum_k x_{i+k} w_{3(k)}^{(i)} + \sum_{j,k} x_{ijk} w_{23(jk)} \right]$$

$$= \sum_i \left[x_{i++} w^{(i)} + \sum_k x_{i+k} w_{3(k)}^{(i)} \right] + \sum_j x_{+j+} w_{2(j)} + \sum_{j,k} x_{+jk} w_{23(jk)}. \qquad (3.3\text{-}20)$$

The minimal sufficient configuration for the last two terms is C_{23} and for the first two is $C_3^{(i)}$. The elements of $C_3^{(i)}$ are $\{x_{i+k}\}$, and so for the complete set of multinomials, $\{C_3^{(i)}\} = C_{13}$. Thus either formulation gives the same sufficient statistics. This equivalence persists for all models that include $u_{1(i)}$.

If we fit a model that does not include variable 1 in the overall formulation equivalent to (3.3-13), the sufficient configurations do not include the sums x_{i++}. Estimates under this model consist of I identical arrays with $\hat{m}_{i++} = N/I$ for all i, which is contrary to the sampling scheme. ∎∎

3.3.6   *Exercises*

1. In five dimensions, if we set $u_{12} = 0$ the hierarchy principle implies that we must set other u-terms equal to zero.
   (i) Which are they?
   (ii) What are the minimal sufficient statistics for this model?
   (Answer: (i) All higher-order relatives, u_{123}, u_{124}, u_{125}, u_{1234}, u_{1235}, u_{1245}, u_{12345}. (ii) C_{1345} and C_{2345}.)

2. In four dimensions, determine which of the following models are (i) comprehensive; (ii) hierarchical.
   (a) $\log m_{ijkl} = u + u_{1(i)} + u_{2(j)} + u_{3(k)} + u_{34(kl)}$,
   (b) $\log m_{ijkl} = u + u_{1(i)} + u_{2(j)} + u_{3(k)} + u_{12(ij)} + u_{13(ik)}$,
   (c) $\log m_{ijkl} = u + u_{1(i)} + u_{2(j)} + u_{3(k)} + u_{4(l)} + u_{123(ijk)}$.

   (Answer: (i) All comprehensive except (b), which does not include variable 4. (ii) (b) is hierarchical, (a) does not include $u_{4(l)}$, and (c) does not include the three two-factor terms implied by the last term.)

3. Suppose that in four dimensions the configuration C_{12} is fixed by the sampling plan.
   (i) If the overall model leads to the sufficient configurations C_{123} and C_{124}, which u-terms have been set equal to zero?
   (ii) If all variables are dichotomous, how many independent multinomials exist?
   (iii) The above overall model is equivalent to the following individual models for each multinomial:

   $$\log m_{kl}^{(ij)} = u^{(ij)} + u_{3(k)}^{(ij)} + u_{4(l)}^{(ij)}.$$

This may be established by equating terms; for instance

$$u = \sum_{i,j} \frac{u^{(ij)}}{IJ}, \qquad u_{1(i)} = \left(\sum_{j} \frac{u^{(ij)}}{J}\right) - u, \qquad u_{2(j)} = \left(\sum_{i} \frac{u^{(ij)}}{I}\right) - u,$$

$$u_{12(ij)} = u^{(ij)} - u_{1(i)} - u_{2(j)} - u.$$

Find the equivalent terms for $u_{4(l)}$, $u_{14(il)}$, $u_{24(jl)}$, and $u_{124(ijl)}$.
(Answer: (i) $u_{1234} = u_{234} = u_{134} = u_{34} = 0$. (ii) $I \times J = 4$. (iii) Relevant sums are obtained starting with

$$\sum_{i,j,k} \log m_{ijkl} = IJK(u + u_{4(l)}) = \sum_{i,j,k} \log m_{kl}^{(ij)} = K\left(IJu + \sum_{i,j} u_{4(l)}^{(ij)}\right),$$

which gives $u_{4(l)} = \sum_{i,j} u_{4(l)}^{(ij)}/IJ$. Other sums give

$$u_{14(il)} = \sum_{j} \frac{u_{4(l)}^{(ij)}}{J} - \sum_{i,j} \frac{u_{4(l)}^{(ij)}}{IJ},$$

$$u_{24(jl)} = \sum_{i} \frac{u_{4(l)}^{(ij)}}{I} - \sum_{i,j} \frac{u_{4(l)}^{(ij)}}{IJ},$$

$$u_{124(ijl)} = u_{4(l)}^{(ij)} - \left(\sum_{i} \frac{u_{4(l)}^{(ij)}}{I} + \sum_{j} \frac{u_{4(l)}^{(ij)}}{J} - \sum_{i,j} \frac{u_{4(l)}^{(ij)}}{IJ}\right).)$$

4. (i) Count the number of log-linear models in four dimensions that are unsaturated, comprehensive, and hierarchical.
(ii) How many are permissible if C_{12} is fixed by the sampling plan?
(Answer: (i) Given by table 3.4-2 as 113. (ii) Derived by reducing the number of
each version to give 67.)

3.4   Methods of Obtaining Maximum Likelihood Estimates

In the preceding section we showed that a unique set of ML estimates for every
elementary cell can be derived from the sufficient statistics alone. Now we consider
methods of deriving these estimates.

For some models, cell estimates may be written directly as functions of the
sufficient statistics. We first give examples in three or four dimensions, then give
a simple procedure for detecting when such direct estimates exist. Proofs establishing the validity of the procedure follow.

When direct estimates do not exist, iterative procedures must be used. We
describe these in Section 3.5 and then give examples of some practical procedures
that have computational advantages.

3.4.1   *Two types of models*

First we consider the well-known estimates for two independent variables in a
framework that enables us to generalize to any dimension. In two dimensions, if
we put $u_{12} = 0$, the sufficient configurations C_1 and C_2 have cells x_{i+} and x_{+j},
respectively, and we know from Birch's first result that $x_{i+} = \hat{m}_{i+}$ and $x_{+j} = \hat{m}_{+j}$.
From our model (3.1-2) we have

$$m_{i+} = e^{u + u_{1(i)}}\left(\sum_j e^{u_{2(j)}}\right),$$

$$m_{+j} = e^{u + u_{2(j)}}\left(\sum_i e^{u_{1(i)}}\right),$$

$$N = e^u\left(\sum_i e^{u_{1(i)}}\right)\left(\sum_j e^{u_{2(j)}}\right),$$

and hence

$$\frac{(m_{i+})(m_{+j})}{N} = e^{u + u_{1(i)} + u_{2(j)}} = m_{ij}.$$

The corresponding function of x_{i+} and x_{+j} gives the elementary cell estimates

$$\hat{m}_{ij} = \frac{x_{i+}x_{+j}}{N}. \tag{3.4-1}$$

A similar procedure enables us to derive the cell estimates directly whenever we
have only one or two configurations and also in some instances when we have
more than two configurations.

When such direct estimates do not exist, we can use an iterative proportional
fitting algorithm to get the expected cell values. This algorithm has the property

that, when used for models that do not require iteration, it yields the direct estimates *at the end of the first cycle*. There is a proviso that the configurations must be
fitted in suitable order for certain models in seven or more dimensions (see
Exercise 1 in Section 3.5.3), otherwise more cycles will be required, but any order
will suffice in lower dimensions.

If we wish to fit a model to a large body of data using a computer, we do not need
to determine whether the model has closed-form estimates; the iterative procedure
always produces the correct estimates. For smaller quantities of data, it may be
useful to make the distinction, as the direct estimates can then be readily computed
by hand, whereas the iterative procedure is sufficiently tedious for recourse to be
had to a computer. The most important reason for distinguishing between the two
models is not, however, related to the ease of computing, but to another property:
when direct estimates exist, the log-likelihood of the fitted model can also be
obtained directly. In Chapter 4 we show how this feature may be exploited in
several ways. We therefore describe closed-form estimates in more detail and give
rules for identifying models that yield direct estimates before describing the general
iterative computing procedure.

3.4.2   *Direct estimates in three and four dimensions*

Let us look first at the three-dimensional model 2 of table 3.3-1 with $u_{12} = u_{123} = 0$.
By examining the model for the cells of the sufficient configurations, we obtain
the direct estimates. The model for the elementary cells is

$$m_{ijk} = \exp(u + u_{1(i)} + u_{2(j)} + u_{3(k)} + u_{13(ik)} + u_{23(jk)}). \tag{3.4-2}$$

The elements x_{i+k} and x_{+jk} of the sufficient configurations C_{13} and C_{23} are estimates of

$$m_{i+k} = \exp(u + u_{1(i)} + u_{3(k)} + u_{13(ik)}) \sum_j \exp(u_{2(j)} + u_{23(jk)}), \tag{3.4-3}$$

$$m_{+jk} = \exp(u + u_{2(j)} + u_{3(k)} + u_{23(jk)}) \sum_i \exp(u_{1(i)} + u_{13(ik)}), \tag{3.4-4}$$

and they have in common the vector C_3 with cells x_{++k}, which are estimates of

$$m_{++k} = \exp(u + u_{3(k)}) \sum_{i,j} \exp(u_{1(i)} + u_{2(j)} + u_{13(ik)} + u_{23(jk)})$$

and may be factored into components

$$m_{++k} = \exp(u + u_{3(k)}) \sum_j \exp(u_{2(j)} + u_{23(jk)}) \sum_i \exp(u_{1(i)} + u_{13(ik)}). \tag{3.4-5}$$

When we divide the product $(m_{i+k})(m_{+jk})$ by m_{++k}, the summed terms cancel, and
we obtain the relationship

$$m_{ijk} = \frac{(m_{i+k})(m_{+jk})}{m_{++k}}. \tag{3.4-6}$$

Thus we can directly obtain the estimates

$$\hat{m}_{ijk} = \frac{x_{i+k} x_{+jk}}{x_{++k}} \tag{3.4-7}$$

for every elementary cell.

In effect, we take each of K two-way tables relating variables 1 and 2 and derive the cell estimates for each table from the one-dimensional margins of sums for the table, in the manner familiar for two-way tables:

$$\frac{(\text{row sum})(\text{column sum})}{\text{table sum}}.$$

Analogous models with one two-factor term and higher-order relatives absent occur frequently in three and higher dimensions, and their cell estimates may be derived in the same way.

We give the form of the estimates for each type of model yielding direct estimates in four dimensions in table 3.4-1. These expressions may be derived by the method of equations (3.4-2)–(3.4-5): (i) write expressions for the cells of each sufficient configuration; (ii) if overlapping subconfigurations exist, write expressions for them and factor them if possible; (iii) examine the expressions in (i) and (ii) to determine whether a direct estimate exists. The general form of the direct estimates is predictable: the numerator has entries from each sufficient configuration; the denominator, entries from redundant configurations caused by overlapping; terms in powers of N appear either in the numerator or in the denominator to ensure the right order of magnitude.

Table 3.4-1   Types of Direct Estimates Possible in Four Dimensions

Reference Number[a]	Sufficient Configurations	Direct Cell Estimate
4	C_{123}, C_{124}	$\dfrac{x_{ijk} + x_{ij+l}}{x_{ij++}}$
7	C_{123}, C_{14}	$\dfrac{x_{ijk} + x_{i++l}}{x_{i+++}}$
8	C_{123}, C_4	$\dfrac{x_{ijk} + x_{+++l}}{N}$
12a	C_{12}, C_{13}, C_{14}	$\dfrac{x_{ij++} + x_{i+k+} + x_{i++l}}{(x_{i+++})^2}$
12b	C_{12}, C_{13}, C_{24}	$\dfrac{x_{ij++} + x_{i+k+} + x_{+j+l}}{x_{i+++} + x_{+j++}}$
13a	C_{12}, C_{34}	$\dfrac{x_{ij++} + x_{++kl}}{N}$
13b	C_{12}, C_{13}, C_4	$\dfrac{x_{ij++} + x_{i+k+} + x_{+++l}}{x_{i+++} N}$
14	C_{12}, C_3, C_4	$\dfrac{x_{ij++} + x_{++k+} + x_{+++l}}{N^2}$
15	C_1, C_2, C_3, C_4	$\dfrac{x_{i+++} + x_{+j++} + x_{++k+} + x_{+++l}}{N^3}$

[a] Refers to table 3.4-2.

Sometimes no combination of the sufficient statistics yields a direct estimate. In expression (3.4-5) for m_{++k}, those terms summed over variable 1 can be separated from those summed over variable 2, and this factorability enables us to cancel out the summed terms of m_{++k} with the summed terms for m_{i+k} and m_{+jk}. In models where the redundant configurations cannot be factored, we cannot find a direct function of the configuration cells that yields an estimate for m_{ijk}. Thus the three-dimensional model with $u_{123} = 0$ (model 1 of table 3.3-1) has no direct estimate.

Closed loops

We can think of two overlapping configurations as being linked to each other. The model $u_{123} = 0$ mentioned above has configurations C_{12}, C_{23}, and C_{13}, and if we link each overlapping pair we have a closed loop. In any dimension, iterative methods are necessary when such closed loops exist, whether all or only some of the configurations are involved in the loop. For example, in five dimensions we can fit C_{12}, C_{23}, C_{34}, and C_{45} directly, but we are forced to use iterative methods if we add a further two-dimensional configuration to these four; for instance, adding C_{13} gives a loop connecting C_{12}, C_{23}, and C_{13}. We can, however, obtain direct estimates from the two configurations C_{123} and C_{345} even though this model has six two-factor effects because the configurations themselves do not make a loop. The rules we give in the next section enable us to detect such loops automatically.

3.4.3   *Rules for detecting existence of direct estimates*

The following rules for examining configurations allow us to determine whether a model may be fitted directly and are applicable in any number of dimensions. We refer to them as the *classifying* rules.

Step 1.   Relabel any group of variables that always appear together as a single variable.

Step 2.   Delete any variable that appears in every configuration.

Step 3.   Delete any variable that only appears in one configuration.

Step 4.   Remove any redundant configurations.

Step 5.   Repeat steps 1–4 until
   (i)   no more than two configurations remain, an indication that a closed-form estimate is available; or
   (ii) no further steps can be taken, an indication that no closed-form estimate exists.

Step 1 is only a simplification to assist in subsequent steps. Step 2 is self-evident if we remember that fixing configurations with s cells in common leads to s independent distributions. Thus if the model can be fitted for any single category of the repeated variable, it can be fitted for every category, and hence we do not have to consider such variables. In step 3 we remove unique variables because they do not appear in the subconfigurations caused by overlapping sufficient configurations, and it is the nonfactorability of subconfigurations that creates the need for indirect estimation.

Step 4 is needed because steps 1–3 may leave some redundant configurations. For instance, in the four-dimensional model 12b of table 3.4-1, the original set is

$$C_{12}, C_{13}, C_{24},$$

and by applying step 3 we have

$$C_{12}, C_{13}, C_2.$$

Now C_2 is redundant, so by applying step 4 we are left with

$$C_{12}, C_{13}.$$

If we were to continue to cycle through steps 1–3 we would apply step 2 and obtain

$$C_2, C_3,$$

and reapplying step 3 would remove these terms. As two configurations can always be removed entirely, it is not necessary to proceed any further when the number has been reduced to 2.

If we cannot get less than three configurations, we must iterate. Consider the three-dimensional model 1 from table 3.3-1 for which the configurations are

$$C_{12}, C_{13}, C_{23}.$$

None of the steps 1–4 can be applied, and the cell estimates can only be obtained by an iterative procedure. We reach the same impasse with

$$C_{123}, C_{234}, C_{14}.$$

This set can only be simplified at step 1 by relabeling 23 as $2'$, to give

$$C_{12'}, C_{2'4}, C_{14},$$

which is another closed loop requiring iteration.

Table 3.4-2   Hierarchical Models in Four Dimensions

	Number of Effects Removed[a]			Degrees of Freedom[b]	Configurations of Each Dimension[a]			Versions	Iterative (I) or Directly Multiplicative (M)
Model	4f	3f	2f		3D	2D	1D		
1	1			1	4			1	I
2	1	1		2	3			4	I
3	1	2		3	2	1		6	I
4	1	2	1	4	2			6	M
5	1	3		4	1	3		4	I
6	1	3	1	5	1	2		12	I
7	1	3	2	6	1	1		12	M
8	1	3	3	7	1		1	4	M
9	1	4		5	6			1	I
10	1	4	1	6	5			6	I
11	1	4	2	7	4			15	I
12a	1	4	3	8	3			4	M
b					3			12	M
c					3	1		4	I
13a	1	4	4	9	2			3	M
b					2	1		12	M
14	1	4	5	10	1	2		6	M
15	1	4	6	11		4		1	M
Total[c]								113	

[a] 4f, 3f, and 2f refer to four-, three-, and two-factor effects, respectively; 3D, 2D, and 1D refer to three-, two-, and one-dimensional configurations, respectively.

[b] Degrees of freedom computed assuming each variable is a dichotomy.

[c] Models that do not involve every variable are not included.

Classification of four-dimensional models

Classification of all the comprehensive hierarchical models in four dimensions in table 3.4-2 shows that for most model types, every version yields estimates that are obtained either by direct multiplication (M) or by an iterative procedure (I). An exception is model type 12, where the terms set equal to zero are the four-factor, all three-factor, and three of the two-factor *u*-terms: only some versions yield direct estimates of the form shown in table 3.4-1; the others, such as the version with sufficient configurations

$$C_{12}, C_{13}, C_{23}, C_4,$$

require iterative fitting.

3.4.4   *Theorems yielding classifying rules*

We now give two theorems, the first of which proves that two configurations always yield direct estimates and the second of which determines when three configurations do not give direct estimates. The rules for determining whether any set of configurations gives direct estimates are a direct outcome of these theorems.

THEOREM 3.4-1   *If the set of minimal sufficient statistics for a model consists of only two configurations, direct estimates exist.*

Proof   The proof is for the three possibilities (i) the configurations are comprehensive and overlapping; (ii) the configurations are comprehensive and nonoverlapping; (iii) the configurations are not comprehensive.

(i)   Let the complete set of subscripts θ be split into three mutually exclusive and comprehensive groups θ_1, θ_2, and θ_3. Then if θ_3 represents variables common to the overlapping configurations C_{φ_1} and C_{φ_2}, we have

$$\varphi_1 = \theta_1 + \theta_3,$$
$$\varphi_2 = \theta_2 + \theta_3. \qquad (3.4-8)$$

Remembering that U_θ represents the sum of all *u*-terms whose subscripts are subsets of θ, we can write the corresponding log-linear model as

$$\log m_\theta = u + U_{\varphi_1} + U_{\varphi_2} - U_{\theta_3}. \qquad (3.4-9)$$

We now proceed, as we did in three dimensions, to derive expressions for the cells of the sufficient configurations and for the cells of the configuration that contains only those variables common to the sufficient configurations. Thus we have

$$m_{\varphi_1} = \sum_{\theta_2} \exp(u + U_{\varphi_1} + U_{\varphi_2} - U_{\theta_3}),$$

where the summation is taken over the variables not present in C_{φ_1}. This gives

$$m_{\varphi_1} = \exp(u + U_{\varphi_1} - U_{\theta_3}) \sum_{\theta_2} \exp(U_{\varphi_2}), \qquad (3.4-10)$$

and similarly,

$$m_{\varphi_2} = \exp(u + U_{\varphi_2} - U_{\theta_3}) \sum_{\theta_1} \exp(U_{\varphi_1}), \qquad (3.4-11)$$

$$m_{\theta_3} = \exp(u - U_{\theta_3}) \sum_{\theta_1} \exp(U_{\varphi_1}) \sum_{\theta_2} \exp(U_{\varphi_2}). \qquad (3.4-12)$$

From equations (3.4-10)–(3.4-12), we find that

$$m_\theta = \frac{m_{\varphi_1} m_{\varphi_2}}{m_{\theta_3}}, \qquad (3.4\text{-}13)$$

and the direct estimates are, correspondingly,

$$\hat{m}_\theta = \frac{x_{\varphi_1} x_{\varphi_2}}{x_{\theta_3}}. \qquad (3.4\text{-}14)$$

(ii)   If the configurations are not overlapping, the set of variables θ_3 does not exist. We may rewrite the model (3.4-10) and derive expressions for m_{φ_1} and m_{φ_2} by deleting the term U_{θ_3} from expressions (3.4-9)–(3.4-11). We can also write an expression for N, the grand total:

$$N = \exp(u) \sum_{\theta_1} \exp(U_{\varphi_1}) \sum_{\theta_2} \exp(U_{\varphi_2}), \qquad (3.4\text{-}15)$$

and hence we have

$$m_\theta = \frac{m_{\varphi_1} m_{\varphi_2}}{N} \qquad (3.4\text{-}16)$$

and the corresponding expression for $\hat{m}_\theta$. In the situation where φ_1 and φ_2 consist of single variables, the estimate is the familiar expression (3.4-1) for two dimensions.

(iii)   If the configurations are not comprehensive, we add another set of subscripts. Thus we have

$$\theta = \theta_1 + \theta_2 + \theta_3 + \theta_4. \qquad (3.4\text{-}17)$$

We define the sufficient statistics φ_1 and φ_2 as in equations (3.4-8) and for convenience define another group

$$\varphi_3 = \theta_1 + \theta_2 + \theta_3. \qquad (3.4\text{-}18)$$

Then we have, from (3.4-14), for a cell of C_{φ_3},

$$\hat{m}_{\varphi_3} = \frac{x_{\varphi_1} x_{\varphi_2}}{x_{\theta_3}}. \qquad (3.4\text{-}19)$$

The model does not contain any of the θ_4 variables when these are not present in the sufficient configurations. These variables therefore do not have any effect on the expected entries in each cell. If L is the total number of cells in C_{θ_4}, we have

$$\hat{m}_\theta = \frac{\hat{m}_{\varphi_3}}{L}. \qquad (3.4\text{-}20)$$

Similarly, if the configurations are not overlapping, we obtain an estimate for m_{φ_3} from (3.4-16) and then use expression (3.4-20). ∎

THEOREM 3.4-2   *If each of three configurations has a different set of variables in common with the other two configurations, direct elementary cell estimates do not exist. Conversely, direct estimation is possible with three configurations if (i) at least two have no variables in common; or (ii) there is one set of variables common to all three, and if these are removed, condition (i) applies.*

Proof   The proof that no direct estimates exist is based on three comprehensive configurations, because extraneous variables were shown in theorem 3.4-1 not to affect the existence of direct estimates.

The complete set of subscripts is divided into four mutually exclusive groups such that $\theta = \theta_1 + \theta_2 + \theta_3 + \theta_4$.

Adding a third configuration C_{φ_3} to the two configurations defined in expression (3.4-8), we have the three sets of subscripts

$$\varphi_1 = \theta_1 + \theta_3,$$

$$\varphi_2 = \theta_2 + \theta_3,$$

$$\varphi_3 = \theta_1 + \theta_2 + \theta_4. \tag{3.4-21}$$

The model may be written as

$$\log m_\theta = u + U_{\varphi_1} + U_{\varphi_2} + U_{\varphi_3} - U_{\theta_1} - U_{\theta_2} - U_{\theta_3}. \tag{3.4-22}$$

To obtain an expression for the cells of C_{φ_1} we must sum over the variables θ_2 and θ_4. Thus, collecting terms that contain these variables, we have

$$m_{\varphi_1} = \exp(u + U_{\varphi_1} - U_{\theta_1} - U_{\theta_3}) \sum_{\theta_2, \theta_4} \exp(U_{\varphi_2} + U_{\varphi_3} - U_{\theta_2}). \tag{3.4-23}$$

We cannot simplify this expression because U_{φ_1} contains not only all the u-terms in U_{θ_1} and U_{θ_3} but also those u-terms whose subscripts contain members of both θ_1 and θ_3, so we cannot subtract U_{θ_1} or U_{θ_3} from U_{φ_1}. Similarly, we cannot sum over U_{φ_2} because it contains u-terms with subscripts from θ_2 and θ_3. By interchanging subscripts 1 and 2, we have an analogous expression for m_{φ_2}:

$$m_{\varphi_2} = \exp(u + U_{\varphi_2} - U_{\theta_2} - U_{\theta_3}) \sum_{\theta_1, \theta_4} \exp(U_{\varphi_1} + U_{\varphi_3} - U_{\theta_1}), \tag{3.4-24}$$

and by summing over variables θ_3 we obtain

$$m_{\varphi_3} = \exp(u + U_{\varphi_3} - U_{\theta_1} - U_{\theta_2}) \sum_{\theta_3} \exp(U_{\varphi_1} + U_{\varphi_2} - U_{\theta_3}). \tag{3.4-25}$$

We can also obtain an expression for the cells of C_{θ_3}, the configuration containing variables common to C_{φ_1} and C_{φ_2}:

$$m_{\theta_3} = \exp(u - U_{\theta_3}) \sum_{\theta_1, \theta_2, \theta_4} \exp(U_{\varphi_1} + U_{\varphi_2} + U_{\varphi_3} - U_{\theta_1} - U_{\theta_2}). \tag{3.4-26}$$

If we define, for convenience,

$$\varphi_4 = \theta_1 + \theta_2 + \theta_3,$$

and proceed to multiply entries from the first two configurations, we find, from expressions (3.4-23), (3.4-24) and (3.4-26), that

$$\frac{m_{\varphi_1} m_{\varphi_2}}{m_{\theta_3}} \neq m_{\varphi_4}.$$

The introduction of the third configuration which contains u-terms with subscripts from θ_1 and θ_2 has the result that in the expression for m_{φ_4} we can no longer separate those terms that are summed over θ_1 from those summed over θ_2. We reach the same impasse if we take any function of the configuration cells, for instance

$$m_\theta \neq \frac{m_{\varphi_1} m_{\varphi_2} m_{\varphi_3}}{m_{\theta_1} m_{\theta_2} m_{\theta_3}}.$$

We can remove the unique set of variables θ_4 or add additional sets of variables to θ_1 and θ_2 or both, but this will not affect the conclusion that three configurations with different pairwise overlapping will not yield direct estimates.

Proof of converse

(i) If any two configurations do not overlap, we can obtain direct estimates from three configurations.

If we redefine C_{φ_3} so that it does not overlap with C_{φ_1}, we have

$$\varphi_1 = \theta_1 + \theta_3,$$

$$\varphi_2 = \theta_2 + \theta_3,$$

$$\varphi_3 = \theta_2 + \theta_4.$$

The model becomes

$$\log m_\theta = u + U_{\varphi_1} + U_{\varphi_2} + U_{\varphi_3} - U_{\theta_2} - U_{\theta_3}, \tag{3.4-27}$$

and we can derive expressions for the cells of the sufficient configurations and for C_{θ_3}, the configuration whose subscripts are common to C_{φ_1} and C_{φ_2}:

$$m_{\varphi_1} = \exp(u + U_{\varphi_1} - U_{\theta_3}) \sum_{\theta_2, \theta_4} \exp(U_{\varphi_2} + U_{\varphi_3} - U_{\theta_2}), \tag{3.4-28}$$

$$m_{\varphi_2} = \exp(u + U_{\varphi_2} - U_{\theta_2} - U_{\theta_3}) \sum_{\theta_1} \exp(U_{\varphi_1}) \sum_{\theta_4} \exp(U_{\varphi_3}), \tag{3.4-29}$$

$$m_{\varphi_3} = \exp(u + U_{\varphi_3} - U_{\theta_2}) \sum_{\theta_1, \theta_3} \exp(U_{\varphi_1} + U_{\varphi_2} - U_{\theta_3}), \tag{3.4-30}$$

$$m_{\theta_3} = \exp(u - U_{\theta_3}) \sum_{\theta_1} \exp(U_{\varphi_1}) \sum_{\theta_2, \theta_4} \exp(U_{\varphi_2} + U_{\varphi_3} - U_{\theta_2}). \tag{3.4-31}$$

Then from (3.4-28), (3.4-29), and (3.4-31), we have

$$\frac{m_{\varphi_1} m_{\varphi_2}}{m_{\theta_3}} = \exp(u + U_{\varphi_1} + U_{\varphi_2} - U_{\theta_2} - U_{\theta_3}) \sum_{\theta_4} \exp(U_{\varphi_3})$$

$$= m_{\varphi_4}, \tag{3.4-32}$$

where $\varphi_4 = \theta_1 + \theta_2 + \theta_3$. We can also derive the relationship

$$m_\theta = \frac{m_{\varphi_1} m_{\varphi_2} m_{\varphi_3}}{m_{\theta_2} m_{\theta_3}}, \tag{3.4-33}$$

and so obtain the direct estimates

$$\hat{m}_\theta = \frac{x_{\varphi_1} x_{\varphi_2} x_{\varphi_3}}{x_{\theta_2} x_{\theta_3}}. \tag{3.4-34}$$

Thus if the third configuration only overlaps with one other configuration, we have direct estimates.

(ii) If removal of a set of variables common to all three configurations gives at least two configurations that do not overlap, we can obtain direct estimates.

Adding a common set of variables to each of the configurations required by model (3.4-27) gives the complete set θ comprising five groups:

$$\theta = \theta_1 + \theta_2 + \theta_3 + \theta_4 + \theta_5,$$

and the sufficient configurations C_{φ_1}, C_{φ_2} and C_{φ_3}, where

$$\varphi_1 = \theta_1 + \theta_3 + \theta_5,$$

$$\varphi_2 = \theta_2 + \theta_3 + \theta_5,$$

$$\varphi_3 = \theta_2 + \theta_4 + \theta_5. \tag{3.4-35}$$

The model is

$$\log m_\theta = u + U_{\varphi_1} + U_{\varphi_2} + U_{\varphi_3} - U_{\theta_2} - U_{\theta_3} - 2U_{\theta_5}, \tag{3.4-36}$$

which differs from model (3.4-27) by having the additional set of u-terms U_{θ_5}. As the proof consists of modifying expressions (3.4-28)–(3.4-30), we leave it to the reader to satisfy himself that

$$\hat{m}_\theta = \frac{x_{\varphi_1} x_{\varphi_2} x_{\varphi_3}}{x_{\varphi_5} x_{\varphi_6}}, \tag{3.4-37}$$

where

$$\varphi_5 = \theta_2 + \theta_5,$$

$$\varphi_6 = \theta_3 + \theta_5. \ \blacksquare$$

Implications of the theorems

Theorems 3.4-1 and 3.4-2 establish when direct estimation is possible for three configurations. There are two important implications. The first is that when direct estimates exist we can compute them in stages; it is this feature that enables us to extend the results to any number of configurations. The other implication is that direct estimates do not exist unless at least one two-factor effect is absent.

a.   *Computation of estimates by stages*

Equations (3.4-32)–(3.4-34) show how direct estimates may be computed in stages. From the first two sufficient statistics C_{φ_1} and C_{φ_2} we get estimates for the cells of the configuration C_{φ_4} that encompass all the variables present in φ_1 and φ_2. At the next stage we introduce the third sufficient configuration C_{φ_3} to get estimates of the elementary cells m_θ.

Whenever direct estimates are available, this computation by stages is possible. The classifying rules consist essentially of successively removing configurations or variables that do not invalidate multiplication by stages. When computing by stages is possible, each intermediate stage yields valid estimates for a configuration, and the final stage gives estimates for the elementary cells. If we can reduce to two configurations in step 5 of the rules, theorem 3.4-1 tells us that the model has direct estimates.

We discuss the convenience of computing by stages in Section 3.7 and give an example. (See also Section 4.5.)

b.   *Absence of two-factor effects*

If a direct estimate exists, then at least one two-factor effect must be absent, because two nonredundant configurations cannot be sufficient for all two-factor effects. This result is a further implication of the stepwise computing of direct estimates.

If direct estimates exist, then the final computation consists of combining two configurations. They can be defined as C_{φ_1} and C_{φ_2}, where

$$\varphi_1 = \theta_1 + \theta_3,$$

$$\varphi_2 = \theta_2 + \theta_3,$$

and the complete set is

$$\theta = \theta_1 + \theta_2 + \theta_3.$$

This definition is completely general if we include the case where θ_3 is an empty set. Removing θ_1 or θ_2 introduces redundancy, and so these must be distinct non-overlapping sets of subscripts.

When θ_1 and θ_2 are two distinct groups, the two configurations are not sufficient for any u-terms including both variables from θ_1 and variables from θ_2. Thus all two-factor terms with members from each group are absent from models yielding direct estimates. Conversely, if all two-factor effects are present, we cannot proceed with stepwise elimination of configurations until only two remain, and we conclude that at least one two-factor effect must be absent for direct estimates to exist.

3.5   Iterative Proportional Fitting of Log-Linear Models

We can obtain MLEs for the elementary cells under any hierarchical model by iterative fitting of the sufficient configurations. The method of successive proportional adjustment that we describe has the following properties:

1. it always converges to the required unique set of maximum likelihood estimates;
2. a stopping rule may be used that ensures accuracy to any desired degree in the elementary cell estimates, instead of a rule that only ensures accuracy in one or more summary statistics;
3. the estimates depend only on the sufficient configurations, and so no special provision need be made for sporadic cells with no observations;
4. any set of starting values may be chosen that conforms to the model being fitted;
5. if direct estimates exist, the procedure yields the exact estimates in one cycle. (For some high-dimensional models this is true only if attention is paid to the order in which the configurations are fitted, but any order will suffice for most models.)

We discuss these features in three dimensions and then in more general terms but first trace briefly the history of methods that have been proposed.

Bartlett [1935] is generally believed to have been the first to describe a method of getting MLEs for a model that does not possess closed-form estimates. He had three dichotomous variables and wished to fit model 1 of table 3.3-1, the model with $u_{123} = 0$. He called this the "no second-order interaction" model.

His development was as follows. Putting $\hat{m}_{111} = x_{111} + \theta$ implies $\hat{m}_{112} = x_{112} - \theta$ because of the marginal constraint $\hat{m}_{11+} = x_{11+}$. Considering all marginal constraints gives

$$|\hat{m}_{ijk} - x_{ijk}| = \theta$$

for all i, j, k. The cross-product condition specified by the model is

$$\frac{\hat{m}_{111}\hat{m}_{221}}{\hat{m}_{121}\hat{m}_{211}} \cdot \frac{\hat{m}_{122}\hat{m}_{212}}{\hat{m}_{112}\hat{m}_{222}} = 1,$$

and this can be written in terms of the observed $\{x_{ijk}\}$ and the unknown deviation θ as

$$\frac{(x_{111} + \theta)(x_{221} + \theta)}{(x_{121} - \theta)(x_{211} - \theta)} \cdot \frac{(x_{122} + \theta)(x_{212} + \theta)}{(x_{112} - \theta)(x_{222} - \theta)} = 1. \qquad (3.5\text{-}1)$$

This gives a cubic equation in θ that can be solved by trial and error. Example 3.5-1 shows our alternative method of analysis applied to Bartlett's original data.

Bartlett's approach was later extended to multiple categories, and methods were proposed for obtaining solutions to the constraining equations (see, e.g., Roy and Kastenbaum [1956] and the review by Goodman [1964b]). Most authors suggested matrix inversion techniques or proposed methods of simplifying the necessary matrix inversion.

More recently several authors have suggested the use of Newton–Raphson techniques and their variants (e.g., see Bock [1970, 1972], and Haberman [1974]). While more efficient numerically, these iterative methods do not in general have properties 3 and 5 listed above. Nevertheless, they represent alternatives that bear consideration, especially for high dimensions.

The method of iterative proportional fitting seems to offer advantages over the earlier techniques proposed. The method is derived from a procedure which we call the "classical" proportional fitting procedure. In the classical procedure the internal cells of an available sample are proportionally adjusted to fit a set of margins derived from another source. This algorithm has a long history and is often associated with Deming and Stephan [1940]. We give examples of the classical use in Section 3.6. The rest of Section 3.5 is devoted to use of the algorithm for fitting unsaturated log-linear models to a single body of data.

3.5.1 *Iterative proportional fitting in three dimensions*

The model with $u_{123} = 0$, considered by Bartlett, is the only hierarchical model in three dimensions that does not have closed-form estimates. We therefore use this model to describe the iterative proportional fitting procedure and its convergence. We then show that this procedure yields estimates identical to those of Bartlett and demonstrate by means of a further example that it is also appropriate when some of the observed cells are empty.

The computing procedure

The MLEs must fit the configurations C_{12}, C_{13}, and C_{23}. Once these configurations have been obtained from the observed cell counts, the observed counts in the elementary cells are discarded and replaced by a set of preliminary expected estimates.

We discuss restrictions on the preliminary estimates in Section 3.5.2, but for the moment denote them as the set $\{\hat{m}_{ijk}^{(0)}\}$. A convenient procedure is to set $\hat{m}_{ijk}^{(0)} = 1$ for every cell; this sets $\log \hat{m}_{ijk}^{(0)} = 0$ and hence sets equal to zero all preliminary u-term estimates.

We then adjust the preliminary estimates to fit successively C_{12}, C_{13}, and C_{23}. Fitting to C_{12} gives

$$\hat{m}_{ijk}^{(1)} = \hat{m}_{ijk}^{(0)} \frac{x_{ij+}}{\hat{m}_{ij+}^{(0)}}. \tag{3.5-2}$$

Subsequent fitting to C_{13} gives

$$\hat{m}_{ijk}^{(2)} = \hat{m}_{ijk}^{(1)} \frac{x_{i+k}}{\hat{m}_{i+k}^{(1)}}, \tag{3.5-3}$$

and similarly after fitting C_{23} we have

$$\hat{m}_{ijk}^{(3)} = \hat{m}_{ijk}^{(2)} \frac{x_{+jk}}{\hat{m}_{+jk}^{(2)}}. \tag{3.5-4}$$

We repeat this three-step cycle until convergence to the desired accuracy is attained.

A satisfactory stopping rule is to decide on a quantity δ (e.g., $\delta = 0.1$ or $\delta = 0.01$) and stop when a complete cycle does not cause any cell to change by more than this amount, that is, when

$$|\hat{m}_{ijk}^{(3r)} - \hat{m}_{ijk}^{(3r-3)}| < \delta \tag{3.5-5}$$

for all i, j, k. Stopping rules are discussed further in Section 3.5.2.

Convergence of the procedure

To show that we can always obtain cell estimates with the desired accuracy we utilize the approach of Brown [1959]. He proved that the likelihood of the $\{\hat{m}_{ijk}\}$ is a monotonic decreasing function and hence converges. We also give a somewhat heuristic proof that the estimates obtained are the required unique set. For other proofs of convergence see Fienberg [1970a], Haberman [1974], Ireland and Kullback [1968a], and Darroch and Ratcliff [1972].

If we define the log-likelihood ratio

$$D_{123}^{(t)} = \sum_{i,j,k} (x_{ijk} \log x_{ijk} - x_{ijk} \log \hat{m}_{ijk}^{(t)}), \tag{3.5-6}$$

then since $\sum_{i,j,k} x_{ijk} \log \hat{m}_{ijk}$ is maximized for $\hat{m}_{ijk} = x_{ijk}$, we know that $D_{123}^{(t)} \geq 0$, provided that $\hat{m}_{ijk}^{(t)} \geq 0$ and $\hat{m}_{+++}^{(t)} = x_{+++}^{(t)}$. If we now proceed to fit the configurations in the order described above, we have

$$\hat{m}_{ijk}^{(3r)} = \hat{m}_{ijk}^{(3r-1)} \frac{x_{+jk}}{\hat{m}_{+jk}^{(3r-1)}},$$

and hence

$$x_{ijk} \log \hat{m}_{ijk}^{(3r)} = x_{ijk} \log \hat{m}_{ijk}^{(3r-1)} + x_{ijk} \log x_{+jk} - x_{ijk} \log \hat{m}_{+jk}^{(3r-1)}. \tag{3.5-7}$$

Using expressions (3.5-6) and (3.5-7), we can write

$$D_{123}^{(3r)} = D_{123}^{(3r-1)} - \sum_{j,k} (x_{+jk} \log x_{+jk} - x_{+jk} \log \hat{m}_{+jk}^{(3r-1)})$$

$$= D_{123}^{(3r-1)} - D_{23}^{(3r-1)}, \tag{3.5-8}$$

where $D_{23}^{(t)}$ is derived from the sets of sums $\{x_{+jk}\}$ and $\{m_{+jk}^{(t)}\}$ and is non-negative, which shows that $D^{(3r)}_{123} \leqq D^{(3r-1)}_{123}$. We similarly derive

$$D^{(3r-1)}_{123} = D^{(3r-2)}_{123} - D^{(3r-2)}_{13}, \tag{3.5-9}$$

$$D^{(3r-2)}_{123} = D^{(3r-3)}_{123} - D^{(3r-3)}_{12}. \tag{3.5-10}$$

Expressions (3.5-8), (3.5-9), and (3.5-10) show that $D^{(s)}_{123}$ is a monotonic decreasing function, i.e., $D^{(3r-3)} \geqq D^{(3r-2)} \geqq D^{(3r-1)} \geqq D^{(3r)}$ for $r = 1, 2, 3,\ldots$ with equality at step t if and only if $\hat{m}^{(t)}_{ijk} = \hat{m}^{(t-1)}_{ijk}$ for all i, j, k. This function is bounded below, and so must converge to this limiting bound, which corresponds to the limiting value for the $\{\hat{m}_{ijk}\}$.

Examining estimates at end of cycle

From equations (3.5-8)–(3.5-10), we also have

$$D^{(3r)}_{123} = D^{(3r-3)}_{123} - (D^{(3r-1)}_{23} + D^{(3r-2)}_{13} + D^{(3r-3)}_{12}), \tag{3.5-11}$$

and so if we compare estimates at the end of each cycle, we have taken into account the divergence between the estimates and the fit of each of the three sufficient configurations. If we establish that $\hat{m}^{(3r)}_{ijk}$ does not differ substantially from $\hat{m}^{(3r-3)}_{ijk}$ and hence that $D^{(3r)}_{123}$ is close to $D^{(3r-3)}_{123}$, we know that each of the terms $D^{(3r-1)}_{23}$, $D^{(3r-2)}_{13}$, and $D^{(3r-3)}_{12}$ is small, and so each of the configurations fits well.

Convergence to correct limits

To show that the estimates exhibit no three-factor effect, we consider the three proportional adjustments

$$\hat{m}^{(3r)}_{ijk} = \hat{m}^{(3r-1)}_{ijk} \frac{x_{+jk}}{\hat{m}^{(3r-1)}_{+jk}}, \tag{3.5-12}$$

$$\hat{m}^{(3r-1)}_{ijk} = \hat{m}^{(3r-2)}_{ijk} \frac{x_{i+k}}{\hat{m}^{(3r-2)}_{i+k}}, \tag{3.5-13}$$

and

$$\hat{m}^{(3r-2)}_{ijk} = \hat{m}^{(3r-3)}_{ijk} \frac{x_{ij+}}{\hat{m}^{(3r-3)}_{ij+}}. \tag{3.5-14}$$

Substituting successively for $\hat{m}^{(3r-1)}_{ijk}$, $\hat{m}^{(3r-2)}_{ijk}$, etc. in expression (3.5-12) gives, after some rearrangement,

$$\hat{m}^{(3r)}_{ijk} = \left[\frac{x_{+jk}}{\hat{m}^{(3r-1)}_{+jk}} \cdot \frac{x_{+jk}}{\hat{m}^{(3r-4)}_{+jk}} \cdots \frac{x_{+jk}}{\hat{m}^{(2)}_{+jk}}\right] \left[\frac{x_{i+k}}{\hat{m}^{(3r-2)}_{i+k}} \cdots \frac{x_{i+k}}{\hat{m}^{(1)}_{i+k}}\right]$$

$$\times \left[\frac{x_{ij+}}{\hat{m}^{(3r-3)}_{ij+}} \cdots \frac{x_{ij+}}{\hat{m}^{(0)}_{ij+}}\right] \hat{m}^{(0)}_{ijk}. \tag{3.5-15}$$

If we set $\hat{m}^{(0)}_{ijk} = 1$, we have

$$\hat{m}^{(3)}_{ijk} = f_1(j, k) f_2(i, k) f_3(i, j),$$

where $f_1(j, k)$ indicates a function depending only on the categories of the second two variables, and similarly for $f_2(i, k)$ and $f_3(i, j)$. Multiplying an estimate by a term that depends only on the categories of two variables cannot introduce a three-factor effect, because the three-factor effect is, by definition, a measure of the difference in two-factor effects attributable to the third variable. The algorithm therefore yields estimates that satisfy the condition $u_{123} = 0$.

Example 3.5-1   Models fitted to Bartlett's data

Bartlett's original data are from an experiment investigating growth of plants following different treatments; thus variable 1, the "response" variable, has categories "alive" and "dead." The "design" variables are again dichotomous, the categories of variable 2 refer to time of planting, and those of variable 3 to length of cutting. The sampling design is such that the configuration C_{23} is fixed and has 240 observations in each of its four cells. We may therefore fit any model that includes u_{23} without violating the structure inherent in this experimental design.

Table 3.5-1   Iterative Proportional Fitting: First Cycle for No-Three-Factor-Effect Model, applied to Bartlett's [1935] Data

Observed Counts			Fitted Estimates		
Cell	x_{ijk}	$\hat{m}_{ijk}^{(0)}$	$\hat{m}_{ijk}^{(1)}$	$\hat{m}_{ijk}^{(2)}$	$\hat{m}_{ijk}^{(3)}$
1, 1, 1	156	1	131.5	166.98	156.26
2, 1, 1	84	1	108.5	89.48	83.74
1, 2, 1	84	1	57.5	73.02	78.40
2, 2, 1	156	1	182.5	150.52	161.60
1, 1, 2	107	1	131.5	96.02	103.09
2, 1, 2	133	1	108.5	127.52	136.91
1, 2, 2	31	1	57.5	41.98	39.29
2, 2, 2	209	1	182.5	214.48	200.71
C_{12}	x_{ij+}	$\hat{m}_{ij+}^{(0)}$			
1, 1, +	263	2			
2, 1, +	217	2			
1, 2, +	115	2			
2, 2, +	365	2			
C_{13}	x_{i+k}		$\hat{m}_{i+k}^{(1)}$		
1, +, 1	240		189.0		
2, +, 1	240		291.0		
1, +, 2	138		189.0		
2, +, 2	342		291.0		
C_{23}	x_{+jk}			$\hat{m}_{+jk}^{(2)}$	
+, 1, 1	240			256.47	
+, 2, 1	240			223.53	
+, 1, 2	240			223.53	
+, 2, 2	240			256.47	

a. First cycle

In table 3.5-1 we show the first cycle of the iterative proportional fitting algorithm for the model with $u_{123} = 0$. We are fitting C_{12}, C_{13}, and C_{23}. If we put $\hat{m}_{ijk}^{(0)} = 1$, the effect of fitting the first configuration C_{12} is to obtain estimates of the form

$$\hat{m}_{ijk}^{(1)} = \frac{x_{ij+}}{2} \qquad (3.5\text{-}16)$$

in every cell. These are the maximum likelihood estimates for the model with u_3 set equal to zero, a model that is not permissible for these data. If in this example we had chosen to fit C_{23} first, we would have obtained estimates appropriate for the hypothesis of equal survival for each of the four treatment groups, i.e., the model with $u_1 = 0$.

In the next column of table 3.5-1, we show the estimates obtained after fitting C_{13}, namely,

$$\hat{m}_{ijk}^{(2)} = \hat{m}_{ijk}^{(1)} \frac{x_{i+k}}{\hat{m}_{i+k}^{(1)}}. \qquad (3.5\text{-}17)$$

From (3.5-16) and (3.5-17) we see that this is equivalent to

$$\hat{m}_{ijk}^{(2)} = \frac{x_{ij+} x_{i+k}}{x_{i++}}.$$

These are the estimates obtained when we fit the model with $u_{23} = u_{123} = 0$. Again, this is a model that is not permissible with this experimental design because the estimates do not yield $\hat{m}_{+jk}^{(2)} = 240$ as required. (We note that it is illegal in table 3.5-3.)

After fitting C_{23} we obtain estimates $\hat{m}_{ijk}^{(3)}$ which satisfy the condition $\hat{m}_{+jk}^{(2)} = 240$, but these estimates no longer fit the configurations C_{12} and C_{13}, and we must repeat the cycle. Our stopping rule would have us examine the estimates at this point and compare with the preceding cycle, in this case the initial estimates. As long as we set $\hat{m}_{ijk}^{(0)} = 1$, we will always have big differences at this point; thus we can assume that for this model we will always need at least two cycles. If we were not convinced of the adequacy of this stopping rule, we could look at the differences $|x_{ijk} - \hat{m}_{ijk}^{(3)}|$, and we would observe that they varied among cells. We know from the discussion of Bartlett's formulation that we must have only one absolute value for all these differences, and so we would again be led to repeating the cycle.

b. *Successive cycles*

We show the cell estimates obtained at the end of each of the first seven cycles in table 3.5-2. We give only $\hat{m}_{1jk}$ because the last configuration fitted in each cycle is C_{23}, which implies that $\hat{m}_{1jk} + \hat{m}_{2jk} = x_{+jk}$. We give the value of $D_{123}^{(t)}$ after each step in each cycle so that we may assess how the changes in cell values are reflected by changes in $D_{123}^{(t)}$. In the first cycle we fit C_{12}, C_{13}, and C_{23} in order, and the value of $D_{123}^{(t)}$ changes rapidly, being 26.72, 3.80, and 1.54. In successive cycles the change is much slower. By the end of the fourth cycle, $D_{123}^{(t)}$ has reached 1.1469 and, to four decimals of accuracy, there is no subsequent change. If we look at the cell estimates we find that if we had required no change greater than 0.1 between successive estimates, we would have stopped at the fifth cycle; for 0.01 and 0.001 we would stop at the sixth and seventh cycles, respectively.

c. *Final estimates*

We give the final estimates obtained after five cycles in table 3.5-3. The absolute difference between observed and expected cell counts is constant, giving $\theta = 5.1$, as Bartlett obtained. Thus our cell values are exactly the same as his. He obtained a X^2 value of 2.27, and this is close to twice our D_{123}, which is 2.29. We conclude that the hypothesis of no three-factor effect is acceptable: there is no evidence that any possible effect of time of planting on growth is affected by the length of cutting, and vice versa.

Table 3.5-2 Estimates Obtained at End of Each Cycle: Model with $u_{123} = 0$ Fitted to Bartlett's Data

| Cycle, r | Cell Estimates | | | | $D_{123}^{(3r-s)}$ | | |
	$\hat{m}_{111}^{(3r)}$	$\hat{m}_{121}^{(3r)}$	$\hat{m}_{112}^{(3r)}$	$\hat{m}_{122}^{(3r)}$	$s = 2$	$s = 1$	$s = 0$
$r = 1$	156.262	78.395	103.090	39.288	26.7202	3.8018	1.5363
$r = 2$	160.179	79.008	101.935	36.706	1.4397	1.2084	1.1610
$r = 3$	160.921	78.921	101.912	36.212	1.1548	1.1492	1.1474
$r = 4$	161.062	78.907	101.905	36.119	1.1472	1.1470	1.1469
$r = 5$	161.089	78.905	101.904	36.101	1.1469	1.1469	1.1469
$r = 6$	161.095	78.904	101.904	36.097	1.1469	1.1469	1.1469
$r = 7$	161.096	78.904	101.904	36.096	1.1469	1.1469	1.1469

Table 3.5-3 Four Models Fitted to Bartlett's Data

| Cell | Observed x | $u_{123} = 0$ | | $u_{23} = u_{123} = 0$[a] | | $u_{13} = u_{123} = 0$ | | $u_{12} = u_{123} = 0$ | |
		Fitted $\hat{m}$	Difference $x - \hat{m}$	Fitted $\hat{m}$	Difference $x - \hat{m}$	Fitted $\hat{m}$	Difference $x - \hat{m}$	Fitted $\hat{m}$	Difference $x - \hat{m}$
1, 1, 1	156	161.1	−5.1	167.0	−11.0	131.5	24.5	120.0	36.0
2, 1, 1	84	78.9	5.1	89.5	−5.5	108.5	−24.5	120.0	−36.0
1, 2, 1	84	78.9	5.1	73.0	11.0	57.5	26.5	120.0	−36.0
2, 2, 1	156	161.1	−5.1	150.5	5.5	182.5	−26.5	120.0	36.0
1, 1, 2	107	101.9	5.1	96.0	11.0	131.5	−24.5	69.0	38.0
2, 1, 2	133	138.1	−5.1	127.5	5.5	108.5	24.5	171.0	−38.0
1, 2, 2	31	36.1	−5.1	42.0	−11.0	57.5	−26.5	69.0	−38.0
2, 2, 2	209	203.9	5.1	214.5	−5.5	182.5	26.5	171.0	38.0
Degrees of Freedom		1		2		2		2	
X^2		2.27		7.42		52.33		101.94	
G^2		2.29		7.63		53.44		105.18	

[a] This model is illegal for these data.

d. *Other models*

Showing that there is no evidence of a significant three-factor effect does not establish that we have the best model for these data. We must investigate whether time of planting and length of cutting have any independent effect on growth. To do this we fit two more models, removing in turn each of the two-factor effects not fixed by the design. We find that neither model fits the data well. In fact, putting $u_{13} = 0$, which stipulates no effect of length of cutting on survival, or putting $u_{12} = 0$, which stipulates no effect of time of planting on survival, we get larger X^2 values than when we fit the illegal model with $u_{23} = 0$.

We conclude that survival is affected by time of planting and by length of cutting but that these two factors operate independently in their influence on survival. ■■

Empty cells

The Bartlett example of three dichotomous variables has observed counts in every elementary cell and hence has entries in every cell of every marginal configuration. With larger tables and sporadic empty elementary cells the configuration cells may or may not all have positive entries.

If a particular *configuration cell entry* is zero, then as soon as this configuration is fitted, all the *elementary cells* that are summed to obtain this configuration cell have estimates of zero. For instance, in any three-dimensional table, if the configuration C_{13} is the second configuration to be fitted and has the entry $x_{2+2} = 0$, we obtain $\hat{m}_{2j2}^{(2)} = 0$ for all j. As subsequent estimates are multiples of $\hat{m}_{2j2}^{(2)}$, they are also zero. Thus occasional empty configuration cells do not invalidate the fitting method, but all those empty cells summing to the empty configuration cells have estimates of zero. We discuss this further when we consider degrees of freedom.

Unless empty configuration cells occur, positive estimates will in general be obtained for every elementary cell, regardless of whether the elementary cell originally contained any observations. Exceptions can occur when the parameters set equal to zero in the model cannot be estimated from the data, as in example 3.3-1. Persistent estimates of zero in elementary cells in the absence of empty configuration cells may thus constitute a warning that the model is inappropriate.

Example 3.5-2   Empty cells in food-poisoning data

We can obtain estimates for the empty cell in the data given in table 3.5-4, but we might wish to keep the cell empty and fit a model only to the remaining cells using the methods described in Chapter 5.

Table 3.5-4   Observed Three-Dimensional Data with a Random Zero

	Food Eaten			
	Crabmeat		No Crabmeat	
Consumer's Illness	Potato Salad		Potato Salad	
	Yes	No	Yes	No
Ill	120	4	22	0
Not Ill	80	31	24	23

Source: Korff, Taback, and Beard [1952].

The data are from an epidemiologic study following an outbreak of food poisoning that occurred at an outing held for the personnel of an insurance company. Questionnaires were completed by 304 of the 320 persons attending. Of the food eaten, interest focused on potato salad and crabmeat [Korff, Taback, and Beard, 1952]. The variables are:

1. presence or absence of illness;
2. potato salad eaten or not eaten;
3. crabmeat eaten or not eaten.

If we consider that entries can occur in any cell, we fit the models (a)–(c) given in table 3.5-5. The first model sets $u_{123} = 0$, and the small X^2 or G^2 values show that the model fits well. We conclude that any association between potato salad and illness is the same for those who ate crabmeat and those who did not.

Table 3.5-5   Models Fitted to Data of Table 3.5-4

| | | Food Eaten | | | | | | |
| | | Crabmeat | | No Crabmeat | | | | |
Model	Consumer's Illness	Potato Salad Yes	No	Potato Salad Yes	No	Degrees of Freedom	X^2	G^2
(a) No Three-Factor Effect	Ill	121.08	2.92	20.92	1.08	1	1.70	2.74
	Not Ill	78.92	32.08	25.09	21.92			
(b) No Interaction between Illness and Potato Salad	Ill	105.53	18.47	14.67	7.33	2	44.34	53.69
	Not Ill	94.47	16.53	31.33	15.67			
(c) No Interaction between Illness and Crabmeat	Ill	115.45	2.41	26.55	1.59	2	5.10	6.48
	Not Ill	84.55	32.59	19.45	21.41			

The second model, with $u_{12} = u_{123} = 0$, gives $X^2 = 44.34$. We conclude that u_{12} is large: potato salad and illness are associated. This two-degree-of-freedom X^2 can be split into two components by summing the contributions for the four cells relating to crab eaters separately from the remaining four cells. The resulting large values, 28.195 and 16.149, confirm our previous conclusion that the association persists whether or not crabmeat was eaten.

Finally we fit the model $u_{13} = u_{123} = 0$ and obtain $X^2 = 5.095$ with two degrees of freedom. The association between illness and crabmeat is not as large as that of illness and potato salad, but it cannot be entirely dismissed.

The observed empty cell indicates that nobody who did not eat either crabmeat or potato salad was ill. If no other food was eaten at the picnic and we were convinced that the illness was caused by the picnic food, we would regard this as a structural zero. Under this approach, if we fit the model $u_{12} = u_{123} = 0$ then we find that all four cells referring to those who did not eat crabmeat are unchanged. The estimates for the remaining cells are identical to those obtained by the previous approach for this model. We previously obtained the one-degree-of-freedom X^2 component 28.195, indicating a high association between potato salad and illness.

These data are a subset of those available. Other food was eaten at the picnic and so the structural zero model is not a good approach. If we did not have such additional knowledge we would need to consider incomplete as well as complete methods for the data. ∎∎

3.5.2   *The general iterative proportional fitting algorithm*

The general procedure for fitting s configurations follows the procedure described for three configurations in three dimensions. If the minimal configurations are C_{θ_q} for $q = 1, 2, \ldots, s$, with cell entries x_{θ_q}, respectively, we choose a set of initial estimates $m_\theta^{(0)}$ and proceed to fit each of these configurations in turn. After r cycles the relations are

$$\hat{m}_\theta^{(sr+1)} = \hat{m}_\theta^{(sr)} \frac{x_{\theta_1}}{\hat{m}_{\theta_1}^{(sr)}},$$

and in general at the tth step we have

$$\hat{m}_\theta^{(t)} = \hat{m}_\theta^{(t-1)} \frac{x_{\theta_q}}{\hat{m}_{\theta_q}^{(t-1)}}, \qquad (3.5\text{-}18)$$

where $t - q$ is a multiple of s. Before giving examples of this general procedure, we will consider in more detail (i) initial estimates and (ii) stopping rules.

(i)   *Initial estimates*

In the description of proportional fitting in three dimensions, we suggested $m_{ijk}^{(0)} = 1$ as a convenient preliminary estimate when fitting a simple log-linear model to a single set of data. For the model with $u_{123} = 0$ we can, however, choose any set of preliminary values that do not exhibit a three-factor effect, and any such set converges to the same maximum likelihood solution. In general, for any model the only restriction on initial values is that the interactions exhibited by the selected values either match or are a subset of those to be fitted. Before proving this, we will demonstrate it by a two-dimensional example.

Example 3.5-3   Effect of changing initial values

Suppose we have added the observed elementary cells to obtain the configurations for the model with $u_{12} = 0$,

				C_2
				25
				75
$C_1 =$	30	30	40	100

If we put

$$\{\hat{m}_{ij}^{(0)}\} = \begin{array}{ccc} 1 & 1 & 1 \\ 1 & 1 & 1 \\ \hline 2 & 2 & 2 \end{array}$$

then after fitting C_1, we obtain

$$\{\hat{m}_{ij}^{(1)}\} = \begin{array}{ccc|c} 15 & 15 & 20 & 50 \\ 15 & 15 & 20 & 50 \end{array}$$

and when we fit C_2, we obtain

$$\{\hat{m}_{ij}^{(2)}\} = \begin{array}{ccc} 7.5 & 7.5 & 10 \\ 22.5 & 22.5 & 30 \end{array}$$

If we had started with any set of estimates which did not exhibit a two-factor effect, for instance

$$\{\hat{m}_{ij}^{(0)}\} = \begin{array}{ccc} 1 & 2 & 3 \\ 2 & 4 & 6 \\ \hline 3 & 6 & 9 \end{array}$$

we would obtain after fitting C_1

$$\{\hat{m}^{(1)}\} = \begin{array}{ccc|c} 10 & 10 & 40/3 & 100/3 \\ 20 & 20 & 80/3 & 200/3 \end{array}$$

and hence after fitting C_2 we have for a single cell

$$\hat{m}_{11}^{(2)} = \frac{(10)(25)}{\left(\dfrac{100}{3}\right)} = 7.5.$$

For this cell and other cells we obtain the same estimates as previously obtained. ∎

THEOREM 3.5-1   *The iterative proportional fitting method converges to the desired limits when we use any initial estimates that only exhibit those interaction terms present in the model.*

Proof   We take the general expression (3.5-18) for $\hat{m}_\theta^{(t)}$ and substitute for $\hat{m}_\theta^{(t-1)}$ from the expression for the preceding step. If we continue to substitute from each preceding step until we reach an expression in $\hat{m}_\theta^{(0)}$, we have, after $r+1$ cycles,

$$\hat{m}_\theta^{(r+1)s} = \hat{m}_\theta^{(0)}\left[\frac{x_{\theta_1}}{\hat{m}_{\theta_1}^{(0)}} \cdot \frac{x_{\theta_1}}{\hat{m}_{\theta_1}^{(s)}} \cdots \frac{x_{\theta_1}}{\hat{m}_{\theta_1}^{(rs)}}\right] \cdots \left[\frac{x_{\theta_i}}{\hat{m}_{\theta_i}^{(i-1)}} \cdots \frac{x_{\theta_i}}{\hat{m}_{\theta_i}^{(rs+i-1)}}\right] \cdots$$

$$\times \left[\frac{x_{\theta_s}}{\hat{m}_{\theta_s}^{(s-1)}} \cdots \frac{x_{\theta_s}}{\hat{m}_{\theta_s}^{(rs+s-1)}}\right].$$

This may be written

$$\hat{m}_\theta^{(t)} = \hat{m}_\theta^{(0)} f_1(\theta_1) f_2(\theta_2) \cdots f_i(\theta_i) \cdots f_s(\theta_s), \tag{3.5-19}$$

where $f_i(\theta_i)$ is a function only of the variables contained in θ_i.

If the initial estimates can be described by a model that includes only the u-terms for which the C_{θ_i} are sufficient statistics, we can write

$$\hat{m}_\theta^{(0)} = g_1(\theta_1) \cdots g_s(\theta_s), \tag{3.5-20}$$

where $g_i(\theta_i)$ is a function only of the variables contained in θ_i. This is a very general formulation and includes the case $\hat{m}_\theta^{(0)} = 1$. We may further define

$$h_i(\theta_i) = f_i(\theta_i) g_i(\theta_i), \tag{3.5-21}$$

where $h_i(\theta_i)$ is a function only of the variables contained in θ_i. From (3.5-19)–(3.5-21), we obtain

$$\hat{m}_\theta^{(t)} = h_1(\theta_1) h_2(\theta_2) \cdots h_s(\theta_s). \tag{3.5-22}$$

Thus at stage t our estimates include only those interrelationships between variables stipulated by the model. If convergence is achieved at this stage the estimates also fit all the sufficient configurations. Birch's [1963] results prove that there is a unique set of estimates that satisfy both these conditions. Thus our estimates converge to the unique solution for any $\{\hat{m}_\theta^{(0)}\}$ that satisfy (3.5-20).

Proof of converse   If the initial estimates exhibit a higher-order interaction than described by the model, convergence to the desired estimates is not achieved. To show this we take expression (3.5-20) and add another term that is a function of a group of variables that do not appear together in the sufficient configurations $g(\theta_i\theta_j)$. We then substitute this value of $\hat{m}_\theta^{(0)}$ in expression (3.5-19), and, using the relation (3.5-21), we obtain

$$\hat{m}_\theta^{(t)} = g(\theta_i\theta_j)h_1(\theta_1)h_2(\theta_2)\cdots h_s(\theta_s). \tag{3.5-23}$$

The term in $g(\theta_i\theta_j)$ is present and unchanged however large t may be.   ∎

Higher-order initial estimates for classical usage
The converse of theorem 3.5-1 brings out the difference between the use of the algorithm for fitting log-linear models to a single set of data and the classical use. In the classical use the set of internal cells that are to be adjusted have higher-order interaction than can be obtained from some set of external configurations from an independent source. When these external configurations are fitted to the internal cells, their original higher-order interactions are retained. We give examples in Section 3.6.

Effect of initial estimates on speed of convergence
As we have a choice, it becomes relevant to consider whether the speed of convergence depends on our choice of initial estimates. We saw from the proof of convergence of the algorithm in three dimensions that the log-likelihood ratio after fitting t configurations,

$$D_\theta^{(t)} = \sum_\theta x_\theta \log x_\theta - \sum_\theta x_\theta \log \hat{m}_\theta^{(t)}, \tag{3.5-24}$$

was a monotonic decreasing function. By taking $\hat{m}_\theta^{(0)} = 1$ we have taken the maximum value of $D_\theta^{(0)}$, and it would seem reasonable that initial values giving a smaller value of $D_\theta^{(t)}$ might be advantageous in speeding convergence.
    When we consider how to choose such estimates, we find, however, that any reasonable estimates are functions of the sufficient configurations, and the simpler estimates are functions not of the minimal set of configurations but of the lower-order redundant configurations. If some of the lower-order configurations are used to give the initial estimates, at the stage t in the iteration procedure when the minimal configurations incorporating these lower-order configurations are fitted, the estimates of $\hat{m}_\theta^{(t)}$ are identical to those obtained by setting $\hat{m}_\theta^{(0)}$ equal to 1. This is best illustrated by an example.

Example 3.5-4   Speed of convergence unchanged by initial estimates
In three dimensions we fit the model $u_{123} = 0$ to an $I \times J \times K$ array, taking the configurations in the order C_{12}, C_{13}, and C_{23} and setting $\hat{m}_{ijk}^{(0)} = 1$. We obtain

$$\hat{m}_{ijk}^{(2)} = \frac{(x_{ij+})(x_{i+k})}{x_{i++}}.$$

If we change the initial estimates to a product,

$$\hat{m}_{ijk}^{(0)} = \lambda(x_{i++})(x_{+jk})(x_{i+k}),$$

where λ is a constant, possibly such that the estimates sum to N, we have

$$\hat{m}_{ijk}^{(1)} = \frac{(x_{++k})(x_{ij+})}{N},$$

and at the next step we have the same estimates for $m_{ijk}^{(2)}$ as we obtained with initial cell estimates of 1. ∎

(ii) *Stopping rules*

In Section 3.5.1 we described a convenient stopping rule based on a cell-by-cell comparison at the end of each cycle. As we are concerned with maximizing the likelihood and our proof of convergence depends on the monotonic change of the likelihood, it is not immediately obvious that cell-by-cell comparisons are equivalent, let alone preferable, to examining a function of the likelihood. We need to consider what function of the likelihood we might use, whether cycle-to-cycle examination is sufficient, and the relative advantages of examining cells or some overall function.

Choosing a function of the likelihood

Our estimates maximize the log-likelihood kernel $\sum x_\theta \log \hat{m}_\theta^{(t)}$, subject to constraints. We know that this statistic increases with t, but although we know the upper bound for the saturated model is $\sum x_\theta \log x_\theta$, we do not know the value attainable for unsaturated models. A criterion for stopping must therefore be based on the rate of change, but examining the log-likelihood kernel itself involves looking for small differences between large numbers, which is computationally undesirable.

This magnitude problem can be overcome by looking instead at changes in

$$D_\theta^{(t)} = \sum_\theta x_\theta \log x_\theta - \sum_\theta x_\theta \log \hat{m}_\theta^{(t)}.$$

We are still faced with the problem of determining how small a difference between successive values can be considered insignificant.

Evaluation at the end of each cycle

As $D_\theta^{(t)}$ is a monotonic decreasing function, we can compare estimates for any two values of t. We know from equation (3.5-11) that cycle-to-cycle comparison takes into account the divergence between the current estimates and the next configuration to be fitted for each of the intermediary steps. So a small cycle-to-cycle change in values of $D_\theta^{(t)}$ ensures that each configuration fits well and is therefore an adequate measure of convergence.

If instead of looking at $D_\theta^{(t)}$ we consider the sum of absolute changes in cell estimates, we no longer have a monotonic function but a periodic function. When cycle-to-cycle changes are small, the periodicity damps out and all step-to-step changes in cell magnitudes are also small, but we have no assurance that the converse holds true. This means that if we look at changes in cell magnitudes from step to step we might find that there is little change between the estimates obtained after fitting a particular consecutive pair of configurations at a stage in the iteration when the cycle-to-cycle changes are still large. It is therefore necessary that we examine differences in cell estimates between cycles, not between steps.

Choosing small changes in cells or in the likelihood
For an s-step cycle, the change in our function of the likelihood between cycles is

$$\Delta D = D_\theta^{(t-s)} - D_\theta^{(t)} = \sum_\theta x_\theta \log \frac{\hat{m}_\theta^{(t)}}{\hat{m}_\theta^{(t-s)}}. \tag{3.5-25}$$

Examination of individual cells requires that we set an upper limit on

$$\Delta m_\theta = |\hat{m}_\theta^{(t)} - \hat{m}_\theta^{(t-s)}|. \tag{3.5-26}$$

Specifying that ΔD is less than δ, for some preset δ, does not ensure that Δm_θ is also less than δ for every cell.

In table 3.5-2 we found that for the Bartlett data ΔD converges to four-decimal accuracy before three-decimal accuracy is attained for Δm_θ. In table 3.5-6 we give $D_\theta^{(t)}$, for $t = 1$ to 18, for the model $u_{123} = 0$ applied to the data of table 3.7-10. We also indicate the cycles for which Δm_θ does not exceed 0.1, 0.01, and 0.001. To four-decimal accuracy $D_\theta^{(t)}$ does not change after the fourth cycle, whereas six cycles are needed before all individual cell changes are less than 0.001.

We conclude that if we are interested in obtaining cell estimates, the best stopping rule is to stipulate that cycle-to-cycle cell changes do not exceed δ. If all that is required is a statistic for hypothesis testing, then it is sufficient to examine changes in $D_\theta^{(t)}$ with possible saving in the number of cycles required.

Table 3.5-6 Value of $D_{123}^{(t)}$ after Each Configuration is Fitted when Model $u_{123} = 0$ is applied to Worcester's [1971] Data

Cycle	C_{12}	C_{13}	C_{23}	Cell Accuracy at End of Cycle
1	26.7155	1.7910	1.3183	
2	1.2953	1.2082	1.1826	
3	1.1824	1.1778	1.1769	
4	1.1769	1.1767	1.1767	0.1
5	1.1767	1.1767	1.1767	0.01
6	1.1767	1.1767	1.1767	0.001

3.5.3 Exercises

1. In seven dimensions show that the model with sufficient configurations C_{123}, $C_{124}, C_{235}, C_{136}, C_{57}$ can be fitted directly. Show also that if fitted in this order, the estimates are obtained at the end of the first cycle of the iterative proportional fitting algorithm. Is this true if they are fitted in the order C_{124}, C_{235}, C_{136}, C_{57}, C_{123}? (Answer: No.)
2. If three comprehensive configurations C_{θ_1}, C_{θ_2}, and C_{θ_3} all have a common subset of variables but do not otherwise overlap, we may define the variables as

$$\theta_1 = \varphi_1 + \varphi_4,$$

$$\theta_2 = \varphi_2 + \varphi_4,$$

$$\theta_3 = \varphi_3 + \varphi_4,$$

where $\varphi_1, \varphi_2, \varphi_3$, and φ_4 are mutually exclusive sets of variables. Show that the model corresponding to these variables may be fitted directly. (Answer: By considering each cell of C_{φ_4} separately, we derive

$$m_\theta = \frac{m_{\theta_1} m_{\theta_2} m_{\theta_3}}{m_{\varphi_4}^2}.)$$

3. In four dimensions if the model $u_{1234} = 0$ is to be fitted, show that the iterative algorithm must be used. Show that the estimates obtained after the second configuration is fitted are identical for either of the following initial estimates

(i) $\hat{m}_{ijkl}^{(0)} = 1$;

(ii) $\hat{m}_{ijkl}^{(0)} = \lambda x_{i+++} x_{+j++} x_{++k+} x_{+++l}$, where λ is any constant.

4. Show by writing out the u-terms that in four dimensions the model that contains all two-factor effects but no higher-order effects cannot be fitted directly.

3.6 Classical Uses of Iterative Proportional Fitting

The iterative proportional fitting method was originally developed not for fitting an unsaturated model to a single body of data but for combining the information from two or more sets of data. We refer to this original purpose as the "classical" use.

In theorem 3.5-1 we showed that, for fitting an unsaturated model from the sufficient configurations, initial estimates containing an interaction effect not present in the model are not permissible. Any such extraneous interaction will still be present when the estimates converge. It is this persistence of high-order interaction that is utilized in the classical approach.

In example 3.6-1 we show how the internal cells of a two-way table can be adjusted to fit more up-to-date margins. In other circumstances the internal cells may be based on a sample of a population and the margins on a complete census. In such updating or scaling to match a known marginal distribution, we are in effect fitting a fully saturated model, using the older or smaller sample to give information about the highest-order effects and the newer or larger data for all the lower-order effects.

The classical method of fitting is also used when we have a simple set of data and wish to scale it to fit hypothetical margins. In example 3.6-2 we show that fitting homogeneous margins to a simple two-way table brings out features of the interaction pattern that are not easily discernible in the raw data. This approach is extended in example 3.6-3, where we fit homogeneous margins to each of two tables describing different populations so that they may be more readily compared.

After these three examples we discuss limitations of the method and describe ways in which it can be used for larger data sets.

Example 3.6-1 Marriage status: updated margins for an old two-way table
Friedlander [1961] used the data of table 3.6-1, which show the distribution of women in England and Wales according to their marital status in mid-1957. He adjusted the cells of this table to fit the official marginal estimates for mid-1958.

We computed the 1958 estimates shown in table 3.6-2 using the iterative proportional fitting algorithm. An accuracy of 0.01 in each cell was obtained after eight cycles. These estimates differ slightly from Friedlander's estimates, presumably because he did not use as many cycles. His procedure, although differently defined, is algebraically equivalent to the proportional adjustment algorithm.

Table 3.6-1   Official 1957 Estimates (in Thousands) of Female Population of England and Wales

Age	Single	Married	Widowed or Divorced	1957 Total
		Marital Status		
15–19	1,306	83	0	1,389
20–24	619	765	3	1,387
25–29	263	1,194	9	1,466
30–34	173	1,372	28	1,573
35–39	171	1,393	51	1,615
40–44	159	1,372	81	1,612
45–49	208	1,350	108	1,666
50 or over	1,116	4,100	2,329	7,545
1957 Total	4,015	11,629	2,609	18,253

Note: Based on Friedlander's [1961] data.

Table 3.6-2   Use of Classical Algorithm to Adjust 1957 Estimates of Table 3.6-1 to Fit 1958 Margins

Age	Single	Married	Widowed or Divorced	1958 Total
		Marital Status		
15–19	1,325.27	86.73	0	1,412
20–24	615.56	783.39	3.05	1,402
25–29	253.94	1,187.18	8.88	1,450
30–34	165.13	1,348.55	27.32	1,541
35–39	173.41	1,454.71	52.87	1,681
40–44	147.21	1,308.12	76.67	1,532
45–49	202.33	1,352.28	107.40	1,662
50 or over	1,105.16	4,181.04	2,357.80	7,644
1958 Total	3,988	11,702	2,634	18,324

We can readily check that the interaction pattern of the original table is preserved by examining the cross-product ratios which, as we saw in Chapter 2, are direct functions of parameters of u_{12}. For instance, for the four cells in the top left-hand corner, we have, for the original data of 1957,

$$\frac{(1,306)(765)}{(619)(83)} = 19.45,$$

and for the 1958 estimates we have

$$\frac{(1,325.27)(783.39)}{(615.56)(86.73)} = 19.45.$$

If instead of using this procedure to retain the interaction information from 1957, we use only the 1958 margins to obtain cell estimates under the hypothesis of independence, we get very different values. For the number of single women over 50, for instance, the cell estimate under independence is

$$\frac{(3,988)\,(7,644)}{18,324} = 1,664 \text{ thousand}$$

which is considerably higher than the estimate of 1,105 thousand obtained when interaction is taken into account. This method does not provide a means of computing an estimate for a cell that is empty in the original table; for example, the cell for the youngest age-group of widowed and divorced remains empty.   ■■

Example 3.6-2 Congressmen's votes: Detecting a diagonal trend by adjusting margins of a single two-way table
The data of table 3.6-3, analyzed by Fienberg [1971], were taken from Benson and Oslick [1969]. Congressmen are classified according to the section of the country they represented and their vote on the 1836 Pinckney Gag rule, a rule that had historical implications in its effect on antislavery petitions. These raw data are difficult to interpret because the margins contain unequal numbers.

Table 3.6-3 Data from Benson and Oslick [1969] on Distribution of Votes by Section

Section Represented	Distribution of Votes			Totals
	Yea	Abstain	Nay	
North	61	12	60	133
Border	17	6	1	24
South	39	22	7	68
Totals	117	40	68	225

Table 3.6-4 Percentage Distribution of Votes from Table 3.6-3

Section Represented	Distribution of Votes			Totals
	Yea	Abstain	Nay	
North	46	9	45	100
Border	71	25	4	100
South	57	32	10	99

Table 3.6-5 Standardized Form of Voting Data from Table 3.6-3

Section Represented	Distribution of Votes			Totals
	Yea	Abstain	Nay	
North	20.1	10.2	69.7	100.0
Border	47.4	42.8	9.8	100.0
South	32.5	47.0	20.5	100.0
Totals	100.0	100.0	100.0	300.0

This two-way table is presented in two other ways that aid interpretation. In table 3.6-4 we have percentages by rows. We observe that relatively fewer northerners voted for the rule but do not gain any further insight. In table 3.6-5, both margins are homogeneous. This adjustment brings out more features of interest; the North-Yea and the South-Nay both show negative interactions of about the same magnitude, and on further inspection we observe a symmetry in the table about the diagonal running from South-Yea to North-Nay which was not apparent previously. Fienberg gives further discussion of these data, extending the procedure to three dimensions. ■■

Example 3.6-3 Family occupation: Comparison of two independent tables by standardizing margins of each

This example is taken directly from Mosteller [1968], who ascribes the analysis to Joel Levine [1967]. Two sets of data, Glass's British and Svalastoga's Danish occupational mobility data, are adjusted to have the same uniform margins so that the interaction patterns can be compared more readily.

The original data are given in table 3.6-6, the standardized data in table 3.6-7, and we see that the adjusted counts not only show much smoother within-category patterns than the original counts, but also show that there is a remarkable similarity between the two countries.

Table 3.6-6   Analysis of Glass's British and Svalastoga's Danish Data: Original Counts

Status Category of Father's Occupation	Status Category of Son's Occupation					
	1	2	3	4	5	Totals
1	50	45	8	18	8	129
	18	17	16	4	2	57
2	28	174	84	154	55	495
	24	105	109	59	21	318
3	11	78	110	223	96	518
	23	84	289	217	95	708
4	14	150	185	714	447	1,510
	8	49	175	348	198	778
5	3	42	72	320	411	848
	6	8	69	201	246	530
Totals	106	489	459	1,429	1,017	
	79	263	658	829	562	

Source: Glass [1954] and Svalastoga [1959].
Note: Upper number in cell is British data, lower is Danish.

By comparing diagonal values we see that, except for category 1, the tendency for fathers and sons to fall into the same category is stronger in Denmark than Britain. By looking across rows we can see, for fathers in each category, in which country the sons are more mobile. ■■

Table 3.6-7 Levine's Analysis of Occupation Data of Table 3.6-6

Status Category of Father's Occupation	Status Category of Son's Occupation					Totals
	1	2	3	4	5	
1	68.5	20.9	4.6	3.7	2.3	
	58.6	25.0	12.0	2.6	1.8	100
2	17.8	37.5	22.5	14.7	7.5	
	21.1	41.6	21.9	10.3	5.1	100
3	8.0	19.2	33.7	24.3	14.9	
	11.7	19.3	33.7	21.9	13.5	100
4	4.1	14.7	22.6	31.1	27.6	
	4.1	11.4	20.7	35.5	28.4	100
5	1.6	7.8	16.6	26.2	47.8	
	4.5	2.7	11.8	29.8	51.2	100
Totals	100	100	100	100	100	

Source: Mosteller [1968].
Note: Upper number in cell is British data, lower is Danish.

Constraints on fitting external configurations

The use of the classical algorithm for combining data sets has limitations. These relate to the compatability of the internal and external cells.

Internally, the persistence of zeros may invalidate use of the method. Externally the configurations to be fitted must be consistent with each other. In example 3.6-2 we could adjust to obtain the total 100 in every marginal cell, but if the number of categories per variable differs we can choose the margins T_{i+} for $i = i, \dots, I$ and T_{+j} for $j = 1, \dots, J$ such that $\sum_i T_{i+} = \sum_j T_{+j}$, but cannot have $T_{i+} = T_{+j}$ for all i, j. In other words, the margins must sum to the same total.

When we have more than two dimensions this consistency requirement refers to any overlap between the configurations to be fitted. For instance, if two configurations C_{12} and C_{23}, each from a different source, are to be fitted to a set of cells $\{x_{ijk}\}$, it is necessary that we obtain identical estimates for $\{\hat{m}_{+j+}\}$ from each configuration. The common vector C_2 must be identical.

When there are more than two overlapping external configurations, even when these configurations are consistent, there may not exist a table which has them as its margins. For details on conditions which consistent external configurations must satisfy in order for the internal cells to exist, see Darroch [1962]. If these conditions are not satisfied the classical algorithm does not converge.

Further uses of classical algorithm

In the examples of this section we have considered adjusting two-way tables to fit new margins. The extension to higher dimensions does not introduce any more constraints, but we may no longer be able to fit a fully saturated model. In example 3.7-3 we consider various methods of combining data from one four-dimensional configuration and several two-way tables, each with a fifth variable and one of the variables in the four-dimensional configuration. In this example we cannot estimate the five-factor effect, or four of the four-factor effects, or indeed any of the effects except those corresponding to the given configurations. As we are no longer dealing with a saturated model we defer the example to the next section, although our purpose is still to fit a model with all the parameters that we can estimate and so is more akin to the classical usage.

3.7 Rearranging Data for Model Fitting

We gave simple steps for detecting models with direct estimates in Section 3.4.3. Whenever any of these steps can be taken, we can perform a corresponding rearrangement of the data. This section illustrates possible advantages of three types of rearrangement, relabeling, partitioning, and condensing so that we fit by stages. We follow this by an example of fitting an incomplete model to a seemingly complete set of data and show how this model can be formulated both in Worcester's notation and the u-term notation of Birch.

3.7.1 Relabeling

Any group of variables that always appear together can be regarded as a single compound variable and relabeled to reduce the dimensionality of the configurations to be fitted. Thus, if in five dimensions we wish to fit

$$C_{124}, C_{1235}, C_{45},$$

we can relabel as

$$C_{1'4}, C_{1'35}, C_{45},$$

where $1'$ represents the $I \times J$ cells of the compound variable 12.

In this example relabeling would enable us to use a computer program written to fit only three-dimensional configurations for fitting a four-dimensional configuration. When we wish to investigate how a particular interaction between two variables, say between 1 and 2, is affected by the other variables present, we are only concerned with models where variables 1 and 2 always appear together in the configurations. Relabeling 12 as $1'$ reduces the dimensionality and ensures that all the between-model comparisons are pertinent to our purpose. We discuss between-model comparisons in Chapter 4 but for the moment turn to an example where relabeling helps us to interpret a complex model.

Example 3.7-1 Survival of breast cancer patients related to three simple and one compound variable

The data of table 3.7-1, kindly supplied by Alan Morrison, are from a study of the survival of breast cancer patients (see Morrison et al. [1973]). The variables and categories are

Variable 1. Degree of chronic inflammatory reaction
      1. Minimal
      2. Moderate–severe

Variable 2. Nuclear grade
      1. Relatively malignant appearance
      2. Relatively benign appearance

Variable 3. Survival for three years
      1. No
      2. Yes

Variable 4. Age of diagnosis
　　　1. Under 50 years
　　　2. 50–69 years
　　　3. 70 or older

Variable 5. Center where patient was diagnosed
　　　1. Tokyo
　　　2. Boston
　　　3. Glamorgan

Table 3.7-1 Three-Year Survival of Breast Cancer Patients According to Two Histologic Criteria, Age and Diagnostic Center

| Diagnostic Center | Age | Survived | Minimal Inflammation | | Greater Inflammation | |
			Malignant Appearance	Benign Appearance	Malignant Appearance	Benign Appearance
Tokyo	Under 50	No	9	7	4	3
		Yes	26	68	25	9
	50–69	No	9	9	11	2
		Yes	20	46	18	5
	70 or over	No	2	3	1	0
		Yes	1	6	5	1
Boston	Under 50	No	6	7	6	0
		Yes	11	24	4	0
	50–69	No	8	20	3	2
		Yes	18	58	10	3
	70 or over	No	9	18	3	0
		Yes	15	26	1	1
Glamorgan	Under 50	No	16	7	3	0
		Yes	16	20	8	1
	50–69	No	14	12	3	0
		Yes	27	39	10	4
	70 or over	No	3	7	3	0
		Yes	12	11	4	1

Source: Morrison et al. [1973].

The investigators wished to know whether the effect of the first two variables on survival differed between centers. They knew that the first two variables were interrelated and together could be regarded as a description of the disease state. Preliminary analysis showed that each of the first two variables were independent of variable 4, age, but the third variable, survival, was related to age. To assess differences in the interaction pattern of the first three variables between centers they fitted model A, based on

$$C_{1235}, C_{34}, C_{45},$$

and compared model A with model B, based on

$$C_{123}, C_{125}, C_{35}, C_{34}, C_{45}.$$

The two models differed in that model A included the effect of variable 5 on the first three variables but model B did not. The reason for choosing this pair of models becomes more apparent when the first two variables are relabeled as the single variable "histology," with categories

1. Malignant appearance, minimal inflammation;
2. Malignant appearance, greater inflammation;
3. Benign appearance, minimal inflammation;
4. Benign appearance, greater inflammation.

The configurations for model A become

$$C_{1'35}, C_{34}, C_{45},$$

and for model B become

$$C_{1'3}, C_{1'5}, C_{34}, C_{35}, C_{45}.$$

In the new notation the two models differ by the term $u_{1'35}$, which is the three-factor term measuring the effect of variable 5 (center) on the interaction between variable 3 (survival) and variable 1' (the compound histology variable).

The estimates obtained under model A shown in table 3.7-2 fitted the data well as judged by the X^2 of 40.59 with 38 degrees of freedom. Model B does not differ greatly; the X^2 statistic of 44.72 with 43 degrees of freedom gives a difference in X^2 statistics of 4.13 between the two models. The difference in degrees of freedom is 5, instead of 6 as would be expected for a measure associated with $u_{1'35}$. One degree of freedom is lost because $C_{1'35}$ has one empty cell, and consequently three of the elementary cells are not estimated. See exercise 2 of Section 3.8.4.

Table 3.7-2   Fitted Values Obtained by Treating Histologic Criteria as a Single Variable and Fitting Model that Includes the Three-Factor Effect (Histology × Survival × Center) and All Two-Factor Effects Except Histology × Age

Diagnostic Center	Age	Survived	Minimal Inflammation		Greater Inflammation	
			Malignant Appearance	Benign Appearance	Malignant Appearance	Benign Appearance
Tokyo	Under 50	No	10.11	9.60	8.09	2.53
		Yes	24.66	62.96	25.18	7.87
	50–69	No	8.16	7.75	6.53	2.04
		Yes	19.52	49.84	19.94	6.23
	70 or over	No	1.74	1.65	1.39	0.43
		Yes	2.82	7.20	2.88	0.90
Boston	Under 50	No	4.81	9.42	2.51	0.42
		Yes	10.51	25.80	3.58	0.96
	50–69	No	10.26	20.08	5.35	0.89
		Yes	21.98	53.95	7.49	2.00
	70 or over	No	7.92	15.50	4.13	0.69
		Yes	11.52	28.27	3.93	1.05
Glamorgan	Under 50	No	9.96	7.84	2.72	0.00
		Yes	18.15	23.10	7.26	1.98
	50–69	No	15.50	12.21	4.23	0.00
		Yes	27.70	35.26	11.08	3.02
	70 or over	No	7.54	5.94	2.06	0.00
		Yes	9.15	11.65	3.66	1.00

We conclude that, when the effect of age is taken into account, the relationship between survival and histology does not differ among centers. This is a complex statement, but is considerably simpler than any statement that can be made if we regard degree of chronic inflammatory reaction and nuclear grade as two different variables rather than one compound variable, histology. ■■

3.7.2 *Partitioning*

When we examine configurations to determine whether direct estimates exist, we delete any variable that appears in every configuration. We are able to do this because if direct estimates exist for one category of the common variable, then they must exist for all categories. Deleting the common variable is thus equivalent to splitting the data into segments according to the categories of the common variable and examining each segment. The same argument permits us to compare estimates separately for each segment. For example, to fit the model with configurations

$$C_{1234}, C_{1235}, C_{1345},$$

we can partition the data according to the categories of variable 1, and for each segment we fit the model with configurations

$$C_{234}, C_{235}, C_{345}.$$

The dimensionality of the configurations to be fitted has again been reduced, as in the preceding example.

Partitioning is particularly useful when variable 1 is a dichotomy, such as "dead" and "alive," or "admitted to hospital" and "at risk but not admitted." In large-scale studies where the number of observations falling into the first category is small compared with the number falling into the second category, the data may be obtained from different sources. The "numerator" data may be a complete census and the "denominator" data a random sample. In such circumstances fitting the two portions of the data separately enables us to correct our measures of goodness of fit for the appropriate sampling fractions.

Example 3.7-2   Halothane study: Deaths and operations fitted separately
Bishop and Mosteller [1969] used the partitioning approach in the National Halothane study of death rates following surgery under various anesthetics during the four-year period 1959–1962. In this study two sets of data were collected from each of 34 institutions. One set consisted of information on all patients dying in the institution within six weeks of surgery following a general anesthetic. The other consisted of a sample of those exposed to this risk, namely, patients undergoing surgical operations. Samples of about the same number of operations were taken at each institution. As the total number of operations varied between institutions, each record of an operation included in the sample had to be weighted by the institution sampling factor before data from different institutions could be combined. These sampling factors varied between 5.36 and 74.59.

We give an example from the analysis of a segment of these data. This segment describes patients who had all undergone the same operation, cholecystectomy. The total number of deaths following cholecystectomy was 672. The total number in the random samples of patients who had undergone cholecystectomy was 1,220. When the random samples are adjusted for the appropriate institution sampling factor, we find that the estimated number of cholecystectomy operations is 27,620. Thus the average sampling factor is 22.64.

Models were fitted to each set of data separately. Thus we split the data according to the categories "died" and "had operation." These are not mutually exclusive categories, such as "died" and "survived," but were most convenient because of the manner in which the data were collected. After the models were fitted, we combined both sets of data to give estimates of death rates.

In table 3.7-3 we give measures of goodness of fit and degrees of freedom for two different models. Ignoring the survival variable leaves us with six other variables: anesthetic agent (5 categories), sex (2 categories), estimate of risk (3 categories), age (3 categories) type of operation (3 categories), and time period (2 categories). This gives a total of 540 cells for each set of data, deaths and operations.

Table 3.7-3   Splitting Data: Estimates of Goodness of Fit for Two Models Applied to National Halothane Study Data

	Degrees of Freedom	Measure of Fit
Model 1, All Two-Factor Effects		
Total Deaths	483	383.65
Estimated Total Operations	483	16,800.25
Sample of Operations (Unweighted)	483	407.25
Model 2, All Two-Factor Effects and Three-Factor Effects		
Total Deaths	471	377.83
Estimated Total Operations	471	16,101.91
Sample of Operations (Unweighted)	471	383.39

Source: Bishop and Mosteller [1969].

Model 1, which includes all two-factor effects among the six variables, is based on the 15 configurations C_{12}, C_{13}, C_{14}, C_{15}, C_{16}, C_{23}, C_{24}, C_{25}, C_{26}, C_{34}, C_{35}, C_{36}, C_{45}, C_{46}, C_{56}. Fitting this model to the two separate sets of 540 cells is equivalent to including the variable "survival" as the seventh variable and fitting the model based on the corresponding fifteen configurations each including variable 7 (C_{127}, C_{137}, etc.) to the complete set of 1,080 cells. Similarly, model 2, which fits all 20 three-factor effects to the two data sets, is equivalent to fitting a selection of the four-factor effects to the seven-variable array.

The advantage of partitioning lies in our ability to compute three different measures of goodness of fit. The first measure of fit refers to the 540 cells containing counts of the number of deaths. The second measure refers to the 540 cells containing estimated counts of the number of operations. As each observed count was multiplied by a sampling factor, we get a very large value for the second fit statistic. The first measure is the sum of squared Freeman–Tukey deviates and, as we describe in Chapter 4, is distributed as χ^2. In each of the models shown, its value for the deaths is less than the degrees of freedom, indicating that either model fits well. The effect of the inflating sampling factor on the measure for estimated operations is difficult to assess.

To determine the effect of the sampling factor we fitted the models again to the 540 cells containing the observed counts of the number of operations before they were inflated by the sampling factor. The measure of goodness of fit is again slightly less than the degrees of freedom. From this test statistic, and from examining its cell-by-cell components, we conclude that the observed counts of operations do not exhibit any marked divergence from either of the models. The large value of the second measure can safely be attributed to the sampling factor, and we can use the resulting estimates to compute "smoothed" rates in each cell. The result of fitting either model is to obtain a set of rates that do not exhibit erratic cell-to-cell changes due solely to sampling variation. Further analysis of these data is given in Chapter 4, example 4.3-1. ■ ■

3.7.3 *Fitting by stages*

If a variable only appears in one configuration, it also affords a means of partitioning the data. Thus, for example, the model with configurations $C_{123}, C_{234}, C_{1245}$ can be fitted by partitioning the data according to the categories of variable 5. For each segment the fitted configurations become $C_{123}, C_{234}, C_{124}$.

A particular instance occurs when the unique variable stands alone. The data can be collapsed by summing over the unique variable and a model fitted to the condensed table. The unique variable can then be reintroduced by proportionally adjusting the fitted cell estimates. Thus, if we wished to fit the configurations $C_{123}, C_{234}, C_{134}, C_5$, we can iterate on the arrays $C_{123}, C_{234}, C_{134}$ and get estimates $\{\hat{m}_{ijkl+}\}$. The final estimates are obtained from these intermediary estimates directly,

$$\hat{m}_{ijklm} = \hat{m}_{ijkl+} \frac{x_{++++m}}{N}.$$

We can extend this stepwise fitting to any number of independent variables, either single or compound. It is applicable both for a single body of data and for combining data from several sources, as in the following example.

Example 3.7-3   Soviet population: Combining sets of data in stages

The data in tables 3.7-4–3.7-7 were provided by Ithiel de Sola Pool and come from Selesnick [1970]. Table 3.7-4, taken from the census, shows the distribution of the population in the Soviet Union for 1959 according to:

1. schooling   4 categories;
2. sex   2 categories;
3. age   3 categories;
4. region   2 categories.

In tables 3.7-5, 3.7-6, and 3.7-7 we have information about a fifth dichotomous variable, party membership, and its two-way cross-classification with schooling, sex, and age, respectively. This information describing the Communist party was obtained from publications and has been slightly adjusted so that the marginal totals are consistent with each other.

Suppose we wish to determine what proportion of the urban population are party members. We can compute simply the percentage of urban dwellers in the total population, 49.73%, and take this proportion of the total party membership. We get an estimate of 4,097,000. If we look at the data more closely and observe the disproportionate sex ratios by region and party membership, we do this computation separately for each sex and get estimates of 3,346,000 males and

Table 3.7-4 Distribution of Soviet Population in 1959: Data from Census by Region, Amount of Schooling, Sex, and Age

Sex	Age	Urban Region				Rural Region				Totals
		Less than 4	Years of Schooling 4–7	8–10	More than 10	Less than 4	Years of Schooling 4–7	8–10	More than 10	
Male	16–29	631,571	4,381,454	7,557,356	915,307	1,719,046	5,266,179	5,718,852	153,669	26,342,934
	30–49	1,176,623	4,174,797	5,384,310	1,055,949	2,386,652	4,297,528	3,050,642	316,853	21,843,354
	50 or over	2,376,214	1,807,879	1,187,537	501,741	5,378,665	1,862,132	376,084	68,963	13,559,215
Female	16–29	664,308	2,843,925	9,180,042	1,274,867	2,536,078	4,246,961	5,886,728	232,464	26,865,373
	30–49	3,825,831	4,481,894	6,873,450	1,070,910	7,181,386	4,653,138	3,089,527	241,507	31,417,643
	50 or over	7,432,953	1,888,644	1,312,793	275,948	13,089,113	1,071,037	207,498	24,283	25,302,269

Source: Selesnick [1970].

790,000 females, for a total of 4,136,000. Neither of these estimates utilizes all the information available. We consider instead three ways of combining the four configurations of tables 3.7-4–3.7-7, C_{1234}, C_{15}, C_{25}, and C_{35}. All three methods lead to the same result.

Table 3.7-5 Distribution of Soviet Population in 1959: Published Data by Party and Schooling

| Party Membership | Years of Schooling | | | | |
	Less than 4	4–7	8–10	More than 10	Totals
Yes	0	2,290,478	4,580,957	1,367,696	8,239,131
No	48,398,440	38,685,090	45,243,862	4,764,765	137,092,157
Percentage Party Members	0.0	5.59	9.19	22.30	5.67

Source: Selesnick [1970].

Table 3.7-6 Distribution of Soviet Population in 1959: Published Data by Party and Sex

| Party Membership | Sex | | Totals |
	Male	Female	
Yes	6,632,500	1,606,631	8,239,131
No	55,113,503	81,978,654	137,092,157
Percentage Party Members	10.74	1.92	5.67

Source: Selesnick [1970].

Table 3.7-7 Distribution of Soviet Population in 1959: Published Data by Party and Age

| Party Membership | Age | | | Totals |
	16–29	30–49	50 or Over	
Yes	1,623,109	4,663,348	1,952,674	8,239,131
No	51,585,698	48,597,649	36,908,810	137,092,135
Percentage Party Members	3.05	8.76	5.02	5.67

Source: Selesnick [1970].

Method 1
We discard all information about non–party members. Then we adjust the census information C_{1234} to fit the new marginal totals C_1, C_2, and C_3 given by the numbers of party members in each category. The classical algorithm fits a fully saturated four-variable model. Thus the interactions from the internal cells of C_{1234} are retained, and the cells are adjusted to fit independently derived margins.

Method 2
We could repeat the classical approach for the non–party members, and by combining party and non–party segments obtain a five-variable array incorporating all the information. To obtain this five-variable array in one step, we no

longer use the classical approach, because we are not fitting a fully saturated five-variable model. Instead we treat the four data sets as if they were derived from a single body of data and fit the corresponding unsaturated model. We accomplish this by setting up a five-dimensional array with provisional cell estimates, such as an entry of 1 in each cell, then fit the four configurations by iterative proportional fitting. The resulting cell estimates fit the model with highest-order interactions u_{1234}, u_{15}, u_{25}, and u_{35}.

Method 3

Variable 4, region, only appears in one configuration. So to minimize computing we sum over this unique variable. Then we obtain cell estimates $\{\hat{m}_{ijk+m}\}$ from C_{123}, C_{15}, C_{25}, and C_{35} and adjust these estimates proportionally to reintroduce variable 4. This procedure differs from method 2 by splitting the computations into two stages, first the iterative procedure on the minimum number of cells and second the reintroduction of the solitary variable which can be fitted directly.

Results

Method 3 is illustrated in table 3.7-8. We give three numbers in each cell, (a) the observed array C_{123}, obtained from C_{1234} by summing over variable 4, (b) half the estimated array C_{1235}, the portion referring to party membership, and (c) the estimated percentage party members derived from (a) and (b).

Table 3.7-8 Result of Combining Four Sets of Data from Tables 3.7-4–3.7-7, (a) Observed Party Plus Non-Party Members, (b) Estimated Party Members, (c) Percent Party Members

Sex	Age		Less than 4	4–7	8–10	More than 10	Totals
				Years of Schooling			
Male	16–29	a	2,350,617	9,647,633	13,276,208	1,068,976	26,343,434
		b	0.0	263,941	912,765	156,054	1,332,760
		c	0.0	2.74	6.88	14.60	5.06
	30–49	a	3,563,275	8,472,325	8,434,952	1,372,802	21,843,354
		b	0.0	962,909	2,124,020	601,249	3,688,178
		c	0.0	11.37	25.18	43.80	16.88
	50 or over	a	7,754,879	3,670,011	1,563,621	570,704	13,559,215
		b	0.0	688,363	589,989	333,213	1,611,565
		c	0.0	18.76	37.73	58.39	11.89
Female	16–29	a	3,200,386	7,090,886	15,066,770	1,507,331	26,865,373
		b	0.0	37,281	206,159	46,910	290,350
		c	0.0	0.53	1.37	3.11	1.08
	30–49	a	11,007,217	9,135,032	9,962,977	1,312,417	31,417,643
		b	0.0	214,929	592,609	167,632	1,002,170
		c	0.0	2.35	5.95	12.77	3.19
	50 or over	a	20,522,066	2,959,681	1,520,291	300,231	25,302,269
		b	0.0	123,058	155,413	62,639	341,110
		c	0.0	4.16	10.22	20.86	1.35

Table 3.7-9 shows the estimated number of urban party members obtained when variable 4 is reintroduced. The introduction is proportional. For instance, for males aged 16–29 with more than 10 years of schooling from table 3.7-4, we have

Table 3.7-9   Estimates of Urban Party Membership from Table 3.7-8

Sex	Age	Less than 4	4–7	8–10	More than 10	Totals
			Years of Schooling			
Male	16–29	0.0	119,868	519,583	133,621	773,072
	30–49	0.0	474,480	1,355,832	462,476	2,292,788
	50 or over	0.0	339,093	448,084	292,948	1,080,125
Female	16–29	0.0	14,952	125,611	39,676	180,239
	30–49	0.0	105,450	408,840	136,785	651,075
	50 or over	0.0	78,526	134,201	57,573	270,300
						5,247,599

the proportion urban as $915,307/(915,307 + 153,669) = 0.856247$. The estimated number of party members, both urban and rural, from table 3.7-8 row (b) is 156,054. The resulting estimate of $156,054 \times 0.856247 = 133,621$ urban party members appears in table 3.7-9. Our final estimate of the total urban party membership, 5,247,599, is considerably higher than either of our two previous estimates. ■ ■

3.7.4   *Fitting incomplete models to seemingly complete data*

We stress the importance of the hierarchy principle, because it defines a set of models for which the sufficient configurations may be readily defined. We can enlarge this set of models by taking subsets of the available cells. When the subsets are incomplete we use the fitting techniques described in Chapter 5. A different formulation of the log-linear model, described by Worcester [1971], leads to definition of the subsets for dichotomous variables, and we describe this with an example. In Chapter 8 we show another way of enlarging the set of models, by rearranging the data.

Worcester's log-linear model

In general the correspondence between the sufficient configurations and the included u-terms does not exist for nonhierarchical models, and it is necessary to treat each case individually. Worcester [1971] formulates her model rather differently. In effect she defines for three-dimensional dichotomies

$$\log m_{ijk} = w + \delta_i w_1 + \delta_j w_2 + \delta_k w_3 + \delta_{ij} w_{12} + \delta_{jk} w_{23} + \delta_{ik} w_{13} + \delta_{ijk} w_{123},$$

where i, j, and k take the value 1 when the attribute is present and 2 when it is absent. When δ has only subscripts of 1 it takes the value 1, thus

$$\delta_1 = \delta_{11} = \delta_{111} = 1,$$

and all other δ take the value zero.

The w-terms appear with single subscripts because they each have a single value. For 2^n tables the u-terms also have a single absolute value but must be doubly subscripted because they are positive or negative in adjacent cells. There is a direct relationship between the w-terms and the absolute value of the u-terms, for instance,

$$w_{123} = 8|u_{123}|,$$
$$w_{12} = 4[|u_{12}| - |u_{123}|],$$

etc.

For the hierarchical models the cell estimates obtained using either notation are identical.

The w-notation is more easily understood than the u-term notation when the variable of prime interest is presence or absence of a response, and the other variables are presence or absence of various stimuli. We reproduce the example that Worcester gives.

Example 3.7-4   Factors affecting thromboembolism: Worcester's synergistic model

The first response variable is presence or absence of thromboembolism. The stimuli variables are variable 2 (use or nonuse of oral contraception) and variable 3 (smoking or nonsmoking). We see from table 3.7-10 that the total number of cases is 58. As this was a case-control study, both this total and twice 58, the total number of controls, is fixed by the study design.

Table 3.7-10 Thromboembolism Data Analyzed by Worcester [1971]

| Type of Patient | Smoker | | Nonsmoker | | |
| | Contraceptive User | | Contraceptive User | | |
	Yes	No	Yes	No	Totals
Thromboembolism	14	7	12	25	58
Control	2	22	8	84	116

The incomplete model

In Worcester's notation, w_{123} is described as the "second-order relative odds" and is a measure of synergism. If this measure is large, it shows that using contraceptives and smoking enhances the risk of disease more than either factor alone. We wish to retain this term and so cannot fit any hierarchical models.

The kernel of the likelihood for the saturated model is

$$Nw + x_{1++}w_1 + x_{+1+}w_2 + x_{++1}w_3 + x_{11+}w_{12} + x_{1+1}w_{13}$$
$$+ x_{+11}w_{23} + x_{111}w_{123}.$$

If we put $w_{13} = w_{23} = 0$ but retain w_{123}, we have the sufficient statistics N, x_{1++}, $x_{+1+}, x_{++1}, x_{11+}$, and x_{111}.

As in the u-term notation, the sufficient statistics are MLEs of their expected values. We have $x_{111} = \hat{m}_{111}$ and can therefore remove this cell from the array of cells to be fitted. Similarly, $x_{11+} = \hat{m}_{11+}$, and as $x_{112} = x_{11+} - x_{111}$ we can also remove x_{112}. We are left with an incomplete array.

From the incomplete array remaining after removal of the two fixed cells, we derive marginal configurations C'_{12} and C'_3, with elements

$$\{x'_{ij+}\} = \begin{Bmatrix} 0 & x_{12+} \\ x_{21+} & x_{22+} \end{Bmatrix}$$

and

$$\{x'_{++k}\} = \begin{Bmatrix} x_{++1} - x_{111} \\ x_{++2} - x_{112} \end{Bmatrix}.$$

These are nonredundant functions of the required sufficient statistics and thus form a set of sufficient statistics.

We use the methods of Chapter 5 to fit an incomplete data array (see table 3.7-11). In this instance the estimates are derived directly by proportionally adjusting C'_{12}; thus, for example,

$$\hat{m}_{121} = x'_{12+} \frac{x'_{++1}}{N'}$$

$$= x_{12+} \left(\frac{x_{++1} - x_{111}}{N - x_{111} - x_{112}} \right).$$

For purposes of comparison, we show in table 3.7-12 estimates obtained under model 2, $w_{123} = 0$. This model is the usual hierarchical model with $u_{123} = 0$.

Table 3.7-11   Model 1 with $w_{13} = w_{23} = 0$ Applied to Thrombo-embolism Data of Table 3.7-10

| | Smoker | | Nonsmoker | | |
| | Contraceptive User | | Contraceptive User | | |
Type of Patient	Yes	No	Yes	No	Totals
Thromboembolism	14	6.703	12	25.297	58
Control	2.095	22.203	7.905	83.797	116

Table 3.7-12   Model 2 with $w_{123} = 0$ Applied to Thromboembolism Data of Table 3.7-10

| | Smoker | | Nonsmoker | | |
| | Contraceptive User | | Contraceptive User | | |
Type of Patient	Yes	No	Yes	No	Totals
Thromboembolism	12.338	8.662	13.662	23.338	58
Control	3.662	20.338	6.338	85.662	116

Comparison of models

We see from table 3.7-13 that the incomplete model, model 1, provides a good fit. The X^2 value with two degrees of freedom is 0.0244. It is a much closer fit than the hierarchical model with $u_{123} = 0$. We conclude that the data are in accordance

Table 3.7-13   Goodness of Fit for Models in Tables 3.7-11 and 3.7-12

	X^2	G^2	Degrees of Freedom
Model 1	0.0244	0.0243	2
Model 2	2.2209	2.3533	1

with the hypothesis that the use of contraceptives is positively associated with thromboembolism, particularly among smokers, but among those who do not use contraceptives, smoking is not associated with the disease.

Worcester discusses other models for these data, but none fit as well as model 1. It corresponds to the u-term model with the constraint that $u_{13} = u_{23} = u_{123}$, but the terms are not all equal to zero. In previous examples of hierarchical models we only considered the possibility of putting u-terms equal to zero.

In the discussion on hierarchical models we stressed the need, if the model with $u_{123} = 0$ is found to be a good fit, for lower-order models to be fitted subsequently. Here we see that if $u_{123} \neq 0$, we may be able to gain more insight by fitting a model that, in effect, holds part of the table constant and fits a hierarchical model to the remaining incomplete table. We analyze these data in a different way in example 4.4-2. ∎∎

3.8   Degrees of Freedom

Most of the statistics that we use to assess the goodness of fit of a model are distributed asymptotically as χ^2. The associated number of degrees of freedom depends on the data structure and the number of independent parameters in our model. We describe two straightforward ways of computing the degrees of freedom. These are suitable for any dimension, but for simplicity we use three-dimensional notation to describe them and also to describe the adjustment needed for empty cells. We then proceed to some general formulas.

3.8.1   *Three dimensions*

Each of the u-terms of the saturated model consists of a set of parameters. The constraints on these parameters determine the number of degrees of freedom associated with each u-term. Once we have computed the number of degrees of freedom associated with each u-term we can readily obtain the number of degrees of freedom for any unsaturated model.

The three-factor effect u_{123} consists of a three-dimensional array of parameters $I \times J \times K$. It can be thought of as K layers of two-way tables, each of size $I \times J$. The marginal totals of each two-way table are constrained to sum to zero. So we have $(I - 1)(J - 1)$ independent parameters in each two-way table. By interchanging dimensions, we find that u_{123} has $(I - 1)(J - 1)(K - 1)$ independent parameters. The number of independent parameters associated with each u-term in three dimensions, given in table 3.8-1, sums to IJK, the total number of cells in the complete three-way array.

We can compute the number of degrees of freedom, V, associated with each unsaturated model by one of two methods.

Method (a). We count the number of independent parameters set equal to zero. For example, for model 2 of table 3.3-1, we have $u_{123} = u_{12} = 0$. Adding the parameters associated with each of these terms gives $(I - 1)(J - 1)(K - 1) + (I - 1)(J - 1)$ degrees of freedom, or more concisely, $V = K(I - 1)(J - 1)$.

Method (b). We count the number of independent parameters estimated, T_p, and subtract this number from the total number of cells estimated, T_e. For example, model 4 of table 3.3-1 fits $u, u_1, u_2,$ and u_3. The number of independent parameters estimated is

$$T_p = 1 + (I - 1) + (J - 1) + (K - 1).$$

Subtracting this sum from $T_e = IJK$ gives

$$V = IJK - (I + J + K) + 2$$

degrees of freedom for models that provide nonzero estimates in every cell. If nonzero estimates are not provided for every cell, the degrees of freedom V computed by these methods must be adjusted.

Table 3.8-1 Degrees of Freedom Associated with Each u-term in Three Dimensions

u-term	Number of Independent Parameters			
u_{123}	$(I-1)(J-1)(K-1) = IJK$	$- [IJ + JK + IK]$	$+ [I + J + K]$	$- 1$
u_{12}	$(I-1)(J-1) =$	$[IJ]$	$- [I + J]$	$+ 1$
u_{13}	$(I-1)(K-1) =$	$[IK]$	$- [I + K]$	$+ 1$
u_{23}	$(J-1)(K-1) =$	$[JK]$	$- [J + K]$	$+ 1$
u_1	$(I-1) =$		I	$- 1$
u_2	$(J-1) =$		J	$- 1$
u_3	$(K-1) =$		K	$- 1$
u	$1 =$			1
Total	IJK			

Adjustment for empty cells

Adjustment is only necessary when empty estimated cells occur. We do not adjust for empty observed cells if the model provides nonzero estimates for these cells. Zero estimates for some cells occur in two ways.

(i)   The u-terms set equal to zero in the model cannot be measured from the data. In this pathological situation the configurations we are fitting may have nonzero estimates in every cell, but some empty elementary cells will remain empty, as in example 3.3-1.

(ii)   The sporadic empty observed cells are so arranged that some of the configuration cells are empty. In this more usual situation, we cannot obtain estimates for the empty elementary cells that sum to the empty configuration cells. As an example, consider a $2 \times 3 \times 2$ array with $x_{111} = x_{112} = 0$ and all other cell entries positive. For C_{12} we have $\hat{m}_{11+} = x_{11+} = 0$, and any model that fits C_{12} will have $\hat{m}_{111} = \hat{m}_{112} = 0$. Although two cells do not contain structural zeros, we cannot obtain nonzero estimates for them if our model includes u_{12}.

The procedure described below gives negative degrees of freedom in situation (i), thus indicating the model is inappropriate for the data, and gives the correct degrees of freedom in situation (ii). The steps are:

1.   compute V by either method (a) or (b);
2.   count the number z_e of elementary cells with zero estimates;
3.   count the number z_p of parameters that cannot be estimated;
4.   get adjusted degrees of freedom V' from the relation

$$V' = V - z_e + z_p. \tag{3.8-1}$$

The general method (b) computes $V = T_e - T_p$, where T_e is the total number of cells, and T_p the number of parameters fitted. In the $2 \times 3 \times 2$ example above, for all models including u_{12} we have zero estimates for two cells; thus $z_e = 2$, and we must correct T_e by subtracting 2, because we fit only ten cells. The configuration C_{12} has one empty cell, $z_p = 1$, and this reduces the number of parameters in u_{12} by 1, so the total number of parameters fitted is $T_p - 1$. The adjustments z_e and z_p are thus corrections to T_e and T_p, and we have

$$V' = (T_e - z_e) - (T_p - z_p), \tag{3.8-2}$$

which is equivalent to expression (3.8-1).

Care must be taken in counting z_p when more than one empty cell appears in any configuration. Suppose the configuration C_q of order s that corresponds to the term u_q has adjacent zeros arranged so that zeros persist in the related configurations of order $s - 1$. Then the zeros of C_q satisfy some of the constraints on the parameters of u_q. Consequently, if we define $z(u_q)$ as the loss of independent parameters from u_q, then $z(u_q)$ will be less than the observed number of zeros. Moreover, the related u-terms of order $s - 1$ also lose parameters when the configurations of order $s - 1$ have zeros.

Thus we must examine the configurations corresponding to all the u-terms of the model, not just the minimal set that is used for fitting. For each configuration we estimate $z(u_q)$, which is less than the observed number of zeros whenever zeros persist in the related configurations of one order lower but otherwise equals the number of zeros. Finally, if the model has N_u subscripted u-terms, we determine

$$z_p = \sum_{q=1}^{N_u} z(u_q).$$

We turn now to an example of adjusting for empty cells before giving an alternative formulation for models with all configurations of the same dimension and a simple method for models with closed-form estimates. Further examples of computing degrees of freedom when empty cells are present are given in Section 5.4.

Example 3.8-1   Recurrence of rheumatic fever: Adjusting degrees of freedom for empty cells

Figure 3.8-1 reproduces a diagram from Spagnuolo, Pasternack, and Taranta [1971]. They followed 393 patients with rheumatic fever and recorded whenever these patients had a streptococcal infection and whether infection was followed by a recurrence of rheumatic fever. Their objective was to find the factors that influence the recurrence rate. The diagram shows the distribution of 711 episodes of infection according to the "control" variables, variables that had to be taken into account when assessing the effect of other variables. We use the diagram only to show how degrees of freedom are adjusted for empty cells.

For this purpose we combine their first two variables, "rise in antibody" and "ASO rise in tube numbers" into a single variable with four categories.

Variable 1.   Laboratory results
                1 = ASO rise of 0 or 1 in presence of antibody rise
                2 = ASO rise of 2 or 3 in presence of antibody rise
                3 = ASO rise of 4, 5, or 6 in presence of antibody rise
                4 = ASO rise of 0 or 1 in absence of antibody rise

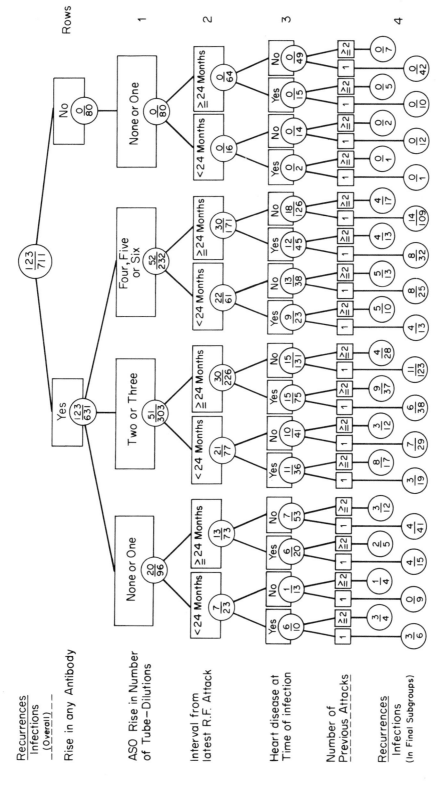

Figure 3.8-1   Data from Spagnuolo, Pasternack, and Taranta [1971].

The remaining variables are dichotomies which we have numbered as follows:

Variable 2.   Interval from last rheumatic fever attack
                 1 = less than 24 months
                 2 = 24 months or more

Variable 3.   Heart disease
                 1 = Yes
                 2 = No

Variable 4.   Number of previous attacks
                 1 = only the initial attack
                 2 = more than the initial attack

Variable 5.   Recurrence of rheumatic fever
                 1 = Yes
                 2 = No.

Structure of the diagram
In each cell of the table a ratio appears,

$$\frac{\text{number with recurrence}}{\text{number with recurrence} + \text{number without recurrence}}.$$

Thus to obtain the cell counts for numbers without recurrences, we must at each level subtract the numerator from the denominator. If we perform this subtraction, the row labeled 1 gives the configuration C_{15}. Similarly, the numbers in the next row give the configuration C_{125}, the third row gives C_{1235}, and the final row gives the elementary cells for the five-dimensional array.

We observe that there are zeros in the numerators of the eight ratios on the right of row 4. Thus the eight cells $(4, j, k, l, 1)$ are empty for $j = 1, 2$; $k = 1, 2$; and $l = 1, 2$. These zeros persist up to row 1. There is one other empty elementary cell, $(1, 1, 2, 1, 1)$, bringing the total to 9.

Degrees of freedom
The eight cells $(4, j, k, l, 1)$ contribute empty cells to all those configurations in which variables 1 and 5 are present. Table 3.8-2 gives the possible number of independent parameters for all three-factor and lower-order terms, together with the number of empty cells that occur in the corresponding configuration and the number of parameters lost.

To obtain degrees of freedom for the model that includes all three-factor effects we count parameters lost, not only in the three-dimensional configurations but also in the configuration C_{15}. The number of parameters fitted is thus not the possible 48, but is reduced by four to 44.

When the model is fitted, the cells that sum to the six empty three-dimensional configuration cells $(4, j, +, +, 1), (4, +, k, +, 1)$, and $(4, +, +, l, 1)$ remain empty. These are again the eight cells $(4, j, k, l, 1)$. The other empty cell $(1, 1, 2, 1, 1)$ has an estimate for this model and is not counted in z_e. Thus, the number of elementary cell estimates obtained is not 64, but is reduced by 8 to 56.

The corrected number of degrees of freedom is the difference between the number of cells estimated, 56, and the number of parameters fitted, 44. So we have twelve

Table 3.8-2   Effect of Empty Cells in Reducing Number of Parameters Estimated when Fitting Models

Three-Factor Effects				Two-Factor Effects				Single-Factor and Main Effects		
Effect	Possible Param- eters	Empty Cells	Param- eters not Estimated	Effect	Possible Param- eters	Empty Cells	Param- eters not Estimated	Effect	Possible Param- eters	Empty Cells
123	3	0	0	12	3	0	0	1	3	0
124	3	0	0	13	3	0	0	2	1	0
125	3	2	1	14	3	0	0	3	1	0
134	3	0	0	15	3	1	1	4	1	0
135	3	2	1	23	1	0	0	5	1	0
145	3	2	1	24	1	0	0	Main	1	0
234	1	0	0	25	1	0	0			
235	1	0	0	34	1	0	0			
245	1	0	0	35	1	0	0			
345	1	0	0	45	1	0	0			
Totals	22	6	3		18	1	1		8	

Note: Based on data from Spagnuolo, Pasternack, and Taranta [1971] (see Fig. 3.8-1).

Table 3.8-3   Adjustment of Degrees of Freedom Using Parameter and Cell Counts from Table 3.8-2

	All Three-Factor Effects Model	All Two-Factor Effects Model
Total Number of Available Cells	$T_e = 64$	64
Total of Possible Parameters	$T_p = 48$	26
Possible Number of Degrees of Freedom	$V = 16$	38
Number of Parameters Not Estimated	$z_p = 4$	1
Number of Cells with Zero Estimates	$z_e = 8$	8
Correct Degrees of Freedom	$V' = 12$	31

degrees of freedom, and V' differs from V by four degrees of freedom, although the number of cells estimated is reduced by 8.

The corresponding computations for the model fitting all two-factor effects are given in table 3.8-3. The same eight cells remain empty because the cell $(4, +, +, +, 1)$ of C_{15} is empty. The number of parameters fitted is, however, only reduced by 1. The result is that V and V' differ by seven degrees of freedom. ■■

3.8.2   *General form for models with all configurations of same dimension*

In any situation we can derive degrees of freedom by taking the number of cells estimated minus the number of independent parameters fitted, but sometimes we can find closed-form expressions. To do this we need to introduce some notation that will cover any number of dimensions, say r.

Notation

We denote the number of categories for each of r variables by $v_1, v_2, \ldots, v_r$, respectively. We also need to refer to groups of variables and for this purpose use double subscripts. Thus the product of the number of categories of any two variables may be written $v_{i_1} v_{i_2}$, where the numbers indicate that we have two different variables and i denotes that we may have any two. With this notation we can define products without specifying which variables are included, as follows:

$$\prod_{j=1}^{2} v_{i_j} = v_{i_1} v_{i_2},$$

$$\prod_{j=1}^{s} v_{i_j} = v_{i_1} v_{i_2} \cdots v_{i_s},$$

where $s \leqq r$. We define the sum of all possible product pairs as

$$P_2 = \sum_i \prod_{j=1}^{2} v_{i_j}$$

and note that the summation is meant to extend over all $\binom{r}{2}$ possible products $v_{i_1} v_{i_2}$. Extending this notation, we have

$$P_0 = 1 = \prod_{j=1}^{0} v_{i_j},$$

$$P_1 = \sum_{i=1}^{r} v_i = \sum_i \prod_{j=1}^{1} v_{i_j},$$

and in general,

$$P_s = \sum_i \prod_{j=1}^{s} v_{i_j},$$

where the sum extends over $\binom{r}{s}$ products each of s terms, for $s < r$. The last term is

$$P_r = \sum_i \prod_{j=1}^{r} v_{i_j} = v_1 v_2 \cdots v_r,$$

which equals the total number of elementary cells.

Terms of same order

For the model that includes all u-terms of order k the sufficient statistics are the $\binom{r}{k}$ configurations of dimension k. The highest-order u-term of the ith configuration has $(v_{i_1} - 1)(v_{i_2} - 1) \cdots (v_{i_k} - 1)$ independent parameters. We can expand and then sum this expression to give

$$\prod_{j=1}^{k} (v_{i_j} - 1) = \prod_j v_{i_j} - \sum_i \prod_j^{k-1} v_{i_j} + \sum_i \prod_j^{k-2} v_{i_j} + \cdots + \prod_j^{0} v_{i_j}$$

$$= \sum_{s=0}^{k} \left[(-1)^{k-s} \sum_i \prod_j^{s} v_{i_j} \right], \qquad (3.8\text{-}3)$$

where the second sum is taken over $\binom{k}{s}$ products, each of s terms.

We have a u-term of order k for each of the $\binom{r}{k}$ configurations. Thus if we compute the number of parameters for all u-terms of order k, we have the sum of $\binom{r}{k}\binom{k}{s}$ products of s terms. Each unique product will, by symmetry, appear the same number of times. We have defined P_s as the sum of the $\binom{r}{s}$ unique products of s-terms. Thus from (3.8-3), the number of independent parameters for all u-terms of dimension k may be written as

$$\sum_{s=0}^{k}\left[(-1)^{k-s}\frac{\binom{r}{k}\binom{k}{s}}{\binom{r}{s}}P_s\right] = \sum_{s=0}^{k}(-1)^{k-s}\binom{r-s}{r-k}P_s. \tag{3.8-4}$$

As all u-terms of lower order must be included (by the hierarchy principle), we can get the number of parameters by summing (3.8-4) over all lower orders. Hence we obtain the degrees of freedom

$$V = P_r - 1 - \sum_{j=1}^{k}\sum_{s=0}^{j}(-1)^{j-s}\binom{r-s}{r-j}P_s. \tag{3.8-5}$$

Special case

The closed form of expression (3.8-5) is particularly useful if all variables are of the same length, say v. Then we have $P_s = \binom{r}{s}v^s$, and the degrees of freedom become

$$P_r - 1 - (r)! \sum_{j=1}^{k}\sum_{s=0}^{j}\frac{(-1)^{j-s}v^s}{(r-j)!(j-s)!(s)!}. \tag{3.8-6}$$

3.8.3   *Models with closed-form cell estimates*

In Section 3.4 we defined a class of models that can be fitted directly. In the simplest examples the estimates for each cell are of the form

$$m_\theta = \frac{\prod_i x_{\theta_i}}{\prod_j x_{\varphi_j}},$$

where the x_{θ_i} are cells of the sufficient configurations and the x_{φ_j} are cells from the common subconfigurations. To obtain the number of independent parameters fitted, we count the total number of cells, T_{θ_i}, in each sufficient configuration C_{θ_i} and the number of cells T_{φ_j} in each redundant configuration C_{φ_j}. We then obtain the degrees of freedom

$$V = T_e - \left(\sum_i T_{\theta_i} - \sum_j T_{\varphi_j}\right). \tag{3.8-7}$$

Computing degrees of freedom for direct models—Illustrative examples

1. In two dimensions the sufficient configurations C_1 and C_2, with members x_{i+} and x_{+j}, have in common the grand total N. Thus the parameters fitted are $(I + J) - 1$, and we have the degrees of freedom

$$V = IJ - I - J + 1 = (I - 1)(J - 1).$$

2. The three-dimensional model (model 2 of table 3.3-1) with $u_{12} = u_{123} = 0$ has sufficient configurations C_{13} and C_{23}. They have C_3 in common. Thus, the

degrees of freedom are

$$V = IJK - (IK + JK - K) = (I - 1)(J - 1)K.$$

Partial association model in general

The preceding examples are both particular cases of a frequently occurring model, the partial association model, in which one two-factor term, say u_{12}, is set equal to zero. Of all models that can be fitted directly, this model has the most parameters. The sufficient statistics consist of two matrices C_{θ_1} and C_{θ_2}, each of dimension $r - 1$. Thus θ_1 includes all variables except variable 1, and θ_2 all variables except variable 2. The total number of cells in the two configurations can be written as a function of the number of elementary cells T_e, namely,

$$\frac{T_e}{v_1} + \frac{T_e}{v_2}. \tag{3.8-8}$$

Similarly, the common configuration C_{θ_3} is of dimension $r - 2$, and contains all variables except 1 and 2, so its cells total to

$$\frac{T_e}{v_1 v_2}. \tag{3.8-9}$$

From expressions (3.8-8) and (3.8-9) we obtain the number of degrees of freedom

$$V = T_e - T_e\left(\frac{1}{v_1} + \frac{1}{v_2} - \frac{1}{v_1 v_2}\right) = T_e\frac{(v_1 - 1)(v_2 - 1)}{v_1 v_2}. \tag{3.8-10}$$

When all variables have v categories, this reduces to

$$V = v^{r-2}(v - 1)^2. \tag{3.8-11}$$

When $v = 2$, this is 2^{r-2} or one-quarter of the total number of elementary cells.

3.8.4   *Exercises*

1. Show from formula (3.8-6) or otherwise that the degrees of freedom for the model that includes all two-factor effects in three dimensions, when each variable has v categories, is $v^3 - 3v^2 + 3v - 1$.

2. In example 3.7-1, when we fitted model A to Morrison's data we had one empty configuration cell resulting in estimates of zero in three elementary cells. Check that the degrees of freedom are adjusted from 40 to 38. Show that when model B is fitted, there are no empty cells and so the degrees of freedom are 46. To compare models we kept three cells empty when we fitted model B to the rest of the array. Check that this incomplete model has 43 degrees of freedom.

3. When the model based on configurations C_{12} and C_{13} is fitted to an $I \times J \times K$ array with no empty cells, we have the degrees of freedom $V = I(J - 1)(K - 1)$. Show that if one category of variable 1 has no observations, the loss of parameters is $J - 1$ for u_{12}, $K - 1$ for u_{13}, and 1 for u_1. Hence show that the number of degrees of freedom corrected for zeros is

$$V' = (I - 1)(J - 1)(K - 1).$$

4 Formal Goodness of Fit: Summary Statistics and Model Selection

4.1 Introduction

This chapter presents plans of campaign appropriate for the following different objectives:

1. We want to fit a model that describes the data. Such a model aids our understanding of the data and the relationship of the underlying variables. We also use fitted or "smoothed" estimates for calibration purposes, for a summary description, or for comparison of two or more data sets. When we fit a log-linear model to multivariate discrete data, the fitted cell estimates provide a smoothed description of the data because the most important structural elements are retained and random sampling fluctuations are damped.

2. We want to obtain summary statistics. Discrete data often have one dichotomous variable, such as "alive-dead," that is of prime interest, and we wish to obtain for each of several subsets of data a comparable measure of this variable. We can achieve this by fitting a model and using the fitted estimates to compute adjusted rates. In other circumstances, fitting a model enables us to determine that some variables can safely be collapsed, and we can confidently use one or more reduced arrays as a summary of the larger original array.

3. We want to detect "outliers." In contingency tables all individuals who are alike form the observed count in an elementary cell. We detect whether some cells differ largely from the general pattern of the others by fitting models and examining the goodness of fit for each cell.

4. We want to test whether two specified variables are associated and to assess the magnitude of the association. We set up the null hypothesis by fitting a model with the two-factor effect of interest set equal to zero, then fit a second model that differs from the first only by including this two-factor effect. The difference in measures of goodness of fit between the two models becomes our test statistic, provided that we select a pair of models such that the second model fits the data well.

Whatever our purpose, we need to measure how well the selected model fits the data. This presupposes that a suitable model has been selected; to some extent the objectives and the sampling scheme define an appropriate subset of models, but within the subset we usually have a choice. The final selection is the outcome of a search procedure. Most search procedures that have been suggested fit successive models and examine summary measures of goodness of fit at each stage. In this chapter we therefore discuss measures of goodness of fit before turning to the problem of search selection. We deal first with summary measures of goodness of fit, then with summary descriptive statistics, and finally with internal measures that relate to individual cells, particular parameters, or specific log contrasts.

4.2 Summary Measures of Goodness of Fit

Throughout this book we use two summary statistics for assessing the overall goodness of fit of our model, Pearson's X^2 and twice the log-likelihood, G^2. The formal properties of these measures are given in Chapter 14. We now consider the practical advantages of these and other suggested measures for overall fit.

4.2.1 *Pearson X^2 compared with likelihood statistic G^2*

As our initial discussion does not depend on the structural dimensions of the data, we use a simple subscript running over all I elementary cells. Thus we have sets of observed values $\{x_i\}$ and fitted values $\{\hat{m}_i\}$.

The Pearson X^2

As will be shown in Chapter 14, the value

$$X^2 = \sum_i \frac{(x_i - \hat{m}_i)^2}{\hat{m}_i} \tag{4.2-1}$$

is asymptotically distributed as χ^2 with the number of degrees of freedom appropriate for the particular set of $\{\hat{m}_i\}$.

This statistic X^2 appears in most elementary statistics books that deal with two-dimensional contingency tables, and its familiarity makes it readily acceptable. In particular, for the 2×2 table $X^2 = z^2$, where z is the standard deviate used to compare two binomial proportions. For 2×2 tables with small cell entries, the well-known Yates [1934] correction is easy to use and often improves the agreement between the distribution of X^2 and its asymptotic χ^2 distribution under the null hypothesis. (See Conover [1968], Grizzle [1967], Mantel and Greenhouse [1968], and Plackett [1964] for discussion of when it may be useful.) As the corrected statistic is only useful for 2×2 tables, we use double subscripts to define it:

$$X_c^2 = \sum_{i,j} \frac{(|x_{ij} - \hat{m}_{ij}| - \frac{1}{2})^2}{\hat{m}_{ij}}$$

for $i = 1, 2$, and $j = 1, 2$.

Example 4.2-1   Responses of lymphoma patients to combination chemotherapy (Skarin et al. [1973])

Table 4.2-1 gives the number of patients suffering from lymphocytic lymphoma who responded to a course of combination chemotherapy. The response rate is 58 % for females but only 28 % for males. Although this seems a large percentage difference, the numbers are small, and fitting the model of independence gives

$$X^2 \text{ without correction} = (2.2)^2 \left(\frac{1}{10.8} + \frac{1}{7.2} + \frac{1}{7.2} + \frac{1}{4.8} \right)$$

$$= 2.80,$$

$$X^2 \text{ with Yates correction} = 1.67.$$

These statistics indicate that we have no reason to reject the hypothesis that the observed sex difference in response rate is the result of chance. The numbers of

Table 4.2-1 Responses of Lymphoma Patients to Combination Chemotherapy: Distribution by Sex

Sex	No Response	Response	Percentage Response
Male	13 (10.8)	5 (7.2)	27.7
Female	5 (7.2)	7 (4.8)	58.3
Total	18	12	40.0

Source: Skarin et al. [1973].

Note: Values obtained under model of independence are given in parentheses.

observations are, however, very small. With such small numbers the observed difference must be about 50% or greater to reach significance at the 5% level. In a sufficiently large sample a difference of 30% is significant, and in the absence of further information we are left wondering whether the observed 30% difference would persist in a larger sample. In this example, other information is available, and when we introduce a further variable the apparent interaction between response rate and sex disappears (see example 4.4-4). ■■

Likelihood statistic G^2

The likelihood ratio statistic is defined as

$$G^2 = -2 \sum_i x_i \log \frac{\hat{m}_i}{x_i}$$

$$= 2 \sum_i x_i \log \frac{x_i}{\hat{m}_i}, \tag{4.2-2}$$

where logarithms are to the base e. The statistic G^2 is also distributed asymptotically as χ^2 with degrees of freedom appropriate for $\{\hat{m}_i\}$ (see Chapter 14, theorem 14.9-4).

The G^2 statistic is less familiar than X^2 for some readers. For desk-top calculations it has the additional disadvantage that we need to refer to a table of logarithms. It does, however, have important properties that X^2 does not possess: it is the statistic that is minimized by MLEs, and it can conveniently be broken into parts in two ways, conditionally and structurally, as we now show.

1. *G^2 appropriate for MLEs*

The statistic G^2 is inherently suitable for MLEs. If we define

$$L(\mathbf{x}) = \sum_i x_i \log x_i, \tag{4.2-3}$$

$$L(\hat{\mathbf{m}}) = \sum_i x_i \log \hat{m}_i,$$

then from (4.2-2) we have

$$G^2 = -2(L(\hat{\mathbf{m}}) - L(\mathbf{x})). \tag{4.2-4}$$

For the saturated model, $\hat{m}_i = x_i$ for all i, and so $L(\hat{\mathbf{m}}) = L(\mathbf{x})$ and $G^2 = 0$. For any unsaturated model, say model (1), with the constraints that $\hat{m}_i^{(1)} > 0$ for all i and $\sum_i \hat{m}_i^{(1)} = \sum_i x_i$, we have

$$L(\hat{\mathbf{m}}^{(1)}) < L(\mathbf{x}). \tag{4.2-5}$$

Our fitted MLEs are obtained by maximizing $L(\hat{\mathbf{m}}^{(1)})$, subject to the constraints above and others specific to the model. Thus we ensure that G^2 is a minimum. In Chapter 10 we discuss other estimators belonging to the "best asymptotically normal" (BAN) class, such as minimum χ^2 estimates, $\{\tilde{m}_i\}$, that are obtained by minimizing the value of X^2. Suppose that for the same model the $\{\tilde{m}_i\}$ and our $\{\hat{m}_i\}$ differ. If we compute both summary statistics for each set of estimates, the $\{\hat{m}_i\}$ give the smaller value of G^2, the $\{\tilde{m}_i\}$ the smaller value of X^2. (See Berkson [1972] for examples.) As MLEs necessarily give minimum values of G^2, it is appropriate to use G^2 as a summary statistic measuring this goodness of fit, although the reader will observe that, in those examples where we compute both X^2 and G^2, the difference in numerical value of the two is seldom large enough to be of practical importance.

2. Conditional breakdown of G^2

When we have two linear models, model (1) and model (2), we say they are *nested* if model (2) contains only a subset of the u-terms contained in model (1). For nested models we can break down G^2 conditionally.

In the simplest conditional breakdown of model (2), the value of $G^2(2)$, which measures the fit of the estimates $\{\hat{m}_i^{(2)}\}$, is broken into two parts:

(i)   a measure of the distance of the estimates $\{\hat{m}_i^{(2)}\}$ from those obtained under model (1), $\{\hat{m}_i^{(1)}\}$;

(ii)   a measure of the distance of the estimates for model (1) from the observations.

From (4.2-3) and (4.2-4), we have

$$\begin{aligned}
G^2(2) &= -2[L(\hat{\mathbf{m}}^{(2)}) - L(\mathbf{x})] \\
&= -2[L(\hat{\mathbf{m}}^{(2)}) - L(\hat{\mathbf{m}}^{(1)})] - 2[L(\hat{\mathbf{m}}^{(1)}) - L(\mathbf{x})] \\
&= G^2[(2)|(1)] + G^2(1),
\end{aligned} \tag{4.2-6}$$

where $G^2[(2)|(1)]$ is the conditional measure for model (2) given model (1). If model (1) is the saturated model, this reduces to the original definition of $G^2(2)$. (Exercise 1 in Section 4.2.4 shows that a comparable breakdown for X^2 does not exist.)

For all hierarchical models,

$$\sum_i x_i \log \hat{m}_i = \sum_i \hat{m}_i \log \hat{m}_i,$$

and when the u-terms of model (2) are a subset of the u-terms of model (1), we also have

$$\sum_i \hat{m}_i^{(1)} \log \hat{m}_i^{(2)} = \sum_i \hat{m}_i^{(2)} \log \hat{m}_i^{(2)}$$

(see exercise 2 in Section 4.2.4). This equivalence enables us to reformulate the conditional G^2 so that its similarity to the unconditional likelihood ratio is more apparent:

$$G^2[(2)|(1)] = -2[L(\hat{\mathbf{m}}^{(2)}) - L(\hat{\mathbf{m}}^{(1)})]$$

$$= -2\sum_i \hat{m}_i^{(1)} \log \frac{\hat{m}_i^{(2)}}{\hat{m}_i^{(1)}}. \tag{4.2-7}$$

As we discuss in Chapter 10, this conditional breakdown forms the basis for the analysis of information tables developed by Kullback, Kupperman, and Ku [1962b]. In Chapter 14, theorem 14.9-8, we show that if $G^2(2)$ and $G^2(1)$ are asymptotically distributed as χ^2 with v_2 and v_1 degrees of freedom, respectively, then $G^2[(2)|(1)]$ is asymptotically distributed as χ^2 with $v_2 - v_1$. The breakdown is particularly valuable when we wish to test for the presence of a single multifactor effect between variables, conditional on the presence of other known effects, as discussed in Section 4.3-3 below. We also exploit this result in Chapter 8 to test for marginal homogeneity in the absence of internal symmetry, although we are unable to obtain cell estimates under this model using iterative proportional fitting techniques. This partitioning result can be extended to more than two models with important implications for model selection (see Section 4.5).

3. Structural breakdown of G^2

When we have direct closed-form estimates, the $\{\hat{m}_i\}$ are products of sums of the $\{x_i\}$, which enables us to write G^2 solely as a function of the $\{x_i\}$. Consequently, we can assess how well a model with direct estimates fits without ever computing the cell estimates. Such an inspection of the goodness of fit of models with direct estimates is a great help in selecting the best-fitting model. Nevertheless to ensure that no important features of the data are overlooked, we recommend inspection of the goodness of fit in each cell for the model selected.

We can readily demonstrate partitioning in three dimensions. To do this we revert to multiple subscripts. We define G^2 for model (s) with observed values x_{ijk} and fitted values $\hat{m}_{ijk}^{(s)}$ as

$$G_{123}^2(s) = -2L_{123}(\hat{\mathbf{m}}^{(s)}) + 2L_{123}(\mathbf{x})$$

$$= -2\left(\sum_{i,j,k} x_{ijk} \log \hat{m}_{ijk}^{(s)} - \sum_{i,j,k} x_{ijk} \log x_{ijk}\right). \tag{4.2-8}$$

Note that the subscripts of L and G^2 refer to the dimensions: Thus $G_{12}^2(s)$ is the corresponding expression for model (s) fitted to the configuration C_{12} with members $\{x_{ij+}\}$. In expression (4.2-8), the only term that changes with a change of model is

$$L_{123}(\hat{\mathbf{m}}^{(s)}) = \sum_{i,j,k} x_{ijk} \log \hat{m}_{ijk}^{(s)}, \tag{4.2-9}$$

and we now evaluate it for a selection of the hierarchical log-linear models that appear differently numbered in table 3.3-1 of Chapter 3. For further examples, see Goodman [1969b].

Model (1). The model with one two-factor effect absent, $u_{12} = 0$, has estimates

$$\hat{m}_{ijk} = \frac{x_{i+k}x_{+jk}}{x_{++k}} \tag{4.2-10}$$

for all i, j, k. From expressions (4.2-9) and (4.2-10), we can write

$$
\begin{aligned}
L_{123}(\hat{\mathbf{m}}^{(1)}) &= \sum_{i,j,k} x_{ijk} \log \frac{x_{i+k} x_{+jk}}{x_{++k}} \\
&= \sum_{i,k} x_{i+k} \log x_{i+k} + \sum_{j,k} x_{+jk} \log x_{+jk} - \sum_{k} x_{++k} \log x_{++k} \qquad (4.2\text{-}11) \\
&= L_{13}(\mathbf{x}) + L_{23}(\mathbf{x}) - L_3(\mathbf{x}),
\end{aligned}
$$

where $L_{13}(\mathbf{x})$ is the likelihood term computed from the observed sufficient configuration C_{13}, and similarly for the other two terms.

Model (2). The model with two two-factor effects absent, $u_{12} = u_{13} = 0$, has sufficient configurations C_{23} and C_1 and cell estimates

$$
\hat{m}_{ijk}^{(2)} = \frac{x_{+jk} x_{i++}}{N},
$$

where $\sum_{i,j,k} x_{ijk} = N$. Following the steps taken for model (1), we write

$$
L_{123}(\hat{\mathbf{m}}^{(2)}) = L_{23}(\mathbf{x}) + L_1(\mathbf{x}) - N \log N. \qquad (4.2\text{-}12)
$$

Model (3). The model of independence fitted to the two-dimensional table of sums $\{x_{i+k}\}$ has cell estimates

$$
\hat{m}_{i+k}^{(3)} = \frac{x_{i++} x_{++k}}{N},
$$

and we write

$$
L_{13}(\hat{\mathbf{m}}^{(3)}) = L_1(\mathbf{x}) + L_3(\mathbf{x}) - N \log N. \qquad (4.2\text{-}13)
$$

Model (4). The no three-factor effect model, $u_{123} = 0$, has sufficient configurations C_{12}, C_{13}, and C_{23}, but direct estimates do not exist. We can, however, compute pseudoestimates $\{y_{ijk}^{(4)}\}$ that are a function of the sufficient statistics, satisfy the no three-factor effect model, and are the right order of magnitude, namely, the estimates reached at the end of the first cycle of iteration (see Chapter 3):

$$
y_{ijk}^{(4)} = \frac{x_{ij+} x_{i+k} x_{+jk}}{x_{i++} \sum_{i} \dfrac{x_{i+k} x_{ij+}}{x_{i++}}}.
$$

Then we have

$$
\begin{aligned}
L_{123}(\mathbf{y}^{(4)}) = L_{12}(\mathbf{x}) + L_{13}(\mathbf{x}) + L_{23}(\mathbf{x}) \\
- L_1(\mathbf{x}) - \sum_{jk} x_{+jk} \log \sum_{i} \frac{(x_{i+k} x_{ij+})}{x_{i++}}. \qquad (4.2\text{-}14)
\end{aligned}
$$

Model (5). If we take the estimates $\{\hat{m}_{ijk}^{(5)}\}$ with $u_{23} = u_{123} = 0$, then by analogy with expression (4.2-11), we have

$$
L_{123}(\hat{\mathbf{m}}^{(5)}) = L_{12}(\mathbf{x}) + L_{13}(\mathbf{x}) - L_1(\mathbf{x}). \qquad (4.2\text{-}15)
$$

We can also define another function of the $\{\hat{m}_{ijk}^{(5)}\}$, which we compute by first summing over the first variable. This function is

$$\bar{L}_{23}(\hat{\mathbf{m}}^{(5)}) = \sum_{j,k} x_{+jk} \log \hat{m}^{(5)}_{+jk}$$

$$= \sum_{j,k} x_{+jk} \log \sum_i \frac{(x_{i+k}x_{ij+})}{x_{i++}}. \tag{4.2-16}$$

We write the left-hand side of (4.2-16) with a bar over the L, because (4.2-16) is unlike L_{23} defined by analogy with (4.2-13). The difference is that the $\hat{\mathbf{m}}$ for (4.2-13) is a log-linear model for the marginal table defined by the subscripts for L, whereas the $\hat{\mathbf{m}}$ in (4.2-15), when collapsed over the first variable does not yield expected values for the original table involving variables 2 and 3, which are log-linear in form.

From expressions (4.2-14), (4.2-15), and (4.2-16), we obtain

$$L_{123}(\mathbf{y}^{(4)}) = L_{123}(\hat{\mathbf{m}}^{(5)}) + L_{23}(\mathbf{x}) - \bar{L}_{23}(\hat{\mathbf{m}}^{(5)}),$$

and hence for the pseudoestimates we have

$$G^2_{123}(\mathbf{y}^{(4)}) = G^2_{123}(5) - \bar{G}^2_{23}(5). \tag{4.2-17}$$

In expression (4.2-17), $\bar{G}^2_{23}(5)$ has the form of a likelihood ratio statistic for the two-way marginal table for variables 2 and 3, but the expected values are a collapsed version of those for model (5). Thus $\bar{G}^2_{23}(5)$ is a pseudo–likelihood ratio statistic. This does not immediately appear useful, but is discussed further in Section 4.5, where we consider strategies for searching for models, and is illustrated in example 4.5-3. For the moment, we only observe that $L_{123}(\mathbf{y}^{(4)}) \leqq L_{123}(\hat{\mathbf{m}}^{(4)})$, and so $L_{123}(\mathbf{y}^{(4)})$ can be regarded as a lower bound for $L_{123}(\hat{\mathbf{m}}^{(4)})$. Thus the value of G^2 computed from the pseudoestimates is an upper bound for the true value.

4.2.2 *Difference between direct models*

It is useful to consider the difference in goodness of fit of different models. From (4.2-8) and (4.2-9), we have

$$G^2_{123}(2) - G^2_{123}(1) = 2[L_{123}(\hat{\mathbf{m}}^{(1)}) - L_{123}(\hat{\mathbf{m}}^{(2)})].$$

When both models have direct estimates, this difference can be computed from the subset of configurations that correspond to the u-terms that are not common to both models. Thus for models (1), (2), and (3) above, we have, from (4.2-11), (4.2-12), and (4.2-13),

$$G^2_{123}(2) - G^2_{123}(1) = 2[L_{13}(\mathbf{x}) - L_3(\mathbf{x}) - L_1(\mathbf{x}) + N \log N]$$

$$= 2[L_{13}(\mathbf{x}) - L_{13}(\hat{\mathbf{m}}^{(3)})]$$

$$= G^2_{13}(3). \tag{4.2-18}$$

The models differ only in that model (1) includes the term u_{13}, whereas this term is set equal to zero in model (2). The relationship (4.2-18) shows that the difference in goodness of fit of the two overall models is the same as the goodness of fit of the model of independence carried out on the marginal sums C_{13}. If models (1) and (2) have v_1 and v_2 degrees of freedom, respectively, then the difference $G^2_{123}(2) - G^2_{123}(1)$ is distributed asymptotically as χ^2 with $v_2 - v_1$ degrees of freedom, as is appropriate for $G^2_{13}(3)$.

If we combine result (4.2-18) with the conditional result (4.2-6), we find that

$$G_{123}^2[(2)|(1)] = G_{13}^2(3), \qquad (4.2-19)$$

which states that the conditional test for model (2) with $u_{12} = u_{13} = 0$, given that model (1) with $u_{12} = 0$ holds, is equivalent to the test for $u_{13} = 0$ carried out on the marginal sums C_{13}. This equivalence should not come as a surprise, because the collapsing theorems of Chapter 2 state that if $u_{12} = 0$ (and $u_{123} = 0$ by the hierarchy principle), we may collapse by summing over variable 2, and the values of the parameters of u_{13} computed from the saturated model applied to C_{13} are identical to those computed from the full array. It follows that we can then test for u_{13} by looking only at C_{13}. The following result is a direct outcome of these considerations.

When two direct models differ only by a single u-term, say u_{θ_1}, the difference in goodness-of-fit statistics G^2 of the two models applied to the entire array may be found from the configuration C_{θ_1} by fitting the model with u_{θ_1} set equal to zero.

This result does not always hold for the difference between two models that do not have direct estimates. See example 4.4-5 and the discussion of partitioning in Section 4.6.

4.2.3   Other summary statistics

In the preceding section we quoted results given in Chapter 14 to the effect that both X^2 and G^2 are asymptotically distributed as χ^2 with appropriate degrees of freedom. In Chapter 14 we will also show that when sample sizes are small, it is not very useful to be told that the distribution of the test statistic is known for large samples. For this reason, some authors have suggested using variance-stabilizing transformations that attempt to make the standardized deviations in each cell closer to normal deviates with zero mean and unit variance. If this can be achieved, then not only is the sum of the squares of these deviates distributed as χ^2, but the individual deviates may be examined to detect aberrant cells and important patterns in the data.

We will discuss the relative merits of various cell deviates in Section 4.4. Among them, we discuss the Freeman–Tukey transformation, which yields deviates

$$z_i = \sqrt{x_i} + \sqrt{x_i + 1} - \sqrt{4\hat{m}_i + 1}. \qquad (4.2-20)$$

The sum of these squared deviates,

$$Z^2 = \sum_i z_i^2, \qquad (4.2-21)$$

is asymptotically distributed as χ^2. As all the summary statistics follow the same distribution, it is not surprising to find that they are usually very close to each other in magnitude.

4.2.4   Exercises

1. (i)   Show that if $\sum_i \hat{m}_i = \sum_i x_i = N$, then $X^2 = \sum (x_i^2/\hat{m}_i) - N$;

    (ii) if $X^2(2) =$ goodness of fit for $\{\hat{m}_i^{(2)}\}$ and $X^2(1) =$ goodness of fit for $\{\hat{m}_i^{(1)}\}$, show that

$$X^2(2) \neq \sum_i \frac{(\hat{m}_i^{(1)} - \hat{m}_i^{(2)})^2}{\hat{m}_i^{(2)}} + X^2(1),$$

and thus the conditional breakdown of G^2 does not apply to X^2.

2. Show by breaking up the likelihoods for any hierarchical model in a three-way table that

$$\sum_{i,j,k} x_{ijk} \log \hat{m}_{ijk} = \sum_{i,j,k} \hat{m}_{ijk} \log \hat{m}_{ijk}.$$

4.3 Standardized Rates

So far we have been concerned with summary statistics that describe the goodness of fit of a particular model. It is also often useful to summarize the findings of the data, and adjusted rates are often used for this purpose. Fitting log-linear models before computing these descriptive summary statistics has two purposes:

1. knowledge of the structure of the data enables us to determine whether the whole array can be described by a simple statistic or whether partitioning the array into smaller segments is appropriate;
2. computing the statistics from the fitted values rather than from the observed counts removes some of the variability due to random fluctuations.

4.3.1   *Two methods of adjustment*

When we have a three-way array $\{x_{ijk}\}$ and the first variable is a dichotomy, it is common practice to present the data in the form of rates. We can think of the subscript j as referring to one of J exposed populations and the subscript k as referring to one of K categories of some stratifying variable, such as age. In this frame of reference, we wish to compare age-adjusted rates for the different populations. We often deal with death rates, and for population j and age-group k we have deaths d_{jk} and persons exposed e_{jk}. Thus we define

$$d_{jk} = x_{1jk},$$
$$e_{jk} = x_{1jk} + x_{2jk} = x_{+jk}, \tag{4.3-1}$$

and hence we have the category-specific rates

$$r_{jk} = \frac{d_{jk}}{e_{jk}} = \frac{x_{1jk}}{x_{+jk}}. \tag{4.3-2}$$

If the age structures of the populations differ, we know that the crude rates d_{j+}/e_{j+} will reflect the age differences. We hope by adjusting for age to be able to determine differences in death rates between the populations that are due to other causes. The two most commonly used methods are often referred to as "direct" and "indirect."

1. *Direct adjustment*

The age-specific rates r_{jk} for a particular population are applied to the counts of those exposed in each age-group of some *standard population* $\{E_k\}$, and the resulting expected numbers of deaths in each age-group are summed. Often the standard population is the sum of the individual populations, and so the expected total number of deaths using rates from population j is

$$\sum_k r_{jk} E_k = \sum_k r_{jk} e_{+k}$$
$$= \sum_k \frac{x_{1jk} x_{++k}}{x_{+jk}}. \tag{4.3-3}$$

We derive the expected death rates for the standard overall population using rates from population j by dividing total deaths (4.3-3) by total exposed:

$$\frac{\sum\limits_k r_{jk}E_k}{E_+} = \frac{\sum\limits_k \dfrac{x_{1jk}x_{++k}}{x_{+jk}}}{x_{+++}}. \tag{4.3-4}$$

The ratio of observed to expected deaths in the overall population can also be written down immediately:

$$\frac{d_{++}}{\sum\limits_k r_{jk}E_k} = \frac{x_{1++}}{\sum\limits_k \dfrac{x_{1jk}x_{++k}}{x_{+jk}}}. \tag{4.3-5}$$

2. Indirect adjustment

A set of *standard rates* $\{R_k\}$ is applied to the counts of those exposed in each age-group of the observed population j, and the expected number of deaths so obtained is compared with the observed number. Often the standard rates are obtained by summing deaths and exposures across populations for each group k. We then readily obtain expressions parallel to (4.3-3)–(4.3-5).

The total expected number of deaths in population j is

$$\sum\limits_k e_{jk}R_k = \sum\limits_k e_{jk}\frac{d_{+k}}{e_{+k}}$$

$$= \sum\limits_k \frac{x_{+jk}x_{1+k}}{x_{++k}}. \tag{4.3-6}$$

The expected death rate for population j is

$$\frac{\sum\limits_k e_{jk}R_k}{e_{j+}} = \frac{\sum\limits_k \dfrac{x_{+jk}x_{1+k}}{x_{++k}}}{x_{+j+}}. \tag{4.3-7}$$

The ratio of observed to expected deaths in population j is

$$\frac{d_{j+}}{\sum\limits_k \dfrac{e_{jk}d_{+k}}{e_{+k}}} = \frac{x_{1j+}}{\sum\limits_k \dfrac{x_{+jk}x_{1+k}}{x_{++k}}}. \tag{4.3-8}$$

This ratio is referred to as the standard mortality ratio (SMR).

There is extensive discussion in the literature of these two methods and their variants (see Yerushalmy [1951], Kitagawa [1964], Kalton [1968], and Goldman [1971]). Much of the discussion is based on the choice of appropriate standards. When either the standard population or the standard rates are selected from an extraneous source instead of being derived (as described above) by summing across populations, the choice of standards becomes a matter of choosing weights appropriate for the particular situation.

4.3.2   Generic form

We confine our discussion to the situation where the standards are derived from the data by summing over all populations. First we expand our original $2 \times J \times K$ array $\{x_{ijk}\}$ to an $2 \times (J + 1) \times K$ array $\{x'_{ijk}\}$, using the category $J + 1$ for totals. Thus we have

$$x'_{ijk} = x_{i+k} \quad \text{for } j = J + 1,$$
$$x'_{ijk} = x_{ijk} \quad \text{otherwise.} \tag{4.3-9}$$

With this notation we can write a generic form for the overall adjusted rate. When the rates observed in population l are applied to the exposed population m, we have the adjusted rate

$$A_{ml} = \frac{\sum_k r_{lk} e_{mk}}{\sum_k e_{mk}}$$
$$= \frac{\sum_k \dfrac{x'_{1lk} x'_{+mk}}{x'_{+lk}}}{\sum_k x'_{+mk}}. \tag{4.3-10}$$

If we put $m = l$, A_{ll} is the crude rate for population l. If $l = J + 1$, we have the form (4.3-7) for indirect adjustment, and if we put $m = J + 1$, we have the form (4.3-4) for direct adjustment, where the total population is used as standard. When $l \neq m$ and $l < J + 1$, $m < J + 1$, we have a variant where a set of rates of one of the observed populations is applied to another observed population. Inspection of the array A_{ml} for a series of observed populations helps us determine not only the relative rates in each, but also whether one population has a structure very different from the others.

   Inspection of the numerator of (4.3-7) shows that we have estimates for each k obtained under the model $u_{12} = u_{123} = 0$, based on the configurations C_{13} and C_{23}. These are the expected values used for the Mantel–Haenszel test (described in Section 4.4.3). It is common knowledge that when the pattern of age-specific rates differs between populations, different adjustments may lead to conflicting conclusions about the relative rates of the populations; this occurs when $u_{123} \neq 0$. The implication is that we should always test the magnitude of u_{123} before computing adjusted rates.

4.3.3   Adjusting for more than one variable

The argument for the appropriateness of computing standardized rates can readily be extended from one to many variables by regarding the third variable as a compound variable. Thus if the data have two categories of sex and four of age, K consists of the eight age $\times$ sex categories. It then follows that if all three-factor effects involving variables 1 and 2 are absent from the structural model, we can summarize the data by computing overall rates that are adjusted for all the underlying variables.

   Fitting a log-linear model to the observed data determines whether the required three-factor effects are absent. In their absence we can compute adjusted rates from the raw data, but if the data are sparse we may prefer to compute adjusted rates from fitted values obtained under a log-linear model. If x_{1jk} is zero for some cell, the corresponding rate r_{jk} is zero, but a well-fitting model may provide a non-zero estimate $\hat{m}_{1jk}$ for this cell. This is particularly true when we have a third compound variable. In the event that x_{1jk} is small in some cells, rates based on $\hat{m}_{1jk}$ can be regarded as more stable than the observed rates. We illustrate by means of an example.

Example 4.3-1   Halothane study: Cholecystectomy rates adjusted for underlying variables

In example 3.7-2 of Chapter 3 we described the method of collecting data used in the National Halothane study and described a subset of these data dealing with cholecystectomies. These data have six variables, but we now reduce to five by

Table 4.3-1   Death Rates Following Cholecystectomy: Observed Rates Standardized for Sex, Risk, Age, and Operation

	Population 1	Population 2	Population 3	Population 4	Population 5	Total Population
Agent 1						
Deaths	148.0	176.7	427.1	167.4	158.0	1,077.2
Base Population	7,930.7	5,065.3	8,044.4	3,702.9	3,549.1	28,292.4
Adjusted Rate	0.0187	0.0349	0.0531	0.0452	0.0445	0.0381
Agent 2						
Deaths	305.2	102.0	445.0	174.6	261.4	1,288.2
Base Population	7,930.7	5,065.3	8,044.4	3,702.9	3,549.1	28,292.4
Adjusted Rate	0.0385	0.0201	0.0553	0.0472	0.0737	0.0455
Agent 3						
Deaths	356.6	135.5	207.0	257.2	177.7	1,134.0
Base Population	7,930.7	5,065.3	8,044.4	3,702.9	3,549.1	28,292.4
Adjusted Rate	0.0450	0.0268	0.0257	0.0694	0.0501	0.0401
Agent 4						
Deaths	896.2	553.7	1,103.2	86.0	598.9	3,238.0
Base Population	7,930.7	5,065.3	8,044.4	3,702.9	3,549.1	28,292.4
Adjusted Rate	0.1131	0.1093	0.1371	0.0232	0.1687	0.1144
Agent 5						
Deaths	706.5	363.6	634.4	252.8	129.0	2,086.3
Base Population	7,930.7	5,065.3	8,044.4	3,702.9	3,549.1	28,292.4
Adjusted Rate	0.0891	0.0718	0.0789	0.0683	0.0363	0.0737
Agent 6						
Deaths	166.2	95.5	209.3	81.9	119.1	672.0
Base Population	7,930.7	5,065.3	8,044.4	3,702.9	3,549.1	28,292.4
Adjusted Rate	0.0210	0.0189	0.0260	0.0221	0.0336	0.0238
SMR	0.8902	1.0681	0.9892	1.0499	1.0833	1.0000

Source: Bishop and Mosteller [1969].

Note: The rates in tables 4.3-1 and 4.3-2 are computed as (deaths)/(death + exposed) instead of the more usual form (death)/(exposed) because of the method of sampling used in this study. Thus the total base population appears as $672 + 27,620.4 = 28,292.4$, and although, as we would expect, the deaths are integers in the diagonal of table 4.3-1, the exposed population values are fractional because they have been weighted by sampling factors.

summing over variable 6. The primary variables are survival and anesthetic agent. The five categories of agent define five different populations of people undergoing surgery. These populations differ with respect to the four underlying variables sex, operative risk, age, and type of operative procedure. In our example they are treated as a compound variable with 54 categories.

In table 4.3-1 rows correspond to rates specific for each agent l, and columns to the population exposed m. In each cell we have three entries:

(i)   observed or expected deaths;
(ii)   exposed base population;
(iii) the adjusted rate A_{ml}.

Thus we have the observed rates on the diagonal, the directly adjusted rates based on the total population in the last column, and the indirectly adjusted rates based on the overall rates in the last row. Below the array we have the SMR, that is, the ratio of the deaths in the diagonal to the deaths in the last row.

The values in table 4.3-1 are based on the observed data. When we compare agents in the same population by looking down the columns, we find a wide range

Table 4.3-2   Death Rates Following Cholecystecomy: Fitted Rates Standardized for Sex, Risk, Age, and Operation[a]

	Population 1	Population 2	Population 3	Population 4	Population 5	Total Population
Agent 1						
Deaths	148.0	82.8	181.8	82.3	109.7	604.6
Base Population	7,930.7	5,065.3	8,044.4	3,702.9	3,549.1	28,292.4
Adjusted Rate	0.0187	0.0164	0.0226	0.0222	0.0309	0.0214
Agent 2						
Deaths	190.7	102.0	231.5	99.6	145.8	769.6
Base Population	7,930.7	5,065.3	8,044.4	3,702.9	3,549.1	28,292.4
Adjusted Rate	0.0240	0.0201	0.0288	0.0269	0.0411	0.0272
Agent 3						
Deaths	192.9	102.5	207.0	95.3	120.7	718.5
Base Population	7,930.7	5,065.3	8,044.4	3,702.9	3,549.1	28,292.4
Adjusted Rate	0.0243	0.0202	0.0257	0.0257	0.0340	0.0254
Agent 4						
Deaths	176.9	89.7	192.5	86.0	108.5	653.6
Base Population	7,930.7	5,065.3	8,044.4	3,702.9	3,549.1	28,292.4
Adjusted Rate	0.0223	0.0177	0.0239	0.0232	0.0306	0.0231
Agent 5						
Deaths	243.7	130.1	253.6	117.0	129.0	873.3
Base Population	7,930.7	5,065.3	8,044.4	3,702.9	3,549.1	28,292.4
Adjusted Rate	0.0307	0.0257	0.0315	0.0316	0.0363	0.0309
Agent 6						
Deaths	174.7	93.4	198.7	89.5	115.8	672.0
Base Population	7,930.7	5,065.3	8,044.4	3,702.9	3,549.1	28,292.4
Adjusted Rate	0.0220	0.0184	0.0247	0.0241	0.0326	0.0238
SMR	0.8472	1.0921	1.0418	0.9614	1.1144	1.0000

Source: Bishop and Mosteller [1969].
[a] See note to table 4.3-2.

of adjusted rates, and if we rank agents according to adjusted rates, we find discrepancies between the columns, i.e., according to which population is used as standard. Log-linear models were fitted to these data, and the model with a selection of two-factor effects (namely, all those involving agent and also type of operation by risk, type of operation by age, and risk by age) was found to fit well. The fitting preserves the total numbers of deaths and exposed in each population but changes the stratum-specific rates. Thus when the fitted values are used to compute adjusted rates, the expected numbers of deaths differ. Adjusted rates computed from the fitted values are given in table 4.3-2. They show much greater agreement between agents, indicating that the previous wide range was largely a result of using stratum-specific rates based on small numbers of deaths. Moreover, if we rank the agents within each population, we find that the ordering of agents by rate is now more consistent between columns. If population 5 is excluded, we have agreement that the lowest rates are obtained with agent 1. For the other agents no population rank differs by more than 1 from the ranks derived from the total population. ■■

4.4   Internal Goodness of Fit

When a multidimensional table has a large number of cells, it is seldom sufficient to compute only a summary measure of goodness of fit; it is often highly desirable to examine the fit in each cell. The reasons for more detailed examination are:

1.   The overall measure for a highly parametized model may be within acceptable limits even when a few cells are poorly fitted. Detection of such cells may be of importance in understanding the data and lead to partitioning of segments of the array;
2.   The pattern of positive and negative cell deviations may indicate that a different model should be selected.

   In addition to examining the goodness of fit in each cell, it is also useful to examine the magnitude of the parameters of the fitted model. Although we obtain fitted cell estimates and measures of goodness of fit for log-linear models without computing the parameters themselves, we can readily derive them from the cell estimates, and inspection of their relative magnitude is useful for interpreting the fitted model. In particular, if we are interested in testing for the presence of a particular multifactor effect, we may learn a great deal by observing how its magnitude changes when we eliminate some effects and add others.

   Both types of examination, cell residuals and model parameters, are best carried out after the measures are suitably standardized. For the cell residuals, several transformations have been suggested, and we proceed next to a discussion of their relative merits. For each model parameter, we compute its variance by means of theorem 14.6-4 of Chapter 14, as we show in example 4.4-3.

4.4.1   *Standardized cell residuals*

In Section 4.2 we described three summary measures of goodness of fit. The corresponding standardized cell residuals are:

1.   Components of X^2

$$z_i^{(1)} = \frac{x_i - \hat{m}_i}{\sqrt{\hat{m}_i}};$$ (4.4-1)

2. Components of G^2

$$z_i^{(2)} = 2x_i \log \frac{x_i}{\hat{m}_i}; \qquad (4.4\text{-}2)$$

3. Freeman–Tukey deviates

$$z_i^{(3)} = \sqrt{x_i} + \sqrt{x_i + 1} - \sqrt{4\hat{m}_i + 1}. \qquad (4.4\text{-}3)$$

This last form is derived by considering the variance-stabilizing transformation

$$y_i = \sqrt{x_i} + \sqrt{x_i + 1}. \qquad (4.4\text{-}4)$$

In the event that x_i follows a Poisson distribution with mean m_i, we use the result that y_i is approximately normally distributed, with approximate mean

$$\sqrt{4m_i + 1} \qquad (4.4\text{-}5)$$

and variance 1 (see Freeman and Tukey [1950], Mosteller and Youtz [1961]). The $z_i^{(1)}$ and $z_i^{(3)}$ are such that $\sum_i z_i^2$ is distributed asymptotically as χ^2 with the number of degrees of freedom v appropriate for the estimated $\{\hat{m}_i\}$. As v is always less than the number I of cells in the models of interest, the average expected value of z_i^2 in each cell is less than 1. As a rough guide to detecting a large deviate, we might take the square root of $\chi^2_{0.05}$, the 5% significance level of the appropriate χ^2 distribution, divided by I, the number of cells, but if we wish to determine more formally whether a cell is an outlier we must fit ancillary models, as we explain below and illustrate in example 4.4-2.

The deviations that are most difficult to evaluate are those in the cells that have small x_i, as the asymptotic theory helps us least when the numbers are small. This is particularly true when $x_i = 0$ and we are unsure whether this is a structural or random zero. Before examining this difficulty further, we give an example of examining standardized deviates for patterns and thereby determining a model that fits well.

Example 4.4-1   Preference for location of training camp for World War II recruits
Stouffer et al. [1949] studied United States soldiers in World War II, and one part of their study consists of data from a sample survey of army recruits, identified by race (black, white), geographic origin (North, South), and location of current training camp (North, South), who were asked to indicate their preference for camp location (see table 4.4-1).

Goodman [1972a] and others have analyzed these data in a slightly different form. First they omit the two undecided columns, and then they merge the "prefer to stay in present camp" category into the remaining two column categories, yielding the arrangement given in table 4.4-2. We refer to race as variable 1, region of origin as variable 2, present location as variable 3, and preferred location as variable 4.

If we fit the model based on the configurations C_{123}, C_{14}, C_{24}, and C_{34} to the data in table 4.4-2, we get $X^2 = 25.73$ and $G^2 = 24.96$ with four degrees of freedom. These values are significant at the 0.001 level when referred to the corresponding chi square distribution, and thus the model does not fit well. We can learn a great deal from an examination of the standardized cell deviates for this model as given in table 4.4-3, or the Freeman–Tukey deviates as given in table 4.4-4. Because of the relatively large cell counts in the original table, the two sets of deviates are almost identical.

Table 4.4-1   Preference of World War II Recruits for Location of Training Camp

| Race | Region of Origin | Location of Present Camp | Prefer to Stay | Prefer to Move to | | | Undecided |
				North	South	Undecided	
Black	North	North	196	191	36	41	52
		South	83	876	167	153	111
	South	North	261	122	270	113	105
		South	924	381	788	353	272
White	North	North	367	588	162	191	162
		South	346	874	164	273	164
	South	North	54	50	176	40	40
		South	481	91	389	91	91

Source: Stouffer et al. [1949].

Table 4.4-2   Data from Table 4.4-1, Reorganized and with "Undecideds" Omitted

| Race | Region of Origin | Location of Present Camp | Number of Soldiers Preferring Camp | |
			in North	in South
Black	North	North	387	36
		South	876	250
	South	North	383	270
		South	381	1,712
White	North	North	955	162
		South	874	510
	South	North	104	176
		South	91	869

Looking at either set of deviates, we notice that the signs associated with the deviates for each level of variable 1 are the same; i.e., if we look at the top part of the table, we have the pattern

$$
\begin{array}{cc}
- & + \\
+ & - \\
\\
+ & - \\
- & +
\end{array}
$$

and this is repeated in the lower part of the table. This pattern of the signs may be indicative of a three-factor effect involving variables 2, 3, and 4. Moreover, the two largest deviates fall in the (1, 1, 1, 2) and (2, 1, 1, 2) cells, i.e., they correspond to the same levels of variables 2, 3, and 4.

Table 4.4-3 Standardized Cell Deviates of the form $(x_{ijkl} - \hat{m}_{ijkl})/\sqrt{\hat{m}_{ijkl}}$ for Model Based on Configurations C_{123}, C_{14}, C_{24}, and C_{34} (as Applied to the Stouffer Data in Table 4.4-2)

Race	Region of Origin	Location of Present Camp	Preferred Camp North	Preferred Camp South
Black	North	North	−0.54	2.14
		South	0.31	−0.56
	South	North	1.77	−1.90
		South	−1.54	0.77
White	North	North	−0.90	2.45
		South	1.03	−1.29
	South	North	0.60	−0.44
		South	−0.77	0.26

Table 4.4-4 Freeman–Tukey Deviates of the form $\sqrt{x_{ijkl}} + \sqrt{x_{ijkl} + 1} - \sqrt{4\hat{m}_{ijkl} + 1}$ for Model Based on Configurations C_{123}, C_{14}, C_{24}, and C_{34} (as Applied to the Stouffer Data in Table 4.4-2)

Race	Region of Origin	Location of Present Camp	Preferred Camp North	Preferred Camp South
Black	North	North	−0.53	1.98
		South	0.31	−0.55
	South	North	1.74	−1.94
		South	−1.56	0.77
White	North	North	−0.90	2.35
		South	1.03	−1.30
	South	North	0.61	−0.42
		South	−0.76	0.27

The model suggested by the examination of the deviates is the one based on the configurations C_{123}, C_{234}, and C_{14}. When we fit this model we get $X^2 = 1.45$ and $G^2 = 1.46$ with three degrees of freedom, indicative of an extremely good fit. What is of interest in these data is how preference of camp location is related to the other three variables. Thus interpreting this fitted model as a logit model (where the logits are taken on variable 4, i.e., preference) is useful (see Sections 2.2.4 and 2.3.5). In the present case, the model based on C_{123}, C_{234}, and C_{14} corresponds to the logit model

$$l_{ijk} = \log\left(\frac{m_{ijk1}}{m_{ijk2}}\right) = w + w_{1(i)} + w_{2(j)} + w_{3(k)} + w_{23(jk)},$$

i.e., the odds for preferring a northern camp depend on race, region of origin, location of present camp, and the interaction between region of origin and location of present camp. ■■

Deviations in cells with no observations

In Chapter 2 we constructed the $\{m_i\}$ for a 2×3 table by selecting values for the u-terms. All the subscripted u-terms are in the range from -0.69 to 3.00, i.e., the limits are $\pm \log 20$, and we find that for a sample size of 3,430 the $\{m_i\}$ range from 2 to 2,880. These u-terms are larger than those we usually encounter, but the example serves to emphasize that for samples of moderate size with large interaction effects, some of the $\{m_i\}$ are so small that the probability for $x_i = 0$ for some i is high. If x_i is an observation from a Poisson distribution with mean m_i, the probability π_i of observing $x_i = 0$ is greater than 0.05 if m_i is less than 3 and greater than 0.5 if m_i is less than 0.7 (see table 4.4-5).

Table 4.4-5   Magnitude of Standardized Deviates When Cell Contains a Random Zero for Fitted Values between 0.2 and 3.0

	Poisson Mean m_i						
	0.2	0.4	0.6	0.8	1.0	2.0	3.0
Probability Π_i of Observing $x_i = 0$	0.82	0.67	0.55	0.45	0.37	0.14	0.05
Normal Deviate with Probability Π_i (Two Tails)	0.23	0.42	0.60	0.76	0.88	1.43	1.96
Cell deviate[a] $|z_i^{(1)}|$	0.45	0.63	0.77	0.89	1.00	1.41	1.73
Cell deviate $|z_i^{(3)}|$	0.34	0.61	0.84	1.05	1.24	2.00	2.61

[a] Cell deviates given are absolute values, although when derived assuming $\hat{m}_i = m_i$, they are all negative.

When the cell contains no observations, we find that

$$z_i^{(1)} = \sqrt{\hat{m}_i},$$

$$z_i^{(2)} = 0,$$

$$z_i^{(3)} = 1 - \sqrt{4\hat{m}_i + 1}. \tag{4.4-6}$$

If we consider the artificial situation where the maximum likelihood estimates are identical to the population values, $\hat{m}_i = m_i$, we can compare the z_i with the standard normal deviates that give appropriate probability levels for Poisson variables. (In table 4.4-5 we give absolute values, but all z_i as defined are negative.) We see that for $m = 2$, the Poisson probability of observing $x_i = 0$ is 0.14, which corresponds to a normal deviate of $+1.43$ or -1.43. The corresponding value for $z_i^{(1)}$ is close, -1.41, but for $z_i^{(3)}$ we obtain the value -2.00. For small values of m_i, both z-transformations give absolute values that are larger than the corresponding normal deviates.

The values in table 4.4-5 show in a somewhat heuristic way that although it is useful to examine the pattern and magnitude of cell deviations, it is not possible to choose a value c such that $|z_i| > c$ indicates, with known probability, that cell (i) is an aberrant cell. For a Monte Carlo exploration of the distribution of various measures in small contingency tables, the reader is referred to Odoroff [1970]. Other references are Craddock and Flood [1970] and Yarnold [1970].

Detecting outliers

If we believe that one cell is an outlier and wish to test this proposition, we can do so by fitting the same model to all cells except the cell in question. If the cell (j)

is to be tested, we treat it as a structural zero and fit the model to the remaining cells using the methods given in Chapter 5. The difference in summary measure of goodness of fit for a model applied to all cells and the same model applied to all cells except (j) is distributed as χ^2 with one degree of freedom. (See exercise 2 in Section 4.4.4.)

Example 4.4-2   Outlier in thromboembolism data

In example 3.7-4 of Chapter 3 we analyzed data on thromboembolism previously analyzed by Worcester [1971]; these data appear in table 3.7-10 of that chapter. If we fit the model based on configurations C_{12} and C_3 to the complete array, we obtain $X^2 = 12.506$ with three degrees of freedom. The cell $(1, 1, 1)$ contributes 7.873 to this X^2, but none of the other cells contribute more than 2.746. If we remove this cell and refit the same model, we get the value $X^2 = 0.024$ with two degrees of freedom, shown in table 3.7-13. The difference of 12.485 between the two X^2 values confirms that this cell is an outlier.   ■■

4.4.2   Standardized parameter estimates

To compare the u-term parameters we must divide each by its standard deviation. To find the appropriate variances we must satisfy ourselves that each parameter is a linear contrast of the logarithms of the expected cell counts. A linear contrast must be of the form

$$u_{\theta(j)} = \sum_i \lambda_i \log m_i, \tag{4.4-7}$$

where $\sum_i \lambda_i = 0$. Theorem 14.6-4 of Chapter 14 shows that the asymptotic variances are then derived by squaring the coefficients for each cell and taking the reciprocals of the cell expected values,

$$\mathrm{Var}(\hat{u}_{\theta(j)}) = \sum_i \frac{\lambda_i^2}{m_i}. \tag{4.4-8}$$

We illustrate for a single parameter $u_{1\,2(ij)}$ of the two-factor effect in two dimensions.

In Chapter 2 the u-term was expressed in a form familiar from ANOVA as a deviation from lower-order terms. We can take this form and rearrange the coefficients so that every cell appears only once. Thus in two dimensions, if $l_{ij} = \log m_{ij}$, we have

$$u_{1\,2(ij)} = l_{ij} - \frac{l_{i+}}{J} - \frac{l_{+j}}{I} + \frac{l_{++}}{IJ}$$

$$= \left(1 - \frac{1}{J} - \frac{1}{I} + \frac{1}{IJ}\right) l_{ij} + \left(-\frac{1}{J} + \frac{1}{IJ}\right) \sum_{s \neq j} l_{is}$$

$$+ \left(-\frac{1}{I} + \frac{1}{IJ}\right) \sum_{r \neq i} l_{rj} + \frac{1}{IJ} \sum_{\substack{r \neq i \\ s \neq j}} l_{rs}$$

$$= \lambda_1 l_{ij} + \lambda_2 \sum_{s \neq j} l_{is} + \lambda_3 \sum_{r \neq i} l_{rj} + \lambda_4 \sum_{\substack{r \neq i \\ s \neq j}} l_{rs}, \tag{4.4-9}$$

where

$$\lambda_1 = \frac{(1 - I)(1 - J)}{IJ},$$

$$\lambda_2 = \frac{(1 - I)}{IJ},$$

$$\lambda_3 = \frac{(1 - J)}{IJ},$$

$$\lambda_4 = \frac{1}{IJ},$$

and $\lambda_1, \lambda_2, \lambda_3$, and λ_4 are the coefficients for $1, (J - 1), (I - 1)$, and $(I - 1)(J - 1)$ cells, respectively. We can readily determine that all cells are included in expression (4.4-9) and that, as required for a linear contrast, the coefficients sum to zero, i.e.,

$$\lambda_1 + (J - 1)\lambda_2 + (I - 1)\lambda_3 + (I - 1)(J - 1)\lambda_4 = 0.$$

We define $\hat{l}_{ij} = \log x_{ij}$ and derive maximum likelihood estimates $\hat{u}_{12}$ for the saturated model by substituting $\hat{l}$ for l in expression (4.4-9). The asymptotic variance of $\hat{u}_{12}$ in the saturated model is

$$\mathrm{Var}(\hat{u}_{12(ij)}) = \lambda_1^2 \frac{1}{m_{ij}} + \lambda_2^2 \sum_{s \neq j} \frac{1}{m_{is}} + \lambda_3^2 \sum_{r \neq i} \frac{1}{m_{rj}} + \lambda_4^2 \sum_{\substack{r \neq i \\ s \neq j}} \frac{1}{m_{rs}}, \qquad (4.4\text{-}10)$$

and we estimate by substituting $\{x_{ij}\}$ for $\{m_{ij}\}$. Some uses of standardized u-terms derived from the saturated model are:

1. ordering interactions according to their magnitude. This is most useful when all variables are dichotomies and a single absolute value describes each u-term;
2. detecting when the number of categories for a variable may be reduced;
3. assessing in advance how well a particular unsaturated model will fit the data.

We can also derive u-terms for unsaturated models. In order to standardize them we must compute their variances. For directly estimable models, we can derive a variance formula for each included u-term from the sufficient configurations. These are typically smaller than the estimates derived for the saturated model, but, as they must be derived ab initio for each individual model, we do not give them here. For general methods, the reader is referred to Section 14.6. For models that can only be derived iteratively, no closed-form variance estimates exist, but some methods of computing the fitted cell estimates yield variance estimates as a byproduct. (See Bock [1970] for the Newton–Raphson method.)

We note that standardized u-terms can only be computed when all cells have positive observed counts. This is the major drawback to using u-terms as a method of selecting appropriate structural models.

Example 4.4-3   Examining parameters, data on detergent preferences (Ries and Smith [1963])

The data of table 4.4-6 were collected by Ries and Smith [1963] and have also

Table 4.4-6   Observed 3×2^3 Data on Detergent Preference and Logarithms of Observed Cell Counts

Variable (1) Water Softness	Variable (4) Brand Preference	Variable (2) Previous Use of M			
		Yes		No	
		Variable (3) Temperature			
		High	Low	High	Low
Soft	X	19	57	29	63
		2.94	4.04	3.37	4.14
	M	29	49	27	53
		3.37	3.89	3.30	3.97
Medium	X	23	47	33	66
		3.14	3.85	3.50	4.19
	M	47	55	23	50
		3.85	4.01	3.14	3.91
Hard	X	24	37	42	68
		3.18	3.61	3.74	4.22
	M	43	52	30	42
		3.76	3.95	3.40	3.74

Source: Ries and Smith [1963].

been analyzed by Cox and Lauh [1967], Ku and Kullback [1968], Goodman [1971b], and Johnson and Koch [1971]. They result from an experiment in which 1,008 people were given two brands of detergent, X and M, and subsequently asked four questions, corresponding to the four variables:

1. water softness: soft, medium, or hard ($I = 3$);
2. previous use of brand M: yes or no ($J = 2$);
3. water temperature: high or low ($K = 2$);
4. brand preference: X or M ($L = 2$).

In addition to the observed counts, table 4.4-6 also gives the logarithms of each count $\log x_{ijkl}$. If we add the six items in each column, that is, sum over variables 1 and 4, we obtain

$$\hat{l}_{+11+} = 20.237,$$

$$\hat{l}_{+12+} = 23.355,$$

$$\hat{l}_{+21+} = 20.434,$$

$$\hat{l}_{+22+} = 24.172.$$

As the addition in the log scale has removed all u-terms involving variables 1 and 4, we can proceed as in equation (4.4-9) to compute for cell $(+, 1, 1, +)$:

$$\hat{u}_{23(11)} = \lambda_1 \hat{l}_{+11+} + \lambda_2 \hat{l}_{+12+} + \lambda_3 \hat{l}_{+21+} + \lambda_4 \hat{l}_{+22+},$$

where

$$\lambda_1 = \lambda_4 = -\lambda_2 = -\lambda_3.$$

This gives $\hat{u}_{23(11)} = 0.0259$. For the variance we have, from (4.4-10),

$$\text{Var}(\hat{u}_{23(11)}) = \lambda_1^2 \sum_{i,k} \frac{1}{x_{i11k}} + \lambda_2^2 \sum_{i,k} \frac{1}{x_{i12k}} + \lambda_3^2 \sum_{i,k} \frac{1}{x_{i21k}} + \lambda_4^2 \sum_{i,k} \frac{1}{x_{i22k}}$$

$$= (0.0336)^2.$$

Thus the standardized value for $\hat{u}_{23(11)}$ is 0.7708. The low-order $\hat{u}$-terms and their standardized values are given in table 4.4-7. We observe that the largest standardized values are $\hat{u}_{3(1)}$ and $\hat{u}_{24(11)}$, which suggests that the minimum hierarchical model is

$$\log m_{ijkl} = u + u_{2(j)} + u_{3(k)} + u_{4(l)} + u_{24(jl)}. \tag{4.4-11}$$

This model has seventeen degrees of freedom, and the summary measures of goodness of fit are $X^2 = 23.54$ and $G^2 = 22.85$. This seems to be a satisfactory fit, but as there is room for a substantial improvement in fit, we might consider including the term u_1 so that all variables appear (model 2), or we might include other two-factor terms selected according to their magnitude in table 4.4-7. The resulting measures of goodness of fit are given in table 4.4-8.

Table 4.4-7   Standardized Parameters Computed from Ries–Smith Data from Table 4.4-6

Main Effects	$\hat{u}_{1(1)}$	$\hat{u}_{1(2)}$	$\hat{u}_{1(3)}$
	-0.047	0.022	0.025
	-0.972	0.469	0.528
	$\hat{u}_{2(1)}$	$\hat{u}_{3(1)}$	$\hat{u}_{4(1)}$
	-0.042	-0.286	-0.015
	-1.259	-8.503	-0.452
Two-factor Terms	$\hat{u}_{12(11)}$	$\hat{u}_{12(21)}$	$\hat{u}_{12(31)}$
	-0.024	0.056	-0.032
	-0.496	1.194	-0.682
	$\hat{u}_{13(11)}$	$\hat{u}_{13(21)}$	$\hat{u}_{13(31)}$
	-0.099	-0.007	0.106
	-2.039	-0.148	2.246
	$\hat{u}_{14(11)}$	$\hat{u}_{14(21)}$	$\hat{u}_{14(31)}$
	0.012	-0.014	0.002
	0.245	-0.294	0.045
	$\hat{u}_{23(11)}$	$\hat{u}_{24(11)}$	$\hat{u}_{34(11)}$
	0.026	-0.157	-0.064
	0.771	-4.673	-1.907

Note: First row gives u-term, second row its standardized value.

We can compare models 1 and 2 and obtain $G^2(1) - G^2(2) = 0.50$ with two degrees of freedom, showing that these models differ little and consequently that variable 1, water softness, has little to contribute. Similar comparisons between other pairs of models confirm the information given by the standardized u-terms that their contribution is not significant.

Table 4.4-8 Goodness of Fit for Models Fitted to Ries–Smith Data from Table 4.4-6 Chosen by Inspecting Standardized u-term Parameters

	Model	Sufficient Statistics	Terms Included	Degrees of Freedom	X^2	G^2
Complete Array	1	C_3, C_{24}	See p. 144	19	23.54	22.85
	2	C_1, C_3, C_{24}	Model 1 + u_1	17	23.13	22.35
	3	C_{13}, C_{24}	Model 2 + u_{13}	15	16.73	16.25
	4	C_{13}, C_{24}, C_{34}	Model 3 + u_{34}	14	11.92	11.89
Reduced Array	1 ⎫			11	144.56	146.67
	2 ⎬	Models as for complete array		10	19.08	18.19
	3 ⎪			9	13.86	13.54
	4 ⎭			8	9.16	9.18

So far we have only used the standardized terms to enable us to determine which u-terms to include in the model. Returning to table 4.4-7, we observe that the first category of u_1, soft water, has a standardized value of -0.97 and differs from the other two categories with values of 0.46 and 0.53. This might suggest that if we were in any doubt about the relative merits of models 1 and 2, instead of collapsing over variable 1 completely we might consider combining the second and third categories. If we do this and again fit model 1, we no longer have a tolerable fit: $X^2 = 144.56$ with eleven degrees of freedom (see table 4.4-8). This large value is understandable if we realize that by excluding variable 1 from the fitted model, we are in effect postulating that the original four-way table can be regarded as three identical three-way tables, one for each category of water softness, whereas if this model is correct when we add two categories, the number of counts in each resulting cell is approximately doubled and so this is no longer a valid hypothesis.

When we fit model 2 to the reduced array, we have $X^2 = 19.08$ with ten degrees of freedom. This is a smaller value than the $X^2 = 23.13$ obtained from the full array for model 2, but the reduction in degrees of freedom is such that $p < 0.05$ for the reduced table, whereas $p > 0.10$ for the original. We conclude that part of the reason why the simple model 2 appears to fit the original array well is that the numbers in the cells are not large. When we look at the reduced array we would probably choose model 4 as the best fit. This example illustrates that although inspecting standardized u-terms helps us in understanding the structure, we cannot set any arbitrary rules, such as "include all u-terms with absolute standardized values greater than 1."

From model 2 we conclude that u_{24}, brand preference versus previous use of M, is large; thus users tend to prefer the brand they know. When we accept model 4 as probably being a better model, we include two more interactions. First, temperature is related to water softness; this is a reflection of the study design, and its inclusion enables us to obtain better estimates of the other associations. Second, temperature is related to brand preference; this indicates that brand M is usually preferred with cold water, brand X with hot water. Knowledge of this tendency may be of importance and is useful in designing future experiments. ■ ■

4.4.3 *Assessing magnitude of an interaction*

Sometimes our main objective in examining a set of data is to test whether a particular interaction is present. The preceding section might suggest that we compute the standardized parameters for the u-term representing this interaction. Certainly this would give us an idea of whether the effect is large or small, but we do not recommend comparing these values with normal deviates as a means of testing the null hypothesis of a zero effect for the following reasons:

1. For variables with more than two categories, the u-terms for a particular interaction have more than one absolute value. We must combine the estimates in some way to form an overall test statistic. Goodman [1971b], in a different context, suggested forming linear and quadratic functions of the parameters. However, unless the categories are ordered, the interpretation of such functions is sometimes difficult.

2. If we compute the u-terms under a fitted model, we obtain different values from those computed under the saturated model. Thus the magnitude of an appropriately standardized u-term depends on which structural model we assume and the other parameters included.

The implicit dependence of the test statistic on the structural model assumed is, unfortunately, often ignored when tests for a particular effect are performed routinely. We consider now one such test that has suffered from much ill use and then outline an alternative approach that avoids the pitfall of ignoring the underlying structure.

Test statistics for a two-factor effect

Most methods proposed for testing a two-factor interaction in more than two dimensions relate to a three-dimensional array. In this context we have K two-way tables of size $I \times J$ relating variables 1 and 2, and we set up the null hypothesis that $u_{12} = 0$. Cochran [1954] computed measures for each of the K 2×2 tables.

$$X_k = \delta_k \sqrt{\sum_{i,j} \frac{(x_{ijk} - \hat{m}_{ijk})^2}{\hat{m}_{ijk}}}, \tag{4.4-12}$$

where δ_k = sign of $x_{11k} - \hat{m}_{11k}$ and

$$\hat{m}_{ijk} = \frac{x_{i+k} x_{+jk}}{x_{++k}}. \tag{4.4-13}$$

Thus $X^2 = \sum_k X_k^2$ is the usual statistic for measuring goodness of fit of the model with $u_{12} = u_{123} = 0$. He then proposed methods of weighting the X_k to form a single summary statistic. Mantel and Haenszel [1959] enlarged on the approach. Birch [1964b] showed their test to be uniformly most powerful *when the interaction was constant in each table*. This proviso led him to recommend prior testing for such uniformity, in other words, testing $u_{123} = 0$ before testing $u_{12} = 0$. We now describe the Mantel–Haenszel statistic and then an alternative based on simplifying the structural model before testing.

The Mantel–Haenszel procedure

For a $2 \times 2 \times K$ table, the Mantel–Haenszel statistic is

$$M_K^2 = \frac{\left(\left| \sum_k A_k - \sum_k \hat{E}(A_k) \right| - \frac{1}{2} \right)^2}{\sum_k \hat{V}(A_k)}, \qquad (4.4\text{-}14)$$

where we have

$$A_k = x_{11k},$$

$$\hat{E}(A_k) = \text{expected value of } A_k \text{ from (4.4-13)},$$

$$\hat{V}(A_k) = \text{estimated variance of } A_k,$$

$$= \frac{x_{1+k} x_{2+k} x_{+1k} x_{+2k}}{x_{++k}^2 (x_{++k} - 1)}.$$

When $K = 1$, expression (4.4-14) reduces to

$$M_1^2 = \frac{\left(|x_{11} x_{22} - x_{12} x_{21}| - \frac{N}{2} \right)^2 (N - 1)}{x_{1+} x_{2+} x_{+1} x_{+2}}, \qquad (4.4\text{-}15)$$

where $N = x_{++}$.

Expression (4.4-15) is the X^2 statistic defined in expression (4.2-1), with continuity correction if the term $N - 1$ is replaced by N.

Examination of the numerator of (4.4-14) shows that we have as the variable component the squared value of

$$\sum_k A_k - \sum_k \hat{E}(A_k) = x_{11+} - \hat{m}_{11+}$$

$$= \sum_k (x_{11k} - \hat{m}_{11k}),$$

where $\hat{m}_{11k}$ is given by (4.4-13). If $x_{11k} - \hat{m}_{11k}$ varies with k, M_K^2 is unrelated to the magnitude of u_{12}, and we see the need to establish that $u_{123} = 0$ before computing the test statistic.

To test for u_{12} in more than three dimensions, we can treat all variables except the first two as a single third composite variable. The Mantel–Haenszel test statistic is then conditional on all higher-order relatives of u_{12} being set equal to zero. In a four-dimensional $I \times J \times K \times L$ array, for instance, we have KL two-way tables corresponding to the joint categories of variables 3 and 4. The statistic M^2 only reflects the magnitude of u_{12} when $u_{123} = u_{124} = u_{1234} = 0$. This must be verified. Thus correct use of the Mantel–Haenszel test statistic requires the following steps for testing u_{12} in four dimensions:

1. Fit a model (1) based on C_{12}, C_{134}, and C_{234}, and compute goodness of fit. If model (1) fits well go to the next step; otherwise, consider different methods of analysis, such as partitioning the data.

2. Fit a model (2) based on C_{134} and C_{234} to obtain estimates $\hat{m}_{ijkl}^{(2)}$.
3. Compute the variance of $\hat{m}_{11kl}$ for all k, l, and hence obtain M_K^2.

A simpler approach that offers advantages over the M^2 statistic is to obtain a measure of goodness of fit for each model. Then the difference $G^2(2) - G^2(1)$ or $X^2(2) - X^2(1)$ is distributed as χ^2 with $(I - 1)(J - 1)$ degrees of freedom and provides a conditional test of the hypothesis $u_{12} = 0$. The conditions are the same for each approach, namely, that u_{12} does not vary among the KL tables. The advantages of comparing a pair of log-linear models are:

1. The same measure of goodness of fit is used for each model. In Section 4.2 we discussed the desirability of matching the test statistic to the method of estimation.
2. Extension to data sets with structural zeros follows directly, using the methodology to be described in Chapter 5.
3. The method is applicable not only for two-factor effects but also for higher-order effects.
4. Further refinement is possible when the structure can be expressed more simply.

These last two items need further development, but before explaining these ideas we give a three-dimensional example of the equivalence of the procedures. For other examples see Bishop [1971]. Although some of the models were incorrectly fitted, the general principles were developed by Kullback et al. [1962b] and later correctly computed by Ku and Kullback [1968] and Goodman [1969b].

Example *4.4-4* *Responses of lymphoma patients to combination chemotherapy, by sex and cell type (Data from Skarin et al. [1973])*
In example 4.2-1 we observed a large sex difference in the percentage of lymphoma patients responding to a course of chemotherapy.

Table 4.4-9 shows that when we introduce a third variable, cell type, the patients with nodular disease show a high response rate and those with diffuse disease a low response rate, regardless of sex. Now we have three variables, we can proceed to test u_{12}, the sex $\times$ response interaction, by fitting two models:

1. model (1) with $u_{123} = 0$ gives $G^2(1) = 0.651$ with one degree of freedom;
2. model (2) with $u_{12} = u_{123} = 0$ gives $G^2(2) = 0.809$ with two degrees of freedom.

Table 4.4-9   Response of Lymphoma Patients to Combination
Chemotherapy: Distribution by Sex and Cell Type

Variable 3	Variable 2	Variable 1 No Response	Response	Percentage Responding
Nodular	Male	1 (1.15)	4 (3.85)	80
	Female	2 (1.85)	6 (6.15)	75
Diffuse	Male	12 (11.47)	1 (1.53)	8
	Female	3 (3.53)	1 (0.47)	25

Source: Skarin et al. [1973].
Note: Values in parentheses obtained under model $u_{123} = u_{12} = 0$.

The difference $G^2(2) - G^2(1) = G^2(2|1) = 0.158$ is distributed as χ^2 with one degree of freedom. The Mantel–Haenszel test statistic without continuity correction is $M^2 = 0.1554$, which agrees closely with $G^2(2|1)$. Both values are less than the X^2 value of 2.80 obtained from the condensed table C_{12} in example 4.2-1, and we conclude that we may put $u_{12} = 0$ when we describe these data. This also means that we can safely collapse over variables 1 or 2 and will not be misled by the configurations C_{23} and C_{13} of table 4.4-10.

Table 4.4-10 Permissible Collapsed Tables for Data of Table 4.4-9

		Variable 1	
Configuration C_{13}	Variable 3	No Response	Response
	Nodular	3	10
	Diffuse	15	2
		Variable 2	
Configuration C_{23}	Variable 3	Male	Female
	Nodular	5	8
	Diffuse	13	4

We proceed to fit further models to evaluate u_{23}, the sex $\times$ cell type effect. To test whether u_{23} is nonzero, we can fit the model with $u_{23} = u_{123} = 0$, based on C_{12} and C_{13}, and compare with model (1). However, we have already concluded that we may put $u_{12} = 0$. It is therefore more reasonable to fit model (3) with $u_{23} = u_{12} = u_{123} = 0$. The test for u_{23} is then the difference $G^2(3) - G^2(2) = 4.808$. Similarly, for assessing the magnitude of u_{13} we have $G^2(4) - G^2(2) = 14.021$, where model (4) has $u_{13} = u_{12} = u_{123} = 0$, as in table 4.4-11.

Table 4.4-11 Fit of Various Models to Data on Lymphoma Patients from Table 4.4-9

Model	Configurations Fitted	Degrees of Freedom	X^2	G^2
1	C_{12}, C_{13}, C_{23}	1	0.67	0.65
2	C_{13}, C_{23}	2	0.93	0.81
3	C_{13}, C_2	3	5.14	5.32
4	C_{23}, C_1	3	13.45	14.83

The test statistics for u_{13} and u_{23} could have been derived directly from the condensed tables C_{13} and C_{23}, respectively (table 4.4-10). This is a result of the collapsing theorems of Chapter 2; if $u_{12} = 0$, then summing over variable 1 does not affect the magnitude of u_{23}. Similarly, summing over variable 2 does not affect u_{13}. An alternative proof follows from the partitioning of G^2 because models (2), (3), and (4) can each be fitted directly.

We conclude that both u_{13} and u_{23} are large; thus cell type is related to both sex and survival. When we take cell type into account, survival is not related to sex, as appeared to be the case when we looked only at the configuration C_{12} in example 4.2-1. ∎

Simplifying structure before testing

As soon as we consider comparing the goodness of fit of two models that only differ by the term u_{12}, we have many choices available. When we use the Mantel–Haenszel statistic in four dimensions, the procedure outlined above depends on models (1) and (2). These models both include configurations C_{134} and C_{234} and hence include the three-factor effects u_{134} and u_{234}. The alternative approach of taking the difference in goodness of fit of an appropriate pair of models gives many other possibilities. For example, if all three-factor effects are absent we could choose the pair:

model (3) based on C_{12}, C_{13}, C_{14}, C_{23}, C_{24}, and C_{34}, i.e., the model with all two-factor effects;
model (4) based on C_{13}, C_{14}, C_{23}, C_{24}, and C_{34}.

Table 4.4-12 (taken from Bishop [1971]) shows the number of pairs of models in four dimensions that differ only by the term u_{12}, assuming that all variables are present in the model. As soon as we set three three-factor terms equal to zero we have eight models with $u_{12} = 0$, and each of these has a matching model that differs only by including u_{12}. These matching pairs appear in the second and third rows of the table in adjacent columns. The total number of matching pairs is 41, but this does not mean that we have 41 different tests for u_{12}.

Table 4.4-12   Number of One-Degree-of-Freedom Tests for a Single Two-Factor Term

Three-Factor Terms Removed	Presence of u_{12}[a]	\multicolumn{7}{c}{Two-Factor Terms Removed}	Totals
		0	1	2	3	4	5	6	
2	Linked	5	5						10
	Alone	1							1
	Absent		1						1[b]
3	Linked	2	6	6	2				16
	Alone	2	4	2	0				8
	Absent		2	4	2				8[b]
4	Alone	1	5	10	10	5	1		32
	Absent		1	5	10	10	5	1	32[b]

Note: The total number of models of a specific type is divided into the number of those with u_{12} present or absent.

[a] C_{123} and C_{124} contain u_{12} linked to other terms; C_{12} contains u_{12} alone.

[b] Each of these models, when combined with the appropriate model from the line above, provides a one-degree-of-freedom test for u_{12}.

From equations (4.2-11)–(4.2-12) we know that for models that have direct estimates we can compute G^2 from the sufficient configurations without obtaining cell estimates. For any two models with direct estimates, the difference in G^2 depends solely on the configurations that differ. When pairs of direct models differ only by u_{12}, the difference in G^2 depends only on C_{12}. We can compute this value by collapsing over all variables except 1 and 2 and then fitting the model of independence to the table of sums, C_{12}. Thus all pairs of direct models yield the same test for u_{12}, and this reduces the number of different tests from 41 to 25.

This is still a large choice. We have no formal proof, but recommend choosing a pair of models such that one member of the pair provides a well-fitting description of the structure of the array. We suggest this course of action to avoid the two extremes of overcondensing and overextending.

a. *Overcondensing extreme*

We look only at the condensed table of sums C_{12}. Testing for independence in this two-way table is equivalent to looking at a pair of models with direct estimates, and is only appropriate because these models permit collapsing. From the collapsing theorems of Chapter 2 we know that this requires specific patterns of independence among variables; one or more two-factor effects other than u_{12} are set equal to zero. The pair of models suffers from not including sufficient terms.

b. *Overextending extreme*

We test that the higher-order relatives of u_{12} are absent and then compute the Mantel–Haenszel statistic. In s dimensions this is equivalent to comparing a pair of models that include all effects of order $s - 1$ that are not relatives of u_{12}. The models involved in this procedure may have many more parameters than are required, and the consequent increase in variance of the estimates may reduce the power of the test. In Chapter 9 we note that even when a low-parameter model leads to slightly biased estimates, we may still have more stable estimates than we would obtain by fitting a model with more parameters.

The following hypothetical situation illustrates the two extremes. Suppose the condensed table C_{12} exhibits a large two-factor effect and has all entries on the diagonal:

$$
\begin{array}{|c|c|}
\hline
a & 0 \\
\hline
0 & b \\
\hline
\end{array}
\qquad (4.4\text{-}16)
$$

If we use the overcondensing method, we look only at C_{12} and conclude that u_{12} is large. When we expand to more dimensions, we have K component 2×2 tables that form this table when collapsed. As K becomes large, many of the component tables are of the form

$$
\begin{array}{|c|c|}
\hline
a'_k & 0 \\
\hline
0 & 0 \\
\hline
\end{array}
\quad \text{or} \quad
\begin{array}{|c|c|}
\hline
0 & 0 \\
\hline
0 & b'_k \\
\hline
\end{array}
\qquad (4.4\text{-}17)
$$

where $a'_k < a$ and $b'_k < b$. Under the overextending method, none of the tables of the form of (4.4-17) contribute to M_K^2, and in the final extreme we have $M_K^2 = 0$ and conclude that $u_{12} = 0$.

Extension to multiple-factor effects

So far we have considered only tests for a two-factor effect, but the procedure of comparing the goodness of fit of a pair of models can readily be extended to three-factor or higher-order effects. All that is required is selection of a pair of models that differ only by the multiple-factor effect under consideration. In example 4.4-5 we examine different pairs of models for each of three three-factor effects and find that the test statistic fluctuates by a factor of 2 according to which pair is chosen. Thus we can illustrate the importance of choosing a suitable pair.

A further advantage of the pairwise comparison approach is that it can be extended to structurally incomplete tables. We can also use it when we have sporadic random zeros, but then a little caution is necessary. If the configuration corresponding to the effect to be tested has a random empty cell, the model of the pair that includes this effect may have more fitted cells with zero counts than the other member of the pair. This can be readily detected because the difference in degrees of freedom for the two models will not correspond to the number of parameters belonging to the u-term being tested. The remedy is to treat these cells as structurally empty cells for both models.

Example 4.4-5   Evaluating magnitude of particular interaction term by comparing pairs of models (Data on basal disc regeneration in hydra from Cohen and Mac-Williams [1974])

A hydra holds fast to its substrate by means of a basal disc or foot. When this foot is cut off the animal can, under certain conditions, regenerate a new foot. To define these conditions, the following experiments were performed. Annuli from donor hydra were grafted to host hydra and observed for foot formation; for each experimental subgroup both the position from which the grafted annulus was taken and the position in which it was implanted were varied. In half the experiments the host's basal disc was removed, while in the other half it was intact. Each pair of annulus positions investigated was repeated 25 times in each half of the experiment. The resulting data form an incomplete contingency table with variables as follows:

Variable 1.   Foot formation
              1 = yes, 2 = no
Variable 2.   Host basal disc
              1 = present, 2 = absent
Variable 3.   Position of donor annulus
              1–7 corresponding to position from foot to head
Variable 4.   Position of host graft
              1–7 corresponding to position from foot to head

We give the observed and fitted values for the $(1, j, k, l)$ cells in table 4.4-13 but do not give the $(2, j, k, l)$ cells, because each experiment was replicated 25 times. This means that C_{234} is fixed by the sampling plan and has 25 in every nonzero cell. The observed values are the first entry in each cell.

Preliminary investigations showed that all two-factor effects must be included in the model. The authors then had a choice between including none, one, two, or all three of the three-factor effects corresponding to configurations C_{123}, C_{124}, and C_{134}. Table 4.4-14 lists the X^2 and G^2 goodness of fit for these possibilities.

Table 4.4-13   Basal Disc Regeneration in Hydra

Host's Disc	Position of Graft in Host	Donor Annulus Position						
		7	6	5	4	3	2	1
Present	7	1	8	22	23	—	—	—
		0.8	8.8	20.7	23.4	—	—	—
	6	—	5	16	21	24	—	—
		—	4.9	17.2	21.6	22.2	—	—
	5	—	—	8	18	18	19	—
		—	—	9.3	15.8	17.2	20.7	—
	4	—	—	—	5	4	15	24
		—	—	—	6.3	7.5	12.0	22.2
	3	—	—	—	—	1	5	19
		—	—	—	—	2.7	5.1	17.3
	2	—	—	—	—	—	0	4
		—	—	—	—	—	0.5	3.5
	1	—	—	—	—	—	—	4
		—	—	—	—	—	—	4.0
Absent	7	2	16	23	—	—	—	—
		2.2	15.3	23.3	—	—	—	—
	6	—	9	21	24	—	—	—
		—	9.4	21.1	23.5	—	—	—
	5	—	—	19	21	22	—	—
		—	—	17.7	21.9	22.5	—	—
	4	—	—	—	20	22	21	—
		—	—	—	19.7	20.6	22.8	—
	3	—	—	—	—	16	19	22
		—	—	—	—	14.3	18.6	24.0
	2	—	—	—	—	—	20	23
		—	—	—	—	—	18.9	24.1
	1	—	—	—	—	—	—	24
		—	—	—	—	—	—	24.0

Source: Cohen and MacWilliams [1974].
Note: The first number is the observed count; the second is the fitted value in each cell.

Table 4.4-14   Goodness of Fit of Various Models to Incomplete Four-Way Array (for Hydra Data from Table 4.4-13)

Model	Configurations[a]	Degrees of Freedom	X^2	G^2
1	C_{12}, C_{13}, C_{14}	26	62.87	68.06
2	C_{123}, C_{14}	20	50.50	48.46
3	C_{134}, C_{12}	17	53.37	56.75
4	C_{124}, C_{13}	20	21.87	21.61
5	C_{123}, C_{134}	11	45.10	35.92
6	C_{124}, C_{134}	11	16.48	15.53
7	C_{123}, C_{124}	14	10.29	10.77
8	$C_{123}, C_{124}, C_{134}$	5	3.46	3.70

[a] All models include C_{234}, which is fixed by design.

The authors selected model 4 with sufficient configurations C_{234}, C_{13}, and C_{124}. The fitted values under this model are the second entry in each cell of table 4.4-13. If we take logits for the dichotomous variable 1, we can write model (4) as

$$\text{logit}(j, k, l) = w + w_{2(j)} + w_{3(k)} + w_{4(l)} + w_{24(jl)} \qquad (4.4\text{-}18)$$

and then compute the parameters of each w-term. The authors inspected these terms and observed that w_4 and w_{24} were approximately linear. They therefore simplified the model further to

$$\text{logit}(j, k, l) = w' + w_{2(j)} + w_{3(k)} + (\lambda_1 + \lambda_2 l)j, \qquad (4.4\text{-}19)$$

where λ_1 and λ_2 are constant weights attached to the category scores l and j, giving the final transplant position and basal disc condition of the host hydra.

The authors were particularly interested in the terms u_{123} and u_{134}. When the model with these terms set equal to zero fitted well, they concluded that the probability of a disc forming depended on the original position of the annulus (variable 3) and independently on two experimental parameters specific to the host (variables 2 and 4). As these terms are of particular interest, we compare all pairs of models that differ only by one of these terms in table 4.4-15. We include the pairs differing by u_{124} for comparison, and note that u_{234} is always included because C_{234} is fixed.

For all model pairs, both u_{123} and u_{134} are associated with small differences in G^2 compared with their degrees of freedom. We note that the smallest value of u_{123} occurs when the preferred model (4) is compared with model (7), and the smallest value of u_{134} when model (4) is compared with model (6). We conclude that neither term differs significantly from zero. However, when we look at pairs that differ by u_{124}, the difference in G^2 is large. The comparison with the preferred model (4) yields the largest value. We agree with the authors' conclusion that the probability of disc formation depends jointly on the site of graft and whether the host's disc is intact.

Table 4.4-15 Comparison of Pairs of Models to Determine Which Three-Factor Effects are Important (for Hydra Data from Table 4.4-13)

Models Forming a Pair	Number of Parameters Common to Both Models	Difference in G^2 for Model Pair	Term by Which Models Differ and Degrees of Freedom
(6), (8)	69	11.83	u_{123}, 6
(3), (5)	63	20.83	
(4), (7)	60	10.84	
(1), (2)	54	19.60	
(7), (8)	66	7.07	u_{134}, 9
(4), (6)	60	6.08	
(2), (5)	60	12.54	
(1), (3)	54	11.31	
(5), (8)	69	32.22	u_{124}, 6
(3), (6)	63	41.22	
(2), (7)	60	37.69	
(1), (4)	54	46.45	

In this example the conclusions reached about each three-factor term do not differ according to which pair of models is examined, but comparisons based on the well-fitting model (4) give the smallest test statistics for the terms that do not contribute significantly and the largest test statistic for the important term. ■ ■

4.4.4 *Exercises*

1. The fitted values of model (4) of table 4.4-14 are given in table 4.4-13 and the logit model in expression (4.4-18). Compute the values of the parameters of w_4 and w_{24} and plot them against the category scores 1–7 for position of host graft. Do they indicate that the authors' simplification of the model in expression (4.4-19) is likely to be a good description of the data?

2. The following hypothetical two-way table has one empty cell but otherwise exactly fits the model of independence:

0	a	a
b	b	b

Find the fitted values for each cell obtained by applying the model of independence to the entire table, and show that

$$X^2 = \frac{a(2a + 3b)}{3(a + b)}.$$

Find values of a and b such that the empty cell would be considered an outlier at the 0.05 level of significance using a χ^2 test.

4.5 Choosing a Model

When we confront a data set we need a strategy that leads us to select an appropriate model with a minimum of computation. In table 3.4-2 in Chapter 3, we show that in four dimensions we have a choice from among 113 hierarchical models that include all four variables. As the number of variables increases, so does the possible number of models. We immediately perceive that it is usually impractical to fit every model and then select the model that is the "best fitting" in the sense that the significance level according to the reference χ^2 is largest.

Most strategies that have been proposed for selecting a model use the significance level as the criterion for choosing between models. The strategies usually have several steps:

1. a routine set of computations is made from the data, and the resulting statistics are used to guide the next step;

2. using as a starting point a model determined in step 1, a series of models is fitted that are close to this model (we consider two models "close" when most of their *u*-terms are common to both);

3. using a model selected in step 2 as baseline, a further series of close models is computed.

In some procedures step 3 is not always necessary; in other procedures it may be followed by further steps, each fitting another series of models adjacent to the model selected previously. The strategies proposed differ mostly in the initial step, and various preliminary computations have been suggested. In Chapter 9 we discuss the difficulties of testing and fitting on the same set of data, with the consequent effect on the significance level of the test statistic, and discuss other criteria for model selection. In this chapter we discuss procedures based on the significance level of the reference χ^2.

4.5.1   *Initial computations*

The choice of initial computation depends on several factors: the size of the data set, whether all the variables are highly interrelated, and finally, our particular objective in analyzing the data. In discussing the contenders we assume for the sake of simplicity that we have a single multinomial. This assumption implies that any hierarchical model is eligible. For stratified sampling schemes, the procedures should be modified so that only models that include the configuration fixed by the sampling design are included. Moreover, if we have a stratified design with unequal sampling fractions, we should consider partitioning the data as described in example 3.7-2 of Chapter 3. We consider four different strategies.

Strategy 1. Standardized u-terms

In example 4.4-3 we demonstrated the use of standardized *u*-terms for model selection, using the four-variable Ries–Smith data on detergents. We could only compute such standardized terms for the saturated model, and then only when we observed no random zeros. In our example we estimated the lower-order *u*-terms and noted that interpretation of terms related to the three-category variable was complex. Examining standardized *u*-terms is useful primarily when we have dichotomous variables with entries in every cell.

In *s* dimensions, if the *s*-factor effect is large, we cannot fit an unsaturated hierarchical log-linear model without first partitioning the data. This partitioning is sometimes equivalent to fitting a nonhierarchical model (see example 3.7-4 of Chapter 3). Prior computation of standardized *u*-terms in this circumstance may alert us to the fact that searching for an unsaturated log-linear model that fits well would be fruitless. To assess whether the highest-order term is significantly large, we must proceed to the next step and compute estimates of the log-linear model that sets only this term equal to zero.

Strategy 2. Fitting models with terms of uniform order

A reasonable first step in the model-selection process is to fit a subset of models. The motivation behind the strategy of fitting models of uniform order is to cover the range of all possible models. Thus for *s* dimensions we fit:

(1)   the complete independence model with only main-effect terms;
(2)   the model with all possible two-factor effects and no higher-order terms;
$\vdots$
$\vdots$

$(s-1)$ the model with all possible terms of order $s-1$, i.e., the model with the *s*-factor effect set equal to zero.

We then examine the goodness of fit of each model in the list and select a consecutive pair such that the lower member of the pair fits poorly and the next higher model fits well. The next step of the search is then confined to the "intervening" models, namely, those models that include all terms of the lower model and a subset of the additional terms of the higher model.

We can readily modify this method to include only the first two or three models of uniform order when we are only interested in lower-order models. Indeed, for large data sets such modification may be essential, as all except the first model must be fit by iterative methods which may prove computationally costly.

The implicit assumptions of this approach are:

1. the goodness of fit of the models selected increases monotonically (in terms of increasing significance levels). For some data sets this assumption is not valid;

2. the data are well described by one of the intervening models. Thus if the model with all terms of order r fits poorly, and the model with all terms of order $r + 1$ fits well, subsequent search is confined to models that include all terms of order r and a subset of the terms of order $r + 1$. As a result, some models are excluded from consideration. In four dimensions, for example, the model based on C_1 and C_{234} will never be examined.

In spite of these drawbacks, the strategy of examining the first three models of uniform order has a great deal of appeal. There are four possible outcomes:

(i)   the model with all three-factor effects does not fit well;

(ii)   the model with all three-factor effects fits well but the model with all two-factor effects does not;

(iii)   the model with all two-factor effects fits well but the model with only main effects does not;

(iv)   the model with only main effects fits well.

Given outcome (i) we should examine the cell residuals. Experience dictates that not only are higher-order effects difficult to interpret, but they typically occur in one of two circumstances. In the circumstance where we have large numbers of counts in every cell, we are well advised to reexamine our motivation in fitting an unsaturated model (for further discussion of this issue, see Section 9.6). The other common circumstance is when the data are uneven: we have some cells with very large counts and other cells with small or zero counts. We may learn more from such data if we partition them into suitable segments. For example, the data collected for the National Halothane study (described in Chapter 3, example 3.7-2) were descriptive of all surgical operations performed in the participating institutions. The death rates varied widely between such procedures as tonsillectomy and brain surgery. After preliminary examination the data were divided for purposes of analysis into three segments corresponding to low, middle, and high death-rate operations. The low and middle segments were then well fit by models that did not include any terms of more than three factors. The high death-rate operations were still so uneven that they required a modified approach, such as that to be proposed in Chapter 12.

Given outcomes (ii) and (iii), we proceed to examine intervening models, taking care to examine cell residuals at every point. If such examination suggests that a model based on configurations of more diverse dimensionality is appropriate, we try fitting this model in addition to the intervening model. As soon as we find a well-fitting model that has one or more two-factor effects set equal to zero, we may profitably consider collapsing the data and continuing our search on the reduced array.

For outcome (iv) we need no further search, and the data are adequately described by the s one-dimensional arrays of sums.

Example 4.5-1   Models with terms of uniform order for Ries–Smith data

In example 4.4-3 we considered the data of Ries and Smith on detergent preference given in table 4.4-6. The data form a $3 \times 2 \times 2 \times 2$ table, and there are only three models with terms of uniform order:

Model	G^2	Degrees of Freedom
(1) Complete Independence	42.9	18
(2) All Two-Factor but No Higher-Order Effects	9.9	9
(3) No Four-Factor Effect	0.7	2

The results of fitting these three models correspond to outcome (iii) described above, i.e., the model with all two-factor effects fits well but the model with only main effects does not. Thus we proceed to examine the intervening models that include all main effects and one or more two-factor effects. ■ ■

Strategy 3. Examining G^2 components for direct models

Although selecting for first examination the models with terms of uniform order has intuitive appeal, it may be costly. All these models except the main effects model must be fitted iteratively. An approach that tries to minimize the amount of computing in the first stage is based on the structural breakdown of G^2 (see Section 4.2.1).

We compute:

1.   The log-likelihood $L_s(\mathbf{x})$ for the observed s-dimensional array;
2.   the log-likelihood $L_{(s-1)}(\mathbf{x}), L_{(s-2)}(\mathbf{x}), \ldots, L_1(\mathbf{x})$ for all the configurations of order $s-1, s-2, \ldots, 1$.

From these log-likelihoods we can obtain G^2 for all possible models with closed-form estimates.

We have only to remember that a model cannot be fit directly if all two-factor effects are included to realize that examining direct models is primarily useful when we believe some of the two-factor effects are absent. A disadvantage of this approach is that when none of the direct models fits well, we have very little guidance as to any next stage of investigation because we do not compute cell estimates and so cannot examine the pattern of residuals.

Example 4.5-2   Direct models for Ries–Smith data

In table 4.5-1 we give the log-likelihood components for the Ries–Smith data on detergent preference from example 4.4-3. Using these values we can construct the log-likelihoods for all direct models, some of which are given in table 4.5-2.

For these data the simplest direct model that fits well and includes all variables is

$$u + u_1 + u_2 + u_3 + u_4 + u_{24}$$

Table 4.5-1 Log-Likelihoods
of All Marginal Configurations
for Data from Table 4.4-6

Component	Value
$L_{1234}(\mathbf{x})$	3,826.88
$L_{123}(\mathbf{x})$	4,509.13
$L_{124}(\mathbf{x})$	4,481.14
$L_{134}(\mathbf{x})$	4,508.61
$L_{234}(\mathbf{x})$	4,926.78
$L_{12}(\mathbf{x})$	5,166.70
$L_{13}(\mathbf{x})$	5,204.86
$L_{14}(\mathbf{x})$	5,165.44
$L_{23}(\mathbf{x})$	5,611.86
$L_{24}(\mathbf{x})$	5,584.95
$L_{34}(\mathbf{x})$	5,612.48
$L_{1}(\mathbf{x})$	5,863.90
$L_{2}(\mathbf{x})$	6,273.32
$L_{3}(\mathbf{x})$	6,308.96
$L_{4}(\mathbf{x})$	6,272.39
$N \log N$	6,971.05

Table 4.5-2 Log-Likelihoods and G^2 Values for Various Direct Models
Computed on Ries–Smith Data from Table 4.4-6

Model	Log-Likelihood	G^2	Degrees of Freedom
C_{12}, C_{13}, C_{14}	$L_{12} + L_{13} + L_{14} - 2L_1 = 3{,}809.20$	35.36	12
C_{12}, C_{23}, C_{24}	$L_{12} + L_{23} + L_{24} - 2L_2 = 3{,}816.87$	20.02	14
C_{13}, C_{23}, C_{34}	$L_{13} + L_{23} + L_{34} - 2L_3 = 3{,}811.28$	31.20	14
C_{14}, C_{24}, C_{34}	$L_{14} + L_{24} + L_{34} - 2L_4 = 3{,}818.09$	17.56	14
C_{13}, C_{24}, C_{34}	$L_{13} + L_{24} + L_{34} - L_3 - L_4 = 3{,}820.94$	11.88	14
C_{123}, C_4	$L_{123} + L_4 - N \log N = 3{,}810.47$	32.82	11
C_{124}, C_3	$L_{124} + L_3 - N \log N = 3{,}819.05$	15.66	11
C_{134}, C_2	$L_{134} + L_2 - N \log N = 3{,}810.88$	32.00	11
C_{234}, C_1	$L_{234} + L_1 - N \log N = 3{,}919.63$	14.50	14
C_{24}, C_1, C_3	$L_{24} + L_1 + L_3 - 2N \log N = 3{,}815.71$	22.34	17
C_1, C_2, C_3, C_4	$L_1 + L_2 + L_3 + L_4 - 3N \log N = 3{,}805.42$	42.92	18

Note: The second decimal place for G^2 may differ slightly from values in other tables because of
roundoff errors introduced by this method.

($G^2 = 22.34$ with seventeen degrees of freedom), and other well-fitting direct
models include it as a special case. In particular, the model based on configurations
C_{13}, C_{24}, and C_{34} (selected as a good model in example 4.4-3) has a smaller value
of G^2, 11.88, than the other models with fourteen degrees of freedom shown in
table 4.5-2. ∎∎

In Section 4.2.1 we showed, using expression (4.2-17), that we can also set bounds
on the value of G^2 for iterative models. Specifically, in expression (4.2-14) we
showed that when fitting the no three-factor effect model to a three-dimensional
table, the estimates obtained at the end of the first cycle can be expressed as the

difference of G^2 values obtained for two other direct models. We know that the likelihood function of the estimates obtained at each step of the iterative procedure are monotonically increasing, and hence the estimates obtained at the end of the first cycle give an upper bound for G^2. For this no three-factor effect model we have three such estimates, according to the order in which the configurations are fit. Thus we can choose the smallest of the three as our upper bound. Unfortunately, this upper bound may still be too large to determine whether or not the model provides a good fit to the data.

Example 4.5-3   Bounds on indirect models for the Stouffer data

In example 4.4-1 we considered a 2^4 table first presented by Stouffer et al. [1949] containing information on soldiers' race (variable 1), region of origin (variable 2), location of present camp (variable 3), and preference as to camp location (variable 4). We saw that the model based on C_{123}, C_{234}, and C_{14} fits the data extremely well. Here we compute an upper bound on the value of G^2 for this model.

First we treat variables 2 and 3 as a single four-category variable, say variable 2'. Thus the configurations to be fit are $C_{12'}$, $C_{2'4}$, and C_{14}. Next we compute the log-likelihood for all the marginal configurations (see table 4.5-3). Finally, we compute the values for G^2 for three models, one with $u_{12'} = 0$, the second with $u_{14} = 0$, and the third with $u_{2'4} = 0$, as well as the values for these models in the appropriate two-way marginal tables.

Table 4.5-3 Log-Likelihoods of All Marginal Configurations for Stouffer Data from Table 4.4.-2

Component	Value
$L_{12'4}(\mathbf{x})$	52,711.61
$L_{12'}(\mathbf{x})$	56,725.74
$L_{14}(\mathbf{x})$	61,155.42
$L_{2'4}(\mathbf{x})$	57,737.95
$L_1(\mathbf{x})$	66,706.18
$L_{2'}(\mathbf{x})$	61,828.40
$L_4(\mathbf{x})$	66,687.33
$N \log N$	72,257.19

Note: Variables 2 and 3 (region and location) are treated as a single combined variable (2').

We have three upper bounds for this value of G^2 for the model based on $C_{12'}$, $C_{2'4}$, and C_{14}, each of the form given in expression (4.2-17):

$$G^2_{12'4}(1) - \bar{G}^2_{12'}(1) = 2{,}162.54 - 740.06 = 1{,}422.48,$$

$$G^2_{12'4}(2) - \bar{G}^2_{14}(2) = 152.65 - 112.34 = 40.31,$$

$$G^2_{12'4}(3) - \bar{G}^2_{2'4}(3) = 3{,}073.26 - 1{,}641.74 = 1{,}431.52.$$

Thus we have 40.31 as an upper bound on the log-likelihood ratio statistic G^2 for the model based on C_{123}, C_{234}, and C_{14}. While this upper bound does indicate that this model provides a significant improvement over the model based on C_{123} and C_{234}, the upper bound does not indicate that the model provides an acceptable fit to the data. ■■

Our rules for detecting whether direct estimates exist also tell us when we can combine G^2 components obtained from the configurations with a G^2 component obtained from iteratively fitted estimates. For instance, in four dimensions, if we wish to fit the model based on C_{12}, C_{13}, C_{23}, and C_{14} and apply the rules, we eliminate the last configuration. We compute $\{\hat{m}_{ijk+}\}$ iteratively from the first three configurations and then compute G^2_{123+} from these estimates. Finally, we use the relationship

$$\hat{m}_{ijkl} = \hat{m}_{ijk+} \frac{x_{i++l}}{x_{i+++}} \tag{4.5-1}$$

to obtain G^2 for the full model as

$$G^2_{1234} = G^2_{123+} - 2L_{14}(\mathbf{x}) + 2L_1(\mathbf{x}). \tag{4.5-2}$$

This technique is especially useful for large data sets and can be combined with the upper bound for G^2_{123+} discussed above. See Section 4.6 for a more detailed discussion of partitioning G^2 for iteratively fitted models.

Strategy 4. Fitting the partial association models
Models with one two-factor effect (and higher-order relatives) set equal to zero that retain all other terms are based on two configurations of order $s - 1$. Such models, termed "partial association models" by Birch [1964b], can always be fit directly. They are, of course, a subset of the models whose goodness of fit can be derived from the sufficient configuration without computing cell estimates. The advantages of computing the cell estimates are that we can examine deviations in each cell, partitions of the goodness-of-fit statistic, and functions of the expected cell estimates.

Suppose we have a four-variable $I \times J \times K \times L$ array and fit the partial association model with $u_{12} = 0$. We obtain cell estimates by fitting C_{134} and C_{234}, and obtain

$$\hat{m}_{ijkl} = \frac{x_{i+kl} x_{+jkl}}{x_{++kl}}.$$

Equivalently, we can partition the data according to the KL combinations of variables 3 and 4, and for each resulting two-way array relating variables 1 and 2 we fit the model of independence. Thus we can partition the overall goodness-of-fit statistic into KL components, each with $(I - 1)(J - 1)$ degrees of freedom, and examine them for consistency. If they vary, the pattern of variation may indicate that u_{123} or u_{124} should be included in the model.

A further advantage of fitting the partial association models is that we can derive information about three-factor terms more formally. For example, if model 2 is the model with u_{12} present but $u_{123} = u_{124} = 0$, then our overall goodness of fit for the partial association model 1 with $u_{12} = 0$ can be split into components

$$G^2_{1234}(1) = G^2_{1234}[(1)|(2)] + G^2_{1234}(2). \tag{4.5-3}$$

In Section 4.4.3 we showed that the Mantel–Haenszel statistic for testing whether $u_{12} = 0$ is similar to the conditional test $G^2[(1)|(2)]$. We can compute the Mantel–Haenszel statistic from our direct estimates obtained under model 1 by summing observed and expected cell estimates to obtain $\{x_{ij++}\}$ and $\{m_{ij++}\}$. Expression (4.5-3) then enables us to compute $G^2_{1234}(2)$ approximately without computing cell estimates for model 2, which would require iterative fitting.

By fitting all partial association models, we can determine whether we must include each two-factor effect in the model, and for each term that we must include, whether its higher-order relatives are likely to be large. A diagrammatic display of all interaction u-terms, such as figure 4.5-1, can help us keep track of terms. Suppose we determine that u_{12} and u_{13} are both large and that the higher-order relatives of u_{12} are also large, but that the higher-order relatives of u_{13} are not. This would be the case if the partial association tests with $u_{12} = 0$, $u_{13} = 0$, $u_{14} = 0$, and $u_{24} = 0$ are significant, while those with $u_{23} = 0$ and $u_{34} = 0$ are not. The relatives of u_{12} that are candidates for inclusion are u_{123}, u_{124}, and u_{1234}, but of these u_{1234} and u_{123} are excluded because they are also relatives of u_{13}. We are left with u_{124} as the only candidate. Alternatively, suppose that the partial association tests for $u_{23} = 0$ and for $u_{14} = 0$ are significant, but those for $u_{12} = 0$, $u_{13} = 0$, $u_{24} = 0$, and $u_{34} = 0$ are not. Then it would appear that none of the three-factor terms must be included in a model and that the only two-factor terms to be included are u_{23} and u_{14}.

This approach is particularly useful when we are not interested in models that include terms of more than three factors, as it then gives us guidelines about both two-factor and three-factor terms. This is a somewhat surprising conclusion, as

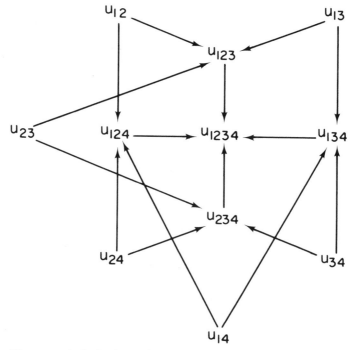

Figure 4.5-1 Diagrammatic display for partial association models in four dimensions. If a two-factor u-term is set equal to zero, all other u-terms which can be reached from it by arrows are also zero.

we are actually fitting a series of models that include two multiple-factor effects of order $s - 1$. When determining whether a particular two-factor effect is small, we must remember that we are assessing its magnitude by a model that includes a large proportion of the possible 2^s u-terms.

Suppose in s dimensions we find that r partial association models fit well as judged by a χ^2 significance level of 5%, say. We next proceed to fit a model that summarizes the findings by setting all the relevant two-factor terms equal to zero. We may find that this summary model can only include main effect terms and the remaining two-factor terms. For large s this is a substantial reduction in parameters compared with each partial association model, and we should not be surprised to find that the summary model does not fit the data well judging by the same 5% criterion. Thus even when we have assessed every two-factor effect individually, we still need a stepwise procedure to enable us to find a model that fits well by any predetermined criterion. We discuss in Chapter 9 the problems involved in interpreting significance levels from our measures of goodness of fit when we use measures from one model to lead us to the next model. For the moment, we assume we have a criterion based on the measure of goodness of fit and associated degrees of freedom and consider possible stepwise procedures, but first give examples of strategy 4.

Example 4.5-4   Partial association models for the Ries–Smith data
Using the log-likelihoods for the Ries–Smith data in table 4.5-1, we can compute the value of the likelihood ratio statistic G^2 for all six partial association models, as in table 4.5-4. The only partial association model that does not provide a good fit to the data has $u_{24} = 0$. For this model the value of G^2 is 28.00 with six degrees of freedom, and the corresponding Mantel–Haenszel statistic is 19.21 with one degree of freedom. Thus the higher-order relatives of u_{24} with five degrees of freedom account for about 8.79 of the G^2 value. This analysis suggests fitting the model

$$u + u_1 + u_2 + u_3 + u_4 + u_{24}.$$

In examples 4.4-3 and 4.5-2 we found that this model fits the data well with $G^2 = 22.35$ and seventeen degrees of freedom. We also found that when we include u_{13} and u_{34} we lose three degrees of freedom but decrease G^2 to 11.88, a significant

Table 4.5-4   Log-Likelihoods and G^2 Values for Partial Association Models Fit to Ries–Smith Data of Table 4.4–5

Model	Log-Likelihood	G^2	Degrees of Freedom	Two-Factor Term Absent
C_{123}, C_{124}	$L_{123} + L_{124} - L_{12} = 3{,}823.57$	6.62	6	u_{34}
C_{123}, C_{134}	$L_{123} + L_{134} - L_{13} = 3{,}812.88$	28.00	6	u_{24}
C_{134}, C_{124}	$L_{134} + L_{124} - L_{14} = 3{,}824.31$	5.14	6	u_{23}
C_{123}, C_{234}	$L_{123} + L_{234} - L_{23} = 3{,}824.05$	5.66	8	u_{14}
C_{124}, C_{234}	$L_{124} + L_{234} - L_{24} = 3{,}822.97$	7.82	8	u_{13}
C_{134}, C_{234}	$L_{134} + L_{234} - L_{34} = 3{,}822.91$	7.94	8	u_{12}

improvement. We note that table 4.5-4 does not help us determine which extra terms are candidates for inclusion. ■■

Example 4.5-5 Structural habitat of lizards from Whitehouse, Jamaica

In table 4.5-5 we give data on the structural habitat of *grahami* and *opalinus* lizards from Whitehouse, Jamaica, taken from Schoener [1970] and analyzed in Fienberg [1970b]. The data consist of observed counts for perch height, perch diameter, insolation, and time-of-day categories for both *grahami* and *opalinus* lizards. The four habitat variables are referred to as variables 1, 2, 3, and 4, respectively, and species is variable 5. The table is of dimension $2 \times 2 \times 2 \times 3 \times 2$.

Table 4.5-5 Counts in Structural Habitat Categories for *Grahami* and *Opalinus* Lizards from Whitehouse, Jamaica

Cell (i, j, k, l, m)	Observed	Cell (i, j, k, l, m)	Observed
1, 1, 1, 1, 1	20	1, 1, 1, 1, 2	2
2, 1, 1, 1, 1	13	2, 1, 1, 1, 2	0
1, 2, 1, 1, 1	8	1, 2, 1, 1, 2	3
2, 2, 1, 1, 1	6	2, 2, 1, 1, 2	0
1, 1, 2, 1, 1	34	1, 1, 2, 1, 2	11
2, 1, 2, 1, 1	31	2, 1, 2, 1, 2	5
1, 2, 2, 1, 1	17	1, 2, 2, 1, 2	15
2, 2, 2, 1, 1	12	2, 2, 2, 1, 2	1
1, 1, 1, 2, 1	8	1, 1, 1, 2, 2	1
2, 1, 1, 2, 1	8	2, 1, 1, 2, 2	0
1, 2, 1, 2, 1	4	1, 2, 1, 2, 2	1
2, 2, 1, 2, 1	0	2, 2, 1, 2, 2	0
1, 1, 2, 2, 1	69	1, 1, 2, 2, 2	20
2, 1, 2, 2, 1	55	2, 1, 2, 2, 2	4
1, 2, 2, 2, 1	60	1, 2, 2, 2, 2	32
2, 2, 2, 2, 1	21	2, 2, 2, 2, 2	5
1, 1, 1, 3, 1	4	1, 1, 1, 3, 2	4
2, 1, 1, 3, 1	12	2, 1, 1, 3, 2	0
1, 2, 1, 3, 1	5	1, 2, 1, 3, 2	3
2, 2, 1, 3, 1	1	2, 2, 1, 3, 2	1
1, 1, 2, 3, 1	18	1, 1, 2, 3, 2	10
2, 1, 2, 3, 1	13	2, 1, 2, 3, 2	3
1, 2, 2, 3, 1	8	1, 2, 2, 3, 2	8
2, 2, 2, 3, 1	4	2, 2, 2, 3, 2	4

Sources: Schoener [1970] and Fienberg [1970b].

Note: Variable 1 is height ($1 = <5'$, $2 = \geqq 5'$), variable 2 is diameter ($1 = \leqq 2''$, $2 = >2''$), variable 3 is insolation ($1 = $ sun, $2 = $ shade), variable 4 is time ($1 = $ early, $2 = $ midday, $3 = $ late), and variable 5 is species ($1 = $ *grahami*, $2 = $ *opalinus*).

We begin our analysis by fitting all ten possible partial association models to the data. The values of G^2 for these models are listed in table 4.5-6. The models not providing an acceptable fit at the 0.05 level are those with $u_{12} = 0$, $u_{15} = 0$, and $u_{34} = 0$, and the model with $u_{25} = 0$ is unacceptable at the 0.10 level. The six partial association models providing an acceptable fit at the 0.10 level are $u_{13} = 0, u_{14} = 0, u_{23} = 0, u_{24} = 0, u_{35} = 0$, and $u_{45} = 0$. Each of these six models sets equal to zero not only the identifying two-factor term but also its higher-order relatives. All three-factor effects except u_{125} are included among the higher-order relatives in at least one of these six models. Thus we only need to assess whether

Table 4.5-6   Partial Association Models Fit to Lizard
Data from Table 4.5-5

Model	G^2	Degrees of Freedom	Two-Factor Term Absent
C_{1345}, C_{2345}	24.5*	12	u_{12}
C_{1245}, C_{2345}	12.8	12	u_{13}
C_{1235}, C_{2345}	19.3	16	u_{14}
C_{1234}, C_{2345}	35.0*	12	u_{15}
C_{1245}, C_{1345}	10.2	12	u_{23}
C_{1235}, C_{1345}	17.5	16	u_{24}
C_{1234}, C_{1345}	20.6†	12	u_{25}
C_{1235}, C_{1245}	61.2*	16	u_{34}
C_{1234}, C_{1245}	14.6	12	u_{35}
C_{1234}, C_{1235}	21.3	16	u_{45}

* Significant at the 0.05 level when referred to the corresponding chi-square distribution on the indicated degree of freedom.
† Significant at the 0.10 level.

u_{125} is large. We do this by computing the Mantel–Haenzsel statistic for u_{12} and obtain $M_K^2 = 8.87$. The difference, 24.5, between M_K^2 and G^2 for the partial association model with $u_{12} = 0$ is large, and we conclude that u_{125} must be included. This leads us to fit the model

$$u + u_1 + u_2 + u_3 + u_4 + u_5 + u_{12} + u_{15} + u_{25} + u_{34} + u_{125}.$$

We can fit this model directly, and we have a value of $G^2 = 48.0$ with 35 degrees of freedom. This model provides an acceptable fit at the 0.05 level, and it implies that the two species differ in their choice of perch height and perch diameter (these being related), but that this choice is independent of insolation and time of day (the latter two being related). ■ ■

4.5.2   Stepwise procedures

We have suggested several approaches for a first look at a large data array. Most of them provide us with two subsets of models. One subset fits poorly. The other subset fits so well that we believe that we can adequately describe the data by a model with fewer parameters. The next step in the examination may be to examine models adjacent to one of the preliminary models or to examine a further subset of models (two models are "adjacent" if they differ only by one u-term).

Adjacent models

If we start by fitting models with terms of uniform order, we know that the model with terms of order $r - 1$ fits poorly but the model with terms of order r fits very well. We can then proceed to find an intermediate model by one of two approaches:

(i) Forward selection. We start with the model containing terms of order $r - 1$ as our base model and add the u-term of order r that gives the greatest improvement in goodness of fit, provided that the test statistic associated with adding this term is significant at a prespecified level. This becomes our new base model. The remaining u-terms of order $r - 1$ are candidates for inclusion at the next stage, and selection is again an automatic procedure governed by the magnitude of improvement in fit compared with the base model. The process continues until no further terms can be added that significantly improve the fit.

(ii) Backward elimination. This procedure is the reverse of the forward selection procedure. We start with the uniform model of order r and first remove the single term of order r that has the least effect on the goodness of fit. Once this term is removed it is never again a candidate for inclusion. On subsequent cycles we remove further terms, one at each cycle, but never reintroduce a term once it has been the choice for removal.

Either adjacent-model method is an automatic procedure and so can readily be machine programmed. Offsetting this advantage, the drawbacks are:
1. many models must be computed at each cycle;
2. the final model may contain a selection of terms that are difficult to interpret;
3. the model ultimately selected is necessarily one based on configurations of order $r - 1$ and order r exclusively;
4. some joint effects of groups of variables are not evaluated. For example, the inclusion of a term in forward selection is never reconsidered when other terms are subsequently added. Thus if we first introduce the term u_{12} and subsequently the related terms u_{13} and u_{23}, we never consider the model that includes u_{13} and u_{23} but not u_{12}. A similar argument applies to backward elimination.

One way to eliminate drawback 4 is to combine features from the forward selection and backward elimination methods, as suggested by Goodman [1971b], yielding either a "stepwise-up" or "stepwise-down" procedure. The key feature of these approaches is that the inclusion (going forward) or the exclusion (going backward) of a given term is not irrevocable. For example, the stepwise-up method follows our description of the forward selection procedure above, except that after we introduce a new term to the model we reassess whether we should keep all of the previously introduced terms. Goodman also relaxes conditions so that the terms eliminated can also be of order less than $r - 1$.

Example 4.5-6   Stepwise selection for Ries–Smith data

In example 4.5-1 we showed that the log-linear model with all two-factor but no higher-order effects fits the Ries–Smith data quite well, but that the model of complete independence does not. Thus we can use a stepwise procedure to find a good model that contains some but not all of the two-factor terms. We arbitrarily choose to work with 0.05 levels of significance.

Using the stepwise-up procedure, we begin with model 1:

(1) $u + u_1 + u_2 + u_3 + u_4$

and add u_{24}, since the difference between the G^2 values for model 1 and for model 2:

(2) $u + u_1 + u_2 + u_3 + u_4 + u_{24}$

is $42.9 - 22.4 = 20.5$ with one degree of freedom (significant at the 0.05 level), and u_{24} is the two-factor term which yields the most significant decrease in G^2 (see exercise 1 in Section 4.5.4). Next we add u_{34}, the most significant of the remaining two-factor u-terms, based on the difference between the G^2 values for model 2 and for model 3:

(3) $u + u_1 + u_2 + u_3 + u_4 + u_{24} + u_{34},$

i.e., $22.4 - 18.0 = 4.4$ with one degree of freedom (see exercise 1 in Section 4.5.4). At this point we are unable to delete any terms. Now we add u_{13}, the next most significant term, based on the difference between the G^2 values for model 3 and for model 4:

(4) $u + u_1 + u_2 + u_3 + u_4 + u_{13} + u_{24} + u_{34},$

i.e., $18.0 - 11.9 = 6.1$ with two degrees of freedom. We can neither delete any two-factor terms (using our prechosen 0.05 level of significance) nor add any further two-factor terms yielding a significant improvement in fit (at the 0.05 level). Thus the stepwise-up procedure leads us to model 4, the "best" model.

Using the stepwise-down procedure we begin with the model with all two-factor but no higher-order effects and delete u_{14}, since the difference between G^2 values for this model and for model 6:

(6) $u + u_1 + u_2 + u_3 + u_4 + u_{12} + u_{13} + u_{23} + u_{24} + u_{34}$

is $G^2 = 10.1 - 9.9 = 0.2$ with two degrees of freedom (not significant at the 0.05 level). Next we drop u_{12}, based on the difference between G^2 values for model 6 and for model 5:

(5) $u + u_1 + u_2 + u_3 + u_4 + u_{13} + u_{23} + u_{24} + u_{34},$

i.e., $11.2 - 10.1 = 1.1$ with two degrees of freedom (again, not significant at the 0.05 level). We are unable to add back in any terms that significantly improve the fit at this stage, and so we proceed to delete u_{23}, based on the difference between G^2 values for model 5 and model 6, i.e., $11.9 - 11.2 = 0.7$ with one degree of freedom. Since we are unable to add back any terms that significantly improve the fit, or to delete any further "insignificant" terms, we end up with the same model as before, model 4. ∎ ∎

Subsets of models

All the preliminary examination methods, except fitting models of uniform order, give us information about individual u-terms. We can usually then divide the terms into three putative categories: (i) those that must be included, (ii) those that can be omitted, and (iii) "doubtfuls." Using this grouping, we can define a limited subset of models based on all members of category (i) and subsets of category (iii) and determine whether members of the subset require iteration. Very often those requiring iteration have a common core of configurations that cannot be combined to give closed-form cell estimates. However, once estimates have been obtained iteratively from this common core, the goodness of fit for some related models can be determined directly. We gave an example in expression (4.5-2) of deriving G^2_{1234} for the model based on C_{12}, C_{13}, C_{23}, and C_{14} by first determining G^2_{123+}. Once the core C_{12}, C_{13}, and C_{23} has been fit, we can assess the effect of adding any one of the following configurations without further iteration: C_{14}, $C_{24}, C_{34}, C_4, C_{124}, C_{134}$, or C_{234}. Thus the same core is common to seven different models.

Determination of the three categories of u-terms cannot be an exact procedure, and so we cannot write a computer program to help us. The fact that we are forced to think at this stage, however, has the great advantage of allowing us to incorporate our knowledge of the subject matter when we select candidates for inclusion. If the first subset of models selected does not yield a good fit, we must revise our preliminary categorization and determine a further subset. This forces us to reexamine the criterion we used initially.

4.5.3   *Selection of strategy*

Our selection of a strategy must depend on our knowledge of the subject matter and on our ultimate purpose. If we have as our objective the computation of adjusted rates, we are well advised to use strategy 4, initially fitting partial association models, because we must partition our data if three-factor or higher-order effects are large. If our primary purpose is to obtain relatively stable estimates in every cell, then a uniform model approach may be quite satisfactory. Computation of standardized u-terms may be a particularly good approach where we have little a priori knowledge of the interrelationships among our variables.

It is not possible to determine a best strategy because it is not possible to give a single criterion for selecting a best model. Even if we were to compute all possible models and then select the model that came closest to some predetermined significance level, we would not necessarily have made the optimum selection. Such a model would probably show no significant difference from many of its close neighbors, and could easily include a combination of terms that are not easy to interpret. These matters are discussed further in Chapter 9. In this chapter we have reviewed a variety of approaches and have emphasized that "cookbook" methods that can be programmed on a machine are usually inferior to approaches that require intelligent input from the researcher.

4.5.4   *Exercises*

1. Using the values of the log-likelihood components from table 4.5-1, show that:

(i)   if model 1 is the complete independence model and model 2 contains the main effects and exactly one two-factor effect, the most significant value of

$$G^2(1) - G^2(2)$$

is given by choosing u_{24} as the two-factor term;

(ii)   if model 2 contains exactly one two-factor effect u_{24} and model 3 contains all the terms of model 2 plus a second two-factor effect, the most significant value of

$$G^2(2) - G^2(3)$$

is given by choosing u_{34} as the extra term.

(*Hint:* The values of $G^2(1) - G^2(2)$ and $G^2(2) - G^2(3)$ can be calculated directly from two-way marginal totals, as described in Section 4.2.2.)

2. Select a model for the lizard data in table 4.5-5, using both the stepwise-up and stepwise-down selection procedures. Be sure to interpret your chosen models and to compare them with the indicated model in example 4.5-5.

3. Recall from example 4.4-1 that, for the Stouffer data in table 4.4-1, the simplest model that fits the data is based on C_{123}, C_{234}, C_{14}. Explain why fitting all possible partial association models cannot point us directly to this model.

4.6   Appendix: Goodman's Partitioning Calculus

4.6.1   *Notation*

In Section 4.2 we discussed the conditional breakdown of the likelihood statistic G^2 and its use in partitioning for a pair of log-linear models, where one is a special case of the other, especially when both models can be fitted directly. We noted that the difference between the G^2 statistics for two such models that can be fitted directly is equivalent to a G^2 statistic for a related model in an appropriate marginal table. We can do this kind of partitioning in a more general setting, and to facilitate our general discussion of partitioning we introduce a special notational representation for log-linear models presented by Goodman [1970, 1971a].

We consider a d-way table and let $Z_1, \ldots, Z_t$ denote mutually exclusive and exhaustive sets of the d variables. We can summarize any hierarchical log-linear model in terms of the corresponding sufficient marginal configurations, e.g., the model of complete independence of the sets of variables in $Z_1, Z_2, \ldots, Z_t$ corresponds to the marginal configurations $C_{Z_1}, C_{Z_2}, \ldots, C_{Z_t}$. In the notation of this section, we denote this model by

$$[\{\bar{Z}_1\}, \{\bar{Z}_2\}, \ldots, \{\bar{Z}_t\}]. \tag{4.6-1}$$

Other models have some sufficient configurations with variables in common, and we represent these models in a similar fashion. For example, the model corresponding to $C_{Z_1 Z_2}, C_{Z_2 Z_3}, \ldots, C_{Z_{t-1} Z_t}$ has the representation

$$[\{\overline{Z_1 Z_2}\}, \{\overline{Z_2 Z_3}\}, \ldots, \{\overline{Z_{t-1} Z_t}\}]. \tag{4.6-2}$$

The basic class of models having direct estimates consists of those models that can be expressed in terms of conditional independence. Goodman refers to these as "elementary models" and suggests that we represent them in a special way. If $t = 2$, we denote by

$$[\bar{Z}_1 \otimes \bar{Z}_2] \tag{4.6-3}.$$

the model of independence of the variables in Z_1 and those in Z_2. When $t > 2$, we use the notation in expression (4.6-3) to denote the model of independence of Z_1 and Z_2 in the marginal table consisting of all the variables in Z_1 and Z_2, i.e., in $Z_1 Z_2$. The symbol "$\otimes$" stands for "independence," and thus we have

$$[\{\bar{Z}_1\}, \{\bar{Z}_2\}, \ldots, \{\bar{Z}_t\}] = [\bar{Z}_1 \otimes \bar{Z}_2 \otimes \cdots \otimes \bar{Z}_t]. \tag{4.6-4}$$

For $t = 3$, we denote by

$$[\bar{Z}_1 \otimes \bar{Z}_2 | \bar{Z}_3] \tag{4.6-5}$$

the model of *conditional* independence of the variables in Z_1 and those in Z_2, *given* the values of the variables in Z_3. Expression (4.6-3) is a special case of (4.6-5), where Z_3 is a null set containing no variables. If $t > 3$, then (4.6-5) refers to the conditional independence of the variables in Z_1 and in Z_2, given those in Z_3, for the marginal table corresponding to the variables in the union of Z_1, Z_2, and Z_3, which we denote by $Z_1 Z_2 Z_3$. In general, we use the vertical slash to indicate that we are conditioning on the variables following the slash. Thus we have the following equivalent representations:

$$[\{\overline{Z_1 Z_2}\}, \{\overline{Z_1 Z_3}\}, \ldots, \{\overline{Z_1 Z_t}\}] = [\bar{Z}_2 \otimes \bar{Z}_3 \otimes \cdots \otimes \bar{Z}_t | \bar{Z}_1],$$

$$[\{\overline{Z_1 Z_2 Z_3}\}, \{\overline{Z_1 Z_3 Z_4}\}, \ldots, \{\overline{Z_1 Z_{t-1} Z_t}\}] = [\{\overline{Z_2 Z_3}\}, \{\overline{Z_3 Z_4}\}, \ldots, \{\overline{Z_{t-1} Z_t}\} | \bar{Z}_1].$$

4.6.2   *Two basic rules for direct models*

Our aim is to show that we can describe certain partitions of the likelihood ratio statistic G^2 for some log-linear model in terms of the "intersections" of other models. For the d-way table, using a structural breakdown similar to that discussed in Section 4.2.1, we can show that

$$G^2[\bar{Z}_2 \bar{Z}_3 \ldots \bar{Z}_{t-1} \otimes \bar{Z}_t | \bar{Z}_1] = G^2[\bar{Z}_2 \otimes \bar{Z}_t | \bar{Z}_1] + G^2[\bar{Z}_3 \otimes \bar{Z}_t | \overline{Z_1 Z_2}]$$

$$+ \cdots + G^2[\bar{Z}_{t-1} \otimes \bar{Z}_t | \overline{Z_1 Z_2 \ldots Z_{t-2}}]. \qquad (4.6\text{-}6)$$

We denote the partitioning described in expression (4.6-6), using the intersection notation of set theory, as

RULE 1 $$[\bar{Z}_2 \bar{Z}_3 \ldots \bar{Z}_{t-1} \otimes \bar{Z}_t | \bar{Z}_1] \equiv [\bar{Z}_2 \otimes \bar{Z}_t | \bar{Z}_1] \cap [\bar{Z}_3 \otimes \bar{Z}_t | \overline{Z_1 Z_2}]$$

$$\cap \cdots \cap [\bar{Z}_{t-1} \otimes \bar{Z}_t | \overline{Z_1 Z_2 \cdots Z_{t-2}}]. \qquad (4.6\text{-}7)$$

For example, suppose we have a three-way table and Z_1 is the null set, $Z_2 = 1$, $Z_3 = 2$, and $Z_4 = 3$. Then (4.6-7) implies that

$$[\overline{12} \otimes \bar{3}] \equiv [\bar{1} \otimes \bar{3}] \cap [\bar{2} \otimes \bar{3} | \bar{1}], \qquad (4.6\text{-}8)$$

as we showed earlier in expression (4.2-14).

Rule 1 follows directly from the equivalence of two sets of probability statements for discrete random variables, i.e., if we think of the $\{Z_i\}$ as groups of random variables:

$$\Pr[Z_2 Z_3 \ldots Z_t | Z_1] = \Pr[Z_2 Z_3 \ldots Z_{t-1} | Z_1] \Pr[Z_t | Z_1]$$

$$\Leftrightarrow \begin{cases} \Pr[Z_2 Z_t | Z_1] = \Pr[Z_2 | Z_1] \Pr[Z_t | Z_1] \\ \Pr[Z_3 Z_t | Z_1 Z_2] = \Pr[Z_3 | Z_1 Z_2] \Pr[Z_t | Z_1 Z_2] \\ \qquad \cdots \\ \Pr[Z_{t-1} Z_t | Z_1 Z_2 \ldots Z_{t-2}] = \Pr[Z_{t-1} | Z_1 Z_2 \ldots Z_{t-2}] \Pr[Z_t | Z_1 Z_2 \ldots Z_{t-2}]. \end{cases} \qquad (4.6\text{-}9)$$

In general, if we can partition the statistic G^2 computed for a particular model into the sum of components, each of which corresponds to a G^2 statistic for another model either for the whole table or for one of the marginal configurations, then we write the original model as the intersection (in the set-theoretic sense) of the models corresponding to all of the components.

We can also deal with models of conditional independence of several sets of variables, e.g., $[\bar{Z}_2 \otimes \bar{Z}_3 \otimes \cdots \otimes \bar{Z}_t | \bar{Z}_1]$. Once again, it is a straightforward exercise in algebra to obtain the second basic rule:

RULE 2 $$[\bar{Z}_2 \otimes \bar{Z}_3 \otimes \cdots \otimes \bar{Z}_t | \bar{Z}_1] \equiv [\bar{Z}_2 \otimes \bar{Z}_3 | \bar{Z}_1] \cap [\overline{Z_2 Z} \otimes \bar{Z}_4 | \bar{Z}_1]$$

$$\cap \cdots \cap [\overline{Z_2 Z_3 \ldots Z_{t-1}} \otimes \bar{Z}_t | \bar{Z}_1]. \qquad (4.6\text{-}10)$$

This rule also follows from the equivalence of the following sets of probability statements that treat the Z_i as groups of random variables:

$$\Pr[Z_2 Z_3 \ldots Z_t | Z_1] = \Pr[Z_2 | Z_1]\Pr[Z_3 | Z_1] \ldots \Pr[Z_t | Z_1]$$

$$\Leftrightarrow \begin{cases} \Pr[Z_2 Z_3 | Z_1] = \Pr[Z_2 | Z_1]\Pr[Z_3 | Z_1] \\ \Pr[Z_2 Z_3 Z_4 | Z_1] = \Pr[Z_2 Z_3 | Z_1]\Pr[Z_4 | Z_1] \\ \ldots \\ \Pr[Z_2 Z_3 \ldots Z_t | Z_1] = \Pr[Z_2 Z_3 \ldots Z_{t-1} | Z_1]\Pr[Z_t | Z_1]. \end{cases} \tag{4.6-11}$$

By applying the partitioning given by (4.6-7) to individual models involved on the right-hand side of (4.6-10), we can get a more detailed partitioning of the likelihood statistic G^2 for the left-hand side of (4.6-10). Continuing with our three-way table example from above, we see that

$$[\bar{1} \otimes \bar{2} \otimes \bar{3}] = [\bar{1} \otimes \bar{2}] \cap [\overline{12} \otimes \bar{3}] \tag{4.6-12}$$
$$= [\bar{1} \otimes \bar{2}] \cap [\bar{1} \otimes \bar{3}] \cap [\bar{2} \otimes \bar{3}|\bar{1}]$$

Furthermore, we can partition the value of G^2 for $[\bar{2} \otimes \bar{3}|\bar{1}]$ into two additive components, one for the no three-factor effect model and the other for the conditional test for $[\bar{2} \otimes \bar{3}|\bar{1}]$, given no three-factor effect, as in (4.2-6) above.

Some further illustrations of the application of (4.6-7) and (4.6-10) for four-way tables are:

$$[\overline{123} \otimes \bar{4}] = [\bar{2} \otimes \bar{4}] \cap [\overline{13} \otimes \bar{4}|\bar{2}],$$
$$[\overline{123} \otimes \bar{4}] = [\bar{1} \otimes \bar{4}] \cap [\overline{23} \otimes \bar{4}|\bar{1}],$$
$$[\bar{2} \otimes \bar{3} \otimes \bar{4}|\bar{1}] = [\bar{2} \otimes \bar{3}|\bar{1}] \cap [\overline{23} \otimes \bar{4}|\bar{1}].$$

4.6.3   *Partitioning for other direct models*

We need not restrict partitioning to direct models of the conditional independence type. For example, suppose we have a model whose sufficient configurations involve the variables in the sets $Z_1 Z_2$, $Z_1 Z_3, \ldots, Z_1 Z_{t-1}$, and Z_t, i.e., $C_{Z_1 Z_2}$, $C_{Z_1 Z_3}, \ldots,$ $C_{Z_1 Z_{t-1}}, C_{Z_t}$:

$$[\{\overline{Z_1 Z_2}\}, \{\overline{Z_1 Z_3}\}, \ldots, \{\overline{Z_1 Z_{t-1}}\}, \{\bar{Z}_t\}]. \tag{4.6-13}$$

We have closed-form estimates for this model, which we denote by $\{\hat{m}_i\}$. Using the notation of Section 4.2,

$$L(\hat{\mathbf{m}}) = \sum_i x_i \log \hat{m}_i$$
$$= L(C_{Z_1 Z_2}) + L(C_{Z_1 Z_3}) + \cdots + L(C_{Z_1 Z_{t-1}}) \tag{4.6-14}$$
$$- (t - 3)L(C_{Z_1}) + L(C_{Z_t}) - N \log N.$$

Because

$$G^2 = -2[L(\hat{\mathbf{m}}) - L(\mathbf{x})], \tag{4.6-15}$$

we can show that

RULE 3 $[\{\overline{Z_1 Z_2}\}, \ldots, \{\overline{Z_1 Z_{t-1}}\}, \{\bar{Z}_t\}] \equiv [\bar{Z}_2 \otimes \bar{Z}_3 \otimes \cdots \otimes \bar{Z}_{t-1} | \bar{Z}_1]$
$$\cap\, [\overline{Z_1 Z_2 Z_3 \ldots Z_{t-1} \otimes \bar{Z}_t}]. \quad (4.6\text{-}16)$$

Similarly, if we replace the last configuration for model (4.6-13) by one involving the variables $Z_2 Z_t$, we can show that

RULE 4 $[\{\overline{Z_1 Z_2}\}, \ldots, \{\overline{Z_1 Z_{t-1}}\}, \{\overline{Z_2 Z_t}\}] \equiv [\bar{Z}_2 \otimes \bar{Z}_3 \otimes \cdots \otimes \bar{Z}_{t-1} | \bar{Z}_1]$
$$\cap\, [\overline{Z_1 Z_3 Z_4 \ldots Z_{t-1} \otimes \bar{Z}_t | \bar{Z}_2}]. \quad (4.6\text{-}17)$$

An alternative partitioning for model (4.6-13) which follows from the combined use of rules 1, 3, and 4 is

$$[\{\overline{Z_1 Z_2}\}, \{\overline{Z_1 Z_3}\}, \ldots, \{\overline{Z_1 Z_{t-1}}\}, \{\bar{Z}_t\}] =$$
$$[\{\overline{Z_1 Z_2}\}, \{\overline{Z_1 Z_3}\}, \ldots, \{\overline{Z_1 Z_{t-1}}\}, \{\overline{Z_2 Z_t}\}] \cap [\bar{Z}_2 \otimes \bar{Z}_t]. \quad (4.6\text{-}18)$$

Applying rules 3 and 4 to a four-way table, we get

$$[\{\overline{12}\}, \{\overline{13}\}, \{\bar{4}\}] = [\bar{2} \otimes \bar{3} | \bar{1}] \cap [\overline{123} \otimes \bar{4}],$$
$$[\{\overline{12}\}, \{\overline{13}\}, \{\overline{24}\}] = [\bar{2} \otimes \bar{3} | \bar{1}] \cap [\overline{13} \otimes \bar{4} | \bar{2}].$$

We can also extend formulas (4.6-16) and (4.6-17) in a conditional way, yielding results such as

$$[\{\overline{Z_1 Z_2 Z_3}\}, \{\overline{Z_1 Z_2 Z_4}\}, \ldots, \{\overline{Z_1 Z_2 Z_{t-1}}\}, \{\overline{Z_1 Z_t}\}]$$
$$\equiv [\{\overline{Z_2 Z_3}\}, \{\overline{Z_2 Z_4}\}, \ldots, \{\overline{Z_2 Z_{t-1}}\}, \{\bar{Z}_t\} | \bar{Z}_1] \quad (4.6\text{-}19)$$
$$\equiv [\bar{Z}_3 \otimes \bar{Z}_4 \otimes \cdots \otimes \bar{Z}_{t-1} | \overline{Z_1 Z_2}] \cap [\overline{Z_2 Z_3 \ldots Z_{t-1} \otimes \bar{Z}_t | \bar{Z}_1}],$$
$$[\{\overline{Z_1 Z_2 Z_3}\}, \ldots, \{\overline{Z_1 Z_2 Z_{t-1}}\}, \{\overline{Z_1 Z_3 Z_t}\}]$$
$$\equiv [\{\overline{Z_2 Z_3}\}, \ldots, \{\overline{Z_2 Z_{t-1}}\}, \{\overline{Z_3 Z_t}\} | \bar{Z}_1] \quad (4.6\text{-}20)$$
$$\equiv [\bar{Z}_3 \otimes \cdots \otimes \bar{Z}_{t-1} | \overline{Z_1 Z_2}] \cap [\overline{Z_2 Z_4 Z_5 \ldots Z_{t-1} \otimes \bar{Z}_t | \overline{Z_1 Z_3}}].$$

Applications of (4.6-19) and (4.6-20) require a table of dimension at least 5.

4.6.4 *Partitioning for indirect models*

We can also set out partitioning results of a form similar to those described above for some (but not all) indirect models for which we do not have closed-form expressions for the estimated expected cell values. Consider a model whose minimal sufficient statistics are

$$C_{Z_1 Z_2 Z_3}, C_{Z_3 Z_4}, C_{Z_4 Z_5}, \ldots, C_{Z_{t-1} Z_t}, C_{Z_t Z_2}. \quad (4.6\text{-}21)$$

From the rules for detecting direct models in Section 3.4, we see that there are no closed-form estimates for the model corresponding to (4.6-21). After deleting Z_1 from the first configuration, we get a closed loop involving $Z_2, Z_3, \ldots, Z_t$. Because Z_1 appears in only the first configuration in (4.6-21), we get the following partitioning:

RULE 5 $[\{\overline{Z_1 Z_2 Z_3}\}, \{\overline{Z_3 Z_4}\}, \{\overline{Z_4 Z_5}\}, \ldots, \{\overline{Z_{t-1} Z_t}\}, \{\overline{Z_t Z_2}\}]$
$$\equiv [\{\overline{Z_2 Z_3}\}, \ldots, \{\overline{Z_{t-1} Z_t}\}, \{\overline{Z_t Z_2}\}] \cap [\bar{Z}_1 \otimes \overline{Z_4 Z_5 \ldots Z_t | Z_2 Z_3}]. \quad (4.6\text{-}22)$$

To illustrate (4.6-22), let us consider a four-way table with $Z_1 = 1, Z_2 = 2, Z_3 = 3$, and $Z_4 = 4$:

$$[\{\overline{123}\}, \{\overline{34}\}, \{\overline{24}\}] = [\{\overline{23}\}, \{\overline{34}\}\{\overline{24}\}] \cap [\overline{1} \otimes \overline{4|23}].$$

No further partitioning is possible here.

In general, the "closed loop" models, determined by applying the rules for detecting direct models, cannot be broken into parts, each of which corresponds to a model. Partitions of the forms (4.6-16)–(4.6-20) and (4.6-22) are the basis for the idea of fitting by stages described in Section 3.7.3.

4.6.5 *Partitioning illustrated on Ries–Smith data*

We now illustrate the use of the partitioning formulas described above for model selection for the Ries–Smith data of table 4.4-6, which we discussed in example 4.4-2. In table 4.5-2 we list seven log-linear models, their degrees of freedom, and the values of G^2 when these models are fitted to the 3×2^3 data. Our plan is to produce the values of G^2 for all hierarchical log-linear models for the marginal configuration involving only variables 2, 3, and 4, using the values given in table 4.6-1.

Table 4.6-1 Various Log-Linear Models Fit to Ries–Smith Data from Table 4.4-6

Configurations	Degrees of Freedom	G^2
(1) C_{123}, C_{234}	8	5.66
(2) C_{123}, C_{34}, C_{24}	9	8.44
(3) C_{123}, C_{24}	10	12.24
(4) C_1, C_{234}	14	14.50
(5) C_1, C_{23}, C_{24}	16	17.99
(6) C_1, C_3, C_{24}	17	22.35
(7) C_1, C_2, C_3, C_4	18	42.93

We begin by partitioning model 3 into parts relating to model 1, and the marginal model of conditional independence of variables 3 and 4 given 2:

$$\overline{[13 \otimes 4|2]} = [\overline{1} \otimes \overline{4|23}] \cap [\overline{3} \otimes \overline{4|2}].$$

$$\text{(model 3)} \qquad \text{(model 1)}$$

Using this partitioning, we have

$$G^2[\overline{3} \otimes \overline{4|2}] = 12.24 - 5.66 = 6.58$$

with two degrees of freedom. Next we consider other pairs of models:

$$[\{\overline{123}\}, \{\overline{34}\}, \{\overline{24}\}] = [\overline{1} \otimes \overline{4|23}] \cap [\{\overline{23}\}, \{\overline{34}\}, \{\overline{24}\}],$$
$$\text{(model 2)} \qquad\qquad \text{(model 1)}$$

$$[\{\overline{1}\}, \{\overline{23}\}, \{\overline{24}\}] = [\overline{1} \otimes \overline{234}] \cap [\overline{2} \otimes \overline{3|4}],$$
$$\text{(model 5)} \qquad\qquad \text{(model 4)}$$

$$[\overline{1} \otimes \overline{3} \otimes \overline{24}] = [\{\overline{1}\}, \{\overline{23}\}, \{\overline{24}\}] \cap [\overline{3} \otimes \overline{4}].$$
$$\text{(model 6)} \qquad\qquad \text{(model 5)}$$

$$[\overline{1} \otimes \overline{2} \otimes \overline{3} \otimes \overline{4}] = [\overline{1} \otimes \overline{234}] \cap [\overline{2} \otimes \overline{3} \otimes \overline{4}].$$
$$\text{(model 7)} \qquad\qquad \text{(model 4)}$$

These four partitions and the values of G^2 in table 4.6-1 yield:

$$G^2[\{\overline{23}\}, \{\overline{34}\}, \{\overline{24}\}] = \quad 8.44 - 5.66 \quad = 2.78 \quad \text{with one degree of freedom,}$$

$$G^2[\overline{2} \otimes \overline{3}|\overline{4}] = 17.99 - 14.50 = 3.49 \quad \text{with two degrees of freedom,}$$

$$G^2[\overline{3} \otimes \overline{4}] = 22.35 - 17.99 = 4.36 \quad \text{with one degree of freedom,}$$

$$G^2[\overline{2} \otimes \overline{3} \otimes \overline{4}] = 42.93 - 14.50 = 28.43 \quad \text{with four degrees of freedom.}$$

Now we make use of partitions for the three-dimensional marginal table involving variables 2, 3, and 4:

$$[\overline{2} \otimes \overline{3} \otimes \overline{4}] = [\overline{2} \otimes \overline{3}|\overline{4}] \cap [\overline{2} \otimes \overline{4}] \cap [\overline{3} \otimes \overline{4}]$$

$$= [\overline{3} \otimes \overline{4}|\overline{2}] \cap [\overline{2} \otimes \overline{4}] \cap [\overline{2} \otimes \overline{3}]$$

$$= [\overline{2} \otimes \overline{4}|\overline{3}] \cap [\overline{2} \otimes \overline{3}] \cap [\overline{3} \otimes \overline{4}]$$

and

$$[\overline{2} \otimes \overline{34}] = [\overline{2} \otimes \overline{3}|\overline{4}] \cap [\overline{2} \otimes \overline{4}]$$

$$[\overline{3} \otimes \overline{24}] = [\overline{3} \otimes \overline{4}|\overline{2}] \cap [\overline{2} \otimes \overline{3}]$$

$$[\overline{4} \otimes \overline{23}] = [\overline{2} \otimes \overline{4}|\overline{3}] \cap [\overline{3} \otimes \overline{4}].$$

Using these partitions and the values of G^2 computed above, we have the information needed to obtain the values of G^2 for all eight hierarchical log-linear models for the three-dimensional marginal table involving variables 2, 3, and 4. We summarize these results in table 4.6-2.

Table 4.6-2 All Hierarchical Log-Linear Models Fit to the Three-Dimensional Marginal Table for Ries–Smith Data from Table 4.4-6 Involving Use, Temperature, and Preference

Model	Configurations	Degrees of Freedom	G^2
$[\{\overline{23}\}, \{\overline{24}\}, \{\overline{34}\}]$	C_{23}, C_{24}, C_{34}	1	2.78
$[\overline{2} \otimes \overline{4}|\overline{3}]$	C_{23}, C_{34}	2	22.80*
$[\overline{3} \otimes \overline{4}|\overline{2}]$	C_{23}, C_{24}	2	6.58*
$[\overline{2} \otimes \overline{3}|\overline{4}]$	C_{24}, C_{34}	2	3.49
$[\overline{23} \otimes \overline{4}]$	C_{23}, C_4	3	27.16*
$[\overline{2} \otimes \overline{34}]$	C_2, C_{34}	3	24.07*
$[\overline{24} \otimes \overline{3}]$	C_3, C_{24}	3	7.84*
$[\overline{2} \otimes \overline{3} \otimes \overline{4}]$	C_2, C_3, C_4	4	28.43*

* G^2 is significant at the 0.05 level when compared to the chi square distribution with the appropriate degrees of freedom.

The simplest model which fits the three-dimensional margin is

$$u + u_2 + u_3 + u_4 + u_{24} + u_{34}.$$

This model is consistent with model 4 in table 4.4-7 for the full four-dimensional array because we can collapse over variable 1 (softness).

4.6.6   *Exercises*

1. For a four-dimensional table, show algebraically that the following partitions for G^2 hold:

$$[\overline{123} \otimes \overline{4}] = [\overline{2} \otimes \overline{4}] \cap [\overline{13} \otimes \overline{4}|\overline{2}];$$

$$[\overline{123} \otimes \overline{4}] = [\overline{1} \otimes \overline{4}] \cap [\overline{23} \otimes \overline{4}|\overline{1}];$$

$$[\overline{2} \otimes \overline{3} \otimes \overline{4}|\overline{1}] = [\overline{2} \otimes \overline{3}|\overline{1}] \cap [\overline{23} \otimes \overline{4}|\overline{1}];$$

$$[\{\overline{12}\}, \{\overline{13}\}, \{\overline{4}\}] = [\overline{2} \otimes \overline{3}|\overline{1}] \cap [\overline{123} \otimes 4];$$

$$[\{\overline{12}\}, \{\overline{13}\}, \{\overline{24}\}] = [\overline{2} \otimes \overline{3}|\overline{1}] \cap [\overline{13} \otimes \overline{4}|\overline{2}];$$

$$[\{\overline{123}\}, \{\overline{34}\}, \{\overline{24}\}] = [\{\overline{23}\}, \{\overline{34}\}, \{\overline{24}\}] \cap [\overline{1} \otimes \overline{4}|\overline{23}].$$

2. For a four-dimensional table, write out at least two complete partitionings of $G^2[\overline{1} \otimes \overline{2} \otimes \overline{3} \otimes \overline{4}]$, including as many tests for marginal components as you can.

5 Maximum Likelihood Estimation for Incomplete Tables

5.1 Introduction

5.1.1 *Sampling versus structural zeros*

When analyzing sample tables of counts, we encounter two types of empty cells. The first of these, which we call a "sampling zero," is due to sampling variability and the relative smallness of the cell probability. At least in principle, by increasing the sample size sufficiently we can make sampling zeros disappear. The other type of empty cell, which we call a "structural zero" cell, is known a priori to have a zero value. Such cells can occur naturally as a feature of the data, or as a result of the structure of the data. For example, when there is an underlying order for the categories in each of two variables and the data represent changes that can only occur in one direction, we have a triangular array.

As we noted in Chapter 3, an incomplete table is one having some cells that contain structural zeros that must remain empty under any fitted model. In this context, we also view any cell that is fixed a priori (even if it has a positive rather than a zero value) to be a structural zero for the purposes of model fitting, since its fitted value is always the same and is known in advance. This type of incompleteness differs substantially from that associated with missing data values for some variables, a problem not discussed in this book.

One problem we have encountered in the application of incomplete table methodology is that many people who have data in the form of incomplete contingency tables fail to recognize them as such. Often the applied researcher either fills in the cells containing structural zeros using some "appropriate values" or collapses the data until the structural zeros have "vanished." Sometimes the occurrence of structural zeros leads the researcher to abandon completely the analysis of his data. Such practices often lead to confusion or to inappropriate conclusions. In addition to presenting the methodology required for the analysis of incomplete tables, we emphasize its application in this chapter.

While we devote a great deal of attention to different patterns of structural zeros and related technical details, in general the log-linear models we fit to incomplete tables are similar to those we use for complete tables. We even use the same iterative proportional fitting algorithm, except that the initial values must reflect the presence of structural zero cells. The main source of difficulty is the computation of degrees of freedom, and to handle the difficulties that occasionally arise we introduce and develop the concept of separability.

5.1.2 *Background*

Structural zero cells occur in several different contexts. Observations for certain cells in a contingency table are often truncated or not reported (Goodman [1968],

Watson [1956]). At other times, certain combinations are impossible, and zero probability is attached to these cells (Kastenbaum [1958]). For example, when there is an underlying order for the categories in each of two or more variables, this ordering may constrain certain cells to be zero a priori (Bishop and Fienberg [1969], Chen et al. [1961], Mantel and Halperin [1963], Waite [1915]). To illustrate, in an analysis of scores in games, if one variable is the winning score and another the losing score, then the cells with losing scores exceeding winning scores are all zero a priori. Finally, we may wish to fit a parametric model to one set of observed cell counts within a table and a second model to the remaining cells (Fienberg [1969a], Goodman [1963a, 1968], Savage and Deutsch [1960]). In such situations, maximum likelihood estimation procedures lead us to treat these sets of expected cell counts as being the nonzero entries in two separate incomplete tables.

For incomplete rectangular tables, we make use of quasi-log-linear models, similar to those used in Chapter 3, and these models are applicable only for those cells not containing structural zeros. We begin, in the next section, by examining two-way incomplete tables and the log-linear model of quasi independence.

The term "quasi independence" was formally introduced into the modern statistical literature by Goodman [1968] for two-way tables, but the use of and work with the model of quasi independence dates back to his earlier work on the mover-stayer model, transaction flows, and the analysis of social mobility tables (see Goodman [1961, 1963a, 1964a, 1965]), and to the work of Caussinus [1965]. The extensions to multiway tables have been dealt with in detail by Fienberg [1972a] and Haberman [1974].

Since the models we use for the analysis of incomplete contingency tables are linear in the logarithmic scale, it is natural to question whether there are related analysis-of-variance schemes for incomplete layouts. The determination of degrees of freedom and the problems regarding the estimability of parameters are closely related to the methods and problems discussed in this chapter.

5.2 Incomplete Two-Way Tables

5.2.1 *Quasi independence*

Let m_{ij} be the expected number of individuals out of a sample size N in the (i, j) cell of an $I \times J$ table, where, as before

$$m_{i+} = \sum_{j=1}^{J} m_{ij} \qquad i = 1, \ldots, I,$$

$$m_{+j} = \sum_{i=1}^{I} m_{ij} \qquad j = 1, \ldots, J,$$

$$m_{++} = \sum_{i=1}^{I} \sum_{j=1}^{J} m_{ij}. \tag{5.2-1}$$

Multiplicative definition
The familiar model of independence of the variables corresponding to rows and

to columns is given by

$$m_{ij} = a_i b_j, \tag{5.2-2}$$

where a_i and b_j are positive constants for $i = 1, \ldots, I$ and $j = 1, \ldots, J$. This definition of independence is equivalent to the more commonly used one, where we let π_{ij} denote the probability that an individual in the $I \times J$ population table will fall in the (i, j) cell. Then the row and column classifications are defined as being independent if the cell probabilities can be written as

$$\pi_{ij} = \pi_{i+} \pi_{+j} \tag{5.2-3}$$

for $i = 1, \ldots, I$ and $j = 1, \ldots, J$. (Here, as before, when we sum a subscripted variable over a subscript, we replace that subscript by a "+".)

Now, let S be the set of cells in an incomplete two-way array that consists of all cells not containing structural zeros, and suppose that (5.2-2) holds for all $m_{ij} \in S$. In such cases we say that the rows and columns of S are "quasi-independent." Since those cells not in S contain structural zeros (i.e., $m_{ij} = 0$ for $(i, j) \notin S$), we can continue to use (5.2-1) to represent the marginal totals of S and thus the marginal totals of the incomplete table.

We can also consider models for the expected cell counts in incomplete tables in the (natural) logarithmic scale, where in general, we write

$$\log m_{ij} = u + u_{1(i)} + u_{2(j)} + u_{12(ij)} \tag{5.2-4}$$

for $(i, j) \in S$, with

$$\sum_{i=1}^{I} \delta_i^{(2)} u_{1(i)} = \sum_{j=1}^{J} \delta_j^{(1)} u_{2(j)} = 0, \tag{5.2-5}$$

$$\sum_{i=1}^{I} \delta_{ij} u_{12(ij)} = \sum_{j=1}^{J} \delta_{ij} u_{12(ij)} = 0, \tag{5.2-6}$$

where

$$\delta_{ij} = \begin{cases} 1 & \text{for } (i, j) \in S \\ 0 & \text{otherwise,} \end{cases} \tag{5.2-7}$$

$$\delta_i^{(2)} = \begin{cases} 1 & \text{if } \delta_{ij} = 1 \text{ for some } j \\ 0 & \text{otherwise,} \end{cases} \tag{5.2-8}$$

and $\delta_j^{(1)}$ is defined in a similar manner. If we put no further restrictions on the u-terms, the m_{ij} remain unrestricted. We then define the model of quasi independence for the subtable S by setting

$$u_{12(ij)} = 0 \tag{5.2-9}$$

for $(i, j) \in S$, so that

$$\log m_{ij} = u + u_{1(i)} + u_{2(j)} \tag{5.2-10}$$

for $(i, j) \in S$ and

$$m_{ij} = e^u e^{u_{1(i)}} e^{u_{2(j)}} \tag{5.2-11}$$

for $(i, j) \in S$. This is the same definition as in (5.2-2), since m_{ij} is a product of a parameter depending on i, another one depending on j, and a third depending on neither i nor j.

Interpretation of model

We can discuss the interpretation of quasi independence in two-way incomplete tables in a manner analogous to the interpretation of independence in complete tables. Quasi independence implies that the relative proportions of individuals in the corresponding cells of any two rows (columns) of the table are the same, provided that we do not consider those columns (rows) which have structural zero entries in one of the cells corresponding to these rows (columns). Thus quasi independence is a form of independence, *conditional* on the restriction of our attention to an incomplete portion of a table.

Suppose, for example, that the $(1, 1)$ cell of a 3×3 table is missing. Thus S consists of the remaining eight cells, and if quasi independence holds for the cells in S, we can rewrite (5.2-10) in multiplicative form, i.e., $m_{ij} = a_i b_j$ for $(i, j) \in S$. The expected entries in the table are then:

—	$a_1 b_2$	$a_1 b_3$
$a_2 b_1$	$a_2 b_2$	$a_2 b_3$
$a_3 b_1$	$a_3 b_2$	$a_3 b_3$

(5.2-12)

If we delete the first row, we are left with

$a_2 b_1$	$a_2 b_2$	$a_2 b_3$
$a_3 b_1$	$a_3 b_2$	$a_3 b_3$

(5.2-13)

which is a 2×3 table satisfying independence. The same thing happens if we delete the first column, where we are left with a 3×2 independent table:

$a_1 b_2$	$a_1 b_3$
$a_2 b_2$	$a_2 b_3$
$a_3 b_2$	$a_3 b_3$

(5.2-14)

There are also some special configurations for the set of cell S which do not allow for this reduction to independence by deletion of rows or columns. If there are missing entries in all three diagonal cells in a 3×3 table and if quasi independence holds, the table of expected values is

—	$a_1 b_2$	$a_1 b_3$
$a_2 b_1$	—	$a_2 b_3$
$a_3 b_1$	$a_3 b_2$	—

$$(5.2\text{-}15)$$

Deletion of rows or columns (or both) from (5.2-15) does not reduce us to a subtable which is independent. Nevertheless, there is still one degree of freedom attached to this incomplete 3×3 table, as we shall see in Section 5.2.4.

Linear contrast definition

A second way to define quasi independence in this log-linear scale is via the notion of interaction contrasts defined by

$$\sum_{i=1}^{I} \sum_{j=1}^{J} \delta_{ij} \beta_{ij} \log m_{ij} \tag{5.2-16}$$

$(\beta_{ij} \neq 0$ for some $i, j)$, where

$$\sum_{i=1}^{I} \delta_{ij} \beta_{ij} = \sum_{j=1}^{J} \delta_{ij} \beta_{ij} = 0. \tag{5.2-17}$$

The δ_{ij}, defined above by (5.2-7), are used in (5.2-16) to restrict attention to contrasts involving only cells in S, and $0 \log 0$ always takes the value 0. Using (5.2-4) and (5.2-5), we can rewrite the interaction contrasts as

$$\sum_{i=1}^{I} \sum_{j=1}^{J} \delta_{ij} \beta_{ij} u_{12(ij)} \tag{5.2-18}$$

$(\beta_{ij} \neq 0$ for some $i, j)$, and setting all such contrasts equal to zero for different values of the $\{\beta_{ij}\}$ is equivalent to (5.2-9). This defines quasi independence as being equivalent to zero values for all interaction contrasts.

Noninteractive cells

Some cells have entries which do not contribute to any of the interaction contrasts. We formally define such cells here, but we defer a more detailed discussion of how to locate them until Section 5.2.6. Now suppose there is a cell (k, l) in the incomplete subtable S for which there does not exist a contrast of the form (5.2-18) with $\beta_{kl} \neq 0$, and so the parameter corresponding to that cell $u_{12(kl)}$ is not involved in any of the interaction contrasts. Then the (k, l) cell is "noninteractive," as it tells us nothing about the values of the interaction contrasts. Consider the 3×3 table with expected cell counts:

m_{11}	m_{12}	—
m_{21}	m_{22}	—
m_{31}	m_{32}	m_{33}

$$(5.2\text{-}19)$$

where $m_{13} = m_{23} = 0$ (the dashes in the table represent structural zeros). S thus consists of all the cells except $(1, 3)$ and $(2, 3)$. Then the $(3, 3)$ cell is noninteractive, and the interaction contrasts are all of the form

$$\sum_{i=1}^{3} \sum_{j=1}^{2} \beta_{ij} u_{12(ij)}. \qquad (5.2\text{-}20)$$

Eliminating noninteractive cells from consideration during the estimation procedure greatly simplifies calculations.

Note that we cannot define interaction contrasts simply in terms of 2×2 subtables (i, j), (i, k), (l, j), and (l, k) for $i \neq l$ and $j \neq k$, as we ordinarily choose to do in complete arrays. For example, consider the 3×3 table given in (5.2-15) with structural zeros down the diagonal. For this table, all definable interaction contrasts are multiples of

$$u_{12(12)} - u_{12(13)} + u_{12(23)} - u_{12(21)} + u_{12(31)} - u_{12(32)}, \qquad (5.2\text{-}21)$$

and there is no 2×2 subtable for which we can define an interaction contrast. The proof of this statement follows from the restriction in (5.2-17) that the β_{ij} must sum to 1 over any row or column.

5.2.2 Connectivity and separability

When we consider estimating the m_{ij} in an incomplete table under the model of quasi independence and determining the associated degrees of freedom, situations arise where we should consider the incomplete subset in separate parts. Such situations are best described in terms of the concepts of connectivity and separability. The main reason for worrying about separability is the degrees of freedom, although reducing a separable incomplete table into its separable components often leads to a reduction in the computations required to produce estimated expected cell values.

In a two-way contingency table two cells are *associated* if they do not contain structural zeros and if they are either in the same row or the same column. A set of non-structural-zero cells is *connected* if every pair of cells can be linked by a chain of cells, any two consecutive members of which must be associated. Finally, an incomplete two-way table is connected if its non-structural-zero cells form a connected set. The notion of a connected design in the analysis of variance is defined in a similar way and is due to Bose [1949]. It is also described in Elston and Bush [1964].

An incomplete table that is not connected is said to be *separable*, since after suitable permutation of rows and columns we can divide the nonempty cells of a separable table into at least two separate subtables (subrectangles), where each of the subtables has no row or column in common with any other such subtable (see Goodman [1968]). We can permute the rows and columns of a separable table so that the nonempty cells are arranged into at least two nonoverlapping blocks running down the diagonal, and so we often refer to such a table as being *block diagonal*. Each block may still contain some structural zeros.

We defined separability in terms of the expected values and structural zeros. When we look at an observed table of counts, we may also have random zeros. For purposes of defining whether the observed table is separable, we make no distinction between random zeros and structural zeros. Thus we speak of the observed table as being separable if the incomplete table defined by the *nonzero observed* counts is not connected. Inseparable observed tables are used in the next section.

Example 5.2-1   Separability and the American bladder nut

Harris [1910] presented the following 4×9 table (see table 5.2-1) giving the relationship between the locular composition (the number of locules of the ovary with odd or even numbers of ovules) and radial symmetry (root mean square deviation of the number of ovules from the mean number in the individual ovary) for the fruit of the American bladder nut, *Staphylea trifolia*, which has three locules per ovary. The locules are connected throughout the greater portion of the length of the fruit (see figure 5.2-1). From four to twelve ovules are located in two series along the inner angle of the locule, and not all of these develop into smooth, bony seeds. This is the first of three series of data (*A, B, C*) recorded by Harris. Ovaries made up of two locules with an even number of ovules and one locule with an odd

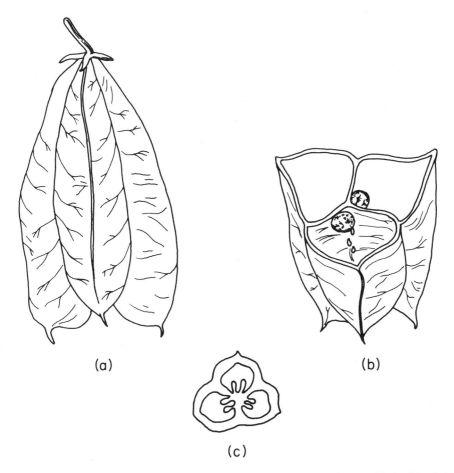

(a)

(b)

(c)

Figure 5.2-1   Diagram showing the external appearance (A) and cross section (B) of fruit of *Staphylea*. Note the one seed and the arrangement of the aborted ovules in the front locule. The cross section of a young ovary (C) shows diagrammatically the disposition of the ovules of each locule in two rows along the inner angle. Reproduced with the kind permission of the Biometrika Trustee from Harris [1910].

Table 5.2-1   Relationship Between Radial Asymmetry and Locular Composition in *Staphylea* (Series A of Harris [1910])

Locular Composition	Coefficient of Radial Asymmetry								
	0.00	0.47	0.82	0.94	1.25	1.41	1.63	1.70	1.89
3 even, 0 odd	462	—	—	130	—	—	2	—	1
2 even, 1 odd	—	614	138	—	21	14	—	1	—
1 even, 2 odd	—	443	95	—	22	8	—	5	—
0 even, 3 odd	103	—	—	35	—	—	1	—	0

Table 5.2-2   Data from Table 5.2-1 after Rearrangement of Rows and Columns

Locular Composition	Coefficient of Radial Asymmetry								
	0.00	0.94	1.63	1.89	0.47	0.82	1.25	1.41	1.70
3 even, 0 odd	462	130	2	1	—	—	—	—	—
0 even, 3 odd	103	35	1	0	—	—	—	—	—
2 even, 1 odd	—	—	—	—	614	138	21	14	1
1 even, 2 odd	—	—	—	—	443	95	22	8	5

number of ovules, or of one locule with an even and two with an odd, cannot have a coefficient of asymmetry of 0.00, since one of the locules must differ from the other two by at least one ovule. Similarly, other combinations, indicated in the table by dashes, are physically impossible, and in our terminology are structural zeros.

Ignoring the apparent order for the row and column categories, we can rearrange these data as in table 5.2-1, and we see that the table is separable with exactly two separable subrectangles, as shown in table 5.2-2. Goodman [1968] gives a detailed analysis of a version of these data, and in exercise 7 in Section 5.2.8 we consider the analysis of the remaining two series of data presented by Harris. We return to a discussion of this example later in the chapter. ■■

Dealing with separable tables

Now suppose that an incomplete table of expected counts is separable with exactly k separable rectangles, so that the set S of nonempty cells consists of exactly k disjoint subsets $S_1, S_2, \ldots, S_k$, each of which contains the nonempty cells of one of the k separable subrectangles. Since the subsets S_v and S_w (for any pair (v, w) with $v \neq w$) have no rows or columns in common, we can write the model of quasi independence given by (5.2-2) in such a way that the expected cell values in S_v have no parameters in common with the expected values in S_w. Thus we have a separate quasi-independence model for each of the subsets S_l, $l = 1, \ldots, k$.

Under the sampling models considered here, the likelihood function for the data from a separable table can be written as a product of components each involving the parameters of only one of the subsets S_l. Thus the maximum likelihood estimates of the expected values in a given subset S_l are based only on the data in that subset. As a result, it is possible to consider each subrectangle of a separable table separately for estimation and testing purposes. In much of what follows, we restrict ourselves to inseparable incomplete tables of expected counts, since we can deal with the inseparable subsets of a separable table one at a time.

5.2.3 *Sampling models*

For complete two-way contingency tables, we know from Chapter 3 that there exist unique maximum likelihood estimates for the expected cell counts under the model of independence when the observed data are generated by any one of the following three sampling schemes:

1. Poisson sampling scheme, with independent Poisson variates for each cell;
2. one common multinomial;
3. product multinomial sampling, i.e., a set of independent multinomials for rows or for columns.

Moreover, these three schemes lead to the same MLEs and are unaffected by the presence of sampling zeros, provided no row or column has an observed zero total.

The other sampling scheme used for two-way tables is multivariate hypergeometric and is based on both row and column totals being fixed (see Chapter 13). The MLEs for hypergeometric sampling are different from those for the other three schemes, and we refer the reader who is interested in a discussion of this alternative to Haberman [1974].

For incomplete tables, just as for complete tables, we can demonstrate the existence of unique nonzero MLEs under the model of quasi independence for the same three types of sampling schemes, provided a further constraint on the observed and expected cell values is satisfied. Once again, the MLEs are the same under each of the sampling schemes. For convenience we now adopt the Poisson sampling scheme described in detail below.

Poisson likelihood

We assume that x_{ij}, the observed number of counts in the (i, j) cell for $(i, j) \in S$, is an observation for a Poisson variate with mean m_{ij}, the expected value for the (i, j) cell, and that these Poisson variates are mutually independent. For notational convenience we set $N = x_{++}$, so that we can always speak of a total sample size N. This notation allows us to slip back to the multinomial scheme more easily. If we let $\prod^*$ denote the product over all cells contained in the incomplete subset S, the likelihood function for this scheme is

$$\prod{}^* m_{ij}^{x_{ij}} \frac{e^{-m_{ij}}}{x_{ij}!}, \tag{5.2-22}$$

and the natural logarithm of the likelihood is

$$L = \sum \delta_{ij}[x_{ij} \log(m_{ij}) - m_{ij} - \log(x_{ij}!)]. \tag{5.2-23}$$

When (5.2-10) holds for S, we can rewrite (5.2-23) as

$$L = x_{++}u + \sum_i x_{i+}u_{1(i)} + \sum_j x_{+j}u_{2(j)} - m_{++} - \sum \delta_{ij} \log(x_{ij}!). \tag{5.2-24}$$

Since the Poisson distribution is a member of the exponential family of probability distributions, we see that the observed marginal totals, x_{i+} for all i and x_{+j} for all j, are complete sufficient statistics, as was the case for complete tables in Chapter 3.

Existence of unique MLEs

Savage [1973] presents necessary and sufficient conditions for the existence of unique nonzero MLEs for the expected cell counts of an inseparable table of expected counts S. These conditions take into account the distribution of observed random zeros. To describe them we require some further notation. Let K be a subset of the set of rows $\{1, \ldots, I\}$ and let K' be a subset of the set of columns $\{1, \ldots, J\}$. Next, let S_K and $S_{K'}$ be the elements in the rows and columns of S corresponding to the elements in K and K', respectively. Finally, if A is a subset of the set S, we let $x(A) = \sum_{(i,j)\in A} x_{ij}$, i.e., $x(A)$ is the observed total for the subset. We can now state Savage's result as follows:

THEOREM 5.2-1 *Necessary and sufficient conditions for the existence and uniqueness of the MLEs for the expected cell counts of an inseparable table S under the model of quasi independence are*:

(i)   *the observed marginal totals x_{i+} for all i and x_{+j} for all j are positive*;
(ii) *if $S_K \subseteq S_{K'}$, then $x(S_K) \leq x(S_{K'})$, or if $S_K \supseteq S_{K'}$, then $x(S_K) \geq x(S_{K'})$, where if the set inclusion is proper the inequality is strict.*

Savage's conditions supersede the somewhat simpler sufficient conditions given by Fienberg [1970c], where condition (ii) is replaced by:

(ii') *the observed table of counts is itself inseparable.*

For the special case where S consists of the off-diagonal cells in a square $I \times I$ table, condition (ii) reduces to

$$x_{i+} + x_{+i} < N$$

for $i = 1, \ldots, I$.

Condition (i) is also required for complete tables if we are to get nonzero estimates. If the condition is violated, we delete the rows with $x_{i+} = 0$ and columns with $x_{+j} = 0$ and analyze only the remainder of the table. Condition (ii) is required to ensure that the likelihood function (5.2-24) is convex and does not assume its maximum on the boundary of the parameter space. Condition (ii') is stronger than (ii) but serves the same kind of purpose. Neither condition (ii) or condition (ii') is satisfied for the following observed table:

2	2	0	—	4
3	7	—	—	10
—	—	4	5	9
—	—	6	7	13
5	9	10	12	26

The only situations where (ii) holds but (ii') does not are those where the shape of S and the positions of the sampling zeros have a special type of symmetry, such as in the following tabulation of observed values:

	3	0	—	0	3
	0	3	0	—	3
Observed	—	0	3	0	3
	0	—	0	3	3
	3	3	3	3	12

In this table, there are no sets K and K' other than the set of all rows and the set of all columns for which $S_K \subseteq S_{K'}$ or $S_K \supseteq S_{K'}$, yet the observed table of counts is separable. The unique nonzero MLEs for this table under quasi independence are:

	1	1	—	1	3
	1	1	1	—	3
Expected	—	1	1	1	3
	1	—	1	1	3
	3	3	3	3	12

When the conditions of theorem 5.2-1 are met, the MLEs are determined by setting the observed marginal totals, x_{i+} for all i and x_{+j} for all j, equal to the MLEs of their expectations, the expected marginal totals $\hat{m}_{i+}$ for all i and $\hat{m}_{+j}$ for all j, i.e.,

$$\begin{aligned} \hat{m}_{i+} &= x_{i+} & i = 1, \dots, I, \\ \hat{m}_{+j} &= x_{+j} & j = 1, \dots, J, \end{aligned} \tag{5.2-25}$$

where the $\{\hat{m}_{ij}\}$ satisfy (5.2-5) and (5.2-10). The actual solution of the maximum likelihood equations (5.2-25) can always be determined by iterative methods, although for certain incomplete configurations, direct closed-form solutions are available. We devote separate sections to the topics of iterative and closed-form solutions.

Although Savage's conditions are both necessary and sufficient, they do not necessarily imply that the MLEs of the $\{m_{ij}\}$ are different from the observed cell values $\{x_{ij}\}$. We discuss such cases in the next section in the context of degrees of freedom, before going on to consider the solution of the maximum likelihood equations.

5.2.4 *Degrees of freedom for quasi independence*

The general rule for computing the degrees of freedom associated with a fitted model is to subtract the number of independent parameters used in the model from the total number of cells to which the model is being fitted. For the model of quasi independence in inseparable incomplete tables satisfying the two conditions of the previous section, the degrees of freedom are $IJ - z_e - [I + J - 1] = (I - 1) \times (J - 1) - z_e$, where z_e is the number of cells in the table containing structural zeros. Here there are $IJ - z_e$ cells to which the model is being fitted and $I + J - 1$ independent parameters in the model (1 for u, $I - 1$ for the $u_{1(i)}$, and $J - 1$ for the $u_{2(j)}$). This computation of the degrees of freedom follows the rule given by expression (3.8-1) of Chapter 3.

The expected values under the model of quasi independence are identical to the observed values whenever the number of independent parameters being fitted is equal to or greater than the total number of cells. For example, the following incomplete table of observed counts satisfies conditions (i) and (ii) above, but there are zero degrees of freedom:

4	6	—	—	—
—	5	7	—	—
—	—	6	3	—
—	—	—	4	5

$$(5.2\text{-}26)$$

i.e., there are eight cells, while the model of quasi independence includes eight independent parameters. Because the MLEs of the m_{ij} are determined uniquely by the marginal totals, by starting with the first column and then alternating on successive rows and columns we can easily see that the expected table is the same as the observed table. For example, since there is only one cell in the first column which is not a structural zero, (5.2-25) implies that $\hat{m}_{11} = 4$; looking next at the first row, we see that (5.2-25) implies $\hat{m}_{11} + \hat{m}_{12} = 10$, and thus $\hat{m}_{12} = 6$ since $\hat{m}_{11} = 4$. In fact, each of the eight cells is noninteractive, and there are no interaction contrasts of the form (5.2-16).

In general, given an inseparable incomplete table satisfying conditions (i) and (ii) from the preceding section, it is impossible to have "negative" degrees of freedom, and there are zero degrees of freedom if and only if all the cells which do not contain structural zeros are noninteractive.

Finally, consider computing degrees of freedom for a separable table, with, say, k separable subrectangles, each of which is itself inseparable. We denote nonempty cells in these subrectangles by $S_l, l = 1, \ldots, k$. Because quasi independence for the whole table implies separate quasi-independence models for each of the sets of cells S_l, we simply consider each such set by itself. When all of the sets S_l satisfy conditions (i) and (ii), the number of degrees of freedom for the model of quasi independence in the whole table is simply equal to the sum of the degrees of freedom for each S_l. In example 5.2-1 we separated the non-structural-zero cells of table 5.2-1 into two inseparable sets, which form two complete subtables of size 2×4 and 2×5, given in table 5.2-2. The degrees of freedom for these subtables under quasi independence are 3 and 4, respectively, and so the number of degrees of freedom for the whole table equals $3 + 4 = 7$. Had the (2, 4) cell of table 5.2-2 been a structural zero rather than simply a sampling zero, the number of degrees of freedom would have equalled $(3 - 1) + 4 = 6$, and the (1, 4) cell would have become noninteractive.

5.2.5   *Iterative solution for MLEs under quasi independence*

We now return to the multiplicative form of the quasi-independence model so that we may demonstrate two equivalent iterative procedures for determining the MLEs for the nonzero expected cell entries. Recall that quasi independence implies that

$$m_{ij} = \delta_{ij} a_i b_j \tag{5.2-27}$$

for $i = 1, \ldots, I; j = 1, \ldots, J$, where

$$\delta_{ij} = \begin{cases} 1 & \text{for cells } (i, j) \text{ in } S \\ 0 & \text{otherwise.} \end{cases} \tag{5.2-28}$$

(This is equivalent to our previous definition.) Then if unique MLEs for the m_{ij} exist, from the previous section we know that they are completely determined by the quasi-independence relationship (5.2-27) and by the $I + J$ maximum likelihood equations

$$\hat{m}_{i+} = x_{i+} \qquad i = 1, \ldots, I,$$
$$\hat{m}_{+j} = x_{+j} \qquad j = 1, \ldots, J. \tag{5.2-29}$$

Estimating expected cell values

We can use a modification of the Deming–Stephan iterative proportional fitting procedure described in Chapter 3 to compute estimated expected cell values. As in Chapter 3, the choice of initial values is optional, provided the initial values satisfy the quasi-independence relationships (5.2-27). For convenience, and to enable us to show the equivalence of the two iterative methods, at the 0th step we set

$$\hat{m}_{ij}^{(0)} = \delta_{ij} \tag{5.2-30}$$

for all i, j. Then at the vth cycle of the iteration we take

$$\hat{m}_{ij}^{(2v-1)} = \frac{\hat{m}_{ij}^{(2v-2)} x_{i+}}{\sum\limits_{k} \hat{m}_{ik}^{(2v-2)}} \tag{5.2-31}$$

for all i, j and

$$\hat{m}_{ij}^{(2v)} = \frac{\hat{m}_{ij}^{(2v-1)} x_{+j}}{\sum\limits_{k} \hat{m}_{kj}^{(2v-1)}} \tag{5.2-32}$$

for all i, j. We continue the iteration for $v = 1, 2, \ldots$ until we achieve the desired accuracy.

Simple modifications to the proof of convergence given in Section 3.5 for the regular Deming–Stephan procedure allow us to prove the convergence of the iterative method in the present context. (See also the proof of convergence presented by Haberman [1974].) Thus, when conditions (i) and (ii) of theorem 5.2-1 hold, our iterative method converges to the unique MLEs for the $\{m_{ij}\}$.

Estimating the multiplicative parameters

Our second iterative procedure produces maximum likelihood estimates for the $\{a_i\}$ and $\{b_j\}$, which in turn yield MLEs for the $\{m_{ij}\}$. The model for quasi independence, when expressed in the multiplicative form (5.2-27), appears to have $I + J$ independent parameters, but we have already seen that there are in fact only $I + J - 1$, and since

$$\sum_{i=1}^{I} \hat{m}_{i+} = \sum_{j=1}^{J} \hat{m}_{+j} = x_{++}, \tag{5.2-33}$$

there are only $I + J - 1$ maximum likelihood equations. However, if we view the $\{a_i\}$ and the $\{b_j\}$ as being unique only up to a multiplicative constant (i.e., if we multiply a_i by λ for all values of i and b_j by $1/\lambda$ for all values of j, the m_{ij} remain unchanged), we only have $I + J - 1$ independent parameters. Thus any solution of (5.2-29) which satisfies (5.2-33) will do.

We begin by rewriting the maximum likelihood equations using the quasi-independence relationship given by (5.2-27):

$$\hat{a}_i \sum_{j=1}^{J} \delta_{ij}\hat{b}_j = x_{i+} \qquad i = 1, \ldots, I,$$

$$\hat{b}_j \sum_{i=1}^{I} \delta_{ij}\hat{a}_i = x_{+j} \qquad j = 1, \ldots, J,$$

(5.2-34)

which can be rewritten as

$$\hat{a}_i = \frac{x_{i+}}{\sum\limits_{j=1}^{J} \delta_{ij}\hat{b}_j} \qquad i = 1, \ldots, I,$$

$$\hat{b}_j = \frac{x_{+j}}{\sum\limits_{i=1}^{I} \delta_{ij}\hat{a}_i} \qquad j = 1, \ldots, J.$$

(5.2-35)

This second representation immediately suggests the following iterative procedure for estimating the $\{a_i\}$ and the $\{b_j\}$. (This method has been extensively used by Goodman [1964a, 1968].)

We begin by setting

$$b_j^{(0)} = 1 \tag{5.2-36}$$

for $j = 1, \ldots, J$, and then continue at the νth cycle of the iteration ($\nu \geq 1$) by setting

$$a_i^{(\nu)} = \frac{x_{i+}}{\sum\limits_{j} \delta_{ij}b_j^{(\nu-1)}} \tag{5.2-37}$$

for $i = 1, \ldots, I$, and

$$b_j^{(\nu)} = \frac{x_{+j}}{\sum\limits_{i} \delta_{ij}a_i^{(\nu)}} \tag{5.2-38}$$

for $j = 1, \ldots, J$. After the νth cycle, the estimates for the m_{ij} are given by

$$\hat{m}_{ij}^{(2\nu)} = \delta_{ij}a_i^{(\nu)}b_j^{(\nu)} \tag{5.2-39}$$

for all (i, j), and the iteration is continued until we obtain the desired accuracy. The resulting $\{\hat{m}_{ij}\}$ are our maximum likelihood estimates. It is now easy to show that equations (5.2-32) and (5.2-39) yield identical values for all i and j, and thus the two procedures are equivalent (see exercise 3 in Section 5.2.8).

This second method is convenient if we wish to estimate the parameters $\{a_i\}$ and $\{b_j\}$, although estimates for these parameters are also available when we use the Deming–Stephan iterative proportional fitting procedure, provided we accumulate the product of the row and column multipliers used at each cycle. Moreover, the second method usually requires fewer computations than the first method, since we need not compute the estimates of the $\{m_{ij}\}$ after each cycle. The first method is easier to program and offers the advantages of allowing us to check for changes in the estimates from cycle to cycle. Thus we can have a stopping rule based on the accuracy of the cell estimates. Also, we can use one general computer program based on the Deming–Stephan algorithm to get estimated expected cell values for both complete and incomplete tables.

Goodness of fit statistics

The usual asymptotic theory for Poisson and multinomial data is applicable here, and we can test the goodness of fit of the quasi-independence model by using either the Pearsonian statistic

$$X^2 = \sum * \frac{(x_{ij} - \hat{m}_{ij})^2}{\hat{m}_{ij}} \qquad (5.2\text{-}40)$$

or the likelihood-ratio statistic

$$G^2 = 2 \sum * x_{ij} \log(x_{ij}/\hat{m}_{ij}), \qquad (5.2\text{-}41)$$

where $\hat{m}_{ij}$ is the MLE of m_{ij} and the summation $\sum *$ is over all cells in an incomplete subtable S. Under the null hypothesis of quasi independence, both X^2 and G^2 have asymptotic central χ^2 distributions, where the degrees of freedom are determined as in the preceding subsection. For example, suppose there are z_e structural zeros in an $I \times J$ table and the set of cells which are not structural zeros is inseparable. Then X^2 and G^2 have asymptotic χ^2 distributions with $(I - 1)(J - 1) - z_e$ degrees of freedom.

Example 5.2-2   Purum marriages

In the sociological literature there are various models for marriages in monogamous societies, where isonymous marriages (i.e., marriages between persons of the same name or clan) are forbidden, or at least strongly discouraged. The Purums, an old and isolated tribe living in the interior of India, are divided into five sibs: Marrim, Makan, Kheyang, Thao, and Parpa. Das [1945], in his study of the Purums, concluded not only that the Purum sib was an exogamous unit, but also that Purum boys and girls could marry only in one or more selected sibs.

In table 5.2-3 we give the data for 128 Purum marriages from a village census

Table 5.2-3   Classification of Purum Marriages

| | | | Sib of Husband | | |
Sib of Wife	Marrim	Makan	Parpa	Thao	Kheyang
Marrim	—	5	17	—	6
Makan	5	—	0	16	2
Parpa	—	2	—	10	11
Thao	10	—	—	—	9
Kheyang	6	20	8	0	1

Source: White [1963], based on data of Das [1945].

reported by White [1963] and based on data collected by Das. There is, in fact, one case of a Kheyang-Kheyang marriage, and this marriage is between a man and woman in different subsibs; however, Das concluded that the Kheyang sib is slowly disintegrating into subsibs, within each of which exogamy is still a rigid norm. Those cells in the table containing dashes denote those combinations which Das concluded are ruled out by Purum traditions. (Note the lack of symmetry regarding the existence of structural zeros.) There are also two sampling zeros corresponding to marriages observed in other sets of data but not in the present set.

A natural question to ask of the data is whether, given that certain marriage combinations are forbidden, we can adequately describe the expected number of marriages of each type by the multiplicative structure of quasi independence. We give the expected values under quasi independence in table 5.2-4 (based on values computed after five cycles of the iterative procedure). The answer to our question is readily apparent when we examine those expected values corresponding to the sampling zeros. The likelihood ratio goodness-of-fit test statistic, G^2 with eight degrees of freedom, has a value of 76.25, and it is more than reasonable to conclude (as have Das and White) that Purums tend to marry into selected sibs, but if a man cannot find a mate in the sib or sibs into which it is customary for his group to marry, he can be expected to turn elsewhere. ■ ■

Table 5.2-4 Estimated Expected Values for Table 5.2-3 under Quasi Independence

			Husband		
Wife	Marrim	Makan	Parpa	Thao	Kheyang
Marrim	—	10.78	10.68	—	6.54
Makan	4.87	—	6.56	7.54	4.02
Parpa	—	8.38	—	9.54	5.09
Thao	10.40	—	—	—	8.60
Kheyang	5.75	7.83	7.75	8.91	4.75

Note: Values are after five cycles of the iterative procedure.

5.2.6 *Direct estimation procedures for quasi independence*

In the preceding section we demonstrated how to get MLEs for the expected cell counts in an incomplete table under the model of quasi independence using two related iterative procedures. We refer to such methods of calculating estimates, and to the resulting estimates themselves, as being "indirect," since we do not write down explicit (direct) formulas for the estimates. For the example at the end of the preceding section, the only way to get MLEs is by indirect estimation. There are some incomplete tables, however, where the configuration of structural zeros is such that direct estimation for the remaining cells is possible. In this section, we discuss several such configurations and show how they lead to maximum likelihood estimates under the model of quasi independence that we can compute directly. The value of these direct estimates lies, of course, in the fact that they can be computed by hand with only a modest amount of arithmetic calculation, while the indirect estimates usually involve so much arithmetic as to make hand calculations unreasonable.

We begin by describing two rules for locating noninteractive cells, since the definition given at the beginning of this chapter does not provide a constructive way of determining whether a cell is noninteractive except by elimination. The net import of these two rules is that the MLE for a noninteractive cell under quasi independence is the observed count for that cell, and if we can recognize such cells immediately, we can remove them from the incomplete subtable being analyzed and replace them when the analysis is completed. The second rule also provides a way to break some incomplete tables into subtables which can be fitted by maximum likelihood estimation separately, even though they are not separable in the sense described previously.

Two further rules allow us to write down direct MLEs for two classes of incomplete tables. By the use of all four rules together we can determine the direct closed-form MLEs for all incomplete configurations which yield direct estimates. The partial use of these rules was implicit in the work of Bishop and Fienberg [1969], and the explicit formulation, which we present here, was suggested in part by the work of Mantel [1970] and Goodman [1968]. Haberman [1974] also considers the problem of closed-form estimates but in a more general and somewhat different framework. Throughout the remainder of this section we must keep in mind that the MLEs under quasi independence must add up to the observed marginal totals, as specified by the maximum likelihood equations (5.2-25).

Before presenting our four rules, we recall that, for inseparable incomplete tables, we set $\delta_{ij} = 1$ for $(i, j) \in S$, the set of cells which do not contain structural zeros, and $\delta_{ij} = 0$ for $(i, j) \notin S$.

RULE 1 (Cell Isolates)

If $\delta_{ij} = 1$ for some cell (i, j) but the remaining δ_{ij} in the same row (same column) are all zero, then $\hat{m}_{ij} = x_{ij}$, since the MLEs are uniquely determined by the preservation of marginal totals. The (i, j) cell is clearly noninteractive, and in such cases we simply delete the ith row (jth column) and continue the estimation procedure.

In the following expected table, the $(3, 3)$ cell is a cell isolate:

m_{11}	m_{12}	m_{13}
m_{21}	m_{22}	—
—	—	m_{33}

When we delete the third row, the $(1, 3)$ cell becomes a cell isolate in the resulting table (it is also noninteractive), and we can then delete the third column as well. This leaves the 2×2 subtable:

m_{11}	m_{12}
m_{21}	m_{22}

whose MLEs can be written out using the usual product-of-margins rule for complete tables.

RULE 2 (Semiseparability)

Mantel [1970] defined an inseparable incomplete table as being semiseparable if it can be made separable, into two or more separate subtables, by the removal of a single row or a single column. Suppose now that a table is semiseparable by removal of a single column, and we partition this table into sets of rows, with each set corresponding to exactly one of the separable subtables which result from the column deletion. Then we can estimate expected cell counts in each of these partitioned sets of rows, under the model of quasi independence, in the same way we would if each set (after the deletion of empty columns) were a separable subtable. Thus we get explicit MLEs for the entire table if and only if we can find explicit MLEs for every set of partitioned rows. This happens because the sum over a partitioned set of rows of the MLEs in the deletable column is equal to the corresponding observed sum.

Example 5.2-3   Use of Rule 2

Consider the following 5×5 incomplete table with expected entries m_{ij}:

m_{11}	m_{12}	—	—	—
m_{21}	m_{22}	—	—	m_{25}
—	—	m_{33}	m_{34}	m_{35}
—	—	m_{43}	m_{44}	—
—	—	m_{53}	m_{54}	m_{55}

The table is semiseparable, since the deletion of column 5 yields two separable subtables. We now partition the table into the first two rows and the last three rows, and consider the reduced subtables:

1

| m_{11} | m_{12} | — |
| m_{21} | m_{22} | m_{25} |

2

m_{33}	m_{34}	m_{35}
m_{43}	m_{44}	—
m_{53}	m_{54}	m_{55}

In the first subtable the cell containing m_{25} is a cell isolate, and the deletion of the corresponding column leaves a 2×2 subtable whose MLEs can be written directly. Note that the $(2, 5)$ cell was not a cell isolate in the original table even though it is noninteractive. The second subtable also admits direct MLEs, using either rule 3 or rule 4 below. We can then piece the direct MLEs back together again to get the direct MLEs for the original 5×5 incomplete table. ∎∎

RULE 3 (Block-Triangular Tables)

We say that an incomplete table is in "block-triangular" form if, after suitable permutation of rows and columns, $\delta_{ij} = 0$ implies $\delta_{kl} = 0$ for all $k \geq i$ and $l \geq j$. We can determine (direct) explicit formulas for the expected cell values in a block-triangular table.

The following are examples of block-triangular tables:

(1)			

m_{11}	m_{12}	m_{13}	m_{14}
m_{21}	m_{22}	m_{23}	—
m_{31}	m_{32}	—	—
m_{41}	—	—	—

(2)			

—	—	m_{13}	m_{14}
—	—	m_{23}	m_{24}
m_{31}	m_{32}	m_{33}	m_{34}
m_{41}	m_{42}	m_{43}	m_{44}

(3)

m_{11}	m_{12}	m_{13}	m_{14}
—	m_{22}	m_{23}	m_{24}
—	m_{32}	m_{33}	m_{34}
—	—	m_{43}	m_{44}
—	—	m_{53}	m_{54}

We call such tables block-triangular because the non-structural-zero cells, after suitable permutation of rows and columns, form a right-angled triangle with blocks of cells lying along the hypotenuse of the triangle. For notational convenience, we always refer to block-triangular tables as having their structural-zero cells in the lower right-hand corner of the table. Below, we illustrate the direct computation of MLEs for a particular class of block-triangular tables, but the method of computation is applicable in general.

MLEs for block-triangular tables

Table 5.2-5 illustrates a block-triangular table with $I = I_1 + I_2 + I_3$ rows and $J = J_1 + J_2 + J_3$ columns. The structural zeros lie in two blocks of cells, completely specified by setting $\delta_{ij} = 0$ for $i \geq I_1 + 1$ and $j \geq J_1 + J_2 + 1$, and for $i \geq I_1 + I_2 + 1$ and $j \geq J_1 + 1$. To write down the MLEs directly for the incomplete subset S consisting of all cells not containing structural zeros, we must write out m_{ij} in some multiplicative form so that it is a function only of marginal totals. We begin by looking at the block on the diagonal defined by $i = I_1 + 1, I_1 + 2,$ $\ldots, I_1 + I_2$ and $j = J_1 + 1, J_1 + 2, \ldots, J_1 + J_2$, and by noting that

$$
\begin{aligned}
m_{i+}m_{+j} &= \left(\sum_{l=1}^{J_1+J_2} e^{u+u_{1(i)}+u_{2(l)}} \right) \left(\sum_{k=1}^{I_1+I_2} e^{u+u_{1(k)}+u_{2(j)}} \right) \\
&= e^{u+u_{1(i)}+u_{2(j)}} \left(\sum_{l=1}^{J_1+J_2} \sum_{k=1}^{I_1+I_2} e^{u+u_{1(k)}+u_{2(l)}} \right) \\
&= m_{ij} \left(\sum_{k=1}^{I_1+I_2} \sum_{l=1}^{J_1+J_2} m_{kl} \right).
\end{aligned}
\tag{5.2-42}
$$

Since

$$
\sum_{k=1}^{I_1+I_2} \sum_{l=1}^{J_1+J_2} m_{kl} = m_{++} - \sum_{i=I_1+I_2+1}^{I_1+I_2+I_3} m_{i+} - \sum_{j=J_1+J_2+1}^{J_1+J_2+J_3} m_{+j},
\tag{5.2-43}
$$

we can write the m_{ij} for this block as direct functions of the marginal totals, and thus the MLEs are given by

$$\hat{m}_{ij} = \frac{\frac{x_{i+} x_{+j}}{I_1 + I_2 \ J_1 + J_2}}{\displaystyle\sum_{k=1} \sum_{l=1} x_{kl}} \tag{5.2-44}$$

for $i = I_1 + 1, \ldots, I_1 + I_2, j = J_1 + 1, \ldots, J_1 + J_2$.

Table 5.2-5　$I \times J$ Block-Triangular Table (with $I = I_1 + I_2 + I_3$, $J = J_1 + J_2 + J_3$)

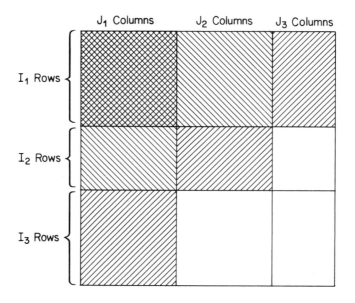

Note Unshaded areas denote blocks of structural zeros.

　　We get the MLEs for the two additional blocks of cells on the diagonal by using formulas analogous to (5.2-42), (5.2-43), and (5.2-44). Since we now have the MLEs for the three diagonal blocks, we can compute revised marginal totals for the remaining as-yet unestimated cells by subtracting the MLEs for the three diagonal blocks from the original marginal totals and proceeding as if the cells already estimated are now structural zeros. This leaves us with a smaller block-triangular table of dimension $(I_1 + I_2) \times (J_1 + J_2)$, where the structural zero cells are completely specified by the condition $\delta_{ij} = 0$ for $i = I_1 + 1$ and $j = J_1 + 1$. We then proceed with the direct estimation using adjusted versions of expressions (5.2-42), (5.2-43), and (5.2-44) along with the revised marginal totals.

Example 5.2-4　Disability scores for patients with strokes
We illustrate direct estimation in block-triangular tables with the data in table 5.2-6 presented originally in Bishop and Fienberg [1969] and collected by Jones and Poskanzer at Massachusetts General Hospital on 121 hospital patients. On admission and on discharge the patients were graded on a five-point scale (A through E) of increasing severity according to their physical disability following a stroke. Since no patient was discharged if he had become worse (except by death),

a patient's score on the second examination could only be the same or better than on the original examination. The ordering of the disability scores, combined with this restriction, produces the block-triangular form of the table. The cross-classification of patients by the results of both examinations is shown in table 5.2-6. Note that our convention of putting the structural zeros in the lower right-hand corner means that the ith row and the ith column do not correspond to the same state, e.g., the (B, B) cell is $(4, 2)$.

Table 5.2-6 Initial and Final Ratings on Disability of Stroke Patients

Initial State	Final State					Totals
	A	B	C	D	E	
E	11	23	12	15	8	69
D	9	10	4	1	—	24
C	6	4	4	—	—	14
B	4	5	—	—	—	9
A	5	—	—	—	—	5
Totals	35	42	20	16	8	121

Source: Bishop and Fienberg [1969].

To compute the expected cell values under quasi independence, we begin with the diagonal cells coresponding to (A, A), (B, B), etc. For example, for the $(4, 2)$ cell, with initial state B and final state B, the expected count is

$$\frac{9 \times 42}{121 - (20 + 16 + 8 + 5)} = 5.25.$$

Similarly, the expected counts for the $(5, 1)$, $(3, 3)$, $(2, 4)$, and $(1, 5)$ cells are 5, 3.37, 4.52, and 8, respectively. Note that the $(5, 1)$ and $(1, 5)$ cells are cell isolates as described in rule 1 above, and so they have their expected value equal to the observed value. We now subtract these diagonal cell estimates from the relevant marginal totals and repeat the procedure.

At the next stage we see that the expected count for the $(3, 2)$ cell is

$$\frac{(14 - 3.37)(42 - 5.25)}{94.86 - (16.63 + 11.48) - 3.75} = 6.20.$$

Similarly, the expected counts for the $(4, 1)$, $(2, 3)$, and $(1, 4)$ cells are 3.75, 4.69, and 11.48, respectively. Continuing this diagonal elimination and marginal adjustment procedure, we arrive at the MLEs given in table 5.2-7.

(We note that Goodman [1968] gives an explicit set of formulas for the MLEs in triangular tables such as the one in this example, and they are equivalent to the ones presented here. Mantel [1970] discusses such tables as well.)

The two goodness-of-fit statistics computed for this table,

$$X^2 = 8.37, \qquad G^2 = 9.60,$$

each with six degrees of freedom, correspond to values around the 80th percentile

Table 5.2-7 Estimated Values for Data on Disability of Stroke Patients, under Quasi-Independence Model

Initial State	Final State A	B	C	D	E	Totals
E	15.66	21.92	11.94	11.48	8.00	69.00
D	6.16	8.63	4.69	4.52	—	24.00
C	4.43	6.20	3.37	—	—	14.00
B	3.75	5.25	—	—	—	9.00
A	5.00	—	—	—	—	5.00
Totals	35.00	42.00	20.00	16.00	8.00	121.00

Source: Bishop and Fienberg [1969].

of the corresponding χ^2 distribution. If we examine the deviations between observed and expected values, we do not find that they exhibit any regular pattern. The largest contribution to X^2 comes from the $(2, 4)$ cell, which has only a single observation in it. The second largest contribution comes from the $(1, 1)$ cell, and this may be considered a more important departure from the quasi-independence model, as it shows that among the patients who started in the worst category fewer than expected finished in the best category. Neither of these contributions is, however, large. The quasi-independence model appears to fit the data fairly well, and in the context of the data implies that, given a patient's initial state and the fact that he cannot get worse, the relative probability of a patient being in an achievable final state is independent of his initial state. ■■

Table 5.2-8 $I \times J$ Block Stairway Incomplete Table (with $I = I_1 + I_2 + I_3 + I_4, J = J_1 + J_2 + J_3 + J_4$)

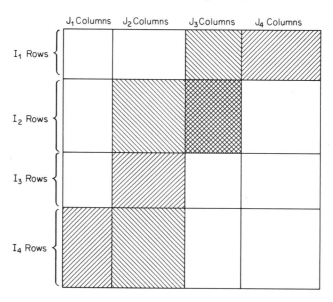

Note: Unshaded areas denote blocks of structural zeros.

RULE 4 (Block-Stairway Tables)

An inseparable incomplete table is said to be a "block-stairway" table if after permutation of rows and columns, we can divide the table into sets of rows, each of which contains one rectangular array of nonzero cells, and each such rectangular array shares columns only with those arrays immediately above and below it. (We give an example of a block-stairway table in table 5.2-8.) All block-stairway incomplete tables have closed-form MLEs for the nonzero expected counts under quasi independence. When one of the rectangular arrays in a block-stairway table has only one row or column, that row or column may have cell isolates or the table may be semiseparable. We now illustrate the algebraic computation of the closed-form MLEs for the block-stairway table illustrated in table 5.2-8.

MLEs for block-stairway tables

We begin by noting that the maximum likelihood marginal constraints (5.2-25) imply that for each of the seven shaded blocks of cells, the sum of the MLEs is equal to the sum of the observed counts. We now give the MLEs for two of these seven blocks of cells, and we leave the calculation of the MLEs for the five remaining blocks, using analogous formulas, as an exercise (see exercise 5 in Section 5.2.8).

First let us consider the heavily shaded block of cells with $i = I_1 + 1, \ldots, I_1 + I_2$ and $j = J_1 + J_2 + 1, \ldots, J_1 + J_2 + J_3$. For the (i, j) cell we first look at the product of the marginal totals, i.e.,

$$
m_{i+}m_{+j} = \left(\sum_{l=J_1+1}^{J_1+J_2+J_3} e^{u+u_{1(i)}+u_{2(l)}} \right) \left(\sum_{k=1}^{I_1+I_2} e^{u+u_{1(k)}+u_{2(j)}} \right)
$$
$$
= e^{u+u_{1(i)}+u_{2(j)}} \left(\sum_{k=1}^{I_1+I_2} \sum_{l=J_1+1}^{J_1+J_2+J_3} e^{u+u_{1(k)}+u_{2(l)}} \right),
\tag{5.2-45}
$$

and thus

$$
m_{ij} = \frac{m_{i+}m_{+j}}{\displaystyle\sum_{k=1}^{I_1+I_2} \sum_{l=J_1+1}^{J_1+J_2+J_3} e^{u+u_{1(k)}+u_{2(l)}}}.
\tag{5.2-46}
$$

Summing over all i, j in the block of cells yields

$$
\frac{\displaystyle\sum_{i=I_1+1}^{I_1+I_2} \sum_{j=J_1+J_2+1}^{J_1+J_2+J_3} m_{i+}m_{+j}}{\displaystyle\sum_{i=I_1+1}^{I_1+I_2} \sum_{j=J_1+J_2+1}^{J_1+J_2+J_3} m_{ij}} = \sum_{k=1}^{I_1+I_2} \sum_{l=J_1+1}^{J_1+J_2+J_3} e^{u+u_{1(k)}+u_{2(l)}}.
\tag{5.2-47}
$$

Finally, substituting (5.2-47) back into (5.2-46) yields an expression for m_{ij} that is a function of the marginal totals $\{m_{i+}\}$ and $\{m_{+j}\}$. Thus we can write the MLEs for this block of cells as

$$
\hat{m}_{ij} = \frac{x_{i+}x_{+j} \left(\displaystyle\sum_{k=I_1+1}^{I_1+I_2} \sum_{l=J_1+J_2+1}^{J_1+J_2+J_3} x_{kl} \right)}{\displaystyle\sum_{k=I_1+1}^{I_1+I_2} \sum_{l=J_1+J_2+1}^{J_1+J_2+J_3} x_{k+}x_{+l}}
\tag{5.2-48}
$$

for $i = I_1 + 1, \ldots, I_1 + I_2;$ $j = J_1 + J_2 + 1, \ldots, J_1 + J_2 + J_3$. Although expression (5.2-48) looks quite formidable, the basic form of the m_{ij} for the block of cells is actually quite simple. First, we take the products of column totals with the row totals, as for a complete table. Since the estimates must sum to the total number of observations in the block, we divide the products of the marginal totals by their sum, and then we multiply by the observed total for the block. Similarly, we can show that

$$
\hat{m}_{ij} = \frac{x_{i+}x_{+j}\left(\displaystyle\sum_{k=I_1+1}^{I_1+I_2}\sum_{l=J_1+1}^{J_1+J_2}x_{kl}\right)}{\displaystyle\sum_{k=I_1+1}^{I_1+I_2}\sum_{l=J_1+1}^{J_1+J_2}x_{k+}x_{+l}} \tag{5.2-49}
$$

for $i = I_1 + 1, \ldots, I_1 + I_2$; $j = J_1 + 1, \ldots, J_1 + J_2$.

Example 5.2-5 Drosophila melanogaster progeny

Novitski and Sandler [1967] studied male *Drosophila melanogaster* carrying a specific translocation between the X chromosome and the minute fourth chromosome, and also carrying a Y chromosome and a normal fourth chromosome. These males produce sperm which carry either the proximal end of the translocation (A) or the Y (A') but not both, and either the distal end of the translocation (B) or the fourth chromosome (B'), but not both. Thus there are four sperm types: AB, $A'B$, AB', $A'B'$.

When such males are mated to attached-X females with a Y chromosome, $A'B$ is lethal. If the Y chromosome is replaced by the proximal segment of the translocation, $A'B$ is recoverable but AB is lethal.

In table 5.2-9 we give data from two experiments conducted by Novitski and Sandler. In the first, a random sample of the translocation males is mated to attached-X females with a Y chromosome, and in the second to attached-X females with section B of the translocation. Thus we have product multinomial sampling. By observing these two types of mating, Novitski and Sandler hoped to determine the relative frequencies of recovery of the four male sperm types. Kastenbaum [1958] and Goodman [1968] have also analyzed these data. Before estimating these relative frequencies (see Kastenbaum [1958] for details), we must first check to see whether the model of quasi independence fits the data. To do this we rearrange the data as shown in table 5.2-10.

Had the (2, 1) and (1, 4) cells been blocks of cells, table 5.2-10 would be a non-degenerate block-stairway table; however, the (2, 1) and (1, 4) cells are isolates, and so their MLEs under quasi independence are equal to their observed counts.

Table 5.2-9 Numbers of Progeny from Mating Translocation-Bearing Males to Attached-X Females

Female Type	Male Sperm Type			
	AB	A′B′	A′B	AB′
Female with a Y Chromosome	1,413	1,029	lethal	2,240
Female with Proximal Segment of the Translocation	lethal	548	346	1,287

Source: Novitski and Sandler [1957].

Table 5.2-10   Data from Table 5.2-9 in Block-Stairway Table Form

		Male Type		
Female Type	A'B	A'B'	AB'	AB
Female with a Y Chromosome	—	1,029	2,240	1,413
Female with Proximal Segment of the Translocation	346	548	1,287	—

In the notation used in rule 4 above, cells $(1, 2)$ and $(1, 3)$ form a block, as do cells $(2, 2)$ and $(2, 3)$. Applying our formula, we get

$$\hat{m}_{12} = \frac{(1,577 \times 4,682) \times 3,269}{(1,577 \times 4,682) + (3,527 \times 4,682)} = \frac{1,577 \times 3,269}{5,104} = 1,010.03,$$

$$\hat{m}_{13} = 2,258.97 \quad \text{(by subtraction)},$$

$$\hat{m}_{22} = \frac{(1,577 \times 2,181) \times 1,835}{(1,577 \times 2,181) + (3,527 \times 2,181)} = \frac{1,577 \times 1,835}{5,104} = 566.97,$$

$$\hat{m}_{23} = 1,268.03.$$

Note that in this example our formulas in (5.2-48) and (5.2-49) reduce to the simple ones given in Goodman [1968].

The value of the Pearson goodness-of-fit statistic is $X^2 = 1.43$ with one degree of freedom. Quasi independence is an acceptable model for the data in the table, and we can use the expected values to compute the relative frequencies of the four male sperm types (see exercise 7 in Section 5.2.8). ■■

5.2.7   Equivalence of closed-form and iterative estimates

We have noted above that the MLEs for expected cell counts in incomplete tables under quasi independence exist and are unique, provided that conditions (i) and (ii) of theorem 5.2-1 hold. Thus if we can write the MLEs for a particular incomplete table in closed form, we know that both the closed-form formulas and the iterative proportional fitting procedure must yield the same set of estimates.

Had we not known about this uniqueness property, we would have considerable trouble convincing ourselves that the closed-form expressions give the same estimates as the iterative procedure, since the latter does not necessarily converge after a finite number of cycles to those estimates produced by the closed-form expressions. This is true even for the simplest of examples. In this respect, incomplete tables differ from complete tables. In Chapter 3 we noted that when a closed-form solution exists, the iterative procedure converges exactly to that solution. It does so in a single cycle as long as the sufficient statistics are fitted in an appropriate order. For incomplete tables, the iterative procedure converges to the closed-form solutions to any specified degree of accuracy, but often requires many cycles to do so.

Example 5.2-6

We illustrate this point using the 2×4 incomplete data of Novitski and Sandler, for which we have already computed expected cell values using closed-form expressions. Table 5.2-9 contains the original data, and we begin by noting that the $(2, 1)$ and $(1, 4)$ cells in the block-stairway version of the table (table 5.2-10), are set equal to their observed counts after the column normalization in each cycle of the iterative procedure. As a result, in table 5.2-11 we give only the iterative estimates for the remaining four nonzero cells at the end of each of the first five cycles. In addition, we give the corresponding estimates computed via the closed-form expressions. At the end of the fifth cycle, the iterative estimates are almost the same as the closed-form estimates to one-decimal accuracy. For this simple 2×4 incomplete configuration, however, the algebraic formulas for the iterative estimates at the end of a finite number of cycles are *not identical* to the algebraic formulas of the closed-form estimates.

Table 5.2-11 Comparison of Iterative and Closed-Form Estimates for Data of Novitski and Sandler [1957]

Cell	Iterative Estimates at End of Cycle Number					Closed-Form Estimates
	1	2	3	4	5	
(1, 2)	963.15	1,000.28	1,007.99	1,009.61	1,009.96	1,010.03
(1, 3)	2,154.11	2,237.14	2,254.39	2,258.00	2,258.76	2,258.97
(2, 2)	613.85	576.72	569.01	567.39	567.04	566.97
(2, 3)	1,372.89	1,289.86	1,272.61	1,269.00	1,268.24	1,268.03

Of course, had we noticed that the (2, 1) and (1, 4) cells are noninteractive and then removed them while fitting the remaining four cells, we would have obtained the closed-form solutions via the iterative procedures at the end of the first complete cycle. It is not very difficult, however, to find more complicated examples where the iterative procedure does not yield the closed-form estimates exactly after a finite number of steps and where the deletion of noninteractive cells does not reduce the table to a complete array, as in the present example. ■■

5.2.8 *Exercises*

1. Show that the following 4×4 incomplete table of expected counts is such that all definable interaction contrasts can be written as a scaled linear combination of two contrasts having coefficients β_{ij} which are all either 1, -1, or 0.

	1	2	3	4
1	—	m_{12}	—	m_{14}
2	m_{21}	—	m_{23}	—
3	m_{31}	m_{32}	—	m_{34}
4	—	m_{42}	m_{43}	—

2. Show that the likelihood equations for an inseparable incomplete two-way table and a multinomial sampling scheme are the same as those given by (5.2-25).
3. Prove that equations (5.2-32) and (5.2-39) yield identical values for all i and j. Since this is true, the two iterative procedures described in Section 5.2.6 are formally equivalent.

4. Ploog [1967] observed the following distribution of genital display among the members of a colony of six squirrel monkeys (labeled as R, S, T, U, V, and W). For each display there is an active and passive participant, but a monkey never displays toward himself.

Active Participant	Passive Participant						Totals
	R	S	T	U	V	W	
R	—	1	5	8	9	0	23
S	29	—	14	46	4	0	93
T	0	0	—	0	0	0	0
U	2	3	1	—	38	2	46
V	0	0	0	0	—	1	1
W	9	25	4	6	13	—	57
Totals	40	29	24	60	64	3	220

(i) Can the MLEs under the model of quasi independence for this table be computed directly? (Answer: No.)

(ii) Determine the degrees of freedom associated with the usual goodness-of-fit statistics when the expected values are computed under quasi independence. (Answer: 15. The obvious answer of 19 is incorrect, since the statistic is examined conditionally, given that the total in the row corresponding to T is 0.)

(iii) Compute the MLEs for this table under quasi independence, and test the goodness of fit of this model using both the likelihood ratio and Pearsonian test statistics. (Answer: $X^2 = 168.05$ and $G^2 = 135.17$, both highly significant. The discrepancy between the values of these statistics is due to the large number of cells with 0 or 1 observed counts. Such cells contribute relatively less to G^2 than to X^2.)

5. Determine algebraic expressions for the MLEs for the five blocks of table 5.2-8 not examined in Section 5.2. Reference to formulas (5.2-48) and (5.2-49) may be of use.

6. In the game of jai alai, as played in Florida, each session consists of twelve games. Game 7 has six competitors, game 10 has seven, and all the rest have eight competitors. Two games (6 and 10) are contested among individual competitors, and the remainder are contested by competitor pairs. For this problem, we draw no distinction between games with competitor pairs and games with individual competitors.

For any particular game, the competitors are assigned unique post positions, which are numbered from 1 up to the total number of competitors in the game. The post positions define a queue. The game begins with competitors 1 and 2 competing for the first point. The winner remains in the game to compete for the next point, and the loser, who is replaced by the player(s) in post position 3, goes to the end of the queue. The game continues in this fashion until some competitor achieves a prespecified number of points. In this problem we are interested in the relative advantages of the different post positions.

The following table gives the distribution of wins across post positions for each of the twelve games separately, as listed in the Tampa jai alai fronton program of Friday, March 26, 1971.

Post Position	Game Number											
	1	2	3	4	5	6	7	8	9	10	11	12
1	10	15	12	8	14	9	12	11	11	14	16	5
2	12	10	10	19	11	11	15	14	13	12	14	14
3	8	14	11	6	12	9	10	11	6	12	10	6
4	10	11	10	7	7	7	9	6	10	11	11	12
5	6	7	6	7	10	7	7	8	7	7	8	12
6	18	11	12	16	8	13	9	14	1	15	9	12
7	8	4	12	8	9	18	—	11	11	12	6	11
8	10	12	10	9	12	9	—	9	9	—	9	10

(i)   Test the hypothesis that the relative proportion of wins by post position remains constant across game numbers (this is the hypothesis of quasi independence or quasi homogeneity of proportions). Note that the MLEs for the expected cell counts can be computed in closed form.

(ii) Given that post position is quasi-independent of game number, determine whether there is any advantage for a randomly chosen pair (individual) to have a particular post position.

7. In example 5.2-1 we discussed the first series A of data from Harris [1910], given in tables 5.2-1 and 5.2-2 on the relationship between radial asymmetry and locular composition in *Staphylea*. Below in tables 5.2-12 and 5.2-13 are series B and C, also taken from Harris.

Table 5.2-12   Relationship between Radial Asymmetry and Locular Composition in *Staphylea* (Series B of Harris [1910])

Locular Composition	Coefficient of Radial Asymmetry										
	0.00	0.47	0.82	0.94	1.25	1.41	1.63	1.70	1.89	2.05	2.16
3 even, 0 odd	610	—	—	130	—	—	0	—	1	—	—
2 even, 1 odd	—	739	146	—	35	9	—	6	—	1	0
1 even, 2 odd	—	451	142	—	28	7	—	2	—	2	1
0 even, 3 odd	110	—	—	42	—	—	3	—	0	—	—

(i)   Analyze each of the three tables separately, in particular, investigating the model of quasi independence. Take care in assessing the degrees of freedom for any test statistics you compute.

(ii) Carry out a single analysis on the three sets of data simultaneously. Be sure to distinguish between those cells you choose to treat as structural zeros and those you choose to treat as sampling zeros.

Table 5.2-13 Relationship between Radial Asymmetry and Locular Composition in *Staphylea* (Series C of Harris [1910])

Locular Composition	Coefficient of Radial Asymmetry						
	0.00	0.47	0.82	0.94	1.25	1.41	1.63
3 even, 0 odd	1,001	—	—	263	—	—	1
2 even, 1 odd	—	744	160	—	20	4	—
1 even, 2 odd	—	340	66	—	11	1	—
0 even, 3 odd	76	—	—	17	—	—	0

8. In example 5.2-5 we considered the data of Novitski and Sandler [1957] on male sperm frequencies, and we decided that the model of quasi independence fit the data in table 5.2-9. The aim of Novitski and Sandler was to estimate the relative frequencies of the four male sperm types.

(i)   Using the closed form formulas for the expected values under quasi independence find the MLEs for these relative frequencies and compute their numerical values.

(Answer:

$$\hat{p}_1 = \frac{x_{11}(x_{22} + x_{24})}{x_{23}(x_{12} + x_{14}) + x_{1+}(x_{22} + x_{24})} = 0.267,$$

$$\hat{p}_3 = \frac{x_{23}(x_{12} + x_{14})}{x_{23}(x_{12} + x_{14}) + x_{1+}(x_{22} + x_{24})} = 0.116,$$

$$\hat{p}_2 = \frac{(x_{12} + x_{22})(1 - \hat{p}_1 - \hat{p}_3)}{(x_{12} + x_{22} + x_{14} + x_{24})} = 0.191,$$

$$\hat{p}_4 = 1 - \hat{p}_1 - \hat{p}_2 - \hat{p}_3 = 0.426.)$$

(ii) Using methods from Chapter 14, compute the asymptotic variances and covariances of the parameter estimates in (i).

9. Vidmar [1972] collected the following data, aimed at discovering the possible effects on decision making of limiting the number of alternatives available to the members of a jury panel:

	Condition						
Alternative	1	2	3	4	5	6	7
First-Degree	11	—	—	2	7	—	2
Second-Degree	—	20	—	22	—	11	15
Manslaughter	—	—	22	—	16	13	5
Not Guilty	13	4	2	0	1	0	2

Larntz [1974] has challenged Vidmar's conclusion "that under conditions of restricted decision alternatives, the more severe the degree of guilt associated with the least severe guilt alternative, the greater were the chances of obtaining a not guilty verdict."

(i) Fit the model of quasi independence to these data. (Answer: $X^2 = 16.5$ with nine degrees of freedom.)

(ii) Interpret the model fitted in (i) and discuss its implications relative to Vidmar's conclusion.

5.3 Incomplete Two-Way Tables for Subsets of Complete Arrays

5.3.1 Composite models

In many complete two-way contingency tables it is clear that the model postulating independence of row and column classifications is inappropriate, either from examination of the data or by analogy with related tables. Alternative ways to describe the expected cell counts in such a table may include composite models, where we fit different models to disjoint subsets of the table. The models for some such disjoint subsets may often correspond to quasi independence in these subsets, while the models for other subsets may be different or even unrestricted.

Example 5.3-1 Social mobility in Britain

Table 5.3-1 presents data collected by Glass and his coworkers [1954] on intergenerational social mobility in Britain. Although these data were actually collected by a version of stratified random sampling, we will proceed as if a form of simple random sampling (either Poisson or multinomial) had been used. This 5×5 table is a condensation of a 7×7 table that uses a more detailed set of status categories. Note that the zero in the (5, 1) cell is a sampling zero. It is quite clear from a superficial examination of this table that the model of independence (which corresponds to perfect social mobility) is highly inappropriate. For example, the expected cell count in the (5, 1) cell under independence is 24.9, while the observed is 0.

Table 5.3-1 British Social Mobility Data

Father's Status	Son's Status				
	1	2	3	4	5
1	50	45	8	18	8
2	28	174	84	154	55
3	11	78	110	223	96
4	14	150	185	714	447
5	0	42	72	320	411

Source: Glass [1954].

Before postulating alternative models to independence we note that there is a one-to-one correspondence between rows and columns, as well as an underlying order for rows and columns. This fact, combined with the observation that the observed diagonal cells (50, 174, 110, 714, 411) are each considerably higher than their corresponding expected values under independence (37.9, 69.2, 65.0, 617.0, 245.7), suggests that we treat the diagonal cells differently from those in the remainder of the table. ■■

Quasi-perfect mobility

These remarks lead us to postulate the following model for the social mobility table, referred to by Goodman [1965, 1969a] as the model of "quasi-perfect mobility":

$$m_{ij} \begin{cases} = a_i b_j & \text{if } i \neq j \\ = m_{ii} & \text{if } i = j. \end{cases} \tag{5.3-1}$$

Here we divide the table into two subsets, S_D (containing the diagonal cells) and S_0 (containing the off-diagonal cells), and for S_0 we postulate the model of quasi independence. A further refinement of the quasi-perfect mobility model given by (5.3-1) is as follows:

$$m_{ij} \begin{cases} = a_i' b_j' & \text{if } i > j, \\ = a_i'' b_j'' & \text{if } i < j, \\ = m_{ii} & \text{if } i = j, \end{cases} \tag{5.3-2}$$

where we have divided the subset S_0 into S_P (the set of cells representing upward mobility with $i - j$ positive) and S_N (the set of cells representing downward mobility with $i - j$ negative).

5.3.2 Breaking apart an $I \times I$ table

Before returning to the Glass data, let us consider the more general problem of maximum likelihood estimation in $I \times I$ tables, under models of the form (5.3-1) and (5.3-2). Once again, we assume that each cell in the table has an observed count x_{ij}, each x_{ij} has a Poisson distribution with mean m_{ij}, and these Poisson variates are mutually independent. We begin by rewriting (5.3-1) and (5.3-2) in log-linear forms:

$$\log m_{ij} \begin{cases} = u + u_{1(i)} + u_{2(j)} & \text{for } (i,j) \in S_0, \\ = v + v_{1(i)} & \text{for } (i,j) \in S_D, \end{cases} \tag{5.3-3}$$

$$\log m_{ij} \begin{cases} = u' + u_{1(i)}' + u_{2(j)}' & \text{for } (i,j) \in S_P, \\ = u'' + u_{1(i)}'' + u_{2(j)}'' & \text{for } (i,j) \in S_N, \\ = v + v_{1(i)} & \text{for } (i,j) \in S_D, \end{cases} \tag{5.3-4}$$

respectively, where as usual, the summation over any subscripted u-term or v-term is zero.

Now, if we set

$$x_{ij}^* = x_{ij} \delta_{ij}, \tag{5.3-5}$$

where

$$\delta_{ij} = \begin{cases} 1 & \text{for } (i,j) \in S_0 \\ 0 & \text{otherwise}, \end{cases} \tag{5.3-6}$$

we can write the natural logarithm for the likelihood function under model (5.3-3) as

$$L = \sum_{i=1}^{I} x_{ii}[v + v_{1(i)}] + x_{++}^* u + \sum_{i=1}^{I} x_{i+}^* u_{1(i)} + \sum_{j=1}^{I} x_{+j}^* u_{2(j)}$$

$$- m_{++} - \sum_{i=1}^{I} \sum_{j=1}^{I} \log(x_{ij}!). \tag{5.3-7}$$

Setting the first partial derivatives of (5.3-7) with respect to the u-terms and the v-terms of (5.3-3) equal to zero, we get the following maximum likelihood equations for the m_{ij}:

$$\hat{m}_{ii} = x_{ii} \quad \text{for } i = 1, \ldots, I, \tag{5.3-8}$$

$$\sum_{j=1}^{r} \hat{m}_{ij} \delta_{ij} = x_{i+}^{*} \quad \text{for } i = 1, \ldots, I,$$

$$\sum_{i=1}^{r} \hat{m}_{ij} \delta_{ij} = x_{+j}^{*} \quad \text{for } j = 1, \ldots, I. \tag{5.3-9}$$

Expression (5.3-8) is immediate, and then (5.3-9) follows either by algebraic manipulation or by a conditional argument which considers the m_{ii} as being structurally fixed. Using arguments similar to those in Section 5.2, or in Haberman [1974], we can show that L has a unique maximum for finite values of the parameters, provided that the incomplete observed subtable formed by the cells in S_0 satisfies conditions (i) and (ii) in theorem 5.2-1.

We now note that (5.3-8) yields the same MLEs as an unrestricted parameter model for S_D, while (5.3-9) yields the MLEs under quasi independence in S_0. Similar arguments show that the MLEs under (5.3-4) are those given by the unrestricted parameter model for S_D and separate quasi-independence models for S_N and S_P. The MLEs here are positive and unique if and only if they are positive and unique for each of the models in each of the subtables. Thus we have at our disposal a technique for using our incomplete table model of quasi independence in our models for complete tables.

Example 5.3-1 *(Continued)*

We now fit models (5.3-1) and (5.3-2) to the Glass social mobility table discussed above. For model (5.3-1) the diagonal cells in the expected table are equal to the observed counts, and we determine the off-diagonal cells by use of the iterative proportional fitting procedure (recall that there are no direct estimates for incomplete tables of the form S_0). These values are given in table 5.3-2.

Table 5.3-2 Expected Values for Glass Social Mobility Data under Model (5.3-1) for Quasi-Perfect Mobility

Father's Status	Son's Status 1	2	3	4	5
1	50.0	9.5	11.0	38.5	20.0
2	6.6	174.0	49.9	174.2	90.3
3	8.5	56.1	110.0	226.2	117.2
4	27.5	181.0	209.0	714.0	378.5
5	10.4	68.4	79.1	276.1	411.0

For model (5.3-2) the expected values in the diagonal cells are also equal to the observed values, and we get the remaining MLEs by considering separately the triangular subtables above and below the diagonal (recall that we can do this using either direct or indirect estimation). This set of expected values is presented in table 5.3-3. The expected values under the model of quasi-perfect mobility given by (5.3-1) are clearly much closer to the observed values than the expected values under independence. The more refined model, with quasi independence in each of the two separate triangular subtables, gives an even closer fit to the observed data.

Table 5.3-3 Expected Values for Glass Social Mobility Data under Model (5.3-2)

Father's Status	Son's Status				
	1	2	3	4	5
1	50.0	45.0	9.6	17.4	7.0
2	28.0	174.0	82.4	150.1	60.4
3	7.7	81.3	110.0	227.5	91.6
4	13.4	122.1	193.5	714.0	447.0
5	4.4	46.6	63.5	320.0	411.0

Table 5.3-4 Summary of Goodness-of-Fit Test Statistics for Data from Table 5.3-1 under Various Models

Model	Incomplete Subset	Degrees of Freedom	X^2	G^2
Independence	Full table	16	1,199.4	811.0
(5.3-1)	S_0	11	328.7	249.4
(5.3-2)	S_P	3	8.5	12.6
	S_N	3	1.3	1.4
	$S_P + S_N$	6	9.9	14.0

Note: The model numbers refer to expressions given in Section 5.3.

We can make a formal assessment of the fit of our models by using the goodness-of-fit criteria (5.2-40) and (5.2-41), and we compute the test statistics for each subtable being fitted by a separate quasi-independence model. Table 5.3-4 summarizes these test statistics for our example. Model (5.3-1) clearly provides a poor fit, even though it is a substantial improvement over the model of independence. Model (5.3-2) can be examined in two parts, and the fit for S_N is remarkably good, while the fit for S_P, as measured by either goodness-of-fit test statistic, is significant at somewhere between the 0.05 and 0.005 level. The combined fit of (5.3-2) corresponds to a level of significance of between 0.10 and 0.025, which, considering the large total sample size, might be acceptable for many purposes. ■■

Goodman [1972b] has considered a general class of social mobility models which includes (5.3-1) and (5.3-2) as special cases. In Section 5.5 on multi-dimensional incomplete tables, we reinterpret (5.3-2) and also discuss other members of this class of multiplicative models in a multidimensional setting.

5.3.3 *Exercise*

1. In table 5.3-5 we present data on intergenerational mobility in Denmark, collected by Svalastoga [1959]. These data are similar to those of Glass [1954] for Britain, analyzed in example 5.3-1 above.

(i) Compute the expected cell values for these data under models (5.3-1) and (5.3-2) as well as for the model of independence.

(ii) Carry out goodness-of-fit tests for the models fitted in (i), displaying your results in a table similar to table 5.3-4.

(iii) Are the characteristics of social mobility represented by our models the same in Britain as in Denmark?

Table 5.3-5 Danish Social Mobility Data

Father's Status	Son's Status				
	1	2	3	4	5
1	18	17	16	4	2
2	24	105	109	59	21
3	23	84	289	217	95
4	8	49	175	348	198
5	6	8	69	201	246

Source: Svalastoga [1959].

5.4 Incomplete Multiway Tables

5.4.1 *Quasi-log-linear models*

Much of the theory from the preceding sections can be generalized to handle the analysis of incomplete multiway tables. These extensions can be used for tables of any number of dimensions, but for simplicity we will discuss the basic ideas in three dimensions. The presentation here follows closely the presentation in Fienberg [1972a].

Here we let S be the set of all cells in an incomplete $I \times J \times K$ three-way array that consists of all cells not containing structural zeros, and m_{ijk} the expected number of individuals in the (i, j, k) cell, where $m_{ijk} = 0$ for $(i, j, k) \notin S$. Tables 5.4-1 through 5.4-6 contain examples of observed incomplete multiway tables. Following the notation in Chapter 3, we can specify the most general log-linear model for those cells in S, i.e.,

$$\log m_{ijk} = u + u_{1(i)} + u_{2(j)} + u_{3(k)}$$
$$+ u_{12(ij)} + u_{13(ik)} + u_{23(jk)}$$
$$+ u_{123(ijk)} \tag{5.4-1}$$

for $(i, j, k) \in S$, where as before the u-terms are deviations and sum to zero over each included variable, e.g.,

$$\sum_i \delta_i^{(23)} u_{1(i)} = \sum_i \delta_{ij}^{(3)} u_{12(ij)} = \sum_i \delta_{ik}^{(2)} u_{13(ik)}$$
$$= \sum_i \delta_{ijk} u_{123(ijk)} = 0, \tag{5.4-2}$$

with

$$\delta_{ijk} = \begin{cases} 1 & \text{if } (i, j, k) \in S \\ 0 & \text{otherwise,} \end{cases} \tag{5.4-3}$$

$$\delta_{ij}^{(3)} = \begin{cases} 1 & \text{if } \delta_{ijk} = 1 \text{ for some } k \\ 0 & \text{otherwise,} \end{cases} \tag{5.4-4}$$

$$\delta_i^{(23)} = \begin{cases} 1 & \text{if } \delta_{ijk} = 1 \text{ for some } (j, k) \\ 0 & \text{otherwise,} \end{cases} \tag{5.4-5}$$

and similar definitions for $\delta_{ik}^{(2)}$, $\delta_{jk}^{(1)}$, $\delta_j^{(13)}$, and $\delta_k^{(12)}$. Note that (5.4-2) includes some u-terms not found in (5.4-1), i.e., those preceded by zero values of δ_{ijk}. We set those u-terms in (5.4-2) which are not included in (5.4-1) equal to an arbitrary finite quantity, so that expression (5.4-2) is well defined. This is a mathematical device which enables us to continue using the same notation as in Chapter 3.

As before, we restrict attention to hierarchical models, where whenever a particular u-term is zero, all of its higher-order relatives must also be zero (e.g., if $u_{12(ij)} = 0$ for all pairs (i, j) for which it is defined in (5.4-1), then $u_{123(ijk)} = 0$ for all $(i, j, k) \in S$).

Unsaturated models and log-linear contrasts

We define unsaturated quasi-log-linear models by setting u-terms in (5.4-1) equal to zero, and the corresponding models can always be described in terms of generalized notions of interaction contrasts. For example, setting $u_{123(ijk)} = 0$ for all $(i, j, k) \in S$ corresponds to setting equal to zero all generalized interaction contrasts of the form

$$\sum_{i=1}^{I} \sum_{j=1}^{J} \sum_{k=1}^{K} \delta_{ijk} \gamma_{ijk} \log m_{ijk} \tag{5.4-6}$$

($\gamma_{ijk} \neq 0$ for some i, j, k), where

$$\sum_{i=1}^{I} \delta_{ijk} \gamma_{ijk} = \sum_{j=1}^{J} \delta_{ijk} \gamma_{ijk} = \sum_{k=1}^{K} \delta_{ijk} \gamma_{ijk} = 0. \tag{5.4-7}$$

Using (5.4-1) through (5.4-5) together, we can rewrite these three-factor interaction contrasts as

$$\sum_{i=1}^{I} \sum_{j=1}^{J} \sum_{k=1}^{K} \delta_{ijk} \gamma_{ijk} u_{123(ijk)} \tag{5.4-8}$$

($\gamma_{ijk} \neq 0$ for some i, j, k). We also have three different sets of two-factor interaction contrasts, which can be expressed in terms of linear contrasts of u-terms, such as

$$\sum_{i=1}^{I} \sum_{j=1}^{J} \delta_{ij}^{(3)} \beta_{ij} u_{12(ij)} \tag{5.4-9}$$

($\beta_{ij} \neq 0$ for some i, j), where

$$\sum_{i=1}^{I} \delta_{ij}^{(3)} \beta_{ij} = \sum_{j=1}^{J} \delta_{ij}^{(3)} \beta_{ij} = 0. \tag{5.4-10}$$

Any unsaturated quasi-log-linear model can be specified by setting equal to zero appropriate sets of interaction contrasts, and a cell is noninteractive with respect to a quasi-log-linear model if it is not included in at least one of these interaction contrasts with nonzero coefficient.

Poisson likelihood function

Suppose now that the observed count x_{ijk} in the (i, j, k) cell, for $(i, j, k) \in S$ has a Poisson distribution with mean m_{ijk}, the expected value for the (i, j, k) cell, and that these Poisson variates are mutually independent. Then the kernel of the log-likelihood is

$$\sum_{ijk} x_{ijk} \log m_{ijk} = x_{+++}u + \sum_i x_{i++}u_{1(i)} + \sum_j x_{+j+}u_{2(j)} + \sum_k x_{++k}u_{3(k)}$$

$$+ \sum_{ij} x_{ij+}u_{12(ij)} + \sum_{ik} x_{i+k}u_{13(ik)} + \sum_{jk} x_{+jk}u_{23(jk)} \qquad (5.4\text{-}11)$$

$$+ \sum_{ijk} x_{ijk}u_{123(ijk)},$$

since $m_{ijk} = 0$ implies $x_{ijk} = 0$. When, for example, $u_{123} = 0$ for all $(i, j, k) \in S$, then the sufficient statistics are once again given by the configurations $C_{12} = \{x_{ij+}\}$, $C_{23} = \{x_{+jk}\}$, and $C_{13} = \{x_{i+k}\}$, just as in the complete table situation.

Other sampling models

Once again, as for complete multiway tables, we could have used either a single multinomial sampling model or a set of multinomials based on exactly one fixed one-dimensional or two-dimensional configuration, as long as the fixed configuration is included among the sufficient statistics. We set the total sample size $x_{+++} = N$ for notational convenience.

As with complete tables, we must take care in interpreting the parameters included in our model under a product multinomial sampling scheme. Suppose that in a three-way incomplete table the configuration $C_{12} = \{x_{ij+}\}$ is fixed, with $x_{ij+} = 20$ for all i and j. Then the u_{12}-terms are included in any quasi-log-linear model for estimation purposes, but when we come to interpret such a fitted model we cannot speak of a row effect, a column effect, or a row by column interaction.

5.4.2   Separability in many dimensions

In two-way tables it was quite simple for us to define the concepts of connectivity and separability because we were dealing with only two dimensions and just one model. In fact, the model, quasi independence, was linked to the definition of separability, since we saw that an alternate way to define the concept was based on the factorization of the likelihood function. Thus it should come as no surprise that the definitions of connectivity and separability in multiway tables depend both on the dimension of the table and on the particular model being fitted. We illustrate the generalization for three-way tables, but as we noted earlier, the basic ideas for higher-way tables are the same, and so our examples are not restricted to three dimensions.

In what follows, we say that the (i, j, k) cell in an $I \times J \times K$ table has first coordinate i, second coordinate j, and third coordinate k. Now if we let D be some nonempty subset of the set of integers $\{1, 2, 3\}$, we say that two cells are D-associated if they do not contain structural zeros and if their coordinates corresponding to the subset D coincide (e.g., if $D = \{1, 3\}$ in a $2 \times 4 \times 2$ table, then cells $(1, 3, 2)$ and $(1, 4, 2)$ are D-associated if they do not contain structural zeros). Next, we say that a set of non-structural-zero cells is (D_1, D_2, D_3)-*connected* if any cell can be linked to any other cell via a chain of cells, any two consecutive members of which must be either D_1-associated, D_2-associated, or D_3-associated. (We define (D_1, D_2)-*connected* in the same way.) Any set of non-structural-zero cells is (D_1, D_2, D_3)-*separable* if it is not (D_1, D_2, D_3)-connected.

Consider a set of non-structural-zero cells which is (D_1, D_2, D_3)-separable. We can divide this set up into subsets, each of which is itself (D_1, D_2, D_3)-connected but no two of which are (D_1, D_2, D_3)-connected when combined. These subsets are referred to as the *separable components* of the original set.

J. E. Cohen [1973] has suggested that the assessment of separability or inseparability of an incomplete multiway table is best done in two steps. The first step deals with the model, whether or not the full array includes structural zeros. Suppose we have a d-way table and that $D = \{1, \ldots, d\}$, the set of integers from 1 to d. Let $D_1, \ldots, D_m$ be the subsets of D which correspond to the highest-order u-terms in the log-linear model. Then the complete d-way array is inseparable with respect to the model if and only if

$$\bigcup_{i=1}^{m} (D - D_i) = D.$$

Thus separability occurs when each of the subsets $\{D_i\}$ includes a common nonzero sub-subset, say D^*, i.e.,

$$\bigcap_{i=1}^{m} D_i = D^*.$$

For example, for a three-way table, the model $u_{23} = u_{123} = 0$ corresponds to the two subsets $D_1 = \{1, 2\}$ and $D_2 = \{1, 3\}$ and is separable, since D_1 and D_2 have $D^* = \{1\}$ in common.

More generally, for any separable model the complete array itself is separable, and it can be broken into subtables, each one of which corresponds to a cell in the cross-classification of the variables in D^*. Then the log-linear model for the full table can be broken down into separate but parallel log-linear models for each subtable. In the assessment of separability for incomplete multiway tables, models for which the complete array is separable should be considered only in terms of the log-linear models for the subtables.

Once we have removed the separable models from consideration, we can proceed as follows. Suppose we look at an unsaturated model; for example, suppose $u_{123} = 0$. Then we let $D_1 = \{1, 2\}$, $D_2 = \{1, 3\}$, and $D_3 = \{2, 3\}$, and the $\{D_i\}$ correspond to the highest-order u-terms in the model for all the subscripts. (If $u_{23} = u_{13} = u_{123} = 0$, we let $D_1 = \{1, 2\}$ and $D_2 = \{3\}$, and we do not need to use D_3.) We have thus linked the model being fitted to the definition of separability. For example, if the model is specified by $u_{123} = 0$ and the non-structural-zero cells are $(\{1, 2\}, \{1, 3\}, \{2, 3\})$-separable, then the table is separable for this specific model into separable components, as defined above.

Finally, we note that if an incomplete three-way table is separable for a particular quasi-log-linear model, say $u_{12} = u_{13} = u_{23} = u_{123} = 0$, then it is separable for all hierarchical models of which it is a special case, e.g., $u_{123} = 0$. The converse is, of course, not true. This follows from the fact that it is more restrictive for two cells to have no coordinates in common than it is for them to have at most one coordinate in common.

Examples of separability in three dimensions

Let us consider the following set of $3 \times 3 \times 3$ tables, where we denote structural zeros by dashes.

(a)

m_{111}	m_{121}	—	m_{112}	m_{122}	—	—	—	—
m_{211}	m_{221}	—	m_{212}	m_{222}	—	—	—	—
—	—	—	—	—	—	—	—	m_{333}

This first table is $(\{1\}, \{2\}, \{3\})$-separable and thus is separable for the quasi-log-linear model specified by $u_{12} = u_{13} = u_{23} = u_{123} = 0$ and for all other quasi-log-linear models which include it as a special case.

(b)

m_{111}	m_{121}	—	m_{112}	m_{122}	—	—	—	—
m_{211}	m_{221}	—	m_{212}	m_{222}	—	—	—	—
—	—	m_{331}	—	—	m_{332}	—	—	m_{333}

This second table is both $(\{1, 3\}, \{2\})$-separable and $(\{2, 3\}, \{1\})$-separable, and thus is separable for both the model specified by $u_{12} = u_{23} = u_{123} = 0$ and for the model given by $u_{12} = u_{13} = u_{123} = 0$. It is not $(\{1, 2\}, \{3\})$-separable nor $(\{1\}, \{2\}, \{3\})$-separable.

(c)

m_{111}	m_{121}	—	m_{112}	m_{122}	—	—	—	m_{133}
m_{211}	m_{221}	—	m_{212}	m_{222}	—	—	—	m_{233}
—	—	m_{331}	—	—	m_{332}	—	—	m_{333}

This table is $(\{1, 3\}, \{2\})$-separable but neither $(\{2, 3\}, \{1\})$- nor $(\{1, 2\}, \{3\})$-separable.

(d)

m_{111}	m_{121}	—	m_{112}	m_{122}	—	—	—	m_{133}
m_{211}	m_{221}	—	m_{212}	m_{222}	—	—	—	m_{233}
—	—	m_{331}	—	—	m_{332}	m_{313}	m_{323}	m_{333}

The final table is only $(\{1, 2\}, \{1, 3\}, \{2, 3\})$-separable, and thus is separable only for the quasi-log-linear model with $u_{123} = 0$, into the two sets of cells $\{m_{111}, m_{112}, m_{121}, m_{122}, m_{211}, m_{212}, m_{221}, m_{222}\}$ and $\{m_{331}, m_{332}, m_{313}, m_{323}, m_{133}, m_{233}, m_{333}\}$. Note that this second subset of seven cells cannot be used to test whether or not $u_{123} = 0$.

Connectedness in experimental designs

As was noted in Section 5.1, the results on connectivity and separability as presented for the analysis of incomplete contingency tables are related to results for analysis-of-variance schemes involving incomplete layouts. Weeks and Williams [1964] discussed the connectedness of multiway cross-classifications for linear analysis-of-variance models without interaction terms. They described a procedure for determining what in the present notation is $(\{1\}, \{2\}, \{3\})$-connectedness by checking for $(\{1, 2\}, \{1, 3\}, \{2, 3\})$-connectedness. As noted above $(\{1, 2\}, \{1, 3\}, \{2, 3\})$-connectedness implies $(\{1\}, \{2\}, \{3\})$-connectedness, but not vice versa. Thus, as Weeks and Williams note in their errata [1965], their conditions are sufficient but not necessary. Other work on this problem is described by Eccleston and Hedayat [1972].

Connectedness versus estimability

Connectedness (or inseparability) is not a sufficient condition for the estimability of the parameters in the log-linear model, since there still may be more parameters to estimate than there are cells without structural zero entries, as in the following table:

m_{111}	—		—	m_{122}		—	—
—	—		m_{212}	—		—	m_{223}

Here the table is $(\{1\}, \{2\}, \{3\})$-connected, but there are only four cell entries to go with the five parameters in the model $u_{12} = u_{23} = u_{13} = u_{123} = 0$. Even when there are more cells than parameters to be estimated, all parameters may not be estimable. For example, the table

m_{111}	—	—	—	m_{122}	m_{132}		—	—	—	
—	—	—	m_{212}	—	—		—	m_{223}	m_{233}	

is $(\{1\}, \{2\}, \{3\})$-connected, and there are six cell entries to go with the six parameters in the model $u_{12} = u_{13} = u_{23} = u_{123} = 0$; however, $u_{1(1)}$ is confounded with $u_{3(2)}$, and $u_{1(2)}$ with $u_{3(3)}$.

Separable components

When an incomplete table is separable for a given quasi-log-linear model, then maximizing the likelihood function for that model turns out to be equivalent to maximizing separately the individual likelihood functions corresponding to each of the separable components or subtables (subsets of cells). This is true because the maximum likelihood equations take the form of linear constraints on the expected cell values (see, for example, (5.4-2) in Section 5.4.3). Separability for a given model can be shown to lead to additional linear constraints on the expected cell values, which lead to a simplification in the ML equations, whereby the equations can be rewritten as the collection of ML equations for each of the separable subtables.

In Section 5.4.3, we discuss the general form of the ML equations under quasi-log-linear models, and we suggest that the reader defer attempts to prove the assertions of the preceding paragraph until a later time (see exercise 1 in Section 5.4.5). The important thing to keep in mind is that separability allows us to consider each separable subset of cells by itself, and such considerations affect the degrees of freedom associated with the overall model for the table, as we shall soon see.

5.4.3   MLEs and degrees of freedom

Because of the difficulty created by the tying of separability to both the dimension of the incomplete table and the quasi-log-linear model being fitted, we are unable to write out a useful set of conditions to ensure the existence of unique nonzero MLEs. Haberman [1973a, 1974] gives general existence theorems for such tables, but the conditions of his theorems are extremely difficult to apply in most situations, and we do not consider them here.

Conditions for existence of MLEs

Although we cannot present general existence conditions for a given table and model, we clearly require that:

(i)   The observed marginal configurations corresponding to the minimal sufficient statistics for the model must have positive entries whenever the corresponding expected configurations have positive entries.

A second condition which we have found useful is:

(ii)   Both the expected and the observed table must be inseparable under the model.

In fact, further conditions are required to prevent patterns of sampling zeros from restricting certain cells to having expected values equal to observed values. For all of the incomplete multiway tables we have encountered in practice, we really have not had to worry about these further conditions. Nevertheless, the iterative procedures described below still converge in these cases, and to the correct MLEs for the unrestricted cells. When the expected value in a cell is restricted so that it must be equal to the observed value, we simply proceed conditionally, treating that cell as if it were structurally fixed, and then we make an appropriate reduction in the degrees of freedom. (For example, any zero total, whether structural or stochastic, leads to a reduction in the degrees of freedom.) These bothersome cases can usually be identified quite easily, and if after several cycles of the iterative procedure the expected values are suspiciously close to the observed values, we can carry out further cycles to see whether the differences in the values for these cells appear to be converging to zero.

When there exist unique nonzero MLEs for the nonzero expected cells of an incomplete multiway table and a particular quasi-log-linear model, these MLEs are uniquely determined by setting expected marginal configurations equal to the observed marginal configurations corresponding to the minimal sufficient statistics. For example, suppose we are fitting the model specified by $u_{123} = 0$ in a three-way table. Then the MLEs are given by the equations

$$\hat{m}_{ij+} = x_{ij+}\,; \qquad \hat{m}_{i+k} = x_{i+k}\,; \qquad \hat{m}_{+jk} = x_{+jk}, \qquad (5.4\text{-}12)$$

where the subscripts in each set of equations range over all sets of values for which the expected marginal values are positive.

Direct versus indirect computation of MLEs

In incomplete two-way tables, for certain configurations of the cells we can write the MLEs for the expected cell values under quasi independence in closed form. In incomplete multiway tables, for certain configurations of the cells we can write closed-form MLEs *for particular quasi-log-linear models*. In a sense, this is like the generalization of the notion of separability to several dimensions, where both the configuration of cells and the model must be considered.

We shall not explore direct estimation procedures for multiway tables in the same exhaustive detail as for two-way tables, in part because for most cell configurations and most quasi-log-linear models no direct estimation is possible. We do give one example in this section where direct estimation is possible, and in Chapter 6 we describe a particular configuration for incomplete 2^k tables and those models for which direct estimation is possible.

Iterative procedure for determining MLEs

Once again, either of two equivalent iterative procedures may be used to obtain the solutions of the maximum likelihood equations. Goodman [1968] suggests one of these procedures, illustrating it for the model with $u_{123} = 0$. The following generalization of the Deming–Stephan iterative proportional fitting procedure is equivalent to it.

1. At the 0th step, let

$$m_{ijk}^{(0)} = \delta_{ijk} \qquad (5.4-13)$$

for all i, j, k, when δ_{ijk} is defined by (5.4-3) above.

2. At each successive cycle of the iteration, there are as many steps as there are configurations of sufficient statistics, and each step consists of a rescaling relative to one of these configurations. For example, in a three-way table for $u_{123} = 0$, at the vth cycle we take

$$m_{ijk}^{(3v-2)} = \frac{m_{ijk}^{(3v-3)} x_{ij+}}{\sum_k m_{ijk}^{(3v-3)}}, \qquad (5.4-14)$$

$$m_{ijk}^{(3v-1)} = \frac{m_{ijk}^{(3v-2)} x_{+jk}}{\sum_i m_{ijk}^{(3v-2)}}, \qquad (5.4-15)$$

$$m_{ijk}^{(3v)} = \frac{m_{ijk}^{(3v-1)} x_{i+k}}{\sum_j m_{ijk}^{(3v-1)}}. \qquad (5.4-16)$$

3. We continue repeating the cycles until desired accuracy is achieved.

This iterative procedure always converges, but the convergence tends to be somewhat slower than for the related procedure for complete multiway tables.

Degrees of freedom

The computation of degrees of freedom for incomplete multiway tables follows the same basic rule as stated earlier: Subtract the number of independent parameters used in the model from the total number of cells to which the model is being fitted.

This is somewhat more complicated in the multiway case, since we may have structural zero totals in the expected marginal configurations associated with various models. Thus we must use the rule given by expression (3.8-1) of Chapter 3:

$$\text{degrees of freedom} = V - z_e + z_p, \tag{5.4-17}$$

where V is the number of degrees of freedom usually associated with the model for the complete table case, z_e is the number of cells with zero expected values, and z_p is the number of zero entries in the expected marginal configurations, adjusted for possible zeros in the marginal totals of these configurations.

As with two-way incomplete tables, we must look at each separable component of a separable table by itself in order to compute the degrees of freedom properly for a given model as applied to the table as a whole. This is the case because separability leads to additional linear constraints on the expected cell values which are equivalent to our fitting extra parameters in the model. These constraints are "additional" in the sense that they do not result directly from the inclusion of particular parameters in the model for a general incomplete table.

Let us illustrate the computation of degrees of freedom with an $I \times J \times K$ incomplete inseparable table for $u_{123} = 0$, containing a total of z_e structural zeros. Let

$$z_{12} = IJ - \sum_{ij} \delta_{ij}^{(3)},$$

$$z_{23} = JK - \sum_{jk} \delta_{jk}^{(1)}, \tag{5.4-18}$$

$$z_{13} = IK - \sum_{ik} \delta_{ik}^{(2)},$$

i.e., z_{12}, z_{23}, and z_{13} are the number of zeros in the expected marginal configurations $C_{12}^* = \{m_{ij+}\}$, $C_{23}^* = \{m_{+jk}\}$, and $C_{13}^* = \{m_{i+k}\}$, respectively. Thus $z_p = z_{12} + z_{23} + z_{13}$. Furthermore, let

$$\sum_i \delta_i^{(23)} = I, \qquad \sum_j \delta_j^{(13)} = J, \qquad \sum_k \delta_k^{(12)} = K. \tag{5.4-19}$$

Then the number of degrees of freedom is:

$$(I - 1)(J - 1)(K - 1) - z_e + (z_{12} + z_{23} + z_{13}). \tag{5.4-20}$$

If the same table is inseparable for the model with $u_{12} = u_{123} = 0$, the number of degrees of freedom for the latter model is then:

$$K(I - 1)(J - 1) - z_e + (z_{23} + z_{13}). \tag{5.4-21}$$

If there is one empty layer in the table and (5.4-19) no longer holds, then further adjustments in the degrees of freedom formulas (5.4-20) and (5.4-21) are necessary. We must add one degree of freedom to (5.4-20) and (5.4-21). On the other hand, if there is one empty row in the table, we add one degree of freedom to (5.4-20) but not to (5.4-21).

Before concluding this section, let us return to an example of an incomplete table which is separable for some model and illustrate the computation of degrees of freedom there. We consider the following table of expected values:

m_{111}	m_{121}	—
m_{211}	m_{221}	—
—	—	m_{331}

m_{112}	m_{122}	—
m_{212}	m_{222}	—
—	—	m_{332}

—	—	m_{133}
—	—	m_{233}
—	—	m_{333}

We have pointed out that this table is $(\{1, 3\}, \{2\})$-separable, i.e., separable for the model

$$u + u_1 + u_2 + u_3 + u_{13}.$$

For this model the number of zeros in the expected marginal configuration $C_{13}^* = \{m_{i+k}\}$ is $z_{13} = 0$, and there are $z_e = 14$ structural zeros. Without considering separability, we compute the degrees of freedom to be

$$IJK - z_e - [1 + (I - 1) + (J - 1) + (K - 1) + (I - 1)(K - 1) - z_{13}] = 2.$$

There are, however, two separable components of the table for this model:

m_{111}	m_{121}
m_{211}	m_{221}

m_{112}	m_{122}
m_{212}	m_{222}

and

—	—	m_{133}
—	—	m_{233}
m_{331}	m_{332}	m_{333}

The first component is a complete 2^3 table and has three degrees of freedom. The second component is two-dimensional (corresponding to one column of the table) and has zero degrees of freedom, since u_{13} is the two-way interaction term for the two-way table. Thus the total number of degrees of freedom is $3 + 0 = 3$, not 2.

Estimability of parameters

As we noted earlier, we must distinguish between the existence of unique MLEs for the expected cell values and the estimability of all of the log-linear model parameters. For the table

m_{111}	—
—	—

—	m_{122}
m_{212}	—

—	—
—	m_{223}

If the appropriate observed counts are nonzero, then there exists a unique set of nonzero expected cell values for the model $u_{12} = u_{13} = u_{23} = u_{123} = 0$ (in fact, the expected equal the observed values), even though not all of the parameters are estimable.

For the table

m_{111}	—	—
—	—	—

—	m_{122}	m_{132}
m_{212}	—	—

—	—	—
—	m_{223}	m_{233}

the estimated cell values under the model $u_{12} = u_{13} = u_{23} = u_{123} = 0$ are:

$$\hat{m}_{111} = x_{111}, \qquad \hat{m}_{212} = x_{212},$$

$$\hat{m}_{122} = \frac{(x_{122} + x_{132})(x_{122} + x_{223})}{x_{122} + x_{132} + x_{223} + x_{233}},$$

$$\hat{m}_{132} = \frac{(x_{122} + x_{132})(x_{132} + x_{233})}{x_{122} + x_{132} + x_{223} + x_{233}}, \qquad (5.4\text{-}22)$$

$$\hat{m}_{223} = \frac{(x_{223} + x_{233})(x_{122} + x_{223})}{x_{122} + x_{132} + x_{223} + x_{233}},$$

$$\hat{m}_{233} = \frac{(x_{223} + x_{233})(x_{132} + x_{233})}{x_{122} + x_{132} + x_{223} + x_{233}}.$$

There is one degree of freedom for testing the goodness of fit of this latter set of expected values and so only one estimable interaction contrast.

5.4.4 Examples of incomplete multiway tables

Example 5.4-1   Taste-bud data

The data for our first example were collected by Frank and Pfaffman and reported in a somewhat reduced form in *Science* (see Frank and Pfaffman [1969]). These authors examined chorda tympani fibers at the posterior of a rat's tongue for sensitivity to four different taste stimuli: N = sodium chloride, H = hydrogen chloride, Q = quinine, S = sucrose. Unfortunately, dead fibers cannot be distinguished from those that respond to none of these stimuli, and so when we think of the responses to the stimuli as forming a four-way table ($2 \times 2 \times 2 \times 2$) with each dimension having two categories (response, no response), there is one cell for which we have no observations. For our purposes, we treat the expected value for this cell as a structural zero. We give the data below in table 5.4-1.

Frank and Pfaffman were interested in whether or not each fiber responds independently to the four different stimuli. The quasi-log-linear model that corresponds to complete independence is given by

$$\log m_{ijkl} = u + u_{1(i)} + u_{2(j)} + u_{3(k)} + u_{4(l)},$$

and the fitted values are given as the second entry in each cell (the third entry is simply observed minus expected value). Note that there are three sampling zeros with nonzero expected values. For the table and model the use of an iterative procedure is required, and below the data in table 5.4-1 we give the maximum deviation of the expected marginal totals corresponding to sufficient statistic configurations from the observed totals at the end of each cycle of the iteration.

Although it most likely would be a mistake to carry out a formal goodness-of-fit test on these data due to the relatively small number of observations per cell, we have computed the value of G^2, which is 12.99 based on $2^4 - 1 - [1 + 1 + 1 + 1 + 1] = 10$ degrees of freedom. Although the data are sparse, the fit of the model for complete quasi independence is reasonably close. Further data are desirable, however, in support of the conclusion that each fiber responds independently to the four different stimuli. ■ ■

Table 5.4-1 Data on Sensitivity of Chorda Tympani Fibers in Rat's Tongue to Four Taste Stimuli

		H Yes	Yes	No	No
		N Yes	No	Yes	No
S	Q				
Yes	Yes	2	0	1	0
		1.32	0.94	1.08	0.77
Yes	No	1	3	3	2
		2.53	1.80	2.07	1.47
No	Yes	3	3	0	1
		1.89	1.34	1.55	1.10
No	No	3	1	4	—
		3.61	2.57	2.95	

Iteration	Maximum Deviation
1	9.000
2	0.423
3	0.062
4	0.004

Source: Frank and Pfaffman [1969], as reported in Fienberg [1972a].
Notes: The first entry in each cell is the observed value, and the second is the MLE of the expected value under the model of complete quasi independence. $G^2 = 12.99$, degrees of freedom = 10.

Example 5.4-2 *Disability scores (Continuation of example 5.2-4)*

The data of Jones and Poskanzer on disability ratings of stroke patients, discussed earlier and presented in table 5.2-6, form a two-way margin of a multiway table. A third dimension of great interest sorted the patients on the basis of whether the stroke corresponded to a right (R) or left (L) lesion of the brain. This version of the data is given in table 5.4-2a, for which the rating scale has only three categories, I, II, and III, corresponding to the original categories in the following way:

$$A \ \& \ B \Leftrightarrow \text{I},$$

$$C \Leftrightarrow \text{II},$$

$$D \ \& \ E \Leftrightarrow \text{III}.$$

We have collapsed the data in this way on the basis of the original definition of disability categories as a result of the sparseness of data in parts of the table.

Based on our findings in the earlier analysis of these data, it is reasonable to begin our new analysis by looking at quasi independence within each layer (i.e., for each of the two types of lesion). This can be done using our closed-form methods for triangular tables, either separately for each type of lesion or by iteration, when we note that the conditional quasi-independence model is given by

$$\log m_{ijk} = u + u_{1(i)} + u_{2(j)} + u_{3(k)} + u_{13(ik)} + u_{23(jk)},$$

i.e.,

[1] $$u_{12} = u_{123} = 0.$$

Table 5.4-2   Three-Way Version of Jones and Poskanzer Data on Stroke Patients

a. Observed Data

| Initial State | R-lesion Final State | | | | L-lesion Final State | | | |
	I	II	III	Totals	I	II	III	Totals
III	17	10	13	40	36	5	10	51
II	7	3	—	10	3	1	—	4
I	6	—	—	6	8	—	—	8
Totals	30	13	13	56	47	6	10	63

b. Expected Values for $u_{12} = u_{123} = 0$

| Initial State | R-lesion Final State | | | | L-lesion Final State | | | |
	I	II	III	Totals	I	II	III	Totals
III	17.51	9.49	13	40.00	35.53	5.47	10	51.00
II	6.49	3.51	—	10.00	3.47	0.53	—	4.00
I	6	—	—	6.00	8	—	—	8.00
Totals	30.00	13.00	13.00	56.00	48.00	7.00	10.00	63.00

c. Expected Values for $u_{12} = u_{13} = u_{123} = 0$

| Initial State | R-lesion Final State | | | | L-lesion Final State | | | |
	I	II	III	Totals	I	II	III	Totals
III	20.35	10.78	13	44.13	31.89	4.98	10	46.87
II	4.19	2.22	—	6.41	6.57	1.02	—	7.59
I	5.46	—	—	5.46	8.54	—	—	8.54
Totals	30.00	13.00	13.00	56.00	47.00	6.00	10.00	63.00

d. Expected Values for $u_{12} = u_{13} = u_{23} = u_{123} = 0$ (Complete Quasi Independence of Initial State, Final State, and Side of Lesion)

| Initial State | R-lesion Final State | | | | L-lesion Final State | | | |
	I	II	III	Totals	I	II	III	Totals
III	24.59	7.41	10.82	42.82	27.66	8.34	12.18	48.18
II	5.06	1.53	—	6.59	5.69	1.72	—	7.41
I	6.59	—	—	6.59	7.41	—	—	7.41
Totals	36.14	8.94	10.82	56.00	40.76	10.06	12.18	63.00

Source: Fienberg [1972a].
Note: Two patients from initial states D and E in table 5.2-6 were omitted from the present table.

The direct fitting is quite simple in this case because the cells corresponding to the states (I, I) and (III, III) are cell isolates, and upon their deletion we simply look at the remaining 2 × 2 table in each layer. There are clearly two degrees of freedom for this model and table (one for each layer), and an examination of the observed and expected values shows an excellent fit, with $G^2 = 0.60$.

We can now look at two simpler quasi-log-linear models:

[2] $$u_{12} = u_{13} = u_{123} = 0,$$

[3] $$u_{12} = u_{13} = u_{23} = u_{123} = 0.$$

The expected values for these models, which can also be written in closed form (see exercise 2 in Section 5.4.5), are given in tables 5.4-2c and 5.4-2d. The corresponding goodness-of-fit statistics are $G^2 = 5.49$ with four degrees of freedom and $G^2 = 11.89$ with six degrees of freedom, respectively, neither of which is significant at the 5% level.

Finally, we note that models [1], [2], and [3] form a nested sequence with [1] as a special case of [2] and [2] as a special case of [3]. For such nested sequences we can partition the likelihood ratio goodness-of-fit statistic G^2 into additive components in a manner analogous to the partitioning for complete tables (see Chapter 4). Thus the statistic for testing the fit of model [3] can be broken into two parts, one for testing the fit of [2], and the other for testing whether [3] is the true model given that [2] is also true. This second component has a value of $11.89 - 5.49 = 6.40$ with two degrees of freedom, which is significant at the 5% level, even though the value of G^2 for model [3] is not. Next we can break the statistic for testing the fit of model [2] also into two parts, one for testing the fit of [1], and the other for testing whether [2] fits given that [1] does. The value of the latter component is $5.49 - 0.60 = 4.89$ with two degrees of freedom, which is not significant at the 5% level.

To summarize, we have a nested sequence of three models, each of which fits the data reasonably well. The difference between [1] and [2] is not significant at the 5% level, and so we opt for model [2] as opposed to [1] because of its simpler form. On the other hand, the difference between [2] and [3] is significant, and this leads us to choose [2] over [3].

We can therefore conclude that model [2], with $u_{12} = u_{13} = u_{123} = 0$, is the appropriate one for the data at hand, and we can say that initial state is quasi-independent of final state and side of lesion jointly. Since the rules for collapsing, described in Chapter 2, are applicable in this situation for collapsing over the categories for side of lesion, we see once again that initial state and final state are quasi-independent. ■ ■

Example 5.4-3 *Color forms for meadow spittlebugs*

The meadow spittlebug *Philaenus spumarius* L. (Homoptera) has received considerable study, at least partly because it possesses a long series of color forms. Many of the forms occur in the females only, but none is confined to the male sex. When data on this species are analyzed, the absence of some color forms in the males gives rise to structural zeros. Table 5.4-3 contains unpublished data of Thompson giving the number of *Philaenus spumarius* by sex and color form for two distinct locales in Wisconsin. The color forms are *populi* (POP), *typica* (TYP), *trilineata* (TRI), *marginella* (MAR), *lateralis* (LAT), *flavicollis* (FLA), and *albomaculata* (ALB).

The three forms MAR, LAT, and FLA are absent in the male sex. An analysis of these data ignoring the structural zeros would lead to the conclusion that a sex-form interaction exists. Fitting the model

$$u + u_{1(i)} + u_{2(j)} + u_{3(k)}$$

of complete quasi independence to these data yields the expected values in table 5.4-4 and a value of $G^2 = 15.82$ with thirteen degrees of freedom. Note that had we included a locale-form interaction in our model, the expected values for MAR, LAT, and FLA would have been the same as the observed values. The fit here is clearly good, and we can conclude that, conditional on certain color forms being absent in males, sex, form, and locale are mutually independent in these data. Further data of this sort for locales in Scandinavia are given by Halkka [1964].

Table 5.4-3   Wisconsin *Philaneus spumarius* Color Forms

Sex	Locale	POP	TYP	TRI	Color Form MAR	LAT	FLA	ALB
M	A	51	123	18	—	—	—	1
M	B	116	303	44	—	—	—	1
F	A	58	91	15	7	5	1	2
F	B	133	289	44	31	13	2	7

Source: V. Thompson, unpublished data, as presented in Fienberg [1972a].

Table 5.4-4   Expected Values for Data from Table 5.4-3, under Model of Complete Quasi Independence

Sex	Locale	POP	TYP	TRI	Color Form MAR	LAT	FLA	ALB
M	A	49.83	112.18	16.84	—	—	—	1.53
M	B	131.66	296.42	44.50	—	—	—	4.05
F	A	48.46	109.10	16.38	10.43	4.94	0.82	1.49
F	B	128.05	288.30	43.28	27.57	13.06	2.18	3.93

Note: Values are after three cycles of iteration.

Table 5.4-5   Breakdown of Data on *Philaneus spumarius* Color Forms for Wisconsin Locale B

Sex	Locale	POP	TYP	TRI	Color Form MAR	LAT	FLA	ALB
M	high	23	55	2	—	—	—	1
M	low	67	186	30	—	—	—	0
M	grassy	24	60	12	—	—	—	0
F	high	28	61	12	11	3	1	3
F	low	63	148	17	9	6	1	3
F	grassy	40	75	15	11	4	0	0

Source: V. Thompson, unpublished data, as presented in Fienberg [1972a].

Thompson gives a further breakdown for most of the data of locale B, based on whether the spittlebug was observed in a high, low, or grassy sublocale. The

Table 5.4-6 Expected Values for Data from Table 5.4-5, under Model with Sex-Locale Two-Factor Effect ($u + u_{1(i)} + u_{2(j)} + u_{3(k)} + u_{23(jk)}$)

Sex	Locale	POP	TYP	TRI	Color Form MAR	LAT	FLA	ALB
M	high	21.45	51.23	7.71	—	—	—	0.61
M	low	74.96	178.98	26.92	—	—	—	2.14
M	grassy	25.42	60.71	9.13	—	—	—	0.73
F	high	28.69	68.48	10.30	7.22	3.03	0.47	0.82
F	low	59.53	142.15	21.38	14.98	6.28	0.97	1.70
F	grassy	34.95	83.45	12.55	8.80	3.69	0.57	1.00

Note: Values are after five cycles of iteration.

sublocale-form marginal totals for these data, presented in table 5.4-5, are somewhat smaller than the observed values for locale B in table 5.4-3 because some of the spittlebugs were not classified by type of sublocale. Note the sampling zero in the last row and column of table 5.4-5.

Fitting the model of complete quasi independence to the data in table 5.4-5 and carrying out a goodness-of-fit test yields a value of $G^2 = 49.23$ with 23 degrees of freedom, a value which is significant at the 0.001 level. When a two-factor effect for sex-locale is included in the fitted model, the expected values are as given in table 5.4-6, with an associated value of $G^2 = 32.17$ with 21 degrees of freedom. This value corresponds to a descriptive level of significance greater than 0.05, and the difference $49.23 - 32.17 = 17.06$ with two degrees of freedom is significant at the 0.001 level. Thus we have found a significant sex-locale effect within site B, even though sex and locale seem to be independent in table 5.4-3. ∎∎

5.4.5 Exercises

1. For all four examples in Section 5.4.2, write out the likelihood function for each of the models discussed, and thus verify the claims made there regarding separability for each model and table combination.

2. Write out closed-form algebraic expressions for the expected values under models [2] and [3] in example 5.4-2 on disability ratings of stroke patients.

5.5 Representation of Two-Way Tables as Incomplete Multiway Arrays

In Section 5.3 we discussed the analysis of complete $I \times I$ tables where two different quasi-independence models were fitted to disjoint subsets of the table, while the model for the remaining cells was unrestricted. Here we reinterpret this model and discuss some related multiplicative models considered by Goodman [1972b].

5.5.1 The model from Section 5.3

We now consider an $I \times I$ table with expected cell values described by the following model:

$$\log m_{ij} = \begin{cases} u' + u'_{1(i)} + u'_{2(j)} & \text{for } (i,j) \in S_P, \\ u'' + u''_{1(i)} + u''_{2(j)} & \text{for } (i,j) \in S_N, \\ v + v_{1(i)} & \text{for } (i,j) \in S_D, \end{cases} \qquad (5.5\text{-}1)$$

where S_D is the set of diagonal cells (i, i), S_P is the set of cells (i, j) with $i - j$ positive, and S_N is the set of cells with $i - j$ negative. In Section 5.3 we saw that fitting this model was equivalent to fitting separate quasi-independent models for S_N and S_P and an unrestricted model to S_D. Thus the number of degrees of freedom associated with this model is $(I - 2)(I - 3)$, since each of the sets S_N and S_P has $(I - 2(I - 3)/2$ degrees of freedom associated with it.

Now let us define a set of three-way expected cell values for an $I \times I \times 3$ table in the following manner:

$$m_{ij1} = \begin{cases} m_{ij} & \text{if } (i, j) \in S_P, \\ 0 & \text{otherwise}; \end{cases} \tag{5.5-2}$$

$$m_{ij2} = \begin{cases} m_{ij} & \text{if } (i, j) \in S_N, \\ 0 & \text{otherwise}; \end{cases} \tag{5.5-3}$$

$$m_{ij3} = \begin{cases} m_{ii} & \text{if } (i, j) \in S_D, \\ 0 & \text{otherwise}, \end{cases} \tag{5.5-4}$$

for all i and j. This resulting table contains $2I^2$ structural zeros and is separable (in the sense of the preceding section) for the model of conditional quasi independence given the level of variable 3, which is just the representation of model (5.5-1) in our three-way table. We illustrate this three-dimensional representation in table 5.5-1 for the Glass social mobility data given in table 5.3-1.

Table 5.5-1   Three-Dimensional Representation of Glass Social Mobility Data (Table 5.3-1), Corresponding to Formula (5.5-1)

Set of Cells	Father's Status	Son's Status 1	2	3	4	5
S_P	1	—	—	—	—	—
	2	28	—	—	—	—
	3	11	78	—	—	—
	4	14	150	185	—	—
	5	0	42	72	320	—
S_N	1	—	45	8	18	8
	2	—	—	84	154	55
	3	—	—	—	223	96
	4	—	—	—	—	445
	5	—	—	—	—	—
S_D	1	50	—	—	—	—
	2	—	174	—	—	—
	3	—	—	110	—	—
	4	—	—	—	714	—
	5	—	—	—	—	411

5.5.2   Other multiplicative models

Goodman [1972b] suggests several multiplicative models for $I \times I$ tables which he applies to the Glass social mobility data discussed in Section 5.3. We will discuss two of these models to show how they can be given a multiway representation. As a result of this representation, the theory of the previous section can be applied directly. Goodman presents what he refers to as the "triangles-parameter" model for the off-diagonal cells given, in our notation, by

$$m_{ij} = \begin{cases} \alpha_i \beta_j \tau_1 & \text{if } (i,j) \in S_P, \\ \alpha_i \beta_j \tau_2 & \text{if } (i,j) \in S_N. \end{cases} \tag{5.5-5}$$

If we use (5.5-2) and (5.5-3) to describe a three-way incomplete $I \times I \times 2$ table, we see that (5.5-5) is simply the quasi-log-linear model for complete quasi independence. Moreover, the three-way table, which contains $I(I + 1)$ structural zeros, is not separable for this model. Applying the methods of the previous section, we find that the number of degrees of freedom for the model is given by

$$2I^2 - (I^2 + I) - [1 + (I - 1) + (I - 1) + 1] = I^2 - 3I \tag{5.5-6}$$

for $I \geq 3$.

A further model considered by Goodman is the "diagonals-parameter" model for off-diagonal cells, given by

$$m_{ij} = \alpha_i \beta_j \delta_k \quad \text{for } (i,j) \in S'_k, \tag{5.5-7}$$

where S'_k is the set of cells (i,j) with $i - j = k$ for $k = \pm 1, \pm 2, \ldots, \pm(I - 1)$. Model (5.5-5) is a special case of (5.5-7), where

$$\delta_k = \begin{cases} \tau_1 & \text{for } k = 1, 2, \ldots, I - 1, \\ \tau_2 & \text{for } k = -1, -2, \ldots, -(I - 1). \end{cases} \tag{5.5-8}$$

Here we define an $I \times I \times 2(I - 1)$ table with expected values given by

$$m_{ijk} = \begin{cases} m_{ij} & \text{if } (i,j) \in S'_k, \\ 0 & \text{otherwise,} \end{cases} \tag{5.5-9}$$

and model (5.5-7) when carried over into this three-way representation becomes the model for complete quasi independence. Expression (5.5-9) yields a three-way table with $I(2I - 1)(I - 1)$ structural zeros, where the $(I, 1, I - 1)$ and $(1, I, -I + 1)$ cells contain cell isolates.

What distinguishes this incomplete table model from all the others in this section is that because $k = i - j$, summing terms like m_{ijk} and $\log m_{ijk}$ over all i and j gives a sum over *all* cells along lines parallel to the diagonal. Thus we get an *additional* constraint on the u-terms $\{u_{3(k)}\}$ besides the usual one, $\sum_k u_{3(k)} = 0$. Thus there is one less parameter to estimate, and we must increase the usual formula for degrees of freedom by 1. The number of degrees of freedom for this model is then given by

$$2(I - 1)I^2 - I(I - 1)(2I - 1) - [1 + (I - 1) + (I - 1) + 2(I - 2)]$$
$$= I^2 - 5I + 5 \tag{5.5-10}$$

for $I \geq 4$.

A special case of (5.5-7) is given by setting

$$\delta_k = \delta_{k*} \tag{5.5-11}$$

for $k* = -k$, with $k = 1, \ldots, I - 1$. The three-way representation of this special case is given via an $I \times I \times (I - 1)$ table, with

$$m_{ijk} = \begin{cases} m_{ij} & \text{if } (i, j) \in S'_k \text{ or } S'_{-k}, \\ 0 & \text{otherwise,} \end{cases} \qquad (5.5\text{-}12)$$

that contains $I(I - 1)^2$ structural zeros. Here the number of degrees of freedom is

$$I^2(I - 1) - I(I - 1)^2 - [1 + (I - 1) + (I - 1) + (I - 2)]$$
$$= I^2 - 4I + 3$$
$$= (I - 3)(I - 1) \qquad (5.5\text{-}13)$$

for $I \geq 4$.

Example 5.5-1   Social mobility in Britain (Continuation of Example 5.3-1)
Table 5.5-2 contains the values of the likelihood ratio goodness-of-fit statistic G^2 for the Glass social mobility table discussed in Section 5.3, using the various models of this section. These values are taken from Goodman [1972b]. An examination of the corresponding descriptive levels of significance suggests that model (5.5-7), possibly with the additional constraints given by (5.5-11), may be a reasonable competitor to model (5.5-1). When Goodman applied these models to a second set of mobility data from Denmark (Svalastoga [1959]), models (5.5-7) and (5.5-7) plus (5.5-11) provided superior fits as compared with the fit of model (5.5-1) (see exercise 1 in Section 5.5.3).

Table 5.5-2   Values of Likelihood Ratio Goodness-of-Fit Statistic for Various Multiplicative Models Applied to the Data from Table 5.3-1

Model	Degrees of Freedom	G^2	Descriptive Level of Significance
(5.5-1)	6	14.0	> 0.025
(5.5-5)	10	242.3	< 0.001
(5.5-7)	5	9.5	0.05
(5.5-7), (5.5-11)	8	19.1	0.025

Note: The model numbers refer to expression given in Section 5.5.

We return to these data and the multiplicative models of this section in Chapter 9, where we discuss more general issues regarding model fitting. The data are also discussed in Chapter 12 from yet another point of view. ■ ■

5.5.3   Exercise

1. Fit the multiplicative models described in this section to the Danish social mobility data in table 5.3-5. Compare the fit of various models with the goodness-of-fit results reported in table 5.5-2.

6 Estimating the Size of a Closed Population

6.1 Introduction

In Chapter 5 we discussed techniques for the analysis of incomplete multiway tables. Here we look at a special application of these techniques to problems of estimating the total number of individuals or objects in a population without the use of complete enumeration. We assume that the population in question is "closed" or fixed, so that there are no changes due to birth, death, emigration, or immigration during the time period when the sampling takes place.

Many problems involving the estimation of the size of a closed population of individuals are handled by means of what is commonly called a "multiple recapture census." Cormack [1968] gives an excellent review of the many techniques currently in use. The technique discussed in this chapter is a generalization of standard multiple recapture census (see Fienberg [1972b]).

The usual multiple recapture census is comprised of a sequence of d samples, lists, or rosters. Those individuals included in the ith sample typically are divided into two groups: those who have been identified, tagged, or marked in preceding samples, and those who have not. The only information provided by traditional multiple recapture data regarding an individual in the ith sample is whether or not he was in some previous sample (i.e., whether or not he was identified, tagged, or marked). For the analysis of such data, one of the crucial assumptions is that the samples are independent of one another.

In this chapter we focus our attention on situations where there is a set of d samples, lists, or rosters, not necessarily ordered sequentially in time. We assume that individuals or objects are uniquely labeled or identifiable so that we can determine whether an individual is present or absent in each of the d samples. As a result, we are able to view the observed data as falling into a 2^d cross-classification, absence or presence in the ith sample or census defining the categories for the ith dimension. This cross-classification has one missing cell, corresponding to absence in all d samples.

The traditional multiple recapture census assumes that the d samples are independent. Under such an assumption, we only need the size of each of the samples and the total number of individuals in the d samples (eliminating overlap) to estimate the population size (see expression (6.5-6) below). Unique tags for individuals are not necessary. Because of this result, some investigators using the traditional multiple recapture formulas may be led to mark newly captured individuals with what Darroch [1958] refers to as "similar" tags. By recording data in a more complete fashion than is possible through the use of similar tags, we can consider possible dependencies among the samples. These dependencies correspond to two-factor and higher-order terms in log-linear models. The analyses we describe here are applicable only when captured individuals are uniquely tagged.

Dependencies among samples arise in several ways. Positive interaction terms might imply that individuals are selectively included in several samples, for reasons of "trap fascination" or the effects of tagging in animal capture-recapture experiments, or for reasons of "social visibility" in other contexts. Negative interaction terms might imply "trap avoidance" in animal tagging experiments, while in social situations they might reflect the stratification of the population according to some latent variable or "social invisibility." Dependencies in the data may also occur because of underenumeration (e.g., the loss of tags) or even overenumeration. The key feature of the estimators of the population size we develop in this chapter is that they change in the direction appropriate for the particular form of interaction observed. Moreover, taking the dependencies into account, we get more accurate confidence bounds for our estimate of population size.

6.1.1 *Practical examples*

Before going into the technical details, let us note some situations where the techniques we describe may be of some use:

1. Estimating the number of children in Massachusetts possessing a specific congenital anomaly. Five sources or lists of names of such children were available, and there were some clear relationships (or dependencies) among the lists (see Wittes [1970], Wittes, Colton, and Sidel [1974], and Fienberg [1972b]). We consider this example in Section 6.5.

2. Estimating the number of volunteer organizations in small cities and towns in Massachusetts (Smith [1973]). In each city or town there were three techniques used to identify individual volunteer organizations. We discuss this example in Section 6.4.

3. Estimating the number of free-roving or feral dogs in a section of the city of Baltimore (Beck [1972]). Pictures of free-roving dogs were taken on nine different dates, and the dogs were uniquely identified by picture. There were nine samples, one for each date. Because of the sparse nature of the data collected by Beck (fewer than 100 dogs were observed altogether), we are unable to make full use of the techniques described below in estimating the size of the free-roving dog population.

4. Estimating the number of drug addicts in the United States. There are five different federal agencies which have registries of drug addicts. Most of the individuals whose names appear in these registries have had their names recorded because of crime-related activities stemming from their involvement with narcotics. The five-sample version of the techniques described here can yield an estimate of the size of the drug addict population which will probably exclude those individuals who have an extremely small probability of being apprehended, either for a narcotics offense or for criminal activity necessitated by the monetary demands of addiction.

5. Estimating the number of crimes committed in a given area. Crime reports are collected by the local, state, and federal police groups. Not all crimes reported are recorded by any one police group. In addition, several local community agencies receive information about neighborhood crimes. A recent national

survey indicated that close to 50% of all felonies (e.g., rapes and robberies), petty thefts, and acts of vandalism go unreported to official police agencies. The techniques described below could be used to combine crime reports from several public and private sources.

6.1.2   *Outline of chapter*

We begin our technical discussion by reconsidering the two-sample capture-recapture problem. The results motivate our general approach. In Section 6.3, we review some results relating conditional and unconditional maximum likelihood estimators which allow us to make direct use of the incomplete table methodology presented in Chapter 5. In Section 6.4, we focus on the three-sample census to keep notational complexities at a minimum. We use example 2 above to illustrate the application of our techniques to the three-sample case. Finally, we discuss the general d-sample census and apply the resulting techniques to example 1 above.

The approach we adopt here consists of several steps. First, we view the observations as falling into a 2^d cross-classification, with absence or presence on the ith census defining the categories for the ith dimension. Next we note that the cell corresponding to absence for all d censuses is missing. Since our object is to estimate the number of individuals in the population who are not observed, and thus correspond to the missing cell, we proceed conditionally, restricting our attention to the 2^d incomplete contingency table which excludes the missing cell. We fit a model with the fewest possible parameters to this incomplete table, allowing for various dependencies among the censuses, in order that the variance of our estimate of the population size be as small as possible. Finally we use the model to predict the content of the unobservable cell. This approach can be conveniently illustrated using the two-sample situation, with which we begin in the next section.

6.2   The Two-Sample Capture-Recapture Problem

6.2.1   *Notation*

Let N be the total number of individuals in the population under consideration, x_{1+} the number of individuals in the first sample, x_{+1} the number in the second sample, and x_{11} the number in both the first and second samples. The number of individuals observed in the second sample but not the first is $x_{21} = x_{+1} - x_{11}$, and the number observed in the first sample but not the second is $x_{12} = x_{1+} - x_{11}$. We can arrange the data in a 2×2 contingency table with one missing cell:

Second Sample

First Sample	Present	Absent	
Present	x_{11}	x_{12}	x_{1+}
Absent	x_{21}	—	

x_{+1}

$$(6.2\text{-}1)$$

As in Chapter 5, we use the symbol "—" to mean that the count for the corresponding cell is missing.

The total number of individuals observed in the two samples is

$$n = x_{11} + x_{12} + x_{21}$$
$$= x_{1+} + x_{+1} - x_{11}. \tag{6.2-2}$$

The number of individuals not observed in either sample is thus $N - n$. We now derive the same estimate of N using two different approaches.

6.2.2 Basic approach

Let p_{11} be the probability of an individual being in both the first and second samples, p_{1+} the probability of being in the first sample, and p_{+1} the probability of being in the second sample. If the two samples are independent, i.e., $p_{11} = p_{1+}p_{+1}$, and if only n (the number of observed individuals) is fixed, then (x_{11}, x_{12}, x_{21}) has the multinomial distribution with probability function

$$\binom{n}{x_{11}, x_{12}, x_{21}} \frac{(p_{1+}p_{+1})^{x_{11}}[p_{1+}(1 - p_{+1})]^{x_{12}}[(1 - p_{1+})p_{+1}]^{x_{21}}}{[1 - (1 - p_{1+})(1 - p_{+1})]^n}. \tag{6.2-3}$$

In (6.2-3) we normalize the probabilities for the three observable cells by dividing each by $1 - (1 - p_{1+})(1 - p_{+1})$, the probability of being observed in at least one of the two samples. The maximum likelihood estimates of p_{1+} and p_{+1}, based on (6.2-3), are

$$\hat{p}_{1+} = \frac{x_{11}}{x_{+1}}, \qquad \hat{p}_{+1} = \frac{x_{11}}{x_{1+}}. \tag{6.2-4}$$

We note that the estimate for p_{1+} is based on information from the second sample, while the estimation of p_{+1} is based on information from the first sample.

If we view n as being binomially distributed, with sample size N and probability $1 - (1 - p_{1+})(1 - p_{+1})$, then the corresponding probability function is

$$\binom{N}{n}[1 - (1 - p_{1+})(1 - p_{+1})]^n[(1 - p_{1+})(1 - p_{+1})]^{N-n}. \tag{6.2-5}$$

Now we find the value of N which maximizes (6.2-5), given the values of p_{1+} and p_{+1}. This is the reverse of the usual binomial estimation problem, where we are given N and wish to estimate the probability of success.

In Chapter 13 we note that the value of N which maximizes (6.2-5) for given values of p_{1+} and p_{+1} is

$$\tilde{N} = \left[\frac{n}{1 - (1 - p_{1+})(1 - p_{+1})} \right], \tag{6.2-6}$$

where the notation $[x]$ in (6.2-6) indicates the greatest integer $\leq x$.

Finally, we put our estimates for p_{1+} and p_{+1} from (6.2-4) into (6.2-6). If we ignore the square brackets, we then get as our estimate of N (see exercise 1 in Section 6.2.4):

$$\hat{N} = \frac{x_{1+}x_{+1}}{x_{11}}. \tag{6.2-7}$$

This is a well-known estimator derived independently for the fixed-sample case (where x_{1+} and x_{+1} are fixed) by Petersen [1896], who was interested in the size of fish populations, and by Lincoln [1930], who was considering banding returns of wildfowl. The case considered above (where x_{1+} and x_{+1} are not fixed) was studied by Geiger and Werner [1924], whose population was the number of flashes on a zinc sulphide screen, and by Sekar and Deming [1949], who wished to assess the birth and death rates in a population using a registration list and a house-to-house canvass.

Sekar and Deming combined the two probability functions, (6.2-3) and (6.2-5), yielding

$$\frac{N!(p_{1+}p_{+1})^{x_{11}}[p_{1+}(1-p_{+1})]^{x_{12}}[(1-p_{1+})p_{+1}]^{x_{21}}[(1-p_{1+})(1-p_{+1})]^{N-n}}{(N-n)!x_{11}!x_{12}!x_{21}!}.$$

$$(6.2\text{-}8)$$

They then showed that (6.2-4) and (6.2-7) give the MLEs of N, p_{1+}, and p_{+1} based on this combined likelihood. They also derived the following estimate of the asymptotic variance of N using the δ method (see Chapter 14):

$$\widehat{\text{Var}}(\hat{N}) = \frac{x_{1+}x_{+1}x_{12}x_{21}}{x_{11}^3}. \qquad (6.2\text{-}9)$$

6.2.3 *Incomplete-table approach*

An equivalent way of viewing the estimation problem is more closely linked to the theory for incomplete tables. If the table of expected values corresponding to the observed table above is

m_{11}	m_{12}
m_{21}	—

$$(6.2\text{-}10)$$

then the MLEs for the m_{ij} under quasi independence are just the corresponding x_{ij}. We can extend this model to the missing cell by noting that its expected value, call it m_{22}^*, under independence of the two samples must satisfy

$$m_{22}^* = \frac{m_{12}m_{21}}{m_{11}}. \qquad (6.2\text{-}11)$$

Thus the MLE for the missing cell is

$$\hat{m}_{22}^* = \frac{\hat{m}_{12}\hat{m}_{21}}{\hat{m}_{11}} = \frac{x_{12}x_{21}}{x_{11}}. \qquad (6.2\text{-}12)$$

If we add this estimate for the missing cell to the total number n of observed individuals, we get as our estimate of N:

$$\hat{N} = x_{11} + x_{12} + x_{21} + \frac{x_{12}x_{21}}{x_{11}}$$

$$= \frac{x_{1+}x_{+1}}{x_{11}}. \qquad (6.2\text{-}13)$$

This is the same estimate as we derived above.

Yet another way to derive $\hat{N}$ is to equate x_{11} to the usual expected value under independence, i.e.,

$$x_{11} = \frac{x_{1+}x_{+1}}{N}. \tag{6.2-14}$$

Solving (6.2-14) for N gives the same estimate $\hat{N}$.

Example 6.2-1 Weevils in a box

Andrewartha [1961] reports data on the number of weevils *Calandra granaria* in a flat box of wheat. About 2,000 weevils were put in a flat box of wheat and allowed to disperse. A sample of size 498 was taken, the members of which were marked with a small spot of paint and replaced. A week later a random sample of 110 was taken, 20 of whose members were found to be marked. The estimate of N is

$$\hat{N} = \frac{498 \times 110}{20} = 2,793.$$

If there were about 2,000 weevils in the box, this example suggests that the variance we should attach to N is relatively large.

The estimate of the asymptotic variance of $\hat{N}$ given by (6.2-9) is

$$\frac{498 \times 110 \times 478 \times 90}{20^3} = 294,079.45,$$

so that an asymptotic 95 % confidence interval is

$$2,739 \pm 1,084.$$

This interval is very large, but it happily includes the known value of N, which was approximately 2,000. ■■

Example 6.2-2 Estimating birth and death rates in India

Sekar and Deming [1949] report the results of an inquiry conducted in February 1947 to estimate birth and death rates in the Singur Health Centre, an area near Calcutta in India. The data cover births and deaths in the years 1945 and 1946 separately.

The headman in each village periodically submits a list of births and deaths to a registrar. The registrar coordinates this information with a second report from each village and a list from the Maternity and Child Welfare Department. We refer to the resulting list as the "registrar's list of births and deaths" (R-list).

The All-India Institute of Hygiene and Public Health undertook a house-to-house canvass of the Singur Health Centre to prepare a list of all births and deaths which occurred in 1945 and 1946 (the I-list).

After removing the illegible, incomplete, and incorrect entries from the R-list, Sekar and Deming applied the two-sample techniques described above. We summarize these corrected data in table 6.2-1.

To illustrate the application of the two-sample techniques, we examine births

Table 6.2-1   Investigators' Report on Comparison of Lists of Singur Health Centre

	Year	Total (x_{1+})	Registrars' Lists		Interviewers' Lists
			Found in Interviewers' Lists (x_{11})	Not found in Interviewers' Lists (x_{12})	Extra in Interviewers' Lists (x_{21})
Births Listed (Excluding	1945	1,504	794	710	741
Nonresident Institutional)	1946	2,242	1,506	736	1,009
Deaths Listed (Excluding	1945	1,083	350	733	372
Nonresident Institutional)	1946	866	439	427	421

Source: Sekar and Deming [1949].

for 1945. We display the data in the form of a 2 × 2 table:

I-list

R-list	Present	Absent	
Present	794	710	1,504
Absent	741	—	

1,535

The estimate for the number of births in 1945 is (to the nearest integer)

$$\hat{N} = \frac{(1,535)(1,504)}{794} = 2,908,$$

and estimated asymptotic standard deviation of $\hat{N}$ is (to the nearest integer)

$$\text{SD} = \sqrt{\frac{1,535 \times 1,504 \times 741 \times 710}{794^3}} = 49.$$

Thus an asymptotic 95 % confidence interval places the number of births in 1945 between 2,810 and 3,006.

Exercise 2 in Section 6.2.4 deals with an analysis of the remaining data.

Sekar and Deming suggest that more accurate estimates might result from subdividing the data into small homogeneous groups. Then, by estimating the number of events for each group separately and adding these together, we can get a new estimate for the overall total. Since the more detailed data required for such stratification are not available, we refer the interested reader to Sekar and Deming [1949] for a discussion of this point. ■ ■

6.2.4   *Exercises*

1. Show that by substituting

$$\hat{p}_{1+} = \frac{x_{11}}{x_{+1}}, \qquad \hat{p}_{+1} = \frac{x_{11}}{x_{1+}}$$

in formula (6.2-6), we get the estimate of N given by (6.2-7).

2. (i) Using the data in table 6.2-1, estimate the total number of births for 1946
       and deaths for 1945 and 1946.
   (ii) Construct 99% asymptotic confidence intervals for your estimates in (i).
   (iii) The total population of the Singur Health Centre was approximately
        64,000 during the time of the study. Was there a change in the birth and
        death rates from 1945 to 1946?

6.3 Conditional Maximum Likelihood Estimation of N

In this section we give a theoretical justification of the two-stage approach for
estimating N illustrated in Section 6.2.

Consider a $(t + 1)$-cell multinomial random variable with cell probabilities $\{p_i\}$
for the first t cells and $p^* = 1 - \sum_{i=1}^{t} p_i$ for the $(t + 1)$st cell. We are interested in
the situation where we are not able to see the observed value for the $(t + 1)$st cell,
i.e., it is missing. Let the total sample for the $(t + 1)$-cell multinomial be N, and
the observed values for the first t cells be $x_i, i = 1, 2, \ldots, t$. Then the missing count
for the $(t + 1)$st cell is $N - n$, where $n = \sum_{i=1}^{t} x_i$.

6.3.1 *Breaking the likelihood into two parts*

The multinomial likelihood function, which involves N as an unknown parameter,
is

$$L(N; \theta) = L(N; \theta|\{x_i\})$$

$$= \frac{N!}{(N - n)! \prod_{i=1}^{t} x_i!} (p^*)^{N-n} \prod_{i=1}^{t} p_i^{x_i}, \tag{6.3-1}$$

where $p_i = p_i(\theta)$ and $p^* = p^*(\theta) = 1 - \sum_{i=1}^{t} p_i(\theta)$. We assume that the dimension
of the vector θ is at most t, so that we have a "reduced" number of parameters to
consider.

Then we can rewrite L in (6.3-1) as a product of two terms:

$$L(N; \theta) = L_1(N; p^*(\theta))L_2(\theta), \tag{6.3-2}$$

where

$$L_1(N; p^*(\theta)) = \binom{N}{n} p^{*N-n}(1 - p^*)^n, \tag{6.3-3}$$

$$L_2(\theta) = n! \prod_{i=1}^{t} \frac{Q_i(\theta)^{x_i}}{x_i!}, \tag{6.3-4}$$

with

$$Q_i(\theta) = \frac{p_i(\theta)}{1 - p^*(\theta)} = \frac{p_i(\theta)}{\sum_j p_j(\theta)}. \tag{6.3-5}$$

L_1 is a binomial likelihood function involving N and the missing individuals.
L_2 is a multinomial likelihood, giving the conditional distribution for the first
t cells.

6.3.2 *Unconditional estimation*

If we compute the *MLE*s of N and the $\{p_i\}$ simultaneously, we have to work
directly with (6.3-1), carrying out the maximization simultaneously. Let us denote
by $\hat{N}_U$ the *MLE* of N determined in this way, and by $\hat{\theta}_U$ the *MLE* of θ (U stands for

"unrestricted"). Although this simultaneous maximization is possible, the algebraic manipulations involved are quite messy. Thus we turn to a simpler approach.

6.3.3   Conditional estimation

A second approach is to compute MLEs for the components of the vector θ using the likelihood $L_2(\theta)$, say $\hat{\theta}_C$ (C stands for "conditional" and reminds us that we first estimated θ using the conditional likelihood (6.3-4)). Then we can estimate N by maximizing $L_1(N; \sum_i p_i(\hat{\theta}_C))$. This approach is much simpler and yields the estimator

$$\hat{N}_C = \left[\frac{n}{\sum\limits_i p_i(\hat{\theta}_C)} \right], \tag{6.3-6}$$

where the notation $[x]$ in (6.3-6) indicates the greatest integer $\leq x$. The first derivation of N for the two-sample census in Section 6.2.2 is based on this approach.

In general, $\hat{N}_C$ and $\hat{N}_U$ need not be equal, although there are special cases of our problem where they are. Sanathanan [1972a] has shown that

$$\text{(i)} \quad \hat{N}_U \leq \hat{N}_C, \tag{6.3-7}$$

and that (ii) under suitable regularity conditions, $(\hat{N}_C, \hat{\theta}_C)$ and $(\hat{N}_U, \hat{\theta}_U)$ are both consistent estimators of (N, θ). It follows from (ii) that $(\hat{N}_C, \hat{\theta}_C)$ and $(\hat{N}_U, \hat{\theta}_U)$ have the same asymptotic multivariate normal distribution. This result is of special importance because the techniques for the analysis of incomplete contingency tables, when we are dealing with log-linear models for the p_i, yield estimates of the form $\hat{\theta}_C$. Thus throughout this chapter we compute $\hat{N}_C$. Other authors have looked at some special cases of the models we consider here and have derived formulas for $\hat{N}_U$ which are identical to our formulas for $\hat{N}_C$ (see Section 6.5.3 for an example). Whether $\hat{N}_C$ and $\hat{N}_U$ are the same for all the parametric models considered here is an as-yet unanswered question.

In the capture-recapture literature, $\hat{N}_U$ and $\hat{N}_C$ are often used to refer to two estimators suggested by Chapman [1951]. The reader should note that the $\hat{N}_U$ and $\hat{N}_C$ of this chapter *do not* refer to the Chapman estimators.

6.4   The Three-Sample Census

6.4.1   Notation

We can arrange the data from a three-sample census in a 2^3 table with one missing cell:

	Third Sample			
	Present		Absent	
	Second Sample		Second Sample	
	Present	Absent	Present	Absent
Present	x_{111}	x_{121}	x_{112}	x_{122}
Absent	x_{211}	x_{221}	x_{212}	—

$$\tag{6.4-1}$$

When individuals are present in a sample, then the subscript on the $\{x_{ijk}\}$ corresponding to that sample is 1. When the individuals are absent in a sample, the corresponding subscript is 2. The (2, 2, 2) cell corresponds to absence in all three samples and contains a structural zero count.

If, as before, we let n be the total number of individuals observed,

$$n = x_{111} + x_{121} + x_{211} + x_{221} + x_{112} + x_{122} + x_{212}$$
$$= \sum{}^* x_{ijk}. \tag{6.4-2}$$

The asterisk reminds us that we do not include the (2, 2, 2) cell in the summation.

If we let m_{ijk} be the expected value for the number of individuals in the (i, j, k) cell of our 2^3 incomplete table, then

$$Q_{ijk} = \frac{m_{ijk}}{n} \tag{6.4-3}$$

is the probability of observing any one of our total sample of size n in the (i, j, k) cell. Thus $Q_{222} = 0$.

Now let p_{ijk} be the underlying probability corresponding to the (i, j, k) cell. For example, p_{111} is the probability of being in all three samples. The probability of being in none of the three samples is p_{222}, and we assume $p_{222} > 0$. Thus we have

$$Q_{ijk} = \frac{p_{ijk}}{1 - p_{222}}, \tag{6.4-4}$$

$$m_{ijk} = p_{ijk}\left(\frac{n}{1 - p_{222}}\right), \tag{6.4-5}$$

for $(i, j, k) \neq (2, 2, 2)$.

6.4.2   Models relating the three samples

As we only have seven cells with observed counts, we cannot fit models containing more than seven parameters. If we examine log-linear models for the expected counts $\{m_{ijk}\}$, we cannot measure a three-factor effect relating the three samples. We therefore assume that there is no three-factor effect.

The log-linear models available for our use and their interpretations are as follows:

1. the three samples are independent:

$$\log m_{ijk} = u + u_{1(i)} + u_{2(j)} + u_{3(k)}; \tag{6.4-6}$$

2. the third sample is independent of the first two:

$$\log m_{ijk} = u + u_{1(i)} + u_{2(j)} + u_{3(k)} + u_{12(ij)} \tag{6.4-7}$$

(there are three versions of this model);
3. two pairs of samples are related:

$$\log m_{ijk} = u + u_{1(i)} + u_{2(j)} + u_{3(k)} + u_{12(ij)} + u_{23(jk)} \tag{6.4-8}$$

(again, there are three versions of this model);

4. all pairwise relationships are present:

$$\log m_{ijk} = u + u_{1(i)} + u_{2(j)} + u_{3(k)} + u_{12(ij)} + u_{23(jk)} + u_{13(ik)}. \qquad (6.4\text{-}9)$$

In Section 6.1 we noted that interaction terms might reflect such effects as trap fascination, trap avoidance, social visibility, or social invisibility.

6.4.3   Choosing a model and estimating N

For the four different log-linear models, (6.4-6), (6.4-7), (6.4-8), and (6.4-9), the MLEs for the expected values as described in Chapter 5 are given by setting the expected values of the marginal totals corresponding to the highest-order u-terms equal to their observed values.

If we have the values $\hat{m}_{ijk}$, we can assess the goodness of fit of our model to the observed data using either

$$X^2 = \frac{\sum^* (x_{ijk} - \hat{m}_{ijk})^2}{\hat{m}_{ijk}} \qquad (6.4\text{-}10)$$

or

$$G^2 = 2 \sum^* x_{ijk} \log\left(\frac{x_{ijk}}{\hat{m}_{ijk}}\right). \qquad (6.4\text{-}11)$$

The associated degrees of freedom for models (6.4-6), (6.4-7), (6.4-8), and (6.4-9) are 3, 2, 1, and 0, respectively. Clearly, model (6.4-9) must fit the data perfectly. By using a model with fewer parameters than (6.4-9), we hope to get a more efficient estimate of N. If the reduced model is correct, we shall see that the asymptotic variance of our estimate of N will be reduced.

Having computed the conditional MLEs of the $\{m_{ijk}\}$, we convert back to the $\{p_{ijk}\}$ using (6.4-5) and then use (6.3-6) to get our estimate of N. This is equivalent to the following procedure. Let m^*_{222} be the number of individuals in the population not included in one of the three samples. Suppose we extend the model for the $\{m_{ijk}\}$ to cover m^*_{222}. Since the Bartlett criterion of no three-factor interaction is met for all four classes of models for the $\{m_{ijk}\}$,

$$m^*_{222} = \frac{m_{111}m_{221}m_{122}m_{212}}{m_{121}m_{211}m_{112}} = \frac{np_{111}p_{221}p_{122}p_{212}}{p_{121}p_{211}p_{112}(1 - p_{222})}$$

$$= \frac{np_{222}}{(1 - p_{222})} = \frac{n}{(1 - p_{222})} - n. \qquad (6.4\text{-}12)$$

Thus

$$\hat{m}^*_{222} = \hat{N}_C - n, \qquad (6.4\text{-}13)$$

$$\hat{N}_C = n + \hat{m}^*_{222}, \qquad (6.4\text{-}14)$$

where

$$\hat{m}^*_{222} = \frac{\hat{m}_{111}\hat{m}_{221}\hat{m}_{122}\hat{m}_{212}}{\hat{m}_{121}\hat{m}_{211}\hat{m}_{112}}. \qquad (6.4\text{-}15)$$

Expression (6.4-15) simplifies for those log-linear models that have common factors in the numerator and denominator.

6.4.4   *Estimates of N and their variances*

We now consider each of the four models for a three-sample census in turn, discussing for each the form of $\hat{N}_C$ and presenting its asymptotic variance. For all but the first model there exist closed-form estimates of the $\{m_{ijk}\}$ and N. Since the techniques required to derive the variances are special cases of those to be discussed in Section 6.5, we present the formulas here without the formal justification.

The three samples are independent

The maximum likelihood equations for the independent samples model (6.4-6), cannot be solved directly, and we suggest solving for the $\{\hat{m}_{ijk}\}$ using the Deming–Stephan iterative proportional fitting procedure described in Chapter 5. This procedure converges to the unique nonzero MLEs of the $\{m_{ijk}\}$ for $(i, j, k) \neq (2, 2, 2)$.

Darroch [1958], Das Gupta [1964], and Wittes [1970, 1974] have all examined the estimation of N for this model. Their approach involves solving the quadratic equation

$$(\hat{N}_C - x_{1++})(\hat{N}_C - x_{+1+})(\hat{N}_C - x_{++1}) = \hat{N}_C^2(\hat{N}_C - n), \qquad (6.4\text{-}16)$$

but does not yield estimates of the $\{m_{ijk}\}$. The iterative proportional fitting approach yields $\hat{N}_C$ and the $\{\hat{m}_{ijk}\}$. Since

$$\hat{p}_{1++} = \frac{x_{1++}}{\hat{N}_C},$$

$$\hat{p}_{+1+} = \frac{x_{+1+}}{\hat{N}_C},$$

$$\hat{p}_{++1} = \frac{x_{++1}}{\hat{N}_C}, \qquad (6.4\text{-}17)$$

and since expression (6.4-15) and the assumption of independence imply that $\hat{m}_{222}^* = \hat{N}_C(1 - \hat{p}_{1++})(1 - \hat{p}_{+1+})(1 - \hat{p}_{++1})$, we see that $\hat{N}_C$ computed using (6.4-16) is identical to $\hat{N}_C$ computed using (6.4-14) and (6.4-15).

No matter which way we choose for getting $\hat{N}_C$ and the $\{\hat{m}_{ijk}\}$, the estimate of the asymptotic variance of $\hat{N}_C$ is

$$\hat{V}(\hat{N}_C) = \frac{\hat{N}_C \hat{m}_{222}^*}{\hat{m}_{112} + \hat{m}_{121} + \hat{m}_{211} + \hat{m}_{111}}. \qquad (6.4\text{-}18)$$

The third sample is independent of the first two

For model (6.4-7), which has only one two-factor term corresponding to the first two samples, the MLEs of the $\{m_{ijk}\}$ have a simple closed form. We let $n' = n - x_{221}$ and $x'_{++1} = x_{++1} - x_{221}$, and the ML equations treat x_{221} as a fixed value and allocate the three sums $\{x_{11+}, x_{12+}, x_{21+}\}$ proportionally according to x'_{++1} and x_{++2}. Thus the estimated expected values are

$$\hat{m}_{221} = x_{221},$$

$$\hat{m}_{111} = \frac{x_{11+}x'_{++1}}{n'}, \qquad \hat{m}_{121} = \frac{x_{12+}x'_{++1}}{n'}, \qquad \hat{m}_{211} = \frac{x_{21+}x'_{++1}}{n'},$$

$$\hat{m}_{112} = \frac{x_{11+}x_{++2}}{n'}, \qquad \hat{m}_{122} = \frac{x_{12+}x_{++2}}{n'}, \qquad \hat{m}_{212} = \frac{x_{21+}x_{++2}}{n'}. \qquad (6.4\text{-}19)$$

Substituting these values in (6.4-15), we get

$$\hat{m}_{222}^* = \frac{x_{221}x_{++2}}{x'_{++1}} = \frac{x_{221}x_{++2}}{x_{++1} - x_{221}}, \tag{6.4-20}$$

and

$$\hat{N}_C = n + \frac{x_{221}x_{++2}}{x_{++1} - x_{221}}. \tag{6.4-21}$$

The estimate of the asymptotic variance of $\hat{N}_C$ is

$$\hat{V}(\hat{N}_C) = (\hat{m}_{222}^*)^2 \left(\frac{1}{(x_{++1} - x_{221})} + \frac{1}{x_{++2}} + \frac{1}{x_{221}} + \frac{1}{\hat{m}_{222}^*} \right). \tag{6.4-22}$$

Two pairs of samples are related
For model (6.4-8), with two two-factor terms (relating 1 and 2, and 2 and 3, respectively), the estimated expected values reduce to

$$\hat{m}_{221} = x_{221}, \qquad \hat{m}_{122} = x_{122}, \qquad \hat{m}_{121} = x_{121},$$

$$\hat{m}_{111} = \frac{x_{11+}x_{+11}}{x_{+1+}}, \qquad \hat{m}_{211} = \frac{x_{21+}x_{+11}}{x_{+1+}},$$

$$\hat{m}_{112} = \frac{x_{11+}x_{+12}}{x_{+1+}}, \qquad \hat{m}_{212} = \frac{x_{21+}x_{+12}}{x_{+1+}}. \tag{6.4-23}$$

Substituting the values from (6.4-23) into (6.4-15) we get

$$\hat{m}_{222}^* = \frac{x_{221}x_{122}}{x_{121}}, \tag{6.4-24}$$

and

$$\hat{N}_C = n + \frac{x_{221}x_{122}}{x_{121}}. \tag{6.4-25}$$

Our estimate of the asymptotic variance of $\hat{N}_C$ is

$$\hat{V}(\hat{N}_C) = (\hat{m}_{222}^*)^2 \left(\frac{1}{x_{221}} + \frac{1}{x_{122}} + \frac{1}{x_{121}} + \frac{x_{121}}{x_{221}x_{122}} \right). \tag{6.4-26}$$

We can rewrite (6.4-26) so that it resembles (6.2-9) more closely, i.e.,

$$\hat{V}(\hat{N}_C) = \frac{x_{221}x_{122}x_{12+}x_{+21}}{x_{121}^3}. \tag{6.4-27}$$

All pairwise relationships are present
For model (6.4-9) there are zero degrees of freedom, and the MLEs of the expected cell counts are the observed values. Thus

$$\hat{m}_{222}^* = \frac{x_{111}x_{221}x_{122}x_{212}}{x_{121}x_{211}x_{112}}, \tag{6.4-28}$$

and

$$\hat{N}_C = n + \frac{x_{111}x_{221}x_{122}x_{221}}{x_{121}x_{211}x_{112}}. \tag{6.4-29}$$

Our estimate of the asymptotic variance of $\hat{N}_C$ is

$$\hat{V}(\hat{N}_C) = (\hat{m}_{222}^*)^2 \left(\frac{1}{x_{111}} + \frac{1}{x_{121}} + \frac{1}{x_{211}} + \frac{1}{x_{221}} + \frac{1}{x_{112}} + \frac{1}{x_{122}} + \frac{1}{x_{212}} + \frac{1}{\hat{m}_{222}^*} \right).$$

(6.4-30)

6.4.5   The simpler the model, the smaller the variance

One of the claims we made in Section 6.1.2 was that the fewer the parameters in the model, the smaller the variance. We offer the following theorem in support of that claim. The proof is straightforward, and we leave it as an exercise.

THEOREM 6.4-1

(i) *Let $\{\hat{m}_{ijk}\}$ be any set of expected cell values, and let m_{222}^* be defined by (6.4-15). Furthermore, let $V_{[1,2,3]}$, $V_{[12,3]}$, $V_{[12,23]}$, and $V_{[12,23,13]}$ be the right-hand sides of variance expressions (6.4-18), (6.4-22), (6.4-26), and (6.4-30), respectively. Then*

$$V_{[12,23,13]} > V_{[12,23]} > V_{[12,3]}$$

when the same values of the $\hat{m}_{ijk}$ and $\hat{m}_{222}^*$ are used to compute the V's.

(ii) *If the $\hat{m}_{ijk}$ satisfy the independent samples model (6.4-6), then we also have*

$$V_{[12,3]} > V_{[1,2,3]}$$

when these values of the $\hat{m}_{ijk}$ are used to compute the V's.

6.4.6   Some examples

Example 6.4-1   Volunteer activity in Massachusetts

Smith [1973], in a study on volunteer activity in Massachusetts towns, attempted to determine the number of formal volunteer organizations (e.g., Girl Scouts, League of Women Voters, Red Cross, etc.) in each town using three different "sampling" techniques:

1.   an organized census using a variety of official and unofficial records;
2.   examination of classified (Yellow Pages) section of the telephone directory;
3.   examination of files, notices, and articles in a sampling of local newspapers.

Since formal volunteer organizations (FVOs) are by definition nonprofit organizations, the only problem in identifying true FVOs, whose members are mainly volunteers, is in distinguishing them from other types of nonprofit organizations (e.g., hospitals), where members are mainly paid employees. For a given town, any FVO may be detected or not by each of the three sampling techniques, and for each town we can estimate the total number of FVOs using the techniques of the three-sample census described above. In table 6.4-1 we present the data for five different towns in Smith's study.

Let us illustrate the techniques of the three-sample census using the data from town IV. Let "telephone" be sample 1, "census" sample 2, and "newspaper" sample 3. The model

$$u + u_{1(i)} + u_{2(j)} + u_{3(k)} + u_{12(ij)} + u_{13(ik)}$$

fits the data perfectly, since

$$\frac{x_{111}x_{122}}{x_{112}x_{121}} = \frac{4 \times 2}{1 \times 8} = 1.$$

Table 6.4-1   Data on Formal Volunteer Organizations in Massachusetts Towns

Town	Telephone	Newspaper	Yes	Yes	No	No
		Census	Yes	No	Yes	No
I	Yes		4	1	5	4
	No		62	35	233	—
II	Yes		15	1	12	5
	No		74	160	259	—
III	Yes		51	2	37	8
	No		139	319	237	—
IV	Yes		4	1	8	2
	No		16	49	113	—
V	Yes		24	0	21	8
	No		92	85	227	—

Source: Smith [1973]

The value of $\hat{m}_{222}^{*}$ under this model is 346.1, and

$$N_C = 193 + 346 = 539.$$

We can then use (6.4-26) to get the following estimate of the asymptotic variance of $\hat{N}_C$:

$$\hat{V}(\hat{N}_C) = 346^2 \left(\frac{1}{16} + \frac{1}{49} + \frac{1}{113} + \frac{1}{346} \right) = 346^2 \times 0.095.$$

Thus the estimated standard deviation is $346 \times \sqrt{0.095} = 106.6$, and an asymptotic 95% confidence interval is [331, 748].

It may be possible to shorten our interval for N by dropping one of the two-factor terms from the model and then reestimating N and the new asymptotic variance. Suppose we drop the term u_{13}. Computing the Pearsonian goodness-of-fit statistic, we get $X^2 = 5.11$ with two degrees of freedom, a value with a descriptive level of significance between 0.10 and 0.05. In this model the estimates of the $\{u_{12(ij)}\}$ indicate a positive interaction between the organized census and the listings in the telephone directory, and this probably reflects the social visibility of some FVOs. This might be a reasonable model, and our new estimate of m_{222}^{*} is $49 \times 123/21 = 287.0$, and

$$\hat{N}_C = 193 + 287 = 480.$$

The new asymptotic variance is

$$287^2 \left(\frac{1}{49} + \frac{1}{123} + \frac{1}{21} + \frac{1}{291} \right) = 287^2 \times 0.079,$$

so that an asymptotic 95% confidence interval for N is [322, 638]. This is still a rather large interval.

Dropping the other two-factor term yields fitted values for the incomplete table that in the Pearsonian goodness-of-fit statistic yield a value significant beyond the 0.001 level. Thus it seems reasonable to stick with our estimate of $\hat{N}_C = 480$ and our asymptotic 95% confidence interval of [322, 638].

Turning to town I and fitting the final model from above for town IV, we get $X^2 = 2.7$ with two degrees of freedom, so that the same model seems quite appropriate again. Here $\hat{m}^*_{222} = 35 \times 242/67 = 126.4$, and $\hat{N}_C = 126 + 342 = 468$, with

$$\hat{V}(\hat{N}_C) = 126^2 \left(\frac{1}{35} + \frac{1}{67} + \frac{1}{242} + \frac{1}{126} \right) = 126^2 \times 0.055.$$

Thus an asymptotic 95% confidence interval for N is $[410, 526]$.

The model

$$u + u_{1(i)} + u_{2(j)} + u_{3(k)} + u_{23(jk)}$$

also provides a good fit for the data from town I. The corresponding value of $\hat{m}^*_{222}$ is 132, with $\hat{N}_C = 474$ and

$$\hat{V}(\hat{N}_C) = 132^2 \left(\frac{1}{330} + \frac{1}{10} + \frac{1}{4} + \frac{1}{132} \right) = 132^2 \times 0.361.$$

The resulting asymptotic 95% confidence interval for N $[319, 629]$ is much larger than the one above. It also includes inadmissible values, since we already know that $N > 342$. The fact that the model does fit the data, however, suggests we look at the independent samples model

$$u + u_{1(i)} + u_{2(j)} + u_{3(k)}.$$

For the latter model we get fitted values of

1.4	0.8		5.0	6.8
63.5	36.7		233.1	—

and $X^2 = 6.6$ with three degrees of freedom. The model fits the data reasonably well. The value of $\hat{N}_C$ for this model is 478, and

$$\hat{V}(\hat{N}_C) = \frac{478 \times 136}{1.4 + 63.5 + 0.8 + 5.0} = 919.5.$$

Thus our final 95% asymptotic confidence interval is $[419, 537]$. Actually, we might choose to use the first interval computed above, $[410, 526]$.

Exercise 2 in Section 6.4.7 deals with the analysis of the data for the remaining three towns presented in table 6.4-1. ■■

Example 6.4-2   Live trapping of cottontail rabbits

At the Rose Lake Wildlife Research Station in Michigan, there was a special project to measure the size of the cottontail rabbit population. For several years the biologists at the station annually trapped rabbits for about three weeks before the hunting season. The trapped rabbits were uniquely marked and then returned to the trapping area. For the purposes of analysis the trapping period was divided in half, and we consider the two halves as separate samples. A final sample of the population was taken by gun, during the hunting season. Thus there are, in effect, three samples:

1. the first half of the live-trapping period;
2. the second half of the live-trapping period;
3. hunting recoveries.

In table 6.4-2 we summarize the data for four selected years from one of the areas under study (Orange Area—980 acres).

For the 1965 data, the independent samples model fits extremely well: $X^2 = 2.2$ or $G^2 = 2.1$ with three degrees of freedom. The estimate of the number of rabbits never seen is $\hat{m}^*_{222} = 241$, and $\hat{N}_C = 241 + 345 = 586$. The estimate of the asymptotic standard error of $\hat{N}_C$ is 45.0.

Table 6.4-2   Live Trapping of Cottontail Rabbits at Rose Lake Wildlife Research Station

	Live Trapping First Half	Present	Present	Absent	Absent
Year	Hunter's Bag Second Half	Present	Absent	Present	Present
1965	Present	8	26	32	189
	Absent	8	41	41	—
1967	Present	9	5	4	28
	Absent	30	17	22	—
1968	Present	9	3	4	41
	Absent	31	41	29	—
1970	Present	4	6	22	59
	Absent	15	17	32	—

Source: Gordon L. Zorb, unpublished data collected under Federal Aid in Wildlife Restoration Michigan Project W-118-R.

For the 1967 data, the independent samples model does not fit well at all: $X^2 = 28.7$ and $G^2 = 28.3$ with three degrees of freedom. The model which allows for the dependence of the two halves of the live-trapping period is

$$u + u_{1(i)} + u_{2(j)} + u_{3(k)} + u_{12(ij)}.$$

This model provides an extremely good fit to the data ($X^2 = 0.6$ and $G^2 = 0.6$ with two degrees of freedom). The estimates of the $\{u_{12(ij)}\}$ here indicate a positive interaction and probably reflect the effects of trapping. The estimate of the number of unseen rabbits under this model for 1967 is $\hat{m}^*_{222} = 107$, and $\hat{N}_C = 222$. Our estimate of the asymptotic standard error of $\hat{N}_C$ is 36.3.

We leave the analysis of the data for the remaining two years for exercise 3 in Section 6.4.7. ■ ■

6.4.7   *Exercises*

1. (i) Prove theorem 6.4-1.

    (ii) Verify part (i) of theorem 6.4-1 using two different sets of expected cell values from example 6.4-1.

2. Analyze the data for towns II, III, and V in table 6.4-1. In particular, for each set of data:

(a) find the simplest model providing an adequate fit, and interpret any interaction terms included in the model;

(b) estimate the total number of formal volunteer organizations;

(c) construct asymptotic 90% confidence intervals for your estimates in (b).

3. (i) Estimate the size of the cottontail rabbit population for 1968 and 1970 using the data in table 6.4-2.

(ii) Taking all four sets of data into account, are animals trapped in the first half of the trapping period more likely to be trapped again in the second half? Does there appear to be any relationship for the rabbits between being trapped and being shot?

4. Das Gupta reports on a model sampling experiment where the three samples are independent with probabilities of detection p_{1++}, p_{+1+}, and p_{++1} equal to 0.4, 0.5, and 0.6, respectively. The true value of N in this sampling experiment is 2,500, and the observed data are:

	Third Sample			
	Present		Absent	
	Second Sample		Second Sample	
First Sample	Present	Absent	Present	Absent
Present	283	297	200	221
Absent	465	435	277	—

(i) Fit the independent samples model to the data and estimate N.

(ii) Compute the estimated asymptotic variance of $\hat{N}_C$ in (i) and find the smallest asymptotic confidence interval containing the true value of N, 2,500.

6.5 The General Multiple Recapture Problem

6.5.1 *Basic framework*

The generalization of the methods described in the preceding section to $d > 3$ samples is straightforward. The data based on a full capture–recapture history can be summarized in the form of a 2^d incomplete cross-classification with one missing cell.

Let $m_{i_1 i_2 \dots i_d}$ be the expected number of individuals in the $(i_1, i_2, \dots, i_d)$ cell of this 2^d table, where i_j (for $j = 1, 2, \dots, d$) takes the value 1 if the individuals corresponding to this cell are present in the jth sample, and 2 if they are absent. Since we cannot observe those individuals absent in all of the d samples, we treat the $(2, 2, \dots, 2)$ cell as a structural zero, i.e., $m_{22 \dots 2} = 0$. Moreover, we assume $m_{i_1 i_2 \dots i_d} \neq 0$ provided $i_j \neq 2$ for some $j = 1, 2, \dots, d$. Then the set S of nonstructural zero cells has $2^d - 1$ elements.

Suppose now that the observed counts $x_{i_1 i_2 \dots i_d}$ for the cells in S have a multinomial distribution and that the total sample size is $n = \sum^* x_{i_1 i_2 \dots i_d}$ where the summation is over all cells in S. The probability associated with the $(i_1, i_2, \dots, i_d)$ cell is $n^{-1} m_{i_1 i_2 \dots i_d}$. We let N be the total number of individuals in the population.

Thus $N - n$ individuals are absent from all d samples.

The most general log-linear model for the cells in S is

$$\log m_{i_1 i_2 \ldots i_d} = u + u_{1(i_1)} + u_{2(i_2)} + \cdots + u_{d(i_d)}$$
$$+ \sum u_{\alpha\beta(i_\alpha i_\beta)} + \cdots + u_{1 2 \ldots d(i_1 i_2 \ldots i_d)}, \qquad (6.5\text{-}1)$$

where the summation of any individual u-term in (6.5-1) over any of its subscripts is zero. Since $(2, 2, \ldots, 2)$ is not in the set S, $u_{1 2 \ldots d(i_1 i_2 \ldots i_d)} = 0$ for all $(i_1, i_2, \ldots, i_d) \in S$. We consider various *unsaturated* log-linear models by setting other u-terms in (6.5-1) equal to zero. As usual, we restrict our attention to hierarchical models (see Chapters 3 and 5).

Darroch [1958], Das Gupta [1964], and Wittes [1970] have studied the problem where the d samples in the multiple recapture census are independent. This problem corresponds to the unsaturated log-linear model, given by

$$\log m_{i_1 i_2 \ldots i_d} = u + \sum_{\alpha=1}^{d} u_{\alpha(i_\alpha)}. \qquad (6.5\text{-}2)$$

Except for $d = 2$, there is no closed-form solution of the likelihood equations for this model.

A special case of the independent samples model (6.5-2) occurs when $u_1 = u_2 = \cdots = u_d$, i.e., the samples are independent *and* there is the same probability of capture associated with each sample. Methods related to those presented in Mantel [1951], Li and Mantel [1968], and Deming and Keyfitz [1965] are appropriate for this model, but we do not consider it here.

For any unsaturated log-linear model, the MLEs for the expected values are given by setting the expected values of the marginal totals, corresponding to the highest-order u-terms in the model, equal to their observed values (see Chapter 5). Once we have computed MLEs for the expected cell counts in the incomplete 2^d table, we can assess the appropriateness of our log-linear model using either of the usual chi square goodness-of-fit statistics.

6.5.2 Formulas for estimating N

For the general d-sample $(d > 3)$ census, once we have chosen a log-linear model for the 2^d incomplete table which provides a good fit to the data, we extend that model to cover the unobserved $(2, 2, \ldots, 2)$ cell, calling the extended expected value $m^*_{22\ldots2}$. The estimate of $m^*_{22\ldots2}$ can be found by writing

$$\hat{m}^*_{22\ldots2} = \hat{M}_{\text{odd}}/\hat{M}_{\text{even}}, \qquad (6.5\text{-}3)$$

where $\hat{M}_{\text{odd}}$ is the product of all the $\hat{m}_{i_1 i_2 \ldots i_d}$ from the incomplete table with $\sum_{j=1}^{d} i_j$ equal to an odd number and $\hat{M}_{\text{even}}$ is the corresponding product of those expected values with $\sum_{j=1}^{d} i_j$ equal to an even number. Expression (6.5-3) implies that there is no dth-order effect in a 2^d table and follows from the natural generalization of Bartlett's condition for $d = 3$. Finally, our estimate of the population size is

$$\hat{N}_C = n + \hat{m}_{22\ldots2}. \qquad (6.5\text{-}4)$$

We note that expression (6.5-3) yields our estimate of $m^*_{22\ldots2}$ no matter with which log-linear model we are working. When d is large and the log-linear model includes only a few interaction terms, the use of expression (6.5-3) may be numerically inefficient even though algebraically correct, due to cancellation of terms in the numerator and denominator.

6.5.3   Independent samples

For the independent samples model (6.5-1) and for general $d \geq 2$, $\hat{N}_C$ must clearly satisfy the relationship:

$$\frac{\hat{m}_{22...2}}{\hat{N}_C} = 1 - \frac{n}{\hat{N}_C} = \left(1 - \frac{x_{1+\,\cdots\,++}}{\hat{N}_C}\right)\left(1 - \frac{x_{+1+\,\cdots\,++}}{\hat{N}_C}\right)\cdots\left(1 - \frac{x_{++\,\cdots\,+1}}{\hat{N}_C}\right), \quad (6.5\text{-}5)$$

which reduces to the approximate equation for the traditional d-sample maximum likelihood estimate (see Cormack [1968] or Darroch [1958]), i.e.,

$$\prod_{i=1}^{d} (\hat{N}_U - n_i) = \hat{N}_U^{d-1}(\hat{N}_U - n), \quad (6.5\text{-}6)$$

where n_i is the number observed in the ith sample. Thus $\hat{N}_C = \hat{N}_U$ in this case.

6.5.4   Models with closed-form expressions for $\hat{N}_C$

There are d classes of hierarchical log-linear models having closed-form MLEs for the expected cell values and thus for $\hat{m}_{22...2}^*$. These classes of models require special definition. The ith class consists of models defined by setting equal to zero exactly i two-factor u-terms ($i = 0, \ldots, d-1$) with one dimension or variable in common (see exercises 1 and 2 in Section 6.5.8 for an elaboration of this point). Recall that if a two-factor term equals zero then so do all its higher-order relatives. For $i = 0$, the model is unrestricted, the expected values equal the observed values, and (6.5-3) applies directly. If exactly one two-factor term (as well as all of its higher-order relatives) equals zero, we get closed-form expected values analogous to (6.4-24). For $u_{12} = 0$,

$$\hat{m}_{22...2}^* = \frac{x_{122...2}x_{212...2}}{x_{112...2}}. \quad (6.5\text{-}7)$$

If exactly two two-factor terms involving a common dimension equal zero, we get closed-form expected values analogous to (6.4-20). For $u_{12} = u_{13} = 0$,

$$\hat{m}_{22...2}^* = \frac{x_{1222...2}x_{2++2...2}}{x_{1++2...2} - x_{1222...2}}. \quad (6.5\text{-}8)$$

If exactly I (less than d) two-factor terms involving a common dimension equal zero, e.g., $u_{12} = u_{13} = \cdots = u_{1,I+1} = 0$, then our estimate of $m_{22...2}$ again is of the form (6.5-8). We replace $x_{j++2...2}$ by the sufficient statistic $x_{j++\cdots+2...2}$, where there are I successive subscripts $+$ and the remaining subscripts are all 2.

6.5.5   Asymptotic variances for $\hat{N}_C$

In analyzing multiple recapture census data it is our aim to fit the incomplete 2^d table by a log-linear model with the fewest possible parameters. The reason for this is readily apparent. The fewer parameters in an "appropriate" model for estimating $m_{22...2}^*$, the smaller the variance of the estimate. Thus it is not good practice simply to use the unrestricted model. On the other hand, if we use a model with too few parameters, we introduce a bias into our estimate of population size that can possibly render our variance formulas meaningless.

There are at least three techniques which in theory can be used to compute asymptotic variances for our population size estimators: the δ method (propagation of error), inversion of the Fisher information matrix, and use of orthogonal projections. Inversion of the Fisher information matrix is practical only for the independent samples model (6.5-2). We can also derive the asymptotic variance for this model using the δ method. Thus we consider only the δ method and the method of orthogonal projections in this chapter.

Using the δ method

By an application of the δ method (see Chapter 14) we can show that

$$E\{(\hat{m}_{22...2}^* - m_{22...2}^*)^2|n\} = n\lambda_1 + \lambda_2 + o(1), \tag{6.5-9}$$

$$E\{(\hat{m}_{22...2}^* - m_{22...2}^*)|n\} = \mu_1 + \frac{\mu_2}{n} + O(n^{-2}), \tag{6.5-10}$$

where λ_1, λ_2, μ_1, and μ_2 are independent of n. Following Darroch [1958], we can use (6.5-9) and (6.5-10) to get an asymptotic variance for $\hat{N}_C = n + \hat{m}_{22...2}^*$, which is unconditional with respect to n, i.e.,

$$\text{Var}(\hat{N}_C) = \lambda_1 E(n) + \left(\frac{n + m_{22...2}^*}{n}\right)^2 \cdot \text{Var}(n) + o(1). \tag{6.5-11}$$

Since n is binomial with parameters $N = n + m_{22...2}^*$ and $p = \{n/(n + m_{22...2}^*)\}$, we get the following estimate of (6.5-11):

$$\frac{\hat{\lambda}_1 n + \hat{m}_{22...2}^* + (\hat{m}_{22...2}^*)^2}{n}. \tag{6.5-12}$$

To apply the δ method to determine λ_1 for any given log-linear model, we must express $\hat{m}_{22...2}$ as an explicit function of the observed cell proportions, or at least we must be able to write out its first partial derivatives with respect to these observed proportions. We can do this only for models where there are closed-form solutions for the MLEs of the expected cell entries, i.e., models such as those described above in Section 6.5.4.

For the unrestricted log-linear model, with $\hat{m}_{22...2}^*$ given directly by (6.5-3), our estimate of the asymptotic variance reduces to

$$(m_{22...2}^*)^2 \left\{ \sum_S x_{i_1 i_2...i_d}^{-1} + (\hat{m}_{22...2}^*)^{-1} \right\}. \tag{6.5-13}$$

The summation in (6.5-13) is over the set S of nonstructural zero cells.

For the log-linear model specified by setting exactly one two-factor term equal to zero (e.g., $u_{12} = 0$), $\hat{m}_{22...2}^*$ is given by (6.5-7), and our estimate for the asymptotic variance of $\hat{N}_C$ is

$$(\hat{m}_{22...2}^*)^2 \left(\frac{1}{x_{122...2}} + \frac{1}{x_{212...2}} + \frac{1}{x_{112...2}} + \frac{1}{\hat{m}_{22...2}^*} \right). \tag{6.5-14}$$

Finally, for the model specified by setting exactly two two-factor terms with a common dimension equal to zero, e.g., $u_{12} = u_{13} = 0$, $\hat{m}_{22...2}^*$ is given by (6.5-8), and our estimate for the asymptotic variance of $\hat{N}_C$ is

$$(\hat{m}_{22...2}^*)^2 \left(\frac{1}{x_{122...2}} + \frac{1}{x_{2++2...2}} + \frac{1}{x_{1++2...2} - x_{122...2}} + \frac{1}{\hat{m}_{22...2}^*} \right). \tag{6.5-15}$$

For the other models with closed-form estimates of $m^*_{22\ldots2}$ we get similar variance formulas.

Darroch [1958] also uses the δ method to compute λ_1 for the log-linear model corresponding to d independent samples. In our notation his estimate of the asymptotic variance of $\hat{N}_C$ is

$$\hat{N}_C \cdot \hat{m}^*_{22\ldots2}(\sum{}^+ \hat{m}_{i_1 i_2 \ldots i_d})^{-1}, \qquad (6.5\text{-}16)$$

where the summation is over all cells in S such that $\sum_{j=1}^d i_j < 2d - 1$, i.e., over all cells corresponding to individuals caught at least twice. Note that when $d = 2$, the independent samples model and the unrestricted model coincide, and expressions (6.5-13) and (6.5-16) are algebraically identical.

Using the method of orthogonal projections

The other technique for computing asymptotic variances is described in detail in Haberman [1974] and is based on a coordinate-free approach which expresses the asymptotic variance in terms of orthogonal projections with respect to various inner products. If we consider the vector of logarithms of the expected cell counts, $\boldsymbol{\mu} = \{\log m_{i_1 i_2 \ldots i_d}\}$, then for any given log-linear model, $\boldsymbol{\mu} \in M$, where M is a linear manifold contained in a $(2^d - 1)$-dimensional vector space and defined by the log-linear model we are considering. Using results from Haberman [1974], we can show that our estimate for the asymptotic variance of $\hat{N}_C$ can always be written as

$$(\hat{m}^*_{2\ldots2})^2 \left[\left(\sum_S \hat{m}^{-1}_{i_1 \ldots i_d} \right) + (\hat{m}^*_{2\ldots2})^{-1} - \left(\sum_S r^2_{i_1 \ldots i_d} \hat{m}_{i_1 \ldots i_d} \right) \right], \qquad (6.5\text{-}17)$$

where

$$\mathbf{r} = \{r_{i_1 \ldots i_d}\} = \left\{ \frac{c_{i_1 \ldots i_d}}{\hat{m}_{i_1 \ldots i_d}} - \mu_{i_1 \ldots i_d} \right\}, \qquad (6.5\text{-}18)$$

$\boldsymbol{\mu} \in M$ is chosen to minimize

$$\sum_S r^2_{i_1 \ldots i_d} \hat{m}_{i_1 \ldots i_d}, \qquad (6.5\text{-}19)$$

and $c_{i_1 \ldots i_d}$ is 1 or -1 depending on whether $\sum_{j=1}^d i_j$ is odd or even.

To demonstrate this minimization, we consider a particular log-linear model, i.e., the one which includes all two-factor terms but no higher-order ones. We wish to minimize

$$\sum_S r^2_{i_1 \ldots i_d} \hat{m}_{i_1 \ldots i_d} = \sum_S \left[\frac{c_{i_1 \ldots i_d}}{\hat{m}_{i_1 \ldots i_d}} - \sum_{j<h} v_{jh(i_j i_h)} \right]^2 \hat{m}_{i_1 \ldots i_d}, \qquad (6.5\text{-}20)$$

where $\mathbf{v} = \{\sum_{j<h} v_{jh(i_j i_h)}\}$ is a typical vector in M, the linear manifold defined by the model. The value of $\mathbf{v}$ which minimizes (6.5-20) is given by the $d(d-1)/2$ sets of normal equations of the form

$$\sum{}^* r_{i_1 \ldots i_d} \hat{m}_{i_1 \ldots i_d} = \sum{}^* \left[\frac{c_{i_1 \ldots i_d}}{\hat{m}_{i_1 \ldots i_d}} - \sum_{j<h} v_{jh(i_j i_h)} \hat{m}_{i_1 \ldots i_d} \right] = 0, \qquad (6.5\text{-}21)$$

where the summation is over all cells in S with *one pair* of fixed subscripts. Were we considering the independent samples model there would be only d sets of normal equations.

We solve the normal equations for the $r_{i_1...i_d}$ using the method of cyclic descent (see Zangwill [1969]). Let our initial values be

$$r^{(0)}_{i_1...i_d} = \frac{c_{i_1...i_d}}{\hat{m}_{i_1...i_d}}. \qquad (6.5\text{-}22)$$

Each cycle of the iterative procedure consists of $d(d-1)/2$ steps, each step corresponding to a particular pair of subscripts. Suppose we have completed l steps and we have come to the equations corresponding to the sth and tth subscripts. Then we let

$$r^{(l+1)}_{i_1...i_d} = r^{(l)}_{i_1...i_d} - \delta^{(l+1)}_{i_s i_t}, \qquad (6.5\text{-}23)$$

where

$$\delta^{(l+1)}_{i_s i_t} = \frac{\sum^* r^{(l)}_{i_1...i_d} \hat{m}_{i_1...i_d}}{\sum^* \hat{m}_{i_1...i_d}} \qquad (6.5\text{-}24)$$

and the summation in (6.5-24) is over all cells in S with the same values (i_s, i_t) for the sth and tth subscripts. The method of cyclic descent as described here always converges (see Zangwill [1969]), and in our applications, we have found that five or six cycles yield values accurate to at least three decimal places.

6.5.6   *Bounds for asymptotic variances*

For many sets of data, relatively tight bounds for the estimate of the asymptotic variance are sufficient for most purposes. One way to get bounds on the variance is to generalize theorem 6.4-1. The generalization follows because we can interpret the second part of (6.5-17) in terms of orthogonal projections with respect to some inner product. As a result, if we use the MLEs for a given model in an asymptotic variance formula for a more general model, we get an upper bound on our estimate of the asymptotic variance for the model in question. For example, if we use MLEs for the model $u_{13} = u_{123} = 0$ in the variance formula for the model $u_{123} = 0$, i.e., in expression (6.4-30), we get an estimate which is larger than the asymptotic variance we get by using expression (6.4-26). Similarly, if we use the MLEs for a given model in an asymptotic variance formula for a model which is a special case of the given model, we get a lower bound. For example, if we use the MLEs for $u_{13} = u_{123} = 0$ in the variance formula for $u_{13} = u_{23} = u_{123} = 0$, i.e., in expression (6.4-22), we get an estimate which is smaller than the asymptotic variance we get by using (6.4-26).

For any given model there may be several upper and lower bounds based on closed-form asymptotic variance formulas, and it may be possible to get relatively tight bounds for our estimate of the asymptotic variance without having to use our iterative method.

To illustrate this point, we consider the following three log-linear models:

1. $u + u_{1(i_1)} + u_{2(i_2)} + \cdots + u_{k(i_k)}$;
2. $u + u_{1(i_1)} + u_{2(i_2)} + \cdots + u_{k(i_k)} + u_{12(i_1 i_2)}$;
3.   the model defined by setting $u_{13(i_1 i_3)} = 0$ for all (i_1, i_3).

Note that model 1 is a special case of model 2, which is in turn a special case of model 3. Let $\{m_{i_1...i_d}\}$ be the expected values under model 2 and $\hat{m}^*_{2...2}$ the corresponding estimate of $m^*_{2...2}$. Then by interpreting the last part of expression

(6.5-17) in terms of orthogonal projections with respect to some inner product, we can show that the estimate of the asymptotic variance for model 2 is bounded above by

$$(\hat{m}_{2\ldots2}^*)^2 \left[\frac{1}{\hat{m}_{1212\ldots2}} + \frac{1}{\hat{m}_{1222\ldots2}} + \frac{1}{\hat{m}_{2212\ldots2}} + \frac{1}{\hat{m}_{2\ldots2}^*} \right] \qquad (6.5\text{-}25)$$

and below by expression (6.5-16), where all the estimates are those computed under model 2. In this example there are actually $(d(d-1)/2) - 1$ upper bounds, each of the form described above and each corresponding to a model defined by setting exactly one two-factor term other than u_{12} equal to zero. There are also other upper bounds corresponding to models defined by setting two two-factor terms (with one variable in common) equal to zero. We use the minimum of the bounds as our upper bound.

A further illustration of the bounding of the asymptotic variance (without the use of the cyclic descent iterative method) is given in the following example.

6.5.7   A five-sample example

Example 6.5-1   Children with a congenital anomaly

Wittes [1970] estimated the number of children possessing a specific congenital anomaly in Massachusetts based on an extensive set of data. Table 6.5-1 contains a subset of these data for children born January 1, 1955, through December 31, 1959, still alive on December 31, 1966. There are five samples (i.e., $d = 5$) corresponding to five different sources of notification of a positive diagnosis: obstetric records, other hospital records, Massachusetts Department of Health, Massachusetts Department of Mental Health, and schools. Henceforth we refer to these five samplings as II, I, V, IV, and III, respectively.

The first step in the analysis of the data is the fitting of log-linear models to the incomplete 2^5 table of counts. For the independent samples model

$$u + u_{1(i_1)} + u_{2(i_2)} + u_{3(i_3)} + u_{4(i_4)} + u_{5(i_5)}, \qquad (6.5\text{-}26)$$

the log-likelihood ratio chi square statistic is $G^2 = 95.5$ with 25 degrees of freedom. This model provides a poor fit to the data.

We then fit each of the ten models with exactly one two-factor term and no higher-order terms, noting which ones give a significant improvement over the independent samples model based on the conditional likelihood ratio test. This leads us to the following model, for which the expected values are given in table 6.5-1.

$$u + u_{1(i_1)} + u_{2(i_2)} + u_{3(i_3)} + u_{4(i_4)} + u_{5(i_5)}$$
$$+ u_{12(i_1 i_2)} + u_{13(i_1 i_3)} + u_{24(i_2 i_4)} + u_{45(i_4 i_5)}. \qquad (6.5\text{-}27)$$

The corresponding value of G^2 is 25.8 with 21 degrees of freedom. This model provides a rather good fit, with a descriptive level of significance between 0.3 and 0.2. We can also look at each of the four models derived from (6.5-27) by dropping exactly one two-factor term. The likelihood ratio statistics for comparing each of these models to (6.5-27), given that (6.5-27) is appropriate for the data, are all highly significant. We proceed, using (6.5-27) as the true model for the data in table 6.5-1.

Table 6.5-1 Children in Massachusetts Possessing a Specific Congenital Anomaly

			I	Yes	Yes	No	No
			II	Yes	No	Yes	No
III	IV	V					
Yes	Yes	Yes		2 / 1.30	0 / 0.50	3 / 1.33	0 / 0.94
Yes	Yes	No		8 / 9.05	23 / 16.96	5 / 9.27	30 / 31.65
Yes	No	Yes		2 / 2.55	0 / 0.99	5 / 3.90	3 / 2.75
Yes	No	No		18 / 17.75	34 / 33.28	36 / 27.14	83 / 92.65
No	Yes	Yes		5 / 3.52	3 / 1.36	1 / 3.60	2 / 2.54
No	Yes	No		25 / 24.46	37 / 45.86	22 / 25.07	97 / 85.59
No	No	Yes		1 / 2.69	1 / 1.04	4 / 4.11	4 / 2.89
No	No	No		19 / 18.69	37 / 35.03	27 / 28.57	—

Source: Wittes [1970].

Note: I, II, III, IV, V correspond to five different methods of ascertainment. The first entry in each cell is the observed value, the second is the expected value under model (6.5-27).

Under (6.5-27), $\hat{m}^*_{22222} = 97.5$, and our estimate of the total population size is $\hat{N}_C = 634.5$, or 634, rounding down to the nearest integer. It is interesting to note that $\hat{N}_C$ is quite close to the estimate under the independent samples model, which was 639.

We can compute an estimate for the asymptotic variance of $\hat{N}_C$ using the method of cyclic descent as described above. As initial values we take

$$r^{(0)}_{i_1\ldots i_5} = \frac{c_{i_1\ldots i_5}}{\hat{m}_{i_1\ldots i_5}}. \tag{6.5-28}$$

The lth step ($l \geq 0$) of the iteration consists of four steps:

$$r^{(4l+1)}_{i_1\ldots i_5} = r^{(4l)}_{i_1\ldots i_5} - \frac{\sum^{i_1 i_2} r^{(4l)}_{i_1\ldots i_5}\hat{m}_{i_1\ldots i_5}}{\sum^{i_1 i_2}\hat{m}_{i_1\ldots i_5}}, \tag{6.5-29}$$

$$r^{(4l+2)}_{i_1\ldots i_5} = r^{(4l+1)}_{i_1\ldots i_5} - \frac{\sum^{i_1 i_3} r^{(4l+1)}_{i_1\ldots i_5}\hat{m}_{i_1\ldots i_5}}{\sum^{i_1 i_3}\hat{m}_{i_1\ldots i_5}}, \tag{6.5-30}$$

$$r^{(4l+3)}_{i_1\ldots i_5} = r^{(4l+2)}_{i_1\ldots i_5} - \frac{\sum^{i_2 i_4} r^{(4l+2)}_{i_1\ldots i_5}\hat{m}_{i_1\ldots i_5}}{\sum^{i_2 i_4}\hat{m}_{i_1\ldots i_5}}, \tag{6.5-31}$$

$$r^{(4(l+1))}_{i_1\ldots i_5} = r^{(4l+3)}_{i_1\ldots i_5} - \frac{\sum^{i_4 i_5} r^{(4l+3)}_{i_1\ldots i_5}\hat{m}_{i_1\ldots i_5}}{\sum^{i_4 i_5}\hat{m}_{i_1\ldots i_5}}. \tag{6.5-32}$$

In (6.5-29) through (6.5-32), $\sum^{i_\alpha i_\beta}$ is the summation over all cells in S with the αth and βth subscripts fixed. After six cycles of the iteration, the maximum deviation between successive values of $r_{i_1...i_5}$ is 0.5×10^{-5}. The resulting estimate of the asymptotic variance is 333.0, yielding a standard deviation of 18.25. Thus a 95% asymptotic confidence interval for N, the total population size, is $[598, 670]$. If model (6.5-26) is appropriate, the asymptotic variance we can compute using expression (6.5-16) for the independent samples model is wrong, and the corresponding 95% asymptotic confidence interval is almost 20% too short, as well as being improperly centered.

Instead of using the cyclic descent iterative procedure, we might try to get bounds for the asymptotic variance or standard deviation. Expression (6.5-16) provides a lower bound of 14.5 on the standard deviation, and the minimum of the upper bounds, 18.4, is provided by the closed-form variance expression associated with the model $u_{14} = u_{15} = 0$, which clearly includes (6.5-27) as a special case. These bounds are relatively tight (especially the upper one), and they allow us to make reasonable inferences regarding the parameter N. ∎∎

6.5.8   *Exercises*

1. Consider a 2^4 table where $m_{2222} = 0$ a priori. Show that the expected values for the following models are expressible in closed form: (i) $u_{12} = 0$; (ii) $u_{12} = u_{13} = 0$; (iii) $u_{12} = u_{13} = u_{14} = 0$. (Hint: For each model one or more of the cells is such that the observed value equals the expected value. If we treat these cells as fixed, then closed-form expressions for the remaining cells are found by analogy with corresponding complete tables.)

2. Consider a 2^d table $(d \geq 3)$ where $m_{22...2} = 0$ a priori. For the log-linear models defined by

   (1) $u_{12} = 0,$
   (2) $u_{12} = u_{13} = 0,$
   $\vdots$
   $(d-1)$ $u_{12} = u_{13} = \cdots = u_{1d} = 0,$

   determine which cells are such that the observed value equals the expected. Show that closed-form estimates are available for these models.

6.6   Discussion

Although the methods described in this chapter lead to a population size estimator based on a model which closely fits the observed data (in the form of a 2^d incomplete table), the estimator can be biased, since we are assuming that the model which describes the observed data also describes the count of the unobserved individuals. We have no way to check this assumption. This situation is analogous to (and has the same dangers as) fitting an arbitrary curve to a series of points (x,y) where $x > 0$, with the intention of estimating y at $x = 0$. Perhaps worse than fitting an arbitrary curve, however, is arbitrarily making the curve a straight line in light of evidence to the contrary, and then estimating the y intercept. The latter is analogous to automatically assuming that the d samples in our problem are independent.

A second, yet related, assumption is that the very same set of 2^d probabilities applies to each individual in the population. In the animal-trapping versions of the multiple-recapture census this assumption is unlikely to be realistic. For example, we would have to consider the possibility that there are many individuals who are extremely wary of all types of traps, where a different type of trap is used for each sampling. Mantel [private communication] has posed the extreme example of a heterogeneous population where for every individual in the population, at least one of the trapping procedures exhibits a fatal fascination. Then every individual would be trapped at least once, but as long as the fatally fascinating traps differ among individuals, we get a nonzero estimate of the number of untrapped individuals.

One way to avoid problems associated with heterogeneous populations is to take advantage of strata of various sorts. If we are fortunate, the strata allow us to divide the population into relatively homogeneous subgroups. In example 6.5-1, the population is stratified by age, and table 6.5-1 contains data for only one stratum. If there are several strata we can treat each separately, projecting a population size for each one independent of the other strata, or we can fit a log-linear model to the set of 2^d incomplete tables which takes into account a stratum effect, as well as possible interactions between strata and the d samplings. In this second approach, once we have an appropriate log-linear model, we can get an estimate of the population size by using (6.5-3) and (6.5-4) for each stratum. Asymptotic variances for each stratum can be computed using the methods described in Section 6.5, but since the population estimates for the strata are not independent, asymptotic covariances must also be computed.

The problem of avoidance learning by individuals who have been trapped (or even those who have not been trapped) can be dealt with by inclusion of appropriate interaction terms in the log-linear model for the 2^d incomplete table. For example, if the effect of being trapped in sample i is in evidence during the collection of sample $i + 1$ but has worn off by the time sample $i + 2$ is gathered, a sensible log-linear model might include only those two-factor terms of the form $u_{i,i+1}$ (for $i = 1, \ldots, d - 1$) and no higher-order terms. Thus we can make an a priori choice of a log-linear model with particular dependencies between samples, but we still have the alternative of seeing whether or not the model fits the observed data well. This type of approach to model-building is only helpful when the samples are taken sequentially in time; however, for the more general situation where we simply have a set of lists, we can still postulate dependencies among lists that are suggested by the problem itself.

The d-sample estimators of N for closed populations, which assume independent samples, do not require the full data on absence and presence in all d samples for each individual caught at least once. For censuses carried out sequentially, we typically need to know how many in the current census were observed in some previous one. Most biological investigators have used these independence estimators exclusively, and as a result have neglected to record the full capture-recapture history for individuals. Thus we have been unable to apply our approach to most of the examples of multiple-recapture data reported in the statistical ecology literature. We hope that the existence of a technique which allows for incorporation of dependencies among samples will lead to new data-recording procedures in multiple-recapture situations.

Sanathanan [1972b, 1973] has considered a different class of models for the d-sample situation which do take advantage of the full information regarding absence or presence in all samples. Her models are related to the latent structure models of Lazarsfeld and have been applied to data from visual scanning experiments in particle physics. There the population is the number of particles on a photograph. The photograph is scanned by several individuals, making up the samples. Interest is focused on the number of undetected particles on the scanning record.

7 Models for Measuring Change

7.1 Introduction

7.1.1 *Contingency tables and Markov models*

Some contingency tables can be viewed as displaying changes in state from one period of time to another, or, more properly, some data sets which display changes in state from one period of time to another can be viewed quite profitably as contingency tables. This is especially true when we use Markov chains to model data where each individual or object in the sample is classified according to a given set of categories at successive points in time.

In this chapter we describe some probability models based on Markov chains and discuss the application of these models to time sequence data of the sort described above. For most of the analyses we discuss we can display the Markov chain data in contingency table form; the analyses of these data are formally equivalent to certain contingency table analyses based on log-linear models. For the analyses we consider, we first present the statistical theory for the Markov chain probability model, then translate these analyses into contingency table notation.

For many sets of time sequence data, the Markov chain analysis is only the first step in the modeling process. As such, the use of Markov chain models helps us to find a convenient way to summarize the data in contingency table form. Once the data are in this form, we turn to other statistical techniques, such as those described in Chapter 8, to summarize or interpret the Markov chain models further.

7.1.2 *Beyond fitting a Markov model*

To bring out some of the salient points in the analysis of time sequence data in contingency table form, we consider a hypothetical example displaying the reported intentions to vote among a panel of 641 adults in two successive polls, one taken in June and the other in August:

June voting intention	August voting intention		Totals	
	Democrat	Republican		
Democrat	142	38	180	(7.1-1)
Republican	32	429	461	
Totals	174	467	641	

For the analysis of the data in (7.1-1), several avenues are open to us. The one we regard as least promising is the usual test for no association or no interaction in a 2×2 table. In a problem of the sort described above, we expect large numbers of individuals to stay with their original category unless some very strange political event has occurred, or unless we are dealing with a special population perhaps suspected of being "changers." Consequently, in carrying out the usual test on such data, we merely compute a large value of chi square and thus verify what we already knew to be true. Exceptions occur in at least two different circumstances. First, we may have several independent 2×2 tables measuring changes of different sorts, in which case we might wish to compare normalized chi square values. Second, in $I \times I$ tables measuring change, examining the residuals from the expected values, which led to the large chi square for the model of independence, often suggests an improved model for the data. This is the way several of the social mobility models, discussed in Chapters 5 and 9, evolved.

A more promising approach to such data raises the question of whether there are an equal number of changers from Democrat to Republican as from Republican to Democrat. This approach arises most simply from the multinomial with population probabilities

| | Second poll | | | |
First poll	D	R		
D	p_{11}	p_{12}	p_{1+}	(7.1-2)
R	p_{21}	p_{22}	p_{2+}	
Totals	p_{+1}	p_{+2}	1	

and table of observed entries

| | Second poll | | | |
First poll	D	R		
D	x_{11}	x_{12}	x_{1+}	(7.1-3)
R	x_{21}	x_{22}	x_{2+}	
Totals	x_{+1}	x_{+2}	N	

corresponding to the example in (7.1-1). The null hypothesis is that $p_{12} = p_{21}$, i.e., the probability of changing from D to R is the same as the probability of going from R to D. This is equivalent in 2×2 tables to the paired test of marginal equality, i.e., $p_{i+} = p_{+i}$, $i = 1, 2$. We can test this hypothesis by using the exact test for the binomial distribution with $p = 1/2$ or by using the chi square approximation with correction (McNemar's test):

$$X^2 = \frac{(|x_{12} - x_{21}| - 1)^2}{x_{12} + x_{21}} = \frac{(|38 - 32| - 1)^2}{38 + 32} = 0.36. \tag{7.1-4}$$

This is quite a small value and suggests that about the same proportion of voters switch from Democrat to Republican as from Republican to Democrat. We discuss estimation and tests for the models of marginal homogeneity and symmetry in further detail in Chapter 8.

The approach we have just described does not necessarily represent the underlying social process in quite the way a political scientist might like when he studies change. He might prefer to think of the outcome of the first poll as fixed and ask about the percentage changing from the first to the second poll. He would reorganize the data as follows:

| | August | | |
June	Same	Different	Totals
Democrat	142	38	180
Republican	429	32	461
Totals	571	70	641

(7.1-5)

which in generic form would be

| | Second poll | | |
First poll	Same	Different	Totals
D	x_{11}	x_{12}	x_{1+}
R	x_{22}	x_{21}	x_{2+}
Totals	$x_{11} + x_{22}$	$x_{12} + x_{21}$	N

(7.1-6)

He would notice that among Democrats as well as among Republicans, 10.9% changed. He could thus regard the difference of 6 ($= 38 - 32$) as being accounted for by the difference in initial sizes of the two political groups. If the two proportions of changers were different, he could apply a test for no association or interaction for (7.1-6). In fact, the observed proportion of Democratic changers is 0.21, while the observed proportion of Republican changers is only 0.07. Testing the revised table (7.1-5) for independence, we get

$$X^2 = \frac{641(|142 \times 32 - 429 \times 38| - 320.5)^2}{180 \times 461 \times 70 \times 571} = 25.3.$$

This is a highly significant value of chi square and suggests that in other communities where the proportion of Democratic voters is somewhat higher, we might expect to find more change from Democrat to Republican than vice versa.

Referring back to (7.1-2), we see that the null hypothesis for this test of no interaction in (7.1-6) can be written as

$$\frac{p_{12}}{p_{1+}} = \frac{p_{21}}{p_{2+}}, \tag{7.1-7}$$

or as

$$p_{11}p_{21} = p_{12}p_{22}. \tag{7.1-8}$$

Thus it is a test of relative symmetry in the process of change as opposed to the test of exact symmetry in a 2×2 table, ($p_{12} = p_{21}$) described above.

If we reparametrize the cell probabilities p_{ij} in log-linear model form, i.e.,

$$\log p_{ij} = u + u_{1(i)} + u_{2(j)} + u_{12(ij)}, \tag{7.1-9}$$

where

$$\sum_{i=1}^{2} u_{1(i)} = \sum_{j=1}^{2} u_{2(j)} = \sum_{i=1}^{2} u_{12(ij)} = \sum_{j=1}^{2} u_{12(ij)} = 0, \tag{7.1-10}$$

then the model of exact symmetry, i.e., $p_{12} = p_{21}$, is defined by

$$u_{1(i)} = u_{2(i)} \qquad i = 1, 2, \tag{7.1-11}$$

whereas the model of relative symmetry, i.e., $p_{11}p_{21} = p_{12}p_{22}$, is defined by

$$u_{2(i)} = 0 \qquad i = 1, 2. \tag{7.1-12}$$

The test of relative symmetry which we have just described does not necessarily assume that row totals are fixed. Thus if only the sample size N is fixed, we might be interested in examining both models (7.1-11) and (7.1-12). On the other hand, if the row totals are fixed, then it rarely makes sense to talk about exact symmetry, and we should focus interest on the model of relative symmetry.

Were we able to carry out a panel study going backward in time, then we could think of the column totals as being fixed, and we could consider a different model of relative symmetry, given by

$$\frac{p_{12}}{p_{+2}} = \frac{p_{21}}{p_{+1}} \quad \text{or} \quad p_{11}p_{12} = p_{21}p_{22}, \tag{7.1-13}$$

which in log-linear form is

$$u_{1(i)} = 0 \qquad i = 1, 2. \tag{7.1-14}$$

This second model of relative symmetry is of interest when the categories for the rows and columns of a contingency table are the same but the two variables occur simultaneously in time, such as in studies of matched pairs of individuals given different treatments. In Chapter 2 we gave a more detailed discussion of the interpretation of log-linear model parameters using the "same-different" approach of expression (7.1-6).

7.1.3   Markov models for aggregate data

Most of the results in the statistical analysis of Markov chain models described in this chapter can be found in a somewhat different form in Anderson [1954], Bush and Mosteller [1955], Anderson and Goodman [1957], and Madansky [1963]. Throughout our presentation we assume that we know the state or category for each member of our sample at every point in time, and we use the observed transitions to fit a Markov chain model.

Often the only information that we have from time-ordered data is the proportion of individuals in each state at every point in time. We refer the reader interested in the analysis of such aggregate data to Miller [1952], Goodman [1953], Madansky [1959], and Lee, Judge, and Zellner [1970].

7.2 First-Order Markov Models

7.2.1 *The basic model*

Measuring change as we have just done suggests that it is convenient to redefine the cell probabilities so that they sum to 1 across rows. Therefore, we set

$$P_{DD} = \frac{p_{11}}{p_{1+}}, \qquad P_{DR} = \frac{p_{12}}{p_{1+}},$$

$$\tag{7.2-1}$$

$$P_{RD} = \frac{p_{21}}{p_{2+}}, \qquad P_{RR} = \frac{p_{22}}{p_{2+}}.$$

The first subscript names the initial state, the second subscript the second state. Then the table in (7.1-3) becomes

| | Second poll | | |
First poll	D	R	Totals
D	P_{DD}	P_{DR}	1
R	P_{RD}	P_{RR}	1

$$\tag{7.2-2}$$

Expression (7.2-2) is a matrix of *transition probabilities*, showing the probabilities of making the transition from one state, D or R, to the other, R or D, or to the same state, D or R, respectively.

7.2.2 *Approach to stationarity*

Now suppose that the pattern of change persists over time. It is natural to ask how the process will end after a long period of time.

To get the expected proportions in the population at the $n + 1$st trial, we need to know the proportions in states D and R for the nth poll. Let D_n be the proportion of people Democrat in the nth poll and R_n the proportion Republican, where $D_n + R_n = 1$. Then at trial $n + 1$ the expected proportions are

$$D_{n+1} = D_n P_{DD} + R_n P_{RD}, \qquad R_{n+1} = D_n P_{DR} + R_n P_{RR}. \tag{7.2-3}$$

The first equation computes the proportion of Democrats in poll $n + 1$, composed of the fraction $D_n P_{DD}$ initially Democratic who stayed Democratic, plus those $R_n P_{RD}$ initially Republican who changed. The overall process under a steady state has a fixed transition matrix $\{P_{ij}\}$, where $i, j = D, R$, and it needs an initial probability vector to get it going, say (D_1, R_1).

What happens to the process represented by equation (7.2-3) after a long time? Rewriting (7.2-3), we have

$$D_{n+1} = D_n(1 - P_{DR}) + R_n P_{RD} = P_{RD} + (1 - P_{DR} - P_{RD})D_n. \tag{7.2-4}$$

Then, assuming $P_{RD} + P_{DR} \neq 0$ and letting $r = 1 - P_{DR} - P_{RD}$, we have

$$D_2 = P_{RD} + rD_1,$$

$$D_3 = P_{RD} + rD_2$$

$$= P_{RD} + r(P_{RD} + rD_1)$$

$$= P_{RD} + rP_{RD} + r^2D_1,$$

$$D_4 = P_{RD}(1 + r + r^2) + r^3D_1, \qquad (7.2\text{-}5)$$

$$\vdots$$

$$D_n = P_{RD}(1 + \cdots + r^{n-2}) + r^{n-1}D_1$$

$$= \frac{P_{RD}}{P_{RD} + P_{DR}}(1 - r^{n-1}) + r^{n-1}D_1.$$

Since $P_{RD} + P_{DR} \neq 0$, then as n grows large r^{n-1} tends to zero, and

$$D_n \to \frac{P_{RD}}{P_{RD} + P_{DR}} \quad \text{as } n \to \infty. \qquad (7.2\text{-}6)$$

Note that the limiting result is independent of the initial probabilities. Furthermore, if any D_n were equal to this limiting value, then D_{n+1} would be too. What we have here is an example of a one-step or *first-order homogeneous Markov chain*. The limiting result (7.2-6) is a special case of more general results discussed by authors such as Bush and Mosteller [1955] and Feller [1968], who give a detailed mathematical treatment of Markov chains.

In many processes the transition probabilities may change very little, or at least very slowly, and then a Markov model like the one just presented is of special interest, since the mathematics tells so much about the flow of the population from one time period to the next. We have just seen that we can regard such a process as having two parts: its transition matrix, which is permanent and central, and its initial probability vector. Since the long-run outcome does not depend on the initial vector except in special cases, the initial vector may for many purposes be regarded as a side issue. The transition probability matrix plays the role of determining interactions in a contingency table, and the types of analyses introduced in the preceding section are of interest. In the following section, we give a summary of the theory for estimating the initial probability vector and the transition probability matrix. (For full details of the theory, see Anderson and Goodman [1957].)

7.2.3   *Estimation of transition probabilities*

Suppose the Markov chain has I possible states (e.g., above $I = 2$, namely D and R), and we observe T sets of successive transitions (the first going from time 0 to time 1). Suppose there are $y_i(0)$ individuals at time 0 in state i, and that the $\{y_i(0)\}$ are multinomially distributed with probabilities η_i and sample size $n = \sum_{i=1}^{I} y_i(0)$. Finally, let $y_{ij}(t)$ be the number of individuals in state i at time $t - 1$ and j at time t, and let $y_i(t)$ be the number of individuals in state i at time t. Note that

$$y_{+j}(t) = y_j(t), \qquad y_{i+}(t) = y_i(t - 1), \qquad (7.2\text{-}7)$$

$$n = \sum_{i=1}^{I} \sum_{j=1}^{I} y_{ij}(t). \qquad (7.2\text{-}8)$$

Expressions (7.2-7) and (7.2-8) imply that $y_{i+}(t) = y_{+i}(t-1)$.

Let $P_{ij}(t)$ be the conditional probability of being in state j at time t, given state i at time $t-1$, so that $\sum_j P_{ij}(t) = 1$ for all i and t. Then for a fixed value of i, the conditional distribution of $y_{ij}(t)$ for $j = 1, \ldots, I$, given $y_{i+}(t) = y_i(t-1)$, is

$$\frac{y_i(t-1)!}{\prod\limits_{j=1}^{I} y_{ij}(t)!} \prod_{j=1}^{I} P_{ij}(t)^{y_{ij}(t)}. \tag{7.2-9}$$

If all the transitions are mutually independent, then the joint probability distribution of the $y_{ij}(t)$ and the $y_i(0)$ is

$$\left[\frac{n!}{\prod\limits_{i=1}^{I} y_i(0)!} \prod_{i=1}^{I} \eta_i^{y_i(0)} \right] \times \prod_{t=1}^{T} \left\{ \prod_{i=1}^{I} \left[\frac{y_i(t-1)!}{\prod\limits_{j=1}^{I} y_{ij}(t)!} \prod_{j=1}^{I} P_{ij}(t)^{y_{ij}(t)} \right] \right\}. \tag{7.2-10}$$

The likelihood function of the $\{\eta_i\}$ and the $\{P_{ij}(t)\}$ is thus

$$\left[\prod_{i=1}^{I} \eta_i^{y_i(0)} \right] \left[\prod_{t=1}^{T} \prod_{i=1}^{I} \prod_{j=1}^{I} P_{ij}(t)^{y_{ij}(t)} \right], \tag{7.2-11}$$

a product of two functions, one involving only the η_i and the other only the $P_{ij}(t)$. The maximum likelihood estimates for the parameters are

$$\hat{\eta}_i = \frac{y_i(0)}{n}, \tag{7.2-12}$$

$$\hat{P}_{ij}(t) = \frac{y_{ij}(t)}{y_i(t-1)} = \frac{y_{ij}(t)}{y_{i+}(t)}. \tag{7.2-13}$$

For a Markov chain with stationary transition probability matrix, $P_{ij}(t) = P_{ij}$ for $t = 1, \ldots, T$, and the second factor of the likelihood function (7.2-11) reduces to

$$\prod_{i=1}^{I} \prod_{j=1}^{I} P_{ij}^{y_{ij}}, \tag{7.2-14}$$

where $y_{ij} = \sum_{t=1}^{T} y_{ij}(t)$. The maximum likelihood estimates of the stationary transition probabilities are thus

$$\hat{P}_{ij} = \frac{y_{ij}}{y_{i+}} = \frac{\sum\limits_{t=1}^{T} y_{ij}(t)}{\sum\limits_{t=1}^{T} y_i(t-1)}. \tag{7.2-15}$$

The maximum likelihood estimates of the initial probabilities remain unchanged.

We can think of these estimates in the following way. In the more general setup, when the transition probabilities are not necessarily stationary, we can represent the data on observed transitions in terms of the sufficient statistics $\{y_{ij}(t)\}$, and we can treat these statistics as if they were the observed counts $x_{ijt} = y_{ij}(t)$ in an $I \times I \times T$ contingency table, where the total for each level of the third variable is fixed equal to the total sample size n. When the transition probabilities are stationary, we can collapse the $I \times I \times T$ table across the third variable, yielding an $I \times I$ table of counts $y_{ij} = x_{ij+}$ whose total equals $N = nT$.

Example 7.2-1   Number-generating experiment

Each of 187 students in a statistics class at Harvard University was asked to close her or his eyes and, at a given signal, to write on a piece of paper one of the numbers 1, 2, or 3 that came to her or his mind. The students were asked to behave like random number generators, or like three-sided dice. The procedure was repeated seven times so that each student wrote down a sequence of seven numbers, each either 1, 2, or 3.

In table 7.2-1 we give the six observed one-step transition matrices, with the corresponding estimated transition probabilities. If the transition probabilities are stationary, then the estimated transition probabilities are those based on the marginal array of counts given in table 7.2-2. Note that the transition probabilities corresponding to the diagonal cells are all less than 0.333. This means

Table 7.2-1   One-Step Transition Data from Number-Generating Experiment (Transition Probabilities in Parentheses)

a.

1st Number	2nd Number			Totals
	1	2	3	
1	10 (0.167)	16 (0.267)	34 (0.567)	60
2	42 (0.575)	10 (0.137)	21 (0.288)	73
3	29 (0.537)	16 (0.296)	9 (0.167)	54
Totals	81 (0.433)	42 (0.225)	64 (0.342)	187

b.

2nd Number	3rd Number			Totals
	1	2	3	
1	23 (0.284)	18 (0.222)	40 (0.494)	81
2	16 (0.381)	11 (0.262)	15 (0.357)	42
3	22 (0.344)	26 (0.406)	16 (0.250)	64
Totals	61 (0.326)	55 (0.294)	71 (0.380)	187

c.

3rd Number	4th Number			Totals
	1	2	3	
1	11 (0.180)	33 (0.541)	17 (0.279)	61
2	22 (0.400)	10 (0.182)	23 (0.418)	55
3	34 (0.479)	27 (0.380)	10 (0.141)	71
Totals	67 (0.358)	70 (0.374)	50 (0.267)	187

d.

4th Number	5th Number			Totals
	1	2	3	
1	20 (0.299)	21 (0.313)	26 (0.388)	67
2	32 (0.457)	18 (0.257)	20 (0.287)	70
3	24 (0.480)	17 (0.340)	9 (0.180)	50
Totals	76 (0.406)	56 (0.299)	55 (0.294)	187

e.

5th Number	6th Number			Totals
	1	2	3	
1	25 (0.329)	26 (0.342)	25 (0.329)	76
2	24 (0.429)	13 (0.232)	19 (0.339)	56
3	13 (0.236)	29 (0.527)	13 (0.236)	55
Totals	62 (0.332)	68 (0.364)	57 (0.305)	187

f.

6th Number	7th Number			Totals
	1	2	3	
1	22 (0.355)	17 (0.274)	23 (0.371)	62
2	29 (0.426)	18 (0.265)	21 (0.309)	68
3	21 (0.368)	25 (0.439)	11 (0.193)	57
Totals	72 (0.385)	60 (0.321)	55 (0.294)	187

Table 7.2-2 Estimated Stationary Transition Probabilities for Data in Table 7.2-1 (Transition Probabilities in Parentheses)

	\(i + 1\)st Number			
\(i\)th Number	1	2	3	Totals
1	111	131	165	407
	(0.273)	(0.322)	(0.405)	
2	165	80	119	364
	(0.453)	(0.220)	(0.327)	
3	143	140	68	351
	(0.407)	(0.399)	(0.194)	
Totals	419	350	352	1,122

that the same number appears less frequently on successive trials than we would expect if the trials were independent of one another.

There appears to be considerable variation among the transition probability matrices in table 7.2-1, and we shall explore whether or not they are stationary. ■■

7.2.4 *Are the transition probabilities stationary?*

To check on the stationarity of the transition probabilities in a first-order Markov chain, we compare the observed transition matrices for the different one-step transitions, treating the data as if they formed an $I \times I \times T$ contingency table as described above. Stationarity of the transition probabilities corresponds to the following log-linear model for the expected cell counts:

$$u + u_{1(i)} + u_{2(j)} + u_{3(k)} + u_{12(ij)} + u_{13(ik)}. \tag{7.2-16}$$

The term $u_{13(ik)}$ is included in the model because, as in (7.2-2), the probabilities in a transition matrix must sum to 1 in each row, i.e., $\sum_j P_{ij}(t) = 1$ for $t = 1, \ldots, T$. Thus we are checking whether or not, given the state at the start of a transition, the state at the end of the transition is independent of the time of transition.

Example 7.2-1 (Continued)

For our number-generating experiment we wish to determine whether or not the one-step transition probabilities are stationary. Testing the goodness of fit of model (7.2-16), we get $X^2 = 54.5$ (and $G^2 = 54.1$) with 30 degrees of freedom. The probability of observing a χ^2 variate with 30 degrees of freedom at least this large is approximately 0.005, so it is reasonable to conclude that the transition probabilities are not stationary. ■■

Example 7.2-2 Voting intentions of the 1940 presidential election

In table 7.2-3, we give data obtained in a panel survey of potential voters in Erie County, Ohio over the course of six months in 1940, as reported by Goodman [1962]. For a description of the survey, see Lazarsfeld, Berelson, and Gaudet [1948], and for detailed analyses of the data see Anderson [1954] and Goodman [1962]. The table summarizes the voting intentions of 445 people who responded to each of the six monthly interviews.

Table 7.2-3 One-Step Transitions in Panel Study of Potential Voters in Erie County, Ohio, 1940

May	June R	June D	June U	Totals	June	July R	July D	July U	Totals
R	125	5	16	146	R	124	3	16	143
D	7	106	15	128	D	6	109	14	129
U	11	18	142	171	U	22	9	142	173
Totals	143	129	173	445	Totals	152	121	172	445

July	August R	August D	August U	Totals	August	September R	September D	September U	Totals
R	146	2	4	152	R	184	1	7	192
D	6	111	4	121	D	4	140	5	149
U	40	36	96	172	U	10	12	82	104
Totals	192	149	104	445	Totals	198	153	94	445

September	October R	October D	October U	Totals
R	192	1	5	198
D	2	146	5	153
U	11	12	71	94
Totals	205	159	81	445

Source: Goodman [1962].
Note: R = Republican, D = Democrat, U = Undecided.

Treating each of the five transition arrays as a layer in a $3 \times 3 \times 5$ table and checking the goodness of fit of model (7.2-16), we get a Pearson chi square value of $X^2 = 103.3$ and a likelihood ratio value of $G^2 = 101.5$. Either of these values is highly significant when compared with a χ^2 variate with 24 degrees of freedom.

Goodman [1962] notes that a cursory inspection of the five transition matrices indicates that the first two are similar, and that they are different from the last two, which also appear to be similar. The third transition matrix appears to be different from the remaining four (it was during this transition that the Democratic convention was held).

In table 7.2-4 we give the goodness-of-fit statistics for comparing the transition

Table 7.2-4 Comparing One-Step Transition Matrices for 1940 Panel Study of Voting Intentions

Transitions	Degrees of Freedom	G^2	X^2
May–June, June–July	6	7.5	7.3
May–June, June–July, July–August	12	68.3	64.7
July–August, August–September, September–October	12	24.7	24.6
August–September, September–October	6	1.5	1.5

probabilities for the first two, first three, last two, and last three periods. The values of these statistics seem to support the explanation given above. In a more detailed look at these data, Goodman [1962] notes that the only significant difference (using a 5% significance level) between the third transition matrix and the fourth and fifth is in the undecided category. Anderson [1954] suggests that the differences are more a matter of time scale, with July–August representing changes at twice the speed of August–September and September–October. ■■

7.3 Higher-Order Markov Models

Consider a sequence of transitions among a set of I states at T fixed points in time. In Section 7.2, we assume that the underlying process of change can be described in terms of one-step transitions, i.e., that the state occupied by an individual at time t depends only on his state at time $t - 1$. Often we find that the current state occupied by an individual depends on his states at time $t - 1, \ldots, t - r$, where $r > 1$. In such situations, we have an rth-order Markov process, and we describe methods for analyzing the data resulting from such a process below.

7.3.1 *Estimating transition probabilities*

Let us concentrate on second-order Markov chains, since results for kth-order chains are similar. As before, we assume there are $y_i(0)$ individuals in state i at time 0 and that the $\{y_i(0)\}$ are multinomially distributed with probabilities η_i and sample size $n = \sum_{i=1}^{I} y_i(0)$. If we let $y_{ijk}(t)$ be the number of individuals in state i at time $t - 2$, state j at time $t - 1$, and state k at time t, then

$$y_{ij+}(t) = y_{ij}(t - 1),$$

$$y_{+jk}(t) = y_{jk}(t),$$

$$y_{i++}(t) = y_i(t - 2),$$

$$y_{+j+}(t) = y_j(t - 1),$$

$$y_{++k}(t) = y_k(t), \tag{7.3-1}$$

for $t = 2, \ldots, T$, where the singly and doubly subscripted quantities on the right-hand sides of (7.3-1) are the counts described in Section 7.2.3 for first-order Markov chains. Thus

$$n = \sum_{i,j,k} y_{ijk}(t). \tag{7.3-2}$$

If the two-step transitions are mutually independent, then the joint probability distribution of the $\{y_{ijk}(t)\}$, the $\{y_{ij}(1)\}$, and the $\{y_i(0)\}$ is

$$\left[\frac{n!}{\prod\limits_{i=1}^{I} y_i(0)!} \prod_{i=1}^{I} \eta_i^{y_i(0)} \right] \times \prod_{i=1}^{I} \left[\frac{y_i(0)!}{\prod\limits_{j=1}^{I} y_{ij}(1)!} \prod_{j=1}^{I} \zeta_{ij}^{y_{ij}(1)} \right]$$

$$\times \prod_{t=2}^{T} \left\{ \prod_{i=1}^{I} \prod_{j=1}^{I} \left[\frac{y_{ij}(t-1)!}{\prod\limits_{k=1}^{I} y_{ijk}(t)!} \prod_{k=1}^{I} P_{ijk}(t)^{y_{ijk}(t)} \right] \right\}, \tag{7.3-3}$$

where ζ_{ij} is the probability of being in state j at time 1, given state i at time 0, and $P_{ijk}(t)$ is the probability of being in state k at t, given state j at $t - 1$ and i at $t - 2$. Thus

$$\sum_j \zeta_{ij} = 1 \qquad i = 1, \ldots, I, \tag{7.3-4}$$

$$\sum_k P_{ijk}(t) = 1 \qquad i, j = 1, \ldots, I; \quad t = 1, \ldots, T. \tag{7.3-5}$$

The likelihood function of the $\{\eta_i\}$, the $\{\zeta_{ij}\}$, and the $\{P_{ijk}(t)\}$ is a product of three factors:

$$\prod_{i=1}^{I} \eta_i^{y_i(0)}, \tag{7.3-6}$$

$$\prod_{i=1}^{I} \prod_{j=1}^{I} \zeta_{ij}^{y_{ij}(1)}, \tag{7.3-7}$$

$$\prod_{t=2}^{T} \prod_{i=1}^{I} \prod_{j=1}^{I} \prod_{k=1}^{I} P_{ijk}(t)^{y_{ijk}(t)}. \tag{7.3-8}$$

The maximum likelihood estimates of the various probabilities are thus

$$\hat{\eta}_i = \frac{y_i(0)}{n}, \tag{7.3-9}$$

$$\hat{\zeta}_{ij} = \frac{y_{ij}(1)}{y_i(0)}, \tag{7.3-10}$$

$$\hat{P}_{ijk}(t) = \frac{y_{ijk}(t)}{y_{ij}(t-1)}. \tag{7.3-11}$$

If the second-order chain has a stationary transition probability matrix $P_{ijk}(t) = P_{ijk}$ for $t = 2, \ldots, T$, then (7.3-8) reduces to

$$\prod_{i=1}^{I} \prod_{j=1}^{I} \prod_{k=1}^{I} P_{ijk}^{y_{ijk}}, \tag{7.3-12}$$

where $y_{ijk} = \sum_{t=2}^{T} y_{ij}(t)$, and the maximum likelihood estimate of P_{ijk} is

$$\hat{P}_{ijk} = \frac{y_{ijk}}{\sum_{k=1}^{I} y_{ijk}}. \tag{7.3-13}$$

Note that $\sum_{k=1}^{I} y_{ijk} \neq y_{ij}$ of the previous section, since here we sum from $t = 2$ to T and there we summed from $t = 1$ to T.

We can treat the observed data $\{y_{ijk}(t)\}$ for the two-step transitions as if they were counts $x_{ijkt} = y_{ijk}(t)$ in an $I \times I \times I \times (T - 1)$ contingency table, where the total for each level of the fourth variable equals n. When the transition probabilities are stationary we can collapse across the fourth variable to get an $I \times I \times I$ table of counts $x'_{ijk} = x_{ijk+}$ whose total equals $n(T - 1)$ (see theorem 2.5-1, Chapter 2).

If we have a stationary transition probability matrix but the Markov chain is really a first-order chain with a stationary transition probability matrix, then

$$(\zeta_{jk} =)P_{ijk} = P_{jk} \qquad i = 1, \ldots, I. \tag{7.3-14}$$

7.3.2 Are the transition probabilities stationary?

Stationarity of the transition probabilities in a second-order chain corresponds to the log-linear model with u_{34} and related higher-order u-terms set equal to zero, i.e., the model

$$u + u_{1(i)} + u_{2(j)} + u_{3(k)} + u_{4(t)} + u_{12(ij)} + u_{13(ik)} + u_{23(jk)}$$
$$+ u_{14(it)} + u_{24(jt)} + u_{123(ijk)} + u_{124(ijt)} \tag{7.3-15}$$

for the $I \times I \times I \times T$ table of counts $x_{ijkt} = y_{ijk}(t)$. We can check for stationarity using the usual chi square goodness-of-fit statistics X^2 and G^2.

7.3.3 Assessing the order of a chain

Suppose we have a Markov chain which has either a first-order or a second-order stationary transition probability matrix. In the latter case, the $I \times I \times I$ table of counts $x'_{ijk} = x_{ijk+} = \sum_{t=2}^{T} y_{ijk}(t)$ satisfies the general log-linear model

$$u + u_{1(i)} + u_{2(j)} + u_{3(k)} + u_{12(ij)} + u_{13(ik)} + u_{23(jk)} + u_{123(ijk)}, \tag{7.3-16}$$

and if we have a first-order chain, the table is described by the model

$$u + u_{1(i)} + u_{2(j)} + u_{3(k)} + u_{12(ij)} + u_{23(jk)}. \tag{7.3-17}$$

Thus to test if the second-order chain is actually a first-order chain, i.e., to check that (7.3-14) holds, we can use the standard goodness-of-fit statistics for the log-linear model (7.3-17).

Although we have described tests for second-order chains in this section, we can start with an rth-order chain, and then by a series of conditional goodness-of-fit tests decide whether the chain is really of order s ($s < r$). We illustrate such a procedure in a related context in the next section.

Example 7.3-1   More on voting intentions and the 1940 presidential election

In example 7.2-2, we assumed that the process of change underlying the panel survey of potential voters in Erie County was a first-order Markov chain. Here we assume that the chain has stationary transition probabilities and attempt to determine if the chain is first or second order. We give the array of counts of second-order transitions in table 7.3-1. Note that the row totals here correspond to the 3×3 sum of the first four transition matrices in table 7.2-3 (May–June, June–July, July–August, August–September).

A quick glance at table 7.3-1 leads one to reject the hypothesis of homogeneity of proportions in the three 3×3 subtables (this is equivalent to the hypothesis that the Markov chain is a first-order chain), and the goodness-of-fit statistics have values $X^2 = 92.1$ and $G^2 = 63.5$, each with twelve degrees of freedom. The unusually large discrepancy between the values of X^2 and G^2 comes from the large sample sizes and the position of the true table of transition probabilities in the parameter space.

In our earlier analysis of these data we concluded that if the chain is of first order, it is not stationary. It is quite conceivable that if we have a second-order

Table 7.3-1   Stationary Two-Step Transitions in Panel Study of
Potential Voters in Erie County, Ohio, 1940

Time $t - 2$	Time $t - 1$	Time t R	D	U	Totals
R	R	557	6	16	579
D	R	18	0	5	23
U	R	71	1	11	83
R	D	3	8	0	11
D	D	9	435	22	466
U	D	6	63	6	75
R	U	17	5	21	43
D	U	4	10	24	38
U	U	62	54	346	462

Source: Goodman [1962].
Note: R = Republican, D = Democrat, U = undecided.

chain, it too is not stationary. Unfortunately, the data required to test stationarity of the second-order chain are unavailable. ∎∎

7.4   Markov Models with a Single Sequence of Transitions

In Sections 7.2 and 7.3 we considered transitions among a set of I states at T fixed points in time for N individuals. The goodness-of-fit statistics for the models we described have the indicated asymptotic χ^2 distributions as $n \to \infty$, while T remains fixed.

Often we are faced with one very long sequence of observations or transitions and would like to investigate its Markov nature, i.e., we have $n = 1$ but T is very large. For such data, we cannot investigate the stationarity properties of the Markov process, since we only have one observation for each transition time. We can assume stationarity, however, and investigate the order of the chain. Bartlett [1951] and Hoel [1954] have shown that the results of Sections 7.3.1 and 7.3.3 on estimating the stationary transition probabilities and assessing the order of chain carry over for the case where $n = 1$ and $T \to \infty$.

When $n = 1$ and T is large and we wish to study a stationary k-step Markov chain (where $U(t) = $ state of the chain at time t) we form the $k + 1$ dimensional table $\{x_{i_1 \ldots i_{k+1}}\}$ where the entry $x_{i_1 \ldots i_{k+1}}$ is the number of values of t for which

$$U(t) = i_1, \quad U(t + 1) = i_2, \quad \ldots, \quad \text{and } U(t + k) = i_{k+1} \qquad (7.4\text{-}1)$$

$t = 1, \ldots, T - k$, and $i_j = 1, \ldots, I$, $j = 1, \ldots, k + 1$.

Example 7.4-1   Pattern of vowels and consonants in biblical text

Newman [1951a] examined the pattern of vowels and consonants in a passage from the New Testament (Mark 4) beginning "And he began again to teach by the sea side...." If we let the letter A stand for a consonant and B for a vowel, then the first part of the biblical passage can be represented as:

BAAABABABABABBAABABBAAABAABABBABAB.

Newman examined a sequence of 1,003 letters from this passage and considered its Markov characteristics. For this sequence we have $n = 1$ and $T = 1,003$. In table 7.4-1 we summarize the matrices of three-step, two-step, and one-step transitions as reported by Newman.

Table 7.4-1   Vowels and Consonants in Biblical Text (Source: Newman [1951a])

a. Three-Step Transitions

		Position t			
		A		B	
		Position $t - 1$		Position $t - 1$	
Position $t - 3$	Position $t - 2$	A	B	A	B
A	A	21	155	69	29
	B	143	56	120	8
B	A	69	108	115	35
	B	41	8	23	0

b. Two-Step Transitions

	Position t			
	A		B	
	Position $t - 1$		Position $t - 1$	
Position $t - 2$	A	B	A	B
A	90	263	184	64
B	184	64	143	8

c. One-Step Transitions

	Position t	
Position $t - 1$	A	B
A	274	327
B	327	72

To test the hypothesis that the chain is of order 1 against the alternative that it is of order 2, we use the data in table 7.4-1b to compute

$$G^2 = 2 \sum_{i,j,k} x_{ijk} \log\left[\frac{x_{ijk}}{x_{ij+} x_{+jk}/x_{+j+}}\right]$$

$$= 36.53.$$

The probability that a χ^2 variate with two degrees of freedom exceeds 36.53 is much less than 0.001, and we can reject the model of a first-order chain in favor of the second-order chain model.

To test the hypothesis that the chain is of order 2 against the alternative that it is of order 3, we use the data in table 7.4-1a to compute

$$G^2 = 2 \sum_{i,j,k,l} x_{ijkl} \log\left[\frac{x_{ijkl}}{x_{ijk+}x_{+jkl}/x_{+jk+}}\right]$$
$$= 13.5.$$

The probability that a χ^2 variate with four degrees of freedom exceeds 13.5 is approximately 0.01, and it appears that the occurrence of a vowel or consonant depends on the preceding three rather than two letters.

Newman [1951b] notes that "the vowel-consonant dichotomy goes to the root of both spoken and written speech. It reveals, for instance, something of the syllabic structure of language...." One way to interpret the third-order Markovity of the vowel-consonant pattern is to note that a very high percentage of monosyllables in the English language have three or four letters, with the sequences ABA, ABAA, and AABA being most common (see Menzerath [1950]). ∎ ∎

Example 7.4-2   Comparing languages by sequences of vowels and consonants
Newman [1951b] used the methods described in the preceding example to compare the vowel-consonant pattern in several languages: basic English, King James English, German, Hebrew, Italian, Latin, Lifu (the language spoken on Lifu, the largest of the Loyalty Islands, a group located in the South Pacific Ocean east of New Caledonia), and Samoan. He first selected a set of passages from the Old and New Testaments, trying to avoid proper names, which tend to be the same in all languages. Each passage was 1,000 letters in length, ending in different places in the text due to differences among the languages. We summarize Newman's results in table 7.4-2.

Table 7.4-2   Pattern of Vowels and Consonants in Different Languages: Log-Likelihood Ratio Goodness-of-Fit Statistics Used to Assess Order of Markov Chains[a]

Language	Goodness-of-Fit Statistics[b]			
	$G^2(1\|2)$	$G^2(2\|3)$	$G^2(3\|4)$	$G^2(4\|5)$
Basic English	66	18	8	30
English	52	16	6	12
German	124	28	2	0[c]
Italian	96	8	6	10
Latin	86	6	8	4
Samoan	4	0	2	2
Hebrew	72	2	26	12
Lifu	36	6	2	10
Degrees of Freedom	2	4	6	8

Note: Raw data from Newman [1951b].
[a] All the numbers are even integers because of the manner in which Newman reports his results.
[b] $G^2(r|u)$ is the statistic for testing the hypothesis that the chain is of order r against the alternative that it is of order $u > r$.
[c] We report zeros for those entries which are negative based on the figures reported by Newman.

Samoan and Lifu are the only languages whose vowel-consonant patterns appear to follow first-order Markov chains. Italian and Latin have patterns which appear to follow second-order chains, while King James English and German appear to follow third-order chains. The nature of the sequences for basic English and Hebrew seems more complicated.

If we view the order of the chains as reflecting syllabic structure of the languages, then Lifu and Samoan are the least complicated and English, German, and Hebrew the most complicated. Newman notes: "It is probably not just a matter of chance that the most primitive languages studied showed the greatest restrictions in terms of patterning, while English showed the least. The very large size of the vocabulary in English is ... a consequence of a complex cultural history." ■■

7.5 Other Models

7.5.1 *Markov models for cross-classified states*

Up to now we have viewed the states of a Markov chain as categories corresponding to a single variable, e.g., political party preference (Democrat, Republican, or undecided). Often we are interested in following the changes among the cells in a cross-classification across time. For example, in panel studies individuals are usually asked several questions, and we are interested not only in the changes in single variables but also in the changes in their interactions.

The simplest Markov model for cross-classified data assumes that the sequences of changes in each classification are independent of the changes in the remaining ones. We illustrate for a first-order stationary Markov chain with two variables.

Let $P_{ij,kl}$ be the transition probabilities of being in state (k, l) at time t given state (i, j) at time $t - 1$, where i and k take values from 1 to I and j and l from 1 to J. Here

$$\sum_{k=1}^{I} \sum_{l=1}^{J} P_{ij,kl} = 1. \tag{7.5-1}$$

If the changes for the two variables are independent, then we can write

$$P_{ij,kl} = P_{ik}^{(1)} P_{jl}^{(2)}, \tag{7.5-2}$$

where

$$\sum_{k} P_{ik}^{(1)} = \sum_{l} P_{jl}^{(2)} = 1. \tag{7.5-3}$$

Thus the $\{P_{ij,kl}\}$ are expressible as products of the transition probabilities $\{P_{ik}^{(1)}\}$ and $\{P_{jl}^{(2)}\}$ of two independent Markov chains. If $\{x_{ij,kl}\}$ are the number of transitions from (i, j) to (k, l), then MLEs of the parameters are

$$\hat{P}_{ik}^{(1)} = \frac{\sum_{j=1}^{J} \sum_{l=1}^{J} x_{ij,kl}}{\sum_{j=1}^{J} \sum_{k=1}^{I} \sum_{l=1}^{J} x_{ij,kl}}, \tag{7.5-4}$$

$$\hat{P}_{jl}^{(2)} = \frac{\sum_{i=1}^{I} \sum_{k=1}^{I} x_{ij,kl}}{\sum_{i=1}^{I} \sum_{k=1}^{I} \sum_{l=1}^{J} x_{ij,kl}}. \tag{7.5-5}$$

If we think of the $\{x_{ij,kl}\}$ as observed values in an $I \times J \times I \times J$ table, then model (7.5-2) corresponds to the log-linear model

$$u + u_{1(i)} + u_{2(j)} + u_{3(k)} + u_{4(l)} + u_{12(ij)} + u_{13(ik)} + u_{24(jl)} \qquad (7.5\text{-}6)$$

for the expected cell values. Note that we include $u_{12(ij)}$ in the model, since the transition probabilities $\{P_{ij,kl}\}$ imply that we fix the corresponding observed totals. The estimated expected values for this model are

$$\hat{m}_{ij,kl} = \frac{x_{+j,+l}x_{i+,k+}x_{ij,++}}{x_{+j,++}x_{i+,++}}. \qquad (7.5\text{-}7)$$

We can test the goodness of fit of this model using either the Pearson or likelihood-ratio statistics. Here the number of degrees of freedom equals $(I-1)(J-1) \times (IJ + I + J)$.

If this model does not fit the data, we can add other u-terms in order to find a model providing an adequate fit.

Example 7.5-1   See–buy

Anderson [1954] considers a two-interview panel study where at each interview, the respondent was asked if he or she had seen an advertisement for a certain product and if he or she had bought that product. We reproduce the data here in table 7.5-1.

Table 7.5-1   See–Buy Transition Data

First Interview		Second Interview			
		See			
		Yes		No	
See	Buy	Buy		Buy	
		Yes	No	Yes	No
Yes	Yes	83	8	35	7
	No	22	68	11	28
No	Yes	25	10	95	15
	No	8	32	6	493

Source: Anderson [1954].

We begin by exploring the goodness of fit of model (7.5-6), which says that the changes in the number of people seeing the advertisement are independent of the changes in the number buying the product. The values of the goodness-of-fit statistics are $X^2 = 150.1$ and $G^2 = 123.9$, and there are eight degrees of freedom. (The value of G^2 reported by Anderson is 37.84, which is clearly incorrect.) It is reasonable to conclude that changes in seeing are related to those in buying.

A logical second step in our analysis of these data is to fit other log-linear models with more terms than (7.5-6). Unfortunately, no unsaturated log-linear model provides an adequate fit to the data. For example, the model with no four-factor effect yields goodness-of-fit statistics $X^2 = 25.5$ and $G^2 = 21.9$, with one degree of freedom. This is not a satisfactory place to end our analysis, and it seems sensible to use alternative models.

Several authors have tried to explain the dependencies between the changes in the two variables in a panel study using models other than those used here. We refer the interested reader to Lazarsfeld [1969] and Yee and Gage [1968]. ■■

7.5.2 *Sequences of unrepeated events*

Often, when observing the behavioral patterns of animals, we can only observe the cases in which a switch occurs from one type of behavior to another. In other cases, each type of behavior pattern is likely to be repeated many times in succession, and the changes from one type to another are of particular interest. If we wish to model the behavior changes using Markov models, we must extend the techniques described above, since successive events cannot be independent in the usual sense. In effect, we wish to determine whether or not the sequence of changes or unrepeated events is like a sequence of events from a Markov chain from which the repeated events have been removed.

To illustrate how the methods from above can be applied in these circumstances, we consider the problem of estimating the order of a chain. In particular, suppose we have a stationary second-order Markov sequence of events from which the repeated events have been removed. Then the matrix $\{x_{ijk}\}$ of observed transitions must have zero entries in those places corresponding to repeated events, i.e., $x_{ijk} \equiv 0$ for $i = j$ or $j = k$ (see Chapter 5 for a more detailed discussion of structural zeros). If the chain is actually of first order, then we can write the two-step transition probabilities as

$$P_{ijk} = P_{jk} \tag{7.5-8}$$

for $i \neq j, j \neq k$. The corresponding model for the expected cell values is

$$\log m_{ijk} = u + u_{1(i)} + u_{2(j)} + u_{3(k)} + u_{12(ij)} + u_{23(jk)} \tag{7.5-9}$$

for $i \neq j, j \neq k$, and the estimated expected values are

$$\hat{m}_{ijk} = \begin{cases} \dfrac{x_{ij+} x_{+jk}}{x_{+j+}} & \text{for } i \neq j, j \neq k \\ 0 & \text{otherwise.} \end{cases} \tag{7.5-10}$$

To test the goodness of fit of the first-order chain model, we can use the usual test statistics, recomputing the degrees of freedom to take into account those cells where we expect zeros a priori. If there are I possible states for the chain, then there are $I(I-1)^2$ cells which are not zero a priori. Moreover, there are I a priori zeros in each of the marginal configurations $C_{12} = \{x_{ij+}\}$ and $C_{23} = \{x_{+jk}\}$. Thus (using results from Chapters 3 and 5), the degrees of freedom are

$$I(I-1)^2 - (2I^2 - I) + 2I = I(I-2)^2. \tag{7.5-11}$$

We must have at least three states to be able to assess the order of such a chain.

Example 7.5-2 Song of the wood pewee

Chatfield and Lemon [1970] discuss a Markov chain analysis of sequences of distinctive units in the song of wood pewee birds. They consider data published by Craig [1943], and we reanalyze one of these sequences (Craig's record 13) which has 342 entries. The bird in question has a repertoire of four distinctive units or phrases. We denote these by A, B, C, and D, and we give the matrix of observed two-step transitions in table 7.5-2.

Table 7.5-2   Triples of Phrases in a Song Sequence of a Wood Pewee

First Phrase	Second Phrase	Third Phrase			
		A	B	C	D
A	A	7	5	0	8
	B	16	2	2	2
	C	2	27	0	0
	D	4	1	0	12
B	A	8	7	6	6
	B	2	0	0	0
	C	24	1	0	0
	D	1	0	0	3
C	A	0	3	23	0
	B	4	0	22	1
	C	0	0	0	0
	D	0	0	0	0
D	A	4	7	0	3
	B	5	0	1	1
	C	0	0	0	0
	D	9	6	0	105

Source: Craig [1943].

First we analyze these data in the usual way, i.e., including the repeated phrases. The values of the goodness-of-fit statistics for testing that the chain is of first order, against the alternative of second order, are $X^2 = 140.5$ and $G^2 = 166.1$. Normally the number of associated degrees of freedom would be 36, but due to zeros in the marginal configurations being fitted, the actual number of degrees of freedom is $36 - 19 + 5 = 21$. At any rate, the fit is not a good one, and we conclude that the chain is at least of second order.

Now we note the extremely large number of triples of the form DDD. Also, there are no triples of the forms BBB and CCC. Actually, the large number of DDDs results from several very long sequences of Ds, one of length 40 and others of lengths 18, 17, and 10. It seems reasonable at this point to delete the repeats from the sequence of phrases and analyze the remaining data. This leaves a sequence of 198 events to analyze, with 36 possible triples (once we have eliminated repeats). We give these two-step transition data in table 7.5-3.

Now we can test that the chain is of first order, against the alternative of second order: $X^2 = 119.7$ and $G^2 = 142.4$, with twelve degrees of freedom (there would normally be sixteen degrees of freedom, but due to zeros in the marginal configuration the actual number of degrees of freedom is $16 - 6 + 2 = 12$). We still do not have a good fit, and so we must conclude that the chain is at least of second order. In this case we do not seem to have gained very much by deleting the repeats from the data. If we had sufficient data to look at the four-step transitions, then the presence or absence of the repeats might be crucial in assessing the order of the chain. ■■

Table 7.5-3 Triples of Phrases in a Song Sequence of a Wood Pewee, with Repeats Deleted

First Phrase	Second Phrase	Third Phrase			
		A	*B*	*C*	*D*
A	*A*	—	—	—	—
	B	19	—	2	2
	C	2	26	—	0
	D	12	5	0	—
B	*A*	—	9	6	12
	B	—	—	—	—
	C	24	1	—	1
	D	1	2	0	—
C	*A*	—	4	22	0
	B	3	—	22	0
	C	—	—	—	—
	D	1	0	0	—
D	*A*	—	11	0	4
	B	5	—	1	1
	C	0	0	—	0
	D	—	—	—	—

Source: Craig [1943].
Note: "—" denotes an unobservable possibility.

7.5.3 *Different chains for different strata*

Up to now we have assumed that the same transition probabilities hold for each individual in the population under study. More often than not, we are studying heterogeneous populations, and the transition probabilities vary from individual to individual depending on certain characteristics, e.g., sex in animal populations. It often makes sense to stratify the population under study according to one or more discrete variables, treating each stratum as a homogeneous population. Then we may investigate whether the transition probabilities, and even the order of the Markov chain, differ among strata.

Suppose we have determined that for each stratum of interest we have an rth-order Markov chain. To test whether the stratification is necessary, we test whether the transition probability matrices for the different strata are the same.

Example 7.5-3 Two-person relationships and sex

Katz and Proctor [1959] report on an analysis of data for 25 eighth-grade students. The students were asked the question "With whom would you like to sit?" at four different times (September, November, January, and May). All pupils made three choices each except in November, when two students gave only two choices each. There are $\binom{25}{2} = 300$ two-person relationships that can be observed at each point in time. A relationship is either mutual (each person picks the other), one-way (only one member of the pair picks the other), or indifferent (neither picks the other).

Since the time gap from January to May is twice that of the other three transition periods, we examine here only the September-November-January data, summarized in table 7.5-4.

Table 7.5-4   Second-Order Transitions in Two-Person Choice Structures

Structure in September	Structure in November	Structure in January Mutual	One-Way	Indifference	Totals
Mutual	Mutual	4	0	3	7
	One-way	2	3	0	5
	Indifference	0	0	3	3
One-way	Mutual	2	0	3	5
	One-way	4	9	6	19
	Indifference	0	6	15	21
Indifference	Mutual	1	0	0	1
	One-way	1	9	12	22
	Indifference	0	20	197	217
Totals		14	47	239	300

Source: Katz and Proctor [1959].

We first ask if the Markov chain of two-person choice structures is of first or second order. Applying the usual formulas for degrees of freedom, we get twelve degrees of freedom, but zero totals in the margins being fitted reduce this number to eight. The goodness-of-fit statistics for the model of a first-order chain have values of $X^2 = 17.1$ and $G^2 = 17.7$. These values are in the 5% upper tail of the χ^2 distribution with eight degrees of freedom and we conclude that the chain may actually be of second order.

Table 7.5-5   Distributions of Types of Transitions in Two-Person Choice Structures (by Three Pairs of Time Points)

Type of Transition	September to November Boy to Boy	Girl to Girl	Cross Sex	November to January Boy to Boy	Girl to Girl	Cross Sex	September to January Boy to Boy	Girl to Girl	Cross Sex
Mutual to									
Mutual	2	5	0	3	4	0	1	5	0
One-way	3	2	0	0	0	0	3	0	0
Indifference	1	2	0	3	3	0	2	4	0
One-way to									
Mutual	4	1	0	0	6	1	2	3	1
Same	2	15	0	8	9	1	2	10	2
Opposite	1	1	0	1	2	0	0	0	1
Indifference	3	11	7	4	13	1	6	15	3
Indifference to									
Mutual	0	1	0	0	0	0	0	2	0
One-way	7	12	3	1	11	14	5	12	12
Indifference	13	70	134	16	72	127	15	69	125
Totals	36	120	144	36	120	144	36	120	144

Source: Katz and Proctor [1959].

Of the 25 students involved in this study, sixteen are girls and nine boys. There are 36 pair relationships among boys, 120 among girls, and 144 across sexes. In table 7.5-5 we show a breakdown of the one-step transitions by the sex of the pairs. Thus there are strata (boy-boy, girl-girl, mixed) for the pairs. Katz and Proctor note that more than 85% of the 144 cross-sex pairs are consistently indifferent, giving this stratum a different appearance from the other two. We can see this difference at a glance in table 7.5-5. We now turn to the two-step transitions for the other two strata. Due to the sparseness of the data, we cannot do much with the boy-boy pairs; however, table 7.5-6 contains the data for the girl-girl pairs.

Table 7.5-6 Second-Order Transitions in Girls' Two-Person Choice Structures

Structure in September	Structure in November	Structure in January			Totals
		Mutual	One-Way	Indifference	
Mutual	Mutual	3	0	2	5
	One-way	2	0	0	2
	Indifference	0	0	2	2
One-Way	Mutual	0	0	1	1
	One-way	3	7	6	16
	Indifference	0	3	8	11
Indifference	Mutual	1	0	0	1
	One-way	1	4	7	12
	Indifference	0	8	62	70
Totals		10	22	87	120

Source: Katz and Proctor [1959].

Testing in table 7.5-6 for first order versus second order, we find that $X^2 = 14.3$ and $G^2 = 13.6$, again with eight degrees of freedom. While these values correspond approximately to the 10% upper tail value for χ^2 with eight degrees of freedom, due to the sparseness of the data it is hazardous for us to conclude that the girl-girl data follow a first-order chain while the combined data follow a second-order chain. At best, we suggest that stratifying future data sets by the sex of the pairs may lead to simpler Markov behavior for the same-sex groups than for the combined heterogeneous mixed-sex group. ■ ■

8 Analysis of Square Tables: Symmetry and Marginal Homogeneity

8.1 Introduction

Two-dimensional tables, with the variable for rows having the same categories as the variable for columns, occur frequently. Such square tables may arise in several different ways:

1. in panel studies where each individual in a sample is classified according to the same criterion at two different points in time;

2. when a sample of pairs of matched individuals, such as husbands and wives or fathers and sons, are classified according to some categorical variable of interest for each member of the pair;

3. when a sample of individuals is cross-classified according to two essentially similar categorical variables (e.g., strength of right hand and strength of left hand);

4. in experiments conducted on matched pairs, where individuals are matched on one or more variables, the members of a pair are subjected to different treatments, and their responses are evaluated.

In all such situations it is natural to compare the proportions in the categories of the variable for rows with those in the categories of the variable for columns, asking if the two sets of underlying marginal probabilities are identical.

When the cross-classification is a double dichotomy (i.e., we have a 2×2 table), the problem of testing for marginal homogeneity is simple and straightforward, since in this special case marginal homogeneity is equivalent to symmetry, i.e., $m_{12} = m_{21}$ (see Chapter 7). For $I \times I$ tables with $I > 2$, complete symmetry is a more restrictive model than marginal homogeneity. The former implies the latter, but not vice versa. In Section 8.2 we examine the models of symmetry and marginal homogeneity in detail. To link the two models we introduce the concept of quasi symmetry, which we also relate to quasi independence (see Chapter 5). After discussing the parametric representation of the models, we derive maximum likelihood estimates for the expected cell values under symmetry and under quasi symmetry. We then present two ways of looking at the notion of marginal homogeneity: one conditional on the model of quasi symmetry being true, and the other unconditional.

There are several ways to generalize the notions of symmetry, quasi symmetry, and marginal homogeneity so that they are useful in the analysis of multidimensional tables. We explore these generalizations in Section 8.3, and we present some new results on estimation and testing for a variety of models. In particular, we discuss tests for the comparison of percentages in matched samples and compare them with the use of Cochran's Q test.

8.2   Two-Dimensional Tables

We now turn to a detailed examination of three models that are useful for two-dimensional square tables: marginal homogeneity, symmetry, and quasi symmetry. Of the three, the model of marginal homogeneity is the only one that we cannot represent in log-linear form (see the discussion in Section 8.2.1).

8.2.1   *Notation and definition of marginal homogeneity*

Let x_{ij} be the observed count in the (i, j) cell of an $I \times I$ table, and let m_{ij} be the expected cell value under some model. We assume that the $\{x_{ij}\}$ are either (i) observations on independent Poisson variates with means $\{m_{ij}\}$ or (ii) an observation on a multinomial variate with cell probabilities $\{p_{ij} = m_{ij}/N\}$, where $\sum m_{ij} = \sum x_{ij} = N$.

If the expected values satisfy the model of marginal homogeneity, then

$$m_{i+} = m_{+i} \qquad i = 1, \ldots, I. \tag{8.2-1}$$

For the 2×2 table, (8.2-1) implies that

$$m_{12} = m_{21}. \tag{8.2-2}$$

The major difficulty encountered in our analysis of marginal homogeneity comes from the fact that we cannot transform (8.2-1) into a simple parametric model for the expected values. For example, equating the main effect terms in the usual log-linear model, i.e., setting $u_{1(i)} = u_{2(i)}$ in

$$\log m_{ij} = u + u_{1(i)} + u_{1(j)} + u_{12(ij)} \tag{8.2-3}$$

(with the usual side constraints), does not imply (8.2-1) unless

$$\sum_j e^{u_{12(ij)}} = \sum_j e^{u_{12(ji)}} \qquad i = 1, \ldots, I. \tag{8.2-4}$$

We return to estimation of expected values under (8.2-1) and associated test statistics below.

8.2.2   *Symmetry*

By symmetry in a two-dimensional square table we mean that

$$m_{ij} = m_{ji} \quad \text{for all } i \neq j. \tag{8.2-5}$$

Clearly, symmetry, as defined by expression (8.2-5) implies marginal homogeneity, as defined by expression (8.2-1). There are at least three ways in which we can represent the model of symmetry as a log-linear model. The first of these is completely straightforward. The other two which we present in this section rely upon our turning the two-dimensional table into a three-dimensional one, either by duplicating all the cells in a particular fashion, or by removing the diagonal and folding the table over onto itself. These three-dimensional representations of symmetry are especially useful in dealing with quasi symmetry and conditional tests for marginal homogeneity.

Regular log-linear model representation

We can easily transform the constraints (8.2-5) into a log-linear model for the $\{m_{ij}\}$:

$$\log m_{ij} = u + u_{1(i)} + u_{1(j)} + u_{12(ij)}, \tag{8.2-6}$$

where

$$u_{12(ij)} = u_{12(ji)}, \tag{8.2-7}$$

$$\sum_i u_{1(i)} = \sum_i u_{12(ij)} = 0. \tag{8.2-8}$$

Note that we have replaced $u_{2(j)}$ by $u_{1(j)}$ in expression (8.2-6).

Assuming a multinomial sampling model, the kernel of the log-likelihood is

$$\sum_{ij} x_{ij} \log m_{ij} = x_{++}u + \sum_i (x_{i+} + x_{+i})u_{1(i)} + \sum_{i,j} \left(\frac{x_{ij} + x_{ji}}{2} \right) u_{12(ij)}. \tag{8.2-9}$$

Maximizing (8.2-9) yields the following estimated expected cell values:

$$\hat{m}_{ij} = \begin{cases} \dfrac{x_{ij} + x_{ji}}{2} & i \neq j \\ \\ x_{ii} & i = j \end{cases} \tag{8.2-10}$$

(see Bowker [1948]). The goodness-of-fit statistics used to test this model of symmetry are

$$G^2 = 2 \sum_{i \neq j} x_{ij} \log \frac{2x_{ij}}{x_{ij} + x_{ji}}, \tag{8.2-11}$$

$$X^2 = \sum_{i > j} \frac{(x_{ij} - x_{ji})^2}{x_{ij} + x_{ji}}. \tag{8.2-12}$$

Both have asymptotic χ^2 distributions with $I(I-1)/2$ degrees of freedom under the null hypothesis of symmetry.

Two three-dimensional representations of symmetry

If we remove the diagonal terms and fold the remainder of the two-dimensional table along the diagonal, we get a three-dimensional table of expected values that resembles a triangular wedge with entries m'_{ijk}, where

$$m'_{ij1} = \begin{cases} m_{ij} & i > j \\ 0 & \text{otherwise} \end{cases} \tag{8.2-13}$$

and

$$m'_{ij2} = \begin{cases} m_{ji} & i > j \\ 0 & \text{otherwise}. \end{cases} \tag{8.2-14}$$

Each triangular layer of side $I - 1$ has $I(I-1)/2$ cells with nonzero entries. In our usual u-term notation for three-dimensional tables, the hypothesis of symmetry consists of fitting the configuration $C'_{12} = \{x'_{ij+}\}$ and thus

$$\log m'_{ijk} = u + u_{1(i)} + u_{2(j)} + u_{12(ij)}. \tag{8.2-15}$$

Note that all terms involving variable 3 have been set equal to zero. This three-dimensional approach yields the same MLEs, goodness-of-fit statistics, and degrees of freedom $I(I-1)/2$ as before.

Using a second three-dimensional approach, we simply take a duplicate of the original table, flip it over, and place it below the original. Thus we have

$$m''_{ijk} = \begin{cases} m_{ij} & k = 1 \\ m_{ji} & k = 2. \end{cases} \qquad (8.2\text{-}16)$$

The hypothesis of symmetry consists of fitting the configuration $C''_{12} = \{x''_{ij+}\}$, and thus (8.2-15) is still our model. The MLEs are again the same as (8.2-10), but since each cell in the original table appears twice in the three-dimensional arrangement, we must divide the associated goodness-of-fit statistics by a factor of 2 in order to get (8.2-11) and (8.2-12). The number of degrees of freedom is $I(I - 1)/2$, as before.

We make use of this second approach again in the next section, where we discuss quasi symmetry. An advantage of the first three-dimensional viewpoint is that we can readily preserve the total number of observations on each side of the diagonal by including the term u_3 in our model. The resulting estimates are

$$\hat{m}'_{ijk} = \frac{x'_{ij+} + x'_{+ + k}}{N}, \qquad (8.2\text{-}17)$$

instead of $x'_{ij+}/2$ as in expression (8.2-10).

Example 8.2-1   Unaided distant vision data

In table 8.2-1 we present unaided distant vision data from case records of eye tests concluded on 7,477 women, aged 30–39, employed in Royal Ordinance factories in Britain. These data have been analyzed earlier by Stuart [1953, 1955], Bowker [1948], Caussinus [1965], Bhapkar [1966], Ireland, Ku, and Kullback [1969], Grizzle, Starmer, and Koch [1969], and Koch and Reinfurt [1971].

Table 8.2-1   Case Records of Eye-Testing of Women (Aged 30–39) Employees in Royal Ordnance Factories

| | Left Eye Grade | | | | |
Right Eye Grade	Highest (1)	Second (2)	Third (3)	Lowest (4)	Totals
Highest (1)	1,520	266	124	66	1,976
Second (2)	234	1,512	432	78	2,256
Third (3)	117	362	1,772	205	2,456
Lowest (4)	36	82	179	492	789
Totals	1,907	2,222	2,507	841	7,477

Source: Stuart [1953].

We begin our analysis of these data by fitting the model of symmetry, and we give the expected values in table 8.2-2. The model does not fit the data well, since $X^2 = 19.11$ and $G^2 = 19.25$, with six degrees of freedom.

Let us reexamine the observed data. We observe that the sum of the diagonal cells is 5,296; thus, of the 7,477 women, most had balanced vision. Of the 2,181 whose vision differed between eyes, 1,171 fall into the upper off-diagonal cells, thus their right eye was better than their left, and 1,010 had better vision in their left eye. We can test whether it was more normal for vision to be better in the right eye or the left using these two triangular off-diagonal sums. The McNemar-like test statistic

$$X^2 = \frac{(b-c)^2}{b+c},$$

where $b = \sum_{i>j} x_{ij}$ and $c = \sum_{i<j} x_{ij}$, for these data has a value of 11.88, with one degree of freedom. We therefore conclude that there is a significant difference between the "right eye better" and "left eye better" groups.

As a result of the preceding analysis, we turn to the three-dimensional representation of these data, in table 8.2-3, where we rearrange the off-diagonal cells to form a three-dimensional array, introducing a third variable with categories "right eye better" and "left eye better." We now fit the model with $u_{13} = u_{23} =$

Table 8.2-2   Expected Values for Data from Table 8.2-1 under Model of Symmetry

Right Eye Grade	Left Eye Grade Highest (1)	Second (2)	Third (3)	Lowest (4)	Totals
Highest (1)	1,520	250	120.5	51	1,941.5
Second (2)	250	1,512	397	80	2,239
Third (3)	120.5	397	1,772	192	2,481.5
Lowest (4)	51	80	192	492	815
Totals	1,941.5	2,239	2,481.5	815	7,477

Table 8.2-3   Three-Dimensional Representation of Data from Table 8.2-1

Grade of Worse Eye	Left Eye Better, $k=1$ Grade of Better Eye (1)	(2)	(3)	Totals	Right Eye Better, $k=2$ Grade of Better Eye (1)	(2)	(3)	Totals
(2)	234	—	—	234	266	—	—	266
(3)	117	362	—	479	124	432	—	556
(4)	36	82	179	297	66	78	205	349
Totals	387	444	179	1,010	456	510	205	1,171

Table 8.2-4   Three-Dimensional Representation of Expected Values of Data from Table 8.2-1 under Model $u_{13} = u_{23} = u_{123} = 0$

Worse Eye	Left Eye Better Better Eye (1)	(2)	(3)	Totals	Right Eye Better Better Eye (1)	(2)	(3)	Totals
(2)	231.5	—	—	231.5	268.5	—	—	268.5
(3)	111.6	367.7	—	479.3	129.4	426.3	—	556.7
(4)	47.2	74.1	177.8	299.1	54.8	85.9	206.2	346.9
Totals	390.3	441.8	177.8	1,009.9	452.7	512.2	206.2	1,171.1

$u_{123} = 0$, which is the hypothesis of symmetry modified to preserve the totals in each triangular array but not the marginal distributions. In other words, we are testing whether not only the pattern of association, but also the distribution between categories is the same whichever eye is better. We see from table 8.2-4 that this model fits well, yielding $X^2 = 7.23$ and $G^2 = 7.35$, with five degrees of freedom. (The largest contributions to X^2, 2.66 and 2.29, come from the cells in each layer where the better eye has grade 1 and the worse eye has grade 4.) Introducing the extra parameter to preserve off-diagonal cells thus leads to a significant improvement in fit over the symmetry model, reducing the value of G^2 by 11.90. This agrees with the hypothesis that when vision is unequal, the right eye is more likely to be the better eye, and that we have symmetry conditional on that occurrence. ∎∎

8.2.3   Quasi symmetry

Quasi symmetry plus marginal homogeneity equals symmetry
We noted above that symmetry implies marginal homogeneity. Algebraically this follows from the fact that expression (8.2-7) implies that (8.2-4) holds. Since we are unable to describe marginal homogeneity in terms of a simple parametric model for the expected cell values, we are forced to examine marginal homogeneity via a different route. We ask: Can we construct a model for the expected cell values which would, with the addition of marginal homogeneity, be the same as symmetry?
One such model is

$$\log m_{ij} = u + u_{1(i)} + u_{2(j)} + u_{12(ij)}, \tag{8.2-18}$$

where

$$u_{12(ij)} = u_{12(ji)}, \tag{8.2-19}$$

$$\sum_i u_{1(i)} = \sum_j u_{2(j)} = \sum_i u_{12(ij)} = 0. \tag{8.2-20}$$

Caussinus [1965] refers to (8.2-18) and (8.2-19) as the model of *quasi symmetry*. Here, if we assume that

$$m_{i+} = m_{+i} \qquad i = 1, \ldots, I, \tag{8.2-21}$$

then

$$e^{u_{1(i)}} = e^{u_{2(i)}} h_i \tag{8.2-22}$$

where

$$h_i = \frac{\sum_k e^{u_{1(k)} + u_{12(ik)}}}{\sum_j e^{u_{2(j)} + u_{12(ij)}}}.$$

Then if we substitute (8.2-22) into the definition of h_i we find that the $\{h_i\}$ satisfy the equations

$$h_i = \sum_k w_{ik} h_k \tag{8.2-23}$$

where

$$w_{ik} = \frac{e^{u_{2(k)} + u_{12(ik)}}}{\sum_j e^{u_{2(j)} + u_{12(ij)}}}.$$

Because $w_{i+} = 1$ and $w_{ik} > 0$ the only solution to (8.2-23) is $h_i = h$ for $i = 1, \ldots, I$. Since $\sum_i u_{2(i)} = \sum_i u_{1(i)} = 0$ we therefore have $u_{1(i)} = u_{2(i)}$. Thus if we let H_S, H_{QS}, and H_{MH} denote the models of symmetry, quasi symmetry, and marginal homogeneity, respectively, we can write

$$H_S \equiv H_{QS} \cap H_{MH}. \tag{8.2-24}$$

Looking at the constraints (8.2-19) and (8.2-20), we see that there are $(I - 1) \times (I - 2)/2$ degrees of freedom associated with the model of quasi symmetry. Subtracting $(I - 1)(I - 2)/2$ from the degrees of freedom $I(I - 1)/2$ associated with the model of symmetry, we get $I - 1$, the degrees of freedom associated with marginal homogeneity.

Caussinus [1965] defined the model of quasi symmetry in a multiplicative form:

$$H_{QS}: m_{ij} = a_i b_j d_{ij}, \tag{8.2-25}$$

where

$$d_{ij} = d_{ji} \qquad i \neq j. \tag{8.2-26}$$

If we let

$$c_{ij} = d_{ij} b_i b_j, \tag{8.2-27}$$

then (8.2-25) implies that

$$m_{ij} = e_i c_{ij}, \tag{8.2-28}$$

where $e_i = a_i/b_i$ and $c_{ij} = c_{ji}$ for $i \neq j$. By eliminating the $\{e_i\}$, we have the alternative definition of quasi symmetry given by Caussinus:

$$\frac{m_{ij}}{m_{ji}} \cdot \frac{m_{jk}}{m_{kj}} \cdot \frac{m_{ki}}{m_{ik}} = 1, \tag{8.2-29}$$

which holds when i, j, and k are all different as well as when any pair of i, j, and k are equal.

Quasi independence implies quasi symmetry

For $I = 2$, the degrees of freedom associated with quasi symmetry equal zero, and the model of quasi symmetry implies no restrictions at all for the $\{m_{ij}\}$. For $I = 3$, (8.2-29) implies that

$$\frac{m_{12} m_{23} m_{31}}{m_{21} m_{32} m_{13}} = 1. \tag{8.2-30}$$

Taking logarithms, we have

$$\log m_{12} + \log m_{23} + \log m_{31} - \log m_{21} - \log m_{32} - \log m_{13} = 0, \tag{8.2-31}$$

which is the log contrast that defines quasi independence in a 3×3 table with $m_{ii} = 0$ for $i = 1, 2, 3$ (see Chapter 5). More generally, since the model of quasi symmetry places no restrictions on the diagonal cells, we have the MLEs

$$\hat{m}_{ii} = x_{ii}, \quad i = 1, \ldots, I, \tag{8.2-32}$$

and we can remove the diagonal cells from the model. Once we have removed these cells, for $I = 3$ quasi symmetry is equivalent to quasi independence. For $I > 3$, with the diagonal cells removed, quasi symmetry is implied by quasi independence. This result follows immediately from the multiplicative definition (8.2-25) of quasi symmetry if we set $d_{ij} = 1$ for $i \neq j$.

Example 8.2-2   A herd of ewes

Tallis [1962] considered the data in table 8.2-5, where 227 Merino ewes are cross-classified by the number of lambs born to them in two consecutive years, 1952 and 1953. A cursory examination of the data reveals that the number of lambs born to a ewe in 1953 is related to the number in 1952 (i.e., they are not independent). Moreover, the table is not symmetric, nor does marginal homogeneity appear to hold.

Table 8.2-5   Cross-Classification of Ewes According to Number of Lambs Born in Consecutive Years

Number of Lambs 1953	Number of Lambs 1952			Totals
	0	1	2	
0	58	52	1	111
1	26	58	3	87
2	8	12	9	29
Totals	92	122	13	227

Source: Tallis [1962].

Turning to the model of quasi symmetry, we compute the expected values for off-diagonal cells using the algorithm from Chapter 5 for the model of quasi independence. We give these expected values in table 8.2-6, and the usual goodness-of-fit statistics take the values $X^2 = 1.31$ and $G^2 = 1.35$, with one degree of freedom. The model clearly fits the data well, and we can interpret it in either of two ways. Most ewes have the same number of lambs in each of the two years, and

Table 8.2-6   Expected Values of Data in Table 8.2-5 under Quasi Symmetry or Quasi Independence

Number of Lambs 1953	Number of Lambs 1952			Totals
	0	1	2	
0	58	50.99	2.01	111
1	27.01	58	1.99	87
2	6.99	13.01	9	29
Totals	92	122	13	227

1. if we condition on only those ewes for which the number of lambs born differed for the two years, then for a given ewe, the number born in 1952 is independent of the number born in 1953;
2. if we take into account the disparity between the marginal distributions for 1952 and 1953, there is symmetry in the table. ∎ ∎

Computing MLEs

Assuming a multinomial sampling model, we write the kernel of the log-likelihood under the model of quasi symmetry as

$$\sum_{i,j} x_{ij} \log m_{ij} = x_{++}u + \sum_i x_{i+}u_{1(i)} + \sum_j x_{+j}u_{2(j)} + \sum_{i,j} \left[\frac{x_{ij} + x_{ji}}{2} \right] u_{12(ij)}. \quad (8.2\text{-}33)$$

The sufficient statistics for this model are the configurations $C_1 = \{x_{i+}\}$, $C_2 = \{x_{+j}\}$, and $S_{12} = \{x_{ij} + x_{ji}\}$, and the MLEs for the expected cell values must satisfy the following equations (see Caussinus [1965]):

$$\hat{m}_{i+} = x_{i+} \qquad i = 1, \ldots, I, \qquad (8.2\text{-}34)$$

$$\hat{m}_{+i} = x_{+i} \qquad i = 1, \ldots, I, \qquad (8.2\text{-}35)$$

$$\hat{m}_{ij} + \hat{m}_{ji} = x_{ij} + x_{ji} \qquad i \neq j. \qquad (8.2\text{-}36)$$

It therefore follows that $\hat{m}_{ii} = x_{ii}$ for $i = 1, \ldots, I$, and we can remove the diagonal cells from our analysis.

Caussinus has suggested removing the diagonal cells from (8.2-34) and (8.2-35) and then using a version of the iterative proportional fitting algorithm (see Chapters 3 and 5), adjusting in turn for each of the three configurations C_1, C_2, and S_{12}. As initial values he takes $\hat{m}_{ij}^{(0)} = 1$ for $i \neq j$, and $\hat{m}_{ii}^{(0)} = 0$. The convergence of this procedure follows directly from the more general results of Darroch and Ratcliff [1972].

Using three-dimensional representations of quasi symmetry

We can transform the iterative proportional fitting approach of Caussinus into our usual algorithm for fitting no three-factor interaction in a three-dimensional table. As in Section 8.2.2, we take a duplicate of the original table, flip it over, and place it below the original, yielding a three-dimensional table of expected values

$$m_{ijk} = \begin{cases} m_{ij} & k = 1 \\ m_{ji} & k = 2, \end{cases} \qquad (8.2\text{-}37)$$

and a three-dimensional table of observed values

$$x_{ijk} = \begin{cases} x_{ij} & k = 1 \\ x_{ji} & k = 2. \end{cases} \qquad (8.2\text{-}38)$$

Then equations (8.2-34) and (8.2-35) imply that

$$\hat{m}_{i+k} = x_{i+k},$$
$$\hat{m}_{+jk} = x_{+jk}, \qquad (8.2\text{-}39)$$

for $i = 1, \ldots, I, j = 1, \ldots, I$, and $k = 1, 2$. Moreover, (8.2-36) implies that

$$\hat{m}_{ij+} = x_{ij+} \tag{8.2-40}$$

for $i = 1, \ldots, I; j = 1, \ldots, I$. Expressions (8.2-39) and (8.2-40) are the ML equations for the no three-factor interaction model $u_{123} = 0$ in our three-dimensional table. Using our usual iterative proportional fitting procedure to solve these equations appears to be equivalent to the use of Caussinus's iterative approach. Before we use this approach, we must verify that $u_{123} = 0$ follows from the quasi-symmetry definition. Taking expression (8.2-29), we translate it into our three-dimensional notation:

$$\frac{m_{ij1}m_{jk1}}{m_{ik1}m_{jj1}} \bigg/ \frac{m_{ij2}m_{jk2}}{m_{ik2}m_{jj2}} = \frac{m_{ij1}m_{jk1}m_{ik2}}{m_{ij2}m_{jk2}m_{ik1}} \cdot \frac{m_{jj2}}{m_{jj1}}$$

$$= \frac{m_{ij}}{m_{ji}} \cdot \frac{m_{jk}}{m_{kj}} \cdot \frac{m_{ki}}{m_{ik}} \tag{8.2-41}$$

$$= 1$$

for all $i \neq j \neq k$. Expression (8.2-41) implies that $u_{123} = 0$ for the three-dimensional table.

Since the second layer in our three-dimensional array is a duplicate of the first (with subscripts on the first two variables interchanged), the estimated expected values in the first level will be duplicated in the second level. Moreover,

$$\hat{m}_{iik} = x_{iik} = x_{ii} \qquad k = 1, 2. \tag{8.2-42}$$

Thus we should divide goodness-of-fit statistics for the three-dimensional array by a factor of 2 to get the correct values for the original data structure. We cannot compute the degrees of freedom for the quasi-symmetry model directly from this three-dimensional representation except by counting the independent u-terms in the model, a process equivalent to our original calculation of the degrees of freedom, where the number of degrees of freedom is $(I - 1)(I - 2)/2$.

As in our discussion of symmetry, we can also deal with the other three-dimensional representation of the data where

$$m'_{ij1} = \begin{cases} m_{ij} & i > j \\ 0 & \text{otherwise,} \end{cases} \tag{8.2-43}$$

$$m'_{ij2} = \begin{cases} m_{ji} & i > j \\ 0 & \text{otherwise.} \end{cases} \tag{8.2-44}$$

By analogy with our quasi-symmetry model in the first three-dimensional approach above, we can fit the model $u_{123} = 0$ to this incomplete structure. This model is similar to the model of quasi symmetry, but it is adjusted to preserve the marginal totals for the two triangular wedges rather than the marginal totals for the original table.

We calculate the degrees of freedom for this model using the usual formulas for incomplete tables:

$$\text{d.f.} = \frac{(I - 2)(I - 3)}{2}. \tag{8.2-45}$$

The formula for degrees of freedom here is different from that for quasi symmetry because we have introduced extra parameters associated with the marginal totals of the triangular wedges. This implies that I must be greater than 3 for us to make use of this model, whereas we can use the model of quasi symmetry if $I = 3$. We list the degrees of freedom associated with each u-term in table 8.2-7.

Table 8.2-7 Degrees of Freedom Associated with Each u-term in a Triangular Wedge of Dimension $(I - 1) \times (I - 1) \times 2$

u-Term	Degrees of Freedom
u	1
u_1	$I - 2$
u_2	$I - 2$
u_3	1
u_{12}	$(I - 2)(I - 3)/2$
u_{13}	$I - 2$
u_{23}	$I - 2$
u_{123}	$(I - 2)(I - 3)/2$
Total	$I(I - 1)$

When we unfold this incomplete table, the estimates for the adjusted quasi-symmetry model are such that there is equality of cross-product ratios on either side of the diagonal, i.e.,

$$\frac{\hat{m}_{ij}\hat{m}_{kl}}{\hat{m}_{kj}\hat{m}_{il}} = \frac{\hat{m}_{ji}\hat{m}_{lk}}{\hat{m}_{jk}\hat{m}_{li}}, \tag{8.2-46}$$

where i, j, k, and l are mutually different and so ordered that all cells on one side of the equation fall on the same side of the diagonal. Expression (8.2-46), however, is not equivalent to the multiplicative definition of quasi symmetry (see expression (8.2-29)) unless we equate two subscripts, for instance putting $k = j$, and we cannot do this, as we have specifically excluded the diagonal cells from our model. The effect of the extra restrictions in the triangular three-dimensional approach (see expressions (8.2-43) and (8.2-44)) is that the estimates for the four cells (1, 2), (2, 1), $(I - 1, I)$, and $(I, I - 1)$ are identical to the original observations. In example 8.2-3, we show how these cells may be related to the others by rearranging the data so that cells in lines parallel to the diagonal become the columns of the new array. Each such arrangement excludes a different four cells from the analysis.

Example 8.2-3 Further analysis of unaided vision data

In example 8.2-1 we examined a set of unaided vision data using the model of symmetry, which we found did not fit the data well. We now consider the model of quasi symmetry for these data. We give the expected values under this model in table 8.2-8. The goodness-of-fit test statistics have values $G^2 = 7.27$ and $X^2 = 7.26$, with three degrees of freedom. These values lie between the upper 5% and 10% tail values of the χ^2 distribution with three degrees of freedom, and so quasi symmetry is an adequate model for the vision data, providing a significant improvement over the symmetry model ($G^2 = 19.25 - 7.27 = 11.98$, with three degrees of freedom).

Table 8.2-8   Expected Values for Unaided Distant Vision Data
from Table 8.2-1 under Quasi-Symmetry Model

Right Eye Grade	Left Eye Grade (1)	(2)	(3)	(4)	Totals
(1)	1,520	263.4	133.6	59.0	1,976
(2)	236.6	1,512	419.0	88.4	2,256
(3)	107.4	375.0	1,772	201.6	2,456
(4)	43.0	71.6	182.4	492	789
Totals	1,907	2,222	2,507	841	7,477

In our earlier analysis we also considered the adjusted symmetry model, which preserves the off-diagonal totals and results from the incomplete table representation of the data. While this model fits the data quite well, for illustrative purposes we now proceed further, using the other models for the incomplete array just developed. Thus we continue with the analysis of the data as displayed in table 8.2-3. We refer to this display as "arrangement A." Fitting the adjusted model of quasi symmetry ($u_{123} = 0$), we get the expected values in table 8.2-9. The goodness-of-fit statistics are $G^2 = 6.79$ and $X^2 = 6.72$, with one degree of freedom. Thus the extra four parameters do not yield a substantially better fitting model than the best one we found in example 8.2-1.

Table 8.2-9   Expected Values for Unaided Vision Data from Table 8.2-4, under
Adjusted Quasi-Symmetry Model

Worse Eye	Left Eye Better Better Eye (1)	(2)	(3)	Totals	Right Eye Better Better Eye (1)	(2)	(3)	Totals
(2)	234	—	—	234	266	—	—	266
(3)	108.2	370.8	—	479	132.8	423.2	—	556
(4)	44.8	73.2	179	297	57.2	86.8	205	349
Totals	387	444	179	1,010	456	510	205	1,171

The single log-linear contrast that corresponds to the adjusted quasi-symmetry model just fitted is rather difficult to interpret, but we can rearrange the triangular arrays as in tables 8.2-10 and 8.2-11 by considering the number of grades of difference between the eyes. In table 8.2-10 we have tabulated the difference according to the grade of the better eye, and in table 8.2-11 according to the grade of the worse eye. Fitting the model with $u_{123} = 0$ to each of these, we obtain $G^2 = 0.44$ and 0.078, respectively. In each case, the linear contrast we are examining is based on a different set of eight cells, but the pairs defined as (1, 4) and (4, 1) in the original table are only included in the model for arrangement A. Thus these two cells are the largest contributors to the possible three-factor effect. If we believed the goodness-of-fit statistics based on arrangement A were sufficiently large to be important, we would conclude that we more frequently encounter the extreme disparity of one eye being grade 1 and the other grade 4 when the right eye is the better eye, even when we take into account the preponderance of women with a better right eye. ∎∎

Table 8.2-10 Second Three-Dimensional Arrangement (Arrangement B) of Vision Data and Estimates Obtained under Model $u_{123} = 0$, for Incomplete Layout in Table 8.2-3

		Left Eye Better				Right Eye Better			
	Grade of Better Eye	Difference in Grade				Difference in Grade			
		(1)	(2)	(3)	Totals	(1)	(2)	(3)	Totals
Observed	(1)	234	117	36	387	266	124	66	456
	(2)	362	82	—	444	432	78	—	510
	(3)	179	—	—	179	205	—	—	205
	Totals	775	199	36	1,010	903	202	66	1,171
Fitted	(1)	231.1	119.9	36	387	268.9	121.1	66	456
	(2)	364.9	79.1	—	444	429.1	80.9	—	510
	(3)	179	—	—	179	205	—	—	205
	Totals	775	199	36	1,010	903	202	66	1,171

Note: $X^2 = 0.4489$, $G^2 = 0.4488$.

Table 8.2-11 Third Three-Dimensional Arrangement (Arrangement C) of Vision Data and Estimates Obtained under Model $u_{123} = 0$, for Incomplete Layout in Table 8.2-3

		Left Eye Better				Right Eye Better			
	Grade of Worse Eye	Difference in Grade				Difference in Grade			
		(1)	(2)	(3)	Totals	(1)	(2)	(3)	Totals
Observed	(2)	234	—	—	234	266	—	—	266
	(3)	362	117	—	479	432	124	—	556
	(4)	179	82	36	297	205	78	66	349
	Totals	775	199	36	1,010	903	202	66	1,171
Fitted	(2)	234	—	—	234	266	—	—	266
	(3)	360.8	118.2	—	479	433.2	122.8	—	556
	(4)	180.2	80.8	36	297	203.8	79.2	66	349
	Totals	775	199	36	1,010	903	202	66	1,171

Note: $X^2 = 0.0784$, $G^2 = 0.0784$.

8.2.4 Marginal homogeneity

Conditional test for marginal homogeneity

In the preceding section we noted that

$$H_S \equiv H_{QS} \cap H_{MH}, \qquad (8.2\text{-}47)$$

where H_S, H_{QS}, and H_{MH} denote the models of symmetry, quasi symmetry, and marginal homogeneity. Thus the difference between the likelihood ratio goodness-of-fit statistics for H_S and H_{QS} is a test statistic for H_{MH}, conditional on H_{QS}

being true, with degrees of freedom equal to

$$\frac{I(I-1)}{2} - \frac{(I-1)(I-2)}{2} = I - 1. \qquad (8.2\text{-}48)$$

Caussinus [1965] was the first to suggest the use of this statistic.

We note that using this conditional test statistic does not yield expected cell estimates for the model of marginal homogeneity.

Example 8.2-4   Marginal homogeneity for vision data

In examples 8.2-1 and 8.2-3 we fitted the models of symmetry and quasi symmetry to the unaided vision data. The quasi-symmetry model provided an adequate fit to the data, so it seems reasonable for us to test for marginal homogeneity conditional on quasi symmetry being true. The goodness-of-fit statistic has a value of $G^2 = G_S^2 - G_{QS}^2 = 19.25 - 7.27 = 11.98$, with three degrees of freedom. This value is just greater than the 1% tail value of a chi square variable with three degrees of freedom, and thus we conclude that the marginal distribution for vision in the right eye is different from the marginal distribution for the left eye. ■■

Unconditional test and cell estimates for marginal homogeneity

We can take a more direct approach to fitting the model of marginal homogeneity to square tables by maximizing the kernel of the log-likelihood function

$$\sum x_{ij} \log m_{ij}, \qquad (8.2\text{-}49)$$

subject to the $I - 1$ constraints

$$m_{i+} = m_{+i} \qquad i = 1, \ldots, I \qquad (8.2\text{-}50)$$

(see Madansky [1963]). Thus we must maximize the Lagrangian

$$L = \sum_{ij} x_{ij} \log m_{ij} - \mu\left(\sum_{ij} m_{ij} - N\right) - \sum_{i=1}^{I} \lambda_i(m_{i+} - m_{+i}) \qquad (8.2\text{-}51)$$

with respect to the $\{m_{ij}\}$, μ, and the $\{\lambda_i\}$. Setting the partial derivatives of L equal to zero, we obtain the I^2 equations:

$$\frac{x_{ij}}{m_{ij}} - \mu - (\lambda_i - \lambda_j) = 0, \qquad (8.2\text{-}52)$$

which we must solve subject to the constraints (8.2-50) and $\sum_{ij} m_{ij} = N$.

If we first rewrite (8.2-52) as

$$x_{ij} = m_{ij}(\mu + \lambda_i - \lambda_j), \qquad (8.2\text{-}53)$$

then summing over all i and j yields $\mu = 1$. Thus we can write the estimated expected values in the form

$$\hat{m}_{ij} = \frac{x_{ij}}{1 + (\lambda_i - \lambda_j)}, \qquad (8.2\text{-}54)$$

subject to the constraints

$$\sum_j \frac{x_{ij}}{1 + (\lambda_i - \lambda_j)} = \sum_j \frac{x_{ji}}{1 + (\lambda_j - \lambda_i)}. \qquad (8.2\text{-}55)$$

In particular, we note that

$$\hat{m}_{ii} = x_{ii} \qquad i = 1, \ldots, I, \tag{8.2-56}$$

and that

$$\frac{\hat{m}_{ij}\hat{m}_{II}}{\hat{m}_{iI}\hat{m}_{Ij}} = \frac{x_{ij}x_{II}}{x_{iI}x_{Ij}} \bigg/ \frac{1 + \lambda_i - \lambda_j}{1 + \lambda_i - \lambda_j + (\lambda_i - \lambda_I)(\lambda_I - \lambda_j)} \tag{8.2-57}$$

for $i \neq I$ and $j \neq I$. Expression (8.2-57) implies that the MLEs of the cross-product ratios under marginal homogeneity are not the same as the observed cross-product ratios unless $x_{i+} = x_{+i}$ for $i = 1, \ldots, I$, in which case $\hat{m}_{ij} = x_{ij}$ for all cells. Moreover,

$$\hat{m}_{ij}\hat{m}_{ji} \geqq x_{ij}x_{ji}, \tag{8.2-58}$$

with equality if and only if $\lambda_i = \lambda_j$.

Since the likelihood equations (8.2-54) and (8.2-55) are nonlinear, we must use some iterative method to solve them. Because the likelihood function (8.2-49) is convex and the constraints (8.2-50) are linear, the maximization problem is a standard one from nonlinear programming (see Zangwill [1969]), and we can solve it using nonlinear programming techniques. Madansky [1963] describes a linear approximation or gradient method for handling the maximization, but because of the complexity of the actual algorithm we do not include it here. Alternatively, we can use the convex simplex method for maximization of (8.2-49) subject to (8.2-50), or one of its variants (again, see Zangwill [1969]).

Once we have computed estimates of the expected cell values under marginal homogeneity, we can use either of the usual goodness-of-fit statistics to check the fit of the model, e.g.,

$$G^2 = 2 \sum_{i \neq j} x_{ij} \log \frac{x_{ij}}{\hat{m}_{ij}}, \tag{8.2-59}$$

with $I - 1$ degrees of freedom.

Ireland, Ku, and Kullback [1969] consider the problem of finding estimated cell values under the model of marginal homogeneity, but they use a modified version of minimum discrimination information estimation. The estimates they derive have the form:

$$\tilde{m}_{ij} = c \left(\frac{a_i}{a_j} \right) x_{ij}, \tag{8.2-60}$$

where the $\{a_i\}$ are determined from the marginal constraints (8.2-50) and c is a normalization constant, i.e.,

$$c = \frac{N}{\sum_{ij} \left(\frac{a_i}{a_j} \right) x_{ij}}. \tag{8.2-61}$$

We note that these estimates are not the same as the MLEs and that they satisfy:

$$\frac{\tilde{m}_{ij}\tilde{m}_{II}}{\tilde{m}_{iI}\tilde{m}_{Ij}} = \frac{x_{ij}x_{II}}{x_{iI}x_{Ij}} \tag{8.2-62}$$

for $i \neq I$ and $j \neq I$, unlike the MLEs. Moreover, $\tilde{m}_{ii} \geq x_{ii}$, whereas $\hat{m}_{ii} = x_{ii}$ for $i = 1, \ldots, N$. (See Chapter 10 for a discussion of the modified minimum discrimination information approach, and for related results regarding the model of symmetry discussed earlier in this chapter.)

Ireland, Ku, and Kullback also consider an alternate model for marginal homogeneity based only on the off-diagonal cells. The expected cell values are again of the form (8.2-60), but with the $\{\tilde{m}_{ii}\}$ equal to the $\{x_{ii}\}$. For both models, they suggest the use of an iterative proportional fitting algorithm to get the estimated expected values.

8.2.5 Zero diagonals in square tables

Many square contingency tables result from situations where it is impossible for the row and column categories to be the same, i.e., we have structural zeros down the diagonal of the table. Since for the models of symmetry, quasi symmetry, and marginal homogeneity we have $\hat{m}_{ii} = x_{ii}$, the entries in the diagonal cells do not affect the expected values in the off-diagonal cells. Thus for square tables with zero diagonals, we can use the same formulas for expected values (for the off-diagonal cells), the same test statistics, and the same degrees of freedom as for square tables with nonzero diagonals.

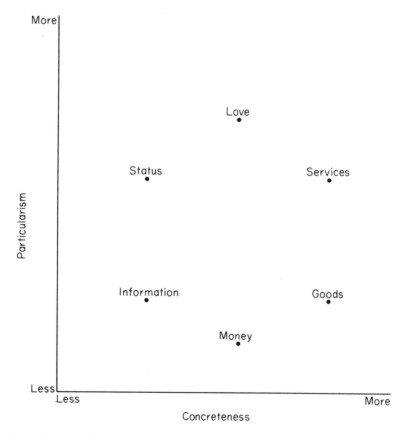

Figure 8.2-1   Position of the Six Resource Classes of Foa [1971] Plotted on the Two Coordinates of Particularism and Concreteness.

Example 8.2-5   Interpersonal behavior

In a study of the exchange of material and psychological resources through interpersonal behavior, Foa [1971] postulated that six resource categories (love, status, information, money, goods, and services) have a cyclic ordering, which results from the positioning of the categories according to the two coordinates of "particularism" and "concreteness," as illustrated in figure 8.2-1. Each resource has two neighbors, one on each side, e.g., love is nearest to status and services while being most distal from money. The cyclic ordering of the categories plays a key role in the model-building for this data set.

Each of a group of 37 subjects received a series of eighteen messages (three for each of the six categories) and was asked to return a message from an attached deck of messages, the one most similar to the message just received. In the deck of messages provided to the subject all classes were represented with the exception of the one to which the message received belonged. Thus the subject was denied the option of returning a message from the same class. The receipt of the message can be viewed as occurring at time T_1 and the return at time T_2. The resulting data are given in table 8.2-12.

Table 8.2-12 Cross-Classification of Resource Returned by Resource Received (Where Resource Returned Was to be Most Similar to Resource Received)

Resource Received	Resource Returned						Totals
	Love	Status	Information	Money	Goods	Services	
Love	—	73	11	0	2	25	111
Status	69	—	22	11	3	6	111
Information	19	38	—	12	27	15	111
Money	0	18	9	—	66	18	111
Goods	7	6	23	61	—	14	111
Services	46	20	8	18	19	—	111
Totals	141	155	73	102	117	78	666

Source: Foa [1971].

We can see from a cursory examination of this table that the model of quasi homogeneity of proportions (i.e., the model of quasi independence) is inappropriate, and the model of symmetry also appears to be inappropriate. Note that we cannot use the usual test statistics to test for symmetry in this table since the row totals are fixed. We give the expected values under quasi symmetry in table 8.2-13. The goodness-of-fit statistics take values $X^2 = 5.2$ and $G^2 = 5.2$, with ten degrees of freedom, and the quasi-symmetry model fits the data extremely well.

(In the preceding analysis, we might wish to decrease the number of degrees of freedom for the model of quasi symmetry by 1. Because we know that $x_{14} = x_{41} = 0$ implies that $\hat{m}_{14} = \hat{m}_{41} = 0$ for both models, if we condition on $\hat{m}_{14} = \hat{m}_{41} = 0$, we decrease the number of cells under consideration by 2 and the number of parameters being fitted by 1.)

Foa's hypothesis for these data is that an individual is more likely to substitute for the missing class a proximal rather than a distal one. This appears to be true in table 8.2-13, but for a given missing class the probabilities of substitution are

Table 8.2-13 Expected Values of Data from Table 8.2-12, under Model of
Quasi Symmetry

Resource Received	Love	Status	Resource Returned Information	Money	Goods	Services	Totals
Love	—	72.80	10.05	0.00	3.42	24.73	111
Status	69.20	—	19.43	10.30	3.31	8.76	111
Information	19.95	40.57	—	11.23	27.42	11.84	111
Money	0.00	18.70	9.77	—	65.24	17.28	111
Goods	5.58	5.69	22.58	61.75	—	15.39	111
Services	46.27	17.24	11.16	18.72	17.61	—	111
Totals	141	155	73	102	117	78	666

not monotonic functions of proximity. The quasi-symmetry model implies that
if we adjust for the heterogeneity of the marginal totals, the table is symmetric.
If we were to make all the marginal totals equal to 111, using iterative proportional
fitting, then we might wish to use the inverse of the entry in the (i, j) cell as a
measure of distance between categories i and j. Then we could use some version
of multidimensional scaling (see Shepard [1962a, 1962b] and Kruskal [1964a,
1964b]) to try to produce an adjusted version of figure 8.2-1. ■■

8.2.6 Exercises

1. In Chapter 5 we considered the analysis of two social mobility tables from
   Britain and Denmark. Reexamine these tables, checking on the appropriateness
   of the models of symmetry, quasi symmetry, and marginal homogeneity.
2. The data in table 8.2-14 represent the cross-classification of a sample of
   individuals according to their socioprofessional category in 1954 and in 1962
   (see Caussinus [1965]). Show that the model of quasi symmetry provides an
   adequate fit to the data.

Table 8.2-14 Cross-Classification of Individuals According
to Socioprofessional Status

Status in 1954	Status in 1962 1	2	3	4	5	6
1	187	13	17	11	3	1
2	4	191	4	9	22	1
3	22	8	182	20	14	3
4	6	6	10	323	7	4
5	1	3	4	2	126	17
6	0	2	2	5	1	153

Source: Caussinus [1965].

3. In table 8.2-15 we give further data from Foa [1971] on interpersonal behavior.
   These data are similar to those discussed in example 8.2-5, but here the resource
   returned was to be most unlike, rather than most like, the resource received.
   Fit the model of quasi symmetry to these data.

Table 8.2-15 Cross-Classification of Resource Returned by Resource Received (Where Resource Returned Was to be Most Unlike the Resource Received)

Resource Received	Resource Returned						Totals
	Love	Status	Information	Money	Goods	Services	
Love	—	5	21	48	29	8	111
Status	4	—	19	27	30	31	111
Information	20	11	—	20	25	35	111
Money	56	10	21	—	4	20	111
Goods	42	18	27	6	—	18	111
Services	12	20	37	26	16	—	111
Totals	134	64	125	127	104	112	666

Source: Foa [1971].

8.3 Three-Dimensional Tables

8.3.1 *Different kinds of symmetry and marginal homogeneity*

As we mentioned in Section 8.1, there are several ways to generalize the models of symmetry, quasi symmetry, and marginal homogeneity when we go from two dimensions to three or more. For example, in a d-dimensional table, we can examine the c-dimensional margins (for $c = 1, \ldots, d - 1$) for homogeneity, and for fixed $c > 1$ there are different kinds of homogeneity to explore. For the $I \times I \times I \times I$ table with expected cell values m_{ijkl}, Kullback [1971] considers the following models of marginal homogeneity:

1. $m_{i+++} = m_{+i++} = m_{++i+} = m_{+++i}$ $\quad i = 1, \ldots, I$;
2. $m_{ij++} = m_{i+j+} = m_{i++j} = m_{+ij+} = m_{+i+j} = m_{++ij}$ $\quad i, j = 1, \ldots, I$;
3. $m_{ijk+} = m_{ij+k} = m_{i+jk} = m_{+ijk}$ $\quad i, j, k = 1, \ldots, I$.

While model 3 implies 2, and 2 implies 1, there are other homogeneity models for the two- and three-dimensional margins, which involve the symmetry of the margins (i.e., interchange of subscripts):

2′. $m_{ij++} = m_{ji++} = m_{i+j+} = m_{i++j} = m_{+ij+} = m_{+i+j} = m_{++ij}$;
3′. $m_{ijk+} = m_{ikj+} = m_{jik+} = m_{ij+k} = m_{i+jk} = m_{+ijk}$;

where model 3′ implies 3 and 2′, and model 2′ implies 2.

In this section we explore various models of symmetry, quasi symmetry, and marginal homogeneity for three-dimensional tables. In doing so we again exploit, wherever possible, log-linear model representations. As with models for two-dimensional marginal homogeneity, we find that models of marginal homogeneity for three-dimensional tables can be handled in our log-linear model framework only in the context of conditional tests involving models for symmetry and quasi symmetry.

8.3.2 *Conditional symmetry and marginal homogeneity*

The simplest models involving symmetry and marginal homogeneity in three dimensions are based directly on the models of Section 8.2 for two-dimensional tables. If we consider an $I \times I \times K$ table then conditional on the level of variable 3 we can use these models directly, i.e., we have the models of:

1. conditional symmetry

$$m_{ijk} = m_{jik} \qquad i, j = 1, \dots, I; \quad k = 1, \dots, K; \qquad (8.3\text{-}1)$$

2. conditional marginal homogeneity

$$m_{i+k} = m_{+ik} \qquad i = 1, \dots, I; \quad k = 1, \dots, K. \qquad (8.3\text{-}2)$$

The analyses of these models are based directly on the analysis of each of the K layers of the table using the methods of Section 8.2. The sampling model may either be multinomial or product multinomial with the layer totals fixed.

The model (8.3-1) of conditional symmetry has $KI(I - 1)/2$ degrees of freedom, and the goodness-of-fit statistics can be partitioned into K components of the form (8.2-11) or (8.2-12) computed separately for each table, each with $I(I - 1)/2$ degrees of freedom. The maximum likelihood estimates for the expected cell values under conditional marginal homogeneity are of the form (8.2-54), and we can get a conditional test statistic with $I - 1$ degrees of freedom for each layer of the table, as in Section 8.2.4.

Example 8.3-1   More on social mobility tables

The two 5×5 social mobility tables from Britain and Denmark can be placed on top of one another to form a $5 \times 5 \times 2$ table with the layer totals fixed by sampling design. All the information needed to test for conditional marginal homogeneity and symmetry is contained in the answer to exercise 1 in Section 8.2.6, and we summarize the information here in table 8.3-1. The overall test for conditional symmetry is based on twenty degrees of freedom, and for these data we get the test statistic values $X^2 = 67.37$ and $G^2 = 71.00$. The overall conditional test for conditional marginal homogeneity is based on eight degrees of freedom, and the test statistics take values $X^2 = 56.39$ and $G^2 = 59.95$. Due to the large values of both sets of test statistics, we conclude that neither conditional symmetry nor conditional marginal homogeneity adequately describes these data. ■■

Table 8.3-1   Summary of Tests for Symmetry and Quasi Symmetry on British and Danish Social Mobility Data

	British			Danish		
	Degree of Freedom	X^2	G^2	Degree of Freedom	X^2	G^2
Symmetry	10	42.95	46.20	10	24.42	24.80
Quasi Symmetry	6	4.67	4.58	6	6.31	6.47
Marginal Homogeneity (Conditional Test)	4	38.28	41.62	4	18.11	18.33

8.3.3   Complete symmetry

The model and degrees of freedom

By "complete symmetry" we mean that a table of cell values is invariant in the permutation of the subscripts indexing the values. For two-dimensional tables, complete symmetry is equivalent to the model of symmetry, i.e., $m_{ij} = m_{ji}$ for all i and j. For three-dimensional $I \times I \times I$ tables, the model of complete symmetry is

$$m_{ijk} = m_{ikj} = m_{jik} = m_{jki} = m_{kji} = m_{kij} \qquad (8.3\text{-}3)$$

for all i, j, and k. Madansky [1963] refers to (8.3-3) as the "interchangeability hypothesis."

We get the degrees of freedom for model (8.3-3) by breaking the I^3 cells into three groups:

(i)   there are I cells with all three subscripts equal, and these remain unaffected by (8.3-3);
(ii) there are $I(I - 1)$ cells with the first two subscripts equal but different from the third, and (8.3-3) matches each of these with one of the $I(I - 1)$ cells with the first and third subscripts equal but different from the second, and with one of the $I(I - 1)$ cells with the second and third subscripts equal but different from the first;
(iii) there are $I(I - 1)(I - 2)$ cells with all subscripts taking different values, and these can be grouped in sets of six, given by the six permutations of the three subscripts.

From (i) we get 0 degrees of freedom, from (ii) we get $2I(I - 1)$, and from (iii) we get $\frac{5}{6}I(I - 1)(I - 2)$. Summing these contributions and rearranging terms, we get

$$\text{d.f.} = \frac{I(I - 1)(5I + 2)}{6}. \qquad (8.3\text{-}4)$$

For $I = 2$, there are no cells in the third category, and there are four degrees of freedom.

We can represent the complete symmetry in terms of the log-linear model

$$\log m_{ijk} = u + u_{1(i)} + u_{2(j)} + u_{3(k)} + u_{12(ij)} + u_{13(ik)} + u_{23(jk)} + u_{123(ijk)} \qquad (8.3\text{-}5)$$

by setting

$$u_{1(i)} = u_{2(i)} = u_{3(i)} \qquad i = 1, \ldots, I, \qquad (8.3\text{-}6a)$$

$$u_{12(ij)} = u_{12(ji)} = u_{13(ij)} = u_{23(ij)} \qquad i, j = 1, \ldots, I, \qquad (8.3\text{-}6b)$$

$$u_{123(ijk)} = u_{123(ikj)} = u_{123(jik)} = u_{123(jki)} = u_{123(kji)} = u_{123(kij)} \qquad i, j, k = 1, \ldots, I.$$
$$(8.3\text{-}6c)$$

Suppose we have observed cell values $\{x_{ijk}\}$ and a multinomial sampling scheme. Then the kernel of the log-likelihood is

$$\sum_{ijk} x_{ijk} \log m_{ijk} = \sum_i (x_{i++} + x_{+i+} + x_{++i}) u_{1(i)}$$

$$+ \sum_{i,j} \left[\frac{x_{ij+} + x_{i+j} + x_{+ij} + x_{ji+} + x_{j+i} + x_{+ji}}{2} \right] u_{12(ij)}$$

$$+ \sum_{i,j,k} \left[\frac{x_{ijk} + x_{ikj} + x_{kji} + x_{jik} + x_{jki} + x_{kij}}{6} \right] u_{123(ijk)}. \qquad (8.3\text{-}7)$$

The minimal sufficient statistics are given by the coefficients of $u_{123(ijk)}$ in (8.3-7), and the MLEs for the $\{m_{ijk}\}$ are

$$\hat{m}_{ijk} = \begin{cases} x_{iii} & i = j = k \\[2ex] \dfrac{x_{iik} + x_{iki} + x_{kii}}{3} & i = j \neq k \\[2ex] \dfrac{x_{ijk} + x_{ikj} + x_{kji} + x_{jik} + x_{jki} + x_{kij}}{6} & i \neq j, i \neq k, j \neq k. \end{cases} \tag{8.3-8}$$

We can test the goodness of fit of this model using the standard statistics.

For the $I \times I \times I$ table we may also be interested in the special case of the model of complete symmetry where there is no three-factor interaction, i.e., the constraints on the log-linear model are given by (8.3-6a), (8.3-6b), and $u_{123(ijk)} = 0$ for all i, j, k. The likelihood function for this model is given by (8.3-7) with $u_{123} = 0$, and the minimal sufficient statistics are the coefficients of the $u_{12(ij)}$ in that expression. This model has $(I - 1)^3 - 2I(I - 1)$ degrees of freedom, and the MLEs for the expected cell values cannot be written in closed form. We consider the computation of the MLEs for this model below.

Four-dimensional representation

We can handle the complete symmetry model as a regular log-linear model without any additional side constraints by creating a fourth dimension for the $I \times I \times I$ table. We let

$$m'_{ijkl} = \begin{cases} m_{ijk} & l = 1 \\ m_{ikj} & l = 2 \\ m_{kji} & l = 3 \\ m_{jik} & l = 4 \\ m_{jki} & l = 5 \\ m_{kij} & l = 6 \end{cases} \tag{8.3-9}$$

(and similarly for the x'_{ijkl}), so that for the second through the sixth categories of variable 4 we simply have a rearrangement of the original three-dimensional table based on a permutation of the indexing subscripts. The configuration

$$C'_{123} = \{x'_{ijk+}\} = \{x_{ijk} + x_{ikj} + x_{kji} + x_{jik} + x_{jki} + x_{kij}\} \tag{8.3-10}$$

corresponds to the minimal sufficient statistics for the model of complete symmetry. Thus the hierarchical log-linear model for the four-dimensional representation which corresponds to complete symmetry is

$$u_{4(l)} = 0 \qquad l = 1, \ldots, 6. \tag{8.3-11}$$

Since all cells are replicated six times in this representation, we must divide the corresponding goodness-of-fit statistics by a factor of 6.

This four-dimensional representation of complete symmetry is especially useful when we consider generalizations of quasi symmetry and marginal homogeneity. We can also use it to get MLEs for the special version of complete symmetry with

$u_{123} = 0$, which in the four-dimensional table representation corresponds to setting $u_{123} = u_4 = 0$.

Complete symmetry implies marginal homogeneity

For $I \times I \times I$ tables we consider two types of marginal homogeneity, the first based on one-dimensional margins, i.e.,

$$m_{i++} = m_{+i+} = m_{++i} \qquad i = 1, \ldots, I, \tag{8.3-12}$$

and the second based on two-dimensional margins, i.e.,

$$m_{ij+} = m_{ji+} = m_{i+j} = m_{+ij} \qquad i, j = 1, \ldots, I. \tag{8.3-13}$$

We note that complete symmetry implies two-dimensional marginal homogeneity (8.3-13), which in turn implies one-dimensional marginal homogeneity (8.3-12). Expression (8.3-12) does not imply (8.3-13), however, nor does (8.3-13) imply complete symmetry.

8.3.4 Quasi symmetry

Two models

Since complete symmetry implies the two forms of marginal homogeneity given by (8.3-12) and (8.3-13), we are interested in models which are similar to complete symmetry but do not satisfy the constraints of marginal homogeneity. We propose two such models here for $I \times I \times I$ tables having all the features of complete symmetry except that in one case (8.3-12) does not hold and in the other case neither (8.3-12) nor (8.3-13) holds. We can define these models in terms of constraints on the basic log-linear model (8.3-5), as follows:

1. quasi symmetry preserving two-dimensional margins. There is complete symmetry only for the three-factor u-terms $u_{123(ijk)}$, i.e.,

$$u_{123(ijk)} = u_{123(ikj)} = u_{123(jik)} = u_{123(jki)}$$

$$= u_{123(kji)} = u_{123(kij)} \qquad i, j, k = 1, \ldots, I; \tag{8.3-14}$$

2. quasi symmetry preserving one-dimensional margins. There is complete symmetry for both the three-factor and all of the two-factor u-terms, i.e., (8.3-14) holds and in addition,

$$u_{12(ij)} = u_{12(ji)} = u_{13(ij)} = u_{23(ij)} \qquad i, j = 1, \ldots, I. \tag{8.3-15}$$

For model 2 we only drop the constraints (8.3-6a) from those in the complete symmetry model, so that there are

$$\frac{I(I-1)(5I+2)}{6} - 2(I-1) = \frac{(I-1)(5I^2+2I-12)}{6} \tag{8.3-16}$$

degrees of freedom for model 2. For model 1 we drop both (8.3-6a) and (8.3-6b), the latter corresponding to $(I-1)(I-2)/2 + 2(I-1)^2$. Thus there are

$$\frac{I(I-1)(5I+2)}{6} - 2(I-1) - \frac{(I-1)(I-2)}{2} - 2(I-1)^2$$

$$= \frac{(I-1)(5I-3)(I-2)}{6} \tag{8.3-17}$$

degrees of freedom for model 1. We can only investigate this model when $I \geqq 3$.

If we assume a multinomial sampling model for the $I \times I \times I$ table of observed counts $\{x_{ijk}\}$, the kernel $\sum_{ijk} x_{ijk} \log m_{ijk}$ of the log-likelihood under model 1 becomes:

$$x_{+++}u + \sum_i x_{i++}u_{1(i)} + \sum_j x_{+j+}u_{2(j)} + \sum_k x_{++k}u_{3(k)}$$
$$+ \sum_{i,j} x_{ij+}u_{12(ij)} + \sum_{i,k} x_{i+k}u_{13(ik)} + \sum_{j,k} x_{+jk}u_{23(jk)}$$
$$+ \sum_{i,j,k} \left[\frac{x_{ijk} + x_{ikj} + x_{kji} + x_{jik} + x_{jki} + x_{kij}}{6} \right] u_{123(ijk)}, \quad (8.3\text{-}18)$$

and under model 2 becomes:

$$x_{+++}u + \sum_i x_{i++}u_{1(i)} + \sum_j x_{+j+}u_{2(j)} + \sum_k x_{++k}u_{3(k)}$$
$$+ \sum_{i,j} \left[\frac{x_{ij+} + x_{i+j} + x_{+ji} + x_{ji+} + x_{j+i} + x_{+ij}}{2} \right] u_{12(ij)}$$
$$+ \sum_{i,j,k} \left[\frac{x_{ijk} + x_{ikj} + x_{kji} + x_{jik} + x_{jki} + x_{kij}}{6} \right] u_{123(ijk)}. \quad (8.3\text{-}19)$$

The minimal sufficient statistics for model 1 are $C_{12} = \{x_{ij+}\}$, $C_{13} = \{x_{i+k}\}$, $C_{23} = \{x_{+jk}\}$, and $S_{123} = \{x_{ijk} + x_{ikj} + x_{kji} + x_{jik} + x_{jki} + x_{kij}\}$, and for model 2 are $C_1 = \{x_{i++}\}$, $C_2 = \{x_{+j+}\}$, $C_3 = \{x_{++k}\}$, and S_{123}, since we get the coefficients of u_{12} in (8.3-19) by summing S_{123} over k. Moreover, the MLEs for the expected cell values must satisfy the following equations for model 1:

$$\hat{m}_{ij+} = x_{ij+},$$
$$\hat{m}_{i+k} = x_{i+k}, \quad (8.3\text{-}20)$$
$$\hat{m}_{+jk} = x_{+jk},$$

$$\hat{m}_{ijk} + \hat{m}_{ikj} + \hat{m}_{kji} + \hat{m}_{jik} + \hat{m}_{jki} + \hat{m}_{kij} = x_{ijk} + x_{ikj} + x_{kji} + x_{jik} + x_{jki} + x_{kij},$$
$$(8.3\text{-}21)$$

and for model 2:

$$\hat{m}_{i++} = x_{i++},$$
$$\hat{m}_{+j+} = x_{+j+}, \quad (8.3\text{-}22)$$
$$\hat{m}_{++k} = x_{++k},$$

and (8.3-21). For both models it follows that $\hat{m}_{iii} = x_{iii}$ for $i = 1, \ldots, I$, and we can delete these I cells from our analysis.

We can solve the ML equations for the two three-dimensional quasi-symmetry models by use of a version of the iterative proportional fitting algorithm, where we adjust in turn for each of the four sufficient configurations (C_{12}, C_{13}, C_{23}, and S_{123} for model 1; C_1, C_2, C_3, and S_{123} for model 2). As initial values we take $\hat{m}_{ijk}^{(0)} = 1$.

Using the four-dimensional representation

An alternative approach to computing the estimated expected cell values as described above is available via the four-dimensional representation of Section 8.3.3, where the fourth dimension corresponds to the six possible permutations of the indexing subscripts. In this representation, fitting model 1 corresponds to fitting

$$u_{1234(ijkl)} = 0 \quad \text{for all } i, j, k, l, \tag{8.3-23}$$

and fitting model 2 corresponds to fitting

$$u_{124(ijl)} = u_{134(ikl)} = u_{234(jkl)} = 0 \quad \text{for all } i, j, k, l. \tag{8.3-24}$$

As before, since all the cells are replicated six times, we must divide the corresponding goodness-of-fit statistics by a factor of 6.

Example 8.3-2   1940 voting intentions

In Chapter 7 we examined data from a panel survey of potential voters in Erie County, Ohio in 1940, and we concluded that the data might reasonably be modeled by a stationary second-order Markov chain. We reproduce the $3 \times 3 \times 3$ array of counts of second-order transitions here in table 8.3-2. It is reasonable to explore this array for various symmetries involving time.

Table 8.3-2   Stationary Two-Step Transitions in Panel Study
of Potential Voters in Erie County, Ohio, 1940

| Time $t-2$ | Time $t-1$ | Time t | | | |
		R	D	U	Totals
R	R	557	6	16	579
D	R	18	0	5	23
U	R	71	1	11	83
R	D	3	8	0	11
D	D	9	435	22	466
U	D	6	63	6	75
R	U	17	5	21	43
D	U	4	10	24	38
U	U	62	54	346	462

Note: See Chapter 7. R = Republican, D = Democrat, U = undecided.

We begin by considering the model of complete symmetry, which we interpret as indicating the interchangeability of evenly spaced points in time with respect to voting intentions. For this model $X^2 = 225.5$, $G^2 = 229.8$, and there are seventeen degrees of freedom, so we can reasonably turn to a consideration of models which exhibit less than complete symmetry.

For the model of quasi symmetry preserving one-dimensional margins (model 2), $X^2 = 95.15$, $G^2 = 110.30$, and there are thirteen degrees of freedom. Even though this model yields a highly significant improvement over complete symmetry, it still provides a poor fit to the data. For the model of quasi symmetry preserving two-dimensional margins (model 1), $X^2 = 2.85$, $G^2 = 3.90$, and there are four degrees of freedom. This model fits the data well, so that we appear to have symmetry for the three-factor interaction term u_{123}. An examination of the estimates of the u-terms suggests that we fit the somewhat more restrictive model $u_{123} = 0$. This latter model also fits the data extremely well, with $X^2 = 4.69$, $G^2 = 6.23$, and eight degrees of freedom. We conclude that there is no three-factor effect needed to describe these second-order transitions, but that there is an apparent lack of symmetry involving the remaining terms in the log-linear model. ■■

8.3.5   *Marginal homogeneity*

MLEs

As we mentioned in Section 8.3.3, we consider two types of marginal homogeneity, the first based on one-dimensional margins, i.e.,

$$m_{i++} = m_{+i+} = m_{++i} \qquad i = 1, \ldots, I, \tag{8.3-25}$$

and the second based on two-dimensional margins, i.e.,

$$m_{ij+} = m_{ji+} = m_{i+j} = m_{+ij} \qquad i, j = 1, \ldots, I, \tag{8.3-26}$$

which also implies (8.3-25). Neither of these models is directly expressible as a log-linear model, but we are still able to get MLEs for the expected cell values under both models by maximizing the multinomial log-likelihood subject to the appropriate constraints.

For the model of homogeneity of one-dimensional marginal totals, the MLEs, based on a multinomial likelihood, are of the form

$$\hat{m}_{ijk} = \frac{x_{ijk}}{1 + (\lambda_i - \lambda_j) + (\mu_i - \mu_k)}, \tag{8.3-27}$$

where the λ_i and the μ_i are Lagrangian multipliers whose values are determined by the marginal constraints (8.3-25). As with the model of marginal homogeneity in two dimensions, the actual determination of the λ_i and the μ_i (and thus the expected values) requires the use of an iterative method other than a version of iterative proportional fitting. Madansky [1963] suggests one such method.

For the model of homogeneity of two-dimensional marginal totals, the MLEs are of the form

$$\hat{m}_{ijk} = \frac{x_{ijk}}{1 + (\lambda_{ij} - \lambda_{ik}) + (\mu_{ij} - \mu_{jk}) + (v_{ij} - v_{ji})}, \tag{8.3-28}$$

where the λ_{ij}, μ_{ij}, and v_{ij} are Lagrangian multipliers whose values are determined by the marginal constraints (8.3-26). Again, we need to use an iterative method to actually compute the $\hat{m}_{ijk}$.

We note that the form of the MLEs under homogeneity of one-dimensional margins, as given by (8.3-27), is different from the form of the minimum discrimination information estimates discussed by Kullback [1971]. Bhapkar [1970], Fryer [1971], and Grizzle, Starmer, and Koch [1969] all consider the model of homogeneity of one-dimensional margins using the minimum modified chi square approach, but they do not actually compute the estimates of the expected cell values. Koch and Reinfurt [1971] refer to all the models discussed here as "mixed categorical data models of order 3."

Conditional tests

While we cannot get MLEs of the expected cell values under the marginal homogeneity models of this section directly from the log-linear model methods found elsewhere in the book, we can construct conditional tests for marginal homogeneity by analogy with the approach in Section 8.2.4.

We let H_{CS}, $H_{1\text{-}QS}$, $H_{2\text{-}QS}$, $H_{1\text{-}MH}$, and $H_{2\text{-}MH}$ denote the models of complete symmetry, quasi symmetry preserving one-dimensional margins, quasi symmetry preserving two-dimensional margins, homogeneity of one-dimensional margins, and homogeneity of two-dimensional margins, respectively. We defined the models of quasi symmetry so that

$$H_{CS} \equiv H_{1\text{-}QS} \cap H_{1\text{-}MH}, \tag{8.3-29}$$

$$H_{CS} \equiv H_{2\text{-}QS} \cap H_{2\text{-}MH}. \tag{8.3-30}$$

Thus the difference between the likelihood ratio goodness-of-fit statistics for H_{CS} and $H_{1\text{-}QS}$ is a test for $H_{1\text{-}MH}$ conditional on $H_{1\text{-}QS}$ being true, and the difference between the statistics for H_{CS} and $H_{2\text{-}QS}$ is a test for $H_{2\text{-}MH}$ conditional on $H_{2\text{-}QS}$ being true. These two conditional tests have $2(I - 1)$ and $(I - 1)(I - 2)/2 + 2(I - 1)^2$ degrees of freedom, respectively.

Example 8.3-3   More on voting intentions

In example 8.3-2 we carried out goodness-of-fit tests for H_{CS}, $H_{1\text{-}QS}$, and $H_{2\text{-}QS}$ on data from the 1940 panel study on voting intentions displayed in the form of a $3 \times 3 \times 3$ table. Model $H_{2\text{-}QS}$ fits the data extremely well, and we now test for homogeneity of two-dimensional margins conditional on $H_{2\text{-}QS}$ being true. The likelihood ratio statistic for this test is based on $G^2 = G^2_{CS} - G^2_{2\text{-}QS} = 229.8 - 3.9 = 225.9$, with nine degrees of freedom, and we conclude that the two-dimensional margins are not homogeneous.

For illustrative purposes we also compute the conditional test for homogeneity of one-dimensional margins conditional on $H_{1\text{-}QS}$ being true, even though the fit for $H_{1\text{-}QS}$ is very poor. The statistic for this test is $G^2 = G^2_{CS} - G^2_{1\text{-}QS} = 229.8 - 110.3 = 119.5$, with four degrees of freedom, indicating that the one-dimensional margins are not homogeneous. ■■

Cochran's Q test

Cochran [1950] proposed the use of a statistic to test the homogeneity of one-dimensional margins, for situations where there are m distinct matched samples and two possible outcomes for each individual. Madansky [1963] generalized Cochran's test to cover the case of an arbitrary number of possible outcomes. We focus here on the two-outcome case with $d = 3$, so that the basic data can be displayed as counts $\{x_{ijk}\}$ forming a $2 \times 2 \times 2$ table.

Cochran's test statistic for the hypothesis $H_{1\text{-}MH}$ in the 2^d table case is

$$Q = \frac{(d - 1) \sum\limits_{j=1}^{d} (T_j - \bar{T})^2}{\sum\limits_{i=1}^{N} S_i \left(1 - \dfrac{S_i}{d}\right)}, \tag{8.3-31}$$

where T_j is the number of individuals falling into the first category for dimension j, $\bar{T} = \sum_{j=1}^{d} T_j/d$, and S_i is a count for the ith individual of the number of dimension for which he falls in the first category. We note that

$$\sum_{i=1}^{N} S_i = \sum_{j=1}^{d} T_j. \tag{8.3-32}$$

For the $2 \times 2 \times 2$ table case, $d = 3$ and we can write (8.3-31) in our notation as

$$Q = \frac{3(x_{1++}^2 + x_{+1+}^2 + x_{++1}^2) - (x_{1++} + x_{+1+} + x_{++1})^2}{x_{+++} - x_{111} - x_{222}}. \tag{8.3-33}$$

Under the model of complete symmetry Q has an asymptotic χ^2 distribution with two degrees of freedom. Thus by using Q we are testing for marginal homogeneity in a restricted sense, just as in the case when we use the conditional tests of homogeneity described above, i.e., Q has an asymptotic noncentral χ^2 distribution for those situations where marginal homogeneity holds but complete symmetry does not.

In expressions (8.3-8) and (8.3-21) we saw that $\hat{m}_{iii} = x_{iii}$ for $i = 1, 2$, in the complete symmetry model and both quasi-symmetry models. Thus in our conditional tests of marginal homogeneity, the size of the counts in these two cells does not affect the value of the test statistic. As Cochran has noted, Q is also not affected by the counts in these cells.

Example 8.3-4   Drug response patterns
Grizzle, Starmer, and Koch [1969] consider an example of data on 46 subjects, each given drugs A, B, and C. Every subject has either a favorable or an unfavorable response to each drug. We summarize the data here in table 8.3-3.

Table 8.3-3   Reponses to Drugs

Response to C	Response to A — Favorable		Response to A — Unfavorable	
	Response to B		Response to B	
	Favorable	Unfavorable	Favorable	Unfavorable
Favorable	6	2	2	6
Unfavorable	16	4	4	6

Source: Grizzle, Starmer, and Koch [1969].

If the three drugs are equally effective, the one-dimensional margins for the table of expected counts are homogeneous, i.e., $H_{1\text{-}MH}$ holds. For the data in table 8.3-3, $Q = 8.47$ with two degrees of freedom. The likelihood ratio goodness-of-fit statistic for the model of quasi symmetry preserving one-dimensional margins is $G^2 = 10.36$, with two degrees of freedom. Even though this statistic is significant at the 0.01 level, we proceed to examine the conditional test statistic for homogeneity of one-dimensional margins, whose value is $G^2 = 18.93 - 10.36 = 8.57$. We refer all three test statistics for marginal homogeneity to a χ^2 distribution with two degrees of freedom, and for each statistic the descriptive label of significance is somewhere between 0.05 and 0.01. Thus we have moderately strong evidence against the model of homogeneity of one-dimensional margins, and looking at the data we see that drugs A and B appear more effective than C. ■ ■

8.4 Summary

Many authors have concerned themselves with models and tests of marginal homogeneity and symmetry for tables of counted data. In this chapter we have presented one approach to such problems, which uses hierarchical log-linear models and the method of iterative proportional fitting to compute estimated expected cell values. Wherever possible we have tried to link our approach to the work of others, and to explain how their results are either equivalent to those presented here (e.g., McNemar's test and Bowker's generalization of it) or are quite similar (e.g., Cochran's Q test).

9 Model Selection and Assessing Closeness of Fit: Practical Aspects

9.1 Introduction

In earlier chapters we described various structural and sampling models for counted data, and we presented a variety of formal tools and techniques for choosing an appropriate model for an observed set of data. Although this formal development is statistically correct, it is not always useful, or indeed applicable, for a given set of data. Our analysis in any practical situation is usually governed by how well we understand the underlying process generating the data, why the data were collected, and how much we choose to report about the data. Since these influencing factors vary so much from one set of data to another, it is extremely difficult to give a formal set of rules for data analysis. Nevertheless, some general remarks may help. This chapter offers a collection of ideas on some practical aspects of assessing goodness of fit that can be regarded as advice sometimes backed by theory, sometimes by experience, sometimes by intuition, but most often by some combination of the three. Only the beginner knows how hard it is to start. These suggestions should help the researcher start work, not stifle his ingenuity. Although we discuss both structure and sampling models, our primary effort deals with structure.

We have three main reasons for fitting models to observed data, especially data in the form of a table of counts. First, the fitted model may help us understand the complex data in the table. To get this understanding, the model need not yield a perfect fit to the data. Indeed, we often pursue the theme that a poor fit may give greater understanding than a close one. Second, we may be interested in assessing the magnitude of a particular interaction in the presence of many other variables. Third, a reasonable model, not necessarily the correct one, may give more precise estimates of expected frequencies than do the original data by themselves. This is especially true of multidimensional tables where data are spread quite thinly over a large number of cells. Detailed proofs are difficult to formulate for several dimensions, so we illustrate this point for $I \times J$ tables.

Our objectives directly affect our attitude toward model-fitting and goodness of fit. Although we rely heavily on the standard Pearson and likelihood-ratio chi square goodness-of-fit statistics throughout this chapter, let us note at the outset that we use such statistics as indices of fit rather than for formal tests of goodness of fit. Consequently, it is often reasonable to look at indices of fit other than those of the usual chi square form, or to normalize the usual Pearson statistic for comparative purposes when the sample size is large. We elaborate on the latter idea in Section 9.6.

In a few problems, we may believe that a specific sampling model yields a reasonable approximation to reality and also that a particular structural model holds or that such a model might well describe our data. Moreover, in this latter situation we may believe that when the model is found, other sets of data will also obey it closely. A good example here is the model for Mendelian inheritance. We believe it to be structurally correct, and we also believe that we know the sampling model whereby data are generated for model verification. In these circumstances, it is appropriate to study how closely the model fits the data, and the methods of Chapter 4 are suitable for such assessments. Nevertheless, Mendel's own data present some surprises (see example 9.5-3).

Other sections of this chapter deal with such topics as the fitting and testing of the same set of data, the handling of confounding and other data anomalies, and the summarizing of large tables using the frequency of frequencies approach. There are many other topics which we might easily have included in this chapter, and much is currently being done on the innovative use of informal data-analytic techniques for counted data. The work of Tukey [1971] is an excellent example of what can be done within a data-analytic framework. A key point is that every new set of data presents us with new models to build and techniques to develop.

9.2 Simplicity in Model Building

9.2.1 General comments

When several equally sensible models provide a reasonable fit to a set of data, the scientist usually chooses the simplest such model. What do we mean by simplest? The simplest model might be the one that (i) has the fewest parameters and (ii) has the least complicated parametric structure. It is difficult, if not impossible, to work with criteria (i) and (ii) simultaneously.

Suppose we have ten models, and suppose that two of these models involve only two parameters, while the remainder involve three or more. Then these two two-parameter models qualify under condition (i). Now if one of the models is a linear function of the two parameters, while the other involves a power series expansion, then most of the time we could agree that the linear model is simpler. If both models are complicated functions of the parameters, the choice of the simplest model is difficult unless we have a natural interpretation for the parameters in one of the models. Many authors have attempted to define the concept of simplicity precisely (see, for example, Kemeny [1953]), but we know of no particular definition that seems natural for every problem.

9.2.2 Why simplicity?

Why do we search for simplicity in model-building? Some reasons were mentioned briefly in the introduction to this chapter, and we elaborate on them at this time.

First, we believe that a simple model often helps us make sense out of a complex set of data, while a more complicated model actually disguises whatever meaning the data have to offer. Consider an experiment which is conceptually equivalent to drawing balls from an urn containing an unknown number of black and white balls. If 495 of the 1,000 drawn balls are white, a simple model postulates that the proportion of white balls in the urn is 1/2, while a more complicated model might postulate a proportion of 495/1,000. Focusing on a proportion of 0.495 may obscure the fact that roughly half the balls appear to be white. When the experiment involves a sample survey instead of balls in an urn, it would be easy to get side-tracked trying to explain the deviation from 50 %.

Second, a model with fewer parameters may improve the precision of the estimates. We have in mind the following:

Suppose we have two models for predicting the cell frequencies in a table of counts, both of which are compatible with the observed data, one model being a special case of and having fewer parameters than the other. Then the "overall variability" of the estimates from the simpler model about the "true" values for the cells is smaller than the "overall variability" for the model with more parameters requiring estimation.

We have no general proof of this theorem, although it is specifically true for many regression problems, and we believe it is true in a rather general way. We can, however, illustrate how the theorem works in a particular situation, with a particular definition of overall variability.

9.2.3   *Precision of estimates in $I \times J$ tables*

Consider the problem of estimating the cell probabilities for an $I \times J$ table with observed counts $\{x_{ij}\}$ following a multinomial distribution with cell probabilities $\{p_{ij}\}$, where $\sum_{ij} x_{ij} = N$ and $\sum p_{ij} = 1$. We wish to compare the maximum likelihood estimator

$$T_{ij}^{(1)} = \frac{x_{i+} x_{+j}}{N^2} \tag{9.2-1}$$

based on the model of independence, i.e., $p_{ij} = p_{i+} p_{+j}$, with the unrestricted maximum likelihood estimator

$$T_{ij}^{(2)} = \frac{x_{ij}}{N}. \tag{9.2-2}$$

In this case, the unrestricted model has more parameters to be estimated than does the independence model.

Now, $T_{ij}^{(2)}$ is always an unbiased estimator of p_{ij}, but

$$E(T_{ij}^{(1)}) = \left(\frac{N-1}{N}\right) p_{i+} p_{+j} + \frac{1}{N} p_{ij}. \tag{9.2-3}$$

Thus $T_{ij}^{(1)}$ is an unbiased estimate of p_{ij} only when $p_{ij} = p_{i+} p_{+j}$, i.e., only under independence. Let us measure the overall variability of our two estimators in terms of the expected mean squared error (risk), i.e.,

$$R_k = R(\mathbf{T}^{(k)}, \mathbf{p}) = \sum_{ij} E(T_{ij}^{(k)} - p_{ij})^2 \quad k = 1, 2, \tag{9.2-4}$$

noting that this index of variability is to some extent arbitrary. For each cell we can write the contribution to the risk as a sum of two components, variance and bias:

$$\text{Var} = E[T_{ij}^{(k)} - E(T_{ij}^{(k)})]^2, \tag{9.2-5}$$

$$\text{Bias} = [E(T_{ij}^{(k)}) - p_{ij}]^2. \tag{9.2-6}$$

For $T_{ij}^{(2)}$, the bias term is always zero, and

$$R_2 = \frac{1}{N}\left(1 - \sum_{ij} p_{ij}^2\right). \tag{9.2-7}$$

For $T_{ij}^{(1)}$, substituting in (9.2-5) and (9.2-6) from (9.2-3), we find that the bias term is

$$\left(\frac{N-1}{N}\right)^2 (p_{i+}p_{+j} - p_{ij})^2, \tag{9.2-8}$$

and the variance term is

$$\frac{(N-1)(6-4N)}{N^3}p_{i+}^2 p_{+j}^2 + \frac{(N-1)(N-4)}{N^3}p_{ij}p_{i+}p_{+j}$$

$$+ \frac{(N-1)(N-2)}{N^3}(p_{i+}p_{+j}^2 + p_{i+}^2 p_{+j}) + \frac{(N-2)}{N^3}p_{ij}^2 \tag{9.2-9}$$

$$+ \frac{(N-1)}{N^3}(2p_{ij}p_{+j} + 2p_{ij}p_{i+} + p_{i+}p_{+j}) + \frac{p_{ij}}{N^3}.$$

When $p_{ij} = p_{i+}p_{+j}$, (9.2-9) simplifies, (9.2-8) becomes zero, and we find that

$$R_1 = R_2\left[1 - \frac{\left(1 - \frac{1}{N}\right)\left(1 - \sum_j p_{+j}^2\right)\left(1 - \sum_i p_{i+}^2\right)}{1 - \left(\sum_j p_{+j}^2\right)\left(\sum_i p_{i+}^2\right)}\right]. \tag{9.2-10}$$

The quantity in square brackets on the right-hand side of (9.2-10) is always less than 1, so we know that $\mathbf{T}^{(1)}$ is always more precise than $\mathbf{T}^{(2)}$ when $p_{ij} = p_{i+}p_{+j}$, i.e., under independence, as intuition suggests. Thus when the simpler model is correct, the overall variability for the corresponding estimates is smaller than for the estimates based on the more complicated model. Our real interest here, however, should be focused on situations when the simpler model is not quite correct.

Now suppose that

$$p_{i+}p_{+j} = p_{ij} + \varepsilon_{ij} \tag{9.2-11}$$

with ε_{ij} small enough that p_{ij} is "close" to $p_{i+}p_{+j}$. After some algebraic manipulation, we get

$$R_1 = \frac{1}{N} - \frac{(N-1)}{N^2}\sum_{ij} p_{ij}^2 + \frac{(N-1)(N-2)(N-3)}{N^3}\sum_{ij} \varepsilon_{ij}^2$$

$$- \frac{2(N-1)(3N-2)}{N^3}\sum_{ij} \varepsilon_{ij}p_{ij} + \frac{(N-1)(N-2)}{N^3}\sum_{ij} \varepsilon_{ij}(p_{i+} + p_{+j}) \tag{9.2-12}$$

$$+ \frac{(N-1)}{N^2}\sum_{ij} p_{ij}(p_{i+} + p_{+j} - 1).$$

For a given sample size N, if $\{p_{ij}\}$ is close enough to $\{p_{i+}p_{+j}\}$, then $R_1 < R_2$, so we still prefer (9.2-1) to the observed proportions. For example, if $I = J = 2$ and

$p_{i+} = p_{+j} = 1/2$, then $\varepsilon_{11} = -\varepsilon_{12} = -\varepsilon_{21} = \varepsilon_{22} = \varepsilon$, and (9.2-12) simplifies to

$$R_1 = \frac{1}{N}\left(1 - \sum_{ij} p_{ij}^2\right) - \frac{(N-1)}{N^2}\sum_{ij} p_{ij}^2 + 4\varepsilon^2 \frac{(N^2 + N + 2)(N-1)}{N^3}$$

$$= R_2 - \frac{(N-1)}{4N^2} + 4\varepsilon^2 \frac{(N^2 + 2)(N-1)}{N^3}.$$

(9.2-13)

Expression (9.2-13) is less than R_2 provided that

$$\varepsilon^2 < \frac{N}{16(N^2 + 2)}.$$

(9.2-14)

For moderately large N, this essentially means that $\{x_{i+}x_{+j}/N^2\}$ has smaller overall variability than $\{x_{ij}/N\}$ if $|\varepsilon| < 1/(4\sqrt{N})$, or if

$$\frac{1}{4}\left(1 - \frac{1}{\sqrt{N}}\right) < p_{ij} < \frac{1}{4}\left(1 + \frac{1}{\sqrt{N}}\right)$$

(9.2-15)

for 2×2 tables with $p_{i+} = p_{+j} = 1/2$. For example, suppose $N = 24$ and the true value of $p_{11} = 0.30$, so that $p_{11} = p_{22} = 0.30$, $p_{12} = p_{21} = 0.20$, and the actual cross-product ratio is $\alpha = 2.25$. Then the independence estimates still have smaller overall variability than the unrestricted ones.

The results described here for the 2×2 table show what can happen if we use two parameters (one for rows and one for columns) instead of three (the previous two plus one for interaction) in cases where the model based on two parameters is "almost" true. For multidimensional tables and log-linear models with potentially large numbers of parameters, the utility of simple log-linear models based on a small number of parameters quite often leads to the same kind of increased precision, even when the simple models are not quite correct.

9.3 Searching for Sampling Models

In analyzing tables of counts, we usually begin by assuming a sampling model, and then we search for a structural model. This approach often results from the unavailability of statistical methodology allowing us to distinguish among sampling models for cross-classified data, and it suggests several research problems of a theoretical nature which go beyond the scope of this book. There are problems, however, in which a search for a sampling model is not only a natural path on which to embark, but also easy to carry out. In the two examples that follow we illustrate the search for an appropriate sampling model to describe the distribution of counts.

Example 9.3-1 Home runs

Let us consider the distribution of the number of men on base when a home run is hit, shown in table 9.3-1, for a particular year in the National Baseball League. The obvious model to fit is the binomial model. Since there can be 3, 2, 1, or 0 men on base, we can have three trials ($N = 3$) and estimate the probability of a success (each occupied base equals one success) from the distribution observed. If we do this, we find that the fit is a poor one from a chi square point of view. The mean probability of a man being on a base, i.e., a success, is $\bar{p} = 0.21$, and the expected counts for the binomial model are shown in table 9.3-1. The contribution to the Pearson chi square statistic from the term for bases loaded, i.e., $x = 3$, is $(21 - 7.3)^2/7.3 = 25.8$, which is already much too large for a good fit with only two degrees of freedom. The binomial assumption gives a poor fit for every number of men on base except 2, and there it is just bearable. The binomial model contains strong assumptions about equal probabilities and independence between trials.

Table 9.3-1   Number of Men on Base when a Home Run Is Hit

	Number of Men on Base				
	0	1	2	3	Totals
Observed Frequency of Home Runs	421	227	96	21	765
Binomial Expected Frequency	377.9	299.6	80.2	7.3	765.0
Poisson Expected Frequency	407.8	256.1	80.9	20.2	765.0
Poisson $(O - E)^2/E$	0.4	3.7	2.4	0.0	6.5

On the other hand, a closely related model is the Poisson model. It has one parameter instead of two, and its reasonableness comes from the idea that having men on base is a somewhat rare event. First we compute the sample mean $\bar{x} = 0.63$, and then we get the expected frequencies using $\bar{x}$ in place of the true mean. Since the Poisson model allows counts well beyond 3, we associate with $x = 3$ the Poisson probability $\Pr(X \geq 3)$. We include these expected Poisson counts in table 9.3-1. Note that the fit is closer for every number of men on base, although the fit for the cell when two men are on base is not much improved. The total chi square is 6.5, just beyond the 5 % level, which is about 6.0. Of course, the pooling which we used to get the fitted value for $x = 3$ does present some problems. Nevertheless, we regard the move from the binomial to the Poisson as a major improvement. Whether we stop here depends on our interest in the problem and the availability of the data. For this set, at least, the Poisson model gives a fair representation. If a person had just one set of such data, he had best remind himself that if the Poisson had not worked he would have tried the negative binomial and various other discrete distributions. ■■

Example 9.3-2   Distribution of word counts

In their study of the authorship of the disputed Federalist papers, Mosteller and Wallace [1964] used the frequencies with which different words, such as "by," "from," "to," "upon," "while," and "whilst" occurred in short articles to decide whether Alexander Hamilton or James Madison had written certain political essays intended to persuade the people of the state of New York to ratify the constitution of the United States. Since one of the methods Mosteller and Wallace use to discriminate between the writings of the two authors is Bayesian, it is important to know the distribution of the number of times an author used a given word in an essay of given length.

It is easy to estimate the mean number of times a word is used by a given author; however, in this case, the actual distribution is needed. Otherwise it is not convenient to calculate the effect on the odds of the writing being by Hamilton, say, when the count of the word is found.

Two sampling models suggested themselves to Mosteller and Wallace. The first is the Poisson distribution, the argument being that the occurrence of a given word in a particular spot is a rare event, and the number of rare events of a given kind in an article are distributed approximately according to a Poisson distribution.

An argument against this possibility is that a writer may not use the same rate for a particular word in each article that he writes. Instead, he may have one rate at one time, another at another time. If this is true, then the distribution of the counts for a given word would have to follow a more complicated distribution than the Poisson. An attractive candidate for this more complicated distribution is the negative binomial, which can arise in many ways. One way to produce the negative binomial distribution is for there to be a different Poisson rate for each essay and for these different rates to be drawn at random from a gamma distribution (see Chapter 13).

By doing an empirical study, Mosteller and Wallace have found that some of the words they use for discriminating between writing by Hamilton and by Madison are distributed closely according to the Poisson distribution, but that other words follow the negative binomial distribution more closely. Since the negative binomial distribution includes the Poisson as a limiting case, they chose the former.

The odds for authorship of a paper depend on whether the Poisson theory or the negative binomial is used. The odds are much less extreme for the negative binomial than for the Poisson. On the average, the logarithm of the odds is about half as large for the negative binomial as for the Poisson, or equivalently, the odds themselves under the Poisson model are approximately the square of those for the negative binomial. Thus if the odds according to the Poisson are 1,000,000 to 1 in Madison's favor, say, under the negative binomial they are only about 1,000 to 1. This shows that, in some contexts, the attention paid to the details of sampling distributions can have substantial effects.

Table 9.3-2 shows some distributions of numbers of words in Federalist papers of known authorship together with the fitting that was done for the Poisson and the negative binomial distributions. The analysis deals with blocks of 200 words each, 247 blocks for Hamilton, 262 for Madison. Of the words presented, "an" for Hamilton has a fairly close fit for either the Poisson or the negative binomial distribution, but for Madison the same word needs the negative binomial, and the word "his" requires the negative binomial for both authors. In the end, "his" and all other pronouns are abandoned by Mosteller and Wallace for discrimination purposes because they seem too contextual, that is, the frequency of their use seems to vary too much depending upon subject matter.

The point of the discussion, then, is that while building a structural model is often primary, in some problems details of the sampling distribution can be most important and require substantial investigation. ■ ■

9.4 Fitting and Testing Using the Same Data

Sometimes an investigator collects a set of data for the purpose of accepting or rejecting one particular hypothesis or well-defined structural model. More typically, we speak, in a loose sense, of wishing to test for the presence of a particular interaction, but such a test has meaning only in the context of a model which reflects the population being sampled and the other variables being considered. Even though we begin our investigation with specific objectives in mind, we only have

Table 9.3-2 Observed and Fitted Poisson and Negative Binomial Distributions for Selected Words

| Word | | Occurrences | | | | | | | |
		0	1	2	3	4	5	6	7+
Alexander Hamilton									
an	Observed	77	89	46	21	9	4	1	0
	Poisson	71.6	88.6	54.9	22.7	7.0	1.7	0.4	0.1
	Negative binomial	81.0	82.7	49.2	22.0	8.2	2.7	1.0	0.2
his	Observed	192	18	17	7	3	2	4	4
	Poisson	131.7	82.7	26.2	5.5	0.9	0.1	0.0	0.0
	Negative binomial	192.2	23.8	11.0	6.4	4.0	2.7	1.9	5.0
James Madison									
an	Observed	122	77	40	14	8	0	1	0
	Poisson	106.1	95.9	43.4	13.1	2.9	0.5	0.1	0.0
	Negative binomial	121.8	80.2	37.2	14.9	5.4	1.9	0.6	0.0
his	Observed	213	21	9	11	2	2	1	3
	Poisson	167.3	74.9	17.0	2.6	0.3	0.0	0.0	0.0
	Negative Binomial	222.7	14.7	8.4	5.2	3.4	2.4	1.3	3.9

Source: Mosteller and Wallace [1964].

vague ideas about most of the structure, and we plan to use our data to find the appropriate model.

In building a model from a set of data, the usual procedure is a sequential one. We postulate a model which we know to be inappropriate and get the fitted values for the data under that model. Then we look at the residuals (observed minus fitted values in some scale, not necessarily that of the original counts), and if necessary, we postulate a new model which we hope accounts for some of the larger residuals. We then fit the revised model and repeat the procedure over and over again. At some stage (often the first), we begin to test the fit of the model to the data using chi square and similar techniques, and this is where our problems usually begin.

1. First, we are building a model on the same set of data that we are using to test the goodness of fit of the model. As the model-building procedure evolves, the models we examine increasingly reflect information gleaned from the data. Should we give up some degrees of freedom for our chi square test statistics because we used the data to suggest the models? If so, how many?

2. Since the first batch of models we use are, in all probability, unlikely to provide a good fit to the data according to any of the standard criteria, how can we compare chi square values for these different models on the same set of data? If the models are nested, then the partitioning of chi square statistics as described in Chapter 4 may help, but not necessarily.

3. How do we compare chi square values when all of the models after a particular one begin to fit reasonably well? This problem is related to those that arise when we get two models, both of which fit the data moderately well and have roughly the same number of parameters, yet have very little overlap (in terms of parameters). For log-linear models and multidimensional contingency tables we have a rule of thumb in such situations: When comparing two models with the same number of parameters, only one of which has some three-factor interaction terms, we almost always prefer the model containing no three-factor effects. Exceptions occur when the three-factor effect has a natural interpretation in the context of data. In the same sense, we prefer a model with no four-factor effects to a model with one or more four-factor effects.

4. Does there come a time after which we cannot profit by further analysis of the same data? Some think that formal statistical analysis beyond this ill-defined point is overanalysis, or more elegantly, that there comes a point at which fitted models, for a given set of data, are only explaining the natural variability that arises since we are dealing with a sample rather than with a population. This is a troublesome matter, since we do not always know the nature of the variability at hand.

5. Should we break the data into pieces and deal with them in different ways? If we do, techniques such as "jackknifing" for measuring internal variability may be of help; however, more theoretical work is needed before we can utilize such methods. (See Mosteller and Tukey [1968] for an elementary discussion of jackknifing and other methods of direct assessment of internal variability.)

6. Are methods available for assessing our sequential fitting procedure rather than the fit of an individual model that arises along the way? In general, the answer is no.

For the types of situations just described, we need a fresh set of data to assess the goodness of fit after the choice of a structural model (although not necessarily the values of the parameters in that model) has been made. This would give an honest measure of the departure of the data from the model. Before illustrating points 1–6 with examples, we indicate at least one potential hazard in this cross-validation approach. Quite often investigators have available several comparable sets of data from different sources which appear to manifest the same general characteristics. If this is the case and we build a model on one of these sets of data, it may be of little help for model verification to test the chosen model on any one of the other sets of data which we have already analyzed and recognize as similar to the one generating the model. On the other hand, if we fit the same model to all sets of data simultaneously, we may be able to produce a measure of internal variability that can be of great value in assessing the fit of the model.

Example 9.4-1   The British and Danish mobility data

Since we have already analyzed the British mobility data of Glass [1954] and the Danish data of Svalastoga [1959] in earlier chapters, we can illustrate several of the points raised above by simply pulling together these previous analyses and adding a few more.

In Chapter 5 we fitted a variety of multiplicative models to the British data. We briefly summarize these here, along with a few additional ones. Following our formal statement of the models, we link them up via a tree diagram. The models (see Goodman [1972b] for details) are defined for $I \times I$ tables.

We begin with the models of independence and quasi independence for off-diagonal cells. Then we add parameters, in a multiplicative fashion, to account for different types of departures from the quasi-independence model. For each model there are obvious (and sometimes not so obvious) parameter constraints not presented here. These constraints have been taken into account in determining the degrees of freedom for the models. By representing the $I \times I$ table in different ways as an incomplete multidimensional table, we can describe many of the models given below in a somewhat more standard log-linear form (see Chapter 5). Because of their multiplicative form, it is relatively easy to derive maximum likelihood estimators for each of the models. By introducing generalized marginal totals which correspond to each set of parameters, we get maximum likelihood equations similar to those developed elsewhere in this book (i.e., the observed totals equal the expected totals) which can be solved using a version of iterative proportional fitting. The details are somewhat lengthy, and we refer the interested reader to Goodman [1972b]. We have discussed models (a), (b), (c), and (e), listed below, in greater detail in Section 5.5.

(a) Independence (IND):

$$m_{ij} = a_i b_j, \tag{9.4-1}$$

with $(I - 1)^2$ degrees of freedom.

(b) Quasi independence for off-diagonal cells (QO):

$$m_{ij} = a_i b_j g_{ij} = \begin{cases} a_i b_j & i \neq j \\ a_i b_j g_{ii} & i = j \end{cases} \tag{9.4-2}$$

(where $g_{ij} = 1$ for $i \neq j$), with $(I - 1)^2 - I = I^2 - 3I + 1$ degrees of freedom. All the models which follow are specializations of the QO model.

(c) Separate quasi independence for upward and downward mobility (QPN):

$$m_{ij} = a_i b_j g_{ij} = \begin{cases} a_i b_j & i > j \\ a_i' b_j' & i < j \\ a_i b_j g_{ii} & i = j, \end{cases} \tag{9.4-3}$$

with $(I - 1)(I - 3)$ degrees of freedom. Note that $g_{ij} = 1$ for $i > j$ and $g_{ij} = (a_i'/a_i)(b_j'/b_j)$ for $i < j$. This model can be broken down into two components, representing upward $(i > j)$ and downward $(i < j)$ mobility. These component models, each with $(I - 2)(I - 3)/2$ degrees of freedom, are referred to as the QP and QN models (where P and N correspond to $i - j$ being positive and negative), respectively. There is a third component consisting of diagonal cells $(i = j)$, but it has zero associated degrees of freedom.

(d) Crossing-parameter model (C):

This model extends the QO model to take into account the differential effects of class changes:

$$m_{ij} = a_i b_j g_{ij} c_{ij}, \tag{9.4-4}$$

where $g_{ij} = 1$ for $i \neq j$ and

$$c_{ij} = \begin{cases} \prod_{u=j}^{i-1} c_u' & i > j \\ 1 & i = j \\ \prod_{u=i}^{j-1} c_u' & i < j. \end{cases} \tag{9.4-5}$$

Here there is a multiplicative factor c_u' for each crossing between adjacent classes (from class u to $u + 1$ or $u + 1$ to u). There are $(I - 2)^2$ degrees of freedom, and this model is a special case of the QPN model, where

$$a_i b_i = a_i' b_i' \quad i = 1, \dots, I. \tag{9.4-6}$$

(e) Diagonals-parameter model (D):

Here we extend the QO model to take into account class changes, this time as a function of the magnitude $i - j$ of the change:

$$m_{ij} = a_i b_j g_{ij} d_{i-j}, \tag{9.4-7}$$

where $d_0 = 1$. There are $I^2 - 5I + 5$ degrees of freedom for this model provided $I \geq 4$. There are also four special cases of the D model which are listed separately below.

(f) Triangles-parameter model (T):

We now consider a special case of the QPN model where $a_i = h_1 a_i'$ and $b_j = h_2 b_j'$. We can describe the expected values for this model by

$$m_{ij} = a_i b_j g_{ij} t_k, \tag{9.4-8}$$

where $k = 1$ if $i > j$, $k = 2$ if $i < j$, and $t_k = 1$ if $i = j$. Note that expression (9.4-8) is a special case of the D model with

$$d_{i-j} = \begin{cases} t_1 & i > j \\ t_2 & i < j. \end{cases} \tag{9.4-9}$$

There are $I(I-3)$ degrees of freedom provided $I > 3$.

(g) Paired diagonals model (DA):

By pairing diagonals above and below the main diagonal in the D model, we get a special case of the quasi-symmetry model discussed in Chapter 8, i.e.,

$$m_{ij} = a_i b_j g_{ij} d_{|i-j|}, \qquad (9.4\text{-}10)$$

with $(I-1)(I-3)$ degrees of freedom provided $I \geq 3$. For $I = 3$, the DA model is the same as the QO model.

(h) Diagonal parameters for upward mobility (DP):

This is the special case of the D model with

$$d_k = 1 \quad k = -1, -2, \ldots, -(I-1), \qquad (9.4\text{-}11)$$

and there are $I^2 - 4I + 2$ degrees of freedom for $I \geq 4$. The analogous model for downward mobility is referred to as the DN model.

We can also consider models formed by taking various combinations of (d) through (h), such as the CT model, combining (c) and (f), and the DACT model, combining (d), (f), and (g). These are defined in the obvious ways.

Figure 9.4-1 presents a treelike structure showing the connections among all of the models described above, as well as all of the related hybrid models. The arrows are used to show how to go from any model to all of its special cases. Beside each model we give the number of degrees of freedom in parentheses and the value of the likelihood-ratio goodness-of-fit test statistic (these values are taken from Goodman

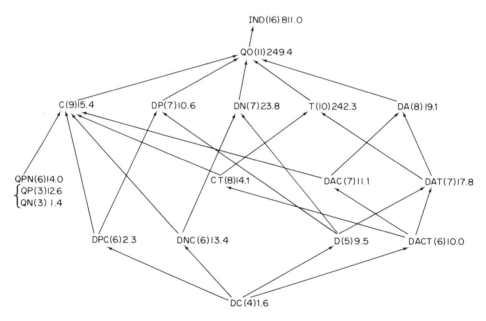

Figure 9.4-1 Hierarchy of multiplicative models for British social mobility data. The number in parentheses beside each model is the associated number of degrees of freedom; the other number is the value of the likelihood ratio statistic.

[1972b] and were computed by techniques similar to those described in Chapter 5). Connected paths of arrows indicate hierarchies of models which can be examined by the partitioning of likelihood-ratio statistics, and starting from any one point on the tree, there are usually several different hierarchies which we can construct.

Had we chosen to look at a single hierarchy of models before examining the data, say IND-QO-T-DAT-D-DC, we could carry out a partitioning of the likelihood-ratio goodness-of-fit test statistic for the IND model. We would find that each of the likelihood-ratio components used for testing a difference between a pair of successive models is significant at the 5 % level, while the component used for testing the goodness of fit of the DC model provides a chi square value smaller than the number of degrees of freedom. Note that the D model when examined by itself also provides a reasonably good fit to the data.

Data analysis, however, does not normally proceed in orderly hierarchies of the sort described above. A more likely path of inquiry would start with the IND model, then go on to the QO model because of big residuals for the cells on the main diagonal, i.e., (1, 1), (2, 2), etc. Following this, we might examine several models simultaneously, say QPN, T, and DA. The relatively good fits for QPN (and in particular QN) and DA might then lead us to examine the DP, C, and D models. By now, the significance levels associated with the goodness-of-fit tests have lost their meaning, and we already have several possible models to describe the data. Nevertheless, we might still proceed to fill all of the numerical values in figure 9.4-1.

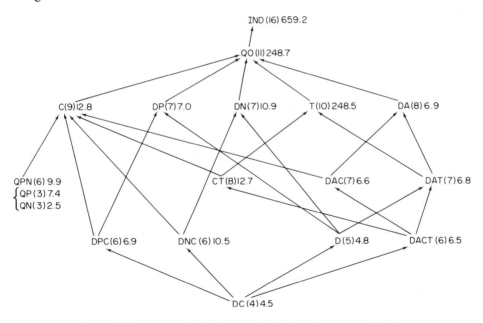

Figure 9.4-2   Hierarchical tree of multiplicative models for Danish social mobility data.

It is reasonable to choose a suitable hierarchy of models, say IND-QO-DP-DPC-DC, and apply these models to a fresh set of data, e.g., the Danish mobility data of Svalastoga. Indeed, we might even choose to compute the statistics for the entire tree of models, yielding figure 9.4-2. The general pattern of likelihood-ratio values for the Danish data is similar to that for the British data, except that all models but IND, QO, and T provide good fits.

Before concluding that we have found the best model(s) for our data, we had better stop and recall two other analyses which we previously carried out. In Chapter 12 we examine the mobility data as a $2 \times 5 \times 5$ table, and we fit the log-linear model which has $u_{123} = 0$. The likelihood-ratio statistic for this model is 39.33 with sixteen degrees of freedom, and although it is significant at the 0.001 level, the relatively small value of the test statistic and the fact that the fitted values are similar to the observed values except for a few cells suggest that the interaction structure in the two 5×5 tables is approximately the same. This is the same conclusion we reached in Chapter 3 by standardizing the marginal totals for the two tables. Now, since the interaction structure (described in terms of cross-product ratios) is similar for both sets of data, we should not be surprised that the same sets of multiplicative models describe both sets of data, as the multiplicative models are simply describing the interaction structure of the tables. Given this conclusion, we can no longer view the Danish data as "fresh," and in order to verify the appropriateness of the multiplicative models described above we must find yet another set of data. ■■

In no way are the foregoing remarks meant to discourage the kinds of explorations and considerations being described. We merely warn against taking the final measure of fit very seriously in the face of selection among models on the basis of examinations of the data.

9.5   Too Good a Fit

It frequently happens that when we assess the goodness of fit of given structural and sampling models to a particular set of data, we get a fit that appears to be extremely good (e.g., X^2 and G^2 may be considerably smaller than the associated degrees of freedom). In such situations we still may wish to question the appropriateness of the underlying models. For example, we may find that the data were culled selectively from a larger set of data (thus our sampling model is grossly untrue) and that when the goodness-of-fit test is applied to the full set of data the fit is no longer close.

We often get too good a fit because we use a model with too many parameters, especially when the parameters are chosen by looking at the data. This problem was examined to some extent in the preceding section. For the DC model applied to the British social mobility data, where we use 21 parameters to describe 25 cells, we should not be astonished to find that the model fits the data extremely well.

Another possibility is that we examine several sets of data which appear homogeneous as if they are actually one single set of data, and we conclude that a specified model fits extremely well. By lumping the sets of data together we may disguise various sources of variability which, when properly considered, lead to quite different tests and conclusions. One of the classic data sets used to illustrate the Poisson distribution falls into this category, and thus it becomes our first example.

Example 9.5-1   Deaths by horsekick

Bortkiewicz [1898] reported on the number of men killed by horsekicks in the Prussian army. Many authors refer to one of the tables reported by Bortkiewicz and use the data therein to show how the Poisson distribution describes the number of men killed by horsekicks. Table 9.5-1 contains this oft-referred-to set of data for the number of men killed per year per army unit, using data for a period of twenty consecutive years for fourteen different army corps. Computing $X^2 = \sum (O - E)^2/E$ for these data yields a value of 2.3 with five* degrees of freedom, a remarkably good fit. Most stop at this point and conclude that the number of deaths by horsekick follows a Poisson distribution.

Table 9.5-1   Observed and Fitted Poisson Distributions for Deaths by Horsekick per Year per Army Unit (for Twenty Years and Fourteen Units)

Number Killed	Observed Frequency	Expected Frequency	Poisson X^2
0	144	139.0	0.2
1	91	97.3	0.4
2	32	34.1	0.1
3	11	8.0	1.1
4	2	1.4	0.3
5+	—	0.2	0.2
Totals	280	280.0	2.3

Source: Winsor [1947], pp. 154–161.

Winsor [1947] points out that "the fact that we have obtained a good fit assuming homogeneity among the corps raises the question as to whether there is any real reason" to present the data as in table 9.5-1. He then goes on to present the full array of data considered by Bortkiewicz, reproduced here as table 9.5-2, and to consider different sources of heterogeneity. If the corps do differ in the number of members killed by horsekicks, then an examination of the corps totals should reveal such differences. As the sum of Poisson variates is also a Poisson variate, if we denote a typical entry in table 9.5-2 as x_{ij} (i for corps and j for year), then the test statistic based on row totals x_{i+},

$$\sum_i \frac{\left(x_{i+} - \dfrac{x_{++}}{14} \right)^2}{\left(\dfrac{x_{++}}{14} \right)}$$

is also a test for the suitability of the Poisson distribution. This statistic has a value of 27.3, and can be compared with the χ^2 distribution with thirteen degrees of freedom, whose 99th percentile is 27.7. Thus there is evidence of heterogeneity among the corps. A similar examination of the yearly totals also suggests significant year-to-year variations in the number of deaths by horsekicks for the Prussian army as a whole.

We point out that Bortkiewicz did not overlook this question regarding the possibility of heterogeneity and was not misled by the occurrence of too good a fit.

* Because we use the mean to estimate λ, the actual reference distribution is between χ_4^2 and χ_5^2 (see p. 523).

Table 9.5-2　Full Array of Horsekick Data (Number Killed by Year [Column] and by Unit [Row])

	75	76	77	78	79	80	81	82	83	84	85	86	87	88	89	90	91	92	93	94	Totals
G		2	2	1			1	1		3		2	1			1		1		1	16
I				2				2				3	1			3		4	1		16
II				2		3			2				2	1				2			12
III	1			1	1	2	2		1	1			1	1	1						12
IV		1		1	1	1	1								1	1		1			8
V					2	1			1		1	1		1	1		1	1	1		11
VI			1		2	1		1	2				3	1	1	1	1	3			17
VII									2	2			2			4			2		12
VIII					1		1	1					1				2			1	7
IX						2	1	1	1		2	1	1		1	1	1	1			13
X			1			1		2	2	1			1		2	1	2	1	1		15
XI			2	1	2	4		1			1	1	2	1	2	1	3		3	1	25
XIV	1	1	1	1	1	3		4		1		3		1		3	2	1		1	24
XV	1	1						1		1	1				2	1					8
Totals	3	5	7	9	10	18	6	14	11	9	5	11	15	6	11	17	12	15	8	4	196

Source: Winsor [1947].

In fact, he viewed the horsekick data as an example of an empirical phenomena which he called the "law of small numbers" (referring to the small number of events). Simply stated, the law of small numbers says that events with low frequency in a large population can be fitted by a Poisson distribution even when the probability of an event varies somewhat among the strata of the population (see Gumbel [1968]). ■■

Example 9.5-2   World cup soccer data

Just as the frequency of deaths by horsekicks in the stables of the Prussian army has been described by a Poisson distribution, so can the number of goals scored per team per game in the 1966 World Cup matches. Table 9.5-3 contains the observed data and the fitted Poisson distribution for the 32 matches played for the 1966 World Cup. Computing $X^2 = \sum (O - E)^2/E$ for this set of data yields a value of 2.46 with six* degrees of freedom, once again a remarkably good fit.

Table 9.5-3   Observed and Fitted Poisson Distributions
(for Goals Scored per Team per Game in
1966 World Cup Soccer Play)

Number of Goals	Observed Frequency	Expected Frequency	Poisson X^2
0	18	15.9	0.28
1	20	22.2	0.22
2	15	15.4	0.01
3	7	7.1	0.00
4	2	2.5	0.10
5	2	0.7	2.41
6+	0	0.2	0.02
Totals	64	64	3.04

Source: *Nature* [1966], 211, p. 670.

Displaying the data as in table 9.5-3, however, ignores potentially important aspects of soccer play. For example, the goal-scoring ability of the two teams playing in a given game may be a function of the opponent's ability and style of play (offensive versus defensive). An interesting way to display the goal data to elucidate this point is in a table giving winner's score versus loser's score, in a manner similar to Mosteller [1970]. We do not have the original soccer data, so we are unable to proceed in this fashion. ■■

Example 9.5-3   Mendel's garden pea data

In 1866 Gregor Mendel reported the results of his experiments on the hybridization of garden peas, illustrating the rules of what we now refer to as Mendelian inheritance (Mendel's paper has been recently reprinted in English; see Mendel [1967]). In studying the offspring of hybrids of garden peas, Mendel used seed plants and pollen plants which differed in the form of the seed and the color of the albumen. Mendel took fifteen plants of pure strain with yellow seed albumen and smooth seeds and pollinated them with seeds from plants with green seed albumen and wrinkled seeds. Following Mendel, we label the seed parents by their genotype

* The actual reference distribution here is between χ_5^2 and χ_6^2 (see p. 523).

as *AABB* (A = smooth; B = albumen yellow) and the pollen parents as *aabb* (a = wrinkled; b = albumen green).

The resulting fertilized seeds were round and yellow like those of the seed parents. The hybrid plants raised from these fertilized seeds, whose genotype is *AaBb*, were crossed with each other and yielded 556 seeds, of which 315 were round and yellow, 101 wrinkled and yellow, 108 round and green, and 32 wrinkled and green.

We can cross-classify the 529 offspring in the next generation (which may have any of the nine possible genotypes) according to form and color of albumen of their seeds, as in table 9.5-4 (see Mosteller and Rourke [1973], p. 201, and Fisher [1936]). For example, 38 offspring plants had only round yellow seeds. Since all the seeds were the same, we conclude that the parental combination was of form *AA* and *BB*. Similarly, 60 had both round yellow and wrinkled yellow seeds, corresponding to a parental combination of *Aa* and *BB*.

Table 9.5-4   Mendel's Data on Garden Peas

Color of Albumen of Hybrid Parents	Form of Hybrid Parents			Totals
	AA	*Aa*	*aa*	
BB	38	60	28	126
Bb	65	138	68	271
bb	35	67	30	132
Totals	138	265	126	529

Source: Mendel [1967], p. 16.

One of the questions we try to answer using these data is whether or not the genetic characteristics form and color are inherited independently of one another. Using the Pearson goodness-of-fit statistic to check for independence, we get $X^2 = 1.85$, with four degrees of freedom. Using the χ^2 distribution with four degrees of freedom, this value corresponds to a descriptive level of significance 0.76.

The theory of Mendelian inheritance also postulates theoretical probabilities for the marginal totals of table 9.5-4, i.e., the true totals for each margin should be in the proportions $(\frac{1}{4}, \frac{1}{2}, \frac{1}{4})$. We can combine these theoretical probabilities with the independence of inherited traits to get a set of predetermined expected values. For example, the expected number of *BBAA* is $N \times 1/4 \times 1/4 = 529/16 = 33.0625$. Using these new expected values, we can test for the adequacy of the Mendelian theory. Now $X^2 = 4.59$ with eight degrees of freedom, corresponding to a descriptive level of significance 0.80, again an extremely good fit.

For this example most biologists accept the Mendelian model as correct, and descriptions of Mendel's experiments suggest that the multinomial sampling model is also correct. While the chi square values given above are small, we should be careful not to conclude immediately that they are too small, even though our suspicions are aroused. We can examine some of the other data sets given in Mendel's paper, as Fisher [1936] has. After a detailed analysis Fisher concludes that in almost all cases the observed values are too close to the expected values under Mendel's theory, i.e., the variability in the reported experimental data is consistently smaller than that which we expect for genuine data. Moreover, Fisher notes that:

One natural cause of bias of this kind is the tendency to give the theory the benefit of doubt when objects such as seeds, which may be deformed or discolored by a variety of causes, are being classified .... Although no explanation can be expected to be satisfactory, it remains a possibility among others that Mendel was deceived by some assistant who knew too well what was expected. This possibility is supported by independent evidence that the data of most, if not all, of the experiments have been falsified so as to agree closely with Mendel's expectations.

Fisher's conclusions are regarded as controversial by some, but the point here is the same as in the previous examples. It is often almost as important to look for small values of chi square as it is to look for big values. While the small values can occur by chance, they indicate to us a need to reexamine various aspects of the data. ■■

9.6   Large Sample Sizes and Chi Square When the Null Model is False

9.6.1   *An overview*

In some sets of data the sample size is so large that almost any model yields a highly significant chi square value, yet we may think that a certain model is roughly, though by no means exactly, true. If we want to compare the chi square values for several tables of this type, having the same number and arrangement of cells but differing in N, then we divide each chi square by its own N. This makes the measures more comparable, since the chi square statistic not only increases as the model fitted increases in distance from the true model, but also increases with N whenever the model fitted is not the true model.

Many different approaches can be used to explore this point. In most of these, we fit one model to the data while a somewhat different model is in fact correct. In this section we focus on the behavior of the Pearson chi square statistic for a k-cell multinomial with known cell probabilities.

9.6.2   *Chi square under the alternative*

We consider the case where there is multinomial variation for k cells, with p_i the hypothesized probability and p_i^* the true probability for the ith cell. If $p_i = p_i^*$ for all i, the expected value for the Pearson chi square statistic

$$X^2 = \sum_{i=1}^{k} \frac{(x_i - Np_i)^2}{Np_i} \tag{9.6-1}$$

is $k - 1$. When $p_i \neq p_i^*$ for some i, some computation similar to that of Section 9.2 involving bias and variance yields the mean value of X^2 as

$$E(X^2|p_i \text{ fitted, } p_i^* \text{ true}) = \sum_{i=1}^{k} \frac{p_i^*(1 - p_i^*)}{p_i} + N \sum_{i=1}^{k} \frac{(p_i^* - p_i)^2}{p_i}$$

$$= k - 1 + \sum_{i=1}^{k} \frac{(p_i^* - p_i)}{p_i} + (N - 1) \sum_{i=1}^{k} \frac{(p_i^* - p_i)^2}{p_i}. \tag{9.6-2}$$

The first term gives the expected value for the case when the fitted probabilities are true ($p_i = p_i^*$), and the next two terms give the change in expected value which results from fitting the wrong model. When N is sufficiently large compared with k, the latter of these terms dominates, and the average value of X^2 increases with N, as is indicated by its noncentral asymptotic χ^2 distribution.

Example 9.6-1 A numerical illustration

We consider a case with many categories and a close fit for all but two of them to illustrate our point. Let $k = 21$ and $p_i = p_i^* = 0.05$ for $i = 1, \ldots, 19$. If we let $\varepsilon = p_{20} - p_{20}^*$, then we can examine $E(X^2)$ for varying ε, p_{20}, and increasing N. If $p_{20} = 0.025$, the second term in (9.6-2) drops out, and the expected value increases directly with N. Table 9.6-1 shows $E(X^2)$ for $p_{20} = 0.02$, for varying ε and N. For N larger than 100, the third term in (9.6-2) dominates the second, but in several cases both of these terms are still of the order of magnitude of the degrees of freedom. ■ ■

Table 9.6-1   Values of $E(X^2 | p_i$ fitted, p_i^* true)

N	0.02	0.01	0.005	ε -0.005	-0.01	-0.02	-0.03
101	23.67	21.00	20.29	20.13	20.67	23.00	27.00
201	27.00	21.83	20.50	20.34	21.50	26.33	34.00
501	27.00	24.33	21.13	20.97	24.00	36.33	57.00
1,001	53.67	28.50	22.18	22.02	28.17	53.00	94.50
2,001	87.00	36.83	24.28	24.12	36.50	86.33	149.50
5,001	187.00	61.83	30.58	30.42	61.50	186.33	374.50

Note: $k = 21$, $p_i = p_i^* = 0.05$ for $i = 1, \ldots, 19$, $p_{20} = 0.02$, $p_{21} = 0.03$, and values of $\varepsilon = p_{20} - p_{20}^*$ vary.

9.6.3   Comparing chi squares

The preceding observations suggest a way for comparing X^2 values from several multinomials with large but different values of N. If we divide the X^2 statistic by N, then for large sample sizes the value of X^2/N approximates

$$\sum_{i=1}^{k} \frac{(p_i^* - p_i)^2}{p_i}, \tag{9.6-3}$$

which is a measure of the closeness of our hypothesized values of the cell probabilities $\{p_i\}$ to the true values $\{p_i^*\}$, like the measure Φ^2 discussed in Section 11.3.

(There are other measures or indices we might wish to consider. For some purposes it is attractive to think of the family of indices of the form

$$\sum_{i} \frac{|p_i^* - p_i|^a}{(p_i^*)^b p_i^c} \tag{9.6-4}$$

for various values of a, b, and c. This family includes (9.6-3) as a special case ($a = 2$, $b = 0$, and $c = 1$).)

To compare several values of X^2, as in example 9.6-2, we may profitably divide X^2 by N, thus getting estimates of our closeness index (9.6-3). In some cases we might even consider refining our estimate, for instance by subtracting $k - 1$ from X^2 and then dividing by N. Such a correction might be of importance when k varies from one X^2 to another.

Example 9.6-2   Contextuality of American English

Mosteller [1968] and Mosteller and Rourke [1973] discuss a set of data gathered by Kučera and Francis [1967] dealing with samples of writing published around 1961. The 500 samples are each approximately 2,000 words long, and each can be categorized as falling into one of fifteen different genres (as indicated in table 9.6-2). The authors wished to know whether or not the 100 most frequently used words occur with the same frequencies from one genre to the next.

Table 9.6-2   Distribution of Writing Samples
by Genre

Genre	Number of Samples
A. Press reportage	44
B. Press editorial	27
C. Press review	17
D. Religion	17
E. Skills and hobbies	36
F. Popular lore	48
G. Belles lettres	75
H. Miscellaneous (governmentese)	30
J. Learned and scientific	80
K. Fiction: general	29
L. Fiction: mystery and detective	24
M. Fiction: science	6
N. Fiction: adventure and western	29
P. Fiction: romance and love	29
R. Humor	9
Total Number of Samples	500

Note: Samples as used in Kučera and Francis [1967] contextuality
study.

The null model for this contextuality problem holds that the expected number of occurrences of a word in genre i is proportional to the number of words in samples from i. For example, if each sample were exactly of length 2,000 then, using the figures in table 9.6-2, we see that the expected proportions in genres C and D are each 17/500. Considering that each sample was approximately 2,000 words long, these are not quite the hypothesized proportions. Since there are fifteen genres we have $k = 15$, and all X^2 values are normally compared with a chi square distribution having fourteen degrees of freedom. Typical values of X^2 found by Kučera and Francis are 688.7 for "the" and 2362.6 for "of." While these values do indicate that the null model is incorrect, modest deviations from the model have been blown up by sample sizes of 69,971 for "the" and 36,411 for "of."

Given the sample sizes in this example, we gain nothing by subtracting 14 from X^2 before dividing by N. In table 9.6-3 we list the twenty words, out of the 100 most frequent words, with the smallest values of X^2/N and the twenty words with the biggest values of X^2/N. The values of X^2/N are multiplied by 1,000.

If the rate of use of a word varies from genre to genre more than the rate of other words, it is referred to as being more "contextual." Thus in table 9.6-3 we list the twenty least contextual and the twenty most contextual words. Mosteller [1968]

Table 9.6-3   Values of $1,000X^2/N$ for Twenty Least
Contextual and Twenty Most Contextual Words among
100 Most Frequent Words

Least		Most	
$1,000\,X^2/N$	Word	$1,000\,X^2/N$	Word
6.2	to	317.0	will
6.9	and	329.5	may
9.8	the	331.4	out
10.3	with	340.6	our
11.5	from	364.1	up
12.6	a	372.2	man
23.4	been	471.3	like
23.8	as	537.4	he
24.7	on	565.8	had
25.0	one	671.0	my
26.1	in	721.6	down
26.9	that	747.6	I
30.0	an	771.5	him
37.0	for	858.3	back
38.8	some	871.9	you
40.2	more	894.5	me
45.7	time	999.2	said
45.7	two	1,141.4	her
45.9	made	1,148.3	your
46.2	at	1,397.7	she

Source: Mosteller and Rourke [1973].

notes that the three least contextual words, "to," "and," and "the," have the
33rd, 43rd, and 79th largest values of X^2; thus unnormalized values of X^2 can be
misleading if we are interested in measuring contextuality. ■■

9.7   Data Anomalies and Suppressing Parameters

In fitting models to data, we often find that the model providing the best fit either
contains a parameter whose estimated value we simply do not believe or, in the
case of log-linear models for multiway contingency tables, includes a higher-order
term which is hard to understand. In many such situations, a reexamination of the
data and the process generating it reveals that the data have been influenced by
factors not controlled by the investigator. Indeed, the unwanted term or parameter
may very well have resulted from the manner in which the data were gathered
rather than from something of intrinsic interest. In this case, if omitting the unwanted term leads to a model providing a reasonable fit to the data (although this
model is significantly different from the best model), it is sometimes wiser to use the
simpler model to describe as well as to report the data.

In observational studies we attempt to measure all variables that might be of
interest; unfortunately, when we use data gathered by others we often encounter
problems with variables not measured. To illustrate this point, let us consider a
group of social psychologists investigating responses to the question "Do you like
to arrange flowers?" at some time in 1974. They carefully note the sex and occupation of the respondents, as well as information on other key variables. One of the

purposes of the study is to compare the results with similar data collected in 1934; but in 1934 information on occupation was not collected. Suppose we focus on the three-way contingency table involving response, sex, and time (1934 versus 1974). We find a three-factor interaction in this table that may reflect actual changes over time of responses by men as opposed to women. On the other hand, were we able to include occupation as a fourth variable in our cross-classification of these data, we might find that the three-factor effect disappears, having been caused by a two-factor effect involving occupation and response and two-factor effects involving occupation-time and occupation-sex. If we have information on occupation shifts over time, the observed sex-time-response effect in the three-way table may be an artifact, depending upon its relative size and other collateral information. This discussion is directly related to the effects of collapsing multidimensional tables (see Sections 2.4 and 2.5).

It is a matter of substantive judgment, in the context of the problem at hand, whether such data anomalies should lead to the suppression of parameters. We must try not to suppress parameters in a fitted model which, if included, may lead to a breakthrough in science; nevertheless, we should use common sense when examining the results of a formal statistical analysis, taking into account auxiliary information not used in the analysis.

Example 9.7-1   Torus Mandibularis in Eskimoid groups

Anthropologists have traditionally used the physical structure of the mouth to study differences among populations and among groups within populations. One often-studied characteristic is the incidence of the morphological trait torus mandibularis, a small protuberance found in the lower jaw at the front of the mouth. Table 9.7-1 contains data reported by Muller and Mayhall [1971] on the

Table 9.7-1   Incidence of Torus Mandibularis in Three Eskimo Populations

Population	Sex	Incidence	1–10	11–20	21–30	31–40	41–50	Over 50	Totals
					Age Groups				
Igloolik	Male	Present	4	8	13	18	10	12	65
		Absent	44	32	21	5	0	1	103
	Female	Present	1	11	19	13	6	10	60
		Absent	42	17	17	5	4	2	87
Hall Beach	Male	Present	2	5	7	5	4	4	27
		Absent	17	10	6	2	2	1	38
	Female	Present	1	3	2	5	4	2	17
		Absent	12	16	6	2	0	0	36
Aleut	Male	Present	4	2	4	7	4	3	24
		Absent	6	13	3	3	5	3	33
	Female	Present	3	1	2	2	2	4	14
		Absent	10	7	12	5	2	1	37
Totals			146	125	112	72	43	43	

Source: T. P. Muller and J. T. Mayhall [1971].

incidence of torus mandibularis, where incidence is cross-classified by age (six classes) and sex for each of three Eskimo groups. The first two groups, Igloolik and Hall Beach, are from a pair of villages in the Foxe Basin area of Canada, and the data for these groups were collected by a different investigator than for the third group, the Aleuts from Western Alaska, with a time difference between investigations of about twenty years. This is an important point, since investigator differences are confounded with population differences and effects involving population differences.

The data were fit by the following series of log-linear models, using the methods described in Chapter 3:

1. $u + u_1 + u_2 + u_3 + u_4 + u_{12} + u_{14} + u_{24}$;
2. model 1 $+ u_{124}$;
3. model 2 $+ u_{23}$;
4. model 3 $+ u_{13} + u_{34} + u_{123} + u_{234}$,

where variable 1 is age, variable 2 is presence or absence of torus mandibularis, variable 3 is sex, and variable 4 is population. This particular hierarchy of models was chosen to reflect information suggested by earlier investigations involving skeletal populations. Initially torus mandibularis was believed to be influenced by age, sex, and population, both separately and jointly (model 4). More recent studies have indicated that the effects due to sex may not be as important as had been previously thought. Thus models 3 and 2 include fewer of the (interaction) effects involving sex. Finally, model 1 involves only two-factor effects not involving sex. We note that none of our models uses the information that age is monotonic.

Table 9.7-2 contains the values of the likelihood-ratio goodness-of-fit statistic G^2 for the four models. We have partitioned the value for model 1 into a set of four additive components to allow for examination of the differences between pairs of successive models. Models 1, 2, and 3 provide an acceptable fit to the data, while model 4 is barely significant at the 5% level. The fact that the most complicated model provides the poorest fit to the data simply demonstrates that the use of a large number of parameters in an injudicious manner does not in itself guarantee a good fit. The differences between models 3 and 4 and between models 2 and 3 are not significant at the 5% level, while the difference between 1 and 2 is, although just barely. This suggests that model 2 is the most appropriate even though model 1 is simpler, i.e., it contains no three-factor effects.

Table 9.7-2 Values of Log-Likelihood Ratio Statistics for Models Fit to Torus Mandibularis Data in Table 9.7-1

	G^2	Degrees of Freedom	Value of χ^2 $p = 0.05$
Model 1	60.57	45	61.7
Difference between 1 and 2	18.71	10	18.3
Model 2	41.06	35	49.8
Difference between 2 and 3	0.80	1	3.8
Model 3	40.26	34	48.6
Difference between 3 and 4	8.09	14	23.7
Model 4	32.17	20	31.4

Table 9.7-3 contains the estimated u-values for models 1 and 2. We note that the addition of u_{124} in model 2 does not change the sign of any of the u-terms, nor does it greatly affect the actual values of the u-terms, except for those involving age category 1–10. Turning to the Freeman–Tukey deviations for models 1 and 2, in tables 9.7-4 and 9.7-5, we are unable to note any substantial change in the pattern of signs or the magnitude of the larger deviations (say those greater than 0.90), with one exception. Three of the four largest deviates for model 1 correspond to age category 1–10, and the extra term in model 2 substantially reduces the size of these deviates. The pattern of the largest deviations for both models suggests a possible age-sex-population effect (i.e., u_{134}). Including this term in either model 1 or model 2 does not provide a significant improvement (at the 0.05 level), and the pattern of the large deviations still persists when this effect is included in either model.

We now recall that investigator differences have been confounded with population differences, and note that the three-factor effect relating incidence to age and population may also result from the investigator differences, i.e., the incidence-age interactions may have changed over time or may be detected in a different manner by different investigative techniques. As a result of this irregularity in the data generation, it might be reasonable to supress the three-factor effect and to report model 1 as the "best" model. ■ ■

Table 9.7-3   Estimated u-Terms for Models 1 and 2 Fitted to Data in Table 9.7-1
(Note: In each case values for model 1 are given in parentheses.)

(i) $\hat{u} = 1.57$ (1.58)

(ii) $\hat{u}_1$

Age	
1–10	0.28(+0.19)
11–20	0.45(+0.47)
21–30	0.45(+0.45)
31–40	0.02(+0.02)
41–50	−0.49(−0.43)
Over 50	−0.71(−0.69)

(iv) $\hat{u}_3$

Sex	
Male	0.07(+0.07)
Female	−0.07(−0.07)

(iii) $\hat{u}_2$

Incidence	
Present	−0.06(−0.08)
Absent	0.06(+0.08)

(v) $\hat{u}_4$

Population	
Igloolik	0.66(+0.58)
Hall Beach	−0.34(−0.37)
Aleut	−0.32(−0.21)

(vi) $\hat{u}_{12}$

Age	Incidence Present	Absent
1–10	−0.93(−1.07)	0.93(+1.07)
11–20	−0.61(−0.55)	0.61(+0.55)
21–30	−0.17(−0.13)	0.17(+0.13)
31–40	0.42(+0.46)	−0.42(−0.46)
41–50	0.50(+0.49)	−0.50(−0.49)
Over 50	0.79(+0.80)	−0.79(−0.80)

(vii) $\hat{u}_{14}$

	Igloolik	Population Hall Beach	Aleut
1–10	−0.09(+0.24)	0.06(+0.16)	0.03(−0.40)
11–20	0.13(−0.01)	0.33(0.28)	−0.46(−0.27)
21–30	0.26(0.17)	−0.01(−0.04)	−0.26(−0.12)
31–40	0.01(−0.05)	−0.07(−0.11)	0.06(+0.15)
41–50	−0.28(−0.33)	−0.02(−0.01)	0.30(+0.33)
Over 50	−0.04(−0.03)	−0.29(0.28)	0.32(0.31)

(viii) $\hat{u}_{24}$

		Incidence	
Population		Present	Absent
Igloolik		0.11(+0.11)	−0.11(−0.11)
Hall Beach		0.09(+0.08)	−0.09(−0.08)
Aleut		−0.20(−0.19)	0.20(+0.19)

(ix) $\hat{u}_{123}$

Population	Incidence	Age 1–10	11–20	21–30	31–40	41–50	Over 50
Igloolik	Present	−0.54	0.09	0.04	0.10	0.15	0.17
	Absent	0.54	−0.09	−0.04	−0.10	−0.15	−0.17
Hall Beach	Present	−0.24	−0.01	−0.01	0.01	0.17	0.08
	Absent	0.24	0.01	0.01	−0.01	−0.17	−0.08
Aleut	Present	0.78	−0.08	−0.03	−0.10	−0.32	−0.25
	Absent	−0.78	0.08	0.03	0.10	0.32	0.25

Table 9.7-4   Freeman–Tukey Deviations
(Based on Data in Table 9.7-1, for Model 1 from Table 9.7-2)

Population	Sex	Incidence	Age Groups 1–10	11–20	21–30	31–40	41–50	Over 50
Igloolik	Male	Present	−0.53	−0.43	−0.92	0.54	0.76	0.28
		Absent	0.14	0.98	0.12	−0.30	−2.49	−0.68
	Female	Present	−2.04	0.95	1.13	−0.16	−0.24	0.16
		Absent	0.75	−1.34	−0.16	0.03	0.97	0.25
Hall Beach	Male	Present	0.28	0.31	0.95	−0.05	0.16	0.57
		Absent	0.47	−0.99	−0.06	0.05	0.53	0.53
	Female	Present	−0.28	−0.35	−1.08	0.26	0.43	−0.29
		Absent	−0.29	1.17	0.29	0.25	−1.46	−0.79
Aleut	Male	Present	2.19	0.14	0.37	0.71	−0.01	−0.57
		Absent	−1.77	0.85	−1.93	−0.23	1.24	1.08
	Female	Present	1.79	−0.41	−0.47	−1.30	−0.82	0.21
		Absent	0.08	−0.58	1.79	0.98	−0.08	−0.08

Table 9.7-5 Freeman–Tukey Deviations
(Based on Data in Table 9.7-1, for Model 2 from Table 9.7-2)

Population	Sex	Incidence	1–10	11–20	21–30	31–40	41–50	Over 50
					Age Groups			
Igloolik	Male	Present	0.81	−0.63	−1.00	0.39	0.54	0.13
		Absent	−0.27	1.10	0.19	−0.05	−2.09	−0.31
	Female	Present	−0.79	0.76	1.06	−0.30	−0.44	0.01
		Absent	0.37	−1.22	−0.09	0.26	1.33	0.58
Hall Beach	Male	Present	0.42	0.42	0.97	−0.05	−0.02	0.51
		Absent	0.42	−1.05	−0.07	0.05	0.85	0.64
	Female	Present	−0.15	−0.25	−1.06	0.26	0.26	−0.34
		Absent	−0.33	1.11	0.27	0.24	−1.17	−0.69
Aleut	Male	Present	0.23	0.42	0.51	0.97	0.51	−0.27
		Absent	−0.85	0.72	−2.03	−0.53	0.68	0.64
	Female	Present	−0.01	−0.15	−0.34	−1.06	−0.34	0.50
		Absent	0.94	−0.70	1.70	0.70	−0.59	−0.49

9.8 Frequency of Frequencies Distribution

9.8.1 *Summarizing data sets*

Many discrete multivariate data sets are extremely large, and while to describe them in terms of a log-linear or similar model would be nice, we are often willing to settle for much less. However, this may be as valuable to us as a log-linear model description. For example, on some occasions we might simply want to standardize the marginal totals of a set of tables and to make "eyeball" comparisons of the standardized tables. We did this in Chapter 3 with the British and Danish occupational mobility tables, where we discovered that the cell values in the standardized tables are very similar. Mosteller [1968] discusses these standardized tables, noting that "it is a great economy of description to find societies having much the same intergenerational mobility, needing only the row and column frequencies to finish the description."

While the form of log-linear models is conceptually easy to work with, as the number of categories per dimension as well as the number of dimensions grows we need to estimate more and more parameters to fit the simplest of models. In such large tables we often have many sparse cells, and an economical summary description of the data is achievable by an examination of the "frequency of frequencies."

9.8.2 *Some basic theory*

Suppose we have N observations distributed over t cells (e.g., we have an $I_1 \times I_2 \times \cdots \times I_p$ contingency table where $t = \prod_{j=1}^{p} I_j$). Now if f_x is the number of cells containing a count of x observations, $x = 0, \ldots, N$, we call f_x the *frequency of the frequency* x in the table. Thus $N = \sum_x x f_x$ and $t = \sum_x f_x$. The frequency of frequencies approach has been used in the studies of literary vocabulary and the occurrence of species of plants or animals (see Good [1953, 1965]), where the distribution of the frequencies of frequencies is used to predict the probabilities of occurrence of new species or new words. Here we propose to fit the frequency of frequencies distribution in order to provide a succinct summary of the data. Such an approach is best illustrated by example, but first we need some statistical theory.

Let $p_i, i = 1, \ldots, t$, be the cell probabilities for the t cells in some table and, as before, let $f_x, x = 0, \ldots, t$, be the number of cells with exactly x observations. If all observations are independent and identically distributed, then

$$E(f_x) = \sum_{i=1}^{t} \binom{N}{x} p_i^x (1 - p_i)^{N-x}. \tag{9.8-1}$$

In fitting the Poisson distribution to the frequencies of frequencies, we assume that $p_i = 1/t$ for $i = 1, \ldots, t$, from which it follows that as both N and t grow with $\lambda = N/t$ remaining constant,

$$E(f_x) \simeq t \frac{e^{-\lambda} \lambda^x}{x!} \tag{9.8-2}$$

for x small relative to N.

For the more general case where the p_i are unequal, following the approach of Good [1953], we assume the existence of a density function $h(p)$, where $h(p)dp$ is the number of cells whose cell probabilities p_i lie between p and $p + dp$. Thus

$$\int_0^1 h(p)\, dp = t, \tag{9.8-3}$$

$$\int_0^1 p h(p)\, dp = 1, \tag{9.8-4}$$

the latter of which is a replacement for $\sum_i p_i = 1$. There is some sleight of hand in this approach, since the use of the density suggests that there are an infinite number of cells, but then we let the density integrate to t, a finite number of cells. Nevertheless we continue, keeping in mind that we will shortly let t get large.

The expected value for the frequency of the frequency x is

$$E(f_x) = \binom{N}{x} \int_0^1 p^x (1 - p)^{N-x} h(p)\, dp, \tag{9.8-5}$$

and if we take $h(p)$ to be concentrated on small values of p where Np is moderate in size,

$$E(f_x) \simeq \frac{1}{x!} \int_0^1 (Np)^x e^{-Np} h(p)\, dp. \tag{9.8-6}$$

Now we take $h(p)$ to be

$$h(p) = \frac{(\alpha t)^{\alpha+1}}{\alpha!} p^{\alpha-1} e^{-pt\alpha} \tag{9.8-7}$$

for $\alpha > 0$, which, while it strictly speaking allows values of p greater than 1, assigns all such p combinations a very small probability when t is large. Using this form for $h(p)$, we get

$$E(f_x) \simeq t \binom{\alpha + x - 1}{\alpha - 1} \left(\frac{\lambda}{\lambda + \alpha}\right)^x \left(\frac{\alpha}{\lambda + \alpha}\right)^\alpha, \tag{9.8-8}$$

where $\lambda = N/t$ as before, i.e., the expected frequency for a negative binomial distribution for a sample of size t.

When the t cells represent a cross-classification of qualitative variables, this approach to the frequency of frequencies distribution ignores all information relating to the contingency table structure. Further research is required on this aspect of the frequency of frequencies approach.

9.8.3   Some applications

Example 9.8-1   Pearson's mobility table

In Chapter 12 we explore the estimates of the cell probabilities in a 14×14 occupational mobility table first presented by Pearson [1904b]. For this table, $t = 14^2 = 196$ and $N = 775$. In table 9.8-1 we give the observed frequency of frequencies distribution for all the cells in the Pearson table, and in table 9.8-2 we give the corresponding distribution for the same table, with the cells along the main diagonal having been deleted, so that $t = 14^2 - 14 = 182$ and $N = 532$. Note that in both cases N and t are large, but $\lambda = N/t$ is moderate, in keeping with the theory we describe above.

Since the sample variance is far greater than the sample mean for both the full and the reduced tables, we do not expect the Poisson distribution to provide a

Table 9.8-1   Frequency of Frequencies for Cells in Pearson [1904b] Occupational Mobility Table

I	II	III	IV
	Observed	Expected	Expected
x	f_x	Poisson	Negative Binomial
0	53	35	62
1	43	14.3	27.8
2	18	28.7	18.9
3	16	38.3	14.1
4	12	38.3	11.2
5	12	30.6	9.0
6	11	20.4	7.4
7	5	11.7	6.3
8	2	5.8	5.3
9	3	2.6	4.5
10	1	1.0	3.7
11	2		3.3
12	3		2.7
13	3		2.4
14	1		2.0
15	2		1.8
16	1		1.6
17	2		1.4
18	1	5.3	1.2
19	1		1.0
20	1		1.0
23	1		
28	1		6.2
51	1		
54	1		

Note: $\bar{x} = 3.90$, $s^2 = 46.2$, $N = 775$, $t = 196$.

Table 9.8-2 Frequency of Frequencies
for Pearson Mobility Table with Main
Diagonal Deleted

I	II	III	IV
	Observed	Expected	Expected
x	f_x	Poisson	Negative Binomial
0	52	10	59.0
1	43	29.1	33.1
2	18	42.1	22.3
3	16	40.1	16.2
4	11	29.5	12.2
5	12	17.1	8.9
6	8	8.3	6.9
7	4	3.4	6.3
8	2		4.0
9	2		3.1
10	1		2.3
11	1		1.8
12	3		1.5
13	3	2.4	1.1
14	1		0.9
15	2		0.7
16	1 ⎫6	3.8	0.5
17	2		0.4
18+	0		1.3

Note: $\bar{x} = 2.92$, $s^2 = 14.4$, $N = 532$, $t = 182$.

good description of the frequencies of frequencies. (For completeness, we include
the expected Poisson frequencies in tables 9.8-1 and 9.8-2.) Turning to the negative
binomial distribution, we give the expected frequencies in column IV of tables
9.8-1 and 9.8-2. We estimate the parameter values using the method of moments
(see Chapter 13). The fit in both cases is quite good ($X^2_{13} = 19.1$ for table 9.8-1 if
we pool frequencies for $x \geq 15$, and $X^2_9 = 8.8$ for table 9.8-2 if we pool frequencies
for $x \geq 11$). The fit in the tails for the full set of cells (including those on the diagonal)
is actually deceptively good, since we have one observed frequency for $x = 51$ and
another for $x = 54$, while the corresponding expected frequencies are extremely
small. Thus it is reasonable to conclude that the negative binomial distribution
adequately describes the frequency of frequencies distribution for the off-diagonal
cells in the table. As a result, we have a two-parameter summary for a 182-cell set of
data, one which may be especially useful if this occupational mobility table is
compared with others having a larger list of occupations and which are also
summarized by a pair of parameters. ∎∎

Example 9.8-2   Similarity of two poems
Eckler [1973] discusses the similarity of two poems, both entitled "Winter Retrospect," by J. A. Lindon and H. W. Bergerson. Bergerson wrote his poem first, then
sent to Lindon an alphabetical list of the words in the poem together with specifica-
tions regarding the stanza sizes and line lengths. Without any knowledge of the
original poem, Lindon then constructed a poem out of these words, using the same

Table 9.8-3   Location of 162 Single-Appearance Words in Lindon's and Bergerson's Poems "Winter Retrospect," by Line

Line in Bergerson	Line in Lindon																							
	1	2	3	4	5	6	7	8	9	10	11	12	13	14	15	16	17	18	19	20	21	22	23	24
1	2	0	1	0	0	1	0	0	0	1	1	1	0	0	0	0	0	2	1	0	0	0	0	0
2	1	1	1	2	0	0	0	0	0	0	0	1	0	0	2	1	0	0	0	0	0	0	1	0
3	0	0	1	0	0	1	0	0	1	0	0	0	0	0	1	0	0	0	0	0	0	0	0	0
4	1	1	1	2	1	0	0	0	0	0	0	0	0	0	0	1	1	0	0	0	0	0	0	0
5	1	0	0	1	0	0	0	0	0	0	0	1	0	1	0	1	1	0	0	0	1	0	1	0
6	0	0	1	0	0	0	1	0	0	0	0	0	0	0	0	0	0	0	1	2	0	1	1	0
7	0	0	0	0	0	0	0	0	0	0	1	1	0	1	1	0	0	0	0	3	4	1	0	0
8	0	0	0	0	0	0	0	0	0	0	0	0	1	2	0	0	0	0	0	0	1	0	0	1
9	0	0	0	1	0	0	0	0	0	1	1	0	0	0	0	1	0	0	0	0	0	0	0	0
10	0	0	0	0	0	1	1	0	0	0	1	0	0	0	1	0	0	0	0	0	1	0	0	0
11	0	0	1	0	0	0	4	2	0	0	0	0	0	0	0	0	1	0	0	0	0	0	0	0
12	0	0	0	0	0	2	0	1	0	1	0	0	0	0	0	0	0	0	0	0	1	0	0	0
13	0	0	1	0	1	0	0	1	2	0	0	0	0	0	0	0	0	0	0	0	0	0	0	0
14	0	0	0	0	0	0	0	0	2	1	0	0	0	0	1	0	0	0	0	0	0	0	0	0
15	0	0	0	1	1	0	0	0	0	0	0	0	0	0	0	0	0	0	0	0	0	0	0	0
16	0	0	0	0	0	0	0	1	0	2	0	0	0	1	0	0	0	0	0	0	0	0	0	0
17	0	0	0	0	1	1	0	0	0	0	1	0	2	0	2	0	0	0	0	0	0	0	0	0
18	1	0	0	0	0	2	2	0	0	0	0	0	2	0	2	0	0	0	0	0	0	1	0	0
19	0	2	0	0	1	0	0	0	1	0	1	0	0	0	0	1	1	1	0	0	0	0	1	0
20	0	0	0	1	1	0	0	1	0	0	0	0	0	0	1	0	0	0	0	0	0	0	0	0
21	0	1	0	0	0	0	1	0	0	0	1	0	0	0	2	0	0	0	0	1	0	0	1	1
22	0	0	0	2	0	0	0	0	1	1	0	0	0	0	0	0	0	0	0	0	1	0	1	0
23	0	0	0	0	0	1	0	0	1	1	1	0	0	0	0	0	0	0	0	1	0	0	1	0
24	0	0	1	0	0	0	0	0	1	0	0	1	1	0	1	0	0	0	0	1	0	0	0	0

Source: Eckler [1973].

stanza sizes and line lengths. Both poems are 24 lines in length and consist of 478 words, of which 162 appear only once. Our intent here is to compare the poems, specifically looking at whether some words cluster together in both.

To compare the two poems, original and reconstruction, we can use the positioning of the 162 single-appearance words in the two poems, and choose as our unit of investigation the line. We give these data in table 9.8-3 in the form of a 24 × 24 table. Since the number of cells in the table, 596, is more than triple the number of observations, 162, we cannot deal with this table in the usual fashion, testing for independence. The frequency of frequencies distribution for these data, however, provides a convenient summary of this table, and we give this distribution in table 9.8-4.

Table 9.8-4 Frequency of Frequencies Distribution for 162 Words as Described in Table 9.8-3

x	Observed f_x	Poisson Expected Values
0	439	435.0
1	117	122.0
2	17	17.3
3	1	1.6
4	2	0.1
Totals	576	576.0

With the exception of the frequency $x = 4$, the observed frequency of frequencies appears to follow a Poisson distribution. Eckler interprets the two counts of 4 (which indicate that the two sets of words—elbow, hand, head, quilt; and bakes, clean, grates, oven—are closely associated) as evidence of similarity between the poems, but notes that, with the exception of these two sets of words, the remaining 154 words appear to be used independently by the authors. Eckler also uses units of investigation other than the single line to measure "distance" between poems, but reaches similar conclusions in these additional analyses. ■ ■

10 Other Methods for Estimation and Testing in Cross-Classifications

10.1 Introduction

In other chapters, we have presented methodology developed specifically for the analysis of categorical (qualitative) data. Moreover, we have focused on a particular technique of estimation (maximum likelihood) and a particular class of parametric models (log-linear models). Here we review alternative methods and models for analyzing qualitative data, and we discuss some approaches for handling mixtures of qualitative and quantitative data.

Since a definitive treatment of all such techniques and approaches is beyond the scope of this book, the reader should use this chapter as a guide to some of the available alternatives to the methods presented in other chapters.

We can always use methods developed for categorical data when measurements are made on both discrete (qualitative) and continuous (quantitative) variables by condensing the quantitative measurements into a few categories. Except for situations where we make implicit use of the ordering of categories within a variable (e.g., see example 5.2-4 in Chapter 5), throughout this book we have treated all variables as if they consisted of discrete categories with no underlying order. Even when considering an essentially continuous variable such as age, we have dealt with age-groups as if they were unordered categories. Such an approach has the potential for loss of information at two stages: when we break the continuous variable into categories, and when we ignore the ordering of the categories.

Other authors have suggested different methods for handling mixtures of qualitative and quantitative variables. Many of these alternatives involve the assignment of scores to the categories of the qualitative variables so that we may include these variables in analyses appropriate for continuous variables. Others involve transformations on rates or proportions, and some use models quite similar to log-linear ones.

If we limit ourselves to cross-classifications whose variables have unordered categories, there are many variations in the methods and procedures:

1. the method of estimation may not be maximum likelihood;
2. the models proposed may not be log-linear;
3. the underlying sampling model for the cell counts may differ from those considered here;
4. the method of choosing a model may differ from ours;
5. no underlying parametric model may be assumed at all.

These differences usually, but not invariably, lead to different cell estimates and different values of test statistics. Some authors primarily concerned with testing hypotheses may even derive test statistics which avoid the computation of individual cell or parameter values.

10.1.1   *Approaches considered*

The early sections of this chapter deal with methods strictly for categorical data. In Section 10.2 we show that the information theory approach usually leads to estimates for the log-linear model identical with those obtained under maximum likelihood. There are, however, some variants of this approach leading to estimates that are not maximum likelihood. Section 10.3 deals first with classical minimum X^2 estimation and then describes variants of this procedure, such as modified minimum X^2, one of the prominent techniques in the recent literature. In Section 10.4 we consider a method due to Lancaster of partitioning X^2 that has been in use for some time, and we discuss a theoretical difficulty. By contrast, in Section 10.5 we discuss "exact" methods that have important theoretical justification but also overwhelming computational disadvantages when used for the analysis of large arrays.

We then turn to methods for mixtures of qualitative and quantitative variables that attempt to utilize all the information rather than debase the continuous measurements. Section 10.6 is devoted to methods based on the logistic model, which is closely related to the use of logits for strictly discrete data. We review other transformations for categorical data in Section 10.7. Some of these require that categories be ordered, and can presumably be used equally well for mixtures of variables.

Finally, we indicate in Section 10.8 the problem areas where more theoretical development is desirable.

10.1.2   *Approaches not covered*

We make no attempt to consider the details of analyses based on sampling models other than the multinomial, product multinomial, or Poisson sampling schemes. Kleinman's approach [1973], referred to in Section 10.7, represents one alternative. Two other sampling models, by Steyn [1959] and by Rudolph [1967], referred to in Section 13.8, are based on the negative multinomial distribution.

There remain several approaches to the analysis of discrete multivariate data, even within the framework of multinomial or Poisson sampling schemes, which we do not discuss. For example, we do not consider the techniques of latent structure analysis. There are relationships between latent structure analysis and the analyses considered in this book, some of which are pointed out by Bock [1970, 1972]. Some additional references for latent structure analysis are Lazarsfeld and Henry [1968], Rasch [1960], and Samejima [1969, 1972].

10.2   The Information-Theoretic Approach

Kullback [1959], Ku and Kullback [1968], Ku, Varner, and Kullback [1971], and others have used the method of minimum discrimination information estimation (MDIE) to estimate multinomial cell probabilities, especially in cases of multi-dimensional cross-classifications.

Let us consider a t-dimensional multinomial with probability vector $\mathbf{p}$, and let $\boldsymbol{\pi}$ be some "null" probability vector (such as $\boldsymbol{\pi} = (1/t, \ldots, 1/t)$). Then we define the discrimination information statistic by

$$I(\mathbf{p}:\boldsymbol{\pi}) = \sum p_i \log \frac{p_i}{\pi_i} = \log \left[\frac{\prod p_i^{p_i}}{\prod \pi_i^{p_i}} \right]. \qquad (10.2\text{-}1)$$

$I(\mathbf{p}:\boldsymbol{\pi})$ is also the Kullback-Liebler distance function between $\mathbf{p}$ and $\boldsymbol{\pi}$, and this distance is measured by comparing $\prod p_i^{p_i}$ with $\prod \pi_i^{p_i}$. Expression (10.2-1) is always nonnegative.

There are two components to the information theory approach, model generation and estimation of parameters. Both make use of the discrimination information statistic (10.2-1).

10.2.1   *Model generation*

One way to specify a parametric model for the multinomial probability vector $\mathbf{p}$ is to require a set of $t - 1$ linear constraints. If we impose fewer constraints, then we require a device to choose among the models which satisfy these constraints.

The method for model generation proposed by Kullback fits into this framework. First we require $\mathbf{p}$ to satisfy certain linear constraints (for multidimensional tables these constraints are typically values for certain marginal totals). Then the *minimum discrimination information model*, denoted by $\mathbf{p}^*$, is that vector of the form $\mathbf{p}$ which most closely resembles $\boldsymbol{\pi}$ in that it minimizes (10.2-1) over all probability vectors $\mathbf{p}$ subject to the linear constraints. Good [1963] refers to this approach of finding $\mathbf{p}^*$ as the "method of maximum entropy."

The equality

$$I(\mathbf{p}:\boldsymbol{\pi}) = I(\mathbf{p}:\mathbf{p}^*) + I(\mathbf{p}^*:\boldsymbol{\pi}) \qquad (10.2\text{-}2)$$

holds for the $\mathbf{p}^*$ which minimizes (10.2-1), where $\mathbf{p}^*$ satisfies the same linear constraints as does $\mathbf{p}$. Both of the terms on the right-hand side of (10.2-2) are positive. Equation (10.2-2) is important because it serves as the basis for the exact partitioning of the goodness-of-fit statistics.

Example 10.2-1   Multiplicative models for two-way tables
Suppose that the multinomial represents an $I \times J$ table, and let $\mathbf{p} = \{p_{ij}\}$ satisfy the marginal constraints $p_{i+} = \sum_j p_{ij}$ and $p_{+j} = \sum_i p_{ij}$. Then the minimum discrimination information model $\mathbf{p}^*$, which minimizes $I(\mathbf{p}, \boldsymbol{\pi})$, is of the form

$$p_{ij}^* = a_i b_j \pi_{ij}, \qquad (10.2\text{-}3)$$

where the a_i and b_j are determined by the marginal constraints (see Ireland and Kullback [1968b]).

If $\boldsymbol{\pi} = (1/IJ, \ldots, 1/IJ)$, then $a_i \propto p_{i+}$, $b_j \propto p_{+j}$, and the model is that of independence, i.e., $p_{ij}^* = p_{i+} p_{+j}$. ■■

On the choice of $\boldsymbol{\pi}$
The simplest model for the probability vector is the one which sets all cell probabilities equal to $1/t$. Thus the simplest "null" probability vector is $\boldsymbol{\pi} = (1/t, \ldots, 1/t)$ and choosing this value of $\boldsymbol{\pi}$ in the model-selection problem is a device for finding the simplest possible model $\mathbf{p}^*$ (i.e., the model with the fewest parameters) which satisfies the linear constraints.

In example 10.2-1 above, where $t = IJ$, we saw that choosing $\pi = (1/IJ, \ldots, 1/IJ)$ reduces the multiplicative model (10.2–3) to the simpler multiplicative model of independence.

Generating log-linear models

We can generate the log-linear models considered in Chapter 3 using the minimum discrimination model-generation technique. This fact is a consequence of the following theorem, due to Good [1963], and its generalizations.

THEOREM 10.2-1   *Let the vector π be uniform, and let $\mathbf{p}$ represent the class of cell probabilities for a $I_1 \times I_2 \times \cdots \times I_k$ contingency table, with all r-dimensional marginal totals fixed, where $r < k$. Suppose further that the constraints are consistent and that there exist probability vectors $\mathbf{p}$ satisfying them with no components equal to zero. Then the model generated by the method of maximum entropy (i.e., the $\mathbf{p}^*$ vector) is the log-linear model with all $(r + 1)$-factor and higher-order effects equal to zero.*

10.2.2   Estimating expected values

Once we have a model for the multinomial probability vector, $\mathbf{p}$, our interest turns to estimating $\mathbf{p}$, or equivalently, the vector of expected cell values $\mathbf{m} = N\mathbf{p}$. The MDIE of $\mathbf{p}$ is defined as that value of $\mathbf{p}$ (subject to the constraints of the model) which minimizes

$$NI\left(\frac{\mathbf{x}}{N}:\mathbf{p}\right) = \sum_i x_i \log \frac{x_i}{Np_i}. \tag{10.2-4}$$

But (10.2-4) when viewed as a function of $\mathbf{p}$ is just a constant minus the kernel of the multinomial log-likelihood,

$$\sum_i x_i \log p_i = \log \prod_i p_i^{x_i}, \tag{10.2-5}$$

and the equivalence of MDIEs and MLEs follows immediately. Kullback suggests the use of the likelihood ratio statistic

$$G^2 = 2NI\left(\frac{\mathbf{x}}{N}:\hat{\mathbf{p}}\right) \tag{10.2-6}$$

to test the goodness of fit of the model under consideration.

Modified MDI estimates

Ireland, Ku, and Kullback [1969] propose an alternate method of estimation (still referred to by the authors as minimum discrimination information). They suggest minimizing

$$NI\left(\mathbf{p}:\frac{\mathbf{x}}{N}\right) = \sum_i Np_i \log \frac{Np_i}{x_i} \tag{10.2-7}$$

with respect to $\mathbf{p}$ (subject to the constraints of the model for $\mathbf{p}$) rather than minimizing $NI(\mathbf{x}/N:\mathbf{p})$. Since the minimum discrimination information statistic is not symmetric, this is a new (and different) minimization problem, and in general does not yield estimates identical to MDIEs and MLEs.

Berkson [1972] notes that the differences between $I(\mathbf{p}:\mathbf{x}/N)$ and $I(\mathbf{x}/N:\mathbf{p})$ are similar to the differences between the chi square statistic of Neyman [1949] and the standard Pearson chi square in that the observed and estimated quantities are interchanged. The test statistic suggested by the minimization of (10.2-7) is

$$G_M^2 = 2 \sum N\hat{p}_i \log \frac{N\hat{p}_i}{x_i}, \tag{10.2-8}$$

which is to the likelihood ratio statistic as the Neyman chi square (see Section 10.3) is to the usual Pearson chi square statistic.

Example 10.2-2   Homogeneity of proportions for J binomials

If we have J independent binomials, we can display the observed data in a $2 \times J$ contingency table with cell observations x_{ij}, where $i = 1, 2$ and $j = 1, \ldots, J$. Each x_{+j} is fixed by the sampling plan, and the observed row proportions are

$$\frac{x_{1j}}{x_{+j}}, \quad \frac{x_{2j}}{x_{+j}} \tag{10.2-9}$$

for $j = 1, \ldots, J$. If we let $m_{ij} = x_{+j} P_{j(i)}$ (where $P_{j(i)} = p_{ij}/p_{+j}$, so that $P_{j(1)} + P_{j(2)} = 1$) be the expected cell count in the (i, j) cell under the model of homogeneity of proportions, then the modified minimum discrimination information estimates (MMDIs) cannot be expressed in closed form, and $\sum_j \hat{m}_{ij} = \sum_j x_{ij}$ only if we impose this additional constraint. If we do impose the constraint, then the estimates are the same as the MLEs. ■■

Example 10.2-3   Symmetry in square tables

Suppose we have an $I \times I$ table where the categories for rows have the same labels as the categories for columns (i.e., the rows are paired with the columns). We use the notation of expected cell values $\{m_{ij}\}$ here in place of cell probabilities $\{p_{ij}\}$, since we would like to consider two versions of this problem, one which includes the diagonal cells in the analysis and one which does not.

First we include the diagonal cells. We are interested in the hypothesis of symmetry denoted by

$$m_{ij} = m_{ji} \quad i \neq j \tag{10.2-10}$$

(see Chapter 8). Using the modified minimum discrimination estimation approach (MMDIE), we must find the values of m_{ij} which minimize

$$\sum_{ij} m_{ij} \log \frac{m_{ij}}{x_{ij}} \tag{10.2-11}$$

subject to the constraints (10.2-10). Ireland, Ku, and Kullback [1969] give these MMDIEs as

$$\hat{m}_{ij} = \hat{m}_{ji} = k\sqrt{x_{ij}x_{ji}} \quad i \neq j,$$
$$\hat{m}_{ii} = kx_{ii}, \tag{10.2-12}$$

where k is such that $\sum \hat{m}_{ij} = N$, the size of the multinomial sample. Since the geometric mean of two unequal numbers is less than the arithmetic mean, we have

$$\sqrt{x_{ij}x_{ji}} < \frac{x_{ij} + x_{ji}}{2} \tag{10.2-13}$$

unless $x_{ij} = x_{ji}$. Thus

$$\sum_{ij} \hat{m}_{ij} = k\left[\sum_{i \neq j} \sqrt{x_{ij}x_{ji}} + \sum_i x_{ii}\right]$$

$$< k\left[\sum_{i \neq j} \frac{x_{ij} + x_{ji}}{2} + \sum_i x_{ii}\right] \tag{10.2-14}$$

$$= k\sum_{ij} x_{ij} = kN$$

(unless symmetry holds exactly in the observed table), and k must be greater than 1. Thus when the diagonal cells are included in the analysis, we get cell estimates greater than the observed counts for the diagonal entries. Such estimates are intuitively unappealing, since there is no reason that a subset of expected values should always be greater than the corresponding observed values.

Now we exclude the diagonal cells from the analysis, and we minimize

$$\sum_{i \neq j} m_{ij} \log \frac{m_{ij}}{x_{ij}} \tag{10.2-15}$$

subject to (10.2-10). The MMDIEs for the off-diagonal cells are now

$$\hat{m}'_{ij} = \hat{m}'_{ji} = \frac{x_{ij} + x_{ji}}{2}, \tag{10.2-16}$$

which are the MLEs (see Chapter 8).

The modified MDI approach for the analysis of square tables does have the advantage that we can obtain cell estimates under the hypothesis of marginal homogeneity, whereas we can only consider such a model conditionally by the maximum likelihood approach (see Ireland, Ku, and Kullback [1969] and Chapter 8) and the iterative proportional fitting techniques. ∎∎

10.3 Minimizing Chi Square, Modified Chi Square, and Logit Chi Square

10.3.1 *Minimizing Pearson's chi square*

If we think of the usual Pearson chi square statistic

$$X_p^2 = \sum \frac{(\text{observed} - \text{expected})^2}{\text{expected}} \tag{10.3-1}$$

as a measure of the distance (or discrepancy) between the observed vector $\mathbf{x}$ of multinomial observations and the expected vector $\mathbf{m} = N\mathbf{p}$, where $N = \sum x_i$, then it is natural for us to consider choosing $\mathbf{m}$ to minimize (10.3-1), subject to certain constraints given by an underlying model for $\mathbf{m}$. Suppose that the underlying cell probabilities p_i are functions of the reduced set of parameters $\boldsymbol{\theta} = (\theta_1, \ldots, \theta_s)$. Then choosing $\boldsymbol{\theta}$ to minimize

$$X_p^2 = \sum_{i=1}^t \frac{(x_i - Np_i(\theta_1, \ldots, \theta_s))^2}{Np_i(\theta_1, \ldots, \theta_s)}, \tag{10.3-2}$$

where $s < t - 1$, yields s equations of the form

$$0 = -\frac{1}{2}\frac{\partial X_p^2}{\partial \theta_j} = \sum_{i=1}^{t}\left(\frac{x_i - Np_i}{p_i} + \frac{(x_i - Np_i)^2}{2Np_i^2}\right)\frac{\partial p_i}{\partial \theta_j}. \tag{10.3-3}$$

Neyman [1949] shows that solving the equations (10.3-3) leads to BAN estimates for the p_i. Even in the simplest cases (such as independence in a two-way contingency table), however, these equations can be difficult to solve.

10.3.2   A modification to minimizing Pearson's chi square

Cramér [1946] suggests a way around these difficulties that can be used when N is large enough to make the influence of the second term on the right-hand side of (10.3-3) negligible. This yields the equations

$$\sum_{i=1}^{t}\left(\frac{x_i - Np_i}{p_i}\right)\frac{\partial p_i}{\partial \theta_j} = 0 \quad j = 1, \ldots, s. \tag{10.3-4}$$

By recalling that $\sum p_i = 1$, we can reduce (10.3-4) to

$$\sum_{i=1}^{t}\frac{x_i}{p_i}\frac{\partial p_i}{\partial \theta_j} = 0 \quad j = 1, \ldots, s, \tag{10.3-5}$$

which are the same as the ML equations for the multinomial likelihood function obtained by maximizing $\sum_i x_i \log p_i$. Ignoring the second term on the right-hand side of (10.3-3) in most problems, however, is only an approximation to minimizing X_p^2.

10.3.3   Minimizing Neyman's chi square

Neyman [1949] suggests an alternative to the method of minimizing Pearson's chi square, where the chi square statistic (10.3-2) is modified to

$$X_N^2 = \sum_{i=1}^{t}\frac{(x_i - Np_i(\theta_1, \ldots, \theta_s))^2}{x_i}. \tag{10.3-6}$$

Choosing $\boldsymbol{\theta}$ to minimize X_N^2 yields the equations

$$\sum_{i=1}^{t}\frac{(x_i - Np_i)}{x_i}\frac{\partial p_i}{\partial \theta_j} = 0 \quad j = 1, \ldots, s, \tag{10.3-7}$$

which resemble (10.3-4), and simplify to

$$\sum_{i=1}^{t}\frac{p_i}{x_i}\frac{\partial p_i}{\partial \theta_j} = 0 \quad j = 1, \ldots, s. \tag{10.3-8}$$

Except in cases where the underlying parameters $\boldsymbol{\theta}$ lead to s independent linear constraints on the $\{p_{ij}\}$, solving equations (10.3-8) is a fairly difficult task. If we denote the solution to equations (10.3-8) by $\{\hat{\theta}_i\}$, we actually compute X_N^2 in (10.3-6) using these values. This modified chi square statistic has an asymptotic χ^2 distribution if the postulated model is true.

Example 10.3-1   Estimation equations for two-way tables

The simplest problem we considered in Chapter 3 is the estimation of cell probabilities in two-way tables under the hypothesis of independence of the variables corresponding to rows and columns. In an $I \times J$ table, when the model is that of independence of rows and columns, $\mathbf{x} = \{x_{ij}\}$, $\mathbf{p} = \{p_{ij}\}$, $\boldsymbol{\theta} = (p_{1+}, \ldots, p_{I+}, p_{+1}, \ldots, p_{+J})$, and (10.3-8) reduces to

$$\frac{\hat{p}_{k+}}{\hat{p}_{I+}} = \frac{\sum_j \dfrac{\hat{p}_{+j}}{x_{Ij}}}{\sum_j \dfrac{\hat{p}_{+j}}{x_{kj}}} \quad k = 1, \ldots, I - 1, \tag{10.3-9}$$

$$\frac{\hat{p}_{+l}}{\hat{p}_{+J}} = \frac{\sum_i \dfrac{\hat{p}_{i+}}{x_{iJ}}}{\sum_i \dfrac{\hat{p}_{i+}}{x_{il}}} \quad l = 1, \ldots, J - 1, \tag{10.3-10}$$

equations which do not have a simple solution in closed form unless $I = J = 2$. For the 2×2 table case, the solution to (10.3-9) and (10.3-10) is

$$\hat{p}_{1+} = \frac{\sqrt{\hat{\alpha}_1}}{1 + \sqrt{\hat{\alpha}_1}}, \qquad \hat{p}_{+1} = \frac{\sqrt{\hat{\alpha}_2}}{1 + \sqrt{\hat{\alpha}_2}}, \tag{10.3-11}$$

where (in the notation of Section 2.2)

$$\hat{\alpha}_2 = \frac{x_{11} x_{21}}{x_{12} x_{22}} \qquad \hat{\alpha}_1 = \frac{x_{11} x_{12}}{x_{21} x_{22}}. \quad \blacksquare\blacksquare \tag{10.3-12}$$

Example 10.3-2   Chevrolets, Fords, and Plymouths
Suppose we take three independent samples of sizes n_1, n_2, and n_3 for a very large population of automobiles, consisting only of Chevrolets, Fords, and Plymouths. The proportion of Chevrolets, Fords, and Plymouths in the population are p_1, p_2, and p_3, respectively, where $p_1 + p_2 + p_3 = 1$. In the first sample we count the number of Chevrolets (y_1) and non-Chevrolets ($n_1 - y_1$); in the second we count the number of Fords (y_2) and non-Fords ($n_2 - y_2$); in the third we count the number of Plymouths (y_3) and non-Plymouths ($n_3 - y_3$). We would like to estimate the proportions p_1, p_2, and $p_3 = 1 - p_1 - p_2$.

The Neyman chi square statistic for the three independent samples in our problem is

$$X_N^2 = \frac{(y_1 - n_1 p_1)^2}{y_1} + \frac{[(n_1 - y_1) - n_1(1 - p_1)]^2}{n_1 - y_1}$$

$$+ \frac{(y_2 - n_2 p_2)^2}{y_2} + \frac{[(n_2 - y_2) - n_2(1 - p_2)]^2}{n_2 - y_2}$$

$$+ \frac{[y_3 - n_3(1 - p_1 - p_2)]^2}{y_3} + \frac{[(n_3 - y_3) - n_3(p_1 + p_2)]^2}{n_3 - y_3}. \tag{10.3-13}$$

Setting the derivatives of X_N^2 (given by (10.3-13)) with respect to p_1 and p_2 equal to zero, we get the following equations:

$$\frac{n_1(p_1 - \bar{p}_1)}{\bar{p}_1(1 - \bar{p}_1)} = \frac{n_2(p_2 - \bar{p}_2)}{\bar{p}_2(1 - \bar{p}_2)} = \frac{n_3(p_3 - \bar{p}_3)}{\bar{p}_3(1 - \bar{p}_3)}, \tag{10.3-14}$$

where $\bar{p}_i = y_i/n_i$, for $i = 1, 2, 3$. Thus each difference $p_i - \bar{p}_i$ is inversely weighted by the observed variance $\bar{p}_i(1 - \bar{p}_i)/n_i$ of $\bar{p}_i$. Equations (10.3-14), when combined with the linear constraint

$$p_1 + p_2 + p_3 = 1, \tag{10.3-15}$$

give us a system of three linear equations in the three parameters to be estimated, the solution to which we can write in closed form.

Using the method of maximum likelihood (or equivalently, using the modification to minimizing Pearson's chi square from Section 10.3.2), we arrive at the equations

$$\frac{n_1(p_1 - \bar{p}_1)}{p_1(1 - p_1)} = \frac{n_2(p_2 - \bar{p}_2)}{p_2(1 - p_2)} = \frac{n_3(p_3 - \bar{p}_3)}{p_3(1 - p_3)}. \tag{10.3-16}$$

Here each difference $p_i - \bar{p}_i$ is inversely weighted by the true variance $p_i(1 - p_i)/n_i$ of $\bar{p}_i$. Equations (10.3-16), while closely resembling (10.3-14), unfortunately turn into a pair of cubic equations in the three unknowns instead of a pair of linear equations. As a result, we are unable to get closed-form estimates of p_1, p_2, and p_3.

If we reduce this example to two classes and two samples, then the pooled estimate of p_1 is

$$\hat{p}_1 = \frac{y_1 + (n_2 - y_2)}{n_1 + n_2},$$

which agrees with (10.3-16), but not with (10.3-14).   ■ ■

10.3.4   *Neyman's chi square with linearized constraints*

Neyman [1949] suggests a second way to view the problem of minimizing X_N^2. The information that the $\{p_i\}$ are known functions of s independent parameters is equivalent to the restrictions on the $t - 1$ independent $\{p_i\}$ imposed by means of $v = t - 1 - s$ equations of the form

$$F_j(\mathbf{p}) = F_j(p_1, \ldots, p_t) = 0 \tag{10.3-17}$$

for $j = 1, \ldots, v$, which are obtainable by eliminating the parameters $\boldsymbol{\theta}$ from the equations $p_i = p_i(\theta_1, \ldots, \theta_s)$. Instead of minimizing (10.3-6) with respect to $\boldsymbol{\theta}$, we now minimize

$$\sum_{i=1}^{t} \frac{(x_i - Np_i)^2}{x_i} + \sum_{j=1}^{v} \gamma_j F_j(\mathbf{p}) \tag{10.3-18}$$

with respect to $\mathbf{p}$ and the Lagrangian multipliers $\{\gamma_j\}$. When the functions $F_j(\mathbf{p})$ are linear, the minimization problem reduces to finding the solution of a system of linear equations.

To get around the difficulty involved in solving nonlinear equations such as (10.3-9) and (10.3-10), Neyman [1949] suggests linearizing the constraints (10.3-17), using the first terms in the Taylor series expansions about the vector $\mathbf{x}/N$ of observed proportions. Thus he replaces (10.3-17) by

$$0 = F_j^* \left(\mathbf{p}, \frac{\mathbf{x}}{N} \right) = F_j \left(\frac{\mathbf{x}}{N} \right) + \sum_{i=1}^{t} \left(p_i - \frac{x_i}{N} \right) \left[\frac{\partial F_j(\mathbf{p})}{\partial p_i} \right]_{\mathbf{p} = \mathbf{x}/N} \tag{10.3-19}$$

and uses the F_j^* in place of the F_j in (10.3-18). The minimization problem again reduces to finding the solution of a system of linear equations, and Neyman proves that this solution also yields BAN estimates of the $\{p_i\}$.

Reducing the minimization problem to finding the solution of a system of linear equations still leaves us with a matrix inversion problem.

Example 10.3-3   Continuation of example 10.3-1

We continue to consider estimation of cell probabilities in an $I \times J$ table under the model of independence of rows and columns. The restrictions on **p** to be used in the minimization problem are

$$F_{ij}(\mathbf{p}) = \log\left(\frac{p_{ij}p_{IJ}}{p_{iJ}p_{Ij}}\right) = 0 \qquad (10.3\text{-}20)$$

for $i = 1, \ldots, I - 1$ and $j = 1, \ldots, J - 1$, and $\sum_{i=1}^{I}\sum_{j=1}^{J} p_{ij} = 1$. Linearizing the constraints (10.3-20) using (10.3-19), we have

$$F_{ij}^*(\mathbf{p}) = \log\left(\frac{x_{ij}x_{IJ}}{x_{iJ}x_{Ij}}\right) + \left(p_{ij} - \frac{x_{ij}}{N}\right) + \left(p_{IJ} - \frac{x_{IJ}}{N}\right)$$

$$- \left(p_{iJ} - \frac{x_{iJ}}{N}\right) - \left(p_{Ij} - \frac{x_{Ij}}{N}\right). \qquad (10.3\text{-}21)$$

Our problem now reduces to minimizing

$$\sum_{i=1}^{I}\sum_{j=1}^{J} \frac{(x_{ij} - Np_{ij})^2}{x_{ij}} + \sum_{i=1}^{I-1}\sum_{j=1}^{J-1} \gamma_{ij}F_{ij}^*(\mathbf{p}) + \lambda \sum_{i=1}^{I}\sum_{j=1}^{J} p_{ij} \qquad (10.3\text{-}22)$$

with respect to **p** and the Lagrangian multipliers γ_{ij}. This leads us to the set of IJ linear equations

$$\sum_{j=1}^{J} \frac{p_{ij}}{x_{ij}} = \sum_{j=1}^{J} \frac{p_{Ij}}{x_{Ij}} \qquad i = 1, \ldots, I - 1,$$

$$\sum_{i=1}^{I} \frac{p_{ij}}{x_{ij}} = \sum_{i=1}^{I} \frac{p_{iJ}}{x_{iJ}} \qquad j = 1, \ldots, J - 1, \qquad (10.3\text{-}23)$$

$$F_{ij}^*(\mathbf{p}) = 0 \qquad i = 1, \ldots, I - 1; j = 1, \ldots, J - 1,$$

$$\sum_{i=1}^{I}\sum_{j=1}^{J} p_{ij} = 1$$

in the IJ unknown parameters p_{ij}. ∎∎

10.3.5   *Equivalence of Neyman chi square and*
weighted least squares approaches

Bhapkar [1961, 1966] points out the equivalence of the X_N^2 statistic (10.3-6) and yet another statistic suggested by Wald. We can express the Wald statistic as a quadratic form which we minimize using a weighted least squares regression algorithm. When the constraints (10.3-17) are linear, the X_N^2 statistic, based on the estimated parameters, is algebraically the same as the minimum sum of squared residuals obtained using a general weighted least squares technique. When the constraints (10.3-17) are nonlinear, the Wald-type quadratic form is algebraically equivalent to the value of X_N^2 based on the estimates derived using the linearized version of the constraints (10.3-19). Thus we can use the well-known computation techniques for weighted least squares to minimize Neyman's chi square statistic with the linearized constraints. We give the details of the weighted least squares approach below.

Bhapkar and Koch [1968], Grizzle, Starmer, and Koch [1969], Theil [1970], Johnson and Koch [1971], and Grizzle and Williams [1972] describe the application of the weighted least squares approach to contingency table problems. The quadratic form of the test statistics used by these authors is similar to the form of the test statistics proposed by Plackett [1962], Gart [1962], and Goodman [1963b, 1964b]. Zellner and Lee [1965] discuss a related generalized least squares estimation approach.

10.3.6   *Weighted least squares estimation*

To describe the weighted least squares approach, we continue with the simple multinomial problem considered in the sections above, and we assume that the model for the vector **p** of cell probabilities is specified by the set of $v = t - 1 - s$ constraints (10.3-17). (The several multinomial problem is similar and is described in detail by Grizzle, Starmer, and Koch [1969]). To cover both linear and non-linear cases, we express the left-hand sides of (10.3-17) as

$$F_j(\mathbf{p}) = \mathbf{a}_j \boldsymbol{\beta} \qquad j = 1, \dots, v, \qquad (10.3\text{-}24)$$

where $\mathbf{a}_j$ is a $1 \times u$ vector of known constants, $\boldsymbol{\beta}$ is a $u \times 1$ vector of functions of the elements of **p**, and $u \leq t$. Letting

$$\mathbf{F}(\mathbf{p}) = \begin{pmatrix} F_1(\mathbf{p}) \\ \vdots \\ F_v(\mathbf{p}) \end{pmatrix}, \qquad (10.3\text{-}25)$$

$$\mathbf{A} = \begin{pmatrix} \mathbf{a}_1 \\ \vdots \\ \mathbf{a}_v \end{pmatrix}, \qquad (10.3\text{-}26)$$

we can rewrite (10.3-24) as

$$\mathbf{F}(\mathbf{p}) = \mathbf{A}\boldsymbol{\beta}. \qquad (10.3\text{-}27)$$

We get an unrestricted estimate of **F** in (10.3-27) by replacing p_i with the observed proportion x_i/N for $i = 1, \dots, t$. Then the weighted least squares estimator of $\boldsymbol{\beta}$ is the vector **b** which minimizes the quadratic form

$$(\hat{\mathbf{F}} - \mathbf{A}\mathbf{b})' \mathbf{S}^{-1} (\hat{\mathbf{F}} - \mathbf{A}\mathbf{b}), \qquad (10.3\text{-}28)$$

where S is a covariance matrix of $\hat{\mathbf{F}}$ estimated from the data. If the functions $F_j(\mathbf{p})$ are linear in the elements of **p**, S is the sample covariance matrix of $\hat{\mathbf{F}}$; if the functions $F_j(\mathbf{p})$ are nonlinear in the elements of **p**, S is the estimated asymptotic covariance matrix of $\hat{\mathbf{F}}$ which we obtain using the δ method (see Chapter 14), i.e., using linearization. Finally, we use the minimum value of the quadratic form (10.3-28) to test the goodness of fit of the model; under the null hypothesis this minimum is distributed as χ^2 with $u - v$ degrees of freedom.

Example 10.3-4   More on two-way tables and independence

Once again we consider an $I \times J$ contingency table with observed entries x_{ij} and underlying cell probabilities p_{ij}, and we take our model to be the independence of the variables corresponding to rows and columns. As we noted in example 10.3-3, the constraints for this model are linear in the logarithms of the p_{ij}, i.e.,

$$F_{ij}(\mathbf{p}) = \log\left(\frac{p_{ij}p_{IJ}}{p_{iJ}p_{Ij}}\right) = \log p_{ij} + \log p_{IJ} - \log p_{iJ} - \log p_{Ij}. \quad (10.3\text{-}29)$$

The covariance matrix $\mathbf{S}$ is the estimated covariance matrix of the logarithms of the observed cross-product ratios $x_{ij}x_{IJ}/x_{iJ}x_{Ij}$, which we derive in Section 14.6, i.e., the components of $\mathbf{S}$ are given by

$$\widehat{\text{Var}}\left[\log\left(\frac{x_{ij}x_{IJ}}{x_{iJ}x_{Ij}}\right)\right] = \frac{1}{x_{ij}} + \frac{1}{x_{IJ}} + \frac{1}{x_{iJ}} + \frac{1}{x_{Ij}}, \quad (10.3\text{-}30)$$

$$\widehat{\text{Cov}}\left[\log\left(\frac{x_{ij}x_{IJ}}{x_{iJ}x_{Ij}}\right), \log\left(\frac{x_{kl}x_{IJ}}{x_{kJ}x_{Il}}\right)\right] = \begin{cases} -\left[\dfrac{1}{x_{IJ}}\right] & \text{if } i \neq k, j \neq l \\[2mm] -\left[\dfrac{1}{x_{IJ}} + \dfrac{1}{x_{iJ}}\right] & \text{if } i = k, j \neq l \\[2mm] -\left[\dfrac{1}{x_{IJ}} + \dfrac{1}{x_{Ij}}\right] & \text{if } i \neq k, j = l, \end{cases} \quad (10.3\text{-}31)$$

for $(i, j) \neq (k, l)$.

Except for the 2×2 table case, the weighted least squares approach for this problem requires our inverting the matrix $\mathbf{S}$, and where any of the observed counts are zero $\mathbf{S}$ has elements which are infinite. ∎ ∎

Advantages and disadvantages

The papers cited above give details for other forms of $\mathbf{F}$ and for the computation of the corresponding chi-square-like test statistics. The emphasis of these authors is on testing rather than estimation of cell probabilities or parameters, although the latter is possible. Even though there are instances where the weighted least squares approach yields estimates without iteration while maximum likelihood estimates require iteration (see the minimum logit χ^2 example below), the reverse situation is usually the case. Nevertheless, this approach has the advantage that we can use the same weighted least squares algorithm to minimize (10.3-28), as we would use for the analysis of linear models with normally distributed errors, and it allows us great latitude in our choice of models.

The weighted least squares approach to estimation and testing suffers from the drawback that when several of the x_i are zero, (10.3-6) and (10.3-28) are not properly defined. The effects of replacing such zero values for observed cell counts by some positive value, as has been advocated by Grizzle, Starmer, and Koch, have not been definitively investigated. Elsewhere in this book we proposed a different approach to the handling of zero counts (see Chapter 12). In Chapter 3 we showed that the presence of sporadic zeros causes no problem when we compute MLEs for log-linear models.

A special case: Minimum logit χ^2

A special case of weighted least squares estimation and testing log-linear models is the minimum logit χ^2 approach suggested by Berkson [1955, 1968] for the analysis of several binomials. We outline this approach below and show that it is equivalent to the weighted least squares approach.

Suppose we have J independent binomial variates. Then we can display the resulting data in a $2 \times J$ table with observed counts x_{ij}, where $x_{+j} = n_j$ are the sample sizes fixed by the sampling plan. The observed and expected logits are defined as

$$l_j = \log\left(\frac{x_{1j}}{x_{2j}}\right),$$

$$L_j = \log\left(\frac{m_{1j}}{m_{2j}}\right),$$

(10.3-32)

where m_{ij} is the expected cell count in the (i, j) cell of the table. Then the logit χ^2, as defined by Berkson [1968], is

$$X_B^2 = \sum_{j=1}^{J} \frac{x_{1j} x_{2j}}{n_j} (l_j - L_j)^2.$$

(10.3-33)

Recalling that $x_{1j} + x_{2j} = n_j$, we can rewrite (10.3-33) as

$$X_B^2 = \sum_{j=1}^{J} \frac{\left(\log\dfrac{x_{1j}}{x_{2j}} - \log\dfrac{m_{1j}}{m_{2j}}\right)^2}{\dfrac{1}{x_{1j}} + \dfrac{1}{x_{2j}}},$$

(10.3-34)

which is a quadratic form of the logits, i.e.,

$$X_B^2 = (\hat{\mathbf{L}} - \mathbf{L})' \mathbf{S}_l^{-1} (\hat{\mathbf{L}} - \mathbf{L}),$$

(10.3-35)

where $\hat{\mathbf{L}} = (l_1, \ldots, l_J)'$, $\mathbf{L} = (L_1, \ldots, L_J)'$, and $\mathbf{S}_l$ is the estimated covariance matrix of $\hat{\mathbf{L}}$. Since we are dealing with J independent binomial variates, the logits $\{l_i\}$ are independent, and inverting $\mathbf{S}_l$ is a trivial matter.

Finally, if we consider linear models for the expected logits $\{L_i\}$, then minimizing X_B^2 with respect to the parameters of the model is equivalent to minimizing (10.3-28), and we can use the minimized value of X_B^2 to test the goodness of fit of the model.

Example 10.3-5   Minimum logit χ^2 for $2 \times 2 \times 2$ tables

We consider a $2 \times 2 \times 2$ table with observed values $\{x_{ijk}\}$, expected values $\{m_{ijk}\}$, and $n_{jk} = \sum_{i=1}^{2} x_{ijk}$ fixed by the sampling plan, for $j, k = 1, 2$. We can specify the log-linear model for no three-factor effect by the constraint on the expected logits given by

$$L_{11} - L_{21} - L_{12} + L_{22} = 0.$$

(10.3-36)

To minimize X_B^2 for this model, we minimize

$$\sum_{j,k} \frac{x_{1jk} x_{2jk}}{n_{jk}} (l_{jk} - L_{jk})^2 + \lambda(L_{11} - L_{21} - L_{12} + L_{22}),$$

which yields estimates

$$\hat{L}_{11} = l_{11} - \frac{\lambda}{2} \frac{1}{w_{11}},$$

$$\hat{L}_{12} = l_{12} + \frac{\lambda}{2} \frac{1}{w_{12}},$$

$$\hat{L}_{21} = l_{21} + \frac{\lambda}{2} \frac{1}{w_{21}}, \qquad (10.3\text{-}37)$$

$$\hat{L}_{22} = l_{22} - \frac{\lambda}{2} \frac{1}{w_{22}},$$

where

$$\frac{\lambda}{2} = \frac{l_{11} - l_{21} - l_{12} + l_{22}}{\dfrac{1}{w_{11}} + \dfrac{1}{w_{21}} + \dfrac{1}{w_{12}} + \dfrac{1}{w_{22}}}, \qquad (10.3\text{-}38)$$

$$w_{jk} = \frac{x_{1jk} x_{2jk}}{n_{jk}}$$

$$= \frac{1}{\dfrac{1}{x_{1jk}} + \dfrac{1}{x_{2jk}}}. \qquad (10.3\text{-}39)$$

We can write these estimates in closed form, whereas to get the MLEs for this model we must iterate or solve the cubic equation of Bartlett. The minimum logit χ^2 statistic used to test the no three-factor effect model becomes

$$X_B^2 = \frac{(l_{11} - l_{21} - l_{12} + l_{22})^2}{\sum_{ijk} x_{ijk}^{-1}}, \qquad (10.3\text{-}40)$$

which is the square of the natural logarithm of the observed three-factor effect divided by the sample estimate of its asymptotic variance. ∎∎

Extensions of minimum logit χ^2

Minimum logit χ^2 is mainly useful in cases with a dichotomous variable and product multinomial sampling (i.e., the marginal totals corresponding to the cross-classification of all the remaining variables are fixed). Extensions to polytomies with product multinomial sampling, although possible, are somewhat complex and lack the immediate appeal of (10.3-33).

Minimum logit χ^2 and marginal constraints

Minimum logit χ^2 estimates clearly satisfy the marginal constraints implied by the product multinomial sampling scheme but in general do not satisfy other marginal constraints, unless these are specified. For example, under the hypothesis of homogeneity of proportions, minimum logit χ^2 does not yield estimates such that $\sum_j \hat{m}_{ij} = \sum_j x_{ij}$; however, $\hat{m}_{+j} = x_{+j}$. The method has the advantage that estimates may be obtained without iteration in some cases where ML estimation requires iteration.

Example 10.3-6   Bartlett's data

Table 10.3-1 contains the data of Bartlett, originally analyzed in Chapter 3, arising from an experiment investigating the growth of plants following different treatments. The data form a $2 \times 2 \times 2$ table with variable 1 acting as a response variable with categories "alive" and "dead," and the margin corresponding to variables 2 and 3 fixed. Table 10.3-1 also gives the MLEs under the no three-factor effect model with $u_{123} = 0$ (these values were computed earlier in Chapter 3). The Pearson and likelihood ratio goodness-of-fit test statistics for this model take the values

$$X^2 = 2.27,$$

$$G^2 = 2.29.$$

The final columns of table 10.3-1 give the minimum logit χ^2 estimates of the expected cell counts under the same model, and

$$X_B^2 = 2.26.$$

Because of the relatively large cell counts and sample sizes, the close agreement among the different test statistics is to be expected. ■ ■

Table 10.3-1 Comparison of Maximum Likelihood and Minimum Logit χ^2 Estimates Computed from Data of Bartlett [1935]

		Observed			MLEs		Minimum Logit χ^2	
Time of Planting	Length of Cutting	Dead	Alive	Total	Dead	Alive	Dead	Alive
At Once	Long	84	156	240	78.904	161.096	79.070	160.930
	Short	133	107	240	138.096	101.904	137.975	102.025
In Spring	Long	156	84	240	161.096	78.904	160.930	79.070
	Short	209	31	240	203.904	36.096	203.647	36.353

(Expected Values for $u_{123} = 0$)

10.4 The Logistic Model and How to Use It

For multivariate problems where one of the variables is a dichotomy, we often wish to represent the dependence of the probabilities for this dichotomy on the remaining variables or groupings of the data. There are three situations here among which we choose to distinguish:

1. the remaining k variables form h categories or groupings of the data;
2. the remaining k variables are all continuous;
3. some of the remaining variables are categorical and some are continuous.

We discuss each situation separately.

10.4.1 *The discrete-variable linear logistic model*

Suppose that in our multivariate problem with one dichotomy, there are h combinations of the remaining variables or h groupings of the data. If θ_i is the probability of a "success" on the dependent variable for the ith grouping, then a simple way to represent the dependence is via a linear model for the logits of the θ_i, i.e.,

$$L_i = \log\left(\frac{\theta_i}{1 - \theta_i}\right) = \boldsymbol{\alpha}_i'\boldsymbol{\beta} \tag{10.4-1}$$

where $\boldsymbol{\alpha}_i$ is a vector of known constants ($\boldsymbol{\alpha}_i'$ is the transpose of $\boldsymbol{\alpha}_i$) and $\boldsymbol{\beta}$ is a vector of unknown parameters. The logit models of Chapter 2 and the preceding section have this general form, where $\boldsymbol{\alpha}_i$ is a "design vector" with entries 0 and 1 and $\boldsymbol{\beta}$ consists of simple functions of the u-terms in our log-linear models. Rewriting (10.4-1), we see that the postulated model for dependence takes the form

$$\theta_i = \frac{1}{1 + e^{-\boldsymbol{\alpha}_i'\boldsymbol{\beta}}}. \tag{10.4-2}$$

Cox [1970] discusses various ways to consider inference for linear logistic models, including a weighted least squares approach similar to the minimum modified χ^2 approach described in the preceding section and an "exact" or conditional approach which will be in part described in Section 10.6.

10.4.2   The continuous-variable linear logistic model

We turn now to the multivariate problem with one dichotomous variable and k continuous variables. If $\theta(y_1, \ldots, y_k)$ is the probability of a success conditional on the k continuous variables $Y_1, \ldots, Y_k$ taking the values $y_1, \ldots, y_k$, then the continuous-variable linear logistic model expresses the dependence of the dichotomy on the remaining k variables as follows:

$$\log\left[\frac{\theta(y_1, \ldots, y_k)}{1 - \theta(y_1, \ldots, y_k)}\right] = \gamma_0 + \sum_{i=1}^{k} \gamma_i y_i. \tag{10.4-3}$$

Berkson [1944, 1953] has considered the problem of estimating the parameters $\gamma' = (\gamma_0, \ldots, \gamma_k)$ using the minimum logit chi square method. Cox [1958] and Hitchcock [1966] have derived tests of hypotheses about one of the regression coefficients, say γ_i ($i = 1, \ldots, k$).

The problems of estimation and testing related to model (10.4-3) become considerably more complicated when we take into account the fact that $y_1, \ldots, y_k$ are observations on k random variables $Y_1, \ldots, Y_k$. If the marginal distribution of the $\{Y_i\}$ is multivariate normal, then the combined joint distribution of the dichotomous variate and the $\{Y_i\}$ is occasionally of a manageable form, e.g., in the discrimination problem described in the next section.

10.4.3   The mixed-variable linear logistic model

The linear logistic models of the form (10.4-1) and (10.4-3) have been used in more general problems, involving mixtures of discrete and continuous independent variables. When investigations involve observing human beings, the association between a response and any specific variable often can be assessed only when many other variables are taken into consideration. This is particularly true of investigations into chronic disease, where the primary objective is often to investigate which of a large number of variables (some continuous, some discrete) are associated with the presence or absence of disease. The Framingham longitudinal study of coronary heart disease described by Dawber, Kannel, and Lyell [1963] is an example of such an investigation. It is the first of four studies in which data have been collected annually (or more often) on many thousands of participants. Portions of these data have been analyzed by many investigators, who have proposed different methods of handling the mixture of discrete and continuous variables.

Truett, Cornfield, and Kannel [1967] and Cornfield [1962] consider the following model for data from the Framingham study. Suppose that the unconditional probability of heart disease is p and that the multivariate frequency distributions of those who will and those who will not develop the disease are known functions of the k variables $y_1, \ldots, y_k$, say, $f_1(y_1, \ldots, y_k)$ and $f_0(y_1, \ldots, y_k)$. Then the conditional probability of disease in a person characterized by the k values $y_1, \ldots, y_k$ is

$$\theta(y_1, \ldots, y_k) = \frac{1}{1 + \frac{(1-p)}{p}\left[\frac{f_0(y_1, \ldots, y_k)}{f_1(y_1, \ldots, y_k)}\right]}. \tag{10.4-4}$$

If the frequency functions f_0 and f_1 in (10.4-4) have a multivariate normal form with different means but the same variances and covariances, then

$$\frac{1-p}{p} \cdot \frac{f_0(y_1, \ldots, y_k)}{f_1(y_1, \ldots, y_k)} = \exp\left[-\alpha - \sum_{i=1}^{k} \beta_i y_i\right]. \tag{10.4-5}$$

Then (10.4-4) can be converted into the linear logistic function

$$\log\left[\frac{p f_1(y_1, \ldots, y_k)}{(1-p) f_0(y_1, \ldots, y_k)}\right] = \alpha + \sum_{i=1}^{k} \beta_i y_i, \tag{10.4-6}$$

and the $\{\beta_i\}$ are the coefficients of the usual linear discriminant function from normal theory used to separate the populations corresponding to f_0 and f_1.

Truett, Cornfield, and Kannel [1967] present some arguments supporting the use of the multivariate normal assumption when departures from it are substantial, and they claim that their approach "seems to promise a more penetrating analysis than can be achieved by contemplation of cross-classifications." They present an example where they have essentially eight y-variables, of which four, serum cholesterol, systolic blood pressure, relative weight, and hemoglobin, are quantitative measurements, and a further two have coded categories (the variable "cigarettes per day" has five coded categories, and "electrocardiagram reading" (ECG) has two). The remaining two variables, age and sex, they use to partition the data into separate segments for analysis. These variables are known to affect the levels of the other variables, and thus partitioning avoids introducing covariate terms (Cox [1970] suggests a similar approach to adjusting for concomitant variables). Nevertheless, the effect of assigning scores to the categorical variables, especially for situations such as in the Framingham study, where several of the y-variables are discrete, needs further investigation. The scoring method of Abelson and Tukey [1963] may be of use here.

Walker and Duncan [1967], using a weighted least squares approach to estimate the logistic function (10.4-6), claim that the logit transformation of the dichotomous variable relating to the presence or absence of disease "submits the zero-one data, not only to the techniques of multiple regression but to its extensions, such as analysis of variance and analysis of covariance." They discuss various iterative computing methods, one of which uses the discriminant function estimates discussed above as initial estimates. Their examples are based on a slightly different subset of the variables available in the Framingham data and again include both quantitative and categorical variables. Although they do not use the multivariate normal assumptions of Truett, Cornfield, and Kannel, Walker and Duncan do use t-tests for various regression coefficients (including coefficients of discrete variables), and they give the related ANOVA table for the overall regression line.

10.4.4   *Multivariate logistic response functions*

All of the models discussed so far in Section 10.4 are designed for use with a single dichotomous response variable. Many of the methods described above can be extended to cover situations involving a multiple-response variable. The multiple responses may be ordered or unordered, and in the latter case they may still have a factorial structure representative of several dichotomous variables. Bock [1970], Cox [1970], and Walker and Duncan [1967] all consider methods for the analysis of multivariate logistic models.

More recently, Nerlove and Press [1973] have considered in detail the case of jointly dependent qualitative variables, and they interpret the resulting conditional probabilities as analogues of structural equations in systems of simultaneous equations, encountered in the econometric literature. Their paper also contains a Fortran listing for a computer program to do such analyses.

10.4.5   *Treelike structures for mixed-variable problems*

A popular method for analyzing data involving large numbers of variables of mixed type (i.e., discrete and continuous) is the AID (automatic interaction detection) technique. According to Sonquist [1969], "the essence of the algorithm is the sequential application of a one-way analysis of variance model." The examples of the AID technique given by Sonquist and Morgan [1964] are primarily concerned with the situation where there is one dependent variable that is quantitative, such as income, and a mixture of independent variables, such as age, occupation, and race, which they treat as categorical variables. Their technique produces a treelike structure which explains as much variability in the dependent variable as possible. Thus AID bears some resemblance to divisive methods of cluster analysis, but Sonquist and Morgan do not discuss their measure of distance in this context.

While AID may be a useful exploratory device for generating models and hypotheses from data, it has serious limitations because it never really takes into account the sampling variability inherent in the data. Thus, the treelike structures (or models) generated by AID tend to be too elaborate (i.e., they have too many branches). Einhorn [1972] has applied AID to sets of independent random deviates which had no relationship at all to the dependent variable. He allowed for minimum group sizes of 10, 25, and 50 in the final "branches" of the "tree," and he found that AID consistently created a treelike structure for the data, where none in fact existed.

10.4.6   *Scoring categorical variables versus categorizing continuous variables*

More investigation is clearly desirable into the similarities and differences between the approach whereby categorical variables are assigned arbitrary values and then combined with continuous variables, and the approach where all variables are categorized and then analyzed without regard to their position in the scale. Categorizing continuous variables and using log-linear models, and the related techniques described in other chapters, offer a method of detecting interactive variables readily and might be a useful first step. Subsequently, methods of analysis which use more of the information given by the continuous variables can be attempted. The effects of changing the categorization in log-linear models and of removing one or more variables was discussed in Chapter 2. The possible effects of changing the categorization should continually be kept in mind for mixed variable problems.

10.5   Testing Via Partitioning of Chi Square

Lancaster [1951] introduced a test for a three-factor effect in three-way contingency tables which differs from all others discussed so far in this book. We describe the formal distinction between the approach given in Chapters 3 and 4, for example, and that of Lancaster in this section because of its theoretical interest.

Consider an $I \times J \times K$ contingency table with observed counts $\{x_{ijk}\}$, having arisen via a multinomial sampling scheme with underlying cell probabilities $\{p_{ijk}\}$ and sample size $N = \sum_{ijk} x_{ijk}$. The Pearson goodness-of-fit statistic for the model of complete independence of the variable corresponding to rows, columns, and layers where the expected values are MLEs is (using the partitioning notation of Section 4.5)

$$X^2[\bar{1} \otimes \bar{2} \otimes \bar{3}] = \sum_{i=1}^{I} \sum_{j=1}^{J} \sum_{k=1}^{K} \frac{(x_{ijk} - x_{i++}x_{+j+}x_{++k}/N^2)^2}{x_{i++}x_{+j+}x_{++k}/N^2}, \qquad (10.5\text{-}1)$$

where the number of related degrees of freedom is $IJK - I - J - K + 2$. (The notation for X^2 and G^2 in this section is a shorthand denoting the model for which we compute expected values.) Next we define the three Pearson statistics for the models of marginal independence of pairs of underlying variables, i.e.,

$$X^2[\bar{1} \otimes \bar{2}] = \sum_{i=1}^{I} \sum_{j=1}^{J} \frac{(x_{ij+} - x_{i++}x_{+j+}/N)^2}{x_{i++}x_{+j+}/N}, \qquad (10.5\text{-}2)$$

$$X^2[\bar{1} \otimes \bar{3}] = \sum_{i=1}^{I} \sum_{k=1}^{K} \frac{(x_{i+k} - x_{i++}x_{++k}/N)^2}{x_{i++}x_{++k}/N}, \qquad (10.5\text{-}3)$$

$$X^2[\bar{2} \otimes \bar{3}] = \sum_{j=1}^{J} \sum_{k=1}^{K} \frac{(x_{+jk} - x_{+j+}x_{++k}/N)^2}{x_{+j+}x_{++k}/N}, \qquad (10.5\text{-}4)$$

with $(I - 1)(J - 1)$, $(I - 1)(K - 1)$, and $(J - 1)(K - 1)$ degrees of freedom, respectively. Lancaster [1951], by analogy with the partitioning of sums of squares in ANOVA, proposed a test of no three-factor interaction given by

$$L^2 = X^2[\bar{1} \otimes \bar{2} \otimes \bar{3}] - X^2[\bar{1} \otimes \bar{2}] - X^2[\bar{1} \otimes \bar{3}] - X^2[\bar{2} \otimes \bar{3}], \qquad (10.5\text{-}5)$$

thus completing the partition of $X^2[\bar{1} \otimes \bar{2} \otimes \bar{3}]$ into four components. L^2 is a chi-square-like quantity, and Lancaster assigns to it the number of degrees of freedom equal to the difference of the degrees of freedom corresponding to the terms on the right-hand side of (10.5-5).

The statistic L^2 is not the same as the statistic we have used to test the no three-factor interaction model defined in log-linear model terms and based on MLEs.

This is most easily seen by looking at the corresponding statistic based on the likelihood ratio test statistics. If we let $G^2[\bar{1} \otimes \bar{2} \otimes \bar{3}]$, $G^2[\bar{1} \otimes \bar{2}]$, $G^2[\bar{1} \otimes \bar{3}]$, and $G^2[\bar{2} \otimes \bar{3}]$ be the statistics corresponding to $X^2[\bar{1} \otimes \bar{2} \otimes \bar{3}]$, $X^2[\bar{1} \otimes \bar{2}]$, $X^2[\bar{1} \otimes \bar{3}]$, and $X^2[\bar{2} \otimes \bar{3}]$, then

$$K^2 = G^2[\bar{1} \otimes \bar{2} \otimes \bar{3}] - G^2[\bar{1} \otimes \bar{2}] - G^2[\bar{1} \otimes \bar{3}] - G^2[\bar{2} \otimes \bar{3}]$$

$$= 2 \sum_{ijk} x_{ijk} \log\left(\frac{x_{ijk} x_{i++} x_{+j+} x_{++k}}{N x_{ij+} x_{i+k} x_{+jk}}\right). \tag{10.5-6}$$

Viewing (10.5-6) as a likelihood-ratio test statistic for the no three-factor interaction model, we see that the estimated cell probabilities under the model are of the form

$$\hat{p}_{ijk} = \frac{x_{ij+} x_{i+k} x_{+jk}}{x_{i++} x_{+j+} x_{++k}}. \tag{10.5-7}$$

This is not a BAN estimate for p_{ijk} under the model $u_{123} = 0$, so that K^2 and L^2, given by (10.5-6) and (10.5-5), do not have asymptotic central χ^2 distributions under this model. Only under the null model of complete independence, i.e., under $u_{12} = u_{13} = u_{23} = u_{123} = 0$, do K^2 and L^2 have central χ^2 distributions with $(I - 1)(J - 1)(K - 1)$ degrees of freedom (see Lancaster [1971]). We illustrate in an example.

Example 10.5-1   Occurrence of smallpox in Boston, 1753

Cassedy [1969] presents data collected in 1753 by Thomas Prince on the occurrence of smallpox in Boston in that year. The data are summarized in table 10.5-1 as a $2 \times 2 \times 2$ table, where the three dimensions are (1) race (white or black), (3) mortality (dead or alive), and (2) type of smallpox (natural or inoculated). The likelihood ratio statistics for these data are:

$$G^2[\bar{1} \otimes \bar{2} \otimes \bar{3}] = 230.47,$$

$$G^2[\bar{1} \otimes \bar{2}] = 10.38,$$

$$G^2[\bar{1} \otimes \bar{3}] = 18.13,$$

$$G^2[\bar{2} \otimes \bar{3}] = 202.96.$$

Thus $K^2 = 230.47 - 10.38 - 18.13 - 202.96 = -1.0$, while the value of the likelihood-ratio goodness-of-fit statistic for no three-factor interaction in the loglinear model is $G^2[\{\overline{12}\}, \{\overline{23}\}, \{\overline{13}\}] = 2.50$.

Table 10.5-1 Comparative Deaths from Inoculation and Natural Smallpox at Boston, in 1753

	Natural		Inoculated	
	Dead	Alive	Dead	Alive
White	24	1,961	470	4,590
Black	6	133	69	416

Source: Cassedy [1969].

Table 10.5-2 Likelihood-Ratio Goodness-of-Fit
Statistics for Various Log-Linear Models Applied to
Data in Table 10.5-1

Model	Degrees of Freedom	G^2
$u + u_1 + u_2 + u_3 + u_{12} + u_{13} + u_{23}$	1	2.50
$u + u_1 + u_2 + u_3 + u_{12} + u_{13}$	2	201.96
$u + u_1 + u_2 + u_3 + u_{12} + u_{23}$	2	9.37
$u + u_1 + u_2 + u_3 + u_{13} + u_{23}$	2	17.12
$u + u_1 + u_2 + u_3 + u_{12}$	3	220.09
$u + u_1 + u_2 + u_3 + u_{13}$	3	212.34
$u + u_1 + u_2 + u_3 + u_{23}$	3	27.51
$u + u_1 + u_2 + u_3$	4	230.47

In table 10.5-2, we give the values of G^2 for all eight hierarchical log-linear models for this set of data. The only model which fits the data is that of no three-factor interaction. We conclude that race, mortality, and type of smallpox are pairwise related, but that there is no three-factor interaction. Our conclusion is thus the same as the one based on the Lancaster–Kullback partition method, but our inference procedure is quite different. ∎∎

Lancaster [1971] notes that L^2 cannot take negative values, as can K^2. In cases where K^2 is negative, Kullback, Kupperman, and Ku [1962a] suggest the use of alternate analyses and alternate partitions of $G^2[\bar{1} \otimes \bar{2} \otimes \bar{3}]$. In a more recent paper, however, Ku, Varner, and Kullback [1971] propose the use of partitioning analyses similar to those we illustrated in Chapter 4.

For many three-way tables, the values of L^2 and $X^2[\overline{12} \otimes \overline{13} \otimes \overline{23}]$ and of K^2 and $G^2[\{\overline{12}\}, \{\overline{23}\}, \{\overline{13}\}]$ are not grossly different. Lancaster [1969] gives several examples. Thus we can view the Lancaster method as an empirical procedure that often closely approximates the ML procedure, where $X^2[\{\overline{12}\}, \{\overline{23}\}, \{\overline{13}\}]$ and L^2 are almost the same. The problem we face is that with present methods we cannot tell how close the approximation will be. The discrepancy may be relatively large, as we illustrate below.

Example 10.5-2 More on Bartlett's data
Based on Bartlett's data in table 10.3-1 (with variable 2 being long-short and variable 3 time of planting), we have

$$X^2[\bar{1} \otimes \bar{2} \otimes \bar{3}] = 141.053,$$

$$X^2[\bar{1} \otimes \bar{2}] = 45.400,$$

$$X^2[\bar{1} \otimes \bar{3}] = 95.583,$$

$$X^2[\bar{2} \otimes \bar{3}] = 0.$$

($X^2[\bar{2} \otimes \bar{3}] = 0$ because the $[2, 3]$-margin was fixed by the experimental design.) Thus $L^2 = 141.053 - 45.400 - 95.583 = 0.070$, whereas the value of $X^2[\{\overline{12}\}, \{\overline{23}\}, \{\overline{13}\}]$ is 2.274 (as given earlier). Here L^2 and $X^2[\{\overline{12}\}, \{\overline{23}\}, \{\overline{13}\}]$ are relatively quite different in value, although if we compare these values to the χ^2 table with one degree of freedom neither is significant. If, however, all the entries in the table are inflated by a factor of 10, the $X^2[\{\overline{12}\}, \{\overline{23}\}, \{\overline{13}\}]$ value of 22.74 indicates a large three-factor effect, whereas the L^2 value of 0.7 does not. ■■

10.6   Exact Theory for Tests Based on Conditional Distributions

10.6.1 2×2 tables

For the analysis of 2×2 contingency tables, an alternative to maximum likelihood estimation and chi square goodness-of-fit tests is the use of Fisher's exact test for independence. The basic idea is that if we have independence (or homogeneity of proportions) in the 2×2 table and if we consider the marginal totals as given, a significance test for the hypothesis of independence is easy to determine, both theoretically and computationally. For example, suppose we are dealing with a pair of binomials with sample sizes x_{1+} and x_{2+}. We observe x_{11} successes out of x_{1+} in the first sample and x_{21} successes out of x_{2+} in the second. Here the observed table is of the form

$$
\begin{array}{|c|c|c}
\hline
x_{11} & x_{12} & x_{1+} \\
\hline
x_{21} & x_{22} & x_{2+} \\
\hline
\end{array}
\tag{10.6-1}
$$

and the row totals are already given. Under the null hypothesis of equality of proportions, the joint probability of observing such an outcome is

$$
\binom{x_{1+}}{x_{11}} p^{x_{11}}(1 - p)^{x_{12}} \binom{x_{2+}}{x_{21}} p^{x_{21}}(1 - p)^{x_{22}} = \binom{x_{1+}}{x_{11}}\binom{x_{2+}}{x_{21}} p^{x_{+1}}(1 - p)^{x_{+2}}, \tag{10.6-2}
$$

where p is the common binomial probability parameter. The conditional probability given the second set of margins (x_{+1}, x_{+2}), which is also governed by a binomial distribution, is

$$
\frac{\binom{x_{1+}}{x_{11}}\binom{x_{2+}}{x_{21}} p^{x_{+1}}(1 - p)^{x_{+2}}}{\binom{x_{++}}{x_{+1}} p^{x_{+1}}(1 - p)^{x_{+2}}} = \frac{\binom{x_{1+}}{x_{11}}\binom{x_{2+}}{x_{21}}}{\binom{x_{++}}{x_{+1}}}, \tag{10.6-3}
$$

i.e., is hypergeometric (see also the discussion in Chapter 3). This leads to the Fisher exact test (see Fisher [1935]) based on the hypergeometric distribution, and the key point is that under the null hypothesis of equality of proportions, p is a "nuisance" parameter. What is of interest is not the common value of p, but the fact that p is the same in both samples. Thus we can be pleased that the conditional distribution does not depend on the nuisance parameter.

More generally, for the two-binomial problem with underlying probabilities p_1 and p_2, we might wish to consider a null hypothesis regarding the odds ratio

$$
\alpha = \frac{p_1(1 - p_2)}{p_2(1 - p_1)}, \tag{10.6-4}
$$

say $\alpha = \alpha_0$. If $\alpha_0 = 1$, this reduces to the problem above. The unconditional probability of the observed table is now

$$\binom{x_{1+}}{x_{11}} p_1^{x_{11}}(1 - p_1)^{x_{12}} \binom{x_{2+}}{x_{21}} p_2^{x_{21}}(1 - p_2)^{x_{22}}, \tag{10.6-5}$$

and the conditional probability given the second set of marginal totals, as Fisher [1935] notes, turns out to be

$$g(x_{11}|x_{+1}, \alpha_0) = \frac{\binom{x_{1+}}{x_{11}}\binom{x_{2+}}{x_{21}}\alpha_0^{x_{11}}}{\displaystyle\sum_{i=0}^{x_{+1}} \binom{x_{1+}}{i}\binom{x_{2+}}{x_{+1}-i}\alpha_0^i}. \tag{10.6-6}$$

Significance tests and confidence intervals for α based on this conditional distribution are described by Cornfield [1956], Cox [1970], and Gart [1971], among others. For example, the significance test of $H_0 : \alpha = \alpha_0$ against $H_A : \alpha > \alpha_0$ involves computing the tail probability as follows:

$$\sum_{i=x_{11}}^{x_{+1}} g(i|x_{+1}, \alpha_0). \tag{10.6-7}$$

The theory related to the probability distribution given by (10.6-6) is said to be "exact" because it does not depend on any nuisance parameters. To describe the probabilities p_1 and p_2 completely, we need to know not only the value of the log odds ratio α, but also the value of some other parameter, say γ, which is related to the overall magnitude of the $\{p_i\}$. Conditioning on the second set of margins (which is a sufficient statistic for the nuisance parameter γ) is the standard Neyman–Pearson approach to the elimination of nuisance parameters and leads to uniformly most powerful tests (see Lehmann [1959], pp. 140–146).

10.6.2   *The combination of 2 × 2 tables*

In recent years, several authors have extended this exact theory to problems dealing with the combination of several 2×2 tables (see Birch [1964b], Cox [1970], Gart [1971], and Zelen [1971]) as well as to inference problems regarding parameters in several $2 \times J$ tables and in one or more multidimensional contingency tables (in addition to the references just cited, see Plackett [1971]).

Extending the notation above, we consider K pairs of independent binomial variates x_{11k} and x_{21k} $(k = 1, \ldots, K)$, with corresponding parameters p_{1k} and p_{2k} and sample sizes x_{1+k} and x_{2+k}. If we assume that the odds ratio for each of the K 2×2 tables is constant, i.e.,

$$\frac{p_{1k}(1 - p_{2k})}{(1 - p_{1k})p_{2k}} = \alpha \quad k = 1, \ldots, K, \tag{10.6-8}$$

then we are working with the log-linear model for the expected cell counts in the corresponding $2 \times 2 \times K$ table, given in our notation by $u_{123} = 0$. To make inferences about the common odds ratio α, we look at the conditional distribution of the $\{x_{11k}\}$ given the $\{x_{+1k}\}$ (and the $\{x_{1+k}\}$, since this marginal configuration is fixed by our design), which is based on the joint conditional distribution of the $\{x_{11k}\}$, i.e., on

$$h(x_{111}, \ldots, x_{11k} | x_{+11}, \ldots, x_{+1k}; \alpha) = \prod_{k=1}^{K} \left[\frac{\binom{x_{1+k}}{x_{11k}} \binom{x_{2+k}}{x_{21k}} \alpha^{x_{11k}}}{\sum_{j=0}^{x_{+1k}} \binom{x_{1+k}}{j} \binom{x_{2+k}}{x_{+1k} - j} \alpha^{j}} \right]. \qquad (10.6\text{-}9)$$

10.6.3 *Practicality of exact theory*

This alternative approach to inference, drawing from exact theory based on conditional distributions, has a considerable intuitive (as well as theoretical) appeal. Unfortunately, except for simple problems such as those dealing with the 2×2 table, the conditional distribution is neither easy to compute nor available in the form of sets of tables. Thus anyone who wishes to consider tests based on the exact theory is forced to use asymptotic normal approximations to the null exact distributions.

Cornfield [1956], Gart [1971], and Plackett [1971] outline a method which gives multivariate normal approximations to the exact theory, and this typically leads back to chi-square-like quadratic forms similar to those described earlier in the chapter and elsewhere in this book.

Zelen [1971] makes a normal approximation to his exact theory approach for testing $H_0: u_{123} = 0$ in $2 \times 2 \times K$ tables, and he ends up with a chi-square-like statistic of the form $X_1^2 - X_2^2$, where

$$X_1^2 = \sum_{k=1}^{K} \frac{(x_{11k} x_{22k} - x_{12k} x_{21k})^2}{x_{1+k} x_{2+k} x_{+1k} x_{+2k} / x_{++k}} \qquad (10.6\text{-}10)$$

(the usual Pearson chi square statistic for testing $u_{12} = u_{123} = 0$), and

$$X_2^2 = \frac{\sum_{k=1}^{K} [(x_{11k} x_{22k} - x_{12k} x_{21k}) / x_{++k}]^2}{\sum_{k=1}^{K} x_{1+k} x_{2+k} x_{+1k} x_{+2k} / x_{++k}^3} \qquad (10.6\text{-}11)$$

(the Birch-Cochran-Mantel-Haenszel conditional chi square statistic for testing $u_{12} = 0$, given $u_{123} = 0$ (see Chapter 4)).

The adequacy of many of these asymptotic approximations, as Gart [1971] points out, remains to be investigated for the small samples typically encountered in practice.

10.7 Analyses Based on Transformed Proportions

Most of the analyses considered throughout this book are based on log-linear or logistic models. We can consider alternative models, especially when we compare several proportions. Suppose we have J binomial variates, where for the ith one we have probability θ_i of observing a success, n_i trials, and an observed value x_i. Here i can represent a multiple cross-classification according to several other variables of interest.

10.7.1   *Two alternatives to the logistic transformation*

The logistic model described in Section 10.4 is based on a linear model for the logits, i.e.,

$$L_i = \log\left(\frac{\theta_i}{1 - \theta_i}\right) = \alpha_i' \beta, \tag{10.7-1}$$

where α_i is a vector of known constants and β is a vector of unknown parameters. Alternatively, we can consider linear models for other transforms of the probabilities $\{\theta_i\}$. Two transformations which have received considerable attention are the integrated normal transform $\Phi^{-1}(\theta_i)$, where $\Phi(t)$ is the value of the cumulative normal curve at the point t, and the angular transform arc sin $\sqrt{\theta_i} = \sin^{-1}\sqrt{\theta_i}$. Thus two alternatives to (10.7-1) are

$$N_i = \Phi^{-1}(\theta_i) = \alpha_i' \beta, \tag{10.7-2}$$

$$A_i = \text{arc sin } \sqrt{\theta_i} = \alpha_i' \beta. \tag{10.7-3}$$

10.7.2   *Probit analysis*

The integrated normal transformation (10.7-2) has received considerable attention in bioassay, and the approach adopted there by Finney [1971] and others is referred to as *probit* analysis. Probit analysis is concerned with the relationship between the dose of a drug and the frequency of a *quantal* (all-or-nothing) response, where the dosage is typically measured in terms of the logarithm of the concentration. If J different doses are applied to independent samples of size n_i $(i = 1, \ldots, J)$ and x_i members of the ith sample are "affected," then the *probit regression line model* is

$$N_i = \Phi^{-1}(\theta_i) = \beta_1 + \beta_2 d_i \quad i = 1, \ldots, J, \tag{10.7-4}$$

where θ_i is the probability of an affected response for the ith dosage level and d_i is the ith log-dose concentration.

We can obtain estimates of the underlying parameters β_1 and β_2 by the method of maximum likelihood (or by some other estimation technique). The log-likelihood function for model (10.7-4) is

$$\sum_{i=1}^{J} \{x_i \log[\Phi(\beta_1 + \beta_2 d_i)] + (n_i - x_i)\log[1 - \Phi(\beta_1 + \beta_2 d_i)]\}, \tag{10.7-5}$$

and Finney [1971] describes an iterative procedure for maximizing (10.7-5), gives asymptotic variance formulas for β_1 and β_2, and discusses the problem of assessing the goodness of fit of the model.

10.7.3   *Arc sin transformation for proportions*

The angular or arc sin transformation goes back to Fisher, and its great advantage stems from its variance-stabilizing property. Suppose $A_i = \text{arc sin } \sqrt{\theta_i}$ and we measure angles in radians. Then $a_i = \text{arc sin } \sqrt{x_i/n_i}$ has an asymptotic normal distribution with mean A_i, and variance $1/(4n_i)$ which is independent of θ_i. Rather than using the $\{a_i\}$ in small samples, we typically use the variant $\bar{a}_i$ suggested by Freeman and Tukey [1950], i.e.,

$$\bar{a}_i = \frac{1}{2}\left(\text{arc sin } \sqrt{\frac{x_i}{n_i + 1}} + \text{arc sin } \sqrt{\frac{x_i + 1}{n_i + 1}}\right), \tag{10.7-6}$$

which has been tabulated for all x and n with $n \leqq 50$ by Mosteller and Youtz [1961].

When the indexing subscript i corresponds to a multiple cross-classification, a model for the angularized proportions which suggests itself is the ANOVA model, especially when the n_i are all equal and we have identical "error variances" (see Bartlett [1947] and Mosteller and Tukey [1968]).

Cox [1970] and others have pointed out that when the probabilities θ_i of success are in a limited range near 0.5, the use of the logit, normal, and angular transforms leads to virtually equivalent results. We illustrate this point below, comparing the logit and angular transform analysis for a set of data.

Example 10.7-1   Evaluation of helping behavior

Rosen [1969], in a study of helping behavior, examined the effects of three independent variables on the probability of accepting help: (i) the value of help offered; (ii) the behavioral constraints imposed on the recipient; and (iii) the alternatives available to the potential recipient. Subjects were presented with a detailed description of a college scholarship of either high or low value, which imposed either highly restrictive or minimal behavioral constraints on the recipient (this variable is referred to as "cost"), in conditions where supplementary financial assistance was either available or not ("alternatives"). Since we have a 2^3 design, the subscript i from above is replaced by the triple subscript (i, j, k), where each subscript takes the value 1 or 2. For each of the eight design conditions a binomial sample of size 20 was questioned, and the resulting data are given in table 10.7-1.

Table 10.7-1   Evaluation of Helping Behavior

	Alternative		No Alternative		
	High Value	Low Value	High Value	Low Value	Totals
High Cost	16	5	13	10	44
Low Cost	18	5	19	17	59
Totals	34	10	32	27	103

Source: Rosen [1969].
Note: The entry in each cell of the table is an observed binomial count based on a sample of size 20, indicating acceptance of scholarship offer.

Using an arc sin transformation on the observed proportions and then computing the usual unweighted ANOVA χ^2 components for a complete 2^3 layout, we arrive at the ANOVA table given in table 10.7-2. Under the asymptotic theory, each of the eight components has a null χ^2 distribution with one degree of freedom, so the suggested model for the transformed proportions is

$$\mu + \alpha_i + \beta_j + \gamma_k + (\alpha\gamma)_{ik} + (\beta\gamma)_{jk} \tag{10.7-7}$$

in the usual ANOVA notation, if we include terms where corresponding χ^2 components are significant at the 0.05 level. We might wish to drop $(\alpha\gamma)_{ik}$ from this model, since the χ^2 component is barely significant.

Viewing the response variable (accept or reject) as a fourth variable, we can analyze the data in table 10.7-1 as a 2^4 contingency table with the configuration

C_{123} fixed. Table 10.7-3 gives a list of hierarchically ordered log-linear models, the degrees of freedom, and the corresponding values of the likelihood-ratio test statistic. Partitioning the statistic for model 4 into various components suggested by the hierarchical order and using a 0.05 level of significance, we arrive at the log-linear model:

$$u_{124} = u_{134} = u_{1234} = 0. \qquad (10.7\text{-}8)$$

Table 10.7-2 Analysis of Variance (ANOVA) for Data in Table 10.7-1 Following an Arc Sin Transformation

Source of Variation	Degrees of Freedom	Normed MS
Cost (A)	1	8.82
Value (B)	1	26.69
Alternative (C)	1	7.02
$A \times B$	1	0.27
$A \times C$	1	4.26
$B \times C$	1	6.56
$A \times B \times C$	1	0.15

Table 10.7-3 Likelihood-Ratio Goodness-of-Fit Test Statistic Values for Log-Linear Models Fitted to the $2 \times 2 \times 2 \times 2$ Table Constructed from the Data in Table 10.7-1

Model	G^2	Degrees of Freedom
(1) $u_{1234} = 0$	0.01	1
(2) $u_{124} = u_{1234} = 0$	0.68	2
(3) $u_{124} = u_{134} = u_{1234} = 0$	4.51	3
(4) $u_{124} = u_{134} = u_{234} = u_{1234} = 0$	13.26	4

Converting this to a model for the logits (see Chapters 2 and 3) we get as our model:

$$w + w_{1(i)} + w_{2(j)} + w_{3(k)} + w_{23(jk)}, \qquad (10.7\text{-}9)$$

which is the same as the model arrived at for the angularized proportions (10.7-7) except for the term $(\alpha\gamma)_{ik}$, which we may or may not have included before.

The conclusion here is that all three of the independent variables, cost, value, and alternative, have an appreciable effect on the proportion of acceptance, with value and alternative interacting in their effect on acceptance. The latter interaction is understandable, since both value and alternative involve monetary considerations which one might expect to be considered jointly before a decision to accept or reject is reached. ∎∎

10.7.4   *Linear models for proportions*

We do not necessarily have to transform proportions before building a linear model, although such transformations are often desirable. Many authors (e.g., Bhat and Kulkarni [1966]) have discussed the use of linear models in the arithmetic scale. An example of the use of linear models is given by Kleinman [1973], who considers the analysis of special concern when we are interested in estimating the expected value of proportions which exhibit greater variation than we would expect under the binomial distribution. Thus we consider a set of observed proportions $\hat{p}_1, \ldots, \hat{p}_k$, with the ith proportion based on n_i observations, which are independent random variables, where the ith one has mean μ and variance $\mu(1 - \mu)/n_i + r\mu(1 - \mu)(1 - (1/n_i))$ (where $1 > \mu > 0$ and $r > 0$), rather than the usual binomial distribution with mean μ and variance $\mu(1 - \mu)/n_i$. We are interested in estimating the mean μ, and gauging the efficiency of weighted estimates of the form

$$\hat{p} = \frac{\displaystyle\sum_i w_i \hat{p}_i}{\displaystyle\sum_i w_i}. \tag{10.7-10}$$

In the case of equal $\{n_i\}$, Kleinman [1973] suggests applying the usual unweighted analysis of variance methods on the observed proportions when there is more than one μ to estimate. In the case of unequal $\{n_i\}$, he proposes an empirical weighting scheme which is like a weighted analysis of variance on the observed proportions.

10.7.5   *Transformations for multiresponse data*

We have emphasized the analysis of transformed proportions here, because of the difficulties which arise when we wish to transform sets of multiresponse (more than two) data. In Section 10.4 we indicated some of the conceptual problems in extending the idea of logits to multiresponse data. For angular-type transformations of multiresponse data the problem is more serious, because to replace the variance-stabilizing property of the arc sin transform, we really desire a multidimensional transformation which stabilizes both variances and covariances. Holland [1973] has shown that for multinomial data, such transformations do not exist; however, the use of arc sin or square-root transformations may still come close to doing the desired job of variance-covariance stabilization.

For the case of integrated normal or probit transform, Aitchison and Silvey [1957] have proposed a generalization of probit analysis to the case of multiple responses. They envision an experiment where random samples of subjects experience J different doses of a stimulus, and as a result of stimulus i $(i = 1, \ldots, J)$ each subject is placed in one of $s + 1$ classes. (If $s = 1$, we get the usual probit analysis situation.) In order for their methods to be of use, Aitchison and Silvey note that three conditions must be met:

1. The classes must be ordered, mutually exclusive, and exhaustive;
2. the reaction of a subject to increasing doses must be systematic, in that if dose x places a subject into a class j, then doses greater than x must place him into a class $j' \geqq j$;

3. the incremental dosages required to boost a subject from the jth to the $(j + 1)$st class, say ξ_j, for $j = 1, \ldots, s$, are independent random variables whose standard deviations are small relative to the means.

This generalized probit analysis assumes that the $\{\xi_j\}$ are normally distributed. Finney [1971] discusses other generalizations of probit analysis.

10.8 Necessary Developments

Lest the reader conclude that the only problem remaining is to make an intelligent choice of existing methods, we now indicate the urgent need for the development of further methodology. In this chapter we have given an overview of current developments, but much remains to be done, particularly when the data base is large and consists of mixed variables, and measurements are unevenly replicated.

The advent of computers has made the collection of large quantities of data more commonplace. To some extent, the use of 80-column punch cards has encouraged investigators with a mixture of categorical and quantitative variables to degrade the continuous measurements to a series of categories. In some areas, the emphasis on more sophisticated analyses has led investigators to distrust data generated by subjective assessment, even when this assessment is based on expert opinion. The wisdom of these tendencies is open to debate. It is certain, however, that in each situation, the decisions should be based on considerations of the problem at hand, and not dictated by the lack of statistical techniques.

When we use methods suitable for discrete variables on continuous variables, a choice of category boundaries arises. In Chapter 2 we showed that different choices of boundaries (based on collapsing of categories) can lead to different conclusions regarding the dependence or independence of variables. Little guidance is available to help the investigator make such choices. Similarly, when scores are assigned to discrete categories prior to an analysis based on methods suitable for continuous data, a choice of scores must be made. Again, the matter needs more attention.

In Chapter 5, we indicated that the incomplete nature of arrays is not always recognizable. All too often when such structural incompleteness occurs, the analysis begins with the replacement of structural zeros by some average value, with consequent possible distortion of the results. There is a need for methods that deal with mixed variables and incomplete arrays.

Still further methodological problems require solution. In this book we have considered the analysis of data using log-linear models analogous to fixed-effects ANOVA models; however, we have not considered the analysis of data using log-linear models in situations corresponding to nesting and random-effects ANOVA models. Several research problems related to this topic require solution.

The problems described above arise in many areas, but are particularly pertinent to large-scale medical trials. We quote from Frei's summations of the 1970 Symposium on Statistical Aspects of Protocol Design conducted by the National Cancer Institute:

> One of the major limitations in the quantitative detailed definitive analysis of large and particularly multi-institute studies is the sheer mass of data accumulated and relatively unsophisticated approaches to data management and analysis. A truly definitive analysis of a study involving several

hundred patients is rarely performed because of the time and skills required and usually the focus is limited to response rate, toxicity, survival, etc. We know that the response rate for many treatments can be influenced in a major way by subtle (and not so subtle) aspects of the natural history. Since the natural history is continuously changing, it is essential that studies be definitively analyzed.... Clearly we need more biostatisticians as well as clinical investigators with the time, interest and talent essential to such analyses.

11 Measures of Association and Agreement

11.1 Introduction

11.1.1 *This chapter is different*

This chapter differs from the others in this book in several ways. It deals exclusively with two-dimensional tables with no structural zeros. Instead of dealing with fitting models to two-dimensional tables, in this chapter we consider single summary numbers that describe relationships between two cross-classified variables. These summary numbers are generally called "measures of association." When properly used, they provide a useful description of the structure displayed by a two-dimensional table.

The single most important series of research papers on measures of association is by Goodman and Kruskal [1954, 1959, 1963, 1972]. The 1954 paper lays down criteria for judging measures of association and introduces several new measures having specific contextual meanings. The 1959 paper serves as a supplement to the earlier one and surveys additional historical and bibliographic material. The 1963 paper derives large-sample standard errors for the sample analogues of population measures of association and presents some numerical results about the adequacy of large-sample normal approximations. The 1972 paper presents a more unified way to derive asymptotic variances for measures of association that can be expressed as ratios of functions of the cell probabilities.

In their much-quoted first paper, Goodman and Kruskal argue that ideally, each research problem should have a measure of association developed for its unique needs. Over the years, scientists have, in fact, developed many different measures. Thus when we look for a measure of association, there is rarely a need to develop a new measure. Rather, the problem is to choose wisely from among the variety of existing measures. It is therefore helpful to summarize existing measures, sort them into general classes, and suggest settings where each measure is applicable. That is the aim of this chapter.

The material in this chapter is limited to two-dimensional tables whose rows and columns are not ordered. For an extended discussion of measures that have been developed for the case when one or both of the cross-classified variables are ordered, we refer the reader to the references by Goodman and Kruskal cited above and Kruskal [1958]. For an excellent discussion of some natural multi-dimensional extensions of the measures presented here, including measures of partial and conditional association, see Davis [1967, 1971].

11.1.2 *Measures of association in general*

The measures that we discuss in this chapter may be grouped into four classes:

1. measures based on the ordinary chi square statistic used to test independence in a complete two-dimensional table;
2. measures based on the cross-product ratio for a 2×2 table;
3. a "proportional reduction of error" measure that indicates the relative value of using the column categories, say, to predict the row categories, formulated by Goodman and Kruskal [1954];
4. a "proportion of explained variance" measure which provides a close analogy for categorical data to the squared correlation coefficient for continuous data, formulated by Light and Margolin [1971].

In the rest of this section we discuss a number of issues that arise in the use and construction of measures of association.

Sample versus population

The point of view we adopt here is that a measure of association is a population parameter that we estimate from a sample. In all cases, the sampling model we assume here is the multinomial (see Chapter 13), and it is this sampling distribution that we use to compute asymptotic variances for each measure. In more formal terms, a measure of association is a function $f(\mathbf{p})$ that maps the space of all possible probability vectors $\mathbf{p} = \{p_{ij}\}$ into an interval on the real line. We estimate $f(\mathbf{p})$ by $\hat{f} = f(\hat{\mathbf{p}})$, where $\hat{\mathbf{p}}$ is the set of observed cell proportions in a sample, and we usually require asymptotic theory (i.e., $N \to \infty$) to obtain an approximate variance of $\hat{f}$ under the multinomial distribution. We denote this asymptotic variance of $\hat{f}$ by

$$\sigma_\infty^2[\hat{f}] = \text{asymptotic variance of } \hat{f}. \tag{11.1-1}$$

In general, $\sigma_\infty^2[\hat{f}]$, like $f(\mathbf{p})$, is a function of the population parameters $\mathbf{p}$. Thus to set confidence intervals for $f(\mathbf{p})$ using $\hat{f}$, we must estimate $\sigma_\infty^2[\hat{f}]$ by $\hat{\sigma}_\infty^2[\hat{f}]$, where we replace $\mathbf{p}$ in $\sigma_\infty^2[\hat{f}]$ by the sample values $\hat{\mathbf{p}}$. Using asymptotic normality, we then may set asymptotic $(1 - \alpha) \cdot 100\%$ confidence intervals for $f(\mathbf{p})$, of the form

$$\hat{f} - z_{1-(\alpha/2)}\hat{\sigma}_\infty[\hat{f}] < f(p) < \hat{f} + z_{1-(\alpha/2)}\hat{\sigma}_\infty[\hat{f}]. \tag{11.1-2}$$

For further discussion of this and related issues, see Section 14.4.

The relationship between measures of association and tests for independence is sometimes confused. We regard the two as separate enterprises. We use tests of independence to determine whether a relationship exists between the two cross-classified variables, but we use a measure of association to help us understand the particular type and extent of this relationship.

Interpretability of measures of association

Measures of association vary considerably in their ease of interpretation. Chi-square-based measures are little more than index numbers that indicate when one table shows a greater degree of association (in a not very interpretable sense) than another table. Contrastingly, the proportional reduction of error and proportion of explained variance measures were designed with ease of interpretation in mind. All the measures of association we consider here have the following property: If the table exhibits perfect independence, i.e.,

$$p_{ij} = p_{i+}p_{+j} \quad \text{for all } i, j, \tag{11.1-3}$$

then the measure exhibits no association. The measures differ in the way in which they define perfect and intermediate stages of association. Except in the 2×2 table case, a single function (and hence a single measure of association) cannot reflect the large variety of ways in which a table can depart from independence. It is this fundamental mathematical fact that leads to the variety of measures and to the difficulty inherent in choosing a single measure on any given occasion.

Association versus marginal distributions

One way to view the discussion on the structure of two-dimensional contingency tables in Chapter 2 is that it separates the information in the set of cell probabilities, **p**, into two parts. The first part consists of the two marginal distributions $(p_{1+}, \ldots, p_{I+})$ and $(p_{+1}, \ldots, p_{+J})$, while the second part is the association between the two cross-classified variables. We can make an analogous distinction for the bivariate normal distribution, in which the means and variances give information about the marginal distributions and the covariance or correlation coefficient gives information relating to the association or degree of dependence between the two variables. However, instead of having a single number, such as the correlation coefficient, to measure association, as we have in the bivariate normal case, it is more profitable to regard association in a contingency table as a multidimensional concept. If we choose to think in terms of degrees of freedom, then there should be $IJ - 1$ degrees of freedom in total, $I - 1$ for the rows, and $J - 1$ for the columns. This leaves $IJ - 1 - (I - 1) - (J - 1) = (I - 1)(J - 1)$ degrees of freedom for association in the table. Thus in general it takes $(I - 1)(J - 1)$ functions of **p** to specify completely the degree of association in a general two-dimensional contingency table. In the 2×2 case, we see that $(I - 1)(J - 1) = 1$, so that indeed, one measure is sufficient to describe association in this case. Various authors have suggested that "association" in a two-dimensional table should be invariant under the transformation

$$p_{ij} \to t_i s_j p_{ij} \tag{11.1-4}$$

for any sets of positive numbers $\{t_i\}$, $\{s_j\}$ that preserve $\sum_{i,j} p_{ij} = 1$ (Yule [1912], Plackett [1965], and Mosteller [1968]). We can use the transformation (11.1-4) to change the given marginal distributions of the p_{ij} into any other set of marginal distributions without changing any of the cross-product ratios in the table. One way we can discriminate between measures is to determine whether they change their values under the transformation (11.1-4) (i.e., we can ask if they are "margin-sensitive"), or whether they are invariant (i.e., we ask if they are "margin-free"). In this sense, a margin-sensitive measure of association mixes marginal information with information about association, whereas a margin-free measure does not.

Norming measures

It is customary to normalize measures of association so that they lie between 0 and 1 or between -1 and 1. All the measures considered here take the value 0 when the $\{p_{ij}\}$ satisfy the hypothesis of independence. The measures may take on the maximum value 1 (or -1, if the measure assigns directionality to association) when the association between the two cross-classified variables is "perfect." What perfect association means, however, varies from measure to measure.

While it is customary to normalize measures, the reasons for doing so remain obscure. Goodman and Kruskal [1954] argue that the importance of the norming convention diminishes as the interpretability or meaningfulness of the measure increases. They also warn that such conventions may lead users of measures to transfer size preconceptions from one normed measure to another in an attempt to clarify the absolute magnitude a measure must attain before it has serious substantive significance.

One positive contribution that norming may make is that we can use it to remove (to some degree) the extraneous effect that the number of rows and columns has on the value of a measure. Removing this effect is useful if we wish to compare measures across a series of different-sized tables and it is the motivation behind the various functions of the chi square measure described in Section 11.3. Although this aim is worthy, we know of no evidence that shows that norming can really accomplish it.

Symmetric versus asymmetric measures

We may have one of two perspectives relative to a two-dimensional table. Either we are interested in the joint distribution of the two cross-classified variables, or we are interested in the conditional distribution of one of the variables given the other. In the first case, we want a measure of association that is symmetric in the way it treats the two variables, while in the second case an asymmetric measure may reflect the situation more appropriately. This second viewpoint is most often associated with experiments in which we control the number of units falling into each category of one margin, but we may also wish to use this format when studying nonexperimental data where only the total number of observations is fixed.

11.1.3   *Organization of this chapter*

In the next section we begin our discussion of measures of association by focusing on 2×2 tables, where there are basically two kinds of measures. In Section 11.3 we turn to the more general case of $I \times J$ tables, where we concentrate on examples of measures falling into the four classes described briefly in Section 11.1.2. In the final section, we take up a special case of association in two-dimensional tables—the problem of measuring agreement or reliability (Goodman and Kruskal [1954, 1959]). In 2×2 tables, agreement can be defined as essentially equivalent to association, but this is not the case for $I \times I$ or $I \times J$ tables. We identify some common difficulties in measuring agreement, such as choosing among alternative definitions, and we suggest one possible solution.

11.2   Measures of Association for 2×2 Tables

11.2.1   *Why deal separately with 2×2 tables?*

As we have noted several times throughout this book (see, for example, Chapter 2) there is one degree of freedom available for measuring association or interaction in 2×2 tables. As a result, something special happens to the array of measures proposed for more general $I \times J$ tables. Almost all standard measures reduce to functions of the cross-product ratio or of the standard Pearson chi square statistic. Moreover, distinctions between measures treating the row and column variables symmetrically and those treating them asymmetrically (for prediction purposes) often vanish for 2×2 tables.

Below, we deal in turn with measures which are functions of the cross-product ratio (or are closely related to such functions), and with measures which are functions of the standard correlation coefficient relating row and column variables. The correlation coefficient and the Pearson chi square statistic are directly related. We then contrast the two types of measures and suggest how to choose between them.

11.2.2 Measures based on the cross-product ratio

Basic properties of the cross-product ratio

In Chapter 2 we considered in great detail the sense in which the cross-product ratio $\alpha = p_{11}p_{22}/p_{12}p_{21}$ measures association in 2×2 tables. As such, α plays a central role in the construction of log-linear models. The basic properties of the cross-product ratio are:

1. α is invariant under the interchange of rows and columns, although an interchange of only rows or only columns changes α into $1/\alpha$;

2. α is invariant under transformation (11.1-4). That is, suppose we multiply the probabilities in row 1 by $t_1 > 0$, row 2 by $t_2 > 0$, column 1 by $s_1 > 0$, and column 2 by $s_2 > 0$, and then we renormalize these values so that they once again add to 1. The normalizing constant cancels out, and we get

$$\frac{(t_1 s_1 p_{11})(t_2 s_2 p_{22})}{(t_1 s_2 p_{12})(t_2 s_1 p_{21})} = \frac{p_{11}p_{22}}{p_{12}p_{21}} = \alpha. \tag{11.2-1}$$

This property was noted by Yule [1900];

3. α has an interpretation. If we think of row totals as fixed, then p_{11}/p_{12} is the odds of being in the first column conditional on being in the first row, and p_{21}/p_{22} is the corresponding odds for the second row. The relative odds for the two rows, or the odds ratio, is then

$$\frac{p_{11}/p_{12}}{p_{21}/p_{22}} = \frac{p_{11}p_{22}}{p_{12}p_{21}} = \alpha. \tag{11.2-2}$$

We get the same odds ratio if we start with column totals fixed.

The quantity α runs from 0 to ∞, but its natural logarithm is symmetric about 0, ranging from $-\infty$ to ∞. The cross-product ratio is symmetric in the sense that α and $1/\alpha$ represent the same degree of association but in opposite directions (i.e., $\log(1/\alpha) = -\log \alpha$).

The observed cross-product ratio $\hat{\alpha} = x_{11}x_{22}/x_{12}x_{21}$ is the maximum likelihood estimate of α under the multinomial sampling model, and under the product binomial model where either the observed row or the observed column totals are fixed. Under both types of sampling models, the approximate large sample variance is

$$\sigma_{\infty}^2(\hat{\alpha}) = \frac{\alpha^2}{N}\left(\frac{1}{p_{11}} + \frac{1}{p_{12}} + \frac{1}{p_{21}} + \frac{1}{p_{22}}\right), \tag{11.2-3}$$

(see Chapter 14 for a detailed discussion) which we estimate by

$$\hat{\sigma}_{\infty}^2(\hat{\alpha}) = \hat{\alpha}^2\left(\frac{1}{x_{11}} + \frac{1}{x_{12}} + \frac{1}{x_{21}} + \frac{1}{x_{22}}\right). \tag{11.2-4}$$

The presence of inverses of the observed counts in our estimated variance implies that expression (11.2-4) is only of use when the x_{ij} are all positive. If one count is zero, we can smooth the table of observed values before computing the variance. (For a discussion of how to smooth tables of counts, see Chapter 12.)

General functions of cross-product ratios

Several other measures of association for 2×2 tables are monotonically increasing or decreasing functions of α. The large number of measures based on the cross-product ratio makes it useful to have a general formula for their asymptotic variances.

Let $f(\alpha)$ be a positive monotonic increasing function of α such that $f(1) = 1$. Then a normalized measure of association based on $f(\alpha)$ whose maximum absolute value is 1 is:

$$g(\alpha) = \frac{f(\alpha) - 1}{f(\alpha) + 1}. \tag{11.2-5}$$

The asymptotic variance of $g(\hat{\alpha})$ is

$$\sigma_\infty^2[g(\hat{\alpha})] = \frac{[1 - g(\alpha)]^4[f'(\alpha)]^2}{4} \times \sigma_\infty^2(\hat{\alpha}), \tag{11.2-6}$$

where $\sigma_\infty^2(\hat{\alpha})$ is the asymptotic variance of $\hat{\alpha}$ given in expression (11.2-3) and $f'(\alpha)$ is the derivative of $f(\alpha)$ with respect to α. We give a special case of (11.2-6) in exercise 10 of Section 14.6.

Specific functions of α as measures of association

Yule proposed the use of two different functions of α of the form (11.2-5). They are his "measure of association" (Yule [1900]):

$$Q = \frac{p_{11}p_{22} - p_{12}p_{21}}{p_{11}p_{22} + p_{12}p_{21}} = \frac{p_{11}p_{22}/p_{12}p_{21} - 1}{p_{11}p_{22}/p_{12}p_{21} + 1} = \frac{\alpha - 1}{\alpha + 1}, \tag{11.2-7}$$

and his "measure of colligation" (Yule [1912]):

$$Y = \frac{\sqrt{p_{11}p_{22}} - \sqrt{p_{12}p_{21}}}{\sqrt{p_{11}p_{22}} + \sqrt{p_{12}p_{21}}} = \frac{\sqrt{\alpha} - 1}{\sqrt{\alpha} + 1}. \tag{11.2-8}$$

For Q, $f(\alpha) = \alpha$, and for Y, $f(\alpha) = \sqrt{\alpha}$. Replacing α by $\hat{\alpha}$, we obtain the MLEs:

$$\hat{Q} = \frac{\hat{\alpha} - 1}{\hat{\alpha} + 1}, \tag{11.2-9}$$

$$\hat{Y} = \frac{\sqrt{\hat{\alpha}} - 1}{\sqrt{\hat{\alpha}} + 1}. \tag{11.2-10}$$

We then get the large-sample estimated standard deviations directly from expression (11.2-6) as

$$\hat{\sigma}_\infty[\hat{Q}] = \frac{1}{2}(1 - \hat{Q}^2)\sqrt{\frac{1}{x_{11}} + \frac{1}{x_{12}} + \frac{1}{x_{21}} + \frac{1}{x_{22}}}, \tag{11.2-11}$$

$$\hat{\sigma}_\infty[\hat{Y}] = \frac{1}{4}(1 - \hat{Y}^2)\sqrt{\frac{1}{x_{11}} + \frac{1}{x_{12}} + \frac{1}{x_{21}} + \frac{1}{x_{22}}}. \tag{11.2-12}$$

For large samples, $\hat{Q}$ and $\hat{Y}$ are normally distributed with means Q and Y, respectively, so we can use $\hat{\sigma}_\infty[\hat{Q}]$ and $\hat{\sigma}_\infty[\hat{Y}]$ to get confidence intervals for Q and Y.

The statistics $\hat{Q}$ and $\hat{Y}$ always give values that are consistent in the sense that if $\hat{Q}_1 > \hat{Q}_2$ for a pair of tables, then $\hat{Y}_1 > \hat{Y}_2$. Both Q and Y have a range of $[-1, 1]$, taking the value 0 when the row and column variables are independent and the value 1 or -1 when there is complete positive or negative association, i.e., when $p_{12} = p_{21} = 0$ or $p_{11} = p_{22} = 0$, respectively. As both measures are simple functions of α, they can also take the values 1 and -1 when only one cell probability is zero. We consider this point in more detail below when we discuss how to choose a measure for 2×2 tables.

We can interpret Yule's Q as the difference between the conditional probabilities of like and unlike "orders," when two individuals are chosen at random from the same population. To illustrate this interpretation, let variable 1 measure the presence or absence of formal political affiliation among husbands and variable 2 the presence or absence among wives. If we write Q as

$$Q = \frac{p_{11}p_{22}}{p_{11}p_{22} + p_{12}p_{21}} - \frac{p_{12}p_{21}}{p_{11}p_{22} + p_{12}p_{21}}, \tag{11.2-13}$$

then Q measures the conditional probability that of two husband-wife pairs chosen at random from the population, both members of one pair are affiliated and both members of the other pair are nonaffiliated (like orderings), minus the conditional probability of one pair with husband affiliated and wife not and the other pair with wife affiliated and husband not (unlike orderings).

To get a simple interpretation of Y, we must do a little work. If we standardize the 2×2 table so that both row and column marginal totals are $(\frac{1}{2}, \frac{1}{2})$ while the cross-product ratio α remains unchanged, we get as our adjusted cell probabilities:

$$p_{11}^* = p_{22}^* = \frac{1}{2}\left[\frac{\sqrt{\alpha}}{\sqrt{\alpha} + 1}\right],$$

$$p_{12}^* = p_{21}^* = \frac{1}{2}\left[\frac{1}{\sqrt{\alpha} + 1}\right] \tag{11.2-14}$$

(provided $\alpha \neq \infty$ and $\alpha \neq 0$). Yule [1912] argued that, as we have removed all of the information about the margins in producing these adjusted totals, a reasonable measure of association is the difference between the probabilities in the diagonal and off-diagonal cells:

$$2(p_{11}^* - p_{12}^*) = \frac{\sqrt{\alpha} - 1}{\sqrt{\alpha} + 1} = Y. \tag{11.2-15}$$

Example 11.2-1   Family structure and political affiliation
Braungart [1971] collected data relating the family structure in which a college student grew up to the student's political affiliation. Table 11.2-1 gives some of Braungart's data, where political affiliation is restricted to Students for a Democratic Society (SDS) and Young Americans for Freedom (YAF), which represented in the 1960s two extremes of the political spectrum. Family structure here is indicated by whether parental decision-making was authoritarian or democratic.

Table 11.2-1 Parental Decision-Making
and Political Affiliation

| Parental | Political Affiliation | | |
Decision-Making	SDS	YAF	Totals
Authoritarian	29	33	62
Democratic	131	78	209
Totals	160	111	271

Source: Braungart [1971].

For these data, $\hat{\alpha} = 0.523$, $\hat{Q} = -0.313$, and $\hat{Y} = -0.161$. ($\hat{Y}$ is always smaller in absolute value than $\hat{Q}$.) The estimated large sample standard deviations of $\hat{Q}$ and $\hat{Y}$ are 0.124 and 0.072, respectively, and the resulting asymptotic 95% confidence intervals are $[-0.561, -0.065]$ for Q and $[-0.305, -0.017]$ for Y. We can conclude that an authoritarian family structure is associated with YAF political affiliation, but that this association is not necessarily very strong (since the 95% asymptotic confidence intervals come quite close to containing zero.) ∎∎

Special cases of more general measures

Goodman and Kruskal [1959] propose a special measure of association γ, applicable in $I \times J$ tables when the categories for both variables are ordered and when we are interested in how much more probable it is to get like as opposed to unlike orders in the two variables for a pair of individuals chosen at random from a population. For 2×2 tables, γ is identical to Yule's Q, and our interpretation of Q described above is based on the interpretation for γ.

In Section 11.3 we discuss Goodman and Kruskal's measures for optimal prediction or proportional reduction in error. If we standardize a 2×2 table so that all the marginal totals are $1/2$ (see expression (11.2-14)), then both the symmetric and asymmetric versions of these optimal prediction measures are equal to the absolute value of Yule's coefficient of colligation Y.

11.2.3   *Measures which are functions of the correlation coefficient*

Correlation coefficient for a 2 × 2 table

Suppose we think of the categories of the variables for a 2×2 table as taking the scores 0 for the first row and column and 1 for the second row and column. Then the means of the row and column margins are, respectively,

$$\mu_r = p_{2+}, \qquad \mu_c = p_{+2}. \tag{11.2-16}$$

The corresponding row and column variances are

$$\sigma_r^2 = \mu_r - \mu_r^2 = \mu_r(1 - \mu_r) = p_{1+}p_{2+},$$
$$\sigma_c^2 = \mu_c - \mu_c^2 = \mu_c(1 - \mu_c) = p_{+1}p_{+2}. \tag{11.2-17}$$

The covariance between the two variables is

$$p_{22} - \mu_r\mu_c = p_{22} - p_{2+}p_{+2}. \tag{11.2-18}$$

Dividing the covariance by the square root of the product of the variances gives the product moment correlation coefficient:

$$\rho = \frac{p_{22} - p_{2+}p_{+2}}{\sqrt{p_{1+}p_{2+}p_{+1}p_{+2}}}$$

$$= \frac{p_{11}p_{22} - p_{21}p_{12}}{\sqrt{p_{1+}p_{2+}p_{+1}p_{+2}}}. \tag{11.2-19}$$

The coefficient ρ is invariant under interchange of both rows and columns, and changes only its sign if we interchange one but not the other. If the row and column variables are independent, $\rho = 0$; if $p_{12} = p_{21} = 0$, $\rho = 1$; if $p_{11} = p_{22} = 0$, $\rho = -1$. Furthermore, as ρ is invariant under positive linear transformations, we get the same formula for ρ if we score the rows and columns by any monotonic increasing function of 0 and 1.

Yule [1912] notes that if we standardize the table (using expression (11.2-14)), then ρ is identical to his coefficient Y discussed above. Thus when we remove the effects of marginal totals from the correlation-like measures of this section, we are led directly back to measures of association based on functions of the cross-product ratio.

If different 2×2 tables have the same cross-product ratio α, the table for which the marginal totals are most disproportionate has the lowest correlation coefficient. To be more explicit, suppose we take a standardized table with $p_{11}^* = p_{22}^* = \frac{1}{2}[\sqrt{\hat{\alpha}}/(\sqrt{\hat{\alpha}} + 1)]$ and $p_{12}^* = p_{21}^* = \frac{1}{2}[1/(\sqrt{\hat{\alpha}} + 1)]$, and then we multiply the second row by a factor $R > 1$, the second column by a factor $C > 1$, and renormalize. This is equivalent to multiplying the other row and column by factors <1. We now have

$$\rho = Y\sqrt{\frac{RC}{(p_{11}^* + p_{12}^*R)(p_{11}^* + p_{12}^*C)(p_{11}^*R + p_{12}^*)(p_{11}^*C + p_{12}^*)}}, \tag{11.2-20}$$

from which it follows that as R and C increase, the reduction of ρ relative to Y increases. Moreover,

$$|\rho| \leqq |Y|, \tag{11.2-21}$$

with equality only when all the margins are equal to 1/2.

The observed correlation coefficient

$$r = \frac{x_{11}x_{22} - x_{12}x_{21}}{\sqrt{x_{1+}x_{2+}x_{+1}x_{+2}}} \tag{11.2-22}$$

is the maximum likelihood estimate of ρ under the multinomial sampling model, and the approximate large sample variance of r (see exercise 12 in Section 14.6.5) is

$$\sigma_\infty^2[r] = \frac{1}{N}\left\{1 - \rho^2 + \left(\rho + \frac{1}{2}\rho^3\right)\frac{(p_{1+} - p_{2+})(p_{+1} - p_{+2})}{\sqrt{p_{1+}p_{2+}p_{+1}p_{+2}}}\right.$$

$$\left. - \frac{3}{4}\rho^2\left[\frac{(p_{1+} - p_{2+})^2}{p_{1+}p_{2+}} + \frac{(p_{+1} - p_{+2})^2}{p_{+1}p_{+2}}\right]\right\}. \tag{11.2-23}$$

Yule [1912] was the first to derive this asymptotic variance. Under the null hypothesis of independence of rows and columns, (11.2-23) reduces to $1/N$. Formula (11.2-23) for the asymptotic variance is appropriate for use in constructing confidence intervals. If the table is completely symmetrical, with $p_{1+} = p_{2+} = p_{+1} = p_{+2} = 1/2$, then (11.2-23) reduces to

$$\frac{1 - \rho^2}{N}. \tag{11.2-24}$$

To estimate the asymptotic variance from sample data, we substitute r for ρ, and the observed marginal proportions for the population values, in (11.2-23). Because r is asymptotically normal with mean ρ and variance (11.2-23), we can use r and $\hat{\sigma}_\infty[r]$ to construct confidence intervals for ρ.

Example 11.2-2   Family structure and political affiliation (continued)
For the data in table 11.2-1, the product moment correlation coefficient is $r = -0.136$, and its estimated large sample standard deviation is $\hat{\sigma}_\infty[r] = 0.061$. Since the marginal totals in the table are unequal, $|r| < |Y|$, as we expected. An asymptotic 95% confidence interval for ρ is $[-0.256, -0.016]$, confirming the result in example 11.2-1: YAF membership is associated with authoritarian parental decision-making. ∎∎

Correlation and chi square
   Many of the generalizations of ρ to the $I \times J$ table situation have resulted from a direct relationship between ρ and the population analogue of the Pearson chi square statistic in the 2×2 case:

$$\Phi^2 = \sum_{i=1}^{2} \sum_{j=1}^{2} \frac{(p_{ij} - p_{i+}p_{+j})^2}{p_{i+}p_{+j}}$$

$$= \frac{(p_{11}p_{22} - p_{21}p_{12})^2}{p_{1+}p_{2+}p_{+1}p_{+2}} = \rho^2. \tag{11.2-25}$$

Thus in 2×2 tables, r^2 is simply X^2/N, i.e., the Pearson chi square test statistic for independence of rows and columns divided by the sample size.

Functions of the correlation coefficient
Various authors have used different functions of ρ as measures of association. For example, Pearson [1904b] proposed the use of the "coefficient of mean square contingency"

$$P = \sqrt{\frac{\rho^2}{\rho^2 + 1}}. \tag{11.2-26}$$

   For a general function $f(\rho)$ of ρ, the asymptotic variance of $f(r)$ is

$$\sigma_\infty^2[f(r)] = [f'(\rho)]^2 \sigma_\infty^2[r], \tag{11.2-27}$$

where $f'(\rho)$ is the derivative of f with respect to ρ. For example, the sample estimate of Pearson's coefficient P of mean square contingency is

$$\hat{P} = \sqrt{\frac{X^2}{X^2 + N}} \tag{11.2-28}$$

(where X^2 is the chi square statistic), and the asymptotic variance of $\hat{P}$ is

$$\frac{P^6}{\rho^6} \sigma_\infty^2[r]. \tag{11.2-29}$$

Under "complete association," $\hat{P}$ cannot attain the upper limit of 1. We discuss this problem in more detail in Section 11.3.1.

11.2.4 *Choosing a measure for 2 × 2 tables*

As we have seen, the choice of a measure for a 2 × 2 table essentially becomes a choice between a function of the cross-product ratio and a function of the product-moment correlation coefficient. For simplicity, in this section we focus on choosing between Q and ρ, because both measures take the value zero under independence, always lie between -1 and $+1$, and have reasonable interpretations.

When should we choose Q rather than ρ, or vice versa? By examining cases where the two measures behave differently, we can get some guidance.

1. The measure Q takes the value 1 or -1 whenever any one of the cell probabilities in a 2 × 2 table is zero, while for $\rho = 1$ or $\rho = -1$ both entries on a diagonal must be zero.

2. In a given table, the marginal totals may constrain the cell entries in such a way that ρ cannot take the value 1 or -1. This is the price we pay for using a margin-sensitive measure. Specifically, $\rho = 1$ or $\rho = -1$ implies the marginals for the first variable are identical to those for the second. The more different the row and column marginals, the lower the upper limit on the absolute value of ρ. We have noted above that Q is not affected by row and column multiplications and can always take its full range of values irrespective of the distribution of the marginal totals.

Thus selection of Q versus ρ depends upon whether we wish our measure to be sensitive to marginal totals and whether we consider an association as complete when only one cell of a 2 × 2 table is zero. Finally, if we plan to incorporate our measure along with other coefficients in a correlation matrix to be used for further analysis (e.g., a factor analysis), then we might prefer to use ρ, since the use of Q could endanger the positive definiteness of the correlation matrix.

11.2.5 *Geometry of measures for 2 × 2 tables*

As any contingency table can be normalized to have entries that sum to 1, there is a natural one-to-one correspondence between the set of 2 × 2 tables whose entries sum to 1 and the points of the three-dimensional simplex (tetrahedron). In Section 2.7 we discuss ideas about 2 × 2 tables in terms of the geometry of the three-dimensional simplex. There we note that all the points corresponding to tables with a given cross-product ratio form a two-dimensional surface within the simplex.

For 2 × 2 tables, different measures of association may correspond to different surfaces within the simplex. As all the measures of interest for the 2 × 2 table are reflected by the behavior of two measures, the cross-product ratio and the correlation coefficient, we need only examine the geometry of these two by comparing their contours within the simplex.

When $\rho = 0$, $\alpha = 1$, and conversely. However, contours of constant ρ do not generally correspond to values of α because ρ is not independent of the marginal totals, while α is.

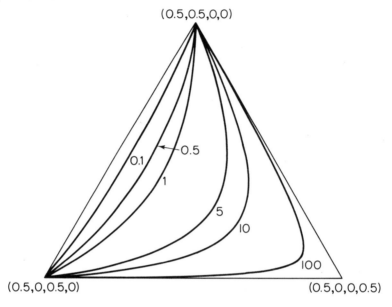

Figure 11.2-1   Contours of constant α for $p_{11} = 0.5$ ($\alpha = 100, 10, 5, 1, 0.5, 0.1$). Reproduced from Fienberg and Gilbert [1970].

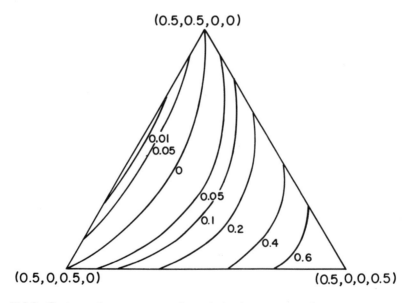

Figure 11.2-2   Contours of constant squared correlation for $p_{11} = 0.5$ ($\rho^2 = 0.6, 0.4, 0.2, 0.1, 0.05, 0, 0.05, 0.1$). Reproduced from Fienberg and Gilbert [1970].

To examine contours corresponding to α and to ρ, we take a slice of the tetra-hedron (i.e., we fix the coordinate corresponding to p_{11}). Figure 11.2-1 contains contours of constant α ($\alpha = 100, 10, 5, 1, 0.5, 0.1$) for p_{11} equal to 0.5, and figure 11.2-2 contains contours of constant ρ^2 ($\rho^2 = 0.6, 0.4, 0.2, 0.1, 0.05, 0, 0.05, 0.1$) for the same value of p_{11}. It is clear that near $\rho^2 = 0$, contours of constant ρ^2 closely approximate contours of constant α, except for points very near the faces of the tetrahedron. Such approximations become less satisfactory as ρ^2 becomes larger.

For $I \times J$ tables, each measure of association corresponds to a manifold of dimension $IJ - 2$ lying within the simplex of dimension $IJ - 1$ used to represent all possible $I \times J$ tables (see Fienberg [1968]). Tables with a fixed cross-product ratio structure (see Section 2.3) correspond to a manifold of dimension $I + J - 2$. This discrepancy indicates that for $I > 2$ or $J > 2$, a single measure of association corresponds at best to a class of different cross-product ratio structures, i.e., we cannot express a vector of cross-product ratios by a single number without using the same number to represent more than one vector.

11.3 Measures of Association for $I \times J$ Tables

We now turn to general measures of association for $I \times J$ tables. We begin with a brief discussion of several measures based on chi square, and then move to other kinds of measures offering clearer interpretations. Finally, we conclude with suggestions for choosing among different groups of measures.

11.3.1 Measures based on chi square

Pearson's coefficient Φ^2 of mean square contingency is defined for 2×2 tables in expression (11.2-25). In general, we have

$$\Phi^2 = \sum_{i=1}^{I} \sum_{j=1}^{J} \frac{(p_{ij} - p_{i+}p_{+j})^2}{p_{i+}p_{+j}}$$

$$= \left(\sum_{ij} \frac{p_{ij}^2}{p_{i+}p_{+j}} \right) - 1, \tag{11.3-1}$$

and if $I > 2$ or $J > 2$, Φ^2 no longer lies between 0 and 1. To overcome this deficiency, Pearson [1904] proposed the use of

$$P = \sqrt{\frac{\Phi^2}{\Phi^2 + 1}}. \tag{11.3-2}$$

The measure P must lie between 0 and 1, but it cannot always attain the upper limit of 1. For example, in an $I \times I$ table with $p_{ij} = 0$ for $i \neq j$, we have $\Phi^2 = I - 1$ and $P = [(I - 1)/I]^{1/2}$. Thus even under complete association, this maximum value of P depends on the number of rows and columns.

Other authors have suggested alternative functions of Φ^2 to avoid the limitations of P. Tschuprow proposed the use of

$$T = \left\{ \frac{\Phi^2}{[(I - 1)(J - 1)]^{1/2}} \right\}^{1/2}, \tag{11.3-3}$$

which achieves the value 1 for the case of complete association in an $I \times I$ table, but not when $I \neq J$. When association is complete with nonzero probabilities only on one of the longest diagonals of an $I \times J$ table, the maximum value of Φ^2 is

$$\min[(I - 1), (J - 1)], \tag{11.3-4}$$

and as

$$[(I - 1)(J - 1)]^{1/2} \geqq \min[(I - 1), (J - 1)], \tag{11.3-5}$$

we find that $T < 1$. Cramér [1946] suggests norming by (11.3-4) to produce a measure that could attain the maximum value of 1 in any $I \times J$ table:

$$V = \left\{ \frac{\Phi^2}{\min[(I - 1), (J - 1)]} \right\}^{1/2} \tag{11.3-6}$$

Because (11.3-5) holds, $V \geq T$ for $I > 2$ or $J > 2$.

The MLE of Φ^2 under multinomial sampling is X^2/N. Thus we get the MLEs of P, T, and V, by substituting X^2/N for Φ^2 in (11.3-2), (11.3-3), and (11.3-6), respectively.

The asymptotic variance of $\hat{\Phi}^2 = X^2/N$ is

$$\sigma_\infty^2[\hat{\Phi}^2] = \frac{1}{N} \left\{ 4 \sum_{ij} \frac{p_{ij}^3}{p_{i+}^2 p_{+j}^2} - 3 \sum_i \frac{1}{p_{i+}} \left(\sum_j \frac{p_{ij}^2}{p_{i+}p_{+j}} \right)^2 - 3 \sum_j \frac{1}{p_{+j}} \left(\sum_i \frac{p_{ij}^2}{p_{i+}p_{+j}} \right)^2 \right.$$

$$\left. + 2 \sum_{ij} \left[\frac{p_{ij}}{p_{i+}p_{+j}} \left(\sum_k \frac{p_{kj}^2}{p_{k+}p_{+j}} \right) \left(\sum_l \frac{p_{il}^2}{p_{i+}p_{+l}} \right) \right] \right\}, \tag{11.3-7}$$

and since

$$\sigma_\infty^2[f(\hat{\Phi}^2)] = [f'(\Phi^2)]^2 \sigma_\infty^2[\hat{\Phi}^2], \tag{11.3-8}$$

we find that (for $\Phi^2 \neq 0$)

$$\sigma_\infty[\hat{P}] = \frac{1}{2\Phi(1 + \Phi^2)^{3/2}} \sigma_\infty[\hat{\Phi}^2],$$

$$\sigma_\infty[\hat{T}] = \frac{1}{2(I - 1)(J - 1)T} \sigma_\infty[\hat{\Phi}^2],$$

$$\sigma_\infty[\hat{V}] = \frac{1}{2\{\min[(I - 1), (J - 1)]\}^{1/2} V} \sigma_\infty[\hat{\Phi}^2]. \tag{11.3-9}$$

The several measures based on chi square can be interpreted in terms of squared departure between observed and expected frequencies on a scale of 0 to 1 (or slightly less than 1). Such an interpretation is only useful for comparing several tables.

Example 11.3-1   Political preference and Cold War allies
Willick [1970] presents survey data for a sample of size 500 from Great Britain. Respondents are cross-classified by party preference and by preferred Cold War ally. We consider these data in the form of a 3×2 table (see table 11.3-1).

First we compute the chi square value, because $\hat{P}$, $\hat{V}$, and $\hat{T}$ are all functions

Table 11.3-1 Party Preference and
Preferred Cold War Ally in Great Britain

Preferred	Party Preference			
Cold War Ally	Right	Center	Left	Totals
U.S.	225	53	206	484
U.S.S.R.	3	1	12	16
Totals	228	54	218	500

Source: Willick [1970].

of chi square. We obtain $X^2 = 6.668$. For the derived measures, we then get

$$\hat{P} = \sqrt{\frac{6.668}{500 + 6.668}} = 0.114,$$

$$\hat{T} = \left[\frac{6.668}{500(1 \cdot 2)^{1/2}}\right]^{1/2} = 0.097,$$

$$\hat{V} = \left[\frac{6.668}{500(1)}\right]^{1/2} = 0.115.$$

Using expression (11.3-7), we get that the estimated asymptotic standard deviation of Φ^2 is $\hat{\sigma}_\infty[\hat{\Phi}^2] = 0.0095$, and the corresponding standard deviations for $\hat{T}$ and $\hat{V}$ are

$$\hat{\sigma}_\infty[\hat{P}] = 0.0405,$$

$$\hat{\sigma}_\infty[\hat{T}] = 0.0246,$$

$$\hat{\sigma}_\infty[\hat{V}] = 0.0415.$$

As usual, we estimate the asymptotic standard errors by substituting sample for population quantities. Both $\hat{T}$ and $\hat{V}$ are within three standard deviations of zero, while $\hat{T}$ is within four. We observe that all of the derived statistics are close in value and indicate a slight association between party preference and ally preference. ■ ■

11.3.2 *Proportional reduction in error measures*

Goodman and Kruskal [1954] point out that it is useful for a measure to possess a clear interpretation in terms of probabilities, an attribute that is lacking in measures based on chi square. They proposed several sets of measures with direct interpretations. One such set is designed for tables where our goal is to predict optimally the category for the row variable from the category of the column variable, or vice versa.

We can predict the category of variable C from the category of variable R by (i) assuming C is statistically independent of R or (ii) assuming C is a function of R. A proportional reduction in error (PRE) strategy relates these two assumptions by taking

$$\text{PRE} = \frac{\text{Probability of error in (i)} - \text{Probability of error in (ii)}}{\text{Probability of error in (i)}}. \quad (11.3\text{-}10)$$

Thus the degree of association shown by a PRE measure is defined as the relative improvement in predicting the C category obtained when the R category is known, as opposed to when the R category is not known.

For an $I \times J$ table, if R corresponds to rows and we wish to predict the category of the column variable C, expression (11.3-10) leads to the following measure:

$$\lambda_{C|R} = \frac{(1 - p_{+m}) - \left(1 - \sum_1^I p_{im}\right)}{1 - p_{+m}} = \frac{\sum_1^I p_{im} - p_{+m}}{1 - p_{+m}}, \tag{11.3-11}$$

where

$$p_{im} = \max(p_{i1}, p_{i2}, \ldots, p_{iJ}) = \max_j(p_{ij}), \tag{11.3-12}$$

$$p_{+m} = \max(p_{+1}, p_{+2}, \ldots, p_{+J}) = \max_j(p_{+j}). \tag{11.3-13}$$

The computing formula for estimating $\lambda_{C|R}$ using the MLEs for the p_{ij} is then

$$\hat{\lambda}_{C|R} = \frac{\sum_1^I x_{im} - x_{+m}}{N - x_{+m}}, \tag{11.3-14}$$

where the cell counts x_{im} and x_{+m} are the maxima for the ith row and for the column totals, respectively. The range for $\hat{\lambda}_{C|R}$ is $[0, 1]$, and the value 1 is attained if and only if each row has a single nonzero cell.

Goodman and Kruskal [1963] show that when $\hat{\lambda}_{C|R}$ is neither 0 nor 1, it has a sampling distribution that is asymptotically normal with mean $\lambda_{C|R}$ and estimated variance

$$\hat{\sigma}_\infty^2[\hat{\lambda}_{C|R}] = \frac{(N - \sum x_{im})(\sum x_{im} + x_{+m} - 2\sum^* x_{im})}{(N - x_{+m})^3}, \tag{11.3-15}$$

where $\sum^* x_{im}$ is the summation of the maximum frequency in a row, taken only over those rows where x_{im} falls in the same column as x_{+m}.

Example 11.3-2   Party preference and preferred ally (continued)
To illustrate the computation of the measure $\hat{\lambda}$, we return to the data from Willick [1970] in example 11.3-1. For predicting party from ally preference, we note that

$$x_{1m} = 225, \qquad x_{2m} = 12, \qquad x_{+m} = 228,$$

and so

$$\hat{\lambda}_{C|R} = \frac{225 + 12 - 228}{500 - 228} = 0.033.$$

Thus the estimated proportional reduction in error when we try to predict party preference from knowledge of preferred ally, compared with basing our prediction only on the marginal distribution of party preference, is 3.3 percent.

The estimated asymptotic variance of $\hat{\lambda}$ is

$$\hat{\sigma}_\infty^2(\hat{\lambda}_{C|R}) = \frac{(500 - 237)(237 + 228 - 2 \times 225)}{(500 - 228)^3}$$

$$= 0.00036,$$

and an asymptotic 95% confidence interval for $\lambda_{C|R}$ is $[-0.004, 0.060]$.

Since only the total of the table is fixed, we may wish to predict the preferred ally using knowledge of party preference. Thus we compute

$$\hat{\lambda}_{R|C} = \frac{225 + 53 + 206 - 484}{500 - 484} = 0.$$

Here the value of $\hat{\lambda}_{R|C}$ tells us there is no proportional reduction in error when we try to predict ally from knowledge of party preference. This example illustrates the asymmetric aspects of $\lambda_{R|C}$ and $\lambda_{C|R}$. ■■

11.3.3 When to use λ

The main attraction of the λ statistics is their PRE interpretability. An asymmetric or directional prediction may be particularly useful when causal or chronological direction between variables is suspected.

Goodman and Kruskal [1954] give several formal properties of $\lambda_{C|R}$:

1. $\lambda_{C|R}$ is indeterminate if and only if the population lies in one column. Otherwise $0 \leqq \lambda_{C|R} \leqq 1$;
2. $\lambda_{C|R}$ is 0 if and only if knowledge of the row category is of no help in predicting the column category;
3. $\lambda_{C|R}$ is 1 if and only if knowledge of an individual's row category completely specifies his column category, i.e., if and only if each row of the table contains at most one nonzero probability;
4. in the case of statistical independence, $\lambda_{C|R}$, when determinate, is zero. The converse need not hold, i.e., $\lambda_{C|R}$ may be zero without statistical independence holding;
5. $\lambda_{C|R}$ is unchanged by permutation of rows or columns.

Property 4, that $\lambda_{C|R}$ can be zero even in the absence of statistical independence, does not imply that it is an unsatisfactory measure. The rationale underlying $\lambda_{C|R}$ is a predictive interpretation of association. When other measures of association, such as those based on chi square, indicate that a particular table shows a substantial departure from independence, a $\lambda_{C|R}$ value of 0 for the same table can occur, and this indicates the absence of predictive association when predicting column categories from row categories. Rather than being a drawback, this predictive interpretation is what differentiates $\lambda_{C|R}$ and $\lambda_{R|C}$ from other measures.

11.3.4 A proportion of explained variance measure

We next consider a different kind of measure of association for two-dimensional tables, $\tau_{R|C}$, which we view as a qualitative analogue to the coefficient of determination for continuous data. It offers a proportion of explained variance interpretation of the relationship between row and column variables. Since this interpretation is frequently used to describe relations among continuous variables, it is useful to have the analogue for qualitative data in contingency tables.

For N categorical responses, each in one and only one of I possible categories, Gini [1912] defined the total variation to be:

$$\text{Variation} = \frac{N}{2} - \frac{1}{2N} \sum_{i=1}^{I} x_i^2, \tag{11.3-16}$$

where x_i is the number of responses in the ith category, $i = 1, \ldots, I$, and $\sum_i x_i = N$. Light and Margolin [1971] illustrate that Gini's measure of variation has two properties expected of a general measure of variation. First, the variation is minimized to a value of zero if and only if all N responses fall into the same category. Second, the variation is maximized when the responses are distributed among the categories as evenly as possible. A third property that follows directly from this definition of variation is that the aggregation of two categories into one new category cannot increase the variation, and in fact decreases it unless one of the categories being combined has not appeared among the responses.

We develop the measure $\tau_{R|C}$ by using Gini's definition of variation. Keeping with analysis of variance terminology, the total sum of squares in an $I \times J$ table is

$$\text{TSS} = \frac{N}{2} - \frac{1}{2N} \sum_{i=1}^{I} x_{i+}^2 . \tag{11.3-17}$$

This variation is then partitioned, in a fashion similar to the standard analysis of variance, into within-group sums of squares (WSS) and between-group sums of squares (BSS):

$$\text{WSS} = \frac{N}{2} - \frac{1}{2} \sum_{j=1}^{J} \frac{1}{x_{+j}} \sum_{i=1}^{I} x_{ij}^2, \tag{11.3-18}$$

$$\text{BSS} = \text{TSS} - \text{WSS}, \tag{11.3-19}$$

or

$$\begin{aligned}
\text{BSS} &= \frac{1}{2} \sum_{j=1}^{J} \frac{1}{x_{+j}} \sum_{i=1}^{I} x_{ij}^2 - \frac{1}{2N} \sum_{i=1}^{I} x_{i+}^2 \\
&= \frac{1}{2} \sum_{i=1}^{I} \sum_{j=1}^{J} \frac{1}{x_{+j}} \left(x_{ij} - \frac{x_{i+} x_{+j}}{N} \right)^2 .
\end{aligned} \tag{11.3-20}$$

Based on these components of variation, a sample measure of the proportion of variation in the row variable attributable to the column variable is defined as the ratio BSS/TSS, which, after some cancellation, is

$$\hat{\tau}_{R|C} = \frac{\sum_j \dfrac{1}{x_{+j}} \sum_i x_{ij}^2 - \dfrac{1}{N} \sum_i x_{i+}^2}{N - \dfrac{1}{N} \sum_i x_{i+}^2} . \tag{11.3-21}$$

We may view $\hat{\tau}_{R|C}$ as a sample estimate of the following population quantity under multinomial sampling:

$$\begin{aligned}
\tau_{R|C} &= \frac{\sum_j \dfrac{1}{p_{+j}} \sum_i p_{ij}^2 - \sum_i p_{i+}^2}{1 - \sum_i p_{i+}^2} \\
&= \frac{\sum_{ij} \dfrac{(p_{ij} - p_{i+} p_{+j})^2}{p_{+j}}}{1 - \sum_i p_{i+}^2} .
\end{aligned} \tag{11.3-22}$$

Goodman and Kruskal [1954] note that we can interpret $\tau_{R|C}$ as the relative decrease in the proportion of incorrect predictions when we go from predicting the row category based only on the row marginal probabilities to predicting the row category based on the conditional proportions p_{ij}/p_{+j}.

The second expression for $\tau_{R|C}$ in (11.3-22) has a form somewhat resembling chi square. When $p_{i+} = 1/I$ for $i = 1, \ldots, I$, i.e., the row totals are equal, $\tau_{R|C}$ reduces to $\Phi^2/(I-1)$. When $I = 2$, $\tau_{R|C} = \Phi^2$ for every value of p_{1+} and p_{2+}.

The measure $\tau_{R|C}$ has the following properties:

(i) If there exists an i such that $p_{i+} = 1$, then all components are zero and $\tau_{R|C}$ is undefined;

(ii) if there does not exist an i such that $p_{i+} = 1$, and if $p_{ij} = p_{i+}p_{+j}$ for all i and j, then $\tau_{R|C} = 0$ (zero association);

(iii) if there does not exist an i such that $p_{i+} = 1$, and if for each j there exists an i (not necessarily unique) such that $p_{ij} = p_{+j}$, then $\tau_{R|C} = 1$ (perfect association);

(iv) if none of (i), (ii), or (iii) occurs, then $0 < \tau_{R|C} < 1$;

(v) $\tau_{R|C}$ is invariant under permutations of rows or columns.

For a detailed study comparing the exact small sample behavior of X^2, $\hat{\lambda}_{R|C}$, and $\hat{\tau}_{R|C}$, see Margolin and Light [1974].

To test the hypothesis of no association, the test statistic

$$U^2 = \frac{(N-1)(I-1)\,\text{BSS}}{\text{TSS}} = (N-1)(I-1)\hat{\tau}_{R|C} \qquad (11.3\text{-}23)$$

is available. Under the null hypothesis, U^2 is asymptotically approximated by a chi square distribution with $(I-1)(J-1)$ degrees of freedom (Light and Margolin [1971]).

Under the multinomial sampling model, the asymptotic variance of $\hat{\tau}_{R|C}$ (Goodman and Kruskal [1972]) is

$$\hat{\sigma}^2_\infty[\hat{\tau}_{R|C}] = \frac{1}{N\delta^4} \sum_{ij} p_{ij} \left[2v \sum_{k \ne i} p_{k+} - \delta \left(2 \sum_{k \ne i} \frac{p_{kj}}{p_{+j}} - \sum_{\substack{k,l \\ k \ne i}} \frac{p_{kl}}{p_{+l}} \right) \right]^2, \qquad (11.3\text{-}24)$$

where v and δ are the numerator and denominator, respectively, of $\tau_{R|C}$ as given in (11.3-22). This formula is applicable in the nonnull case where $\tau_{R|C} \ne 0$.

Example 11.3-3   Party preference and preferred ally (continued)
Again we examine the Willick [1970] data from Great Britain in table 11.3-1. To test the null hypothesis that the sample data come from a population with no association, we compute

$$U^2 = (N-1)(I-1)\hat{\tau}_{R|C} = 6.665.$$

For two degrees of freedom, the critical value of chi square at the 0.05 level is 5.991. Thus we reject the null hypothesis of no association in the population. For predicting preferred ally from party preference, we get the estimate

$$\hat{\tau}_{R|C} = \frac{469.437 - 469.024}{30.976} = 0.013.$$

We conclude that 1.3% of the variation in choice of preferred ally for the British sample is explained by knowing a person's party preference.

The estimated nonnull asymptotic standard error of $\hat{\tau}_{R|C}$ here is 0.135, and a 95% asymptotic confidence interval for $\tau_{R|C}$ is $[-0.256, 0.282]$. As with the measures based on X^2 for this example, our asymptotic 95% confidence interval contains the value zero because the asymptotic variance formula used is appropriate for the nonnull case. This example also illustrates how a statistically significant association may nevertheless involve a very small amount of explained variation. ■■

11.3.5 Measures insensitive to marginal totals

In Section 11.1 we noted that measures of association that are not affected by the transformation (11.1-4) must be functions only of the cross-product ratios for all 2×2 subtables in an $I \times J$ table. We now show by means of an example that none of the measures discussed so far in this section has this property.

Example 11.3-4 Measures not invariant under scale transformation

Using the raw data in table 11.3-1, we have already computed the sample measures $\hat{T}, \hat{V}, \hat{\lambda}_{C|R}, \hat{\lambda}_{R|C}$, and $\hat{\tau}_{R|C}$ in examples 11.3-1, 11.3-2, and 11.3-3. We standardize the raw data by the iterative proportional fitting procedure (see Section 3.6) so that the new row totals are 0.500 and the new column totals are 0.333 (see table 11.3-2).

Table 11.3-2 Standardized Version of Willick Data in Table 11.3-1 (with Equal Row and Column Totals)

Preferred Cold War Ally	Party Preference			Totals
	Right	Center	Left	
U.S.	0.215	0.187	0.098	0.500
U.S.S.R.	0.118	0.147	0.235	0.500
Totals	0.333	0.334	0.333	1.000

This procedure transforms the observed proportions in the manner of expression (11.1-4).

We can now recompute the measures of association using the transformed table and compare these new values with those we obtained originally.

	New Value	Original Value
$\hat{T}$	0.254	0.097
$\hat{V}$	0.300	0.115
$\hat{\lambda}_{C|R}$	0.177	0.033
$\hat{\lambda}_{R|C}$	0.280	0
$\hat{\tau}_{R|C}$	0.090	0.013

All measures now have different values. In particular, we observe that $\hat{\lambda}_{R|C}$ now shows substantial predictive power, whereas for the raw data it showed no predictive power. The explanation is that when the modal row is identical for all columns (as in the Willick raw data), $\hat{\lambda}_{R|C}$, being a measure of predictivity, is zero even if the data do not show independence. ∎∎

Altham [1970a] gives several measures of association for $I \times J$ tables which are functions only of the cross-product ratios

$$\alpha_{ij} = \frac{p_{ij}p_{IJ}}{p_{iJ}p_{Ij}} \qquad (11.3\text{-}25)$$

for $i = 1, \ldots, I - 1; j = 1, \ldots, J - 1$. An example of the measures she discusses is:

$$\left\{ \sum_{ij} |\log \alpha_{ij}|^s \right\}^{1/s} \quad \text{for } s \geqq 1. \qquad (11.3\text{-}26)$$

This measure has several interesting and desirable properties. Its interpretation for the nonmathematician, however, is somewhat difficult, and even with $s = 2$ in (11.3-26) the observed table of counts must be smoothed before the measure can be estimated whenever sampling zeros occur.

For these reasons we do not consider Altham's measures in subsequent discussion. Nevertheless, they represent a potentially important class of measures that deserve further attention. For details of the measures and their extensions, we refer the reader to Altham [1970a, 1970b].

11.3.6 *Choosing among measures for $I \times J$ tables*

Let us summarize when each measure of association is most useful. The measures based on chi square, P, T, and V, are symmetric and have the convenience of providing a built-in test of significance. The major difficulty in their use is their lack of clear interpretation. The only reasonable interpretation is that each measure departs from zero as the data in a table depart from independence. Of the three measures, Cramér's V is preferred on the grounds of its being normed to a $(0, 1)$ scale for all $I \times J$ tables.

The PRE measure $\lambda_{R|C}$ is useful for studying a different kind of association: predictability. Although two variables are not independent, each may still have zero predictability for the other. It is also possible that one variable has some predictive value for the second, while the reverse is not true. Thus it is not uncommon for a chi-square-based measure of association for an $I \times J$ table to be positive, while a PRE measure for the same table is zero. The measure $\tau_{R|C}$ offers still another interpretation; it focuses on explained variation, although it also has a proportional prediction interpretation.

No single measure is better than all others in every circumstance. Different measures have different purposes, and our selection must depend on our objective in studying a set of data. If the focus is on departures from bivariate independence, P, T, and V are useful, while $\lambda_{R|C}$ or $\tau_{R|C}$ may mislead. If the focus is on directional prediction, the reverse is true, and we may profitably choose $\lambda_{R|C}$ or $\tau_{R|C}$.

11.4 Agreement as a Special Case of Association

11.4.1 *The problem*

Suppose two observers independently categorize items or responses among the same set of nominal categories, and we wish to develop a measure of agreement for these observers. Goodman and Kruskal [1954] have described this problem as one of measuring the reliability between two observers. In example 2.2-1 of Chapter 2 we showed that in a 2×2 table we may wish to measure both association and two measures of agreement, sensitivity and specificity.

Agreement can be regarded as a special case of association. Examining the display in table 11.4-1 helps us to see why. Assume that each of two observers independently assigns each of N items to one of I categories. The $I \times I$ contingency table in table 11.4-1 is the general form for such a setting. Notice that the layout for measuring agreement requires a higher level of specificity for the table structure than is required for measuring association. Not only is the number of rows equal to the number of columns, but as Goodman and Kruskal point out, the categories for the rows must appear in the same order as the categories for the columns. Thus, two rows cannot be permuted unless the corresponding columns are similarly permuted. This gives meaning to the main diagonal of any agreement table.

Table 11.4-1   Data Format for Measuring Agreement

Observer 1	Observer 2				Totals
	1	2	$\cdots$	I	
1	x_{11}	x_{12}	$\cdots$	x_{1I}	x_{1+}
2	x_{21}			$\cdot$	x_{2+}
$\cdot$	$\cdot$			$\cdot$	$\cdot$
$\cdot$	$\cdot$			$\cdot$	$\cdot$
$\cdot$				$\cdot$	$\cdot$
I	x_{I1}	$\cdot$	$\cdot \quad \cdots$	x_{II}	x_{I+}
Totals	x_{+1}	x_{+2}	$\cdot \quad \cdots$	x_{+I}	N

We can give the layout in table 11.4-1 two possible interpretations. In the first interpretation, we view the data as based on N observers of one type (e.g., mothers) paired with N of a second type (e.g., fathers), where each of the $2N$ observers assigns an item to one of I categories. An alternative interpretation is based on two observers, each categorizing N items, with table 11.4-1 reporting the results of the paired assignments of each item.

The distinction between agreement and association for nominal data is that for two responses to agree, they must fall into the identical category, while for two responses to be perfectly associated we only require that we can predict the category of one response from the category of the other response. Thus, a table may exhibit high association along with either low or high agreement. To illustrate this distinction, we consider a table of hypothetical paired responses from 30 mother-father pairs, as displayed in table 11.4-2. Each of the 60 people answers a multiple-choice question with alternatives a_1, a_2, and a_3.

Since responses on the main diagonal indicate agreement, the table displays no agreement; yet there is perfect association in a predictive sense. Given the father's responses, the mother's responses are perfectly predictable, and vice versa. Thus while most measures of association discussed earlier have a value of 1.0, this value reflects predictability, and not agreement.

Table 11.4-2 Hypothetical
Illustration of Full Association
but No Agreement

		Mothers		
Fathers	a_1	a_2	a_3	Totals
a_1	0	10	0	10
a_2	0	0	10	10
a_3	10	0	0	10
Totals	10	10	10	30

11.4.2 *A measure of agreement*

The simplest measure of agreement is the proportion of the population whose categorization in the two variables is identical, i.e., $\sum_i p_{ii}$. The possible values this simple proportion can take are affected by the marginal totals, and the effects of not taking the marginal totals into account are suggested by the contrived example in table 11.4-3. Mother-father pairs in two different countries are contrasted by their responses to a dichotomous question with categories A and B. In country 1, 60% of the pairs agree, while in country 2, only 40% agree. Yet given the constraints of the marginal totals, the pairs of country 1 display the minimum possible agreement, while those of country 2 display the maximum possible agreement. This problem of disentangling absolute versus relative agreement has led to several measures that take into account information regarding the marginal probabilities.

Cohen [1960] and others suggest comparing the actual agreement $\theta_1 = \sum_i p_{ii}$ with the "chance agreement" $\theta_2 = \sum_i p_{i+} p_{+i}$, that occurs if the row variable is independent of the column variable, i.e., they look at

$$\theta_1 - \theta_2 = \sum_i p_{ii} - \sum_i p_{i+} p_{+i}. \tag{11.4-1}$$

They then normalize (11.4-1) by its maximum possible value for the given marginal totals $\{p_{i+}\}$ and $\{p_{+i}\}$. This leads to the measure of agreement:

$$K = \frac{\sum\limits_i p_{ii} - \sum\limits_i p_{i+} p_{+i}}{1 - \sum\limits_i p_{i+} p_{+i}} = \frac{\theta_1 - \theta_2}{1 - \theta_2}. \tag{11.4-2}$$

Table 11.4-3 Hypothetical Agreement Data for Two Countries

| | Country 1 | | | | Country 2 | | |
| | Mothers | | | | Mothers | | |
Fathers	A	B	Totals	Fathers	A	B	Totals
A	60	20	80	A	20	60	80
B	20	0	20	B	0	20	20
Totals	80	20	100	Totals	20	80	100

For an extensive discussion on the use of K, see Cohen [1968], Fleiss, Cohen, and Everett [1969], and Light [1971]. Note that the use of K does not require an assumption of identical marginal probabilities for the two observers, as does a measure suggested by Scott [1955].

For a multinomial sampling model, where only the total N is fixed, we get the maximum likelihood estimate of K by substituting observed proportions $\hat{p}_{ij} = x_{ij}/N$ for cell probabilities p_{ij}:

$$\hat{K} = \frac{N \sum\limits_{i=1}^{I} x_{ii} - \sum\limits_{i} x_{i+} x_{+i}}{N^2 - \sum\limits_{i} x_{i+} x_{+i}}. \qquad (11.4\text{-}3)$$

The approximate large sample variance of $\hat{K}$, determined by using the δ method (see exercises 13 and 14 in Section 14.6.5) is

$$\sigma_\infty^2[\hat{K}] = \frac{1}{N} \left\{ \frac{\theta_1(1 - \theta_1)}{(1 - \theta_2)^2} + \frac{2(1 - \theta_1)(2\theta_1\theta_2 - \theta_3)}{(1 - \theta_2)^3} \right.$$
$$\left. + \frac{(1 - \theta_1)^2(\theta_4 - 4\theta_2^2)}{(1 - \theta_2)^4} \right\}, \qquad (11.4\text{-}4)$$

where $\theta_1 = \sum_i p_{ii}$ and $\theta_2 = \sum_i p_{i+} p_{+i}$ as above, and

$$\theta_3 = \sum_i p_{ii}(p_{i+} + p_{+i}), \qquad (11.4\text{-}5)$$

$$\theta_4 = \sum_{i,j} p_{ij}(p_{j+} + p_{+i})^2. \qquad (11.4\text{-}6)$$

Under the null model of independence, the asymptotic variance (11.4-4) reduces to

$$\frac{\left[\theta_2 + \theta_2^2 - \sum\limits_{i} p_{i+} p_{+i}(p_{i+} + p_{+i}) \right]}{N(1 - \theta_2)^2}, \qquad (11.4\text{-}7)$$

a formula given by Fleiss [1973] for testing the hypothesis $K = 0$ (under the assumption of independence). It is important for us to note that (11.4-4) is not an expression for the asymptotic variance conditional on the marginal totals.

We can estimate the asymptotic variances (11.4-4) and (11.4-7) by substituting observed proportions $\hat{p}_{ij}$ for cell probabilities p_{ij}. We refer to the estimate of $\sigma_\infty^2[\hat{K}]$ as $\hat{\sigma}_\infty^2[\hat{K}]$. Since $\hat{K}$ is asymptotically normal, we can use $\hat{\sigma}_\infty[\hat{K}]$ to construct a confidence interval for the true value K.

Example 11.4-1   Ratings of student teachers' classroom style
Gross [1971] has collected data based on two supervisors who were asked to rate independently the classroom style of 72 student teachers as authoritarian, democratic, or permissive. We give the agreement table for the two supervisors in table 11.4-4.

The estimated measure of agreement for these data is

$$\hat{K} = \frac{0.583 - 0.347}{1 - 0.347} = 0.362,$$

Table 11.4-4 Student Teachers Rated by Supervisors

Rating by Supervisor 1	Rating by Supervisor 2			
	Authoritarian	Democratic	Permissive	Totals
Authoritarian	17	4	8	29
Democratic	5	12	0	17
Permissive	10	3	13	26
Totals	32	19	21	72

Source: Gross [1971].

and its estimated asymptotic variance is

$$\hat{\sigma}^2_\infty[\hat{K}] = \frac{1}{72}\left\{ \frac{0.583 \times 0.417}{0.653^2} + \frac{2 \times 0.417 \times (0.405 - 0.401)}{0.653^3} \right.$$
$$\left. + \frac{0.417^2 \times (0.505 - 0.481)}{0.653^4} \right\}$$

$$= \frac{1}{72}(0.570 + 0.011 + 0.023)$$

$$= 0.0084.$$

A 95% confidence interval for the true value of K is thus [0.182, 0.542], so we conclude that the supervisors agree to a moderate extent, and more than they would by chance. ∎∎

11.4.3 *Measuring conditional agreement*

We can also look at the agreement between two observers conditional on one of the observers' categorization. For example, we may want a measure of agreement between the observers for only those items which the first observer placed into the ith category. Coleman [1966] and Light [1969] define a measure of agreement between two observers for those items which the first observer (appearing on the rows of the table) assigns to the ith specific category; it is

$$K_i = \frac{\left(\dfrac{p_{ii}}{p_{i+}} - p_{+i} \right)}{1 - p_{+i}} = \frac{p_{ii} - p_{i+}p_{+i}}{p_{i+} - p_{i+}p_{+i}}. \tag{11.4-8}$$

If we sum the numerators and denominators of the K_i separately over all I categories, we get the numerator and denominator, respectively, of K.

While the population quantity K_i is based on a conditional argument, the sampling model for the data in the agreement table remains the same, i.e., multinomial, with only the total fixed, not product multinomial with the row totals fixed. The maximum likelihood estimate of K_i under multinomial sampling is

$$\hat{K}_i = \frac{Nx_{ii} - x_{i+}x_{+i}}{Nx_{i+} - x_{i+}x_{+i}}, \tag{11.4-9}$$

and its asymptotic variance is

$$
\sigma^2_\infty[\hat{K}_i] = \frac{1}{N} \frac{(p_{i+} - p_{ii})}{p^3_{i+}(1 - p_{+i})^3} [(p_{i+} - p_{ii})(p_{i+}p_{+i} - p_{ii})
$$

$$
+ p_{ii}(1 - p_{i+} - p_{+i} + p_{ii})]. \tag{11.4-10}
$$

Under the null model of independence, (11.4-10) reduces to

$$
\frac{1}{N} \frac{p_{+i}(1 - p_{i+})}{p_{i+}(1 - p_{+i})}. \tag{11.4-11}
$$

As usual, we can estimate the asymptotic variances by substituting observed proportions for cell probabilities.

Example 11.4-2   Student teachers' classroom style (continued)

In example 11.4-1 we concluded that the supervisors agreed more often than they would by chance, but the value of $\hat{K}$ we computed was an omnibus measure of agreement. Using the conditional measures K_i, we can try to localize the agreement. Suppose we think of supervisor 1 as a standard against whose judgement we wish to compare the judgement of supervisor 2. Using expression (11.4-9), we have

$$
\hat{K}_1 = 0.255, \qquad \hat{K}_2 = 0.600, \qquad \hat{K}_3 = 0.294.
$$

We see that supervisor 2 agrees most with supervisor 1 when the latter assigns students to the "democratic" category. Setting 95 % confidence bounds separately on each K_i, we get

$$
K_1 : [-0.001, 0.511],
$$

$$
K_2 : [0.329, 0.871],
$$

$$
K_3 : [0.078, 0.510]. \quad \blacksquare\blacksquare
$$

11.4.4   *Studying disagreement*

While studying agreement we have focused our attention on the diagonal cells of $I \times I$ tables. Often we already expect high agreement, and so we may turn our attention to the disagreement, as expressed by the $I(I - 1)$ off-diagonal cells. The study of disagreement is also of special interest when we compare two tables with similar levels of agreement. For example, in table 11.4-5 we present two hypo-thetical tables with the same marginal totals and the same diagonal cell probabili-ties. For both tables $p_{ii} = p_{i+}p_{+i}$ for $i = 1, 2, 3$ and $K = 0$. The difference between the tables is due to the entries in the off-diagonal cells, with table (a) having rows independent of columns and table (b) illustrating substantial departures from independence. Thus it would be useful for us to have a way to focus on these off-diagonal cells.

One method of studying disagreement utilizes the model of quasi independence for the off-diagonal cells in a manner similar to its use in the study of social mobility data in Section 5.3. The postulated model is then

$$
p_{ij} = \begin{cases} p_{ii} & i = j \\ a_i b_j & i \neq j, \end{cases} \tag{11.4-12}
$$

which resembles the model of independence when attention is directed only to the off-diagonal cells. Under this model the maximum likelihood estimates of the diagonal cells (assuming a multinomial sampling scheme) are the observed proportions, i.e., $\hat{p}_{ii} = x_{ii}/N$. To compute the MLEs $\hat{p}_{ij}$ for the off-diagonal cell probabilities we take the raw data, in the form of table 11.4-1, and replace the diagonal entries x_{ii} by zeros, adjusting the observed marginal totals appropriately. Then we apply iterative proportional fitting for the model of quasi independence to this adjusted table, assuming that the diagonal cells are structural zeros (see Chapter 5 for specific examples of these computations).

Table 11.4-5   Different Disagreement Patterns with Identical Values of K

(a)	Observer 2				(b)	Observer 2			
Observer 1	a_1	a_2	a_3	Totals	Observer 1	a_1	a_2	a_3	Totals
a_1	0.08	0.04	0.08	0.20	a_1	0.08	0.10	0.02	0.20
a_2	0.16	0.08	0.16	0.40	a_2	0.10	0.08	0.22	0.40
a_3	0.16	0.08	0.16	0.40	a_3	0.22	0.02	0.16	0.40
Totals	0.40	0.20	0.40	1.00	Totals	0.40	0.20	0.40	1.00

We can assess the disagreement in our table in terms of the departure of the observed off-diagonal proportions from the MLEs of the cell probabilities under model (11.4-12), either by looking at the fit in individual cells or by computing some kind of distance measure, like

$$\sum_{i \neq j} \frac{\left(\dfrac{x_{ij}}{N} - \hat{p}_{ij}\right)^2}{\hat{p}_{ij}}. \tag{11.4-13}$$

We use the model of quasi independence as a baseline here, just as we used the model of independence as a baseline for measuring agreement.

Example 11.4-3   Victimization survey data

As part of an ongoing methodological inquiry into the self-reporting of crimes, the Law Enforcement Assistance Administration [1972] conducted a series of surveys in San Jose, California. In table 11.4-6 we present data from one of the San Jose surveys on the cross-classification of original police descriptions of a sample of crimes versus the victims' categorization of the crimes based on recall.

Clearly, most of the observations in table 11.4-6 lie along the diagonal, and we have strong agreement between police classification and victims' recall, reflected

Table 11.4-6   Law Enforcement Assistance Administration Data

Police Classification	Victim's Recall					
	Assault	Burglary	Larceny	Robbery	Rape	Totals
Assault	33	0	0	5	1	39
Burglary	0	91	2	0	0	93
Larceny	0	12	56	0	0	68
Robbery	0	0	6	54	0	60
Rape	5	0	0	0	25	30
Totals	38	103	64	59	26	290

by our estimate of $\hat{K} = 0.86$. Given this high degree of agreement, we now turn our attention to the disagreement. As only six of the twenty off-diagonal cells here contain nonzero counts, we do not really need to use the quasi-independence model as a baseline from which to study departure: we simply look at the six cells. Some (13.2%) of the crimes that victims classify as assault, the police classify as rape; of the crimes the police classify as assault, the victims classify 12.8% as robbery and 2.6% as rape. The nature of the disagreement in these three cells reflects the social stigma attached to rape and assault, and perhaps the victims' reluctance to admit to having been raped. The remaining three nonzero off-diagonal cells illustrate the fine dividing lines between burglary and larceny and between larceny and robbery. ∎∎

11.4.5 *Agreement among more than two observers*

The measure of agreement K for two observers is essentially based on pairs of responses, e.g., if both members of a mother-father pair choose alternative a_1 on a multiple-choice question, they agree. But when three or more observers are studied, several alternative definitions of agreement can be used. One option is to define agreement as a pairwise occurrence, so that a measure of overall agreement becomes an average of a set of pairwise agreements. Fleiss [1971] suggests this strategy, presents an illustration involving six observers, and defines his agreement statistic as the mean of the $\binom{6}{2}$ or fifteen pairwise $\hat{K}$'s.

While the pairwise agreement approach is useful in some data analysis settings, it is not directly applicable to a situation where agreement among a group of observers must be stronger than a set of pairwise agreements. For example, consider six psychiatrists classifying patients by illness. If the illnesses are severe, some patients may have to be institutionalized and given special treatment. In this case, we may want to require that at least five, say, of the psychiatrists agree on the identical diagnosis. Thus, we define the agreement of interest here as a five-way agreement. If two psychiatrists diagnose a particular patient as depressive, two diagnose the same patient as having a personality disorder, and two diagnose him as being schizophrenic, there are three two-way agreements but no five-way agreements. Marx [1973] and Marx and Light [1973] consider measuring r-way agreement among m observers. A major problem in measuring agreement among several observers is determination of the baseline for norming the observed agreement; this difficulty exists however we choose to define agreement. Different baselines would seem to be appropriate for different problems; e.g., for some problems we might wish to use chance agreement as implied by the complete independence among all the observers, while for others a baseline model allowing pairwise dependencies among observers would be more appropriate. A fruitful approach here may be to tie together measures of multiple agreement with log-linear models (see Chapter 3). Lin [1974] explores this approach as well as other multidimensional agreement problems.

For a discussion of measures of disagreement among m observers in recording the presence or absence of a clinical sign, using no baselines at all, see Bennett [1972] and Armitage, Blendis, and Smyllie [1966].

12 Pseudo-Bayes Estimates of Cell Probabilities

12.1 Introduction

When analyzing contingency tables, we frequently wish to provide a table of expected cell frequencies that can be used for other purposes, such as creating standardized rates or discriminating among treatments. All too often, the observed table of counts provides an unsatisfactory table of estimated values since there are many cells and few observations per cell. Such observed tables usually contain a large number of sampling zeros, and these are troublesome because, paradoxically, some zeros are smaller than others, especially when we are computing rates. For example, it may be misleading to report both 0/5 and 0/500 as being equal to zero, since as rates they carry quite different information. A main thrust of the analyses in the National Halothane study is to distinguish such zero proportions from one another (see Bishop and Mosteller [1969]). In this sense, the observed table of counts seems too abrupt, and we want to smooth the observed counts.

The basic problem here is one of simultaneously estimating a large number of parameters (the expected cell frequencies). One way to provide such estimates is to assume an underlying parametric model for the expected frequencies, where the number of parameters in the model is typically much smaller than the total number of cells, and then to estimate the parameters. This approach, explored in Chapters 3 and 5, involves not only the estimation problem but also the problem of model selection. In this chapter we take a different approach, which, although it draws on the results of Chapters 3 and 5, does not involve the problem of model selection as such.

A striking conclusion of the asymptotic analyses in Section 12.3 is that the method we propose is clearly superior to the generally accepted practice of adding 1/2 to the count in each cell of a large, sparse table.

12.1.1 *Central formulas*

The pseudo-Bayes estimates that are derived in Sections 12.2 and 12.5 can be described succinctly by a pair of formulas. To obtain pseudo-Bayes estimates for an observed array $\{x_{ij}\}$ with $N = \sum_{i,j} x_{ij}$, the procedural steps are:

1. select a prior array of probabilities $\{\lambda_{ij}\}$, which may be based on external information or the data themselves;
2. compute the weighting factor

$$\hat{K} = \frac{N^2 - \sum_{i,j} x_{ij}^2}{\sum_{i,j} (x_{ij} - N\lambda_{ij})^2} \qquad (12.1\text{-}1)$$

using the computing form of expression (12.2-26) of this chapter;
3. compute the cell estimates

$$m_{ij}^* = N p_{ij}^* = \frac{N}{N + \hat{K}}(x_{ij} + \hat{K}\lambda_{ij}). \tag{12.1-2}$$

Example 12.1-1   Supervisors rating student teachers

In Chapter 11 we considered the following data of Gross [1971] on the agreement of two supervisors asked independently to rate 72 student teachers' classroom style:

Supervisor 1	Supervisor 2			
	Authoritarian	Democratic	Permissive	Totals
Authoritarian	17	4	8	29
Democratic	5	12	0	17
Permissive	10	3	13	26
Totals	32	19	21	72

Suppose we select as our estimate of the cell probabilities the values of λ_{ij} given by

$$
\begin{array}{ccc}
3/15 & 1/15 & 1/15 \\
1/15 & 3/15 & 1/15 \\
1/15 & 1/15 & 3/15
\end{array}
$$

These values reflect the fact that the data table is about agreement, and we expect clumping on the main diagonal.

The weighting factor $\hat{K}$, computed using (12.2-26), is

$$\hat{K} = \frac{5{,}184 - 816}{78.7199} = 55.49.$$

The resulting pseudo-Bayes estimates of the cell frequencies, computed using (12.1-2), are

15.9	4.4	6.6	26.9
4.9	13.0	2.1	20.0
7.7	3.8	13.6	25.1
28.5	21.2	22.3	72.0

The major change in the table from the original is that the zero has been changed to 2.1. ∎∎

At this point, the nonmathematical reader may wish to skip to the more detailed examples of Sections 12.5 and 12.6, which illustrate the computation and use of pseudo-Bayes cell estimates. Following this, he can return to our technical derivation and discussion of the properties of pseudo-Bayes estimators.

12.1.2 *Historical background*

We refer to the class of estimators proposed here as pseudo-Bayes estimators because their form is similar to that of the standard Bayes estimators discussed by Good [1965, 1967] except that the parameters in the prior distribution are replaced by "estimates" based on the data themselves. Other names that have been used to describe our approach are "semi-Bayes" and "bootstrap-Bayes." There is also an analogy here to the empirical Bayes approach of Robbins [1955]. In Section 12.2, we show how our estimators might arise from the use of an argument due to Good [1967] and a two-stage prior distribution which differs markedly from the one-stage approach in Altham [1969, 1971], Bloch and Watson [1967], and Lindley [1964].

We view the ideas on pseudo-Bayes estimators presented here as being in the spirit of recent work by several authors, and we describe some of this work briefly. The basic idea is to assume that the parameters we wish to estimate can be grouped in some way, and that we can improve their estimation by the use of additional parameters underlying the grouping. For contingency tables, cells naturally group together to form marginal totals, for example, and various functions of the marginal totals may be used to strengthen estimators of the entire array $\mathbf{p}$ of cell probabilities. This idea is pursued in detail in Sections 12.2, 12.5, and 12.6.

Stein and others have studied the simultaneous estimation of the parameters composing the mean vector of the multivariate normal distribution, where the number t of dimensions is at least 3. Given $\mathbf{X} \sim N(\boldsymbol{\theta}, I)$, the MLE for $\boldsymbol{\theta} = (\theta_1, \ldots, \theta_t)'$ is $\hat{\boldsymbol{\theta}} = \mathbf{X}$, and under the squared-error loss function for the general estimator $\mathbf{T}$,

$$L(\boldsymbol{\theta}, \mathbf{T}) = \|\mathbf{T} - \boldsymbol{\theta}\|^2 = \sum_{i=1}^{t} (T_i - \theta_i)^2, \qquad (12.1\text{-}3)$$

$\hat{\boldsymbol{\theta}}$ has expected loss (risk) equal to t. James and Stein [1961] have shown that an estimator

$$\boldsymbol{\theta}^*(\mathbf{X}) = \left(1 - \frac{t-2}{\mathbf{X}'\mathbf{X}}\right)\mathbf{X} \qquad (12.1\text{-}4)$$

has smaller risk than $\hat{\boldsymbol{\theta}}$, uniformly in $\boldsymbol{\theta}$. Lindley [1962], in his discussion of a paper by Stein [1962], used a Bayesian argument to justify Stein's estimator. Similar arguments can be found in Efron and Morris [1971, 1972] and in Zellner and Vandaele [1974]. Basically, these arguments assume a common univariate prior distribution for the components of $\boldsymbol{\theta}$, and this grouping of parameters (one big group) yields a strengthened estimator of $\boldsymbol{\theta}$.

Cornfield [1970] was concerned with smoothing data from higher-order cross-classifications, and he wished to apply the Lindley–Stein results. He converted his data into mortality rates and then performed an analysis of variance on the rates, using the Lindley–Stein adjustments to improve upon the estimators for each of the ANOVA components.

Lindley [1971] and Lindley and Smith [1972] have extended this approach for the multivariate normal problem using multistage groupings and a multistage Bayesian analysis. Empirical evidence supporting Lindley's work is given by Novick, et al. [1971], and Novick [1970] refers to this as a "Bayesian Model II" (variance-components) method. Novick, Lewis, and Jackson [1973] and Leonard

[1972] have applied the method to the estimation of binomial proportions, following arc sine and logit transformations of the proportions. The multistage Bayes approach is analytically feasible in the multivariate normal problem because of the linearity of the normal models and the fact that the normal distribution is its own natural conjugate. These properties do not carry over to the multinomial problem, and so Lindley's method is not directly applicable.

Dickey [1968a, 1968b, 1969] has proposed some smoothed estimators of multinomial cell probabilities which, while different from the estimators proposed here, have the property that the data themselves are used to estimate some unknown prior parameters, which are related to weak stationarity conditions he introduces.

12.1.3   *Organization of chapter*

In the next section we introduce our class of pseudo-Bayes estimators for the multinomial probability vector, using both a geometric argument of Stein [1962] and a Bayesian argument. The presentation closely follows that of Fienberg and Holland [1970, 1973].

In Section 12.3 we give two kinds of asymptotic results that help clarify the type and extent of the improvement in the estimation of **p** that is possible by use of the biased estimator developed above. In Section 12.4 we give some exact comparisons of the risk functions of our proposed pseudo-Bayes estimators and the usual maximum likelihood estimators, in small samples and small dimensions. In Section 12.5 we consider some pseudo-Bayes estimators which also rely on the idea of gaining strength in estimation from grouping parameters.

In Section 12.6 we illustrate the use of the pseudo-Bayes estimators on the pair of two-way social mobility tables considered elsewhere in this book. Finally, in Section 12.7 we discuss some further development of pseudo-Bayes estimators, and we advise the reader which pseudo-Bayes estimators to use for particular sets of data.

12.2   Bayes and Pseudo-Bayes Estimators

In this section we develop Bayes and pseudo-Bayes estimators for the vector of cell probabilities of a multinomially distributed random vector (or set of counts). After setting up our notation, we give the Bayes estimators of **p** under both Dirichlet and compound Dirichlet (two-stage) prior distributions. We then discuss the risk function of estimators of **p** and introduce an easily computed class of pseudo-Bayes estimators. Finally, we illustrate the various estimators discussed in this section on the numerical example given in example 12.1-1. Throughout this chapter we make explicit the distinction between a random vector **X** and a particular value **x** it may take on.

12.2.1   *Bayes estimators*

Let $\mathbf{X} = (X_1, \ldots, X_t)$ have the multinomial distribution with parameters $N = \sum_{i=1}^{t} X_i$ and $\mathbf{p} = (p_1, \ldots, p_t)$. We observe a vector of values $\mathbf{x} = (x_1, \ldots, x_t)$, where x_i is the observed count in the ith category and $\sum_{i=1}^{t} x_i = N$. The vector **p** takes values in the parameter space $\mathscr{S}_t$, where

$$\mathscr{S}_t = \left\{ \mathbf{p} = (p_1, \ldots, p_t) : p_i \geqq 0 \quad \text{and} \quad \sum_{i=1}^{t} p_i = 1 \right\}, \tag{12.2-1}$$

and we denote the "center" of $\mathscr{S}_t$ by $\mathbf{c} = (t^{-1}, \ldots, t^{-1})$.

The kernel of the likelihood function for this multinomial distribution is

$$l(\mathbf{p}|\mathbf{x}) = l(p_1, \ldots, p_t|x_1, \ldots, x_t) = \prod_{i=1}^{t} p_i^{x_i}. \tag{12.2-2}$$

Dirichlet prior distribution

The natural conjugate family of prior distributions for this likelihood is the Dirichlet, whose densities have the form

$$f(\mathbf{p}|\boldsymbol{\beta}) = \Gamma\left(\sum_{i=1}^{t} \beta_i\right) \prod_{i=1}^{t} \frac{p_i^{\beta_i - 1}}{\Gamma(\beta_i)}, \tag{12.2-3}$$

where $\beta_i > 0$ for all i and $\Gamma(y)$ is the gamma function given by $\Gamma(y) = \int_0^\infty e^{-z} z^{y-1} \, dz$. When the prior distribution is Dirichlet with parameters $\boldsymbol{\beta} = (\beta_1, \ldots, \beta_t)$, the posterior distribution is also Dirichlet with parameters $\boldsymbol{\beta} + \mathbf{x} = (\beta_1 + x_1, \ldots, \beta_t + x_t)$.

If we let

$$D(\beta_1, \beta_2, \ldots, \beta_t) = \frac{\Gamma\left(\sum_{i=1}^{t} \beta_i\right)}{\prod_{i=1}^{t} \Gamma(\beta_i)}, \tag{12.2-4}$$

then the moments of the Dirichlet distribution are given by

$$E\left(\prod_{i=1}^{t} p_i^{a_i} \bigg| \boldsymbol{\beta}\right) = \frac{D(\beta_1, \ldots, \beta_t)}{D(\beta_1 + a_1, \ldots, \beta_t + a_t)} \tag{12.2-5}$$

(see, for example, Wilks [1962] or Watson [1965]). If we set

$$K = \sum_{i=1}^{t} \beta_i, \qquad \lambda_i = \frac{\beta_i}{K} \tag{12.2-6}$$

we see that the prior and posterior means of p_i are given by

$$E(p_i|K, \lambda) = \lambda_i \quad \text{(prior mean)}, \tag{12.2-7}$$

$$E(p_i|K, \lambda, \mathbf{x}) = \frac{x_i + K\lambda_i}{N + K} \quad \text{(posterior mean)}. \tag{12.2-8}$$

We can rewrite (12.2-8) in vector notation as

$$E(\mathbf{p}|K, \lambda, \mathbf{x}) = \frac{N}{N + K}(\mathbf{x}/N) + \frac{K}{N + K}\lambda. \tag{12.2-9}$$

For a geometric interpretation of K when $\lambda = \mathbf{c}$, see Fienberg and Holland [1972].

The posterior mean is a Bayesian point estimate of $\mathbf{p}$. When the prior distribution is Dirichlet with parameters K and λ, this Bayesian point estimate is given by (12.2-9).

Compound Dirichlet prior distributions (two-stage Bayesian models)

A wider class of prior distributions for $\mathbf{p}$ is given by a two-stage Bayesian model used by Good [1967]. As before, the distribution of $\mathbf{X}$ given $\mathbf{p}$ is multinomial with parameters N and $\mathbf{p}$, and that of $\mathbf{p}$ given (K, λ) is Dirichlet; however, we now let (K, λ) have a prior distribution with density function $\varphi(K, \lambda)$. This two-stage approach results in a prior for $\mathbf{p}$ that is a mixture of Dirichlet distributions (a compound Dirichlet distribution). If $\varphi(K, \lambda)$ is a degenerate prior distribution concentrated on a single value of (K, λ), then the compound Dirichlet distribution reduces to the ordinary Dirichlet distribution discussed earlier.

Following an argument given in Good [1967], we may show that the posterior mean of $\mathbf{p}$ with a compound Dirichlet prior is

$$E(\mathbf{p}|\varphi, \mathbf{x}) = w(\mathbf{x})\hat{\mathbf{p}} + (1 - w(\mathbf{x}))\lambda(\mathbf{x}), \tag{12.2-10}$$

where

$$\hat{\mathbf{p}} = \mathbf{x}/N,$$

$$w(\mathbf{x}) = \frac{N}{N + K(\mathbf{x})}, \tag{12.2-11}$$

$$K(\mathbf{x}) = \frac{\int \dfrac{K}{N + K} H(\mathbf{x}, K, \lambda)\varphi(K, \lambda) \, dK \, d\lambda}{\int \dfrac{1}{N + K} H(\mathbf{x}, K, \lambda)\varphi(K, \lambda) \, dK \, d\lambda}, \tag{12.2-12}$$

$$\lambda_i(\mathbf{x}) = \frac{\int \lambda_i \dfrac{K}{N + K} H(\mathbf{x}, K, \lambda)\varphi(K, \lambda) \, dK \, d\lambda}{\int \dfrac{K}{N + K} H(\mathbf{x}, K, \lambda)\varphi(K, \lambda) \, dK \, d\lambda}, \tag{12.2-13}$$

and $H(\mathbf{x}, K, \lambda)$ is the Bayes factor

$$H(\mathbf{x}, K, \lambda) = \frac{\Gamma(K)}{\Gamma(N + K)} \prod_i \frac{\Gamma(x_i + K\lambda_i)}{\Gamma(K\lambda_i)}. \tag{12.2-14}$$

12.2.2 *The risk criterion*

We shall adopt the expected value of the squared distance from an estimator $\mathbf{T}$ to $\mathbf{p}$ as our risk criterion for comparing different choices of $\mathbf{T}$, i.e., our risk function is

$$R(\mathbf{T}, \mathbf{p}) = NE\|\mathbf{T} - \mathbf{p}\|^2 = N \sum_{i=1}^{t} E(T_i - p_i)^2. \tag{12.2-15}$$

The risk function of the usual estimator

If we consider the usual estimator of $\mathbf{p}$ given by

$$\hat{\mathbf{p}} = \frac{\mathbf{X}}{N}, \tag{12.2-16}$$

then we can easily calculate its risk function $R(\hat{\mathbf{p}}, \mathbf{p})$:

$$R(\hat{\mathbf{p}}, \mathbf{p}) = 1 - \|\mathbf{p}\|^2 = 1 - \sum_{i=1}^{t} p_i^2. \tag{12.2-17}$$

Each coordinate $\hat{p}_i$ of $\hat{\mathbf{p}}$ is well-known to be the unique minimum variance unbiased estimate of p_i. From this fact it follows that if $\mathbf{T}$ is any unbiased estimator of $\mathbf{p}$ (i.e., $E(\mathbf{T}) = \mathbf{p}$), then the risk function of $\mathbf{T}$ is never smaller than the risk function of $\hat{\mathbf{p}}$ for any $\mathbf{p} \in \mathscr{S}_t$. Thus no improvement over $\hat{\mathbf{p}}$ can be achieved unless we leave the class of unbiased estimators. Furthermore, Johnson [1971] has shown that $\hat{\mathbf{p}}$ is an admissible estimator of $\mathbf{p}$ with respect to the risk function (12.2-15), so that there exists no biased estimator whose risk function is uniformly smaller than that of $\hat{\mathbf{p}}$. Nevertheless, as Johnson points out, the reason $\hat{\mathbf{p}}$ is admissible is not because it has small risk everywhere; rather, it is due to the smallness of its risk on the boundary of the parameter space. The risk of $\hat{\mathbf{p}}$ is smallest when $\mathbf{p}$ has one component near unity. Hence we would expect to be able to improve on $\hat{\mathbf{p}}$ for those values of $\mathbf{p}$ that are not so extreme. We show in Sections 12.3 and 12.4 that pseudo-Bayes estimators provide a substantial improvement over $\hat{\mathbf{p}}$ away from the boundary, and that the region of improvement in the parameter space increases as t becomes large.

The risk function of some Bayes estimators

We denote the random variable version of the Bayes estimator given in (12.2-9) by

$$\hat{\mathbf{q}} = \hat{\mathbf{q}}(K, \lambda) = \frac{N}{N + K}(\mathbf{X}/N) + \frac{K}{N + K}\lambda. \qquad (12.2\text{-}18)$$

Since K and λ are constants, we can easily compute the risk function of $\hat{\mathbf{q}}$:

$$R(\hat{\mathbf{q}}, \mathbf{p}) = \left(\frac{N}{N + K}\right)^2 (1 - \|\mathbf{p}\|^2) + \left(\frac{K}{N + K}\right)^2 N \|\mathbf{p} - \lambda\|^2. \qquad (12.2\text{-}19)$$

Various choices of K in (12.2-18) have appeared in the literature (e.g., see the list in Fienberg and Holland [1972]). Two examples are

$$K = \tfrac{1}{2}t, \qquad (12.2\text{-}20)$$

$$K = \sqrt{N}. \qquad (12.2\text{-}21)$$

When $\lambda = \mathbf{c}$, (12.2-20) corresponds to adding a fake count of $\tfrac{1}{2}$ to each cell. This method is often used to eliminate zeros in contingency tables. The risk function of $\hat{\mathbf{q}}(\tfrac{1}{2}t, \mathbf{c})$ is obtained by substituting the appropriate values into equation (12.2-19). This yields

$$R(\hat{\mathbf{q}}(\tfrac{1}{2}t, \mathbf{c}), \mathbf{p}) = \left(\frac{2\delta}{2\delta + 1}\right)^2 (1 - \|\mathbf{p}\|^2) + \left(\frac{1}{2\delta + 1}\right)^2 N\left(\|\mathbf{p}\|^2 - \frac{1}{t}\right), \qquad (12.2\text{-}21a)$$

where $\delta = N/t$. The choice $K = \sqrt{N}$ given in (12.2-21) yields the unique, constant risk, minimax estimator of $\mathbf{p}$ (Trybula, [1958]) when $\lambda = \mathbf{c}$. We denote the minimax estimator by $\mathbf{p}_M$, i.e.,

$$\mathbf{p}_M = \hat{\mathbf{q}}(\sqrt{N}, \mathbf{c}). \qquad (12.2\text{-}22)$$

The constant risk of $\mathbf{p}_M$ is

$$R(\mathbf{p}_M, \mathbf{p}) = \left(\frac{\sqrt{N}}{1 + \sqrt{N}}\right)^2 (1 - t^{-1}). \qquad (12.2\text{-}23)$$

12.2.3   *Pseudo-Bayes estimator of* **p**

In order to use the Bayesian estimator $\hat{\mathbf{q}}(K, \lambda)$ given in (12.2-18), we need to know the values of K and λ. Typically, the assessment of these prior parameters is a difficult task. In this section we discuss a way of choosing K so that it depends on the data and the choice of λ.

If λ is regarded as fixed, then we can find the value of K that minimizes the risk $R(\hat{\mathbf{q}}(K, \lambda), \mathbf{p})$ by differentiating (12.2-19) in K and solving the resulting equation. This yields

$$K = K(\mathbf{p}, \lambda) = \frac{1 - \|\mathbf{p}\|^2}{\|\mathbf{p} - \lambda\|^2}. \tag{12.2-24}$$

This optimal value of K depends on the unknown value of **p**. We may obtain an estimate of this unknown optimal value of K by replacing **p** by $\hat{\mathbf{p}} = \mathbf{X}/N$, yielding

$$\hat{K} = K(\hat{\mathbf{p}}, \lambda) = \frac{1 - \|\hat{\mathbf{p}}\|^2}{\|\hat{\mathbf{p}} - \lambda\|^2}, \tag{12.2-25}$$

or, in terms of **x**, the observed value of the random variable **X**,

$$\hat{K} = \frac{N^2 - \sum_{i=1}^{t} x_i^2}{\sum_{i=1}^{t} x_i^2 - 2N \sum_{i=1}^{t} x_i \lambda_i + N^2 \sum_{i=1}^{t} \lambda_i^2}. \tag{12.2-26}$$

A pseudo-Bayes estimator of **p** is then

$$\mathbf{p}^* = \hat{\mathbf{q}}(\hat{K}, \lambda) = \left(\frac{N}{N + \hat{K}}\right)\hat{\mathbf{p}} + \frac{\hat{K}}{N + \hat{K}}\lambda, \tag{12.2-27}$$

where $\hat{K}$ is given in (12.2-26). Other pseudo-Bayes estimators of **p** are possible, and they correspond to alternative ways of estimating the optimal value of K (see Section 12.7).

When $\lambda = \mathbf{c}$, (12.2-26) may be written as

$$\hat{K} = \frac{N^2 - \sum x_i^2}{\sum x_i^2 - \frac{N^2}{t}}. \tag{12.2-28}$$

Because $\hat{K}$ is a function of the data **x**, it is not an easy matter to calculate the risk function of $\mathbf{p}^*$ and to compare it with the risk of $\hat{\mathbf{p}}$ and the other estimators we have mentioned so far. In Section 12.3, we give some asymptotic approximations to the risk function of $\mathbf{p}^*$ and in Section 12.4 we show some exact comparisons of the risk functions for small samples.

12.2.4   *Geometric motivation for the use of Bayes and pseudo-Bayes estimators*

All of the estimators discussed in this section ((12.2-9), (12.2-10), (12.2-16), (12.2-18), (12.2-22), and (12.2-27)) share a common feature. They are all weighted averages of $\hat{\mathbf{p}}$ and some value λ, i.e.,

$$\mathbf{T} = w\hat{\mathbf{p}} + (1 - w)\lambda \quad \text{for some } 0 \leq w \leq 1. \tag{12.2-29}$$

To provide additional motivation for the estimators of the form (12.2-29), we make use of a heuristic geometric argument adapted from one originally given by Stein [1962]. As above, we let λ denote a fixed vector of probabilities, i.e., $\lambda \in \mathscr{S}_t$. We let λ be our choice of "origin" within the parameter space; often we take $\lambda = \mathbf{c}$, but in this development we let λ denote a general choice of the origin. Now consider the triangle whose vertices are the three vectors λ, $\mathbf{p}$, and $\hat{\mathbf{p}}$. Let θ denote the angle between the vectors $\hat{\mathbf{p}} - \mathbf{p}$ and $\mathbf{p} - \lambda$. The inner product of these two vectors has expected value equal to zero, so that on the average, θ is a right angle. Also, if $\mathbf{p}$ is constrained to lie away from the boundary of $\mathscr{S}_t$, then Sutherland [1974] shows that

$$E(\cos^2 \theta) = O(t^{-1}) \quad \text{as } t \to \infty. \tag{12.2-30}$$

Thus for large t, θ is very nearly a right angle (with high probability), and hence figure 12.2-1 is representative of the typical relationship between λ, $\mathbf{p}$, and $\hat{\mathbf{p}}$.

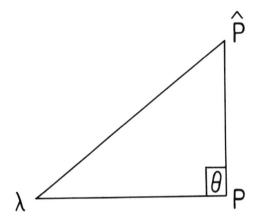

Figure 12.2-1   Geometric relationship among $\mathbf{p}, \hat{\mathbf{p}}$, and λ.

From figure 12.2-1 we see that there are points along the line connecting $\hat{\mathbf{p}}$ and λ that are closer to $\mathbf{p}$ than is $\hat{\mathbf{p}}$. This fact leads us to consider estimates of $\mathbf{p}$ that are formed by shrinking $\hat{\mathbf{p}}$ towards the origin λ. Any point along the line connecting $\hat{\mathbf{p}}$ and λ may be represented as in (12.2-29). Thus we see that both Bayes and pseudo-Bayes estimators can provide improvements over $\hat{\mathbf{p}}$ as estimators of $\mathbf{p}$, especially when the dimension t is large. In Sections 12.3 and 12.4 we investigate the nature and extent of these improvements.

Table 12.2-1   The Value of Three Different Choices of K (for the Data Described in Example 12.1-1)

Choice of K	K	$\dfrac{N}{N + K}$
$\frac{1}{2}t$	4.50	0.941
$\sqrt{N}$	8.49	0.895
$\hat{K}$ from (12.2–28)	18.20	0.798

12.2.5 *A comparison of some choices of K in an example*

Example 12.2-1   Supervisors rating student teachers (continued)
For the data described in example 12.1-1 we have

$$N = 72, \qquad t = 3 \times 3 = 9.$$

Suppose we choose $\lambda = \mathbf{c} = (\frac{1}{9}, \ldots, \frac{1}{9})$. Then table 12.2-1 gives the values of three different choices of K, as well as the weight the corresponding estimators give to $\hat{\mathbf{p}}$. ∎∎

12.3   Asymptotic Results for Pseudo-Bayes Estimators

We are unable to write a closed-form expression for the risk function of the pseudo-Bayes estimator $\mathbf{p}^*$. By the use of asymptotic approximations, however, we can compare the risk of $\mathbf{p}^*$ with the risk of the unrestricted maximum likelihood estimator $\hat{\mathbf{p}} = \mathbf{X}/N$.

The usual asymptotic approach to multinomial problems (such as that taken in Chapter 14) holds the dimension t fixed and lets the sample size N tend to infinity, and we consider this technique in Section 12.3.2.

A different approach lets both N and t tend to infinity, but at the same rate so that N/t remains constant. One reason for looking at this special type of asymptotics comes from practical considerations. Typically, multinomial data arrive in the form of a cross-classification of discrete variables. In many situations there are a large number of variables which can be used to cross-classify each observation, and if all of these variables are used the data would be spread too thinly over the cells in the resulting multidimensional contingency table. Thus if the investigator uses a subset of the variables to keep the average number of observations from becoming too small, he is in effect choosing t so that N/t is moderate. If N and t are both large, then he is in the special type of asymptotic situation described in detail in Section 12.3.1. In another context, Roscoe and Byars [1971] also advocate describing multinomial variables in terms of N/t.

12.3.1   *Special asymptotics for sparse multinomials*

Under our special asymptotic setup, which lets the dimension t of the multinomial tend to infinity at the same rate as the sample size N, we show below that the risk of the pseudo-Bayes estimator $\mathbf{p}^*$ is uniformly smaller than the risk of the usual estimator $\hat{\mathbf{p}}$. In addition, the risk of $\mathbf{p}^*$ is uniformly smaller than the risk of the estimator formed by adding 1/2 to each cell count.

A simplifying device
The asymptotic setup that describes a sparse multinomial distribution lets $t \to \infty$ and sets $N = \delta t$, where $\delta = N/t$ is a constant. The dimension t varies in this asymptotic setup so that the parameter space, $\mathscr{S}_t$, also varies with t. Instead of having a single fixed probability vector, we must consider an infinite sequence of probability vectors whose dimensions increase without bound. This type of asymptotic setup has been treated before, e.g., by Morris [1969]. We choose to simplify the structure of this situation by relating the elements of this sequence of probability vectors through the following device. Let $p(\cdot)$ denote a probability

density function on $[0, 1]$. For each value of t, we let

$$p_i = \frac{1}{t} p\left(\frac{i - \frac{1}{2}}{t}\right) \quad i = 1, \ldots, t. \tag{12.3-1}$$

Strictly speaking, p_i defined in (12.3-1) should depend explicitly on t (i.e., should be written $p_{i,t}$), but this excessive notation will not be used here. Furthermore, the vector $\mathbf{p} = (p_1, \ldots, p_t)$ defined from (12.3-1) is not necessarily an element of $\mathscr{S}_t$, since $\sum_{i=1}^{t} p_i$ need not be unity. However, if $p(\cdot)$ is sufficiently smooth (for example, if $p(\cdot)$ has a continuous second derivative), then standard results in numerical integration (Davis and Rabinowitz [1967]) show that

$$\sum_{i=1}^{t} p_i = \sum_{i=1}^{t} \frac{1}{t} p\left(\frac{i - \frac{1}{2}}{t}\right) = \int_0^1 p(x) \, dx + o(t^{-1}) = 1 + o(t^{-1}). \tag{12.3-2}$$

Hence $\sum_{i=1}^{t} p_i$ converges to unity as $t \to \infty$ with an error that is small compared with t^{-1}. As will become evident, this is sufficient for our purposes.

Let $\lambda(\cdot)$ be a second probability density on $[0, 1]$, and set

$$\lambda_i = \frac{1}{t} \lambda\left(\frac{i - \frac{1}{2}}{t}\right) \quad i = 1, \ldots, t. \tag{12.3-3}$$

By assuming that both $p(\cdot)$ and $\lambda(\cdot)$ have continuous second derivatives, for all $\alpha, \beta \geq 0$ we have

$$\sum_{i=1}^{t} p_i^\alpha \lambda_i^\beta = \left(\frac{1}{t}\right)^{\alpha + \beta - 1} \int_0^1 p^\alpha(x) \lambda^\beta(x) \, dx + o(t^{-\alpha - \beta}). \tag{12.3-4}$$

Thus we can replace summations involving p_i and λ_i by integrals involving $p(\cdot)$ and $\lambda(\cdot)$.

Some illustrations of asymptotic formulas

Using the device described in (12.3-1) to (12.3-4) we express the risk for $\hat{\mathbf{p}}$ as follows:

$$R(\hat{\mathbf{p}}, \mathbf{p}) = \sum_{i=1}^{t} p_i - \sum_{i=1}^{t} p_i^2$$

$$= \int_0^1 p(x) \, dx + o(t^{-1}) - \frac{1}{t} \int_0^1 p^2(x) \, dx + o(t^{-2}), \tag{12.3-5}$$

or

$$R(\hat{\mathbf{p}}, \mathbf{p}) = 1 - \frac{1}{t} \int p^2 + o(t^{-1}). \tag{12.3-6}$$

Henceforth, as in (12.3-6), we omit references to x, dx, and the limits of integration in all expressions using integrals.

Equation (12.3-6) is prototypic of our expansions of risk functions in this chapter. The corresponding expansion for the risk function (12.2-19) of $\hat{\mathbf{q}}(K, \lambda)$ depends on how K behaves as a function of t. If $K = \frac{1}{2}t$ as in (12.2-20), then w is constant, and we can express (12.2-17) as

$$R(\hat{\mathbf{q}}(\tfrac{1}{2}t, \lambda), \mathbf{p}) = w_0^2 + (1 - w_0)^2 D - \frac{1}{t} w_0^2 \int p^2 + o(t^{-1}), \tag{12.3-7}$$

where w_0 and D are given by

$$w_0 = \frac{\delta}{\delta + \frac{1}{2}}, \quad D = \delta \int (\lambda - p)^2, \tag{12.3-8}$$

and $\delta = N/t$. The reader should note that as p and λ vary over all possible density functions on $[0, 1]$, D varies from 0 to ∞. We can also express the risk of the minimax estimator $\mathbf{p}_M$ given by (12.2-23) as:

$$R(\mathbf{p}_M, \mathbf{p}) = 1 - \frac{2}{\sqrt{N}} + \frac{1}{t}\left(\frac{3}{\delta} - 1\right) + o(t^{-1}). \tag{12.3-9}$$

Comparison of some estimators

For fixed $p(\cdot)$ and δ, the three expansions (12.3-6), (12.3-7), and (12.3-9) give the risk functions of three estimators of $\mathbf{p}$ out to order t^{-1}. We propose to compare estimators of $\mathbf{p}$ on the basis of the leading term of these expansions. For example, from (12.3-6) and (12.3-9) it follows that for each fixed choice of $p(\cdot)$ and δ, the risk of the minimax estimator is smaller than that of $\hat{\mathbf{p}}$ if t is large enough. The estimator $\hat{\mathbf{q}}(\frac{1}{2}t, \lambda)$ does not possess this property. If D is large enough, the risk of $\hat{\mathbf{q}}(\frac{1}{2}t, \lambda)$ exceeds that of $\hat{\mathbf{p}}$ for large t. The implication here is that if D is large, $K = \frac{1}{2}t$ is too big and $\sqrt{N}$ is preferable.

As stated above, our main purpose in developing the asymptotics of sparse multinomial distributions is to approximate the risk function of $\mathbf{p}^*$. Our presentation of (12.3-6), (12.3-7), and (12.3-9) serves mainly to introduce this type of approximation in situations where we know the functions being approximated. These expressions may also be compared to our expansion of the risk of $\mathbf{p}^*$, which we now give. The derivation here is considerably more difficult, and we give the results without proof. Holland and Sutherland [1971a] have shown that, to order t^{-1}, the risk function of the pseudo-Bayes estimator $\mathbf{p}^*$ is

$$R(\mathbf{p}^*, \mathbf{p}) = S_0(D) + \frac{1}{N}S_1(D) + \frac{1}{t}\left(\int p^2\right)S_2(D) + \frac{\delta}{t}S_3(D)\sigma_{p-\lambda}^2 + o(t^{-1}), \tag{12.3-10}$$

where

$$S_0(D) = \frac{D^2 + 3D + 1}{(D + 2)^2}, \quad S_1(D) = \frac{2(D + 1)}{(D + 2)^3}, \tag{12.3-11}$$

$$S_2(D) = \frac{-(D^4 + 6D^3 + 7D^2 - 6D - 2)}{(D + 2)^4}, \quad S_3(D) = \frac{4(D^2 + 3D - 1)}{(D + 2)^4},$$

$$\sigma_{p-\lambda}^2 = \int (p - \lambda - \mu_{p-\lambda})^2 p, \tag{12.3-12}$$

$$\mu_{p-\lambda} = \int (p - \lambda)p, \tag{12.3-13}$$

and D is defined in (12.3-8).

For comparison we collect all of the leading terms of the risk function expansions given in this section in table 12.3-1.

It is evident that $S_0(D)$ in (12.3-11) satisfies the inequality

$$\tfrac{1}{4} \leqq S_0(D) < 1 \tag{12.3-14}$$

for all $D \geq 0$. Hence this first-order analysis shows that $\mathbf{p}^*$ has a risk function whose leading term is uniformly smaller than that of $\hat{\mathbf{p}}$ for all $p(\cdot)$. This implies that for all fixed $p(\cdot)$ and δ, if t is large enough, then the risk of $\mathbf{p}^*$ is less than that of $\hat{\mathbf{p}}$. In this sense, $\mathbf{p}^*$ is an improvement over $\hat{\mathbf{p}}$ for a large portion of the parameter space.

Table 12.3-1 Leading Term in Expansion of Risk Function for Four Estimators of $\mathbf{p}$

Estimator	Expression Number	Leading Term
$\hat{\mathbf{p}}$	(12.2–16)	1
$\hat{\mathbf{p}}_M = \hat{\mathbf{q}}(\sqrt{N}, \mathbf{c})$	(12.2–18), (12.2–21)	$1 - 2/\sqrt{N}$
$\hat{\mathbf{q}}(\frac{1}{2}t, \lambda)$	(12.2–18), (12.2–20)	$w_0^2 + (1 - w_0)^2 D$
$\mathbf{p}^*$	(12.2–26), (12.2–27)	$(D^2 + 3D + 1)/(D + 2)^2$

In figure 12.3-1 we graph the leading terms for the various estimators given in table 12.3-1 for $\delta = 5$ and $N = 100$. Here $\mathbf{p}^*$, our pseudo-Bayes estimator, has smaller risk than $\hat{\mathbf{q}}(\frac{1}{2}t, \lambda)$, the estimator formed by adding $1/2$ to each cell. Thus the use of $\mathbf{p}^*$ is preferable to the common practice of adding $1/2$ to each cell in large sparse multinomials.

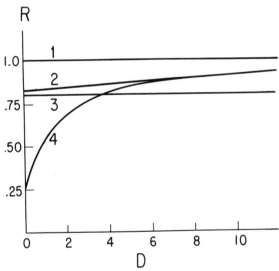

Figure 12.3-1   Leading terms of risk functions ($\delta = 5, N = 100$) for four estimators of $\mathbf{p}$. $1 = \hat{\mathbf{p}}$, $2 = \hat{\mathbf{q}}(\frac{1}{2}t, \lambda), 3 = \hat{\mathbf{q}}(\sqrt{N}, \mathbf{c}) = \mathbf{p}_M, 4 = \mathbf{p}^*$. Reproduced from Fienberg and Holland [1973].

12.3.2   Standard asymptotics for multinomials

We now turn to more standard asymptotic theory, where we keep the dimension of the parameter space fixed and let the sample size tend to infinity. This type of asymptotics helps us complete the picture of the behavior of pseudo-Bayes estimators in situations where there is a large amount of data per cell.

The results of this section indicate that for large samples (with fixed dimensionality), the pseudo-Bayes estimator $\mathbf{p}^*$ is equivalent to the usual estimator $\hat{\mathbf{p}}$, except when $\mathbf{p} = \boldsymbol{\lambda}$. In the latter situation, we form $\mathbf{p}^*$ by shrinking $\hat{\mathbf{p}}$ to the true value of $\mathbf{p}$, and so $\mathbf{p}^*$ has smaller variability about $\mathbf{p}$ than does $\hat{\mathbf{p}}$.

Notation

Here we regard t as fixed, and study $\mathbf{p}^*$ and $\hat{\mathbf{p}}$ as $N \to \infty$. In order to allow for the situation described in Section 12.5, we modify our notation for $\boldsymbol{\lambda}$ somewhat. We let $\boldsymbol{\lambda}$ depend on $\mathbf{X}$ and denote this by $\hat{\boldsymbol{\lambda}} = \boldsymbol{\lambda}(\mathbf{X})$. Furthermore, we assume that there is a function $\boldsymbol{\lambda}^* = \boldsymbol{\lambda}^*(\mathbf{p})$ of $\mathbf{p}$ such that $\sqrt{N}(\hat{\boldsymbol{\lambda}} - \boldsymbol{\lambda}^*)$ has an asymptotic, possibly degenerate, multivariate normal distribution with mean zero as $N \to \infty$. In Section 12.5 the choice of $\hat{\boldsymbol{\lambda}}$ and $\boldsymbol{\lambda}^*$ that receives the most attention is

$$\hat{\lambda}_{ij} = \frac{X_{i+} X_{+j}}{N^2},$$

$$\lambda^*_{ij}(\mathbf{p}) = p_{i+} p_{+j}, \qquad (12.3\text{-}15)$$

where $t = IJ$. It is well known in this case that $\sqrt{N}(\hat{\boldsymbol{\lambda}} - \boldsymbol{\lambda}^*)$ does have an asymptotic multivariate normal distribution. If $\hat{\boldsymbol{\lambda}}$ is constant (i.e., nonrandom), then $\sqrt{N}(\hat{\boldsymbol{\lambda}} - \boldsymbol{\lambda}^*) \equiv 0$, which we interpret as degenerate asymptotic normality. Thus our present notation does not conflict with our previous assumption that $\boldsymbol{\lambda}$ is nonrandom.

Some theorems

Let $\hat{w}_N = N/(N + \hat{K})$, where $\hat{K}$ is given by (12.2-26). Since $\hat{\boldsymbol{\lambda}}$ converges in probability to $\boldsymbol{\lambda}^*$, we have the following theorem, given here without proof.

THEOREM 12.3-1   *If $\mathbf{p} \neq \boldsymbol{\lambda}^*(\mathbf{p})$, then $\hat{K}$ converges in probability to $(1 - \|\mathbf{p}\|^2)/\|\boldsymbol{\lambda}^* - \mathbf{p}\|^2$, and consequently $1 - \hat{w}_N = O_p(N^{-1})$.*

Now let $\mathbf{U}_N = \sqrt{N}(\hat{\boldsymbol{\lambda}} - \boldsymbol{\lambda}^*)$, $\mathbf{V}_N = \sqrt{N}(\hat{\mathbf{p}} - \mathbf{p})$, and assume that $(\mathbf{U}_N, \mathbf{V}_N)$ converges in distribution to $(\mathbf{U}, \mathbf{V})$, which has a possibly degenerate multivariate normal distribution with zero mean. The next theorem complements theorem 12.3-1 when $\mathbf{p} = \boldsymbol{\lambda}^*$.

THEOREM 12.3-2   *If $\mathbf{p} = \boldsymbol{\lambda}^*(\mathbf{p})$, then $\hat{w}_N$ converges in distribution to the random variable*

$$w = \frac{\|\mathbf{U} - \mathbf{V}\|^2}{1 - \|\mathbf{p}\|^2 + \|\mathbf{U} - \mathbf{V}\|^2}. \qquad (12.3\text{-}16)$$

We can now use theorems 12.3-1 and 12.3-2 to show that if $\mathbf{p} \neq \boldsymbol{\lambda}^*$, then $\mathbf{p}^*$ is a consistent estimator of $\mathbf{p}$ as $N \to \infty$. The proof of this result is given in Fienberg and Holland [1970], and we give formal statement of the result in the following theorem.

THEOREM 12.3-3   (i) *If $\mathbf{p} \neq \boldsymbol{\lambda}^*$, then $\sqrt{N}(\mathbf{p}^* - \hat{\mathbf{p}}) = O_p(N^{-1})$; (ii) if $\mathbf{p} = \boldsymbol{\lambda}^*$, then $\sqrt{N}(\mathbf{p}^* - \hat{\mathbf{p}})$ converges in distribution to $(1 - w)(\mathbf{U} - \mathbf{V})$, where w is given by* (12.3-16).

COROLLARY   *The vector $\mathbf{p}^*$ is a consistent estimator of $\mathbf{p}$ as $N \to \infty$.*

Limiting values of risk ratios

From (i) of theorem 12.3-3 we see that if $\mathbf{p} \neq \boldsymbol{\lambda}^*$, then $\mathbf{p}^*$ is asymptotically equivalent to $\hat{\mathbf{p}}$ in the sense that they have the same asymptotic distributions. Hence the ratio of the risk of $\mathbf{p}^*$ to that of $\hat{\mathbf{p}}$ approaches unity as $N \to \infty$. The situation is more complicated when $\mathbf{p} = \boldsymbol{\lambda}^*$. From (ii) of the theorem it follows that if $\mathbf{p} = \boldsymbol{\lambda}^*$, then the ratio of the risk converges to the value of

$$\lim_{N \to \infty} \frac{R(\mathbf{p}^*, \mathbf{p})}{R(\hat{\mathbf{p}}, \mathbf{p})} = (1 - \|\mathbf{p}\|^2)^{-1} E \|w\mathbf{V} + (1 - w)\mathbf{U}\|^2. \qquad (12.3\text{-}17)$$

To see what (12.3-17) looks like in a special case, we set $\hat{\boldsymbol{\lambda}} \equiv \boldsymbol{\lambda}^* = \mathbf{p} = \mathbf{c}$. Now $\mathbf{U} \equiv 0$, and (12.3-17) becomes

$$\lim_{N \to \infty} \frac{R(\mathbf{p}^*, \mathbf{p})}{R(\hat{\mathbf{p}}, \mathbf{p})} = \left(1 - \frac{1}{t}\right)^{-1} E \left\| \left(\frac{\|\mathbf{V}\|^2}{\|\mathbf{V}\|^2 + 1 - \frac{1}{t}} \right)\mathbf{V} \right\|^2 \qquad (12.3\text{-}18)$$

$$= E\left[\left(\frac{t\|\mathbf{V}\|^2}{t\|\mathbf{V}\|^2 + t - 1} \right)^2 \frac{t\|\mathbf{V}\|^2}{t - 1} \right].$$

In (12.3-18), $\mathbf{V}$ has a multivariate normal distribution with mean zero and a covariance matrix whose (i, j) component is

$$\text{Cov}(V_i, V_j) = \begin{cases} -t^{-2} & \text{if } i \neq j \\ t^{-1}(1 - t^{-1}) & \text{if } i = j. \end{cases} \qquad (12.3\text{-}19)$$

Using a standard orthogonality argument, we can show that (12.3-19) implies that (12.3-18) equals

$$E\left[\left(\frac{C^2}{C^2 + t - 1} \right)^2 \frac{C^2}{t - 1} \right], \qquad (12.3\text{-}20)$$

where C^2 has a chi square distribution with $t - 1$ degrees of freedom. We may expand the expectation in (12.3-20) directly to order t^{-1}, yielding

$$E\left[\left(\frac{C^2}{C^2 + t - 1} \right)^2 \frac{C^2}{t - 1} \right] = \frac{1}{4} + \frac{3}{8t} + o(t^{-1}). \qquad (12.3\text{-}21)$$

Alternatively, we can use Jensen's inequality to show that

$$E\left[\left(\frac{C^2}{C^2 + t - 1} \right)^2 \frac{C^2}{t - 1} \right] \geq \frac{1}{4}. \qquad (12.3\text{-}22)$$

R. R. Bahadur [personal communication] has pointed out that both upper and lower bounds for (12.3-20) are given by

$$\frac{1}{4} + \frac{3}{8(t - 1)} - \frac{1}{2(t - 1)^2} + \theta \frac{9}{4(t - 1)^2}\left[1 + \frac{2}{t - 1}\right], \qquad (12.3\text{-}23)$$

where $0 < \theta < 1$. To order t^{-1}, (12.3-23) is consistent with (12.3-21).

If we take the expansion of $R(\mathbf{p}^*, \mathbf{p})$ given by (12.3-10) for our special asymptotics and divide by $R(\hat{\mathbf{p}}, \mathbf{p})$, we also obtain the expansion (12.3-21). To see this we set D equal to zero, set λ equal to 1, and let δ approach infinity.

Finally, from (12.3-21) and (12.3-22), we see that when $\mathbf{p} = \boldsymbol{\lambda} = \mathbf{c}$, the asymptotic risk ratio converges to $1/4$ from above as t gets large. The actual rate of this convergence is indicated in Section 12.4.3.

Our conclusion is that for fixed t, as $N \to \infty$, the risk of the pseudo-Bayes estimator $\mathbf{p}^*$ is approximately equal to the risk of the usual estimator $\hat{\mathbf{p}}$, except when $\mathbf{p} = \boldsymbol{\lambda}^*$. In the latter situation, $\mathbf{p}^*$ is superior to $\hat{\mathbf{p}}$, with the ratio of their risk functions at $\mathbf{p} = \boldsymbol{\lambda}^*$ tending to $1/4$ for very large t.

12.4   Small-Sample Results

As noted in Section 12.2, the risk of $\mathbf{p}^*$ does not have a convenient algebraic closed form due to the dependence of $\hat{K}$ on $\mathbf{X}$. We have already noted that as $N \to \infty$ and $t \to \infty$ with N/t fixed, $\mathbf{p}^*$ provides an almost uniform improvement over $\hat{\mathbf{p}}$. Two questions remain.

1.   Does this large-sample improvement carry over to small samples in any way?
2.   How large do N and t have to be before the special asymptotics are meaningful?

Here we give some exact comparisons of the risk functions of $\hat{\mathbf{p}}$ and $\mathbf{p}^*$ in an attempt to answer these questions. In all our examples we take $\boldsymbol{\lambda} = \mathbf{c}$. The exact values of $R(\mathbf{p}^*, \mathbf{p})$ were evaluated numerically by a computer.

12.4.1   *Binomial case*

Figure 12.4-1 shows the risk of $\mathbf{p}^*$, $\hat{\mathbf{p}}$, and the minimax estimator $\mathbf{p}_M$ for $t = 2$ (the binomial case) and $N = 15$. Here $\mathbf{p}^*$ has smaller risk than $\hat{\mathbf{p}}$ for p_1 (and thus p_2) between 0.33 and 0.67 and has larger risk elsewhere. The minimax estimator $\mathbf{p}_M$ is superior to $\hat{\mathbf{p}}$ between 0.2 and 0.8, but is dominated by $\mathbf{p}^*$ between 0.42 and 0.58 and near the boundary, where p_1 is near 0 or 1.

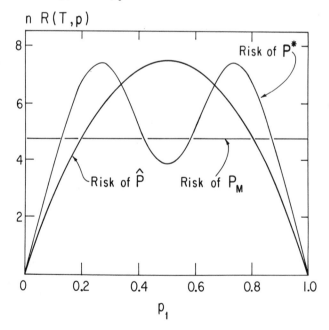

Figure 12.4-1   Risk of $\hat{\mathbf{p}}$, $\mathbf{p}_M$, and $\mathbf{p}^*$ for $t = 2$ and $N = 15$. Reproduced from Fienberg and Holland [1973].

The behavior of $\mathbf{p}_M$ and $\mathbf{p}^*$ near the boundary deserves further comment. We find that $\hat{\mathbf{p}}$ dominates $\mathbf{p}^*$ near the boundary, but both of their risks tend to 0 as p_1 approaches 0 or 1. Thus the behavior of $\mathbf{p}^*$ near the boundary is satisfactory. On the other hand, the ratio of the risk of $\mathbf{p}_M$ to either that of $\mathbf{p}^*$ or that of $\hat{\mathbf{p}}$ tends to ∞ as we near the boundary, and for some purposes this may be unacceptable. This behavior of $\mathbf{p}_M$ near the boundary is typical of all true Bayes estimators of the form (12.2-10), where w is constant. We do not consider $\mathbf{p}_M$ in the subsequent examples.

Finally, we note that as $N \rightarrow \infty$ with $t = 2$ fixed, the "ears" of the risk function of $\mathbf{p}^*$ rising above the risk function of $\hat{\mathbf{p}}$ move toward each other, and the difference between the functions disappears, except at $\mathbf{p} = (1/2, 1/2)$, where the risk of $\mathbf{p}^*$ is strictly less than the risk of $\hat{\mathbf{p}}$.

12.4.2 *Trinomial case*

A three-dimensional picture is required to dispay the risk functions of $\mathbf{p}^*$ and $\hat{\mathbf{p}}$ for $t = 3$. In Fienberg and Holland [1970], rather than looking at these functions as they sit over the two-dimensional probability simplex $\mathscr{S}_3$, the authors present sections along the two lines in $\mathscr{S}_3$ defined by $[p_1, (1 - p_1)/2, (1 - p_1)/2]$ and $[p_1, 1 - p_1, 0]$. Here we consider the ratio of the risks of $\mathbf{p}^*$ and $\hat{\mathbf{p}}$ and plot contours of constant risk ratio, i.e., contours for which

$$\rho(\mathbf{p}) = \frac{R(\mathbf{p}^*, \mathbf{p})}{R(\hat{\mathbf{p}}, \mathbf{p})} \qquad (12.4\text{-}1)$$

is constant. When this ratio is less than 1, $\mathbf{p}^*$ is superior to $\hat{\mathbf{p}}$.

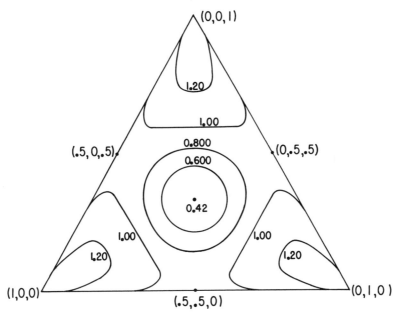

Figure 12.4-2   Contours of constant risk ratio ($\mathbf{p}^*$ over $\hat{\mathbf{p}}$) for $t = 3$ and $N = 15$. Reproduced from Fienberg and Holland [1973].

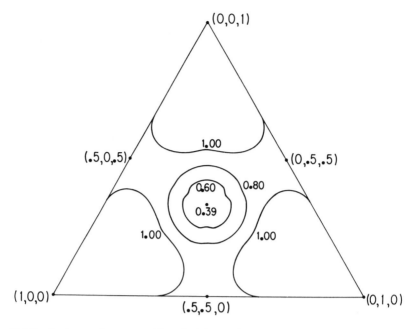

Figure 12.4-3   Contours of constant risk ratio (**p*** over **p̂**) for $t = 3$ and $N = 30$. Reproduced from
Fienberg and Holland [1973].

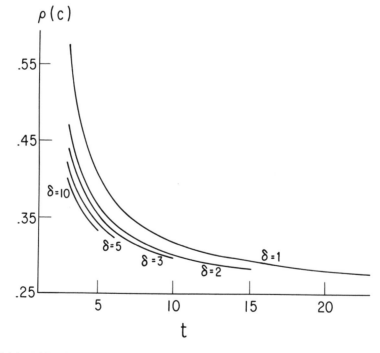

Figure 12.4-4   Risk ratio (**p*** over **p̂**) at center of simplex for various values of t and δ. Reproduced from
Fienberg and Holland [1973].

Figures 12.4-2 and 12.4-3 give contours of $\rho(\mathbf{p})$ in $\mathcal{S}_3$ for $N = 15$ and $N = 30$, respectively. As before, the greatest improvement of $\mathbf{p}^*$ over $\hat{\mathbf{p}}$ occurs at the point $\mathbf{p} = \lambda = (\frac{1}{3}, \frac{1}{3}, \frac{1}{3})$, where the value of $\rho(\mathbf{p})$ equals 0.42 and 0.39 for $N = 15$ and $N = 30$. Over a large section of the parameter space, $\mathbf{p}^*$ has smaller risk than $\hat{\mathbf{p}}$. Although $\hat{\mathbf{p}}$ has smaller risk near the vertices of $\mathcal{S}_3$, the value of $\rho(\mathbf{p})$ near the vertices reaches maxima of 1.25 and 1.17 for $N = 15$ and $N = 30$ and then decreases. Interestingly, $\mathbf{p}^*$ dominates $\hat{\mathbf{p}}$ for sizable segments of the one-dimensional boundaries.

The exact comparisons above indicate that the asymptotic improvement provided by $\mathbf{p}^*$ in the estimation of $\mathbf{p}$ carries over in large measure to cases where both N and t are very small. Yet it is clear in the examples above that the magnitude of the improvement suggested by the asymptotic results is not being realized, e.g., the lower bound of $1/4$ given in (12.3-14) is not reached here.

12.4.3   *Small samples and special asymptotics*

To check on the rate of approach to the asymptotic behavior of $\rho(\mathbf{p})$ described in Section 12.3, we focus on $\mathbf{p} = \lambda = \mathbf{c}$ and examine values of $\rho(\mathbf{c})$ as t increases for various fixed values of δ. Figure 12.4-4 displays exact computations of $\rho(\mathbf{c})$ for $\delta = 1, 2, 3, 5, 10$. By the time t reaches 13, $\rho(\mathbf{c})$ is less than 0.30 for all integral values of δ, and we have come close to the asymptotic value of $1/4$. The larger the value of δ, the faster $\rho(\mathbf{c})$ approaches the asymptote for increasing t. We conjecture that similar rates of approach to asymptotic values occur for other values of $\mathbf{p}$.

12.5   **Data-Dependent λ's**

Returning to figure 12.2-1, we recall that we were led to consider weighted-average estimates of the form (12.2-29) because the angle θ between $\lambda - \mathbf{p}$ and $\hat{\mathbf{p}} - \mathbf{p}$ was close to 90° for large t. The optimal estimator of this form, with w given by $N/(N + K)$ and K given by (12.2-24), is approximately the point at the base of the perpendicular from $\mathbf{p}$ to the hypotenuse in figure 12.2-1. The squared length of this perpendicular is roughly the risk of this optimal estimator, and since θ is close to a right angle, the value of the risk function decreases as we let λ approach $\mathbf{p}$, so that the improvement of the optimal estimator over $\hat{\mathbf{p}}$ increases. This can be seen algebraically, since the risk of the optimal estimator (with K as a function of $\mathbf{p}$) is

$$(1 - \|\mathbf{p}\|^2)\left[\frac{N}{N + K(\mathbf{p}, \lambda)}\right]. \tag{12.5-1}$$

Substituting in (12.5-1) the value of $K(\mathbf{p}, \lambda)$ from (12.2-24), we get

$$\left(\frac{1}{1 - \|\mathbf{p}\|^2} + \frac{1}{N\|\lambda - \mathbf{p}\|^2}\right)^{-1}. \tag{12.5-2}$$

Expression (12.5-2) clearly decreases as λ approaches $\mathbf{p}$.

Since we estimate the optimal values of K by expression (12.2-26), the geometric argument described above is only a rough description, in an asymptotic sense. Nevertheless, it is clear that choosing λ close to $\mathbf{p}$ produces a better estimator $\mathbf{p}^*$ than choosing λ far from $\mathbf{p}$. In the latter case, the maximum improvement of $\mathbf{p}^*$ over $\mathbf{p}$ that we can expect is small. The Bayesian statistician picks λ to reflect his prior information about $\mathbf{p}$ and so hopes to minimize the distance $\|\lambda - \mathbf{p}\|$. Here we are evaluating the performance of $\mathbf{p}^*$ relative to $\hat{\mathbf{p}}$ for all points in the parameter space $\mathscr{S}_t$. Thus a sensible choice of λ might be one where λ is a function of the data and so varies with $\mathbf{p}$. If our choice of such a random λ is sensible, then λ may remain close to $\mathbf{p}$ for a large portion of $\mathscr{S}_t$, and particularly for those points away from the boundary where $\hat{\mathbf{p}}$ gains its admissibility.

For multinomial data without a cross-classification structure, it is natural for us to choose $\lambda = \mathbf{c}$, by symmetry considerations. When the multinomial represents cross-classified data, such a choice is not the most natural, and it is quite reasonable for us to find a λ which either (i) reflects some prior knowledge about the cross-classification structure, or (ii) represents some special cross-classification structure which can serve as a "null" model for $\mathbf{p}$, often characterized by symmetry considerations.

In this section we consider an $I \times J$ cross-classification (i.e., a contingency table with I rows and J columns). Although probabilities and other quantities for two-way contingency tables are normally doubly subscripted, they can still be strung out in vector form, so we need not change the notation established in earlier sections, except that we replace single subscripts by double ones.

We continue to work with an estimator of the form

$$\mathbf{p}^* = \left(\frac{N}{N + \hat{K}}\right)\hat{\mathbf{p}} + \left(\frac{\hat{K}}{N + \hat{K}}\right)\lambda, \tag{12.5-3}$$

$$\hat{K} = K(\hat{\mathbf{p}}, \lambda) = \frac{N^2 - \sum_{i,j} x_{ij}^2}{\sum_{i,j} x_{ij}^2 - 2N \sum_{i,j} x_{ij}\lambda_{ij} + N^2 \sum_{i,j} \lambda_{ij}^2}. \tag{12.5-4}$$

Formula (12.5-4) is the same as (12.2-26) except for the double subscripts.

12.5.1 Decomposing λ

Throughout this book we have pointed out that it is natural to decompose a cross-classified table of probabilities or proportions p_{ij} into three pieces:

1. the row totals p_{i+};
2. the column totals p_{+j};
3. the cross-product ratios, for example

$$\alpha_{ij} = \frac{p_{ij} p_{i+1,j+1}}{p_{i,j+1} p_{i+1,j}}, \tag{12.5-5}$$

for $i = 1, \ldots, I - 1; j = 1, \ldots, J - 1$ (see, for example, Chapter 2). We can perform the same kind of decomposition for λ. Our emphasis in this section is on a choice for the cross-product ratios of the λ_{ij}, since we are usually interested in improving on the interaction structure of the data. As a result, it is natural for us to choose the row and column totals for λ as

$$\lambda_{i+} = \frac{X_{i+}}{N} \qquad i = 1, \ldots, I,$$

$$\lambda_{+j} = \frac{X_{+j}}{N} \qquad j = 1, \ldots, J. \tag{12.5-6}$$

Our first choice for the null values of the cross-product ratios of λ is

$$\frac{\lambda_{ij}\lambda_{i+1,j+1}}{\lambda_{i,j+1}\lambda_{i+1,j}} = 1 \tag{12.5-7}$$

$i = 1, \ldots, I - 1; j = 1, \ldots, J - 1$. This represents the null hypothesis, postulating independence of the variables corresponding to rows and columns, and leads to the usual expected values for computing chi square, i.e.,

$$N\lambda_{ij} = \frac{X_{i+}X_{+j}}{N}. \tag{12.5-8}$$

We do not have to restrict ourselves to vectors λ which express no interaction, as defined by (12.5-8). We can choose any set of cross-product ratios for λ and then construct some table which has these same cross-product ratios. Since this is often a tedious job, an alternative approach is to find another table which has a "relevant" cross-product ratio structure for the data at hand. Then we can give that table the margins specified by (12.5-6), by a simple application of the Deming–Stephen iterative proportional fitting procedure, used elsewhere in this book (see Chapter 3). This produces our new λ, and we then use formula (12.5-4) to produce a new value of $\hat{K}$. This value of $\hat{K}$ and the new λ are then used in (12.5-3) to give a different set of smoothed estimators. We illustrate this approach in our example in Section 12.6.

12.5.2   Small-sample results

We now report on some exact comparisons of $\hat{\mathbf{p}}$ and $\tilde{\mathbf{p}}^*$, where $\tilde{\mathbf{p}}^*$ is the estimate formed by using a random λ with components given by (12.5-8) for $I = J = 2$ (i.e., for the 2×2 table) and for $N = 20$. The parameter space $\mathscr{S}_4$ is now a tetrahedron, and we choose to view it as being composed of surfaces on which the cross-product ratio $\alpha = p_{11}p_{22}/p_{12}p_{21}$ is constant (see Chapter 2). Because of symmetry considerations, it suffices to look only at values of $\alpha \geq 1$. Each of the surfaces of constant α can be mapped one-to-one onto a square (see Fienberg [1970a]) for which the coordinates of any point are given by the marginal totals of the probabilities in the 2×2 table.

Figures 12.5-1, 12.5-2, and 12.5-3 give contours of constant risk ratio

$$\tilde{\rho}(\mathbf{p}) = \frac{R(\tilde{\mathbf{p}}^*, \mathbf{p})}{R(\hat{\mathbf{p}}, \mathbf{p})} \tag{12.5-9}$$

for $\alpha = 1, 3$, and 5, respectively. Recall that $\mathbf{p}^*$ has smaller risk than $\hat{\mathbf{p}}$ when $\tilde{\rho}(\mathbf{p})$ is less than unity. For $\alpha = 1$ (figure 12.5-1), we see that $\tilde{\rho}(\mathbf{p})$ has a minimum value of 0.76 when $p_{1+} = p_{+1} = 0.5$, and nowhere is it greater than 1. For $\alpha = 3$ (figure 12.5-2), $\tilde{\rho}(\mathbf{p})$ is still less than unity everywhere, but the minimum value is approximately 0.92 at four symmetrically placed points, corresponding roughly to $p_{1+} = 0.50$ and $p_{+1} = 0.15$. For $\alpha = 5$ (figure 12.5-3), $\hat{\mathbf{p}}$ begins to show some superiority over $\mathbf{p}^*$ near the center of the surface, although there still remain values of $\mathbf{p}$ for which $\tilde{\rho}(\mathbf{p}) < 1$ and $\mathbf{p}^*$ is superior to $\hat{\mathbf{p}}$. For other values of $\alpha > 5$ (not shown here), $\tilde{\rho}(\mathbf{p})$ has a maximum value of slightly more than 1.22, and for surfaces corresponding to $\alpha > 20$, the maximum value of $\tilde{\rho}(\mathbf{p})$ decreases, tending to 1 as $\alpha \to \infty$.

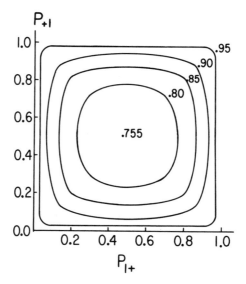

Figure 12.5-1   Contours of constant risk ratio ($\tilde{\mathbf{p}}^*$ over $\hat{\mathbf{p}}$) in a 2 × 2 table. $N = 20, \alpha = 1$. Reproduced
from Fienberg and Holland [1973].

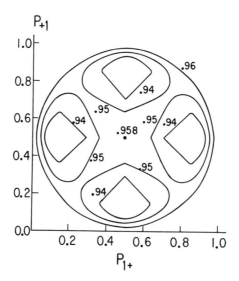

Figure 12.5-2   Contours of constant risk ratio ($\tilde{\mathbf{p}}^*$ over $\hat{\mathbf{p}}$) in a 2 × 2 table. $N = 20, \alpha = 3$. Reproduced
from Fienberg and Holland [1973].

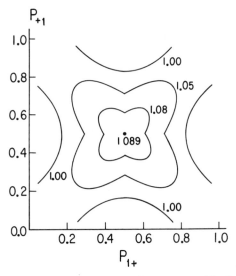

Figure 12.5-3 Contours of constant risk ratio (p̃* over p̂) in a 2 × 2 table. $N = 20, \alpha = 5$. Reproduced from Fienberg and Holland [1973].

For a sizable amount of the tetrahedron surrounding the surface of independence ($\alpha = 1$), our weighted biased estimator $\mathbf{p}^*$ has smaller risk than $\hat{\mathbf{p}}$, the unrestricted maximum likelihood estimator. As the dimensions of the table increase so that IJ/N remains constant, we conjecture that the hypervolume of the region in which $\mathbf{p}^*$ is superior to $\hat{\mathbf{p}}$ increases relative to the hypervolume of the entire simplex.

Table 12.5-1 Contingency between the Occupations of Fathers and Sons

Occupations of Fathers	(i)	(ii)	(iii)	(iv)	(v)	(vi)	(vii)	(viii)	(ix)	(x)	(xi)	(xii)	(xiii)	(xiv)	Totals
(i)	28	0	4	0	0	0	1	3	3	0	3	1	5	2	50
(ii)	2	51	1	1	2	0	0	1	2	0	0	0	1	1	62
(iii)	6	5	7	0	9	1	3	6	4	2	1	1	2	7	54
(iv)	0	12	0	6	5	0	0	1	7	1	2	0	0	10	44
(v)	5	5	2	1	54	0	0	6	9	4	12	3	1	13	115
(vi)	0	2	3	0	3	0	0	1	4	1	4	2	1	5	26
(vii)	17	1	4	0	14	0	6	11	4	1	3	3	17	7	88
(viii)	3	5	6	0	6	0	2	18	13	1	1	1	8	5	69
(ix)	0	1	1	0	4	0	0	1	4	0	2	1	1	4	19
(x)	12	16	4	1	15	0	0	5	13	11	6	1	7	15	106
(xi)	0	4	2	0	1	0	0	0	3	0	20	0	5	6	41
(xii)	1	3	1	0	0	0	1	0	1	1	1	6	2	1	18
(xiii)	5	0	2	0	3	0	1.	8	1	2	2	3	23	1	51
(xiv)	5	3	0	2	6	0	1	3	1	0	0	1	1	9	32
Totals	84	108	37	11	122	1	15	64	69	24	57	23	74	86	775

Source: Karl Pearson [1904b].

Note: The labels (i) to (xiv) have the following significance: (i) army, (ii) art, (iii) teaching, clerical work, civil service, (iv) crafts, (v) divinity, (vi) agriculture, (vii) landownership, (viii) law, (ix) literature, (x) commerce, (xi) medicine, (xii) navy, (xiii) politics and court, (xiv) scholarship and science.

Table 12.5-2   Pseudo-Bayesian Estimates for Expected Counts Corresponding to Table 12.5-1 (Rounded to Two Decimal Places)

Occupations of Fathers	Occupations of Sons													
	(i)	(ii)	(iii)	(iv)	(v)	(vi)	(vii)	(viii)	(ix)	(x)	(xi)	(xii)	(xiii)	(xiv)
(i)	25.53	0.76	3.82	0.08	0.86	0.01	1.00	3.12	3.16	0.17	3.07	1.05	4.98	2.39
(ii)	2.52	46.37	1.21	0.99	2.85	0.01	0.13	1.45	2.38	0.21	0.50	0.20	1.54	1.64
(iii)	5.98	5.28	6.52	0.08	8.95	0.90	2.79	5.83	4.09	1.96	1.32	1.07	2.35	6.89
(iv)	0.52	11.36	0.23	5.41	5.21	0.01	0.09	1.29	6.66	1.04	2.14	0.14	0.46	9.44
(v)	5.82	6.21	2.38	1.07	50.07	0.02	0.24	6.38	9.14	3.95	11.61	3.05	2.09	12.97
(vi)	0.31	2.18	2.81	0.04	3.12	0.00	0.06	1.13	3.82	0.98	3.77	1.87	1.16	4.77
(vii)	16.18	2.23	4.02	0.14	13.98	0.01	5.53	10.59	4.42	1.19	3.38	2.96	16.06	7.30
(viii)	3.49	5.50	5.70	0.11	6.53	0.01	1.93	16.65	12.25	1.12	1.45	1.11	7.85	5.29
(ix)	0.23	1.18	0.99	0.03	3.89	0.00	0.04	1.06	3.75	0.06	1.93	0.95	1.09	3.79
(x)	11.94	15.87	4.12	1.06	15.18	0.01	0.22	5.41	12.61	10.16	6.20	1.23	7.34	14.65
(xi)	0.49	4.19	2.00	0.06	1.60	0.01	0.09	0.37	3.07	0.14	18.14	0.13	4.88	5.84
(xii)	1.10	2.95	0.98	0.03	0.31	0.00	0.93	0.16	1.07	0.95	1.04	5.40	1.97	1.11
(xiii)	5.06	0.78	2.05	0.08	3.55	0.01	1.00	7.59	1.39	1.95	2.19	2.84	21.02	1.51
(xiv)	4.83	3.16	0.17	1.83	5.89	0.00	0.96	2.96	1.20	0.11	0.26	0.99	1.22	8.40

12.5.3 λ and $\hat{K}$ at work

Example 12.5-1   Occupational mobility in England

Table 12.5-1 is a cross-classification of the occupations of fathers and sons in England, presented by Pearson [1904b] and later analyzed by Good [1956]. Fienberg [1969] also worked with these data, and he required nonzero cell entries for the plotting of log contrasts.

If we shrink these data to the constant, i.e., $\lambda_{ij} = (IJ)^{-1} = 1/196$, we get a value of $\hat{K} = 65.29$, which is less than 10% of $N = 775$. The pseudo-Bayes estimator (12.5-3), using these values of $\hat{K}$ and λ, replaces the sampling zeros in table 12.5-1 by 0.31 and does not greatly effect the remaining entries (e.g., cells with entries of 1 are changed to 1.23 and cells with entries of 2 are changed to 2.15).

Next we shrink the data toward independence using the value of λ given by (12.5-8). This leads to $\hat{K} = 95.15$ and a set of smoothed cell estimates given in table 12.5-2, where there is differential smoothing for cells with the same observed entries but different marginal totals. ∎∎

Example 12.5-2   Intelligence for fraternal pairs

Pearson [1904a] examined the relationship of intelligence within fraternal pairs using the data in table 12.5-3. For $\lambda_{ij} = 1/4$, $\hat{K} = 26.3$, and for $\lambda_{ij} = x_{i+}x_{+j}/N^2$, $\hat{K} = 32.5$. Neither of these choices of λ does very much to the data, in the sense that $\mathbf{p}^*$ in both cases is almost identical to $\hat{\mathbf{p}}$. Now suppose we have surmised a positive relationship (as we very well might have) between brothers with regard to intelligence. In particular, let us suppose that we have chosen

$$\frac{\lambda_{11}\lambda_{22}}{\lambda_{12}\lambda_{21}} = 3,$$

and that $N\lambda$ has the margins of the original data (see table 12.5-4). Then the corresponding value of $\hat{K}$ is 2,562.3, and the resulting pseudo-Bayes estimator is given in table 12.5-5. Choosing a cross-product ratio for the λ_{ij} so that $\mathbf{x}$ is reasonably close to $N\lambda$ yields a larger value of $\hat{K}$ than does either of the choices above; however, a large value of $\hat{K}$ does not necessarily yield a great amount of smoothing of the observed table. ∎∎

Table 12.5-3   Correlation of Intelligence between Brothers

	First Brother		
Second Brother	Quick Intelligent and Intelligent	Others	Totals
Quick Intelligent and Intelligent	526	324	850
Others	324	694	1,018
Totals	850	1,018	1,868

Source: Karl Pearson [1904a].

Table 12.5-4   Table with the Margins of
Table 12.5-3 and Cross-Product Ratio of 3

Second Brother	First Brother Quick Intelligent and Intelligent	Others
Quick Intelligent and Intelligent	510.32	339.68
Others	339.68	678.32

Table 12.5-5   Pseudo-Bayesian Estimates for
Table 12.5-3 with $N\lambda$ Given by Table 12.5-4

Second Brother	First Brother Quick Intelligent and Intelligent	Others
Quick Intelligent and Intelligent	516.93	333.07
Others	333.07	684.93

12.6   Another Example: Two Social Mobility Tables

In several other chapters we have fitted models to social mobility tables for Denmark (Svalastoga [1959]) and Britain (Glass [1954]). Suppose we are simply interested in getting smoothed estimates for, say, the British table. Since we have already learned that the two tables have similar interaction structures, it is reasonable to use the Danish data to supply a value of λ toward which we can shrink the British data. Throughout this example we choose to display tables of cell estimates $N\mathbf{p}^*$ rather than cell probability estimates $\mathbf{p}^*$. This choice gives a somewhat more interesting perspective on the changes in $\mathbf{p}^* = \hat{\mathbf{q}}(\hat{K}, \lambda)$ for various choices of λ.

Tables 12.6-1 and 12.6-2 contain the British and Danish data, respectively. Our first task is to construct a new table with the margins of the British data but the cross-product ratios of the Danish data. This table (table 12.6-3), which serves as our new λ, looks remarkably similar to the original British data, and suggests that although we may get a large value of $\hat{K}$, the resulting table of smoothed estimates will not look much different from the original British data. In fact, our new $\hat{K} = 1,051.0$, and the estimates given by (12.5-3) are indeed similar to the original table (see table 12.6-4). Although there are changes of more than ten counts in some cells, the changes are all extremely small relative to the size of the original counts in the cells.

We can also get improved estimators for the Danish data by shrinking them toward the British data. Giving the British data the margins of the Danish data yields table 12.6-5, and it serves as our λ for the Danish data. Here $\hat{K} = 1,097$, and the resulting smoothed estimates are given in table 12.6-6. The two tables of smoothed estimates (tables 12.6-4 and 12.6-6) for the British and Danish data may be more suitable than the original tables of counts for building sociological models for mobility flow.

Table 12.6-1 British Social Mobility Data

50	45	8	18	8
28	174	84	154	55
11	78	110	223	96
14	150	185	714	447
3	42	72	320	411

Source: Glass [1954].

Table 12.6-2 Danish Social Mobility Data

18	17	16	4	2
24	105	109	59	21
23	84	289	217	95
8	49	175	348	198
6	8	69	201	246

Source: Svalastoga [1959].

Table 12.6-3 Danish Data with British Margins

37.5	53.5	20.6	11.5	5.9
32.9	217.2	92.5	111.5	41.0
15.6	86.1	121.4	203.1	91.8
12.7	117.3	171.8	761.2	446.9
7.4	14.9	52.6	341.6	431.4

Table 12.6-4 Pseudo-Bayesian Estimators for British Data with $N\lambda$ Given by Table 12.6-3

47.1	47.0	10.9	16.5	7.5
29.1	184.0	86.0	144.2	51.8
12.1	79.9	112.6	218.4	95.0
13.7	142.5	182.0	724.9	447.0
4.0	35.7	67.5	325.0	415.7

Table 12.6-5 British Data with Danish Margins

26.7	14.8	6.4	6.3	2.8
22.3	85.3	100.6	81.1	28.6
17.9	78.2	269.5	240.2	102.2
9.5	62.5	188.5	319.7	197.8
2.6	22.2	93.0	181.7	230.6

Table 12.6-6 Pseudo-Bayesian Estimators for Danish Data with $N\lambda$ Given by Table 12.6-5

20.7	16.3	13.0	4.7	2.2
23.5	98.8	106.4	66.0	23.4
21.4	82.2	282.9	224.3	97.3
8.5	53.3	179.2	339.1	197.9
4.9	12.5	76.5	194.9	241.2

An alternative to the estimates above comes to mind if we view tables 12.6-1 and 12.6-2 together as a $2 \times 5 \times 5$ (three-way) contingency table. There are several ways to generalize pseudo-Bayes estimation for multiway tables, the most straightforward of which sets $N\lambda$ equal to the table of expected values under some log-linear model, which has been computed using the methods of Chapter 3. Unlike the two-way situation, however, we have a large number of log-linear models from which to choose. If the model chosen fits the data either too well or too poorly, $N\mathbf{p}^*$ will be almost identical to $\mathbf{x}$. Ideally, we should pick a model which almost fits the data, so that $N\mathbf{p}^*$ will differ from $\mathbf{x}$ although not by too much.

It is clear that if a multinomial model is appropriate for each of the two-way tables separately (as we assumed above), then we cannot assume that the $2 \times 5 \times 5$ table has a multinomial sampling model. Nevertheless, the generalization of the pseudo-Bayes estimators to several multinomials (in our case, two) is quite simple, and as long as we work with cell counts instead of proportions, the methods already derived are appropriate.

Table 12.6-7 contains the pair of social mobility tables displayed as a $2 \times 5 \times 5$ table. We have already seen that there is considerable interaction between occupation of father and occupation of son within countries, and that occupational distributions differ in Britain and Denmark. This suggests that we fit the model of no three-factor interaction to the data. The expected values for this model are given in table 12.6-8, and the value of the likelihood-ratio chi square goodness-of-fit statistic for this model is 39.33, with sixteen degrees of freedom. Although the value of the statistic is significant at the 0.001 level, the fitted values are quite similar to the observed values except in a few of the cells, e.g., the $(2, 5, 2)$ and $(1, 5, 3)$ cells.

Using the expected values under the model of no three-factor interaction as our table $N\lambda$, we get $\hat{K} = 11{,}108.3$ and pseudo-Bayesian estimates given in table 12.6-9. The enormous value of $\hat{K}$ reflects the excellent fit of the no three-factor interaction model in most of the cells, even though the chi square statistic for this model is significant. (Had the model fit perfectly, $\hat{K}$ would have been ∞.) The pseudo-Bayesian values for the cells with large residuals from this model (i.e., large values of $x_{ijk} - N\lambda_{ijk}$) are closer to the expected values of table 12.6-8 than to the observed values; however, they are not as close as one might have expected with such a large $\hat{K}$.

Table 12.6-7 British and Danish Social Mobility Data as a $2 \times 5 \times 5$ Table

Country		I	II	III	IV	V
Britain	I	50	45	8	18	8
	II	28	174	84	154	55
	III	11	78	110	223	96
	VI	14	150	185	714	447
	V	3	42	72	320	411
Denmark	I	18	17	16	4	2
	II	24	105	109	59	21
	III	23	84	289	217	95
	IV	8	49	175	348	198
	V	6	8	69	201	246

Table 12.6-8   Expected Values for Table 12.6-7 under Model of No Three-Factor Interaction

Country		I	II	III	IV	V
Britain	I	45.0	46.8	13.4	16.2	7.4
	II	29.5	187.7	88.4	139.4	50.1
	III	13.3	81.5	117.3	212.5	93.3
	IV	13.3	140.4	178.6	730.9	446.9
	V	4.9	32.6	61.3	329.9	419.3
Denmark	I	23.0	15.2	10.6	5.7	2.6
	II	22.5	91.3	104.6	73.6	25.9
	III	20.7	80.5	281.7	227.5	97.7
	IV	8.7	58.6	181.4	331.1	198.1
	V	4.1	17.4	79.7	191.1	237.7

Table 12.6-9   Pseudo-Bayesian Estimators for Data in Table 12.6-7 with $N\lambda$ Given by Table 12.6-8

Country		I	II	III	IV	V
Britain	I	46.8	46.2	11.5	16.9	7.6
	II	29.0	182.9	86.9	144.5	51.8
	III	12.5	80.3	114.8	216.1	94.2
	IV	13.5	143.7	180.8	725.0	446.9
	V	4.2	35.9	65.0	326.5	416.4
Denmark	I	21.2	15.8	12.5	5.1	2.4
	II	23.0	96.0	106.1	68.5	24.2
	III	21.5	81.7	284.2	223.9	96.8
	IV	8.5	55.3	179.2	337.0	198.1
	V	4.8	14.2	76.0	194.5	240.6

12.7   Recent Results and Some Advice

In this section we discuss some recent work that extends and amplifies the material covered earlier in this chapter. We also give some advice to potential users of pseudo-Bayes estimators.

12.7.1   *Further results when λ is fixed a priori*

Sutherland [1974] and Sutherland, Holland, and Fienberg [1974] have used the asymptotic framework of Section 12.3.1 to approximate the risk functions of a large class of estimators of the form (12.1-2) in which K depends on $\mathbf{X}$ and λ is fixed a priori. This class of estimators includes the one given by (12.2-26) and (12.2-27), but not the minimax estimator given by (12.2-21) and (12.2-22).

The approximations to the risk function that Sutherland obtains have the following general form:

$$R(\mathbf{T}, \mathbf{p}) = S_0 + \frac{1}{t}S_1 + o\left(\frac{1}{t}\right), \tag{12.7-1}$$

where S_0 and S_1 depend on $p(\cdot), \lambda(\cdot), \delta$ and various parameters in his family of shrinking constants K. He is able to show that for his class of estimators, the dominant term S_0 of the risk function is uniformly bounded from below by

$$\frac{D}{D+1},$$

(12.7-2)

where D is defined in (12.3-8).

Finally, he shows that for a particular member of his class of estimators, S_0 is identically equal to the lower bound (12.7-2). The estimator which attains this lower bound is $\hat{\mathbf{q}}(\hat{K}^*, \lambda)$ where

$$\hat{K}^* = \frac{N^2 - \sum_i x_i^2}{\sum_i x_i^2 - 2(N-1)\sum_i x_i \lambda_i + N(N-1)\sum_i \lambda_i^2 - N}.$$

(12.7-3)

When $\lambda = \mathbf{c}$, $\hat{K}^*$ reduces to

$$\hat{K}^*(\mathbf{c}) = \frac{N^2 - \sum_i x_i^2}{\sum x_i^2 - N - \left[\dfrac{N(N-1)}{t}\right]}.$$

(12.7-4)

While $\hat{K}^*$ is similar to $\hat{K}$ defined in (12.2-26), the term $-N$ in the denominator suggests that, generally speaking, $\hat{K}^*$ will be larger than $\hat{K}$.

The form of $\hat{K}^*$ may be motivated in the same way as $\hat{K}$ in Section 12.2. In Section 12.2, the value of K that minimized the risk function of $\hat{\mathbf{q}}(K, \lambda)$ for a given choice of λ and value of $\mathbf{p}$ was shown to be

$$K(\mathbf{p}, \lambda) = \frac{1 - \Sigma p_i^2}{\Sigma p_i^2 - 2\Sigma \lambda_i p_i + \Sigma \lambda_i^2}.$$

(12.7-5)

Now we define the two composite parameters θ_1 and θ_2 by

$$\theta_1 = \theta_1(\mathbf{p}) = \sum_i p_i^2,$$

(12.7-6)

$$\theta_2 = \theta_2(\mathbf{p}) = \sum_i \lambda_i p_i.$$

(12.7-7)

Then $K(\mathbf{p}, \lambda)$ is given by

$$K(\mathbf{p}, \lambda) = \frac{1 - \theta_1}{\theta_1 - 2\theta_2 + \Sigma \lambda_i^2}.$$

(12.7-8)

Unbiased estimates of θ_1 and θ_2 are easily shown to be

$$S_1 = (N(N-1))^{-1} \sum_i X_i(X_i - 1),$$

(12.7-9)

$$S_2 = N^{-1} \sum_i X_i \lambda_i,$$

(12.7-10)

so that

$$E(S_1) = \theta_1, \qquad E(S_2) = \theta_2.$$

(12.7-11)

If we estimate the optimal value $K(\mathbf{p}, \lambda)$ by the ratio of the unbiased estimators of its numerator and denominator, we obtain

$$\hat{K}^* = \frac{1 - S_1}{S_1 - 2S_2 + \sum_i \lambda_i^2}.$$

(12.7-12)

It is easy to show that the right-hand side of (12.7-12) is algebraically identical to the right-hand side of (12.7-3). Thus $\hat{K}^*$ can be described as the "ratio unbiased" estimator of $K(\mathbf{p}, \lambda)$, just as $\hat{K}$ is the maximum likelihood estimator of $K(\mathbf{p}, \lambda)$.

As of this writing there is no small-sample analysis for $\hat{K}^*$ comparable to that given for $\hat{K}$ in Section 12.4. Therefore, while the asymptotic theory suggests that $\hat{K}^*$ is an improvement on $\hat{K}$, we lack the crucial small-sample information to guide our choice between them. Since the asymptotic and small-sample theories suggest that either $\hat{\mathbf{q}}(\hat{K}, \lambda)$ or $\hat{\mathbf{q}}(\hat{K}^*, \lambda)$ is preferable to $\hat{\mathbf{p}}$, we are not seriously concerned with our uncertainty in the choice between them. Nevertheless, small-sample analysis of these pseudo-Bayes estimators is an important piece of research that still remains to be done.

12.7.2   *Further work when λ is data dependent*

Sutherland [1974] also considers the choice of K when λ is not a fixed a priori, but depends on the data $\mathbf{X}$. He suggests the following general method for deriving a data-dependent K in this situation. When $\lambda = \hat{\lambda}$ depends on $\mathbf{X}$ and K is fixed, the risk of $\hat{\mathbf{q}} = \hat{\mathbf{q}}(K, \hat{\lambda})$ is given by

$$R(\hat{\mathbf{q}}, \mathbf{p}) = E \left\| \frac{N}{N + K} \hat{\mathbf{p}} + \frac{K}{N + K} \hat{\lambda} - \mathbf{p} \right\|^2 \tag{12.7-13}$$

$$= \sum_i E \left(\frac{N}{N + K} (\hat{p}_i - \hat{\lambda}_i) + (\hat{\lambda}_i - p_i) \right)^2. \tag{12.7-14}$$

Differentiating (12.7-14) in K and setting the result equal to zero yields the following optimal value of K, which we denote by $K(\mathbf{p})$:

$$K(\mathbf{p}) = N \frac{\sum_i E(\hat{p}_i - p_i)(\hat{p}_i - \hat{\lambda}_i)}{\sum_i E(\hat{p}_i - \hat{\lambda}_i)^2}. \tag{12.7-15}$$

Expression (12.7-15) reduces to (12.7-5), as it should if $\hat{\lambda}$ is a constant not depending on $\mathbf{X}$.

In general, the value of $K(\mathbf{p})$ in (12.7-15) is difficult to compute, but when $\mathbf{X}$ is a two-way table and

$$\hat{\lambda}_{ij} = N^{-2} X_{i+} X_{+j}, \tag{12.7-16}$$

as in Section 12.5.1, Sutherland shows that $K(\mathbf{p})$ is

$$K_{\text{two-way}}(\mathbf{p}) = \frac{(1 - N^{-1})(1 - \Delta^2 + \Sigma_R^2 \Sigma_C^2 - \Sigma_R^2 - \Sigma_C^2)}{\Delta^2 - \frac{2}{N} A + \frac{1}{N^2} B - \frac{2}{N^3} A}, \tag{12.7-17}$$

where

$$\Delta^2 = \sum_{i,j} (p_{ij} - p_{i+} p_{+j})^2,$$

$$\Sigma_R^2 = \sum_i p_{i+}^2, \qquad \Sigma_C^2 = \sum_j p_{+j}^2, \qquad \Sigma^2 = \sum_{i,j} p_{ij}^2,$$

$$A = 2\Delta^2 + \Sigma_R^2 \Sigma_C^2 - \Sigma^2, \qquad B = 7\Delta^2 + 4\Sigma_R^2 \Sigma_C^2 - 4\Sigma^2.$$

Sutherland suggests that $K_{\text{two-way}}(\mathbf{p})$ be estimated by its ratio unbiased estimator by anology with his results for the fixed λ case. This approach yields the following formula for $K^*_{\text{two-way}}$:

$$K^*_{\text{two-way}} = \frac{A_N + B_N Q_0 - C_N(Q_1 + Q_2 + Q_3)}{D_N Q_3 - E_N Q_0 + F_N Q_1 Q_2 + G_N(Q_1 + Q_2) - (N-1)}, \quad (12.7\text{-}18)$$

where

$$Q_0 = \sum_{i,j} X_{ij} X_{i+} X_{+j}, \qquad Q_1 = \sum_i X_{i+}^2, \qquad Q_2 = \sum_j X_{+j}^2, \qquad Q_3 = \sum_{i,j} X_{ij}^2,$$

$$A_N = N(N-1),$$

$$B_N = \frac{2N-1}{N^2},$$

$$C_N = \frac{N-1}{N},$$

$$D_N = \frac{N^5 - 5N^4 + 7N^3 - 3N^2 + 24}{N^4(N-3)},$$

$$E_N = \frac{2(N-1)^2}{N^3},$$

$$F_N = \frac{(N-1)(N-2)}{N^4},$$

$$G_N = \frac{N^3 - 12N^2 + 17N - 8}{N^3(N-3)}.$$

Sutherland's choice of K, given by (12.7-18), needs to be compared to the estimator used in Section 12.5.2. In that section we introduced the estimator $\tilde{\mathbf{p}}^*$, where

$$\tilde{p}^*_{ij} = \frac{N}{N + \tilde{K}} \hat{p}_{ij} + \frac{\tilde{K}}{N + \tilde{K}} \hat{p}_{i+} \hat{p}_{+j}, \quad (12.7\text{-}19)$$

$$\tilde{K} = \frac{N^2 - \sum x_{ij}^2}{\sum\limits_{i,j} x_{ij}^2 - \frac{2}{N} \sum\limits_{i,j} x_{ij} x_{i+} x_{+j} + \frac{1}{N^2} \sum\limits_i x_{i+}^2 \sum\limits_j x_{+j}^2}. \quad (12.7\text{-}20)$$

While $K^*_{\text{two-way}}$ and $\tilde{K}$ are similar, a comparison of their risk functions for small and moderate samples would be very useful. $\tilde{K}$ is quite a bit simpler to compute by hand, but if the analysis is done by computer, as it often is, this advantage would be offset if $K^*_{\text{two-way}}$ turns out to yield a significant improvement in the estimation of $\mathbf{p}$. $\tilde{K}$ or $K^*_{\text{two-way}}$ is expected to be preferable to $\hat{p}$ when there are a large number of cells and $\delta = N/t$ is not large.

Example 12.7-1   Occupational mobility in England (continued)

To compare $\tilde{K}$ and $K^*_{\text{two-way}}$ on a set of real data, we compute both for the occupational mobility data given in table 12.5-1 and discussed earlier in example 12.5-1. For these data, $\tilde{K} = 95.2$ and $K^*_{\text{two-way}} = 87.3$. The similarity of these two values suggests that the $\tilde{K}$ may generally give results similar to those we get from the use of $K^*_{\text{two-way}}$. Additional discussion of these and other choices of K for two-way tables is given in Sutherland, Holland, and Fienberg [1974]. ■■

12.7.3 *Some advice about pseudo-Bayes estimators*

Pseudo-Bayes estimators have two primary uses. First, if we are truly interested in the cell probabilities themselves, then most pseudo-Bayes estimators are superior to the raw cell proportions $\hat{p}_i$. Much of this chapter deals with demonstrations of this superiority. If we report the actual value of K, the choice of λ, and the smooth pseudo-Bayes estimates $\hat{q}_i(\hat{K}, \lambda)$, then we can always recover the raw cell counts or proportions if we really need them.

Second, pseudo-Bayes estimators provide an all-purpose method for removing the zeros in an observed frequency distribution or contingency table, so that we may proceed with other analyses of the data that are hampered by the presence of zero counts (e.g., model-fitting via the method of weighted least squares, as described in Chapter 10).

In situations where one has reasonable grounds for choosing an a priori assignment of probabilities λ_i, we recommend the use of the estimators based on expressions (12.2-26) and (12.7-3). If (12.7-3) is negative, then it should be replaced by zero, so that $\hat{\mathbf{q}} = \hat{\mathbf{p}}$. We advocate the use of these estimators in lieu of the traditional practice of adding $1/2$ to all the counts in the table, especially when the number of cells is large and the number N/t of observations per cell is not large. We base this recommendation on the small-sample results of Section 12.4 and on the risk-function calculations of Section 12.3, where we show that the "adding $1/2$" estimator possesses a uniformly larger dominant term in its risk function than either of the recommended pseudo-Bayes estimators.

When one is prepared to use the structure of the data to build a data-dependent assignment of probabilities $\lambda_i(\mathbf{X})$, then alternative pseudo-Bayes estimators are available for use. For two-way contingency tables, we recommend the use of estimators like those based on expressions (12.7-18) and (12.7-20), where we chose $\lambda_{ij}(\mathbf{X}) = X_{i+}X_{+j}/N$. We base this recommendation on the small-sample results of Section 12.4 for 2×2 tables and on the analogy with the results for the fixed λ case.

13 Sampling Models for Discrete Data

13.1 Introduction

In this chapter we describe and relate several discrete distributions whose properties we often use. Since we expect this chapter to be used mainly for reference, we assume that its readers have some familiarity with standard statistical techniques and terminology. For example, we assume that the reader knows what maximum likelihood estimates are and what the method of moments is. Rather than work through this chapter from start to finish, from time to time the reader might read a section or two to explicate material in other chapters.

We give most of the results without derivation. The major exceptions are in Section 13.4, namely:

1. the derivation of the unrestricted maximum likelihood estimator for a multinomial probability vector;
2. the derivation of the relationship between Poisson and multinomial estimators.

In the following sections we discuss separately the binomial, Poisson, multinomial, hypergeometric, multivariate hypergeometric, negative binomial, and negative multinomial distributions. The first three of these distributions are used throughout the book. The hypergeometric distribution and its multivariate generalization are mentioned at various points when conditional arguments are being used or conditional tests are being considered. We include a brief section on the negative binomial distribution partly for completeness and partly because it is a natural generalization of the Poisson distribution. As such, it appears in the examples discussed in Chapter 9. Finally, the negative multinomial distribution can also be used to generate tables of counts. We do not, however, make much use of this distribution in our presentation.

The material in this chapter is not original, but is a summary of ideas to be found elsewhere in the statistical literature. For more details we refer the reader to the books on discrete distributions by Haight [1967], Johnson and Kotz [1969], and Patil and Joshii [1968].

13.2 The Binomial Distribution

The binomial distribution is usually discussed in terms of a sequence of N independent trials. In each trial the probability is p that a particular outcome $\mathscr{E}$ will occur. The number of times $\mathscr{E}$ occurs in this sequence of N trials is represented by the random variable X, which is said to have the binomial distribution with parameters N and p given by

$$\Pr[X = x] = \binom{N}{x} p^x (1 - p)^{N-x} \quad x = 0, \ldots, N, \qquad (13.2\text{-}1)$$

where $1 > p > 0$.

Extensive tables are available containing values of individual binomial probabilities as well as cumulative values. Among these are the tables of the National Bureau of Standards [1950], Romig [1953], the U.S. Office of the Chief of Ordnance [1952], and Harvard University Computation Laboratory [1955]. A briefer table for some values of p and $N = 1(1)30$ is given in Mosteller, Rourke, and Thomas [1970], and it is of particular interest because, unlike the other tables, it uses a full range of p from 0 to 1, rather than a range from 0 to 0.5.

13.2.1 Moments

The characteristic function of the binomial distribution is

$$E[e^{isX}] = (q + p\,e^{is})^N, \qquad (13.2\text{-}2)$$

where $q + p = 1$; the moment-generating function is

$$E[e^{sX}] = (q + p\,e^{s})^N. \qquad (13.2\text{-}3)$$

Expressions (13.2-2) and (13.2-3) can be used to derive formulas for the moments of the binomial distribution; however, a rather pretty formula for the rth *factorial moment* can be found directly, and other moment formulas can be derived from it. The rth factorial moment ($r \le N$) is (using the factorial notation):

$$\mu_{(r)}(X) = E[X^{(r)}] = E[X(X - 1)(X - 2)\ldots(X - r + 1)]$$

$$= \sum_{x=0}^{N} x^{(r)} \frac{N!}{x!(N - x)!} p^x q^{N-x}$$

$$= \sum_{x=r}^{N} x^{(r)} \frac{N!}{x!(N - x)!} p^x q^{N-x}$$

$$= \sum_{x=r}^{N} \frac{N^{(r)}(N - r)!}{(x - r)!(N - x)!} p^r p^{x-r} q^{N-x}$$

$$= N^{(r)} p^r \sum_{y=0}^{N-r} \binom{N - r}{y} p^y q^{N-r-y}$$

$$= N^{(r)} p^r = N(N - 1)(N - 2)\ldots(N - r + 1)p^r$$

$$= \frac{N!}{(N - r)!} p^r. \qquad (13.2\text{-}4)$$

We present the actual derivation of $\mu_{(r)}(X)$ here, because a simple extension of it yields a useful formula for the mixed factorial moments of the multinomial distribution, considered in a subsequent section.

We can go on to compute other moment formulas using (13.2-4) and the following relationships between descending factorial moments and moments about zero:

$$\mu'_1 = \mu_{(1)},$$

$$\mu'_2 = \mu_{(2)} + \mu_{(1)},$$

$$\mu'_3 = \mu_{(3)} + 3\mu_{(2)} + \mu_{(1)},$$

$$\mu'_4 = \mu_{(4)} + 6\mu_{(3)} + 7\mu_{(2)} + \mu_{(1)}, \qquad (13.2\text{-}5)$$

where $\mu'_r = E[X^r]$. This leads to the following formulas for the first four moments about zero:

$$\mu'_1(x) = Np,$$

$$\mu'_2(X) = N(N - 1)p^2 + Np = N^{(2)}p^2 + Np,$$

$$\mu'_3(X) = N^{(3)}p^3 + 3N^{(2)}p^2 + Np,$$

$$\mu'_4(X) = N^{(4)}p^4 + 6N^{(3)}p^3 + 7N^{(2)}p^2 + Np. \qquad (13.2\text{-}6)$$

The lower-order central moments are related to the moments about zero by:

$$\mu_2 = \text{Var}(X) = \mu'_2 - \mu'^2_1,$$

$$\mu_3 = \mu'_3 - 3\mu'_2\mu'_1 + 2\mu'^3_1,$$

$$\mu_4 = \mu'_4 - 4\mu'_3\mu'_1 + 6\mu'_2\mu'^2_1 - 3\mu'^4_1, \qquad (13.2\text{-}7)$$

yielding for the binomial:

$$\mu_2(X) = \text{Var}(X) = Npq,$$

$$\mu_3(X) = Npq(q - p),$$

$$\mu_4(X) = 3(Npq)^2 + Npq(1 - 6pq). \qquad (13.2\text{-}8)$$

13.2.2   Point estimation of parameters

If X is a binomial random variable with parameters N and p, then the maximum likelihood estimate of p is the observed frequency

$$\hat{p} = \frac{X}{N}. \qquad (13.2\text{-}9)$$

Since the sum of independent binomial random variables $X_1, \ldots, X_n$ with a common p and parameters $N_1, \ldots, N_n$, respectively, is itself binomial with $N = \sum_{i=1}^{n} N_i$ and the same p, we only have to consider the estimation problem for a single binomial variate.

The estimator $\hat{p}$ is also the one based on the method of moments, and it is minimum variance unbiased, with variance

$$\frac{pq}{N}. \qquad (13.2\text{-}10)$$

When p is known and N is not (but is fixed), the natural estimator X/p for N is unbiased, with variance Nq/p. The maximum likelihood estimate of N given p is $[X/p]$, where $[x]$ is the greatest integer $\leq x$. We used these estimators for N in Chapter 6, where the binomial likelihood function comes out of a multinomial problem. When both N and p are unknown, we need more than one observation to estimate them (see Johnson and Kotz [1969]).

In Chapter 12 we considered a class of biased estimates of p that provides smaller mean squared error for a large part of the parameter space.

13.2.3 *Relationship to other distributions*

The binomial is perhaps the simplest of the distributions discussed in this chapter, and at the same time it is directly related to many of the other distributions. For example, if we let $N \to \infty$ and $p \to 0$, keeping $Np = m$ fixed, then we obtain in the limit the Poisson distribution.

If X_1 and X_2 are independent binomial random variables with parameters (N_1, p) and (N_2, p), respectively, then $X = X_1 + X_2$ also has a binomial distribution with parameters $N_1 + N_2$ and p, and the distribution of X_1 given the value of $X = X_1 + X_2$, say x, is given by the *hypergeometric*:

$$\Pr[X_1 = x_1 | X_1 + X_2 = x] = \frac{\binom{N_1}{x_1}\binom{N_2}{x - x_1}}{\binom{N_1 + N_2}{x}}, \qquad (13.2\text{-}11)$$

where $\max(0, x - N_2) \leqq x_1 \leqq \min(N_1, x)$.

Looking at relationships in the reverse direction, the binomial can be derived from the Poisson distribution using a conditionality argument similar to that for the binomial to the hypergeometric above. We give such a derivation in Section 13.3.

The binomial naturally appears as a one-dimensional marginal distribution for the multinomial, and we discuss this property in Section 13.5.

Finally, the negative binomial (see Section 13.7) is also related, being based on a sequence of independent trials, and in each the probability is p that the event $\mathscr{E}$ occurs. Instead of the number of trials N being fixed, the number of times $\mathscr{E}$ occurs is fixed, leading to the expression "inverse sampling."

13.3 The Poisson Distribution

The random variable X has a Poisson distribution with parameter $m > 0$ if

$$\Pr[X = x] = \frac{e^{-m}m^x}{x!} \quad x = 0, 1, \dots. \qquad (13.3\text{-}1)$$

The Poisson distribution arises in a variety of different ways. As mentioned in the preceding section, it is the limit of a sequence of binomial distributions in which $N \to \infty$ and $p \to 0$, while $Np = m$ is fixed (see Feller [1968]). In this sense, since $p \to 0$ the Poisson is said to describe the distribution of rare events.

Suppose independent items having a common exponential lifetime are used consecutively, i.e., an item is replaced by a fresh one as soon as it fails. Then the distribution of the number of failures in a time period of fixed length is Poisson.

The Poisson distribution also describes events (again, see Feller [1968]) which are said to occur randomly and independently in time. The generating process for such events, usually referred to as a Poisson process, is characterized by the following assumptions:

(i) the probability that during an interval of time $(t, t + h)$ there is exactly one occurrence of an event $\mathscr{E}$ is $\lambda h + o(h)$, independent of the number of occurrences of $\mathscr{E}$ in $(0, t)$;

(ii) the probability of more than one occurrence of $\mathscr{E}$ in $(t, t + h)$ is $o(h)$.

Based on these assumptions, we can show that the number of occurrences of the event $\mathscr{E}$ in a period of time of length t is given by a Poisson random variable with parameter λt.

It is usually in this latter context that the Poisson distribution is used in connection with tables of counts. If the number of occurrences of some event for each of the cells in a contingency table is governed by an independent Poisson process, and if this set of processes is observed for a fixed period of time t, then the observations in the cells have independent Poisson distributions. Since t is common to the parameters of all of the processes, we usually refer to the parameter of the ith process as m_i instead of $\lambda_i t$.

The relationship between the Poisson and χ^2 distributions (see Johnson and Kotz [1969], p. 98) allows us to use tables of χ^2 to compute cumulative Poisson probabilities. Some of the more useful tables are found in Molina [1942], Owen [1962], and Pearson and Hartley [1958].

13.3.1 *Moments*

The characteristic function of the Poisson random variable X with parameter m is

$$E[e^{isX}] = \exp\{m(e^{is} - 1)\}, \tag{13.3-2}$$

and the moment-generating function is

$$E[e^{sX}] = \exp\{m(e^{s} - 1)\}. \tag{13.3-3}$$

The logarithm of the moment-generating function

$$\log E[e^{sX}] = m(e^{s} - 1) \tag{13.3-4}$$

is the cumulant-generating function, and the coefficient of $s^r/r!$ in its Taylor series expansion is the rth cumulant κ_r. The first four cumulants are related to the central moments by:

$$\kappa_1 = E[X] = \mu_1',$$

$$\kappa_2 = \text{Var}(X) = \mu_2,$$

$$\kappa_3 = \mu_3,$$

$$\kappa_4 = \mu_4 - 3\mu_2^2. \tag{13.3-5}$$

The rth factorial moment of the Poisson random variable X with parameter m is

$$E[X^{(r)}] = m^r, \tag{13.3-6}$$

and we can use the same technique to derive this formula as we did in the derivation of (13.2-4). It then follows, using (13.3-6), that the first four moments about zero are

$$\mu_1'(X) = E[X] = m,$$

$$\mu_2'(X) = m + m^2,$$

$$\mu_3'(X) = m + 3m^2 + m^3,$$

$$\mu_4'(X) = m + 7m^2 + 6m^3 + m^4. \tag{13.3-7}$$

We use the letter m for the parameter of the Poisson because the mean of X is m, and when we speak of expected values in a table of counts, we use the notation m_i, m_{ij}, etc. This use of subscripted letters m is in keeping with the generation of a table of counts via a set of independent Poisson processes.

Formula (13.3-4) for the cumulant-generating function implies that

$$\kappa_r(X) = m \tag{13.3-8}$$

for all $r \geq 1$, and then using (13.3-8), we get the following formulas for the lower-order moments of the Poisson:

$$\mu_2(X) = \text{Var}(X) = m,$$

$$\mu_3(X) = m,$$

$$\mu_4(X) = m + 3m^2. \tag{13.3-9}$$

13.3.2   Sums of independent Poissons

If X_1 and X_2 are independent random variables, each having a Poisson distribution with parameters m_1 and m_2, respectively, then $X = X_1 + X_2$ also has a Poisson distribution with parameter $m = m_1 + m_2$. This result is of considerable importance and has been used several times in earlier parts of this book. The result follows immediately when we note that the characteristic function of X, using (13.3-2), is

$$E[e^{isX}] = E[e^{is(X_1 + X_2)}] = E[e^{isX_1}]E[e^{isX_2}]$$

$$= \exp\{m_1(e^{is} - 1)\} \exp\{m_2(e^{is} - 1)\}$$

$$= \exp\{(m_1 + m_2)(e^{is} - 1)\}, \tag{13.3-10}$$

which is the characteristic function of a Poisson variate with parameter $m = m_1 + m_2$.

13.3.3   Point estimation of parameters

If X is a Poisson random variable with parameter m, then the maximum likelihood estimator of m is

$$\hat{m} = X. \tag{13.3-11}$$

Now if $X_1, \ldots, X_n$ are independent Poisson variates with a common parameter m, then $\sum X_i$ is a Poisson variate with parameter nm, and the MLE of m is $\hat{m} = \sum_{i=1}^n X_i/n$ (from (13.3-11)) and $\text{Var}(\hat{m}) = m/n$. This estimator $\hat{m}$ is the minimum variance unbiased estimator of m, and it is also a complete sufficient statistic.

13.3.4   A conditioning property

Let X_1 and X_2 be Poisson variates with parameters m_1 and m_2, respectively. Then the conditional distribution of X_1 given the value of $X = X_1 + X_2$ is binomial with parameters that are the observed values of X and $m_1/(m_1 + m_2)$, i.e.,

$$\Pr[X_1 = x_1 | X_1 + X_2 = x] = \frac{\dfrac{e^{-m_1}m_1^{x_1}}{x_1!} \cdot \dfrac{e^{-m_2}m_2^{x-x_1}}{(x-x_1)!}}{\dfrac{e^{-(m_1+m_2)}(m_1+m_2)^x}{x!}}$$

$$= \frac{x!}{x_1!(x-x_1)!}\left(\frac{m_1}{m_1+m_2}\right)^{x_1}\left(\frac{m_2}{m_1+m_2}\right)^{x-x_1}$$

$$= \frac{x!}{x_1!(x-x_1)!}p^{x_1}(1-p)^{x-x_1}, \qquad (13.3\text{-}12)$$

where $p = m_1/(m_1 + m_2)$. Moreover, if X_1 and X_2 are independent nonnegative integer-valued random variables and (13.3-12) holds, then X_1 and X_2 must be Poisson variates with parameters in the ratio m_1 to m_2.

We can extend this result to n independent Poisson variates $X_1, \ldots, X_n$ with parameters $m_1, \ldots, m_n$, respectively, and we find that the conditional distribution of $X_1, \ldots, X_n$ given $\sum_{i=1}^{n} X_i$ is multinomial. This property is examined in more detail in the next section.

13.4 The Multinomial Distribution

$\mathbf{X} = (X_1, \ldots, X_t)$ is a *multinomial* random variable with probability vector $\mathbf{p} = (p_1, \ldots, p_t)$ provided the X_i take nonnegative integral values, sum to N (fixed), and

$$\Pr[X_1 = x_1, \ldots, X_t = x_t] = N! \prod_{i=1}^{t} \frac{p_i^{x_i}}{x_i!}, \qquad (13.4\text{-}1)$$

where $p_i > 0$ and $\sum_{i=1}^{t} p_i = 1$. We denote this multinomial by $\mathcal{M}_t(N, \mathbf{p})$. When $t = 2$, this distribution reduces to the binomial, which would be labeled as $\mathcal{M}_2(N, \mathbf{p})$ with $\mathbf{p} = (p, q)$.

The multinomial distribution can be generated via a series of N independent trials, in each of which one of t mutually independent events $\mathscr{E}_1, \ldots, \mathscr{E}_t$ occurs, and where the probability of $\mathscr{E}_i$ occurring is p_i, which is constant across trials. As mentioned in the preceding section, the multinomial can also be generated from a set of independent Poisson variates by conditioning on their sum.

13.4.1 *Moments*

The characteristic function of the multinomial is

$$(p_1 e^{is_1} + p_2 e^{is_2} + \cdots + p_t e^{is_t})^N, \qquad (13.4\text{-}2)$$

and the moment-generating function is

$$(p_1 e^{s_1} + p_2 e^{s_2} + \cdots + p_t e^{s_t})^N. \qquad (13.4\text{-}3)$$

The simplest extension of the moment formulas for the binomial involves looking at *mixed factorial moments,* and we find that

$$\mu_{(r_1,\ldots,r_t)} = E[X_1^{(r_1)}X_2^{(r_2)}\ldots X_t^{(r_t)}]$$

$$= E\left[\frac{X_1!}{(X_1 - r_1)!}\frac{X_2!}{(X_2 - r_2)!}\cdots\frac{X_t!}{(X_t - r_t)!}\right]$$

$$= N^{(\Sigma r_i)}p_1^{r_1}p_2^{r_2}\ldots p_t^{r_t}$$

$$= \frac{N!}{\left(N - \sum_{i=1}^{t} r_i\right)!}p_1^{r_1}p_2^{r_2}\ldots p_t^{r_t}, \qquad (13.4-4)$$

where $r_1,\ldots,r_t$ are nonnegative integers summing to less than $N + 1$. Haldane [1937] proved a special case of (13.4-4).

Since, as we show in the following section, the marginal distribution of the X_i $(i = 1,\ldots,t)$ is binomial with parameters N and p_i, the central moments of X_i are simply those of a binomial, e.g.,

$$E[X_i - Np_i]^2 = Np_iq_i \qquad q_i = 1 - p_i,$$

$$E[X_i - Np_i]^3 = Np_iq_i(q_i - p_i),$$

$$E[X_i - Np_i]^4 = Np_iq_i\{1 + 3(N - 2)p_iq_i\}, \qquad (13.4-5)$$

$$\text{Var}[(X_i - Np_i)^2] = Np_iq_i\{(p_i - q_i)^2 + 2(N - 1)p_iq_i\}. \qquad (13.4-6)$$

The simple form of (13.4-4) makes it convenient for us to work with factorial moments, and as a result we often are forced to handle products of factorial powers of N. The basic formula for such products, for $i \leq j$, is

$$N^{(i)}N^{(j)} = \sum_{k=0}^{i} \frac{i!j!}{k!(i - k)!(j - k)!}N^{(i+j-k)}. \qquad (13.4-7)$$

Table 13.4-1 gives the products of $N^{(i)}$ with $N^{(j)}$ for $i, j \leq 4$, and table 13.4-2 gives $N^{(i)}$ as a polynomial in N for $i = 1,\ldots,6$. These tables are of considerable use in Chapter 12.

From formula (13.4-4) and table 13.4-1, we get the following covariance formulas (see Morris [1966]):

$$\text{Cov}(X_i, X_j) = \begin{cases} -Np_ip_j & i \neq j, \\ Np_iq_i & i = j, \end{cases} \qquad (13.4-8)$$

$$\text{Cov}(X_i, X_j^{(2)}) = \begin{cases} -2N^{(2)}p_ip_j^2 & i \neq j, \\ 2N^{(2)}q_ip_i^2 & i = j, \end{cases} \qquad (13.4-9)$$

$$\text{Cov}(X_i^{(2)}, X_j^{(2)}) = \begin{cases} -(4N^{(3)} + 2N^{(2)})p_i^2p_j^2 & i \neq j \\ (4N^{(3)}p_i + 2N^{(2)})p_i^2 - (4N^{(3)} + 2N^{(2)})p_i^4 & i = j. \end{cases} \qquad (13.4-10)$$

Table 13.4-1 Products of $N^{(i)}$ with $N^{(j)}$ (for $i, j \leqq 4$)

	N	$N^{(2)}$	$N^{(3)}$	$N^{(4)}$
N	$N^{(2)} + N$	—	—	—
$N^{(2)}$	$N^{(3)} + 2N^{(2)}$	$N^{(4)} + 4N^{(3)} + 2N^{(2)}$	—	—
$N^{(3)}$	$N^{(4)} + 3N^{(3)}$	$N^{(5)} + 6N^{(4)} + 6N^{(3)}$	$N^{(6)} + 9N^{(5)} + 18N^{(4)} + 6N^{(3)}$	—
$N^{(4)}$	$N^{(5)} + 4N^{(4)}$	$N^{(6)} + 8N^{(5)} + 12N^{(4)}$	$N^{(7)} + 12N^{(6)} + 36N^{(5)} + 24N^{(4)}$	$N^{(8)} + 16N^{(7)} + 72N^{(6)} + 96N^{(5)} + 24N^{(4)}$

Table 13.4-2 $N^{(i)}$ as a Polynomial in N,
and N^i as a Linear Function of $N^{(j)}$
$(i = 1, \ldots, 6)$

a.

i	$N^{(i)}$
1	N
2	$N^2 - N$
3	$N^3 - 3N^2 + 2N$
4	$N^4 - 6N^3 + 11N^2 - 6N$
5	$N^5 - 10N^4 + 35N^3 - 50N^2 + 24N$
6	$N^6 - 15N^5 + 85N^4 - 225N^3 + 274N^2 - 120N$

b.

i	N^i
1	N
2	$N^{(2)} + N^{(1)}$
3	$N^{(3)} + 3N^{(2)} + N^{(1)}$
4	$N^{(4)} + 6N^{(3)} + 7N^{(2)} + N^{(1)}$
5	$N^{(5)} + 10N^{(4)} + 25N^{(3)} + 15N^{(2)} + N^{(1)}$
6	$N^{(6)} + 15N^{(5)} + 65N^{(4)} + 90N^{(3)} + 31N^{(2)} + N^{(1)}$

13.4-2 Marginal and conditional distributions

Let

$$X'_1 = \sum_{i=1}^{\alpha_1} X_i,$$

$$X'_2 = \sum_{i=\alpha_1+1}^{\alpha_2} X_i,$$

$$\vdots$$

$$X'_s = \sum_{i=\alpha_{s-1}+1}^{\alpha_s} X_i, \tag{13.4-11}$$

where $(X_1, \ldots, X_t)$ is a multinomial $\mathcal{M}_t(N, \mathbf{p})$, $1 \leq \alpha_1 < \alpha_2 < \cdots < \alpha_{s-1} < \alpha_s = t$, and $s < t$. Then $\mathbf{X}' = (X'_1, \ldots, X'_s)$ also has a multinomial distribution with parameters N and $(p'_1, \ldots, p'_s)$, where

$$p'_1 = \sum_{i=1}^{\alpha_1} p_i, \quad \ldots, \quad p'_s = \sum_{i=\alpha_{s-1}+1}^{\alpha_s} p_i, \tag{13.4-12}$$

i.e., $\mathbf{X}' \sim \mathcal{M}_s(N, \mathbf{p}')$. This is the *marginal distribution for sums of the multinomial component variates*, and the result says that the marginal distribution is also multinomial.

The foregoing result is of great importance, and we use it throughout this book when we deal with the distribution of the marginal totals of a cross-classification when the distribution of the entire table is multinomial. For example, suppose we have an $I \times J$ contingency table whose entries $\mathbf{X} = (X_{11}, X_{12}, \ldots, X_{IJ})$ are random variables following a multinomial distribution $\mathcal{M}_{IJ}(N, \mathbf{p})$, where $\mathbf{p} =$

$(p_{11}, \ldots, p_{IJ})$. Then the result for the marginal distribution of multinomial component variates implies that the row totals

$$\mathbf{X}^{(R)} = (X_{1+}, \ldots, X_{I+})$$

have a multinomial distribution with the same sample size N and cell probabilities

$$\mathbf{p}^{(R)} = (p_{1+}, \ldots, p_{I+}).$$

Similarly, the column totals $\mathbf{X}^{(C)} = (X_{+1}, \ldots, X_{+J})$ have a multinomial distribution with the same sample size N and the corresponding column probabilities $\mathbf{p}^{(C)} = (p_{+1}, \ldots, p_{+J})$. These two marginal multinomial random variables are independent if and only if the variable for the row categories is independent of the variable for the column categories.

The conditional distribution of a subset of the components of $\mathbf{X}$, i.e., $\mathbf{X}^* = (X_1^*, \ldots, X_s^*) = (X_{\alpha_1}, \ldots, X_{\alpha_s})$ for $s < t$, given the values of the remaining components of $\mathbf{X}$, is also multinomial, and depends on these remaining components only through their observed sum $S = N - \sum_{i=1}^{s} X_i^*$. If we let $N^* = N - S$ and $p_i^* = p_{\alpha_i}/\sum_{j=1}^{s} p_{\alpha_j}$, then the distribution of X^* is $\mathscr{M}_s(N^*, \mathbf{p}^*)$. This conditional distribution result allows us to break up the multinomial likelihood function into pieces, each of which is a multinomial likelihood, and we make use of this fact in Chapters 4 and 5, as well as in Section 13.4.4 below.

To illustrate the conditional breakdown of the multinomial, we again consider an $I \times J$ table whose entries follow a multinomial distribution $\mathscr{M}_{IJ}(N, \mathbf{p})$. If the observed values of the row totals $\mathbf{x}^{(R)} = (x_{1+}, \ldots, x_{I+})$ are given, then the conditional distribution of the entries in each row is multinomial, where for the ith row the sample size is x_{i+} and the cell probabilities are $(p_{i1}/p_{i+}, \ldots, p_{iJ}/p_{i+})$. Similarly, the conditional distribution of the entries for each column, given the observed column total, is also multinomial. Thus the multinomial for the $I \times J$ table breaks up into either $I + 1$ or $J + 1$ multinomial parts: a marginal multinomial for the row totals and I conditional multinomials for the I rows, or a marginal multinomial for the column totals and J conditional multinomials for J columns, i.e., breaking the multinomial into $I + 1$ parts, we get

$$\frac{N!}{\prod_{i,j} x_{ij}!} \prod_{i,j} p_{ij}^{x_{ij}} = \prod_i \left[\frac{x_{i+}!}{\prod_j x_{ij}!} \prod_j \left(\frac{p_{ij}}{p_{i+}} \right)^{x_{ij}} \right] \times \left[\frac{N!}{\prod_i x_{i+}!} \prod_i p_{i+}^{x_{i+}} \right].$$

Combining the conditional and marginal distribution results in another way, we find that

$$E[X_i' | X_{\alpha_1}', \ldots, X_{\alpha_l}'] = \left(N - \sum_{j=1}^{l} X_{\alpha_j}' \right) p_i' \left(1 - \sum_{j=1}^{l} p_{\alpha_j}' \right)^{-1} \qquad (13.4\text{-}13)$$

$(\alpha_j \neq i$ and $l < s)$, which says that the *multiple regression* of X_i' on $X_{\alpha_1}', \ldots, X_{\alpha_l}'$ is *linear*.

13.4.3 *Estimation of parameters*

If $\mathbf{X}$ has the multinomial distribution $\mathscr{M}_t(N, \mathbf{p})$, then we get the maximum likelihood estimator $\hat{\mathbf{p}}$ of $\mathbf{p}$ by maximizing the logarithm

$$\sum_{i=1}^{t} x_i \log p_i \qquad (13.4\text{-}14)$$

of the likelihood function with respect to the p_i, subject to the constraint that they are nonnegative and sum to unity, i.e., $p_i \geq 0$ for $i = 1, \ldots, t$ and $\sum_{i=1}^{t} p_i = 1$. We incorporate the constraint using a Lagrangian multiplier, and so we maximize the function

$$L = \sum_{i=1}^{t} x_i \log p_i + \lambda \left(\sum_{i=1}^{t} p_i - 1 \right) \qquad (13.4\text{-}15)$$

with respect to $\mathbf{p}$ and λ. Setting the first partial derivatives of (13.4-15) equal to zero, we get

$$\frac{\partial L}{\partial p_i} = \frac{x_i}{p_i} + \lambda = 0 \quad i = 1, \ldots, t, \qquad (13.4\text{-}16)$$

$$\frac{\partial L}{\partial \lambda} = \sum_{i=1}^{t} p_i - 1 = 0. \qquad (13.4\text{-}17)$$

The solution to (13.4-16) and (13.4-17) is

$$\hat{p}_i = \frac{x_i}{N} \quad i = 1, \ldots, t, \qquad (13.4\text{-}18)$$

and $\hat{\lambda} = -N$. Taking the second partial derivative of (13.4-15) and substituting these values, we find that they do yield the maximum of L, and thus (13.4-18) gives the maximum likelihood estimator of $\mathbf{p}$.

When the p_i are functions of a reduced number of parameters $\theta_1, \ldots, \theta_s$ ($s < t - 1$), the maximum likelihood estimation is more complex. We deal with various facets of this problem in Chapters 3 and 5, and in the next section we discuss a result that is of considerable use in such estimation problems.

13.4.4 Relationship between Poisson and multinomial estimators

In the preceding section, we pointed out that if $X_1, \ldots, X_t$ are t independent Poisson variates with parameters $m_1, \ldots, m_t$, respectively, then the conditional distribution of $\mathbf{X} = (X_1, \ldots, X_t)$ given $\sum_{i=1}^{t} X_i = N$ is multinomial, $\mathcal{M}_t(N, \mathbf{p})$, where $p_i = m_i / \sum_{i=1}^{t} m_i$. This result and the fact that certain conditional distributions based on the multinomial are also multinomial (see the preceding section) underly the following result on maximum likelihood estimation. Our presentation follows closely that of Birch [1963].

Our plan is to compare two different sampling experiments and their associated MLEs.

(i) There are t independent Poisson variates with parameters $m_i(\theta)$, $i = 1, \ldots, t$, and observed frequencies x_i, $i = 1, \ldots, t$, respectively. Here θ can be a single parameter or a vector of parameters.

(ii) There are s independent multinomials, and for the jth one we have a fixed sample size of

$$N_j = \sum_{i=\alpha_{j-1}+1}^{\alpha_j} x_i, \qquad (13.4\text{-}19)$$

where $\alpha_0 = 0$ and $\alpha_s \leq t$. The observed values in this jth multinomial are $x_{\alpha_{j-1}+1}, \ldots, x_{\alpha_j}$ in the 1st, $\ldots$, $(\alpha_j - \alpha_{j-1})$st categories, while the probability of falling into the γth category is $m_{\alpha_{j-1}+\gamma}(\theta)/N_j$, where

$$\sum_{i=\alpha_{j-1}+1}^{\alpha_j} m_i(\theta) = N_j \quad j = 1, \ldots, s. \tag{13.4-20}$$

In addition, we have $t - \alpha_s$ independent Poisson values with parameters m_i for $i = \alpha_s + 1, \ldots, t$. Moreover, $m_i(\theta)$ for $i = 1, \ldots, t$ is the same function of θ in both sampling experiments.

THEOREM 13.4-1   *Denote the MLE of θ in (i) by $\hat{\theta}$ and in (ii) by $\hat{\theta}^*$, and let $\hat{m}_i$ and $\hat{m}_i^*$ be abbreviations for $m_i(\hat{\theta})$ and $m_i(\hat{\theta}^*)$, respectively, $i = 1, \ldots, t$. Then $\hat{\theta} = \hat{\theta}^*$ and $\hat{m}_i = \hat{m}_i^*$ for $i = 1, \ldots, t$ (the MLEs are the same for the two sampling schemes) if and only if*

$$\sum_{i=\alpha_{j-1}+1}^{\alpha_j} \hat{m}_i = N_j \quad j = 1, \ldots, s. \tag{13.4-21}$$

Proof   Expression (13.4-20) implies that

$$\sum_{i=\alpha_{j-1}+1}^{\alpha_j} \hat{m}_i^* = N_j \quad j = 1, \ldots, s, \tag{13.4-22}$$

and so the equivalence of the MLEs implies (13.4-21).

Conversely, to prove that (13.4-21) implies the equivalence of the MLEs we consider the likelihood functions. For experiment (i), we have the likelihood

$$l_1 = \prod_{i=1}^{t} \frac{m_i^{x_i}}{x_i!} e^{-m_i} \tag{13.4-23}$$

and for experiment (ii),

$$l_2 = \prod_{j=1}^{s} \left\{ N_j! \prod_{i=\alpha_{j-1}+1}^{\alpha_j} \frac{\left(\dfrac{m_i}{N_j}\right)^{x_i}}{x_i!} \right\} \prod_{k=\alpha_s+1}^{t} \frac{m_k^{x_k}}{x_k!} e^{-m_k}. \tag{13.4-24}$$

Expressions (13.4-19) and (13.4-20) allow us to rewrite (13.4-24) as

$$\prod_{j=1}^{s} \left\{ \frac{N_j! \, e^{N_j}}{N_j^{N_j}} \prod_{i=\alpha_{j-1}+1}^{\alpha_j} \left[\frac{m_i^{x_i} e^{-m_i}}{x_i!} \right] \right\} \prod_{k=\alpha_s+1}^{t} \frac{m_k^{x_k} e^{-m_k}}{x_k!}$$

$$= \prod_{j=1}^{s} \frac{N_j! \, e^{N_j}}{N_j^{N_j}} \prod_{i=1}^{t} \frac{m_i^{x_i} e^{-m_i}}{x_i!}. \tag{13.4-25}$$

Since (13.4-21) implies that N_j ($j = 1, \ldots, s$) does not depend on θ, (13.4-25) is a constant multiple of (13.4-23), and the result follows. ∎

The only difference between the two sampling experiments described above for the purposes of maximum likelihood estimation is that for (ii) we have s additional constraints on $\mathbf{p}$, given by (13.4-20). When (13.4-21) is true, these constraints are also effectively operative in (i), and so the MLEs for (i) and (ii) are the same. Experiment (ii) in effect describes a class of sampling schemes for different choices of α_j ($j = 1, \ldots, s$) and s. We can prove a theorem similar to theorem 13.4-1 demonstrating the equivalence of two sets of MLEs based on two experiments from this class. In this more general theorem, (13.4-21) is replaced by a pair of conditions requiring the MLEs under each experiment to satisfy the explicit constraints imposed by the other. This more general result contains theorem 13.4-1 as a special case, since we can choose $s = 0$ for one sampling scheme, thus yielding experiment (i). If we take $s = 1$ and $\alpha_s = t$, we get a single multinomial, and the MLEs under it are equivalent to those under (i), provided

$$\sum_{i=1}^{t} \hat{m}_i = N. \tag{13.4-26}$$

Because of the results just described, we often get the same maximum likelihood estimators for the expected cell entries in a contingency table under a variety of Poisson-multinomial sampling schemes, such as experiment (ii) above. Each such sampling scheme implies one set of *constraints* on the expected cell entries, and in these constraints any particular cell entry appears just once. Thus if the basic Poisson sampling scheme in (i) implies that the MLEs of the $\{m_i\}$ satisfy the set of constraints implied by each of several experiments of the form (ii), the MLEs under each of the experiments are the same as those under the Poisson scheme. For example, if there are IJ m's cross-classified in an $I \times J$ table, theorem 13.4-1 implies that under the independence model ($m_{ij} = a_i b_j$, $i = 1, \ldots, I$; $j = 1, \ldots, J$), the MLEs for the expected cell values under the Poisson scheme are the same as those under one multinominal for all IJ cells, or I independent multinomials, one for each row, or J independent multinomials, one for each column.

13.5   The Hypergeometric Distribution

Consider a population of N objects, of which M are of a particular type, say type 1. If we take a sample of size n from this population, then the number X of type 1 objects has the hypergeometric distribution

$$\Pr[X = x] = \frac{\binom{M}{x}\binom{N-M}{n-x}}{\binom{N}{n}} \tag{13.5-1}$$

for $\max(0, n + M - N) \leq x \leq \min(n, M)$. The distribution gets its name because the hypergeometric probabilities are scaled terms from the series expansion of the hypergeometric function

$$F(\alpha, \beta; \gamma; \delta) = 1 + \frac{\alpha\beta}{\gamma} \cdot \frac{\delta}{1!} + \frac{\alpha(\alpha+1)\beta(\beta+1)}{\gamma(\gamma+1)} \cdot \frac{\delta^2}{2!} + \cdots, \tag{13.5-2}$$

i.e., the right-hand sides of (13.5-1) are the terms in the expansion of

$$\frac{(N-n)!(N-M)!}{N!(N-M-n)!} F(-n, -M; N-M-n+1; 1). \tag{13.5-3}$$

For example, the first term is

$$\frac{(N-n)!(N-M)!}{N!(N-M-n)!} = \frac{\binom{N-M}{n}}{\binom{N}{n}}, \tag{13.5-4}$$

and the second term is

$$\frac{(N - n)!(N - M)!}{N!(N - M - n)!} \times \frac{(-n)(-M)(1)}{(N - M - n + 1)} = M\frac{\binom{N - M}{n - 1}}{\binom{N}{n}}. \quad (13.5\text{-}5)$$

If we sample sequentially, returning each object to the population after its selection, the probability of drawing a type 1 object at each step is M/N, and the distribution of X is binomial, with parameters n and M/N. The reason the distribution is not binomial is that we are sampling without replacement.

Lieberman and Owen [1961] present an extensive set of tables of the individual hypergeometric terms (13.5-1) and cumulative values.

13.5.1 *Moments*

The moment-generating function of the hypergeometric is

$$E(e^{sX}) = \frac{(N - n)!(N - M)!}{N!}F(-n, -M; N - M - n + 1; e^s), \quad (13.5\text{-}6)$$

where F is the hypergeometric function defined by (13.5-2). This is not a useful function for finding moments.

Once again, the factorial moments are quite easy to compute:

$$\mu_{(r)} = E(X^{(r)}) = \frac{n^{(r)}M^{(r)}}{N^{(r)}} = \frac{(N - r)!n!M!}{(n - r)!(M - r)!N!}. \quad (13.5\text{-}7)$$

From (13.5-7) we get the following formulas for the first three central moments:

$$\mu = E(X) = n\left(\frac{M}{N}\right),$$

$$\mu_2 = \text{Var}(X) = n\left(\frac{N - n}{N - 1}\right)\left(\frac{M}{N}\right)\left(1 - \frac{M}{N}\right), \quad (13.5\text{-}8)$$

$$\mu_3 = n\left(\frac{N - n}{N - 1}\right)\left(\frac{N - 2n}{N - 2}\right)\left(\frac{M}{N}\right)\left(1 - \frac{M}{N}\right)\left(1 - \frac{2M}{N}\right).$$

If we think of M/N as the proportion of objects of type 1, then the moments given by (13.5-8) resemble those for the binomial, except for multiplicative terms in the formulas for μ_2 and μ_3 which occur because we are sampling without replacement.

13.5.2 *Estimation of parameters*

The hypergeometric distribution has three parameters, N, M, and n. Most often we know N and n, and then the maximum likelihood estimator of M is

$$\hat{M} = \left[\frac{x(N + 1)}{n}\right], \quad (13.5\text{-}9)$$

where $[a]$ denotes the greatest integer not exceeding a. When $x(N + 1)/n$ is an integer, $\hat{M}$ equals either $x(N + 1)/n - 1$ or $x(N + 1)/n$.

If M and n are known, the MLE of N is

$$\hat{N} = \left[\frac{nM}{x}\right], \tag{13.5-10}$$

where again $[a]$ is the greatest integer not exceeding a. If nM/x is an integer, nM/x and $nM/x - 1$ are both MLEs. The positive moments of nM/x are infinite unless $n > N - M$. This is because x can take the value 0. The estimators

$$\frac{(n + 1)(M + 1)}{(x + 1)} - 1, \qquad \frac{(n + 2)(M + 2)}{(x + 2)} \tag{13.5-11}$$

do not have infinite moments because the denominators of the fractions are never zero. Alternatively, we can replace x in (13.5-10) by some nonzero value whenever x is zero.

13.5.3   *Relationships to other distributions*

Let X_1 and X_2 be binomial variates with parameters (N_1, p) and (N_2, p), respectively. Then the distribution of X_1 conditional on $X_1 + X_2 = n$ is

$$\Pr[X_1 = x | X_1 + X_2 = n] = \frac{\binom{N_1}{x}p^x q^{N_1 - x}\binom{N_2}{n - x}p^{n - x}q^{N_2 - (n - x)}}{\binom{N_1 + N_2}{n}p^n q^{N_1 + N_2 - n}}$$

$$= \frac{\binom{N_1}{x}\binom{N_2}{n - x}}{\binom{N_1 + N_2}{n}}, \tag{13.5-12}$$

i.e., the hypergeometric with $M = N_1$ and $N = N_1 + N_2$. Since we can get the binomial as the conditional distribution of a pair of independent Poisson variates given their sum, we can get the hypergeometric from a set of four independent Poisson variates conditional on three different sums. Thus if X_1, X_2, X_3, and X_4 are independent Poisson variates with parameters m_1, m_2, m_3, and m_4, respectively, with $m_1/(m_1 + m_2) = m_3/(m_3 + m_4)$, the distribution of X_1 conditional on $X_1 + X_2 = N_1$, $X_3 + X_4 = N_2$, and $X_1 + X_3 = n$ is hypergeometric with probabilities as in expression (13.5-12).

13.6   The Multivariate Hypergeometric Distribution

Consider a population of N objects, of which N_i are of type $i, i = 1, \ldots, t$, with $\sum_{j=1}^{t} N_j = N$. If we take a sample of size n from this population (without replacement), then the joint distribution of the random variables $X_1, \ldots, X_t$ giving the numbers of objects of types $1, \ldots, t$, respectively, in the sample is

$$\Pr[X_1 = x_1, X_2 = x_2, \ldots, X_t = x_t] = \frac{\left\{\prod_{i=1}^{t}\binom{N_i}{x_i}\right\}}{\binom{N}{n}}, \tag{13.6-1}$$

where $\sum_{i=1}^{t} x_i = n$ and $0 \le x_i \le N_i$ for $i = 1, \ldots, t$. The distribution gets its name because the multivariate hypergeometric probabilities are scaled terms from the series expansion of the multivariate hypergeometric series (see Steyn [1955]). This is more complicated than the expansion given by (13.5-2), and we omit the details. When $t = 2$, the distribution reduces to the regular univariate hypergeometric.

13.6.1 *Moments*

As with the hypergeometric distribution, the moment-generating function of the multivariate hypergeometric is of little use for finding moments, but the factorial moments are quite easy to compute:

$$\mu_{(r_1, r_2, \ldots, r_t)} = E\left[\prod_{j=1}^{t} X_j^{(r_j)} \right]$$

$$= \left[\frac{n^{(\Sigma r_j)}}{N^{(\Sigma r_j)}} \right] \prod_{i=1}^{t} N_i^{(r_i)}. \tag{13.6-2}$$

Expression (13.6-2) yields the following formulas for first and second moments:

$$E[X_i] = \frac{nN_i}{N}, \tag{13.6-3}$$

$$\mathrm{Var}(X_i) = n\left(\frac{N-n}{N-1} \right)\left(\frac{N_i}{N} \right)\left(1 - \frac{N_i}{N} \right), \tag{13.6-4}$$

and for $i \ne j$,

$$\mathrm{Cov}(X_i, X_j) = -\frac{nN_iN_j}{N^2}\left(\frac{N-n}{N-1} \right). \tag{13.6-5}$$

Note the similarity between (13.6-4) and (13.6-5) and the corresponding variance-covariance formulas for the multinomial distribution (13.4-8). Expressions (13.6-3) and (13.6-4) for the mean and variance also follow from the fact that the marginal distribution X_i is hypergeometric with parameters n, N_i, and N (see the following section).

13.6.2 *Marginal and conditional distributions*

The results for marginal and conditional distributions here parallel those for the multinomial distribution. Let

$$X_1' = \sum_{i=1}^{\alpha_1} X_i,$$

$$X_2' = \sum_{i=\alpha_1+1}^{\alpha_2} X_i, \tag{13.6-6}$$

$$\vdots$$

$$X_s' = \sum_{i=\alpha_{s-1}+1}^{\alpha_s} X_i,$$

where $\mathbf{X} = (X_1, \ldots, X_t)$ has a multivariate hypergeometric distribution with parameters n and $(N_1, \ldots, N_t)$, and $1 \leq \alpha_1 < \cdots < \alpha_{s-1} < t$ with $s < t$. Then $\mathbf{X}' = (X'_1, \ldots, X'_s)$ also has a multivariate hypergeometric distribution with parameters n and $(N'_1, \ldots, N'_s)$, where $N'_j = \sum_{i=\alpha_{j-1}+1}^{\alpha_j} N_i$ for $j = 1, \ldots, s$. This is the *marginal distribution* for sums of the multivariate hypergeometric component variates, and when $s = 2$, we get the univariate hypergeometric distribution.

The conditional distribution of a subset of the components of $\mathbf{X}$, i.e., $\mathbf{X}^* = (X_{\alpha_1}, \ldots, X_{\alpha_s})$ for $s < t$, given the values of the remaining components of $\mathbf{X}$ is also multivariate hypergeometric and depends on these remaining components only through their observed sum $S = N - \sum_{i=1}^{s} X_{\alpha_i}$.

13.6.3 *Relationships to other distributions*

If $X_1, \ldots, X_t$ are binomial variates with sample sizes $N_1, \ldots, N_t$, respectively and a common p, then the distribution of $\mathbf{X} = (X_1, \ldots, X_t)$ conditional on $\sum_{i=1}^{t} X_i = n$ is multivariate hypergeometric with parameters n, N, and $(N_1, \ldots, N_t)$. This conditional distribution has been used by many authors for the analysis of a $2 \times t$ contingency table with fixed marginal totals (e.g., see Van Eeden [1965]). Haberman [1974] and Zelen [1971], among others, have considered generalizations of the multivariate hypergeometric. Haberman refers to these generalizations as "conditional Poisson distributions," since they can arise from a set of independent Poisson variates subject to a set of linear constraints.

As N tends to infinity with $N_i/N = p_i$ constant, the multivariate hypergeometric tends to the multinomial distribution $\mathcal{M}_t(N, \mathbf{p})$.

13.7 The Negative Binomial Distribution

The random variable X has a negative binomial distribution with positive parameters k and m if

$$\Pr[X = x] = \binom{k + x - 1}{x} \left(\frac{m}{m + k}\right)^x \left(\frac{k}{m + k}\right)^k \qquad (13.7\text{-}1)$$

for $x = 0, 1, \ldots$. The parameters k and m need not be integers.

The negative binomial distribution arises in several different ways. If a proportion p of individuals in a population possesses a certain characteristic, then the number of observations x in excess of k (here k is an integer) that must be taken to obtain just k individuals with the characteristic has a negative binomial distribution, with $p = k/(m + k)$. The process just described is often referred to as inverse sampling.

A second way to generate the negative binomial starts with a Poisson variate with parameter $\lambda > 0$ and gives λ the gamma distribution with probability density function

$$f(\lambda) = [\Gamma(k)]^{-1} \left(\frac{k}{m}\right)^k \lambda^{k-1} e^{-\lambda k/m} \qquad \lambda > 0, k > 0, m > 0. \qquad (13.7\text{-}2)$$

Averaging over this distribution for λ gives us the negative binomial.

When individuals come in groups, where the number of groups has a Poisson distribution and the number of individuals in any group follows a common logarithmic distribution, the number of individuals follows a negative binomial distribution. Thus, the negative binomial is a simple example of a "contagious" distribution.

The negative binomial also occurs as an equilibrium distribution for birth and death processes.

13.7.1   *Moments*

For the negative binomial random variable X with parameters k and m, the moment-generating function is

$$E(e^{sX}) = k^k(m + k - me^s)^{-k}. \tag{13.7-3}$$

The factorial moments of X are given by

$$\mu_{(r)} = E(X^{(r)}) = (k + r - 1)^{(r)} \left(\frac{m}{k}\right)^r. \tag{13.7-4}$$

from which it follows that

$$E(X) = m, \tag{13.7-5}$$

$$\mu_2 = \text{Var}(X) = \frac{m(m + k)}{k}, \tag{13.7-6}$$

$$\mu_3 = \frac{m(m + k)(2m + k)}{k^2}. \tag{13.7-7}$$

13.7.2   *Estimation of parameters*

Suppose $X_1, \ldots, X_n$ are independent identically distributed negative binomial variates with parameters m and k. Then the maximum likelihood estimate of m is

$$\hat{m} = \overline{X} = \frac{1}{n} \sum_{i=1}^{n} X_i, \tag{13.7-8}$$

and the MLE of k is a root of the following equation:

$$n \log\left(1 + \frac{\bar{x}}{k}\right) = \sum_{j=1}^{\infty} f_j\left(\frac{1}{k} + \frac{1}{k + 1} + \cdots + \frac{1}{k + j - 1}\right), \tag{13.7-9}$$

where $\bar{x}$ is the average of the observed values $x_1, \ldots, x_n$ and f_j is the number of x_i taking the value j. For $s^2 = (1/(n - 1)) \sum_{i=1}^{n} (x_i - \bar{x})^2 \leq \bar{x}$, the two sides of (13.7-9) tend to equality as $k \to \infty$, so setting $\hat{k} = \infty$ seems appropriate. For $s^2 > \bar{x}$, the equation always has at least one root (see Fisher [1941]). The derivation of (13.7-8) and (13.7-9) requires extensive algebraic manipulation, which we do not include here.

Instead of maximum likelihood, we can use the method of moments to get simpler estimators of m and k:

$$m^* = \hat{m} = \bar{x}, \tag{13.7-10}$$

$$k^* = \frac{\bar{x}^2}{s^2 - \bar{x}}. \tag{13.7-11}$$

The variances of the two sets of estimators above are:

$$\text{Var}(\hat{m}) = \text{Var}(m^*) = \frac{1}{n}\left(m + \frac{m^2}{k}\right), \tag{13.7-12}$$

$$\text{Var}(\hat{k}) = \frac{2k(k+1)}{n\left(\dfrac{m}{m+k}\right)^2}\bigg/\left[1 + \frac{4\left(\dfrac{m}{m+k}\right)}{3(k+2)} + \frac{3\left(\dfrac{m}{m+k}\right)^2}{(k+2)(k+3)} + \cdots\right], \tag{13.7-13}$$

$$\text{Var}(k^*) = \frac{2k(k+1)}{n\left(\dfrac{m}{m+k}\right)^2}. \tag{13.7-14}$$

The variance of $\hat{m}$ is the sum of two parts: the first part is the variance of the MLE of m when it is a parameter of a Poisson distribution, and the second part is the added variation when the Poisson parameter has a distribution with parameters k and m. Anscombe [1950] indicates that k^* is at least 90% efficient relative to $\hat{k}$ whenever

$$\frac{(k+m)(k+2)}{m} \geqq 15. \tag{13.7-15}$$

13.7.3 *Relationship to other distributions*

If X is a negative binomial variate with parameters k and m, the distribution of X tends to that of a Poisson variate with mean m as $k \to \infty$. Thus the Poisson distribution is a special case of the negative binomial, and the negative binomial as a result is often a first choice as an alternative to the Poisson whenever the observed variation appears to be too large for Poisson variation (i.e., $s^2 \gg \bar{x}$).

13.8 The Negative Multinomial Distribution

The random variable $\mathbf{X} = (X_1, \ldots, X_t)$ has a negative multinomial distribution with positive parameters $k, m_1, \ldots, m_t$ if

$$\Pr[X_1 = x_1, \ldots, X_t = x_t]$$

$$= \binom{k + \sum_{i=1}^{t} x_i - 1}{x_1, \ldots, x_t, k - 1}\left(\frac{k}{\sum_{i=1}^{t} m_i + k}\right)^k \prod_{i=1}^{t}\left(\frac{m_i}{\sum_{i=1}^{t} m_i + k}\right)^{x_i}$$

$$= \frac{\left(k + \sum_{i=1}^{t} x_i - 1\right)!}{x_1! \ldots x_t!(k-1)!}\left(\frac{k}{\sum_{i=1}^{t} m_i + k}\right)^k \prod_{i=1}^{t}\left(\frac{m_i}{\sum_{i=1}^{t} m_i + k}\right)^{x_i} \tag{13.8-1}$$

for $x_i = 0, 1, \ldots$ ($i = 1, \ldots, t$). For $t = 1$, (13.8-1) reduces to (13.7-1), and we get the negative binomial distribution. The parameter k need not be an integer.

Neyman [1965] describes the historical development of the negative multinomial. Steyn [1955, 1959] uses the negative multinomial as a sampling distribution for the cells in a two-way contingency table. He considers situations where observations are taken until a fixed number of observations, say five, is observed in a given cell, say the (1, 1) cell. Here the count in the (1, 1) cell is fixed a priori but the total number of observations in the table is not. In such situations we can no longer use the usual chi square test for independence, and Steyn derives an alternate test statistic which does have an asymptotic χ^2 distribution.

Rudolf [1967], extending the work of Steyn, considers a sampling model which is part multinomial and part negative multinomial. For a two-way contingency table, this model arises by fixing one of the row or column marginal totals instead of the total sample size (as for the multinomial case) or one of the cell counts (as in the negative multinomial case). Thus the distribution of one set of marginal totals is negative multinomial. Rudolf shows that both the exact and the asymptotic chi square test statistics for independence under this sampling model are identical to the usual ones under the multinomial model. We do not consider the work of Steyn and Rudolf elsewhere in this book.

13.8.1   *Moments*

The moment-generating function of the negative multinomial is

$$k^k \left(\sum_{i=1}^{t} m_i + k - \sum_{i=1}^{t} m_i e^{s_i} \right)^{-k}, \tag{13.8-2}$$

and the mixed factorial moments are

$$\mu_{(r_1, r_2, \ldots, r_t)} = E[X_1^{(r_1)} X_2^{(r_2)} \ldots X_t^{(r_t)}]$$

$$= \left(k + \sum_{i=1}^{t} r_i - 1 \right)^{\left(\sum_{i=1}^{t} r_i \right)} \prod_{i=1}^{t} \left(\frac{m_i}{k} \right)^{r_i} \tag{13.8-3}$$

$$= \frac{\left(k + \sum_{i=1}^{t} r_i - 1 \right)!}{(k-1)!} \prod_{i=1}^{t} \left(\frac{m_i}{k} \right)^{r_i}.$$

Since the marginal distribution of X_i $(i = 1, \ldots, t)$ is negative binomial with parameters m_i and k, the moments of X_i are simply those of the negative binomial as given by (13.7-4), with m and X replaced by m_i and X_i.

13.8.2   *Marginal and conditional distributions*

Let

$$X_1' = \sum_{i=1}^{\alpha_1} X_i,$$

$$X_2' = \sum_{i=\alpha_1+1}^{\alpha_2} X_i, \tag{13.8-4}$$

$$\vdots$$

$$X_s' = \sum_{i=\alpha_{s-1}+1}^{\alpha_s} X_i,$$

where **X** is negative multinomial with parameters k and **m**, and $1 \leq \alpha_1 < \cdots < \alpha_{s-1} < \alpha_s = t$. Then the marginal distribution of $\mathbf{X}' = (X_1', \ldots, X_s')$ is also negative multinomial with parameters k and $\mathbf{m}' = (m_1', \ldots, m_s')$, where

$$m_1' = \sum_{i=1}^{\alpha_1} m_i, \quad \ldots, \quad m_s' = \sum_{i=\alpha_{s-1}+1}^{\alpha_s} m_i. \tag{13.8-5}$$

When $s = 2$, the marginal distributions are negative binomial.

The conditional distribution of a subset of the components of $\mathbf{X}$, i.e., $\mathbf{X}^* = (X_1^*, \ldots, X_s^*) = (X_{\alpha_1}, \ldots, X_{\alpha_s})$ for $s < t$, given the values of the remaining components of $\mathbf{X}$, is also negative multinomial with parameters $\mathbf{m}_\alpha = (m_{\alpha_1}, \ldots, m_{\alpha_s})$ and $k_\alpha = k + \sum_{i=1}^t X_i - \sum_{i=1}^s X_{\alpha_i}$ if we condition on $\sum_{i=1}^t X_i - \sum_{i=1}^s X_{\alpha_i} = \sum_{i=1}^t m_i - \sum_{i=1}^s m_{\alpha_i}$.

13.8.3 Estimation of parameters

Suppose $\mathbf{X}$ has a negative multinomial distribution with parameters k and $\mathbf{m} = (m_1, \ldots, m_t)$. The likelihood equations yield:

$$\hat{m}_i = X_i \quad i = 1, \ldots, t, \tag{13.8-6}$$

$$\sum_{j=1}^n \left(\frac{1}{\hat{k} + j - 1} \right) = \log \left(1 + \frac{n}{\hat{k}} \right), \tag{13.8-7}$$

where

$$n = \sum_{i=1}^t X_i. \tag{13.8-8}$$

Equation (13.8-7) has no solution for a positive value of $\hat{k}$, and this is to be expected since there are $t + 1$ parameters to be estimated from a t-dimensional vector of observations. Johnson and Kotz [1969] give the details of the estimation of k and $\mathbf{m}$ when there is a sample of observations on $\mathbf{X}$. The MLE for $\mathbf{m}$ is

$$\hat{\mathbf{m}} = \overline{\mathbf{X}}, \tag{13.8-9}$$

and $\hat{k}$ is a root of an equation analogous to (13.7-9).

14 Asymptotic Methods

14.1 Introduction

Many common statistical methods for discrete data make use of asymptotic results, primarily because of the difficulty of making exact calculations in the discrete case. For example, to compute the exact tail probabilities for a distribution as elementary as the binomial, one must compute sums of the form

$$\sum_{j=x}^{n} \frac{n!}{j!(n-j)!} p^j (1-p)^{n-j} \tag{14.1-1}$$

for various values of n, p, and x. However, while these computations are often complicated, it is also true that for most statistical problems involving discrete random variables, some form of the central limit theorem is available. To continue the above example, the sum (14.1-1) may be approximated by the expression

$$1 - \Phi\left(\frac{x - \frac{1}{2} - np}{\sqrt{np(1-p)}}\right), \tag{14.1-2}$$

where $\Phi(t)$ is the cumulative distribution function of the standard normal distribution. How closely (14.1-2) approximates (14.1-1) depends on n, p, and x, but generally speaking, if n is large, p not near 1 or 0, and x not too far from np relative to $\sqrt{np(1-p)}$, then the approximation is usually adequate for statistical applications (see Pratt and Peizer [1968]).

The difficulty of exact calculations coupled with the availability of normal approximations leads to the almost automatic computation of asymptotic distributions and moments for discrete random variables. Three questions may be asked by a potential user of these asymptotic calculations:

1. How does one make them? What are the formulas and techniques for getting the answers? How is one lead to a formula like (14.1-2)?
2. How does one justify them? What conditions are needed to ensure that these formulas and techniques actually produce valid asymptotic results? What limit theorem justifies the use of (14.1-2)?
3. How does one relate asymptotic results to pre-asymptotic situations? How close are the answers given by an asymptotic formula to the actual cases of interest involving finite samples? For what values of n, p, and x does (14.1-2) provide a good approximation to (14.1-1)?

These questions differ vastly in the ease with which they may be answered. The answer to (1) usually requires mathematics at the level of elementary calculus.

Question (2) is rarely answered carefully, and is typically tossed aside by a remark of the form "... assuming that higher-order terms may be ignored...." Rigorous answers to question (2) require some of the deepest results in mathematical probability theory, but we can go very far with the relatively simple mathematical machinery described in the next few sections of this chapter. Question (3) is the most important, the most difficult, and consequently the least answered. Analytic answers to question (3) are usually very difficult, and it is more common to see reported the result of a simulation or a few isolated numerical calculations rather than an exhaustive answer.

This chapter brings together a variety of useful asymptotic methods and their applications in a self-contained discussion. Primarily we concentrate on questions 1 and 2. Our aim is to catalogue the important definitions, theorems, and methods and to give numerous examples and applications of them. We omit or abbreviate most proofs and cite references in their place. Examples and corollaries serve to clarify concepts and complicated formulas. The sections of this chapter fall naturally into two groups. Sections 14.2–14.5 are concerned with the basic mathematical tools needed to study asymptotic distributions, while Sections 14.6–14.9 apply these tools to finding asymptotic means and variances (Section 14.6) and asymptotic distributions of important statistical quantities that arise in many of the other chapters of this book (Sections 14.7, 14.8, and 14.9).

The results in this chapter that are most commonly used in practice are theorems 14.6-1, 14.6-2, and 14.6-3 on the δ method, theorem 14.8-3 on the limiting distribution of the multinomial maximum likelihood estimator, and theorems 14.9-4 and 14.9-8 on the limiting distribution of chi square goodness-of-fit statistics.

We use vector and matrix notation on occasion and make the convention that all vectors are *row vectors*. We use the notation $\mathcal{N}(0, 1)$ to mean the unit normal distribution, while $\mathcal{N}_T(\mu, \Sigma)$ means the T-dimensional multivariate normal distribution with mean vector μ and covariance matrix Σ.

14.2 The O, o Notation

The essence of asymptotic methods is approximation. We approximate functions, random variables, probability distributions, means, variances, and covariances. Careful use of approximations requires attention to their accuracy or *order*. A notation that is especially useful for keeping track of the order of an approximation is the "big O, little o" notation. In this section, we introduce and review this notation for the reader who may be unfamiliar with it. More advanced readers may wish to skim the material here.

There are really two big O, little o notations: one for nonstochastic variables, denoted O and o, and one for stochastic variables, denoted O_p and o_p. In this section we define and exemplify the O and o notation for nonstochastic variables. In the next section we generalize this to O_p and o_p for stochastic variables.

If $\{a_n\}$ and $\{b_n\}$ are two sequences of real numbers, then the following two formal definitions define the expressions $a_n = O(b_n)$ and $a_n = o(b_n)$, which are basic for understanding the O and o notation.

DEFINITION 14.2-1 $a_n = O(b_n)$ (Read: a_n is big O of b_n) if the ratio $|a_n/b_n|$ is bounded for large n; in detail, if there exists a number K and an integer $n(K)$ such that if n exceeds $n(K)$ then $|a_n| < K|b_n|$.

DEFINITION 14.2-2 $a_n = o(b_n)$ *(Read: a_n is little o of b_n) if the ratio $|a_n/b_n|$ converges to zero; in detail, if for any $\varepsilon > 0$, there exists an integer $n(\varepsilon)$ such that if n exceeds $n(\varepsilon)$ then $|a_n| < \varepsilon|b_n|$.*

The idea behind these two definitions is the comparison of the approximate size or *order of magnitude* of $\{a_n\}$ to that of $\{b_n\}$. In most applications, $\{a_n\}$ is the sequence of interest while $\{b_n\}$ is a comparison sequence. Some important examples of $\{b_n\}$ are: $b_n = n^{-1}$, $b_n = n^{-1/2}$, $b_n = n$, $b_n = n \log n$.

The interpretation of $a_n = O(b_n)$ is that the sequence of interest $\{a_n\}$ is of roughly the same size or order of magnitude as the comparison sequence $\{b_n\}$. The interpretation of $a_n = o(b_n)$ is that $\{a_n\}$ is of a smaller order of magnitude than is $\{b_n\}$. Of course, both of these comparisons refer to "large n" properties of the sequences and are unaffected by their initial behavior.

14.2.1   *Conventions in the use of the O, o notation*

A number of conventions are observed in the use of the O, o notation. We list here some of the more important ones.

(i)   Definitions 14.2-1 and 14.2-2 above are unaffected if we allow a_n to be infinite or even undefined for a finite number of values of n. In a number of statistical applications this is a convenience, and so we will always assume this slight extension of the definitions.

(ii)   Definitions 14.2-1 and 14.2-2 above may easily be extended to apply to a sequence $\{\mathbf{a}_n\}$ of vectors. If $\|\mathbf{a}_n\|$ denotes the length of the vector $\mathbf{a}_n$, i.e.

$$\|\mathbf{a}_n\| = \sqrt{\sum_i a_{ni}^2},$$

then $\mathbf{a}_n = O(b_n)$ means $\|\mathbf{a}_n\| = O(b_n)$, and $\mathbf{a}_n = o(b_n)$ means $\|\mathbf{a}_n\| = o(b_n)$.

(iii)   The expressions $a_n = O(b_n)$ and $a_n = O(cb_n)$ are equivalent if c is a nonzero constant. The same is true for o. Hence multiplicative constants are ignored in the argument of O and o. For example, $o(2n^{-1})$ is written $o(n^{-1})$. The sign, positive or negative, of an O or o term is always ignored.

(iv)   The expression $a_n = o(1)$ is used to signify that $a_n \to 0$, while $a_n = O(1)$ means that $|a_n| \leq K$ for some constant K if n is large enough, i.e., that $\{a_n\}$ is eventually bounded.

(v)   We always have $a_n = O(a_n)$.

(vi)   Products of O and o factors obey these easily proved rules:

(P1) $O(a_n)O(b_n) = O(a_n b_n)$,
(P2) $O(a_n)o(b_n) = o(a_n b_n)$,
(P3) $o(a_n)o(b_n) = o(a_n b_n)$.

(vii)   It is often necessary to add together several O and o expressions to obtain a single order-of-magnitude term. The rule is that the order of magnitude of a sum is the largest order of magnitude of the summands. For example,

$$o(1) + O(n^{-1/2}) + O(n^{-1}) = o(1).$$

This rule is not necessarily correct if the number of terms in the summation depends also on n. For example, if we have the n terms

$$\frac{1}{n} + \frac{1}{n} + \cdots + \frac{1}{n} = 1,$$

then the largest order of magnitude of the summands is n^{-1}, but this is not the order of magnitude of the summation.

The O and o notation also appears with a continuous variable in the argument rather than a sequence, especially in expansions of functions. For example, the first-order Taylor expansion of a function $f(\cdot)$ about the value a is stated as:

$$f(x) = f(c) + (x - c)f'(c) + o(|x - c|) \quad \text{as } x \to c. \tag{14.2-1}$$

In this example, the little o means that if x_n is any sequence such that $x_n \to c$ and if a_n and b_n are defined by

$$a_n = f(x_n) - f(c) - (x_n - c)f'(c), \tag{14.2-2}$$

$$b_n = x_n - c, \tag{14.2-3}$$

then

$$a_n = o(b_n). \tag{14.2-4}$$

In order for (14.2-1) to be true, (14.2-4) must be true for any choice of x_n such that $x_n \to c$. The following two definitions formalize the use of the o, O notation with a continuous argument in terms of definitions 14.2-1 and 14.2-2.

DEFINITION 14.2-3 $a(x) = O(b(x))$ as $x \to L$ if for any sequence $\{x_n\}$ such that $x_n \to L$, we have $a(x_n) = O(b(x_n))$ in the sense of definition 14.2-1.

DEFINITION 14.2-4 $a(x) = o(b(x))$ as $x \to L$ if for any sequence $\{x_n\}$ such that $x_n \to L$, we have $a(x_n) = o(b(x_n))$ in the sense of definition 14.2-2.

In the applications of definitions 14.2-3 and 14.2-4, we observe that when there is no ambiguity the condition "as $x \to L$" is not always explicitly stated. The value of L may be any real number, $+\infty$, or $-\infty$.

In the following section we go through an illustrative example that allows the reader to see the use of the O, o notation in a natural setting.

14.2.2   An example of the use of the O, o notation

Example 14.2-1   Approximating e

We begin by considering the sequence $\{e_n\}$ given by

$$e_n = (1 + n^{-1})^n. \tag{14.2-5}$$

Let $\log(x)$ denote the natural logarithm of x (to the base $e = 2.7182818\ldots$). We may find the limit of $\{e_n\}$ by first finding the limit of $\log(e_n) = n\log(1 + n^{-1})$ and then taking antilogs. We let

$$f(t) = \log(1 + t). \tag{14.2-6}$$

From elementary calculus, we recall that

$$f(0) = 0, \qquad f'(t) = (1 + t)^{-1}, \qquad f'(0) = 1. \tag{14.2-7}$$

Hence the Taylor expansion of f about $t = 0$ may be expressed as

$$f(t) = f(0) + tf'(0) + o(t)$$

$$= t + o(t) \quad \text{as } t \to 0. \tag{14.2-8}$$

Now we apply definition 14.2-3 to (14.2-8) with $x_n = n^{-1}$, and we obtain

$$\log(1 + n^{-1}) = n^{-1} + o(n^{-1}), \tag{14.2-9}$$

so that

$$\log(e_n) = n \log(1 + n^{-1}) = 1 + no(n^{-1}) = 1 + o(1) \tag{14.2-10}$$

and hence

$$\log(e_n) \to 1. \tag{14.2-11}$$

From (14.2-11) it follows that the limit we seek is

$$\lim_{n \to \infty} (1 + n^{-1})^n = \lim_{n \to \infty} e^{\log(e_n)} = e^1 = e. \tag{14.2-12}$$

The convergence of $\{e_n\}$ to e is often too slow for applications, but by considering higher-order Taylor series expansions of $f(t)$, we may find simple "corrections" to e_n that markedly improve the convergence to e. The remainder of this example is devoted to finding some of these corrections as further illustrations of the use of the O, o notation.

Define the sequence x_n by

$$x_n = \log(1 + n^{-1}). \tag{14.2-13}$$

Now we look at the sequence $\{(n + c)x_n\}$, where c is a constant to be determined. The second-order Taylor series expansion of $f(t) = \log(1 + t)$ is given by

$$f(t) = t - \tfrac{1}{2}t^2 + o(t^2) \quad \text{as } t \to 0. \tag{14.2-14}$$

Hence

$$x_n = f(n^{-1}) = n^{-1} - \tfrac{1}{2}n^{-2} + o(n^{-2}), \tag{14.2-15}$$

so that

$$(n + c)x_n = (n + c)(n^{-1} - \tfrac{1}{2}n^{-2} + o(n^{-2}))$$

$$= 1 + (c - \tfrac{1}{2})n^{-1} - \tfrac{1}{2}cn^{-2} + no(n^{-2}) + co(n^{-2})$$

$$= 1 + (c - \tfrac{1}{2})n^{-1} + o(n^{-1}). \tag{14.2-16}$$

By choosing $c = 1/2$, the order of the convergence of $(n + c)x_n$ to 1 improves from $o(1)$ to $o(n^{-1})$. Thus we define a new sequence e_n^* by

$$e_n^* = (1 + n^{-1})^{n + \frac{1}{2}}. \tag{14.2-17}$$

The distinction between e_n and e_n^* is that

$$e_n = e^{1 + o(1)} \quad \text{while} \quad e_n^* = e^{1 + o(n^{-1})}. \tag{14.2-18}$$

We need not stop here. We can find higher-order approximations by the same device. For example, consider the sequence $\{(n + c + dn^{-1})x_n\}$, for constants

c and d to be determined. As an exercise the reader is invited to use the third-order Taylor expansion of $f(t) = \log(1 + t)$ about $t = 0$ to show that if $c = 1/2$ and $d = -1/12$ then

$$(n + c + dn^{-1})x_n = 1 + o(n^{-2}). \tag{14.2-19}$$

This leads us to define e_n^{**} by

$$e_n^{**} = (1 + n^{-1})^{n + \frac{1}{2} - \frac{1}{12n}}, \tag{14.2-20}$$

which converges to e at a faster rate than either e_n or e_n^*.

Accuracy of the approximations to e

The careful reader will have one question that gnaws at him through the above manipulation of little o terms. He remembers that

$$10^6 n^{-2} = o(n^{-1})$$

and realizes that the multiplicative constant 10^6, while not affecting the order of magnitude of the convergence, can have a tremendous effect on the actual degree of the approximation for the values of n that concern him. In other words, after e_n^* and e_n^{**} have been invented it is still important to find out if the actual values of these quantities are really nearer e than e_n is, for small values of n. To answer this question, table 14.2-1 gives values of n, e_n, e_n^*, and e_n^{**} for $n = 1$ to 20. It is evident that for many purposes e_n is a totally inadequate approximation to e for these values of n. On the other hand, two- and three-decimal accuracy is quickly attained by e_n^* and e_n^{**}. ∎∎

Table 14.2-1

n	$\left(1 + \dfrac{1}{n}\right)^n$	$\left(1 + \dfrac{1}{n}\right)^{n + \frac{1}{2}}$	$\left(1 + \dfrac{1}{n}\right)^{n + \frac{1}{2} - \frac{1}{12n}}$
1	2	2.8284	2.6697
2	2.25	2.7557	2.7095
3	2.37	2.7371	2.7153
4	2.44	2.7296	2.7169
5	2.49	2.7258	2.7175
6	2.52	2.7237	2.7178
7	2.55	2.7223	2.7179986
8	2.57	2.7214	2.7180886
9	2.58	2.7207	2.7181441
10	2.59	2.7203	2.7181803
11	2.60	2.719997	2.7182048
12	2.61	2.719733	2.7182220
13	2.62	2.719526	2.7182345
14	2.627	2.719360	2.7182437
15	2.633	2.719225	2.7182507
16	2.638	2.719114	2.7182560
17	2.642	2.719022	2.7182602
18	2.646	2.718944	2.7182636
19	2.650	2.718878	2.7182663
20	2.653	2.718821	2.7182684
∞	2.718...	2.718281...	2.7182818...

14.2.3   *Exercises*

1. Show that $(2n^2 + 3n + 4) \sin n = O(n^2)$.
2. Show that $\log n = o(n^\alpha)$ for any power $\alpha > 0$.
3. Show that $\log x = o(x^{-1})$ as $x \to 0$.
4. Show that $\sin x = O(x)$ as $x \to 0$.
5. Show that if $c = 1/2$ and $d = -1/12$, then (14.2-19) is valid.
6. Prove the multiplicative rules for o and O given in (P1), (P2), and (P3).
7. Expand $(x + yn^{-1/2} + zn^{-1})^2$ to order n^{-1} by applying the rule for multiplying and adding o and O terms.
8. Show that we could have defined $O(b_n)$ and $o(b_n)$ by first defining $a_n = O(1)$ and $a_n = o(1)$ appropriately, and then defining $O(b_n) = b_n O(1)$ and $o(b_n) = b_n o(1)$.
9. Show that $(1 + (\lambda/n))^n \to e^\lambda$ for all λ.
10. Show that $(1 + \lambda/n + o(1/n))^n \to e^\lambda$ for all λ.
11. Does $f(t) = \sqrt{t}$ have a first-order Taylor expansion at $t = 0$?

14.3   Convergence of Stochastic Sequences

There are a large number of senses in which a sequence of stochastic variables can converge to a limit (see, for example, Loeve [1963] and Lehmann [1959]). Two notions of convergence, however, play an especially important role in most statistical applications—convergence in distribution and convergence in probability. Related to these is the notion of convergence of the moments of a stochastic sequence. In this section, we define both convergence in distribution and convergence in probability, and discuss some aspects of convergence in distribution. We delay an extended discussion of convergence in probability until Section 14.4, where we put it into a more general setting.

14.3.1   *Convergence in distribution*

One of the most important statistical applications of the asymptotic methods discussed in this chapter is the approximation of probabilities for computing significance levels and setting confidence intervals. Convergence in distribution is the technical tool that justifies these approximations. We begin with a sequence of random variables which we denote by $\{X_n\}$. Each of these random variables has a probability distribution that may be described in a variety of ways. The description that is important here is the *distribution function* of each random variable X_n. If X_n is a univariate random variable, then the distribution function of X_n is

$$F_n(x) = P\{X_n \leq x\}, \tag{14.3-1}$$

i.e., $F_n(x)$ is the probability that X_n falls on or below the value x. If X_n is a discrete random variable, then $F_n(x)$ is a right-continuous step function that has a point of increase only at the possible values of X_n. When $\mathbf{X}_n = (X_{n1}, \ldots, X_{nT})$ is a random vector, then the distribution function of $\mathbf{X}_n$ is

$$F_n(x_1, \ldots, x_T) = P\{X_{n1} \leq x_1, \ldots, X_{nT} \leq x_T\}. \tag{14.3-2}$$

In this case, F_n is a function of T variables.

By a *stochastic sequence* or a *sequence of stochastic variables* we mean either the sequence $\{F_n\}$ of distribution functions, or the sequence $\{X_n\}$ of random variables. Throughout this chapter, except when we state explicitly to the contrary, we are interested only in the marginal distributions of the random variables X_n and never in their joint distributions. For us, talking about a sequence $\{X_n\}$ of random variables is a convenient way of referring to the sequence $\{F_n\}$ of distribution functions. Some results in probability theory emphasize the joint probabilistic behavior of the entire sequence of random variables (see Feller [1966] and Loeve [1963]). We do not provide such an emphasis.

The next definition gives the basic notion of convergence in distribution.

DEFINITION 14.3-1   *Let X_n have distribution function F_n and X have distribution function F. Then X_n converges in distribution to X if*

$$\lim_{n \to \infty} F_n(x) = F(x)$$

for all x which are continuity points of F.

Convergence in distribution is sometimes called "convergence of probability laws" and accordingly we denote it by

$$\mathscr{L}[X_n] \to \mathscr{L}[X]. \tag{14.3-3}$$

In many applications of convergence in distribution, the limiting distribution F is continuous, so that the condition in definition 14.3-1 that x be a continuity point of F is automatically satisfied by every value of x.

Example 14.3-1   Normal approximations to the binomial distribution
If X_n has the binomial distribution $\mathscr{B}(n, p)$, then probabilities involving X_n can be approximated by the standard normal distribution as follows. Let Z_n be given by

$$Z_n = \frac{X_n - np}{\sqrt{np(1 - p)}}. \tag{14.3-4}$$

Then it is shown in example 14.3-3 that

$$\mathscr{L}[Z_n] \to \mathscr{L}[Z],$$

where Z has the standard normal distribution $\mathscr{N}(0, 1)$. The distribution function of Z is denoted by $\Phi(x)$ and may be expressed as the integral

$$\Phi(x) = (2\pi)^{-1/2} \int_{-\infty}^{x} e^{-u^2/2} \, du. \tag{14.3-5}$$

$\Phi(x)$ is the area under the density function

$$\varphi(u) = (2\pi)^{-1/2} e^{-u^2/2} \tag{14.3-6}$$

from $-\infty$ up to x. Denote the distribution function of Z_n by $F_n(x)$, i.e.,

$$F_n(x) = P\{Z_n \leq x\}. \tag{14.3-7}$$

Since $\mathscr{L}[Z_n] \to \mathscr{L}[Z]$ and $\Phi(x)$ is a continuous function, we have

$$\lim_{n \to \infty} F_n(x) = \Phi(x) \quad \text{for all } x. \tag{14.3-8}$$

To approximate probabilities involving X_n, we note that

$$P\{X_n \leq x\} = P\{X_n - np \leq x - np\}$$

$$= P\left\{\frac{X_n - np}{\sqrt{np(1 - p)}} \leq \frac{x - np}{\sqrt{np(1 - p)}}\right\}$$

$$= P\left\{Z_n \leq \frac{x - np}{\sqrt{np(1 - p)}}\right\}$$

$$= F_n\left(\frac{x - np}{\sqrt{np(1 - p)}}\right). \tag{14.3-9}$$

In the o, O notation, this last quantity may be expressed as

$$\Phi\left(\frac{x - np}{\sqrt{np(1 - p)}}\right) + o(1) \quad \text{as } n \to \infty, \tag{14.3-10}$$

so that we have the approximation

$$P\{X_n \leq x\} = \sum_{j=0}^{x} \binom{n}{j} p^j (1 - p)^{n-j} = \Phi\left(\frac{x - np}{\sqrt{np(1 - p)}}\right) + o(1). \tag{14.3-11}$$

Expression (14.3-11) approximates the lower tail of the binomial distribution. The upper tail is approximated by

$$P\{X_n \geq x\} = \sum_{j=x}^{n} \binom{n}{j} p^j (1 - p)^{n-j} = 1 - \Phi\left(\frac{x - np}{\sqrt{np(1 - p)}}\right) + o(1). \tag{14.3-12}$$

These approximations (14.3-11) and (14.3-12) can be improved by use of "continuity corrections" which take the following forms. For (14.3-11) we use

$$P\{X_n \leq x\} = \Phi\left(\frac{x + 0.5 - np}{\sqrt{np(1 - p)}}\right) + o(1), \tag{14.3-11a}$$

and for (14.3-12) we use

$$P\{X_n \geq x\} = 1 - \Phi\left(\frac{x - 0.5 - np}{\sqrt{np(1 - p)}}\right) + o(1). \ \blacksquare\blacksquare \tag{14.3-12a}$$

14.3.2 *Convergence in probability to a constant*

In this section we define convergence in probability to a constant and discuss its relation to convergence in distribution. A more detailed discussion of convergence in probability to a constant is given in Section 14.4.

A numerical constant c can always be viewed as a degenerate random variable C whose distribution has all of its probability concentrated on the single value c. The distribution function $F_c(x)$ of C is easily shown to be

$$F_c(x) = \begin{cases} 0 & \text{if } x < c \\ 1 & \text{if } x \geq c. \end{cases} \tag{14.3-13}$$

F_c is a right-continuous step function with a single point of discontinuity at $x = c$. What does it mean for a stochastic sequence $\{X_n\}$ to converge in distribution to C? It ought to mean that the distributions of the X_n become more and more concentrated about the point $x = c$ as $n \to \infty$. To show that this is indeed exactly what happens, we calculate the probability that X_n lies between $c - \varepsilon$ and $c + \varepsilon$ for any fixed small positive number ε:

$$
\begin{aligned}
P\{|X_n - c| \leq \varepsilon\} &= P\{c - \varepsilon \leq X_n \leq c + \varepsilon\} \\
&= P\{c - \varepsilon < X_n \leq c + \varepsilon\} + P\{X_n = c - \varepsilon\} \\
&= P\{X_n \leq c + \varepsilon\} - P\{X_n \leq c - \varepsilon\} + P\{X_n = c - \varepsilon\} \\
&= F_n(c + \varepsilon) - F_n(c - \varepsilon) + P\{X_n = c - \varepsilon\}.
\end{aligned}
\tag{14.3-14}
$$

But since $\mathscr{L}[X_n] \to \mathscr{L}[C]$ has been assumed, we have, by definition 14.3-1,

$$
\lim_{n \to \infty} F_n(x) = F_c(x) \quad \text{for all } x \neq c.
\tag{14.3-15}
$$

The convergence of $F_n(c)$ to $F_c(c)$ is not required by the definition because c is a discontinuity point of F_c. From (14.3-15), we have

$$
\lim_{n \to \infty} F_n(x) = \begin{cases} 0 & \text{if } x < c \\ 1 & \text{if } x > c, \end{cases}
\tag{14.3-16}
$$

and hence we conclude that

$$
\lim_{n \to \infty} F_n(c + \varepsilon) = 1,
$$

$$
\lim_{n \to \infty} F_n(c - \varepsilon) = 0.
\tag{14.3-17}
$$

Finally we observe that

$$
0 \leq \lim_{n \to \infty} P\{X_n = c - \varepsilon\} \leq \lim_{n \to \infty} P\{X_n \leq c - \varepsilon\} = \lim_{n \to \infty} F_n(c - \varepsilon) = 0,
$$

from which we conclude that

$$
\lim_{n \to \infty} P\{X_n = c - \varepsilon\} = 0.
\tag{14.3-18}
$$

Putting (14.3-17) and (14.3-18) together with (14.3-14) yields the following fact: for every $\varepsilon > 0$,

$$
\lim_{n \to \infty} P\{|X_n - c| \leq \varepsilon\} = 1.
\tag{14.3-19}
$$

The situation described by (14.3-19) is usually given as the definition of *convergence in probability to c*.

DEFINITION 14.3-2 $\{X_n\}$ *converges in probability to c, denoted* $X_n \underset{P}{\to} c$, *if for every* $\varepsilon > 0$,

$$
\lim_{n \to \infty} P\{|X_n - c| \leq \varepsilon\} = 1.
\tag{14.3-20}
$$

If $\mathbf{X}_n$ is a sequence of T-dimensional random vectors and $\mathbf{c}$ is a constant vector in T dimensions, then we extend definition 14.3-2 as follows.

DEFINITION 14.3-3 $\{\mathbf{X}_n\}$ *converges in probability to* $\mathbf{c}$, *denoted* $\mathbf{X}_n \underset{P}{\to} \mathbf{c}$, *if for every* $\varepsilon > 0$,

$$
\lim_{n \to \infty} P\{\|\mathbf{X}_n - \mathbf{c}\| \leq \varepsilon\} = 1.
\tag{14.3-21}
$$

The earlier discussion showed that if **C** is a degenerate random variable such that

$$P\{\mathbf{C} = \mathbf{c}\} = 1, \tag{14.3-22}$$

then $\mathscr{L}[\mathbf{X}_n] \to \mathscr{L}[\mathbf{C}]$ implies that $\mathbf{X}_n \xrightarrow{p} \mathbf{c}$. The converse of this is also true and is left as an exercise (exercise 1 in Section 14.3.4). This equivalence yields the following theorem.

THEOREM 14.3-1   *If* **C** *is a degenerate random variable satisfying* (14.3-22), *then* $\mathscr{L}[\mathbf{X}_n] \to \mathscr{L}[\mathbf{C}]$ *is equivalent to* $\mathbf{X}_n \xrightarrow{p} \mathbf{c}$.

One way to interpret definition 14.3-3, theorem 14.3-1, and definition 14.3-1 is that convergence in distribution is a generalization of convergence in probability to a constant.

14.3.3   *Methods of demonstrating convergence in distribution*

There are several ways of showing that a stochastic sequence converges in distribution to a particular probability law. In this section we discuss and give examples of the more common methods.

Direct verification that $F_n \to F$

In some cases it is possible to show directly that the conditions of definition 14.3-1 are met. We are required to show that $F_n(x) \to F(x)$ for all continuity points of F. We now consider an example where direct verification is possible.

Example 14.3-2   The distribution of long waiting times

Discrete waiting times are sometimes modeled by the geometric distribution, i.e., the distribution of the first "success" in a sequence of independent Bernoulli trials with "success" probability p. When p is small, the waiting time is large, since the expected waiting time is p^{-1}. We study geometric waiting times for $p = p_n = \lambda n^{-1}, \lambda > 0$, as $n \to \infty$.

Let X_n have the geometric distribution

$$P\{X_n = k\} = (1 - p_n)^{k-1}p_n \quad k = 1, 2, \ldots.$$

Then

$$P\{X_n \leq l\} = 1 - (1 - p_n)^l \quad l = 0, 1, \ldots. \tag{14.3-23}$$

Since X_n has a geometric distribution, its distribution function $G_n(x)$ is a step function that increases at each positive integer and is constant between them. The following notational device is useful to describe $G_n(x)$.

For any nonnegative real number x, let $[x]$ denote the largest nonnegative integer that is less than or equal to x. Then we may express $G_n(x)$ as

$$G_n(x) = P\{X_n \leq x\} = \begin{cases} 1 - (1 - p_n)^{[x]} & x > 0 \\ 0 & x \leq 0. \end{cases} \tag{14.3-24}$$

Now $E(X_n) = \lambda^{-1}n$, so that as $n \to \infty$, X_n wanders off toward $+\infty$. To avoid this situation, we study a standardized variable

$$Y_n = n^{-1}X_n. \tag{14.3-25}$$

The distribution function of Y_n is

$$F_n(x) = P\{Y_n \le x\} = P\{X_n \le nx\} = G_n(nx),$$

or

$$F_n(x) = \begin{cases} 1 - (1 - \lambda n^{-1})^{[nx]} & x > 0 \\ 0 & x \le 0. \end{cases} \tag{14.3-26}$$

Our aim is to find the limit of $F_n(x)$. For $x \le 0$, (14.3-26) implies that $F_n(x) \to 0$, so we need only consider $x > 0$. From (14.3-26), we have

$$F_n(x) = 1 - (1 - \lambda n^{-1})^{nx}(1 - \lambda n^{-1})^{[nx]-nx}. \tag{14.3-27}$$

But the reader can verify that

$$(1 - \lambda n^{-1})^{-1} \ge (1 - \lambda n^{-1})^{[nx]-nx} \ge 1, \tag{14.3-28}$$

so that for any $x > 0$,

$$(1 - \lambda n^{-1})^{[nx]-nx} \to 1,$$

or

$$(1 - \lambda n^{-1})^{[nx]-nx} = 1 + o(1) \tag{14.3-29}$$

as $n \to \infty$. Furthermore, a well-known extension of (14.2-12) (see exercise 9 in Section 14.2.3) is

$$(1 + \lambda n^{-1})^n \to e^\lambda,$$

so that

$$(1 - \lambda n^{-1})^{nx} \to e^{-\lambda x}. \tag{14.3-30}$$

For $x > 0$, from (14.3-27), (14.3-29), and (14.3-30) we have

$$F_n(x) = 1 - (e^{-\lambda x} + o(1))(1 + o(1)) \quad \text{as } n \to \infty,$$

or

$$F_n(x) \to 1 - e^{-\lambda x}. \tag{14.3-30a}$$

However, if we define $F(x)$ by

$$F(x) = \begin{cases} 0 & x \le 0 \\ 1 - e^{-\lambda x} & x > 0, \end{cases} \tag{14.3-31}$$

then $F(x)$ is the distribution function of the well-known exponential distribution whose density function is

$$f(x) = \begin{cases} 0 & x < 0 \\ \lambda e^{-\lambda x} & x \ge 0. \end{cases} \tag{14.3-32}$$

If Y is a random variable with the exponential density function (14.3-32), then we have shown that

$$\mathscr{L}[Y_n] \to \mathscr{L}[Y]. \quad \blacksquare\blacksquare$$

Indirect verification using moment-generating functions

The moment-generating function (MGF) of a random variable or vector is one of the most important tools for establishing convergence in distribution for discrete distributions. If $\mathbf{X} = (X_1, \ldots, X_T)$ is a random vector, then the MGF of $\mathbf{X}$ is given by

$$M_{\mathbf{X}}(t_1, \ldots, t_T) = E[e^{t_1 X_1 + \cdots + t_T X_T}]. \tag{14.3-33}$$

When $T = 1$, (14.3-33) becomes

$$M_{\mathbf{X}}(t) = E[e^{tX}]. \tag{14.3-34}$$

The moment-generating function derives its name from its ability to produce the moments of $\mathbf{X}$, as indicated in the next theorem.

THEOREM 14.3-2 *If the expectation given in (14.3-33) is finite for all $(t_1, \ldots, t_T)$ in a neighborhood of $\mathbf{0} = (0, \ldots, 0)$, then*

$$E(X_1^{a_1} \ldots X_T^{a_T}) = \frac{\partial^{a_1 + \cdots + a_T}}{\partial t_1^{a_1} \ldots \partial t_T^{a_T}} M_{\mathbf{X}}(t_1, \ldots, t_T)\bigg|_{t = 0}.$$

In particular,

$$E(X_i) = \frac{\partial}{\partial t_i} M_{\mathbf{X}}(0, \ldots, 0),$$

$$E(X_i^2) = \frac{\partial^2}{\partial t_i^2} M_{\mathbf{X}}(0, \ldots, 0),$$

$$E(X_i X_j) = \frac{\partial^2}{\partial t_i \partial t_j} M_{\mathbf{X}}(0, \ldots, 0),$$

Exercises 2 and 3 in Section 14.3.4 apply theorem 14.3-2 to the multinomial distribution.

The next theorem states the relationship between convergence of MGFs and convergence in distribution.

THEOREM 14.3-3 *If $\mathbf{X}_n$ has MGF $M_n(\mathbf{t})$ and $\mathbf{X}$ has MGF $M(\mathbf{t})$, and if $M_n(\mathbf{t}) \to M(\mathbf{t})$ for all $\mathbf{t}$ in a neighborhood of $\mathbf{0} = (0, \ldots, 0)$, then $\mathscr{L}[\mathbf{X}_n] \to \mathscr{L}[\mathbf{X}]$.*

For a proof of this theorem, we refer the advanced reader to Loeve [1963] and Feller [1966]. Theorem 14.3-3 is an important tool for establishing convergence in distribution. We use moment-generating functions rather than characteristic functions (see the above references) because they are sufficiently powerful for the applications in this book. In particular, the distributions encountered in this book always satisfy the provision in theorem 14.3-2 regarding the existence of the expectation (14.3-33).

We now apply theorem 14.3-3 to two very important examples involving the multinomial distribution.

Example 14.3-3 Asymptotic normality of the multinomial distribution

Let $\mathbf{X}_n = (X_{n1}, \ldots, X_{nT})$ have the multinomial distribution $\mathscr{M}(n, \mathbf{p})$, where $\mathbf{p} = (p_1, \ldots, p_T)$. From exercise 3 in Section 14.3.4, we see that (in vector notation)

$$E(\mathbf{X}_n) = n\mathbf{p}, \tag{14.3-35}$$

$$\text{Cov}(\mathbf{X}_n) = n(\mathbf{D}_p - \mathbf{p}'\mathbf{p}), \tag{14.3-36}$$

when $\mathbf{D}_p$ denotes the diagonal matrix based on $\mathbf{p}$.

Let $\hat{\mathbf{p}} = n^{-1}\mathbf{X}_n$ be the vector of sample proportions, and set $\mathbf{U}_n = \sqrt{n}(\hat{\mathbf{p}} - \mathbf{p})$. Then

$$E(\mathbf{U}_n) = \mathbf{0}, \tag{14.3-37}$$

$$\mathrm{Cov}(\mathbf{U}_n) = \mathbf{D}_p - \mathbf{p}'\mathbf{p}. \tag{14.3-38}$$

THEOREM 14.3-4 $\mathscr{L}[\mathbf{U}_n] \to \mathscr{L}[\mathbf{U}]$, where $\mathbf{U}$ has the multivariate normal distribution with mean vector $\mathbf{0}$ and covariance matrix $\mathbf{D}_p - \mathbf{p}'\mathbf{p}$.

Proof We compute the MGF of $\mathbf{U}_n$ and show that it converges to that of $\mathbf{U}$:

$$M_{\mathbf{U}_n}(\mathbf{t}) = E[e^{t_1 U_{n1} + \cdots + t_T U_{nT}}] = E[e^{\mathbf{t}\mathbf{U}_n'}]$$

$$= E[e^{\mathbf{t}n^{-1/2}\mathbf{X}_n' - \sqrt{n}\mathbf{t}\mathbf{p}'}] = e^{-\sqrt{n}\mathbf{t}\mathbf{p}'}M_{\mathbf{X}_n}(n^{-1/2}\mathbf{t}).$$

From exercise 2 in Section 14.3.4, we have

$$M_{\mathbf{X}_n}(\mathbf{t}) = \left(\sum_j p_j e^{t_j}\right)^n,$$

so that

$$M_{\mathbf{U}_n}(\mathbf{t}) = e^{-\sqrt{n}\mathbf{t}\mathbf{p}'}\left(\sum_j p_j e^{n^{-1/2}t_j}\right)^n = \left(\sum_j p_j e^{n^{-1/2}(t_j - \mathbf{t}\mathbf{p}')}\right)^n.$$

Since

$$e^x = 1 + x + \tfrac{1}{2}x^2 + o(x^2) \quad \text{as } x \to 0,$$

we have

$$M_{\mathbf{U}_n}(\mathbf{t}) = \left(\sum_j p_j(1 + n^{-1/2}(t_j - \mathbf{t}\mathbf{p}') + (2n)^{-1}(t_j - \mathbf{t}\mathbf{p}')^2 + o(n^{-1}))\right)^n$$

$$= \left(1 + n^{-1/2}\sum_j p_j(t_j - \mathbf{t}\mathbf{p}') + (2n)^{-1}\sum_j p_j(t_j - \mathbf{t}\mathbf{p}')^2 + o(n^{-1})\right)^n.$$

Now

$$\sum_j p_j(t_j - \mathbf{t}\mathbf{p}') = 0,$$

and

$$\sum_j p_j(t_j - \mathbf{t}\mathbf{p}')^2 = \mathbf{t}(\mathbf{D}_p - \mathbf{p}'\mathbf{p})\mathbf{t}',$$

so that

$$M_{\mathbf{U}_n}(\mathbf{t}) = (1 + n^{-1}\tfrac{1}{2}\mathbf{t}(\mathbf{D}_p - \mathbf{p}'\mathbf{p})\mathbf{t}' + o(n^{-1}))^n. \tag{14.3-39}$$

Hence, from exercise 3 in Section 14.3.4 and (14.3-39), we conclude that

$$M_{\mathbf{U}_n}(\mathbf{t}) \to M(\mathbf{t}) = e^{\frac{1}{2}\mathbf{t}(\mathbf{D}_p - \mathbf{p}'\mathbf{p})\mathbf{t}'}. \tag{14.3-40}$$

But $M(\mathbf{t})$ defined in (14.3-40) is the MGF of the multivariate normal distribution with mean vector $\mathbf{0}$ and covariance matrix $\mathbf{D}_p - \mathbf{p}'\mathbf{p}$. ∎

The limiting distribution of $\hat{\mathbf{p}}$ given in theorem 14.3-1 is sometimes described by saying that $\hat{\mathbf{p}}$ has an approximate multivariate normal distribution with mean $\mathbf{p}$ and covariance matrix

$$n^{-1}(\mathbf{D}_p - \mathbf{p}'\mathbf{p}). \tag{14.3-41}$$

The matrix in (14.3-41) is singular, i.e., its determinant is zero, because $\hat{\mathbf{p}}$ satisfies the linear constraint

$$\sum \hat{p}_i = 1. \quad \blacksquare\blacksquare \tag{14.3-42}$$

Example 14.3-4 Asymptotic normality of the multinomial distribution with $\mathbf{p} = \boldsymbol{\pi} + \boldsymbol{\mu} n^{-1/2}$

Continuing with the notation of example 14.3-3, we let

$$\mathbf{p} = \mathbf{p}_n = \boldsymbol{\pi} + \boldsymbol{\mu} n^{-1/2}. \tag{14.3-43}$$

This case is useful in the study of goodness-of-fit tests when the model being tested is wrong but not far wrong. (See example 14.3-6 for an application of the idea to the chi square statistic and Section 14.9.3 for further discussion.) In this case,

$$E(\mathbf{X}_n) = n\boldsymbol{\pi} + \sqrt{n}\boldsymbol{\mu},$$
$$\text{Cov}(\mathbf{X}_n) = n(\mathbf{D}_\pi - \boldsymbol{\pi}'\boldsymbol{\pi}) + \sqrt{n}(\mathbf{D}_\mu - 2\boldsymbol{\pi}'\boldsymbol{\mu}) + \boldsymbol{\mu}'\boldsymbol{\mu}. \tag{14.3-43a}$$

The coordinates of $\boldsymbol{\pi}$ and $\mathbf{p}$ both sum to 1 so that $\boldsymbol{\mu}$ satisfies the condition

$$\sum \mu_i = 0. \tag{14.3-44}$$

In this example, $\boldsymbol{\mu}$ acts as a noncentrality parameter. When $\boldsymbol{\mu} = \mathbf{0}$, we are in the situation of example 14.3-3. This time we let

$$\mathbf{U}_n = \sqrt{n}(\hat{\mathbf{p}} - \boldsymbol{\pi}), \tag{14.3-45}$$

so that

$$E(\mathbf{U}_n) = \boldsymbol{\mu},$$

and

$$\text{Cov}(\mathbf{U}_n) = n^{-1}\,\text{Cov}(\mathbf{X}_n)$$
$$= \mathbf{D}_\pi - \boldsymbol{\pi}'\boldsymbol{\pi} + n^{-1/2}(\mathbf{D}_\mu - 2\boldsymbol{\pi}'\boldsymbol{\mu}) + n^{-1}\boldsymbol{\mu}'\boldsymbol{\mu}. \tag{14.3-46}$$

THEOREM 14.3-5 $\mathcal{L}[\mathbf{U}_n] \to \mathcal{L}[\mathbf{U}]$, *where* $\mathbf{U}$ *has the multivariate normal distribution with mean vector* $\boldsymbol{\mu}$ *and covariance matrix* $\mathbf{D}_\pi - \boldsymbol{\pi}'\boldsymbol{\pi}$.

Proof We compute the MGF of $\mathbf{U}_n$ and show that it converges to that of $\mathbf{U}$:

$$M_{\mathbf{U}_n}(\mathbf{t}) = E[e^{\mathbf{t}\mathbf{U}_n'}] = E[e^{(\mathbf{t}n^{-1/2}\mathbf{X}_n' - \sqrt{n}\mathbf{t}\boldsymbol{\pi}')}]$$

$$= e^{-\sqrt{n}\mathbf{t}\boldsymbol{\pi}'}M_{\mathbf{X}_n}(n^{-1/2}\mathbf{t}) = e^{-\sqrt{n}\mathbf{t}\boldsymbol{\pi}'}\left(\sum_i (\pi_i + n^{-1/2}\mu_i)\,e^{n^{-1/2}t_i}\right)^n$$

$$= \left(\sum_i (\pi_i + n^{-1/2}\mu_i)\,e^{n^{-1/2}(t_i - \mathbf{t}\boldsymbol{\pi}')}\right)^n$$

$$= \left(\sum_i (\pi_i + n^{-1/2}\mu_i)(1 + n^{-1/2}(t_i - \mathbf{t}\boldsymbol{\pi}') + (2n)^{-1}(t_i - \mathbf{t}\boldsymbol{\pi}')^2 + o(n^{-1}))\right)^n$$

$$= \left(1 + n^{-1}\left(\sum_i \mu_i t_i + \tfrac{1}{2}\sum_i \pi_i(t_i - \mathbf{t}\boldsymbol{\pi}')^2\right) + o(n^{-1})\right)^n.$$

But

$$\sum \mu_i t_i = \mu t',$$

$$\sum \pi_i (t_i - t\pi')^2 = t(\mathbf{D}_\pi - \pi'\pi)t',$$

so that we have

$$M_{\mathbf{U}_n}(\mathbf{t}) = (1 + n^{-1}(\mu\mathbf{t}' + \tfrac{1}{2}\mathbf{t}(\mathbf{D}_\pi - \pi'\pi)\mathbf{t}') + o(n^{-1}))^n, \qquad (14.3\text{-}47)$$

and hence

$$M_{\mathbf{U}_n}(\mathbf{t}) \to M(\mathbf{t}) = e^{\mu\mathbf{t} + \tfrac{1}{2}\mathbf{t}(\mathbf{D}_\pi - \pi'\pi)\mathbf{t}'}. \qquad (14.3\text{-}48)$$

But $M(\mathbf{t})$ defined in (14.3-48) is the MGF of the multivariate normal distribution with mean vector μ and covariance matrix $\mathbf{D}_\pi - \pi'\pi$. ∎ ∎∎

Convergence in distribution of a function of a random variable

We have illustrated ways of showing that $\mathscr{L}[X_n] \to \mathscr{L}[X]$. The result we now turn to is useful when we can express the random variable of interest as a function of a more easily analyzed random variable. The essential idea is that if by some means we can establish that $\mathscr{L}[X_n] \to \mathscr{L}[X]$, then we can express the limiting distribution of *functions* of X_n in terms of the distribution of X. This powerful approach allows us to break up complicated distribution problems into more easily solved smaller pieces. The next theorem states the crux of the matter.

THEOREM 14.3-6 *Let* $\mathscr{L}[X_n] \to \mathscr{L}[X]$ *and let* g *be a continuous function. Then* $\mathscr{L}[g(X_n)] \to \mathscr{L}[g(X)]$.

For a proof of this result, see Rao [1965], p. 104.

In this theorem, $\{X_n\}$ may be a sequence of random vectors and g may be a continuous function mapping vectors of dimension c into vectors of a possibly different dimension d. In fact, continuity of g is not necessary. All we really require is that the limiting distribution X assign zero probability to the set of discontinuities of g. For example, if X has a normal distribution and g has at most a finite number of discontinuities, then the conclusion of theorem 14.3-6 applies. The next two examples are important applications of theorem 14.3-6.

Example 14.3-5 Distribution of the Pearson chi square statistic for a simple hypothesis

We continue using the notation of example 14.3-3. For the simple hypothesis

$$H_0 : \mathbf{p} = \pi \quad (\pi \text{ a fixed value})$$

the Pearson chi square test is: reject H_0 if X^2 is too large, where

$$X^2 = \sum_i \frac{(X_i - n\pi_i)^2}{n\pi_i}.$$

Now, using matrix notation, X^2 can be written as

$$X^2 = \mathbf{U}_n \mathbf{D}_\pi^{-1} \mathbf{U}_n', \qquad (14.3\text{-}49)$$

where

$$\mathbf{U}_n = \sqrt{n}(\hat{\mathbf{p}} - \pi), \qquad \hat{\mathbf{p}} = n^{-1}\mathbf{X}_n.$$

Let $g(\mathbf{x}) = \mathbf{x}\mathbf{D}_\pi^{-1}\mathbf{x}'$ for $\mathbf{x} = (x_1, \ldots, x_T)$. Evidently, g is a continuous function of $\mathbf{x}$. From theorem 14.3-4, when H_0 is true, then $\mathscr{L}[\mathbf{U}_n] \to \mathscr{L}[\mathbf{U}]$, where $\mathbf{U}$ has the multivariate normal distribution $\mathscr{N}(0, \mathbf{D}_\pi - \pi'\pi)$. By theorem 14.3-6 and expression (14.3-49), we have

$$\mathscr{L}[X^2] = \mathscr{L}[\mathbf{U}_n\mathbf{D}_\pi^{-1}\mathbf{U}_n'] \to \mathscr{L}[\mathbf{U}\mathbf{D}_\pi^{-1}\mathbf{U}'].$$

Thus the asymptotic distribution of X^2 under H_0 is the distribution of $\mathbf{U}\mathbf{D}_\pi^{-1}\mathbf{U}'$, where $\mathbf{U}$ has the $\mathscr{N}(0, \mathbf{D}_\pi - \pi'\pi)$ distribution. This reduces the problem to finding the distribution of a quadratic form of a multivariate normal random vector. We state without proof the following general result on the distribution of a quadratic form of a multivariate normal random variable.

THEOREM 14.3-7   *If* $\mathbf{X} = (X_1, \ldots, X_T)$ *has the multivariate normal distribution* $\mathscr{N}(0, \mathbf{\Sigma})$ *and* $Y = \mathbf{X}\mathbf{A}\mathbf{X}'$ *for some symmetric matrix* $\mathbf{A}$, *then* $\mathscr{L}[Y] = \mathscr{L}[\sum_{i=1}^T \lambda_i Z_i^2]$, *where* $Z_1^2, \ldots, Z_T^2$ *are independent chi square variables with one degree of freedom each and* $\lambda_1, \ldots, \lambda_T$ *are the eigenvalues of* $\mathbf{A}^{1/2}\mathbf{\Sigma}(\mathbf{A}^{1/2})'$.

For a proof see Rao [1965], p. 149.

If we apply theorem 14.3-7 to the present example, we see that $\mathscr{L}[\mathbf{U}\mathbf{D}_\pi^{-1}\mathbf{U}'] = \mathscr{L}[\sum_{i=1}^T \lambda_i Z_i^2]$, where the λ_i are the eigenvalues of

$$\mathbf{B} = \mathbf{D}_\pi^{-1/2}(\mathbf{D}_\pi - \pi'\pi)\mathbf{D}_\pi^{-1/2} = \mathbf{I} - \sqrt{\pi}'\sqrt{\pi}, \qquad (14.3\text{-}50)$$

with $\sqrt{\pi} = (\sqrt{\pi_1}, \ldots, \sqrt{\pi_T})$.

In this example, the eigenvalues of $\mathbf{B}$ may be found as follows. First observe that

$$\mathbf{B}^2 = \mathbf{I} - 2\sqrt{\pi}'\sqrt{\pi} + \sqrt{\pi}'\sqrt{\pi}\sqrt{\pi}'\sqrt{\pi} = \mathbf{I} - \sqrt{\pi}'\sqrt{\pi} = \mathbf{B},$$

so that $\mathbf{B}$ is *idempotent*. $\mathbf{B}$ is also symmetric. We now invoke the following fact about symmetric idempotent matrices.

THEOREM 14.3-8   *If* $\mathbf{B}^2 = \mathbf{B}$ *and* $\mathbf{B}' = \mathbf{B}$, *then the eigenvalues of* $\mathbf{B}$ *are all either* 0 *or* 1. *The number of eigenvalues equal to* 1 *is equal to the trace of* $\mathbf{B}$, *where*

$$\text{tr}(\mathbf{B}) = \sum_i b_{ii}. \qquad (14.3\text{-}51)$$

For a proof see Rao [1965], p. 150.

Applying theorem 14.3-8 to $\mathbf{B}$ defined in (14.3-50) gives the following result:

$$\text{tr}(\mathbf{B}) = \text{tr}(\mathbf{I} - \sqrt{\pi}'\sqrt{\pi}) = \text{tr}(\mathbf{I}) - \text{tr}(\sqrt{\pi}'\sqrt{\pi})$$
$$= T - \text{tr}(\sqrt{\pi}\sqrt{\pi}') = T - \text{tr}(1) = T - 1.$$

Hence we see that $T - 1$ of the eigenvalues of $\mathbf{B}$ equal 1, and one of them equals 0. Therefore we see that $\mathscr{L}[\mathbf{U}\mathbf{D}_\pi^{-1}\mathbf{U}] = \mathscr{L}[\sum_{i=1}^{T-1} Z_i^2] = \chi_{T-1}^2$, i.e., we have established the celebrated result that under the simple hypothesis H_0, Pearson's chi square statistic X^2 has an asymptotic chi square distribution with $T - 1$ degrees of freedom. ∎

Example 14.3-6   Noncentral distribution of Pearson's chi square
Here we briefly indicate the extension of the result in the previous example for the asymptotic setup described in example 14.3-4. We omit most of the details.

Suppose that we are testing

$$H_0 : \mathbf{p} = \boldsymbol{\pi}$$

with the test statistic X^2 of (14.3-49), but instead of H_0 being true, suppose that $\mathbf{p}$ is given by

$$\mathbf{p} = \boldsymbol{\pi} + n^{-1/2}\boldsymbol{\mu},$$

as in example 14.3-4. The reader can show that

$$\mathscr{L}[X^2] \to \mathscr{L}[\mathbf{U}\mathbf{D}_\pi^{-1}\mathbf{U}'], \tag{14.3-52}$$

where $\mathbf{U}$ has the multivariate normal distribution $\mathscr{N}(\boldsymbol{\mu}, \mathbf{D}_\pi - \boldsymbol{\pi}'\boldsymbol{\pi})$. If $Y = \mathbf{U}\mathbf{D}_\pi^{-1}\mathbf{U}'$, then it can be shown that Y has a noncentral chi square distribution with $T - 1$ degrees of freedom and noncentrality parameter

$$\psi^2 = \boldsymbol{\mu}\mathbf{D}_\pi^{-1}\boldsymbol{\mu}'. \tag{14.3-53}$$

If we write $\boldsymbol{\mu}$ as

$$\boldsymbol{\mu} = \sqrt{n}(\mathbf{p} - \boldsymbol{\pi}),$$

then this result is sometimes stated with the noncentrality parameter given as

$$\psi^2 = n(\mathbf{p} - \boldsymbol{\pi})\mathbf{D}_\pi^{-1}(\mathbf{p} - \boldsymbol{\pi})'. \quad \blacksquare\blacksquare \tag{14.3-54}$$

The effect of a small remainder term on convergence in distribution

The technique discussed in this section is actually a special case of theorem 14.3-6, but it is so important that we discuss it separately. The technique is based on the following theorem.

THEOREM 14.3-9 *If $\mathscr{L}[X_n] \to \mathscr{L}[X]$ and if $Y_n \xrightarrow{p} 0$, then $\mathscr{L}[X_n + Y_n] \to \mathscr{L}[X]$.*

For a proof see Rao [1965], p. 104.

The idea behind theorem 14.3-9 is as follows. Suppose we are interested in the limiting distribution of Z_n and we are able to express Z_n as $Z_n = X_n + Y_n$, where $\{X_n\}$ is a stochastic sequence with an asymptotic distribution we can find by some method and Y_n is a small remainder term that converges to 0 in probability, i.e., $Y_n \xrightarrow{p} 0$. Then the limiting distribution of Z_n is the same as that of X_n. We delay examples of this approach until Section 14.4, 14.6, 14.8, and 14.9, where it is applied extensively.

14.3.4   *Exercises*

1. Show that if $\mathbf{X}_n \xrightarrow{p} \mathbf{c}$, then $\mathscr{L}[\mathbf{X}_n] \to \mathscr{L}[\mathbf{C}]$, where $\mathbf{C}$ is a degenerate random variable concentrated at $\mathbf{c}$.

2. Show that if $\mathbf{X} = (X_1, \ldots, X_T)$ has the multinomial distribution $\mathscr{M}(n, \mathbf{p})$, where $\mathbf{p} = (p_1, \ldots, p_T)$, then

$$M_\mathbf{X}(\mathbf{t}) = \left(\sum_i p_i e^{t_i}\right)^n.$$

3. Use the result of problem 2 to show that $E(X_i) = np_i$, $\mathrm{Var}(X_i) = np_i(1 - p_i)$, and $\mathrm{Cov}(X_i, X_j) = -np_ip_j$, so that

$$E(\mathbf{X}) = n\mathbf{p},$$

$$\mathrm{Cov}(\mathbf{X}) = n(\mathbf{D}_p - \mathbf{p}'\mathbf{p}),$$

where $\mathbf{D}_p = \mathrm{diag}(\mathbf{p})$.

4. Fill in the details of example 14.3-6.
5. Use theorem 14.3-6 to prove theorem 14.3-9.
6. Let X have the negative binomial distribution given by

$$P\{X = x\} = \binom{r + x - 1}{r - 1} p^r (1 - p)^x \quad x = 0, 1, \ldots .$$

(i) Find the moment-generating function of X.

(ii) Find the mean and variance of X. (See Section 13.7.1.)

(iii) If $p = p_n = \lambda/n$ and r remains fixed, find the limiting distribution of $n^{-1}X$.

14.4 The O_p, o_p Notation for Stochastic Sequences

We now develop the O_p, o_p notation for stochastic sequences, which generalizes the O, o notation for nonstochastic sequences. In the first four sections we describe the notation and its elementary properties. In the fifth section, we deal with a more general version of this theory due to Chernoff and Pratt.

14.4.1 Definition of $X_n = o_p(b_n)$

We follow up the idea expressed in exercise 8 of Section 14.2.3, by defining $o_p(1)$ first and then setting $o_p(b_n) = b_n o_p(1)$. The notation $X_n = o_p(1)$ means exactly the same thing as $X_n \xrightarrow{p} 0$. More formally, we have:

DEFINITION 14.4-1 $X_n = o_p(1)$ *if for every* $\varepsilon > 0$,

$$\lim_{n \to \infty} P\{|X_n| \leq \varepsilon\} = 1. \tag{14.4-1}$$

If $\mathbf{X}_n$ is a vector, we say that $\mathbf{X}_n = o_p(1)$ if $\|\mathbf{X}_n\| = o_p(1)$. Then we define the general o_p notation by:

DEFINITION 14.4-2 $X_n = o_p(b_n)$ *if* $X_n/b_n = o_p(1)$, *or equivalently,* $X_n = b_n o_p(1)$.

Again, if $\mathbf{X}_n$ is a vector, we say that $\mathbf{X}_n = o_p(b_n)$ if $\|\mathbf{X}_n\| = o_p(b_n)$.

14.4.2 Definition of $X_n = O_p(b_n)$

We continue the approach of Section 14.4.1 and define $X_n = O_p(1)$ first and then set $O_p(b_n) = b_n O_p(1)$. However, to motivate the definition of $O_p(1)$, it is useful to begin by amplifying definition 14.4-1 somewhat. Since this amplification is nothing more than a restatement of definition 14.4-1 with more technical detail, we call it definition 14.4-1a.

DEFINITION 14.4-1a $X_n = o_p(1)$ *if for every* $\varepsilon > 0$ *and every* $\eta > 0$ *there exists an integer* $n(\varepsilon, \eta)$ *such that if* $n \geq n(\varepsilon, \eta)$, *then*

$$P\{|X_n| < \varepsilon\} \geq 1 - \eta. \tag{14.4-2}$$

Informally, this definition means that with arbitrarily high probability (i.e., $\geq 1 - \eta$), $|X_n| = o(1)$. Taking our cue from this informal description, we want $X_n = O_p(1)$ to mean that with arbitrarily high probability, $|X_n| = O(1)$. In terms of a formal definition this becomes:

DEFINITION 14.4-3 $X_n = O_p(1)$ *if for every $\eta > 0$ there exist a constant $K(\eta)$ and an integer $n(\eta)$ such that if $n \geq n(\eta)$, then*

$$P\{|X_n| \leq K(\eta)\} \geq 1 - \eta. \tag{14.4-3}$$

If $\mathbf{X}_n$ is a vector, we say $\mathbf{X}_n = O_p(1)$ if $\|\mathbf{X}_n\| = O_p(1)$. Finally, we define the general O_p notation by:

DEFINITION 14.4-4 $X_n = O_p(b_n)$ *if $X_n/b_n = O_p(1)$, or equivalently, $X_n = b_n O_p(1)$.* As before, if $\mathbf{X}_n$ is a vector, we say $\mathbf{X}_n = O_p(b_n)$ if $\|\mathbf{X}_n\| = O_p(b_n)$.

If we compare definitions 14.4-1a and 14.4-3 with their counterparts in the O, o notation (definitions 14.2-1 and 14.2-2, with $b_n = 1$) we see a number of parallels. In both O_p and o_p, events are required to hold with a probability that exceeds a preassigned limit $1 - \eta$, replacing the "certainty" of the O, o definitions. In both $o(1)$ and $o_p(1)$, the sequences are required to be less than any prescribed small positive ε if n is sufficiently large. In both $O(1)$ and $O_p(1)$, the sequences are required to be bounded by some constant K if n is sufficiently large.

It is sometimes useful to refer to $X_n = O_p(1)$ by saying that X_n is "bounded in probability," and to $X_n = o_p(1)$ by saying that X_n "converges to zero in probability." The information in the O_p, o_p notation is often referred to as the "stochastic order" of X_n.

14.4.3   *Methods for determining the stochastic order of a sequence*
In this section we discuss two easily applied methods for determining the stochastic order of a sequence of random variables $\{X_n\}$.

You are only as big as your standard deviation
The standard deviation of a random variable is often used as an index of the size or order of magnitude of the typical departure of the random variable from its expected value. This use of the standard deviation to measure typical deviations is usually motivated by the fact that for the normal distribution, the probability that an observation will lie within one standard deviation from the mean is approximately 0.66. However, even for nonnormal distributions, the standard deviation gives the order of magnitude of typical deviations. For example, Tchebychev's inequality asserts that if X is a random variable with mean μ and variance $\sigma^2 < \infty$ and h is any positive number, then

$$P\{|X - \mu| \leq h\sigma\} \geq 1 - h^{-2}. \tag{14.4-4}$$

We use Tchebychev's inequality to connect the O_p, o_p notion of stochastic order of magnitude with the standard deviation as an index of the order of magnitude of deviations from the expected value.

THEOREM 14.4-1   *If $\{X_n\}$ is a stochastic sequence with $\mu_n = E(X_n)$ and $\sigma_n^2 = \mathrm{Var}(X_n) < \infty$, then*

$$X_n - \mu_n = O_p(\sigma_n).$$

Proof   If in (14.4-4) we set $h = \eta^{-1/2}$ for any $0 < \eta < 1$ and apply (14.4-4) to X_n, μ_n, and σ_n, then we have

$$P\left\{\frac{|X_n - \mu_n|}{\sigma_n} < \eta^{-1/2}\right\} \geq 1 - \eta. \tag{14.4-5}$$

Thus (14.4-5) holds for $n = 1, 2, \ldots$. Setting $K(\eta) = \eta^{-1/2}$, we apply definition 14.4-3 and conclude that

$$\frac{X_n - \mu_n}{\sigma_n} = O_p(1),$$

from which the desired result immediately follows. ∎

Example 14.4-1   The order of a binomial variable
If X_n has the binomial distribution $\mathscr{B}(n, p)$, then $E(X_n) = np$, $\mathrm{Var}(X_n) = np(1 - p)$. Hence $\sigma_n = \sqrt{np(1 - p)} = O(\sqrt{n})$, from which we conclude that $X_n - np = O_p(\sqrt{n})$, or, as it is usually written,

$$X_n = np + O_p(\sqrt{n}). \tag{14.4-6}$$

Actually, this example is a special case of the more general result that if X_n is a sum of n independent and identically distributed random variables with mean μ and variance σ^2, then

$$X_n = n\mu + O_p(\sqrt{n}) \tag{14.4-7}$$

(see exercise 2 in Section 14.4.6). ∎∎

A sequence that converges in distribution is also bounded in probability
The subject of this short section is a theorem that is the most common tool used for showing that $X_n = O_p(1)$. This theorem also illustrates one of the many connections between convergence in distribution and the O_p, o_p notation.

THEOREM 14.4-2   *If* $\mathscr{L}[X_n] \to \mathscr{L}[X]$, *then* $X_n = O_p(1)$.

The proof of this result is left for exercise 3 in Section 14.4.6. The reader may wish to apply this result to example 14.4-1. The next example gives a further application of this theorem.

Example 14.4-2   The order of the minimum of exponential variables
Let $Y_1, \ldots, Y_n$ be independent random variables, all identically distributed with the exponential distribution

$$P\{Y_i \leq t\} = F(t) = 1 - e^{-\lambda t}$$

for $t \geq 0$. Now if $X_n = \min\{Y_1, \ldots, Y_n\}$, then the distribution function of X_n is

$$\begin{aligned} G_n(x) = P\{X_n \leq x\} &= 1 - P\{X_n > x\} \\ &= 1 - P\{Y_1 > x, Y_2 > x, \ldots, Y_n > x\} \\ &= 1 - P\{Y_1 > x\}P\{Y_2 > x\} \ldots P\{Y_n > x\} \\ &= 1 - (1 - F(x))^n = 1 - e^{-n\lambda x}. \end{aligned}$$

We conclude that X_n has the exponential distribution with parameter $n\lambda$. From this it follows that nX_n has the same distribution as Y_1, from which we conclude that

$$\mathscr{L}[nX_n] \to \mathscr{L}[Y_1]. \tag{14.4-8}$$

Applying theorem 14.4-2 to (14.4-8), we obtain

$$nX_n = O_p(1), \qquad (14.4\text{-}9)$$

or

$$X_n = O_p(n^{-1}). \quad \blacksquare\blacksquare \qquad (14.4\text{-}10)$$

14.4.4   Convergence in distribution and $o_p(1)$

This section extends the discussion in Section 14.3.3 on the effect of a small remainder term on convergence in distribution. In the O_p, o_p notation, theorem 14.3-9 can be restated as

THEOREM 14.4-3   *If* $\mathcal{L}[X_n] \to \mathcal{L}[X]$, *then* $\mathcal{L}[X_n + o_p(1)] \to \mathcal{L}[X]$.

This theorem, in conjunction with the various tools we have for determining the stochastic order of a sequence, allows us to find the limiting distributions of fairly complicated random variables. We illustrate this point with two examples involving the binomial distribution.

Example 14.4-3   Asymptotic confidence intervals for p

We continue using the notation of example 14.4-1. From example 14.3-3 it follows that if $\hat{p} = n^{-1}X_n$, then

$$\mathcal{L}[\sqrt{n}(\hat{p} - p)] \to \mathcal{L}[W], \qquad (14.4\text{-}11)$$

where W has the normal distribution $\mathcal{N}(0, p(1 - p))$. Define Z_n by

$$Z_n = \frac{\sqrt{n}(\hat{p} - p)}{\sqrt{p(1 - p)}}. \qquad (14.4\text{-}12)$$

By theorem 14.3-6, $\mathcal{L}[Z_n] \to \mathcal{L}[Z]$, where Z has the unit normal distribution $\mathcal{N}(0, 1)$.

From (14.4-11) and theorem 14.4-2, we deduce that

$$\hat{p} = p + O_p(n^{-1/2}). \qquad (14.4\text{-}13)$$

But if $Y_n = O_p(n^{-1/2})$, then $Y_n = o_p(1)$ (see exercise 6 in Section 14.4.6), so that

$$\hat{p} = p + o_p(1) \quad \text{as } n \to \infty.$$

Applying theorems 14.3-1 and 14.3-6, we deduce that if $p \neq 0$, and $p \neq 1$, then

$$\left(\frac{p(1 - p)}{\hat{p}(1 - \hat{p})}\right)^{1/2} = 1 + o_p(1). \qquad (14.4\text{-}14)$$

Now define V_n by

$$V_n = \frac{\sqrt{n}(\hat{p} - p)}{\sqrt{\hat{p}(1 - \hat{p})}} = Z_n\left(\frac{p(1 - p)}{\hat{p}(1 - \hat{p})}\right)^{1/2} = Z_n(1 + o_p(1))$$

$$= Z_n + Z_n o_p(1) = Z_n + O_p(1)o_p(1)$$

$$= Z_n + o_p(1).$$

We conclude from theorem 14.3-9 that

$$\mathcal{L}[V_n] \to \mathcal{L}[Z]. \qquad (14.4\text{-}15)$$

The result (14.4-15) is often used to set asymptotic confidence intervals for p as follows. If n is large enough, (14.4-15) says that

$$P\{-a < V_n < a\} \approx P\{-a < Z < a\}.$$

If a is chosen so that

$$P\{-a < Z < a\} = 1 - \alpha,$$

then inverting the inequality $-a < V_n < a$, we obtain

$$P\left\{\hat{p} - \frac{a}{\sqrt{n}}\sqrt{\hat{p}(1 - \hat{p})} < p < \hat{p} + \frac{a}{\sqrt{n}}\sqrt{\hat{p}(1 - \hat{p})}\right\} \approx 1 - \alpha$$

and so obtain an asymptotic $1 - \alpha$ level confidence interval for p. ∎∎

Example 14.4-4 *The square of a binomial proportion*
We continue using the notation of the example 14.4-3. Our aim now is to find the asymptotic distribution of $\hat{p}^2$. We begin by observing that

$$\hat{p}^2 = (p + (\hat{p} - p))^2 = p^2 + 2p(\hat{p} - p) + (\hat{p} - p)^2,$$

or

$$\sqrt{n}(\hat{p}^2 - p^2) = 2p\sqrt{n}(\hat{p} - p) + \sqrt{n}(\hat{p} - p)^2$$
$$= 2pW_n + n^{-1/2}W_n^2,$$

where we set $W_n = \sqrt{n}(\hat{p} - p)$. We know that W_n converges in distribution, so from theorem 14.3-6 we know that W_n^2 also converges in distribution. Then from theorem 14.4-2 we know that $W_n^2 = O_p(1)$, and hence $n^{-1/2}W_n^2 = O_p(n^{-1/2}) = o_p(1)$. These facts give us

$$\sqrt{n}(\hat{p}^2 - p^2) = 2pW_n + o_p(1).$$

Since $\mathscr{L}[W_n] \to \mathscr{L}[W]$, where W has the normal distribution $\mathscr{N}(0, p(1 - p))$, we now have

$$\mathscr{L}[\sqrt{n}(\hat{p}^2 - p^2)] \to \mathscr{L}[2pW].$$

But $2pW$ has the $\mathscr{N}(0, 4p^3(1 - p))$ distribution, from which we conclude that $\hat{p}^2$ has an approximate normal distribution with mean p^2 and variance $n^{-1}4p^3(1 - p)$. ∎∎

14.4.5 *Chernoff–Pratt theory of stochastic order*
The similarity of the O, o and O_p, o_p notation suggests that they be used in tandem. For instance, in example 14.4-1 we uncritically assumed that $O_p(O(\sqrt{n})) = O_p(\sqrt{n})$. Similarly, in using Taylor expansion arguments we often make use of the fact that

$$o(O_p(n^{-1/2})) = o_p(n^{-1/2}). \tag{14.4-16}$$

Also, if $f(x) = o(x)$ as $x \to 0$, and if $X_n = O_p(n^{-1/2})$, we would like to be sure that

$$f(X_n) = o_p(n^{-1/2}). \tag{14.4-17}$$

Furthermore, in example 14.4-3 we made uncritical use of the fact that $O_p(1)o_p(1) = o_p(1)$.

The approach to o_p and O_p discussed in this section is based on the work of Chernoff, as extended by Pratt [1959]. It requires a little attention to the formal structure that underlies a stochastic sequence; however, once this machinery has been set up, it provides a simple and mathematically rigorous way of turning nonstochastic results involving o and O into parallel stochastic results involving o_p and O_p.

Formal structure of a stochastic sequence

For each $n = 1, 2, \ldots$, let (Ω_n, P_n) be a probability space, where Ω_n is a sample space and P_n is a probability measure on the subsets of Ω_n. (The measurability of all subsets encountered is assumed but not explicitly stated.) Let ω_n be an abstract random variable taking values in Ω_n and distributed according to P_n. Next, suppose H_n is a (measurable) function mapping Ω_n into c-dimensional Euclidean space. Finally, suppose that the sequence of random vectors of interest is given by

$$\xi_n = H_n(\omega_n). \tag{14.4-18}$$

We let w_n denote the elements of Ω_n, i.e., possible values of ω_n.

Example 14.4-5   The multinomial distribution

The important examples in later sections of this chapter concern the multinomial distribution. We give the formal structure here that is used in the later sections. We let

$$\mathscr{S}_T = \left\{ \mathbf{p} = (p_1, \ldots, p_T) : p_i \geqq 0 \quad \text{and} \quad \sum_{i=1}^{T} p_i = 1 \right\}.$$

Then $\mathscr{S}_T$ is the set of all T-dimensional probability vectors. First, we set

$$\Omega_n = \mathscr{S}_T \quad n = 1, 2, \ldots. \tag{14.4-19}$$

In $\mathscr{S}_T$, the multinomial distribution is the distribution of the vector of cell proportions rather than the vector of cell counts. For the $\mathscr{M}(n, \boldsymbol{\pi})$ distribution, the cell proportions are constrained to lie in the subset of $\mathscr{S}_T$ given by

$$T_n = \{\mathbf{p} \in \mathscr{S}_T : np_i \text{ is an integer for } i = 1, \ldots, T\}. \tag{14.4-20}$$

Hence if ω_n has the $\mathscr{M}(n, \boldsymbol{\pi})$ distribution, then P_n is given by

$$P_n\{B\} = \sum_{\mathbf{p} \in B \cap T_n} \binom{n}{np_1, \ldots, np_T} \pi_1^{np_1} \ldots \pi_T^{np_T}, \tag{14.4-21}$$

where B is any subset of $\mathscr{S}_T$.

There are several functions of the ω_n-sequence that arise in later sections of this chapter.

1. If $H_n(\mathbf{w}_n) = n\mathbf{w}_n$ for $\mathbf{w}_n \in \mathscr{S}_T$, then $\xi_n = H_n(\omega_n) = n\omega_n$, so that $\{\xi_n\}$ is the ordinary sequence of multinomial distributions.

2. If

$$H_n(\mathbf{w}_n) = n(\mathbf{w}_n - \boldsymbol{\pi})\mathbf{D}_\pi^{-1}(\mathbf{w}_n - \boldsymbol{\pi})', \tag{14.4-22}$$

where $\mathbf{D}_\pi$ is the diagonal matrix based on $\boldsymbol{\pi}$, then $\{\xi_n\}$ is the sequence of chi square statistics used to test the hypothesis that $\boldsymbol{\pi}$ is the vector of true cell probabilities.

3.   If $H_n(\mathbf{w}_n)$ is any value of a parameter $\boldsymbol{\theta}$ which maximizes $\pi_1(\boldsymbol{\theta})^{nw_{n1}} \ldots \pi_T(\boldsymbol{\theta})^{nw_{nT}}$, then $\xi_n = H_n(\omega_n)$ is the maximum likelihood estimate of $\boldsymbol{\theta}$.

Chernoff–Pratt definitions of o_p and O_p

The following two theorems form the basis of the Chernoff–Pratt approach to the O_p, o_p notation. They can be viewed as providing alternative definitions of o_p and O_p to those given by definitions 14.4-2 and 14.4-4. These theorems make use of the notation developed in the previous section.

THEOREM 14.4-4   *We have $\xi_n = H_n(\omega_n) = o_p(b_n)$ if and only if for every $\eta > 0$, (measurable) subsets $S_n \subseteq \Omega_n$ can be found such that*

(i) $P_n\{\omega_n \in S_n\} \geq 1 - \eta$ *for all n;*

(ii) *if $\{w_n\}$ is a nonstochastic sequence such that $w_n \in S_n$ for every n, then $x_n = H_n(w_n) = o(b_n)$.*

THEOREM 14.4-5   *We have $\xi_n = H_n(\omega_n) = O_p(b_n)$ if and only if for every $\eta > 0$, (measurable) subsets $S_n \subseteq \Omega_n$ can be found such that*

(i) $P_n\{\omega_n \in S_n\} \geq 1 - \eta$ *for all n;*

(ii) *if w_n is a nonstochastic sequence such that $w_n \in S_n$ for every n, then $x_n = H_n(w_n) = O(b_n)$.*

See Pratt [1959] for a proof of these results. Exercises 7, 8, and 9 in Section 14.4.6 constitute proofs of theorems 14.4-4 and 14.4-5.

We note two implications of these new definitions of o_p and O_p. First, the two definitions of O_p and o_p are completely parallel. Second, the definitions separate out two distinct problems to be solved in assessing the stochastic order or ξ_n. One is stochastic, namely, finding events $S_n \subseteq \Omega_n$ with arbitrarily large P_n-probability (i.e., $\geq 1 - \eta$). The other is analytic, namely, establishing the order of magnitude of certain nonstochastic sequences $\{w_n\}$. We illustrate the power of these results in an initial application of the δ method which we discuss more extensively in Section 14.6.

Example 14.4-6   Delta method for the binomial distribution

We continue using the notation of example 14.4-4, in which we found the asymptotic distribution of the square $\hat{p}^2$ of a binomial proportion. We now consider finding the asymptotic distribution of an arbitrary function $g(\hat{p})$ of $\hat{p}$, where g has a first-order Taylor expansion about p, i.e.,

$$g(w) = g(p) + (w - p)g'(p) + o(w - p) \qquad (14.4\text{-}23)$$

as $w \to p$. From (14.4-23) it follows that if w_n is any sequence of numbers such that

$$w_n = p + O(n^{-1/2}), \qquad (14.4\text{-}24)$$

then

$$g(w_n) - g(p) - (w_n - p)g'(p) = o(n^{-1/2}). \qquad (14.4\text{-}25)$$

Note that (14.4-24) and (14.4-25) are statements about nonstochastic sequences and that (14.4-25) can be verified by referring to the definitions and properties of the o, O notation.

Now set $\omega_n = \hat{p}$. From (14.4-13) we know that

$$\omega_n - p = O_p(n^{-1/2}).\tag{14.4-26}$$

Applying theorem 14.4-5 to (14.4-26) with

$$\xi_n^{(1)} = H_n^{(1)}(\omega_n) = \omega_n - p,\tag{14.4-27}$$

we deduce the existence of (measurable) subsets $S_n \subseteq \Omega_n$ such that

(i) $P\{\omega_n \in S_n\} \geq 1 - \eta$ for all n;

(ii)   if w_n is a nonstochastic sequence such that $w_n \in S_n$ for every n, then $x^{(1)} = H_n^{(1)}(w_n) = O(n^{-1/2})$.

From (14.4-24) and (14.4-25), however, we see that these same subsets S_n also have the property that if $w_n \in S_n$ for all n, then (14.4-25) holds. If we set

$$\xi_n^{(2)} = H_n^{(2)}(\omega_n) = g(\omega_n) - g(p) - (\omega_n - p)g'(p),$$

then, applying the sufficiency of the conditions in theorem 14.4-4 to the S_n and $\xi_n^{(2)}$, we obtain

$$\xi_n^{(2)} = o_p(n^{-1/2}).\tag{14.4-28}$$

Expressing (14.4-28) in terms of $\hat{p}$ and g, we have the basic result:

$$g(\hat{p}) - g(p) = (\hat{p} - p)g'(p) + o_p(n^{-1/2}).\tag{14.4-29}$$

Multiplying both sides of (14.4-29) by $\sqrt{n}$ and setting $U_n = \sqrt{n}(\hat{p} - p)$, we have

$$\sqrt{n}(g(\hat{p}) - g(p)) = g'(p)U_n + o_p(1).\tag{14.4-30}$$

Now we can apply theorems 14.3-6 and 14.4-3 to (14.4-30), and we find that

$$\mathscr{L}[\sqrt{n}(g(\hat{p}) - g(p))] \to \mathscr{L}[V],\tag{14.4-31}$$

where V has the normal distribution $\mathscr{N}(0, (g'(p))^2 p(1 - p))$. Setting $g(p) = p^2$ and differentiating, we see that the variance of the asymptotic normal distribution is $4p^3(1 - p)$, which agrees with the result of example 14.4-4. ∎∎

The separation of the stochastic and analytic problems of o_p and O_p made in theorems 14.4-4 and 14.4-5 allows us to use more explicitly the O, o convention (described in (i) of Section 14.2.1) that for a finite number of values of n, the sequence may be infinite or undefined.

Pratt's theory of occurrence in probability

In this section we briefly describe a general theory in which certain aspects of a stochastic sequence "occur in probability." Occurrence in probability is a generalized notion that arises from the form of theorems 14.4-4 and 14.4-5. We use the notation of the earlier parts of this section.

We would like to describe all sequences $\{w_n\}$ such that $w_n \in \Omega_n$ for every n. Let Ω denote the set of such sequences. Then Ω can be expressed as

$$\Omega = \Omega_1 \times \Omega_2 \times \cdots \times \Omega_n \times \cdots = \mathop{\vartimes}_{n=1}^{\infty} \Omega_n,\tag{14.4-32}$$

i.e., the infinite Cartesian product of the Ω_n. Similarly, if S_n is any subset of Ω_n, then

$$S = S_1 \times S_2 \times \cdots \times S_n \times \cdots = \overset{\infty}{\underset{n=1}{\times}} S_n \qquad (14.4\text{-}33)$$

is a subset of Ω. There are subsets of Ω that cannot be described as in (14.4-33). When we describe the convergence properties of a sequence $\{w_n\}$, where $w_n \in \Omega_n$, we are defining a subset of Ω. Pratt [1959] gives the following general definition of occurrence in probability.

DEFINITION 14.4-5 $S \subseteq \Omega$ *occurs in probability if for every $\eta > 0$, (measurable) subsets $S_n(\eta) \subseteq \Omega_n$ can be found such that*

(i) $P\{\omega_n \in S_n\} \geq 1 - \eta$ *for every n*;

(ii) $\overset{\infty}{\underset{n=1}{\times}} S_n \subseteq S$.

This definition is clearly motivated by the form of theorems 14.4-4 and 14.4-5. We may easily show the relation between occurrence in probability and the O_p, o_p notation by constructing the proper subsets of Ω. We define

$$S^{(1)} = \{\{w_n\}: \ w_n \in \Omega_n \quad \text{and} \quad x_n = H_n(w_n) = o(b_n)\},$$

$$S^{(2)} = \{\{w_n\}: \ w_n \in \Omega_n \quad \text{and} \quad x_n H_n(w_n) = O(b_n)\}.$$

Theorem 14.4-4 is equivalent to saying that $\xi_n = o_p(b_n)$ if and only if $S^{(1)}$ occurs in probability. Theorem 14.4-5 is equivalent to saying that $\xi_n = O_p(b_n)$ if and only if $S^{(2)}$ occurs in probability.

The use of the occurrence in probability device is enhanced by the following two theorems.

THEOREM 14.4-6   *If S occurs in probability and T is another subset of Ω such that $S \subseteq T$, then T also occurs in probability.*

THEOREM 14.4-7   *Let $S^{(1)}, S^{(2)}, \ldots$ all be subsets of Ω. $S^{(1)}, S^{(2)}, \ldots$ all occur in probability if and only if their intersection $\bigcap_{j=1}^{\infty} S^{(j)}$ occurs in probability.*

Theorem 14.4-6 can be used to simplify the discussion in example 14.4-6. Theorem 14.4-7 is useful when the analytic parts of theorems 14.4-4 and 14.4-5 require the combination of a number of O, o terms. For example, a simple application of theorems 14.4-6 and 14.4-7 proves the following complicated theorem involving the simultaneous use of the O, o and O_p, o_p notation.

THEOREM 14.4-8   *Let $f_n^{(j)}(\cdot)$ $(j = 1, \ldots, J)$, $g_n^{(k)}(\cdot)$ $(k = 1, \ldots, K)$, and $h_n(\cdot)$ be functions such that if $\{x_n\}$ is a nonstochastic sequence and*

(i) $f_n^{(j)}(x_n) = O(r_n^{(j)})$ *for $j = 1, \ldots, J$,*   (ii) $g_n^{(k)}(x_n) = o(s_n^{(k)})$ *for $k = 1, \ldots, K$,*

then $h_n(x_n) = O(t_n)$ (or $= o(t_n)$). Moreover, let $X_1, X_2, \ldots$ be a stochastic sequence such that

(i) $f_n^{(j)}(X_n) = O_p(r_n^{(j)})$ *for $j = 1, \ldots, J$,*   (ii) $g_n^{(k)}(X_n) = o_p(s_n^{(k)})$ *for $k = 1, \ldots, K$.*

Then $h_n(X_n) = O_p(t_n)$ (or $= o_p(t_n)$).

While this result is sufficiently general to include many applications, it is generally easier in any specific case to apply theorems 14.4-6 and 14.4-7 directly.

14.4.6   *Exercises*

1. Show that o_p and O_p obey the same rules for products as do o, O.

   (P1) $O_p(a_n)O_p(b_n) = O_p(a_n b_n)$,

   (P2) $O_p(a_n)o_p(b_n) = o_p(a_n b_n)$,

   (P3) $o_p(a_n)o_p(b_n) = o_p(a_n b_n)$.

2. Show that if X_n is the sum of n independent and identically distributed random variables, each with mean μ and variance $\sigma^2 < \infty$, then $X_n = n\mu + O_p(\sqrt{n})$.

3. Prove theorem 14.4-2.

4. Prove theorem 14.4-6.

5. Prove theorem 14.4-7.

6. Show that if $Y_n = O_p(n^{-1/2})$, then $Y_n = o_p(1)$.

7. Show that $\xi_n = H_n(\omega_n) = O_p(b_n)$ if and only if for any $\eta > 0$ there is a sequence of (extended real) numbers $d_n = d_n(\eta)$ such that

   (i) $P_n\{\|\xi_n\| \le d_n\} \ge 1 - \eta$ for all n;

   (ii) $d_n = O(b_n)$.

8. Show that $\xi_n = H_n(\omega_n) = O_p(b_n)$ if and only if for any $\eta > 0$ there is a sequence of Borel sets (in c-dimensional Euclidean space) $T_n = T_n(\eta)$ such that

   (i) $P_n\{\xi_n \in T_n\} \ge 1 - \eta$ for all n;

   (ii)   if $x_n \in T_n$ for every n, then $x_n = O(b_n)$.

9. Show that $\xi_n = H_n(\omega_n) = O_p(b_n)$ if and only if for any $\eta > 0$ there is a sequence of (measurable) sets $S_n = S_n(\eta) \subseteq \Omega_n$ such that

   (i) $P_n\{\omega_n \in S_n\} \ge 1 - \eta$ for all n;

   (ii)   if $w_n \in S_n$ for every n, then $x_n = H_n(w_n) = O(b_n)$.

10. Show that the results of problems 7, 8, and 9 remain true if O and O_p are replaced throughout by o and o_p.

14.5   Convergence of Moments

If $\mathcal{L}[X_n] \to \mathcal{L}[X]$, we may also want to know if

$$E(X_n^r) \to E(X^r) \qquad\qquad (14.5\text{-}1)$$

for various choices of r, usually 1 and 2. It should be noted that convergence in distribution does not in general entail (14.5-1) for any value of $r \ne 0$. In most of the applications in this book, however, (14.5-1) does hold if $\mathcal{L}[X_n] \to \mathcal{L}[X]$. The remainder of this section is devoted to a short discussion of this convergence problem.

14.5.1   Limits of moments and asymptotic moments

If $\mathscr{L}[X_n] \to \mathscr{L}[X]$, then we refer to

$$E(X^r) \tag{14.5-2}$$

as the *asymptotic rth moment* of X_n, while

$$\lim_{n \to \infty} E(X_n^r) \tag{14.5-3}$$

is the *limit of the rth moment* of X_n, if it exists. Since (14.5-3) and (14.5-2) may be unequal, we have separate terminology for them. The next example gives a case when they are equal.

Example 14.5-1   Square of a binomial proportion

If X_n has the binomial distribution $\mathscr{B}(n, p)$ and $\hat{p} = n^{-1}X_n$, then from example 14.4-4 we have $\mathscr{L}[\sqrt{n}(\hat{p}^2 - p^2)] \to \mathscr{L}[V]$, where V has the normal distribution $\mathscr{N}(0, 4p^3(1 - p))$. We calculate that the actual variance of $\sqrt{n}(\hat{p}^2 - p^2)$ is:

$$\begin{aligned}
\mathrm{Var}(\sqrt{n}(\hat{p}^2 - p^2)) &= n\,\mathrm{Var}(\hat{p}^2) \\
&= 4p^3(1 - p) + n^{-1}p^2(10p^2 - 16p + 6) \\
&\quad - n^{-2}p(6p^3 - 12p^2 + 7p - 1) \\
&= 4p^3(1 - p) + O(n^{-1}).
\end{aligned} \tag{14.5-4}$$

Hence we have

$$\lim_{n \to \infty} \mathrm{Var}(\sqrt{n}(\hat{p}^2 - p^2)) = \mathrm{Var}(V). \quad \blacksquare\blacksquare$$

The general relationship between the limit of the variances and the asymptotic variance is given in the next theorem.

THEOREM 14.5-1   *If $\mathscr{L}[X_n] \to \mathscr{L}[X]$ and if we let $\mathrm{Var}(X_n)$ denote the variance of X_n when it exists and set it equal to $+\infty$ otherwise, then*

$$\liminf_{n \to \infty} \mathrm{Var}(X_n) \geqq \mathrm{Var}(X). \tag{14.5-5}$$

For a proof, see Zacks [1971], pp. 252–253.

From this theorem we see that the asymptotic variance of X_n can be no larger than the limit of the variances. Examples can be given to show that it can be strictly smaller (see exercise 1 of Section 14.5.3).

14.5.2   The order of moments and o_p

In example 14.5-1, we saw that

$$E(\hat{p}^2) = p^2 + o(1), \tag{14.5-6}$$

$$\mathrm{Var}(\hat{p}^2) = n^{-1}4p^3(1 - p) + o(n^{-1}). \tag{14.5-7}$$

However, from the δ method discussed in Section 14.6 and example 14.4-4, we know that

$$\mathscr{L}[\sqrt{n}((\hat{p})^2 - p^2)] \to \mathscr{N}(0, 4p^3(1 - p)),$$

and hence that

$$\hat{p}^2 = p^2 + o_p(1),\tag{14.5-8}$$

$$\hat{p}^2 = p^2 + n^{-1/2}V + o_p(n^{-1/2}),\tag{14.5-9}$$

where V is defined in example 14.5-1.

We might be tempted to take expectations of both sides of (14.5-8) and variances of both sides of (14.5-9) in order to derive (14.5-6) and (14.5-7). As a general rule, this approach needs justification, since it is easy to construct examples where

$$E(o_p(1)) \neq o(1)\tag{14.5-10}$$

(see exercise 1 in Section 14.5.3). However, for many of the applications in this book, the following theorem, due to Cramér [1946, p. 354], justifies taking expected values over stochastic order-of-magnitude terms.

THEOREM 14.5-2   Let $\overline{\mathbf{X}}_n$ be the average of n independent and identically distributed random vectors with a moment-generating function that exists in a neighborhood of 0, and let $\boldsymbol{\mu} = E(\overline{\mathbf{X}}_n)$, $n^{-1}\boldsymbol{\Sigma} = \mathrm{Cov}(\overline{\mathbf{X}}_n)$. Let $\mathbf{H}(\cdot)$ be any function defined on the sample space of $\overline{\mathbf{X}}_n$ such that

(i) $\mathbf{H}(\mathbf{x})$ has continuous first and second derivatives at $\mathbf{x} = \boldsymbol{\mu}$;

(ii) $\|\mathbf{H}(\overline{\mathbf{x}}_n)\| \leqq Kn^p$ for some fixed K and p, where $\overline{\mathbf{x}}_n$ runs over all of the possible values of $\overline{\mathbf{X}}_n$.

Then

$$E[\mathbf{H}(\overline{\mathbf{X}}_n)] = \mathbf{H}(\boldsymbol{\mu}) + O(n^{-1}),\tag{14.5-11}$$

$$\mathrm{Cov}[\mathbf{H}(\overline{\mathbf{X}}_n)] = n^{-1}\left(\frac{\partial \mathbf{H}}{\partial \boldsymbol{\mu}}\right)\boldsymbol{\Sigma}\left(\frac{\partial \mathbf{H}}{\partial \boldsymbol{\mu}}\right)' + O(n^{-3/2}).\tag{14.5-12}$$

We observe that theorem 14.5-2 is tailor-made for multinomial problems where the moment-generating function always exists and very often the bound in (ii) is easy to compute. This theorem is less useful for continuous variables, where (ii) often just means that H is bounded.

14.5.3   Exercise

1. Let X_n be a random variable with a distribution given by

$$X_n = \begin{cases} 2^n & \text{with probability } 2^{-n} \\ 0 & \text{with probability } 1 - 2^{-n}. \end{cases}$$

Show that $X_n \xrightarrow{p} 0$, but that $EX_n^r \nrightarrow 0$ if $r \geqq 1$. What happens when $r < 1$?

14.6   The δ Method for Calculating Asymptotic Distributions

We have already used the δ method in examples 14.4-4 and 14.4-6. There we applied it to derive the asymptotic distribution of functions of a binomial proportion. However, the δ method is an important general technique for calculating asymptotic distributions and thereby deducing asymptotic means, variances, and covariances. In this section we discuss the details of this general method.

14.6.1   *The one-dimensional version of the δ method*

The δ method requires two ingredients: first, a random variable (which we denote here by $\hat{\theta}_n$) whose distribution depends on a real-valued parameter θ in such a way that

$$\mathscr{L}[\sqrt{n}(\hat{\theta}_n - \theta)] \to \mathscr{N}(0, \sigma^2(\theta)) \tag{14.6-1}$$

(in example 14.4-6, $\hat{\theta}_n = \hat{p}, \theta = p,$ and $\sigma^2(\theta) = \theta(1 - \theta)$); and second, a function $f(x)$ that can be differentiated at $x = \theta$ so that it possesses the following expansion about θ:

$$f(x) = f(\theta) + (x - \theta)f'(\theta) + o(|x - \theta|) \quad \text{as } x \to \theta. \tag{14.6-2}$$

The δ method for finding approximate means and variances of a function of a random variable is justified by the following theorem.

THEOREM 14.6-1 (*The one-dimensional δ method.*)   *If $\hat{\theta}_n$ is a real-valued random variable and θ is a real-valued parameter such that (14.6-1) holds, and if f is a function satisfying (14.6-2), then the asymptotic distribution of $f(\hat{\theta}_n)$ is given by*:

$$\mathscr{L}[\sqrt{n}(f(\hat{\theta}_n) - f(\theta))] \to \mathscr{N}(0, \sigma^2(\theta)[f'(\theta)]^2). \tag{14.6-3}$$

(This result can also be interpreted as saying that for large n, $f(\hat{\theta}_n)$ has an approximate normal distribution with mean $f(\theta)$ and variance $n^{-1}\sigma^2(\theta)[f'(\theta)]^2$. The following proof of theorem 14.6-1 indicates how the tools developed in Sections 14.2, 14.3, and 14.4 are applied in this problem.)

Proof   We set up the underlying structure for the Chernoff–Pratt definitions as follows: $\Omega_n = R^1 =$ the real numbers, P_n is the probability distribution of $\hat{\theta}_n$ on R^1. As in Section 14.4.5, Ω is the set of all sequences $\{t_n\}$ such that $t_n \in \Omega_n$. We define two subsets of Ω:

$$S = \{\{t_n\} \in \Omega : t_n - \theta = O(n^{-1/2})\},$$

$$T = \{\{t_n\} \in \Omega : f(t_n) - f(\theta) - (t_n - \theta)f'(\theta) = o(n^{-1/2})\}.$$

The only subtlety here is the fact that if f satisfies the expansion (14.6-2), then $S \subseteq T$, i.e., any sequence in S is also a member of T. This is because if $\{t_n\} \in S$, then

$$t_n - \theta = O(n^{-1/2}),$$

and hence from (14.6-2) we have

$$f(t_n) - f(\theta) - f'(\theta)(t_n - \theta) = o(|t_n - \theta|)$$
$$= o(O(n^{-1/2}))$$
$$= o(n^{-1/2}).$$

Thus $\{t_n\}$ is also a member of T. The next fact to observe is that since (14.6-1) holds, we have

$$n^{1/2}(\hat{\theta}_n - \theta) = O_p(1), \tag{14.6-4}$$

and hence

$$\hat{\theta}_n - \theta = O_p(n^{-1/2}). \tag{14.6-5}$$

Now we use theorem 14.4-5 and definition 14.4-5 to deduce from (14.6-5) that S occurs in probability. However, since $S \subseteq T$, it follows from theorem 14.4-6 that T must also occur in probability. Finally, from theorem 14.4-4 we conclude that

$$f(\hat{\theta}_n) - f(\theta) - f'(\theta)(\hat{\theta}_n - \theta) = o_p(n^{-1/2}). \tag{14.6-6}$$

Rewriting (14.6-5), we obtain

$$\sqrt{n}(f(\hat{\theta}_n) - f(\theta)) = \sqrt{n}(\hat{\theta}_n - \theta)f'(\theta) + o_p(1). \tag{14.6-7}$$

Now let

$$V_n = \sqrt{n}(f(\hat{\theta}_n) - f(\theta)),$$
$$U_n = \sqrt{n}(\hat{\theta}_n - \theta),$$
$$g(x) = xf'(\theta) \quad \text{for all real numbers } x.$$

Then (14.6-7) may be written as

$$V_n = g(U_n) + o_p(1). \tag{14.6-8}$$

We may express the assumptions (14.6-1) as $\mathcal{L}[U_n] \to \mathcal{L}[U]$, where $U \sim \mathcal{N}(0, \sigma^2(\theta))$. Hence from theorems 14.3-6 and 14.3-9, we may conclude that

$$\mathcal{L}[V_n] \to \mathcal{L}[g(U)], \tag{14.6-9}$$

where

$$g(U) \sim \mathcal{N}(0, \sigma^2(\theta)[f'(\theta)]^2).$$

Since (14.6-9) is equivalent to (14.6-3), our proof is complete. ∎

Example 14.6-1   Asymptotic distribution of powers of a binomial proportion
We continue using the notation of example 14.4-3. Suppose we want to find the asymptotic distribution for a power of $\hat{p}$, i.e.,

$$f(\hat{p}) = \hat{p}^m \quad \text{for } m \neq 0. \tag{14.6-10}$$

Evidently the assumptions of theorem 14.6-1 are met if $0 < p < 1$, and since $f'(p) = mp^{m-1}$, we conclude that

$$\mathcal{L}[\sqrt{n}(\hat{p}^m - p^m)] \to \mathcal{N}(0, m^2 p^{2m-1}(1 - p)).$$

Hence for large n, $\hat{p}^m$ has an approximate normal distribution with mean p^m and variance $n^{-1}m^2 p^{2m-1}(1 - p)$. Setting $m = 2$ yields the result obtained in example 14.4-4. Furthermore, when $m > 0$, using direct calculations on the distribution of $\hat{p}$ we can show that

$$E(\hat{p}^m) = p^m + O(n^{-1}), \tag{14.6-11}$$

$$\text{Var}(\hat{p}^m) = n^{-1}m^2 p^{2m-1}(1 - p) + O(n^{-2}). \tag{14.6-12}$$

When $m < 0$, however, this example illustrates an important problem that occurs often in discrete distributions. If $m < 0$, then $\hat{p}^m$ has positive probability of being infinite, and therefore neither $E(\hat{p}^m)$ nor $\text{Var}(\hat{p}^m)$ is finite. However, since the probability that $\hat{p}^m = \infty$ goes to zero very quickly, for large n the distribution of $\hat{p}^m$ looks very much like the $\mathcal{N}(p^m, n^{-1}m^2 p^{2m-1}(1 - p))$. ∎∎

Example 14.6-2   The asymptotic distribution of an estimator based on Hansen frequencies

A scoring technique frequently used in the study of animal behavior is to record whether a behavior pattern of interest occurs at least once or not at all during each one of n nonoverlapping time intervals of constant length T. The resulting data are often referred to as "Hansen frequencies," from the work of E. W. Hansen [1966] and his students, although such a scoring system was used as early as 1929 by Olson [1929, p. 97]. Fisher [1953] considers a similar problem in dilution experiments.

Assuming that the behavior of interest has a temporal distribution governed by a Poisson process, the Hansen frequencies then follow a binomial distribution. Let p_0 denote the probability of no occurrence of the behavior in an interval of length T:

$$p_0 = e^{-\lambda T}, \tag{14.6-13}$$

where λ is the mean rate of occurrence in a unit time interval. Let N_0 denote the number of intervals in which the behavior did not occur. Then $N_0 \sim \mathcal{B}(n, p_0)$, and $\hat{p}_0 = N_0/n$ is a binomial proportion. From (14.6-13) we have

$$\lambda = -T^{-1} \log_e(p_0), \tag{14.6-14}$$

so that the maximum likelihood estimate of λ based on N_0 is

$$\lambda^* = -T^{-1} \log_e(\hat{p}_0). \tag{14.6-15}$$

We can use the δ method to find the asymptotic distribution of λ^*. First, we have

$$\mathcal{L}[\sqrt{n}(\hat{p}_0 - p_0)] \to \mathcal{N}(0, p_0(1 - p_0))$$

as $n \to \infty$. Second, if we let $f(p_0) = -T^{-1} \log_e(p_0)$, then $f'(p_0) = -T^{-1} p_0^{-1}$, so that by theorem 14.6-1 we have

$$\mathcal{L}[\sqrt{n}(f(\hat{p}_0) - f(p_0))] \to \mathcal{N}(0, T^{-2} p_0^{-1}(1 - p_0)). \tag{14.6-16}$$

Thus for large n, λ^* has an approximate $\mathcal{N}(\lambda, n^{-1} T^{-2}(e^{\lambda T} - 1))$ distribution. Note that, as in the previous example, the moments of λ^* are not finite, since λ^* has positive probability of being infinite. As in the previous example, the moments of the asymptotic distribution of λ^* are finite. ∎∎

Example 14.6-3   Asymptotic distribution of powers of a single Poisson variate

Here we illustrate another use of the Chernoff–Pratt theory to find the asymptotic distribution of a function of a random variable. The result given here is related to the δ method of theorem 14.6-1, but it differs from it because the random variable X_n is wandering off to $+\infty$ rather than converging to a point.

Suppose X_n has a Poisson distribution with mean λ_n, where $\lambda_n \to \infty$ as $n \to \infty$. Using moment-generating functions, we can show that

$$\mathcal{L}[\lambda_n^{-1/2}(X_n - \lambda_n)] \to \mathcal{N}(0, 1). \tag{14.6-17}$$

Thus for large λ_n, X_n has an approximate $\mathcal{N}(\lambda_n, \lambda_n)$ distribution. In this example we find the asymptotic distribution of $(X_n)^m$ for $m \neq 0$.

We set up the following underlying structure for the Chernoff–Pratt theory (see Section 14.4-5 for definitions and details): $\Omega_n = \{0, 1, \dots\}$ = the nonnegative integers, X_n is such that

$$P_n(X_n \in E) = \sum_{j \in E} e^{-\lambda_n} \frac{\lambda_n^j}{j!}.$$

As in Section 14.4.5, we let Ω denote the set of all sequences $\{t_n\}$ such that $t_n \in \Omega_n$. Define the following two subsets of Ω:

$$S = \{\{t_n\} \in \Omega : t_n - \lambda_n = O(\sqrt{\lambda_n})\},$$

$$T = \left\{ \{t_n\} \in \Omega : \frac{x_n^m - \lambda_n^m}{m\lambda_n^{m-1}} - (t_n - \lambda_n) = o(\sqrt{\lambda_n}) \right\}.$$

As in the proof of theorem 14.6-1, our aim is to show that $S \subseteq T$. To do this, we suppose $\{x_n\} \in S$ and observe that

$$\frac{t_n}{\lambda_n} - 1 = O(\lambda_n^{-1/2}), \tag{14.6-18}$$

so that t_n/λ_n converges to unity. Now

$$\begin{aligned}
\frac{t_n^m - \lambda_n^m}{m\lambda_n^{m-1}} &= \frac{\lambda_n}{m}\left[\left(\frac{t_n}{\lambda_n}\right)^m - 1 \right] \\
&= \frac{\lambda_n}{m}\left[\left(1 + \left(\frac{t_n}{\lambda_n} - 1\right)\right)^m - 1 \right] \\
&= \frac{\lambda_n}{m}\left[1 + m\left(\frac{t_n}{\lambda_n} - 1\right) + o\left(\left|\frac{t_n}{\lambda_n} - 1\right|\right) - 1 \right] \\
&= (t_n - \lambda_n) + o(|t_n - \lambda_n|) \\
&= (t_n - \lambda_n) + o(O(\sqrt{\lambda_n})) \\
&= (t_n - \lambda_n) + o(\sqrt{\lambda_n}).
\end{aligned}$$

Thus we have shown that

$$\frac{t_n^m - \lambda_n^m}{m\lambda_n^{m-1}} - (t_n - \lambda_n) = o(\sqrt{\lambda_n}),$$

or in other words, $\{t_n\} \in T$, and hence we have $S \subseteq T$. The next step is to use (14.6-17) to deduce that S occurs in probability. Because $S \subseteq T$, it follows that T also occurs in probability, and we conclude that

$$\frac{(X_n)^m - (\lambda_n)^m}{\sqrt{\lambda_n} m \lambda_n^{m-1}} = \frac{X_n - \lambda_n}{\sqrt{\lambda_n}} + o_p(1). \tag{14.6-19}$$

Now from (14.6-19) and theorems 14.3-6 and 14.3-9, we conclude that

$$\mathscr{L}\left[\frac{X_n^m - \lambda_n^m}{m\lambda_n^{m-1/2}} \right] \to \mathscr{N}(0, 1). \tag{14.6-20}$$

Hence for large λ_n, X_n^m has an approximate $\mathscr{N}((\lambda_n)^m, m^2\lambda_n^{2m-1})$ distribution. In the exercises for this section, the reader is invited to show that the methods used here for $(X_n)^m$ also work for $\log(X_n)$ but fail for e^{X_n}. This failure points up an essential difference between this problem and the δ method. ∎

14.6.2 *Variance-stabilizing transformations*

Suppose $\hat{\theta}_n$ is a real-valued random variable whose distribution depends on a real parameter θ in such a way that

$$\mathscr{L}[\sqrt{n}(\hat{\theta}_n - \theta)] \to \mathscr{N}(0, \sigma^2(\theta)). \qquad (14.6\text{-}21)$$

We assume that $\sigma^2(\theta) > 0$ for θ in an interval D. The variable $\hat{\theta}_n$ has an approximate $\mathscr{N}(\theta, n^{-1}\sigma^2(\theta))$ distribution, so the variance of $\hat{\theta}_n$ depends on both n and θ. The problem of variance stabilization is to find a one-to-one function $f: D \to R^1$ such that the variance of $f(\hat{\theta}_n)$ is proportional to n^{-1} and does not depend on θ. In general, the exact stabilization of the variance of $f(\hat{\theta}_n)$ is not possible for finite n, but an exact asymptotic solution is always possible and provides a useful first-order approximate solution for finite n. To find the asymptotic solution, we begin by assuming that there exists a function $f: D \to R^1$ such that

(i) f is one-to-one and has a derivative for all $\theta \in D$;

(ii) $\mathscr{L}[\sqrt{n}(f(\hat{\theta}_n) - f(\theta))] \to \mathscr{N}(0, 1)$.

Now, applying the δ method to f and (14.6-21), we have

$$\mathscr{L}[\sqrt{n}(f(\hat{\theta}_n) - f(\theta))] \to \mathscr{N}(0, \sigma^2(\theta)[f'(\theta)]^2). \qquad (14.6\text{-}22)$$

Equating the asymptotic variances in (14.6-22) and (ii), we see that f must also satisfy the differential equation

$$[f'(\theta)]^2 = \frac{1}{\sigma^2(\theta)} \qquad (14.6\text{-}23)$$

for all $\theta \in D$. Now suppose $\gamma(\theta)$ is any function that only takes on the values -1 and $+1$, i.e., $\gamma^2(\theta) \equiv 1$. The nonlinear differential equation (14.6-23) is equivalent to the following infinity of linear differential equations for each possible choice of $\gamma(\theta)$:

$$f'(\theta) = \frac{\gamma(\theta)}{\sigma(\theta)} \qquad (14.6\text{-}24)$$

for $\theta \in D$. If $\gamma(\theta)$ is not chosen to be identically $+1$ or identically -1 for $\theta \in D$, then the solution to (14.6-24) has a derivative that changes sign, so that it is not one-to-one. Consequently the desired solution may be found by solving the equation:

$$f'(\theta) = \frac{1}{\sigma(\theta)} \qquad (14.6\text{-}25)$$

for $\theta \in D$. In general, the solution to (14.6-25) is given by

$$f(\theta) = \int_{\theta_0}^{\theta} \frac{dt}{\sigma(t)}. \qquad (14.6\text{-}26)$$

In some important examples, this integration may be carried out explicitly.

Example 14.6-4   Binomial proportion

Using the notation of example 14.6-1, we set $\hat{\theta}_n = \hat{p}$ (a binomial proportion). In this case, $\theta = p$, D is the interval $(0, 1)$, and $\sigma^2(\theta) = \theta(1 - \theta)$. Equation (14.6-25) becomes

$$\frac{df}{d\theta} = \frac{1}{\sqrt{\theta(1-\theta)}},$$

which may be shown to have the celebrated angular transformation as its solution, i.e.,

$$f(p) = 2\sin^{-1}\sqrt{p} + \text{constant}. \qquad (14.6\text{-}27)$$

Thus for large n, $\sin^{-1}\sqrt{\hat{p}}$ has an approximate $\mathcal{N}(\sin^{-1}\sqrt{p}, (4n)^{-1})$ distribution if $\sin^{-1}$ is measured in radians. When $\sin^{-1}$ is measured in degrees, $\sin^{-1}\sqrt{\hat{p}}$ has an approximate $\mathcal{N}(\sin^{-1}\sqrt{p}, (180/\pi)^2(4n)^{-1})$ distribution. Note that $(180/\pi)^2(4n)^{-1} = (820.7)/n$. Freeman and Tukey [1950] suggest the following modification of the angular transformation in order to stabilize the variance for small n:

$$\frac{1}{2}\left\{\sin^{-1}\sqrt{\frac{n}{n+1}\hat{p}} + \sin^{-1}\sqrt{\frac{n}{n+1}\hat{p} + \frac{1}{n+1}}\right\}. \qquad (14.6\text{-}28)$$

Mosteller and Youtz [1961] indicate how the variance of (14.6-28) behaves for $0 \leq p \leq 1$ and selected values of n. They also give tables to facilitate computing the expression (14.6-28). ∎∎

Example 14.6-5   A Poisson variate

In example 14.6-3 we derived the tools we now need to find the variance-stabilizing transformation for a single Poisson variate. We saw there that if X has a Poisson distribution with mean λ, then as $\lambda \to \infty$, X^m has an approximate $\mathcal{N}(\lambda^m, m^2\lambda^{2m-1})$ distribution. If we take $m = \frac{1}{2}$, then we see that $\sqrt{X}$ has an approximate $\mathcal{N}(\sqrt{\lambda}, \frac{1}{4})$ distribution, or $2\sqrt{X}$ has an approximate $\mathcal{N}(2\sqrt{\lambda}, 1)$ distribution. Freeman and Tukey [1950] suggest the following modification of the square root transformation for Poisson variates in order to stabilize the variance even for small values of λ:

$$\sqrt{X} + \sqrt{X+1}. \qquad (14.6\text{-}29)$$

Mosteller and Youtz [1961] indicate how the variance of (14.6-29) behaves for $\lambda > 0$ and give tables that facilitate computing expression (14.6-29).

The expectation of $\sqrt{X} + \sqrt{X+1}$ is well approximated by $\sqrt{4\lambda+1}$ for a wide range of values of λ (note that it is exact for $\lambda = 0$). Thus if λ is estimated by a fitted value, say $\hat{\lambda}$, then we use (14.6-29) to form the Freeman–Tukey deviate

$$\sqrt{X} + \sqrt{X+1} - \sqrt{4\hat{\lambda}+1}.$$

Freeman–Tukey deviates are discussed further in sections 14.7 and 14.9 and have also been discussed in Chapter 4. ∎∎

In the exercises in Section 14.6.5, the reader can use the δ method to find other examples of variance-stabilizing transformations.

14.6.3   Multivariate versions of the δ method

The one-dimensional version of the δ method given in theorem 14.6-1 can be generalized to allow the random variable, the function of the random variable, and the parameter to be vector-valued. We give the details of this generalization now. Let $\hat{\boldsymbol{\vartheta}}_n$ be a T-dimensional random vector: $\hat{\boldsymbol{\theta}}_n = (\hat{\theta}_{n1}, \ldots, \hat{\theta}_{nT})$, and let $\boldsymbol{\theta}$ be a

T-dimensional vector parameter: $\boldsymbol{\theta} = (\theta_1, \ldots, \theta_T)$. We assume that $\hat{\boldsymbol{\theta}}_n$ has an asymptotic normal distribution in the sense that

$$\mathscr{L}[\sqrt{n}(\hat{\boldsymbol{\theta}}_n - \boldsymbol{\theta})] \to \mathscr{N}(0, \boldsymbol{\Sigma}(\boldsymbol{\theta})). \tag{14.6-30}$$

$\boldsymbol{\Sigma}(\boldsymbol{\theta})$ is the $T \times T$ asymptotic covariance matrix of $\hat{\boldsymbol{\theta}}_n$ and is a singular covariance matrix if $\hat{\boldsymbol{\theta}}_n$ has a distribution that is concentrated on a subspace of T-dimensional space. For large n, $\hat{\boldsymbol{\theta}}_n$ has an approximate $\mathscr{N}(\boldsymbol{\theta}, n^{-1}\boldsymbol{\Sigma}(\boldsymbol{\theta}))$ distribution.

Now suppose $\mathbf{f}$ is a function defined on an open subset of T-dimensional space and taking values in R-dimensional space, i.e.,

$$\mathbf{f}(\boldsymbol{\theta}) = (f_1(\boldsymbol{\theta}), \ldots, f_R(\boldsymbol{\theta})).$$

We assume that $\mathbf{f}$ has a differential at $\boldsymbol{\theta}$, i.e., that $\mathbf{f}$ has the following expansion as $\mathbf{x} \to \boldsymbol{\theta}$:

$$f_i(\mathbf{x}) = f_i(\boldsymbol{\theta}) + \sum_{j=1}^{T} (x_j - \theta_j) \frac{\partial f_i}{\partial x_j}\bigg|_{\mathbf{x}=\boldsymbol{\theta}} + o(\|\mathbf{x} - \boldsymbol{\theta}\|) \tag{14.6-31}$$

for $i = 1, \ldots, R$. If we let $(\partial \mathbf{f}/\partial \boldsymbol{\theta})$ denote the $R \times T$ matrix whose (i, j) entry is the partial derivative of f_i with respect to the jth coordinate of $\mathbf{x} = (x_1, \ldots, x_T)$ evaluated at $\mathbf{x} = \boldsymbol{\theta}$, i.e.,

$$\left(\frac{\partial \mathbf{f}}{\partial \boldsymbol{\theta}}\right)_{ij} = \frac{\partial f_i}{\partial x_j}\bigg|_{\mathbf{x}=\boldsymbol{\theta}},$$

then (14.6-31) can be expressed neatly in matrix notation as

$$\mathbf{f}(\mathbf{x}) = \mathbf{f}(\boldsymbol{\theta}) + (\mathbf{x} - \boldsymbol{\theta})\left(\frac{\partial \mathbf{f}}{\partial \boldsymbol{\theta}}\right)' + o(\|\mathbf{x} - \boldsymbol{\theta}\|) \tag{14.6-32}$$

as $\mathbf{x} \to \boldsymbol{\theta}$. Within this framework, the δ method can be stated as follows:

THEOREM 14.6-2 (*Multivariate δ method*)   *Let $\hat{\boldsymbol{\theta}}_n$, $\boldsymbol{\theta}$, and $\mathbf{f}$ be as described above and suppose (14.6-30) and (14.6-32) hold. Then the asymptotic distribution of $\mathbf{f}(\hat{\boldsymbol{\theta}}_n)$ is given by:*

$$\mathscr{L}[\sqrt{n}(\mathbf{f}(\hat{\boldsymbol{\theta}}_n) - \mathbf{f}(\boldsymbol{\theta}))] \to \mathscr{N}\left(0, \left(\frac{\partial \mathbf{f}}{\partial \boldsymbol{\theta}}\right)\boldsymbol{\Sigma}(\boldsymbol{\theta})\left(\frac{\partial \mathbf{f}}{\partial \boldsymbol{\theta}}\right)'\right). \tag{14.6-33}$$

The proof of theorem 14.6-2 is a simple multivariate extension of that given for theorem 14.6-1, and the details are left as an exercise. In terms of the individual coordinates of $\mathbf{f}(\hat{\boldsymbol{\theta}}_n)$, theorem 14.6-2 implies that the asymptotic variance of $f_i(\hat{\boldsymbol{\theta}}_n)$ is

$$n^{-1} \sum_{k, k'=1}^{T} \sigma_{kk'}(\boldsymbol{\theta})\left(\frac{\partial f_i}{\partial \theta_k}\right)\left(\frac{\partial f_i}{\partial \theta_{k'}}\right)$$

and that the asymptotic covariance of $f_i(\hat{\boldsymbol{\theta}}_n)$ and $f_j(\hat{\boldsymbol{\theta}}_n)$ is

$$n^{-1} \sum_{k, k'=1}^{T} \sigma_{kk'}(\boldsymbol{\theta})\left(\frac{\partial f_i}{\partial \theta_k}\right)\left(\frac{\partial f_j}{\partial \theta_{k'}}\right).$$

Holland [1973] uses theorem 14.6-2 and related results to study the possibility of multivariate extensions of the variance-stabilizing transformations discussed above. He gives a complicated condition that must be satisfied in order for a covariance stabilizing transformation to exist and shows that the trinomial distribution does not satisfy it.

As in the one-dimensional case, (14.6-30) can be weakened in various ways. The essential features of (14.6-30) are that $\hat{\boldsymbol{\theta}}_n$ converges to $\boldsymbol{\theta}$, and that for some sequence $\{r_n\}$ of real numbers such that $r_n \to \infty$, the quantity $r_n(\hat{\boldsymbol{\theta}}_n - \boldsymbol{\theta})$ has a limiting distribution. In order for the δ method to work, it is not necessary for $r_n = \sqrt{n}$, nor does the limiting distribution have to be normal; however, (14.6-30) is sufficiently general for most of the common applications. Now suppose that $\boldsymbol{\theta} = \boldsymbol{\theta}(\boldsymbol{\varphi})$, where $\boldsymbol{\varphi}$ is an s-dimensional parameter $\boldsymbol{\varphi} = (\varphi_1, \ldots, \varphi_s)$. Also, suppose that the asymptotic covariance matrix in (14.6-30) depends on $\boldsymbol{\varphi}$, i.e., that

$$\mathscr{L}[\sqrt{n}(\hat{\boldsymbol{\theta}}_n - \boldsymbol{\theta}(\boldsymbol{\varphi}))] \to \mathscr{N}(0, \boldsymbol{\Sigma}(\boldsymbol{\varphi})). \tag{14.6-30'}$$

Finally, let $(\partial \mathbf{f}/\partial \boldsymbol{\theta})$ be the $R \times T$ matrix whose (i, j) entry is the partial derivative of f_i with respect to the jth coordinate of $\mathbf{x} = (x_1, \ldots, x_T)$ evaluated at $\mathbf{x} = \boldsymbol{\theta}(\boldsymbol{\varphi})$. We give the following slightly more general result than theorem 14.6-2, as it is sometimes needed.

THEOREM 14.6-3   *If $\hat{\boldsymbol{\theta}}_n$, $\boldsymbol{\theta}(\boldsymbol{\varphi})$, $\boldsymbol{\varphi}$, and $\mathbf{f}$ are as described above and if (14.6-30') and (14.6-32) hold for $\boldsymbol{\theta} = \boldsymbol{\theta}(\boldsymbol{\varphi})$, then the asymptotic distribution of $\mathbf{f}(\hat{\boldsymbol{\theta}}_n)$ is given by:*

$$\mathscr{L}[\sqrt{n}\mathbf{f}(\hat{\boldsymbol{\theta}}_n) - \mathbf{f}(\boldsymbol{\theta}(\boldsymbol{\varphi}))] \to \mathscr{N}\left(0, \left(\frac{\partial \mathbf{f}}{\partial \boldsymbol{\theta}}\right) \boldsymbol{\Sigma}(\boldsymbol{\varphi}) \left(\frac{\partial \mathbf{f}}{\partial \boldsymbol{\theta}}\right)'\right). \tag{14.6-34}$$

Example 14.6-6   Linear combinations of the logarithms of multinomial counts
We now apply theorem 14.6-2 to find the asymptotic variances and covariances of linear combinations of the logarithms of multinomial counts. These quantities are important because in log-linear models for contingency tables, the maximum likelihood estimates for the u-terms are linear combinations of the logarithms of the cell counts. We first give the general theory for any set of linear combinations of the log counts and then apply this to some specific linear combinations of interest in the analysis of contingency tables. Let $\mathbf{X} = (X_1, \ldots, X_T)$ have the $\mathscr{M}_T(N, \mathbf{p})$ distribution, where $\mathbf{p} = (p_1, \ldots, p_T)$ and $N = \sum_{i=1}^{T} X_i$. Let $\hat{\mathbf{p}} = N^{-1}\mathbf{X}$. Then

$$E(\hat{\mathbf{p}}) = \mathbf{p},$$

$$\mathrm{Cov}(\hat{\mathbf{p}}, \hat{\mathbf{p}}) = N^{-1}(\mathbf{D}_p - \mathbf{p}'\mathbf{p}) = N^{-1}\boldsymbol{\Lambda}(\mathbf{p}), \tag{14.6-35}$$

$$\mathscr{L}[\sqrt{N}(\hat{\mathbf{p}} - \mathbf{p})] \to \mathscr{N}(0, \boldsymbol{\Lambda}(\mathbf{p})),$$

where $\mathbf{D}_p$ is the diagonal matrix with the vector $\mathbf{p}$ along its main diagonal and $\boldsymbol{\Lambda}(\mathbf{p}) = \mathbf{D}_p - \mathbf{p}'\mathbf{p}$. Comparing (14.6-35) with (14.6-30), we see that in this application of theorem 14.6-2, $\hat{\mathbf{p}}$, $\mathbf{p}$, and $\boldsymbol{\Lambda}(\mathbf{p})$ play the roles of $\hat{\boldsymbol{\theta}}_n$, $\boldsymbol{\theta}$, and $\boldsymbol{\Sigma}(\boldsymbol{\theta})$, respectively. Now suppose $\mathbf{x} = (x_1, \ldots, x_T)$, and that $f_1(\mathbf{x}), \ldots, f_R(\mathbf{x})$ are defined by

$$f_i(\mathbf{x}) = \sum_{k=1}^{T} c_{ik} \log x_k \qquad i = 1, \ldots, R.$$

We are interested in the asymptotic distribution of $\mathbf{f}(\hat{\mathbf{p}}) = (f_1(\hat{\mathbf{p}}), \ldots, f_R(\hat{\mathbf{p}}))$ as $N \to \infty$. To apply theorem 14.6-2, we need to compute the partial derivatives

$$\left.\frac{\partial f_i}{\partial x_j}\right|_{\mathbf{x}=\mathbf{p}}. \tag{14.6-36}$$

A straightforward calculation shows that

$$\frac{\partial f_i}{\partial x_j} = c_{ij}(x_j)^{-1}. \tag{14.6-37}$$

In matrix notation (14.6-37) becomes

$$\left(\frac{\partial \mathbf{f}}{\partial \mathbf{p}}\right) = \mathbf{C}\mathbf{D}_p^{-1}, \tag{14.6-38}$$

where $\mathbf{C} = (c_{ij})$. Applying theorem 14.6-2, we obtain the following result for the asymptotic distribution of a set of linear combinations of log counts.

THEOREM 14.6-4   *If* $\mathbf{f}(\hat{\mathbf{p}})$ *is the set of linear combinations of the logarithms of multinomial counts in* (14.6-35), *then*

$$\mathscr{L}[\sqrt{N}(\mathbf{f}(\hat{\mathbf{p}}) - \mathbf{f}(\mathbf{p}))] \to \mathscr{N}(0, \mathbf{C}\mathbf{D}_p^{-1}\mathbf{C}' - (\mathbf{e}\mathbf{C}')'\mathbf{e}\mathbf{C}'), \tag{14.6-39}$$

where $\mathbf{e} = (1, \ldots, 1)$ *is a* $1 \times T$ *vector.*

If we let $c_{i+} = \sum_{k=1}^{T} c_{ik}$, then, in terms of the individual coordinates of $\mathbf{f}(\hat{\mathbf{p}})$, theorem 14.6-4 implies that for large N the asymptotic variance of $f_i(\hat{\mathbf{p}}) = \sum_{k=1}^{T} c_{ik} \log(\hat{p}_k)$ is

$$N^{-1}\left[\sum_{k=1}^{T} c_{ik}^2 p_k^{-1} - c_{i+}^2\right], \tag{14.6-40}$$

and that the asymptotic covariance of $f_i(\hat{\mathbf{p}})$ and $f_j(\hat{\mathbf{p}})$ is

$$N^{-1}\left[\sum_{k=1}^{T} c_{ik}c_{jk}p_k^{-1} - c_{i+}c_{j+}\right]. \tag{14.6-41}$$

Theorem 14.6-4 and expressions (14.6-37), (14.6-38), (14.6-40), and (14.6-41) all presuppose that the logarithms used are to the natural base $e = 2.7182818 \cdots$. If instead the base 10 is used, then the variances and covariances given in (14.6-40) and (14.6-41) must be multiplied by $(\log_{10} e)^2 = (0.4343)^2 = 0.1886$.

An important application of theorem 14.6-4 is to find the asymptotic variance of the log odds ratio used to measure departures from independence in 2×2 tables (see Chapters 2 and 3). In the notation of Chapter 2, let

$$\alpha = \frac{p_{11}p_{22}}{p_{21}p_{12}} \tag{14.6-42}$$

be the cross-product ratio (odds ratio) for a multinomially distributed 2×2 table with observed entries X_{ij} ($i = 1, 2; j = 1, 2$). Note that the single subscript notation used so far in this chapter to denote the coordinates of a random vector is now being altered for this example to agree with standard two-dimensional contingency table double subscript notation. Let θ be given by

$$\theta = \log \alpha = \log p_{11} + \log p_{22} - \log p_{12} - \log p_{21}. \tag{14.6-43}$$

The maximum likelihood estimate of θ is $\hat{\theta}$ given by

$$\hat{\theta} = \log \hat{p}_{11} + \log \hat{p}_{22} - \log \hat{p}_{12} - \log \hat{p}_{21}. \tag{14.6-44}$$

From (14.6-44) we see that $\hat{\theta}$ is a linear combination of the logarithms of the cell proportions, and thus from (14.6-40) the asymptotic variance of $\hat{\theta}$ is

$$N^{-1}\left[\frac{1}{p_{11}} + \frac{1}{p_{22}} + \frac{1}{p_{12}} + \frac{1}{p_{21}}\right]$$

$$= (Np_{11})^{-1} + (Np_{22})^{-1} + (Np_{12})^{-1} + (Np_{21})^{-1}. \qquad (14.6\text{-}45)$$

Note that in this case $c_{i+} = 1 + 1 - 1 - 1 = 0$.

If we consider two log-odds ratios formed from two 2×2 subtables of a larger two-way table, we can use (14.6-41) to calculate the asymptotic covariance between them. Let α_1 and α_2 be given by

$$\alpha_1 = \frac{p_{ij}p_{i+1,j+1}}{p_{i+1,j}p_{i,j+1}},$$

$$\alpha_2 = \frac{p_{lm}p_{l+1,m+1}}{p_{l+1,m}p_{l,m+1}},$$

and set $\theta_k = \log \alpha_k$ ($k = 1, 2$). Let $\hat{\theta}_k$ be the corresponding maximum likelihood estimate of θ_k formed by replacing $\mathbf{p}$ by $\hat{\mathbf{p}}$. Using (14.6-41), we see that the asymptotic covariance between $\hat{\theta}_1$ and $\hat{\theta}_2$ depends on the amount of overlap between the two 2×2 subtables. If there are two cells that overlap (say $(l, m) = (i, j + 1)$), then the asymptotic covariance between $\hat{\theta}_1$ and $\hat{\theta}_2$ is

$$-(Np_{i,j+1})^{-1} - (Np_{i+1,j+1})^{-1}. \qquad (14.6\text{-}46)$$

If there is only one overlapping cell (say $(l, m) = (i + 1, j + 1)$), then the asymptotic covariance is

$$(Np_{i+1,j+1})^{-1}. \qquad (14.6\text{-}47)$$

Finally, if there is no overlap between the two 2×2 subtables, then the asymptotic covariance between $\hat{\theta}_1$ and $\hat{\theta}_2$ is 0. Fienberg [1969] exploits the asymptotic independence of the log-odds ratio for nonoverlapping 2×2 subtables in his suggestion for a preliminary half-normal plot to identify those cells to omit when fitting quasi-independence models.

As a final example of the use of linear combinations of log counts, we consider the general $I \times J$ table and its u-term parametization (see Chapters 2 and 3 for explanations of the notation). Let p_{ij} be the probability of the (i, j) cell, and set

$$\log p_{ij} = u + u_{1(i)} + u_{2(j)} + u_{12(ij)}, \qquad (14.6\text{-}48)$$

where $u_{1(+)} = u_{2(+)} = u_{12(i+)} = u_{12(+j)} = 0$ for $i = 1, \ldots, I; j = 1, \ldots, J$. Solving for the u-terms in terms of the log p_{ij} yields

$$u = \frac{1}{IJ}\sum_{l,m} \log p_{lm},$$

$$u_{1(i)} = \frac{1}{J}\sum_m \log p_{im} - u,$$

$$u_{2(j)} = \frac{1}{I}\sum_l \log p_{lj} - u,$$

$$u_{12(ij)} = \log p_{ij} - u - u_{1(i)} - u_{2(j)}. \qquad (14.6\text{-}49)$$

We can express the right-hand side of (14.6-49) completely in terms of the $\log p_{ij}$ as follows:

$$u = \sum_{l,m} \left(\frac{1}{IJ} \right) \log p_{lm}$$

$$u_{1(i)} = \sum_{\substack{l \neq i \\ m}} \left(\frac{-1}{IJ} \right) \log p_{lm} + \sum_m \frac{1}{J} \left(1 - \frac{1}{I} \right) \log p_{im}$$

$$u_{2(j)} = \sum_{\substack{l \\ m \neq j}} \left(\frac{-1}{IJ} \right) \log p_{lm} + \sum_l \frac{1}{I} \left(1 - \frac{1}{J} \right) \log p_{lj}$$

$$u_{12(ij)} = \left(1 - \frac{1}{I} \right)\left(1 - \frac{1}{J} \right) \log p_{ij} + \sum_{\substack{l = i \\ m \neq j}} \frac{-1}{J} \left(1 - \frac{1}{I} \right) \log p_{lm}$$

$$+ \sum_{\substack{l \neq i \\ m = j}} \frac{-1}{I} \left(1 - \frac{1}{J} \right) \log p_{lm} + \sum_{\substack{l \neq i \\ m \neq j}} \frac{1}{IJ} \log p_{lm}. \tag{14.6-50}$$

If we now substitute $\hat{p}_{ij}$ for p_{ij} in these expressions for the u-terms, we obtain their maximum likelihood estimates under the saturated multinomial model; call these $\hat{u}$, $\hat{u}_{1(i)}$, $\hat{u}_{2(j)}$, and $\hat{u}_{12(ij)}$. These estimates of the u-terms are linear combinations of the log counts, and as such, their asymptotic variances can be obtained from theorem 6-4:

$$\text{Var}_\infty(\hat{u}) = \left(\frac{1}{IJ} \right)^2 \sum_{l,m} (Np_{lm})^{-1} - N^{-1},$$

$$\text{Var}_\infty(\hat{u}_{1(i)}) = \left(\frac{1}{IJ} \right)^2 \sum_{l,m} (Np_{lm})^{-1} + \left(\frac{I - 2}{IJ^2} \right) \sum_m (Np_{im})^{-1},$$

$$\text{Var}_\infty(\hat{u}_{2(j)}) = \left(\frac{1}{IJ} \right)^2 \sum_{l,m} (Np_{lm})^{-1} + \left(\frac{J - 2}{IJ^2} \right) \sum_l (Np_{lj})^{-1},$$

$$\text{Var}_\infty(\hat{u}_{12(ij)}) = \left(\frac{1}{IJ} \right)^2 \sum_{l,m} (Np_{lm})^{-1} + \left(\frac{I - 2}{IJ^2} \right) \sum_m (Np_{im})^{-1}$$

$$+ \left(\frac{J - 2}{JI^2} \right) \sum_l (Np_{lj})^{-1} + \left(\frac{(I - 2)(J - 2)}{IJ} \right) (Np_{ij})^{-1}. \tag{14.6-51}$$

The asymptotic covariances can be found by a similar application of theorem 14.6-4. As the number of different covariances among the u-terms is rather large, we terminate this application of theorem 14.6-4 by calculating the asymptotic covariance between $\hat{u}_{1(i)}$ and $\hat{u}_{2(j)}$. Expression (14.6-50) gives us the coefficients of these two linear combinations of log counts:

$$\text{Cov}_\infty(\hat{u}_{1(1)}, \hat{u}_{1(2)}) = \left(\frac{1}{IJ} \right)^2 \sum_{l,m} (Np_{lm})^{-1} - \left(\frac{1}{IJ^2} \right) \sum_{\substack{l = 1,2, \\ m}} (Np_{lm})^{-1}. \tag{14.6-52}$$

These calculations can be extended to log-linear models for *multidimensional* tables. ■■

14.6.4   *Some uses for asymptotic variances and covariances*

We need to find approximate variances and covariances for a number of reasons, and we now discuss some of the more common ways these quantities are put to use.

Hypothesis testing

Suppose $\hat{\theta}_n$ is an estimate of a real-valued parameter θ, and we wish to test the hypothesis that $\theta = \theta_0$. If

$$\mathscr{L}[\sqrt{n}(\hat{\theta}_n - \theta)] \to \mathscr{N}(0, \sigma^2(\theta)) \tag{14.6-53}$$

for each value of θ in the parameter space and we let

$$Z_n = \frac{\sqrt{n}(\hat{\theta}_n - \theta_0)}{\sigma(\theta_0)}, \tag{14.6-54}$$

we have

(i)   if $\theta = \theta_0$, then $\mathscr{L}[Z_n] \to \mathscr{N}(0, 1)$;

(ii)   if $\theta \neq \theta_0$, then $\lim_{n \to \infty} P\{|Z_n| > K\} = 1$.

From (i) we can set the critical value for the test using the $\mathscr{N}(0, 1)$ tables, and from (ii) we see that tests of the form:

$$\text{reject } \theta = \theta_0 \quad \text{if } |Z_n| > K$$

achieve any level of power if n is large enough.

Another type of testing problem that arises involves estimates of two parameters θ_1 and θ_2. We suppose that the vector $\hat{\boldsymbol{\theta}}_n = (\hat{\theta}_n^{(1)}, \hat{\theta}_n^{(2)})$ has the asymptotic distribution:

$$\mathscr{L}[\sqrt{n}(\hat{\boldsymbol{\theta}}_n - \boldsymbol{\theta})] \to \mathscr{N}\left(0, \begin{pmatrix} \sigma_1^2(\boldsymbol{\theta}) & \sigma_{12}(\boldsymbol{\theta}) \\ \sigma_{12}(\boldsymbol{\theta}) & \sigma_2^2(\boldsymbol{\theta}) \end{pmatrix}\right), \tag{14.6-55}$$

where $\boldsymbol{\theta} = (\theta_1, \theta_2)$.

In some cases $\hat{\theta}_n^{(1)}$ and $\hat{\theta}_n^{(2)}$ are independent and $\sigma_{12}^2(\boldsymbol{\theta}) = 0$, while in other cases $\sigma_i^2(\boldsymbol{\theta})$ only depends on θ_i, i.e., $\sigma_i^2(\boldsymbol{\theta}) = \sigma_i^2(\theta_i)$. In any case we may wish to test the hypothesis that $\theta_1 = \theta_2$. From (14.6-55) it follows that

$$\mathscr{L}[\sqrt{n}(\hat{\theta}_n^{(1)} - \hat{\theta}_n^{(2)}) - (\theta_1 - \theta_2)] \to \mathscr{N}(0, \sigma_1^2(\boldsymbol{\theta}) - 2\sigma_{12}(\boldsymbol{\theta}) + \sigma_2^2(\boldsymbol{\theta})), \tag{14.6-56}$$

and hence we can base a test of $\theta_1 = \theta_2$ on

$$Z_n(\varphi) = \frac{\sqrt{n}(\hat{\theta}_1 - \hat{\theta}_2)}{\sigma(\varphi)}, \tag{14.6-57}$$

where φ is the unknown common value of θ_1 and θ_2 under the hypothesis of equality and

$$\sigma^2(\varphi) = \sigma_1^2(\varphi, \varphi) - 2\sigma_{12}(\varphi, \varphi) + \sigma_2^2(\varphi, \varphi). \tag{14.6-58}$$

To use $Z_n(\varphi)$, we need some estimate of φ. Various candidates are possible. For example, we might use

$$\hat{\varphi} = \frac{\hat{\theta}_n^{(1)} + \hat{\theta}_n^{(2)}}{2}, \tag{14.6-59}$$

but this estimate ignores the possible differences in $\sigma_1^2(\varphi)$ and $\sigma_2^2(\varphi)$. Weighting $\hat{\theta}_n^{(1)}$ and $\hat{\theta}_n^{(2)}$ inversely proportional to their estimated variances is another possibility. This can be done by setting

$$\hat{\varphi} = \hat{w}\hat{\theta}_n^{(1)} + (1 - \hat{w})\hat{\theta}_n^{(2)}, \tag{14.6-60}$$

where

$$\hat{w} = \frac{\sigma_1^{-2}(\hat{\mathbf{\theta}}_n)}{\sigma_1^{-2}(\hat{\mathbf{\theta}}_n) + \sigma_2^{-2}(\hat{\mathbf{\theta}}_n)}. \tag{14.6-61}$$

When $\sigma_{12}^2 = 0$ this procedure is reasonable, but when $\sigma_{12}^2 \neq 0$, (14.6-60) ignores the correlation between the two estimators. If this correlation is significant, then it is better to use an estimate that takes account of it, such as

$$\hat{\varphi} = \frac{\hat{\mathbf{\theta}}_n \hat{\mathbf{\Sigma}}^{-1} \mathbf{e}'}{\mathbf{e} \hat{\mathbf{\Sigma}}^{-1} \mathbf{e}'}, \tag{14.6-62}$$

where $e = (1, 1)$ and $\hat{\mathbf{\Sigma}}$ is formed by substituting $\hat{\mathbf{\theta}}_n$ for $\mathbf{\theta}$ everywhere in the asymptotic covariance matrix given in (14.6-55).

Assuming that φ is estimated by $\hat{\varphi}$, then from (14.6-57) and the further assumption that $\sigma(\varphi)$ is continuous, it follows that if $\theta_1 = \theta_2$, then

$$\mathscr{L}[Z_n(\hat{\varphi})] \to \mathscr{N}(0, 1). \tag{14.6-63}$$

We can use (14.6-63) to set critical values for tests of the form

$$\text{reject } \theta_1 = \theta_2 \quad \text{if } |Z_n(\hat{\varphi})| > K.$$

Confidence intervals

While hypothesis testing has its place, we are often more interested in setting confidence intervals on a parameter value or on the difference between two parameter values. When (14.6-53) holds, then we can set asymptotic $(1 - \alpha)$-level confidence intervals for θ in the following way. Let $Z_{\alpha/2}$ denote the upper $\alpha/2$ probability point for the $\mathscr{N}(0, 1)$ distribution. Then as $n \to \infty$, we have, for any value of θ:

$$P\{\hat{\theta}_n - n^{-1/2}\sigma(\hat{\theta}_n)Z_{\alpha/2} < \theta < \hat{\theta}_n + n^{-1/2}\sigma(\hat{\theta}_n)Z_{\alpha/2}\} \approx 1 - \alpha. \tag{14.6-64}$$

Strictly speaking, in order for (14.6-64) to hold, $\sigma^2(\theta)$ must be continuous in θ.

When (14.6-55) holds, we may be interested in an asymptotic confidence interval for $\theta_1 - \theta_2$. We let $\sigma^2(\mathbf{\theta})$ denote the asymptotic variance in (14.6-56), i.e.,

$$\sigma^2(\mathbf{\theta}) = \sigma_1^2(\mathbf{\theta}) - 2\sigma_{12}(\mathbf{\theta}) + \sigma_2^2(\mathbf{\theta}). \tag{14.6-65}$$

Then as $n \to \infty$, we have, for any value of $\mathbf{\theta}$:

$$P\{\hat{\theta}_n^{(1)} - \hat{\theta}_n^{(2)} - n^{-1/2}\sigma(\hat{\mathbf{\theta}}_n)Z_{\alpha/2} < \theta_1 - \theta_2 < \hat{\theta}_n^{(1)} - \hat{\theta}_n^{(2)} + n^{-1/2}\sigma(\hat{\mathbf{\theta}}_n)Z_{\alpha/2}\} \approx 1 - \alpha. \tag{14.6-66}$$

We note that the main difference between testing hypotheses and setting confidence intervals is in the choice of the way we estimate σ^2. In the hypothesis testing case we use the null hypothesis to generate an estimate of σ^2, while in the confidence interval case we do not use the null hypothesis and estimate $\sigma^2(\theta)$ by $\sigma^2(\hat{\theta}_n)$ rather than $\sigma^2(\theta_0)$. Using $\sigma^2(\theta_0)$ increases the power of the test, while using $\sigma^2(\hat{\theta}_n)$ ensures asymptotic $(1 - \alpha)$-level confidence intervals for all values of the true parameter $\boldsymbol{\theta}$. In Chapter 11 we emphasized the confidence interval use of asymptotic variances.

14.6.5   *Exercises*

1. If X is Poisson distributed with mean λ, show that for large λ, $\log X$ has an approximate $\mathcal{N}(\log \lambda, \lambda^{-1})$ distribution.

2. If X is Poisson distributed with mean λ, show that it is not true that as $\lambda \to \infty$,

$$\mathscr{L}\left[\frac{e^X - e^\lambda}{\sqrt{\lambda} e^\lambda}\right] \to \mathcal{N}(0, 1).$$

What is the behavior of e^X for large λ?

3. *Variance stabilization of the correlation coefficient.* Let $\hat{\rho}$ be the sample correlation coefficient for a sample of n from the bivariate normal distribution with correlation ρ. Use the fact that

$$\mathscr{L}[\sqrt{n}(\hat{\rho} - \rho)] \to \mathcal{N}(0, (1 - \rho^2)^2)$$

to show that

$$f(\rho) = \frac{1}{2} \log\left(\frac{1 + \rho}{1 - \rho}\right)$$

is an asymptotic variance-stabilizing transformation for $\hat{\rho}$. What is the asymptotic variance of $f(\hat{\rho})$?

4. *Variance stabilization for the negative binomial.* Let X have the negative binomial distribution given by

$$P\{X = x\} = \binom{r + x - 1}{r - 1} p^r (1 - p)^x \quad x = 0, 1, \ldots.$$

Then $E(X) = m = r((1 - p)/p)$ and $\mathrm{Var}(X) = r((1 - p)/p^2)$.

(i) If p is known and r is unknown, show that as $m \to \infty$, the asymptotic variance-stabilizing transformation for X is proportional to $\sqrt{X}$. What is the constant of proportionality?

(ii) If r is known and p is unknown, show that as $m \to \infty$, the asymptotic variance-stabilizing transformation is proportional to $\sinh^{-1}((1/r)\sqrt{X})$. What is the constant of proportionality? (Hint: In this example, r remains fixed as $m \to \infty$, and the limiting distribution is not the normal. How does this affect the application of the δ method?)

(iii) If r and p are both unknown, what candidates do you suggest to stabilize the variance of X?

5. In a $2 \times 2 \times 2$ contingency table, suppose the observations follow a multinomial distribution with cell probabilities p_{ijk}. Let

$$\gamma = \left(\frac{p_{111} p_{221}}{p_{121} p_{211}} \middle/ \frac{p_{112} p_{222}}{p_{122} p_{212}}\right),$$

and set $\theta = \log \gamma$. Find the asymptotic variance of the maximum likelihood estimate $\hat{\theta}$ of θ.

6. Define a suitable underlying formal structure and apply the Chernoff–Pratt definitions to fill in the details of the proof of theorem 14.6-2.

7. If $\mathscr{L}[\sqrt{n}(\hat{\theta}_n - \theta)] \to \mathscr{N}(0, \sigma^2(\theta))$ and $\log(\hat{\theta}_n)$ is the asymptotic variance-stabilizing transformation for $\hat{\theta}_n$, what must be the form of $\sigma^2(\theta)$? Give an example of a distribution with this property.

8. Find the asymptotic covariance of $\hat{u}_{1(i)}$ and $\hat{u}_{2(j)}$ for the $I \times J$ multinomial distribution by applying theorem 14.6-4.

9. Find the asymptotic covariance of $\hat{u}_{12(ij)}$ and $\hat{u}_{12(kl)}$ for all choices of i, j, k, and l by applying theorem 14.6-4 for an $I \times J$ multinomial.

10. Various measures of association (see Chapter 11) for the 2×2 table can be expressed as functions of the log odds ratio $\theta = \log(p_{11}p_{22}/p_{12}p_{21})$. One class of these measures is given by $\mu_\beta = f(\beta\theta)$ for $\beta \neq 0$, where $f(x) = (e^x - 1)/(e^x + 1)$. For example, if $\beta = 1$, then

$$\mu_1 = \frac{\alpha - 1}{\alpha + 1} = \frac{p_{11}p_{22} - p_{12}p_{21}}{p_{11}p_{22} + p_{12}p_{21}},$$

so that μ_1 is Yule's Q (see Chapter 11). Show that if m_β is the maximum likelihood estimate of μ_β (i.e., $\hat{p}_{ij}$ substituted for p_{ij} in μ_β), then

$$\mathscr{L}[\sqrt{N}(m_\beta - \mu_\beta)] \to \mathscr{N}\left(0, \left(\frac{\beta}{2}\right)^2 (1 - \mu_\beta^2)^2 \sigma^2\right),$$

where

$$\sigma^2 = \frac{1}{p_{11}} + \frac{1}{p_{12}} + \frac{1}{p_{21}} + \frac{1}{p_{22}}.$$

Use the result to suggest a method for setting approximate confidence intervals for μ_β.

11. Consider a 2×2 table (X_{ij}). In (14.6-45) we gave the asymptotic variance of the log of the sample odds ratio, $\hat{\theta} = \log(X_{11}X_{22}/X_{12}X_{21})$, under multinomial sampling. Now suppose that the sampling fixes one of the margins, so that $X_{11} \sim \mathscr{B}(N_{1+}, p_1)$, $X_{21} \sim \mathscr{B}(N_{2+}, p_2)$, and X_{11} and X_{21} are independent (see Chapter 13). Show that the asymptotic variance of $\hat{\theta}$ is

$$\frac{1}{N_{1+}p_1} + \frac{1}{N_{1+}(1 - p_1)} + \frac{1}{N_{2+}p_2} + \frac{1}{N_{2+}(1 - p_2)}.$$

Use this result to conclude that the usual estimate of the asymptotic variance $\hat{\theta}$ is the same under both sampling schemes.

12. The correlation coefficient for a 2×2 table is

$$\rho = \frac{p_{11}p_{22} - p_{12}p_{21}}{\sqrt{p_{1+}p_{+1}p_{2+}p_{+2}}}.$$

Find the asymptotic variance of the maximum likelihood estimate of ρ under multinomial sampling.

13. In Chapter 11 we discussed the agreement measure $K(\mathbf{p})$ for an $I \times I$ table:

$$K(\mathbf{p}) = \frac{\theta_1 - \theta_2}{1 - \theta_2},$$

where $\theta_1(\mathbf{p}) = \sum_i p_{ii}$ and $\theta_2(\mathbf{p}) = \sum_i p_{i+}p_{+i}$. Apply the δ method to show that

$$\mathscr{L}[\sqrt{N}(K(\hat{\mathbf{p}}) - K(\mathbf{p}))] \to \mathcal{N}(0, \sigma^2(\mathbf{p})),$$

where

$$\sigma^2(\mathbf{p}) = \frac{\theta_1(1 - \theta_1)}{(1 - \theta_2)^2} + \frac{(1 - \theta_1)^2(\theta_4 - 4\theta_2^2)}{(1 - \theta_2)^4} + \frac{2(1 - \theta_1)(2\theta_1\theta_2 - \theta_3)}{(1 - \theta_2)^3},$$

with θ_1, θ_2 as above and

$$\theta_3(\mathbf{p}) = \sum_i p_{ii}(p_{i+} + p_{+i}), \qquad \theta_4(\mathbf{p}) = \sum_{i,j} p_{ij}(p_{+i} + p_{j+})^2.$$

14. Goodman and Kruskal [1972] consider measures of association for two-way tables of the form $\mu(\mathbf{p}) = v(\mathbf{p})/\delta(\mathbf{p})$. Use the δ method to show that

$$\mathscr{L}[\sqrt{N}(\mu(\hat{\mathbf{p}}) - \mu(\mathbf{p}))] \to \mathcal{N}(0, \sigma^2(\mathbf{p})),$$

where

$$\sigma^2(\mathbf{p}) = \delta^{-4}\left[\sum_{i,j} p_{ij}\left(\delta\frac{\partial v}{\partial p_{ij}} - v\frac{\partial \delta}{\partial p_{ij}}\right)^2 - \left(\sum_{i,j} p_{ij}\left(\delta\frac{\partial v}{\partial p_{ij}} - v\frac{\partial \delta}{\partial p_{ij}}\right)\right)^2\right].$$

15. Apply the result of exercise 14 to $K(\hat{\mathbf{p}})$ given in exercise 13.

16. Let $\mathbf{p}$ be the doubly subscripted probability vector (p_{ij}) for a multinomially distributed $I \times J$ contingency table $\mathbf{X} = (X_{ij})$. Let $\theta_1(\mathbf{p})$ and $\theta_2(\mathbf{p})$ be two functions of $\mathbf{p}$, and let $\mu(\mathbf{p})$ be a measure of association of the form $\mu(\mathbf{p}) = (\theta_1 - \theta_2)/(1 - \theta_2)$. Use the δ method to find a formula for the limiting distribution of $\mu(\hat{\mathbf{p}})$.

17. Continuing exercise 16, let $I = J$ and define an agreement measure by

$$K_i(\mathbf{p}) = \frac{p_{ii} - p_{i+}p_{+i}}{1 - p_{i+}p_{+i}}.$$

(i) Find the asymptotic variance of $K_i(\hat{\mathbf{p}})$ using your general formula from exercise 16. (See Chapter 11.)

(ii) Find the asymptotic covariance between $K_i(\hat{p})$ and $K_j(\hat{p})$. To what use could we put this limiting covariance?

14.7 General Framework for Multinomial Estimation and Testing

In this section we develop a general notation useful for discussing a variety of issues that arise in testing and estimation for the multinomial distribution. We first set up the notation and then show how it relates first to estimation and then to goodness-of-fit tests.

14.7.1 Notation

Let $\mathbf{X} = (X_1, \ldots, X_T)$ be a T-dimensional random vector with the multinomial distribution $\mathbf{X} \sim \mathcal{M}_T(N, \boldsymbol{\pi})$, where $\boldsymbol{\pi} = (\pi_1, \ldots, \pi_T)$ is the vector of true cell probabilities and $N = \sum_{i=1}^{T} X_i$. We let $\mathscr{S}_T$ be the set of all possible T-dimensional probability vectors, i.e.,

$$\mathscr{S}_T = \left\{ \mathbf{p} : p_i \geq 0 \quad \text{and} \quad \sum_{i=1}^{T} p_i = 1 \right\}. \tag{14.7-1}$$

We also let $\mathbf{p}$ be a generic point in $\mathscr{S}_T$, while $\boldsymbol{\pi}$ is the special point in $\mathscr{S}_T$ that gives the true cell probabilities generating the random vector $\mathbf{X}$. The vector of observed cell proportions,

$$\hat{\mathbf{p}} = N^{-1}\mathbf{X}, \tag{14.7-2}$$

is also a point in $\mathscr{S}_T$. As we showed in Chapter 13, $\hat{\mathbf{p}}$ is the unconstrained maximum likelihood estimate of $\boldsymbol{\pi}$, and $\hat{\mathbf{p}}$ converges to $\boldsymbol{\pi}$ in probability (as well as in various other senses) as $N \to \infty$.

Most of the models (log-linear or otherwise) that appear in this book can be described as follows. There exists a function $\mathbf{f}(\boldsymbol{\theta})$ that maps each value of a vector $\boldsymbol{\theta} = (\theta_1, \ldots, \theta_s)$ into a point in $\mathscr{S}_T$. The vector $\boldsymbol{\theta}$ is a vector of s parameters and ranges over a subset Θ of s-dimensional Euclidean space. Θ is called the "parameter space," and we call $\boldsymbol{\theta}$ the "vector of parameters" or sometimes just "the parameter." As $\boldsymbol{\theta}$ ranges over the values of Θ, $\mathbf{f}(\boldsymbol{\theta})$ ranges over a subset M of $\mathscr{S}_T$. Any multinomial model is defined by either the subset M or the pair $(\mathbf{f}, \Theta)$. When we assume that a given model is "correct," we are really just assuming that $\boldsymbol{\pi}$ lies in M, or equivalently, that there is a parameter value $\boldsymbol{\varphi}$ in Θ such that $\boldsymbol{\pi} = \mathbf{f}(\boldsymbol{\varphi})$. If the model is incorrect, then $\boldsymbol{\pi}$ does not lie in M, and $\boldsymbol{\varphi}$ does not exist.

To help clarify the notation, we now consider an example.

Example 14.7-1   Truncated Poisson variates

Suppose that a set of N independent and identically distributed Poisson variates with mean θ are observed, but that the observations are truncated at $x = 2$. Let X_1, X_2, and X_3 be the number of observations taking on the values 0, 1, and 2 or more, respectively. Then $\mathbf{X} = (X_1, X_2, X_3)$ has the trinomial distribution $\mathscr{M}_3(N, \mathbf{f}(\theta))$, where

$$\mathbf{f}(\theta) = (e^{-\theta}, \theta e^{-\theta}, 1 - (1 + \theta) e^{-\theta}) \tag{14.7-3}$$

for $\theta > 0$. In the notation we have developed in this section, $T = 3$, $s = 1$, and Θ is the positive real axis. As θ varies over Θ, $\mathbf{f}(\theta)$ traces out a one-dimensional curve in $\mathscr{S}_3$. This curve is the subset M. When $\theta \to 0$, $\mathbf{f}(\theta) \to (1, 0, 0)$, and when $\theta \to \infty$, $\mathbf{f}(\theta) \to (0, 0, 1)$. Thus the boundary points of Θ in this example correspond to boundary points of $\mathscr{S}_3$. Table 14.7-1 gives values of $\mathbf{f}(\theta)$ for various values of $\theta > 0$.

Table 14.7-1   Values of $\mathbf{f}(\theta) = (e^{-\theta}, \theta e^{-\theta}, 1 - (1 + \theta)e^{-\theta})$ for Selected Values of θ

θ	$e^{-\theta}$	$\theta e^{-\theta}$	$1 - (1 + \theta)e^{-\theta}$
0.00	1.000	0.000	0.000
0.25	0.779	0.195	0.026
0.50	0.607	0.304	0.089
0.75	0.472	0.354	0.174
1.00	0.368	0.368	0.264
1.15	0.317	0.365	0.318
2.00	0.135	0.270	0.595
3.00	0.050	0.150	0.800
4.00	0.018	0.072	0.910
∞	0.000	0.000	1.000

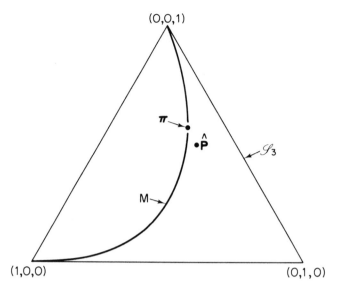

Figure 14.7-1.

Now suppose we observe **X**, and the vector of observed proportions is

$$\hat{\mathbf{p}} = (0.2, 0.3, 0.5). \tag{14.7-4}$$

Suppose further that the "true" parameter φ is $\varphi = 2$. Then $\boldsymbol{\pi} = \mathbf{f}(2)$ equals $(0.135, 0.270, 0.595)$.

Figure 14.7-1 shows the relationships between $\mathscr{S}_3$, M, $\boldsymbol{\pi}$, and $\hat{\mathbf{p}}$ in this example. If the Poisson model is incorrect, then the true value of $\boldsymbol{\pi}$ does not generally lie on the curve M, although in principle it can. Because of the discreteness of the multinomial distribution, it often happens that $\hat{\mathbf{p}}$ does not lie on M (as is the case in figure 14.7-1). ■■

14.7.2   Estimation in multinomial models

Using the notation developed in section 14.7.1 we can describe estimation in a multinomial model as choosing a value $\hat{\boldsymbol{\pi}}$ in M that depends on the observation **X**. Usually we want $\hat{\boldsymbol{\pi}}$ to be close in some sense to $\hat{\mathbf{p}}$. Alternatively, we can choose a value $\hat{\boldsymbol{\theta}}$ in Θ and set $\hat{\boldsymbol{\pi}} = \mathbf{f}(\hat{\boldsymbol{\theta}})$. The values $\hat{\pi}_i$ are called the *fitted values* (sometimes the $N\hat{\pi}_i$ are called the fitted values), and the $\hat{\theta}_i$ are called the *parameter estimates*. There are two important classes of methods for choosing $\hat{\boldsymbol{\pi}}$ or $\hat{\boldsymbol{\theta}}$. These are minimum distance methods and the method of moments. We discuss these briefly in turn.

Minimum distance methods

Many familiar methods of estimation in a multinomial model choose $\hat{\boldsymbol{\pi}}$ or $\hat{\boldsymbol{\theta}}$ to minimize some measure of the distance from the data point $\hat{\mathbf{p}}$ to the model M. We let $K(\mathbf{x}, \mathbf{y})$ be a function such that for all $\mathbf{x}, \mathbf{y}$ in $\mathscr{S}_T$, we have

1. $K(\mathbf{x}, \mathbf{y}) \geqq 0$;
2. $K(\mathbf{x}, \mathbf{y}) = 0$ if and only if $\mathbf{x} = \mathbf{y}$;
3. K is a distance function in the sense that if $\|\mathbf{x} - \mathbf{y}\|$ is increased sufficiently, then $K(\mathbf{x}, \mathbf{y})$ is also increased. (This condition is left vague since it plays no essential role in the rest of the chapter, and the term "distance function" is used here as a heuristic device.)

The minimum K-distance estimate of φ is $\hat{\theta}$, where $K(\hat{p}, f(\hat{\theta})) = \min_{\theta \in \Theta} K(\hat{p}, f(\theta))$. Regularity conditions such as those given in Section 14.8.1 are needed on f and Θ to ensure that $\hat{\theta}$ exists.

The most obvious method of this type is the minimum chi square method of estimation. In this case, $\hat{\theta}$ is chosen to minimize $X^2(\hat{p}, f(\theta))$, where $X^2(\mathbf{x}, \mathbf{y})$ is defined by

$$X^2(\mathbf{x}, \mathbf{y}) = \sum_{i=1}^{T} \frac{(x_i - y_i)^2}{y_i} \qquad (14.7\text{-}5)$$

for two points $\mathbf{x}$ and $\mathbf{y}$ in $\mathscr{S}_T$.

Maximum likelihood estimation for a multinomial model also corresponds to a minimum distance method of estimation. In this case, $\hat{\theta}$ is chosen to minimize $G^2(\hat{p}, f(\theta))$, where $G^2(\mathbf{x}, \mathbf{y})$ is defined by

$$G^2(\mathbf{x}, \mathbf{y}) = 2 \sum_{i=1}^{T} x_i \log\left(\frac{x_i}{y_i}\right) \qquad (14.7\text{-}6)$$

for two points $\mathbf{x}$ and $\mathbf{y}$ in $\mathscr{S}_T$. These and other distance functions for multinomial estimation are studied by Rao [1962].

Generally speaking, the distance functions used for multinomial estimation are not true distances in the sense of being metrics or even monotone functions of a metric. However, if $\mathbf{x}$ and $\mathbf{y}$ are close together and not too near the boundary of $\mathscr{S}_T$, then $X^2(\mathbf{x}, \mathbf{y})$ and $G^2(\mathbf{x}, \mathbf{y})$, given by (14.7-5) and (14.7-6), behave very much like the squares of a metric, so that for many purposes it is quite reasonable to regard them as measures of distance.

Method of moments

Method-of-moments estimators can be described in general terms as follows. We select s functions of $\hat{p}$, $H_1(\hat{p}), \ldots, H_s(\hat{p})$, and compute their theoretical expectations under the model:

$$h_i(\theta) = E_\theta H_i(\hat{p}). \qquad (14.7\text{-}7)$$

We solve the system of s equations in s unknowns:

$$h_1(\hat{\theta}_1, \ldots, \hat{\theta}_s) = H_1(\hat{p}),$$
$$\vdots \qquad\qquad\qquad (14.7\text{-}8)$$
$$h_s(\hat{\theta}_1, \ldots, \hat{\theta}_s) = H_s(\hat{p}),$$

and the solution $\hat{\theta}$ is the estimator of θ. We interpret (14.7-8) by saying that the sample moments $H_i(\hat{p})$ and the population moments $h_i(\theta)$ are matched at $\hat{\theta}$.

In terms of the geometry of $\mathscr{S}_T$ and M, the method of moments can be described as follows. If $\mathbf{p}$ is a general point in $\mathscr{S}_T$, then the set

$$\mathscr{H}(\hat{p}) = \{\mathbf{p} : H_i(\mathbf{p}) = H_i(\hat{p}) \quad \text{for } i = 1, \ldots, s\} \qquad (14.7\text{-}9)$$

gives all the values of $\mathbf{p}$ whose H_i-values agree with those of the observed point $\hat{p}$. Requiring $\hat{\theta}$ to be the solution to the system of equations (14.7-8) is the same as requiring $\hat{\pi}$ to lie simultaneously in both the model M and the set of admissible $\mathbf{p}$-values $\mathscr{H}(\hat{p})$. In general, zero, one, or more than one point of $\mathscr{S}_T$ may be in both M and $\mathscr{H}(\hat{p})$. In well-behaved cases, M and $\mathscr{H}(\hat{p})$ have exactly one point in common, so that $\hat{\pi}$ (and therefore $\hat{\theta}$) is well defined.

Example 14.7-2   Truncated Poisson variates (Continued)

Continuing with example 14.7-1, if we minimize (14.7-5), we obtain

$$X^2(\hat{\mathbf{p}}, \mathbf{f}(\hat{\boldsymbol{\theta}})) = 0.001353, \qquad \hat{\theta} = 1.659. \tag{14.7-10}$$

If we minimize (14.7-6), we obtain

$$G^2(\hat{\mathbf{p}}, \mathbf{f}(\hat{\boldsymbol{\theta}})) = 0.001356, \qquad \hat{\theta} = 1.661. \tag{14.7-11}$$

We see that the two estimates of $\boldsymbol{\theta}$ are very close. This reflects the fact that for small distances, G^2 and X^2 are very nearly identical.

In this example we get a simple method-of-moments estimator as follows. Set

$$H_1(\hat{\mathbf{p}}) = \hat{p}_1. \tag{14.7-12}$$

Then

$$h_1(\theta) = E_\theta(\hat{p}_1) = e^{-\theta}. \tag{14.7-13}$$

Thus equation (14.7-8) becomes

$$e^{-\hat{\theta}} = \hat{p}_1,$$

or

$$\hat{\theta} = -\log(\hat{p}_1)$$
$$= -\log(0.2) = 1.609. \tag{14.7-14}$$

In this example, $\mathcal{H}(\hat{\mathbf{p}})$ is a straight line through $\hat{\mathbf{p}}$ parallel to the boundary of $\mathcal{S}_3$ in figure 14.7-1 that connects $(0, 1, 0)$ to $(0, 0, 1)$. We see that $\mathcal{H}(\hat{\mathbf{p}})$ and M intersect in exactly one point no matter what the value of $\hat{\mathbf{p}}$. ∎∎

When the model is correct and $\mathbf{f}$ satisfies some regularity conditions (see Section 14.8.1), any reasonable estimate of $\boldsymbol{\varphi}$ (e.g., maximum likelihood, minimum chi square, or any appropriate method-of-moments estimator) converges in probability to $\boldsymbol{\varphi}$ as $N \to \infty$ (i.e., is a consistent estimator). The only essential difference among these estimators is the rate at which they converge to $\boldsymbol{\varphi}$ (i.e., their efficiency). However, if the model is incorrect and no $\boldsymbol{\varphi}$ exists such that $\boldsymbol{\pi} = \mathbf{f}(\boldsymbol{\varphi})$, we can still compute an estimate $\hat{\boldsymbol{\theta}}$ and examine its behavior for large N. The behavior of $\hat{\boldsymbol{\theta}}$ when the model is incorrect depends on the type of estimation method used, and while we cannot go into the problem in any depth here, the notation we have developed so far allows us to give a simple qualitative description of both minimum distance estimators and method-of-moments estimators in this situation.

For minimum distance estimators we must extend the notion of the true parameter $\boldsymbol{\varphi}$ to that of the "minimum K-distance parameter".

Thus $\boldsymbol{\varphi}_K$ is defined by

$$K(\boldsymbol{\pi}, \mathbf{f}(\boldsymbol{\varphi}_K)) = \min_{\boldsymbol{\theta} \in \Theta} K(\boldsymbol{\pi}, \mathbf{f}(\boldsymbol{\theta})). \tag{14.7-15}$$

We observe that when the model is correct, $\boldsymbol{\varphi} = \boldsymbol{\varphi}_K$ for any distance function K. In general, regularity conditions are needed to ensure that $\boldsymbol{\varphi}_K$ exists and is unique. When the model is incorrect, the minimum K-distance estimator $\hat{\boldsymbol{\theta}}_K$ converges in probability to $\boldsymbol{\varphi}_K$. Thus the behavior of different minimum K-distance estimators can be quite different when the model is wrong, and generally, the more wrong the model is, the more different are the values of different choices of $\hat{\boldsymbol{\theta}}_K$.

For method-of-moment estimators we must also extend the notion of the true parameter $\boldsymbol{\varphi}$. This time, we define the "matched-moments parameter value" $\boldsymbol{\varphi}_H$

as the solution (assuming it exists and is unique) of the systems of equations
(14.7-8) with $H_i(\pi)$ replacing $H_i(\hat{\mathbf{p}})$. Again, if the model is correct, then $\boldsymbol{\varphi}$ and $\boldsymbol{\varphi}_H$
coincide. When the model is incorrect, the method-of-moments estimator $\hat{\boldsymbol{\theta}}_H$
defined by $H_1, \ldots, H_s$ converges to $\boldsymbol{\varphi}_H$.

Example 14.7-3   Truncated Poisson variates (Continued)
Continuing with examples 14.7-1 and 14.7-2, we now suppose that the model is
incorrect and that $\boldsymbol{\pi} = (0.05, 0.8, 0.15)$. Then the minimum G^2-distance parameter
is $\varphi_{G^2} = 1.17$, and $\mathbf{f}(\varphi_{G^2}) = (0.310, 0.363, 0.327)$. The minimum X^2-distance parameter is $\varphi_{X^2} = 1.057$, and $\mathbf{f}(\varphi_{X^2}) = (0.347, 0.367, 0.285)$. The matched-moments
parameter, using the H given by (14.7-12), is $\varphi_H = 2.996$, and $\mathbf{f}(\varphi_H) = (0.05, 0.15,$

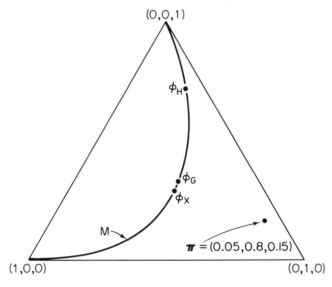

Figure 14.7-2.

0.80). (Figure 14.7-2 shows the relationship of these parameter values to $\boldsymbol{\pi}$.) Thus in
this example, for large N the parameter estimates $\hat{\theta}_{X^2}$, $\hat{\theta}_{G^2}$, and $\hat{\theta}_H$ do not converge
to the same value of θ. ■■

14.7.3   Multinomial goodness-of-fit tests
Continuing with the notation developed in the previous two sections, we now
give a general discussion of multinomial goodness-of-fit statistics. The observed
point $\hat{\mathbf{p}}$ is a natural estimate of $\mathbf{p}$ when we do not restrict ourselves to the model M.
A goodness-of-fit statistic is a measure of the "distance" from $\hat{\mathbf{p}}$ to M. Thus
there is a relationship between goodness-of-fit statistics and minimum distance
methods of estimating a multinomial parameter. There are two general types of
goodness-of-fit statistics for multinomial data. Both begin with a distance function
$K(\mathbf{x}, \mathbf{y})$.

The first type of goodness-of-fit statistic is the minimum value $K_{\min}$ of $K(\hat{\mathbf{p}}, \mathbf{f}(\boldsymbol{\theta}))$.
Thus if $\hat{\boldsymbol{\theta}}$ denotes the minimum K-distance estimate of $\boldsymbol{\theta}$, then

$$K_{\min} = \min_{\boldsymbol{\theta} \in \Theta} K(\hat{\mathbf{p}}, \mathbf{f}(\boldsymbol{\theta})) = K(\hat{\mathbf{p}}, \mathbf{f}(\hat{\boldsymbol{\theta}})) = K(\hat{\mathbf{p}}, \hat{\boldsymbol{\pi}}). \qquad (14.7\text{-}16)$$

Examples of this are the log-likelihood ratio statistic

$$G^2 = NG^2(\hat{\mathbf{p}}, \hat{\boldsymbol{\pi}}),\qquad(14.7\text{-}17)$$

where $\hat{\boldsymbol{\pi}} = \mathbf{f}(\hat{\boldsymbol{\theta}})$ is the maximum likelihood estimate of $\boldsymbol{\pi} = \mathbf{f}(\boldsymbol{\varphi})$, and the Pearson minimum chi square statistic

$$X^2 = NX^2(\hat{\mathbf{p}}, \hat{\boldsymbol{\pi}}),\qquad(14.7\text{-}18)$$

where $\hat{\boldsymbol{\pi}} = \mathbf{f}(\hat{\boldsymbol{\theta}})$ is the minimum chi square estimate of $\boldsymbol{\pi} = \mathbf{f}(\boldsymbol{\varphi})$.

The second type of goodness-of-fit statistic begins with an expression like (14.7-16), but instead of choosing $\hat{\boldsymbol{\pi}} = \mathbf{f}(\hat{\boldsymbol{\theta}})$ to minimize the distance function K, we use some other estimate $\boldsymbol{\theta}^*$ of $\boldsymbol{\theta}$ and set $\boldsymbol{\pi}^* = \mathbf{f}(\boldsymbol{\theta}^*)$. Then

$$K^* = K(\hat{\mathbf{p}}, \boldsymbol{\pi}^*) = K(\hat{\mathbf{p}}, \mathbf{f}(\boldsymbol{\theta}^*))\qquad(14.7\text{-}19)$$

is the K-distance from $\hat{\mathbf{p}}$ to $\boldsymbol{\pi}^*$, but not necessarily the minimum K-distance from $\hat{\mathbf{p}}$ to M.

An example of the latter goodness-of-fit measure is the sum of squared Freeman–Tukey deviates, given by

$$FT^2 = \sum_{i=1}^{T} (\sqrt{X_i} + \sqrt{X_i + 1} - \sqrt{4N\hat{\pi}_i + 1})^2,\qquad(14.7\text{-}20)$$

where the $\hat{\pi}_i$ are usually the fitted values found by maximum likelihood. If we omit terms of order N^{-1}, then we can express (14.7-20) in a form more like the other goodness-of-fit measures. Set

$$K_{FT}^2(\mathbf{x}, \mathbf{y}) = 4 \sum_{i=1}^{T} (\sqrt{x_i} - \sqrt{y_i})^2.\qquad(14.7\text{-}21)$$

Then, ignoring order N^{-1} terms, we have

$$FT^2 = NK_{FT}^2(\hat{\mathbf{p}}, \hat{\boldsymbol{\pi}}).\qquad(14.7\text{-}22)$$

Having computed the distance from $\hat{\mathbf{p}}$ to M using either method, the fit of the model is assessed as good or bad depending on the size of $K(\hat{\mathbf{p}}, \hat{\boldsymbol{\pi}})$. If $K(\hat{\mathbf{p}}, \hat{\boldsymbol{\pi}})$ is large, then the fit is poor, while if it is small, the fit of the model M may be acceptable (see Chapter 9 for more detailed discussion of the interpretation of the size of goodness-of-fit statistics). In order to assess what "large" and "small" mean, we use the sampling distribution of $K(\hat{\mathbf{p}}, \hat{\boldsymbol{\pi}})$ under the hypothesis that $\boldsymbol{\pi}$ is in M. In general, the sampling distribution of $K(\hat{\mathbf{p}}, \hat{\boldsymbol{\pi}})$ depends on the following factors: (i) the sample size N; (ii) the true value of $\boldsymbol{\pi}$; (iii) the model M, especially its dimension (i.e., the number of parameters needed to specify it); (iv) the form of $K(\mathbf{x}, \mathbf{y})$; (v) the method used to determine $\hat{\boldsymbol{\pi}}$. While all of these factors may make a difference in small samples, we will see that when N is large and $\boldsymbol{\pi}$ is in M (i.e., when the model is correct), then for the distance functions (14.7-5), (14.7-6), and (14.7-21), only the dimension of M and the method of determining $\hat{\boldsymbol{\pi}}$ influence the distribution of $K(\hat{\mathbf{p}}, \hat{\boldsymbol{\pi}})$. When N is large and $\boldsymbol{\pi}$ is not in M (i.e., when the model is not correct), then all of the factors (i)–(v) can have a substantial influence on the distribution of $K(\hat{\mathbf{p}}, \hat{\boldsymbol{\pi}})$.

14.7.4 Exercise

1. Show that if $H_1(\hat{\mathbf{p}}) = \hat{p}_2$ in example 14.7-2, then there may be zero, one, or two values of $\mathbf{p}$ in $\mathcal{H}(\hat{\mathbf{p}}) \cap M$, depending on the value of $\hat{\mathbf{p}}$.

14.8 Asymptotic Behavior of Multinomial Maximum Likelihood Estimators

In this section we continue to use the notation of Section 14.7 and develop further the discussion of maximum likelihood estimation for multinomial models given in Section 14.7.2.

We let $\overline{\Theta}$ denote the closure of Θ. Then $\overline{\Theta}$ is the union of Θ and all of the boundary points of Θ. If Θ is unbounded (as it is in the log-linear models discussed elsewhere in this book), we include a point at infinity in $\overline{\Theta}$.

If the model is correct, then the likelihood function is given by

$$L(\theta, \mathbf{X}) = \binom{N}{X_1, \ldots, X_T} \prod_{i=1}^{T} (f_i(\theta))^{X_i}. \tag{14.8-1}$$

Therefore, we have

$$-2 \log(L(\theta, \mathbf{X})) = 2NG^2(\hat{\mathbf{p}}, \mathbf{f}(\theta)) - 2 \log \left(\frac{N!}{X_1! \ldots X_T!} \right)$$

$$- 2N \sum_{i=1}^{T} \hat{p}_i \log(\hat{p}_i), \tag{14.8-2}$$

where $G^2(\mathbf{x}, \mathbf{y})$ is defined in (14.7-6). Hence, as we asserted in Section 14.7.2, we maximize the likelihood function if we minimize the distance measure

$$G^2(\hat{\mathbf{p}}, \mathbf{f}(\theta)) = \sum_{i=1}^{T} \hat{p}_i \log \left(\frac{\hat{p}_i}{f_i(\theta)} \right). \tag{14.8-3}$$

Now we let $\mathbf{p}$ denote a generic point in $\mathscr{S}_T$ and consider the purely analytic problem of minimizing $G^2(\mathbf{p}, \mathbf{f}(\theta))$ for $\theta \in \Theta$. We define $\theta(\mathbf{p})$ to be any point in $\overline{\Theta}$ for which

$$G^2(\mathbf{p}, \mathbf{f}(\theta(\mathbf{p}))) = \inf_{\theta \in \Theta} G^2(\mathbf{p}, \mathbf{f}(\theta)). \tag{14.8-4}$$

Thus $\theta(\mathbf{p})$ minimizes $G^2(\mathbf{p}, \mathbf{f}(\theta))$ for θ in $\overline{\Theta}$. If $\mathbf{f}(\theta)$ is continuous in Θ, then $\theta(\mathbf{p})$ is defined in $\overline{\Theta}$. The only problem is that there may be more than one choice of $\theta(\mathbf{p})$ for some values of $\mathbf{p}$. This ambiguity is not serious and vanishes if $\mathbf{p}$ is sufficiently close to M. When the model is correct, $\hat{\mathbf{p}}$ gets arbitrarily close to M as $N \to \infty$, so that we can define the maximum likelihood estimator as any choice of

$$\hat{\theta} = \theta(\hat{\mathbf{p}}), \tag{14.8-5}$$

and in large samples the problem of ambiguity vanishes. Because $\theta(\hat{\mathbf{p}})$ is in $\overline{\Theta}$, we may have $\hat{\theta} = \infty$, but again, in large samples, under the regularity conditions assumed in the Section 14.8.1, the probability of this happening goes to zero as $N \to \infty$. Birch [1964a] gives a more general definition of maximum likelihood estimators for multinomial models, but the one given here is sufficiently general for our purposes.

14.8.1   *The regularity conditions*

The regularity conditions 1–6 come from Birch [1964a] and are usually satisfied by the models that appear elsewhere in this book. We assume the model is correct, so that $\pi = \mathbf{f}(\varphi)$. Throughout, we assume $s < T - 1$.

1. The point $\boldsymbol{\varphi}$ is an interior point of Θ, so that $\boldsymbol{\varphi}$ is not on the boundary of Θ and there is an s-dimensional neighborhood of $\boldsymbol{\varphi}$ that is completely contained in Θ.

2. $\pi_i = f_i(\boldsymbol{\varphi}) > 0$ for $i = 1, \ldots, T$. Thus $\boldsymbol{\pi}$ is an interior point of $\mathscr{S}_T$ and does not lie on the boundary of $\mathscr{S}_T$.

3. The mapping $f : \Theta \to \mathscr{S}_T$ is totally differentiable at $\boldsymbol{\varphi}$, so that the partial derivatives of f_i with respect to each θ_j exist at $\boldsymbol{\varphi}$ and $\mathbf{f}(\boldsymbol{\theta})$ has a linear approximation at $\boldsymbol{\varphi}$ given by

$$f_i(\boldsymbol{\theta}) = f_i(\boldsymbol{\varphi}) + \sum_{j=1}^{s} (\theta_j - \varphi_j) \frac{\partial f_i(\boldsymbol{\varphi})}{\partial \theta_j} + o(\|\boldsymbol{\theta} - \boldsymbol{\varphi}\|) \qquad (14.8\text{-}6)$$

    as $\boldsymbol{\theta} \to \boldsymbol{\varphi}$.

4. The Jacobian matrix $(\partial \mathbf{f}/\partial \boldsymbol{\theta})$, whose (i, j) element is $\partial f_i(\boldsymbol{\varphi})/\partial \theta_j$, is of full rank (i.e., rank s). Thus $\mathbf{f}(\boldsymbol{\theta})$ maps a small neighborhood of $\boldsymbol{\varphi}$ (in Θ) into a small s-dimensional neighborhood of $\mathbf{f}(\boldsymbol{\varphi})$ in M.

5. The inverse mapping $f^{-1} : M \to \Theta$ is continuous at $f(\boldsymbol{\varphi}) = \boldsymbol{\pi}$. In particular, for every $\varepsilon > 0$ there exists a $\delta > 0$ such that if $\|\boldsymbol{\theta} - \boldsymbol{\varphi}\| \geqq \varepsilon$, then $\|\mathbf{f}(\boldsymbol{\theta}) - \mathbf{f}(\boldsymbol{\varphi})\| \geqq \delta$.

6. The mapping $f : \Theta \to \mathscr{S}_T$ is continuous at every point $\boldsymbol{\theta}$ in Θ.

Discussion of the regularity conditions

The condition 1 that $\boldsymbol{\varphi}$ is not on the boundary of Θ is usually not a serious restriction on the models considered in this book. In the usual parametrizations of these models, the boundary points of Θ (including a point at infinity if necessary) correspond to boundary points of $\mathscr{S}_T$, and thus condition 2 usually implies condition 1. Condition 2 merely ensures that there really are T cells in the multinomial and not fewer. In models for incomplete contingency tables, the structural zero cells are not counted in computing T.

Condition 3 is a smoothness condition on $\mathbf{f}$ and can be replaced by the slightly stronger and more easily verified condition that $\mathbf{f}(\boldsymbol{\theta})$ has continuous partial derivatives in a neighborhood of $\boldsymbol{\varphi}$. Conditions 1, 3, and 4 combine to ensure that the model really does have s parameters and not fewer. Condition 1 ensures that Θ contains a "full" s-dimensional neighborhood about $\boldsymbol{\varphi}$, while 3 and 4 ensure that this neighborhood gets mapped via $\mathbf{f}$ into a full s-dimensional neighborhood about $\boldsymbol{\pi}(\boldsymbol{\varphi})$ in M. If condition 4 were not satisfied, then we could reparametrize the model so that it had fewer than s parameters.

Condition 5 is an additional smoothness condition on $\mathbf{f}$, but it can also be interpreted as a "strong identifiability" condition on the parameters $\boldsymbol{\theta}$ of the model. It ensures that if $\boldsymbol{\theta}$ is kept away from $\boldsymbol{\varphi}$, then $\mathbf{f}(\boldsymbol{\theta})$ cannot get arbitrarily close to $\boldsymbol{\pi} = \mathbf{f}(\boldsymbol{\varphi})$. A weaker identifiability condition is that if $\boldsymbol{\theta} \neq \boldsymbol{\varphi}$, then $\mathbf{f}(\boldsymbol{\theta}) \neq \mathbf{f}(\boldsymbol{\varphi})$. Clearly, if 5 is satisfied, then the weaker form of identifiability is also satisfied. While condition 5 may appear to be a local condition on properties of the mapping $\mathbf{f}$ in a neighborhood of $\boldsymbol{\varphi}$, as are conditions 1–4, it is actually a global condition on $\mathbf{f}$ and is not a consequence of any smoothness conditions assumed on $\mathbf{f}$ near the value of the true parameter $\boldsymbol{\varphi}$. Condition 5 ensures that M does not bend back on itself and is also sufficient to ensure the consistency of the maximum likelihood estimates of $\boldsymbol{\varphi}$.

Condition 6 is not really necessary for the local properties of the maximum likelihood estimates of $\boldsymbol{\varphi}$, but it is virtually always satisfied by the models in this book, and it simplifies our definition of maximum likelihood estimates. Furthermore, in all the models that usually arise in applications, condition 3 is satisfied not merely at one particular point φ of Θ but at all points in Θ. When this happens, condition 6 is automatically satisfied.

14.8.2 *Expansion of the maximum likelihood estimator*

Our aim in this section is to establish the important asymptotic expansion of the MLE $\hat{\theta}$ about the true parameter φ when the model is correct. To use the Chernoff–Pratt machinery developed in Section 14.4, we need to separate the purely analytic parts from the purely stochastic parts of the problem. Thus we first concentrate on the function $\theta(\mathbf{p})$ defined in (14.8-4). Birch [1964a] proves the following key result under assumptions 1–6.

THEOREM 14.8-1   *Let* $\mathbf{A}$ *be the* $T \times s$ *matrix whose* (i, j) *element is* $\pi_i^{-1/2}(\partial f_i(\varphi)/\partial\theta_j)$ *and let* $\theta(\mathbf{p})$ *be defined by* (14.8-4). *Then as* $\mathbf{p} \to \pi$ *in* $\mathscr{S}_T$,

$$\theta(\mathbf{p}) = \varphi + (\mathbf{p} - \pi)\mathbf{D}_\pi^{-1/2}\mathbf{A}(\mathbf{A}'\mathbf{A})^{-1} + o(\|\mathbf{p} - \pi\|). \qquad (14.8\text{-}7)$$

Observe that if $\mathbf{p} = \pi + O(N^{-1/2})$, the o term in (14.8-7) becomes $o(N^{-1/2})$. Theorem 14.8-1 is the crucial analytic result we need, and we refer the reader to Birch's paper for his detailed proof. The next theorem summarizes the necessary stochastic facts we need.

THEOREM 14.8-2   *If* $\mathbf{X} \sim \mathscr{M}_T(N, \pi)$ *and* $\hat{\mathbf{p}} = N^{-1}\mathbf{X}$, *then*

$$\mathscr{L}[\sqrt{N}(\hat{\mathbf{p}} - \pi)] \to \mathscr{N}(0, \mathbf{D}_\pi - \pi'\pi), \qquad (14.8\text{-}8)$$

$$\hat{\mathbf{p}} = \pi + O_p(N^{-1/2}). \qquad (14.8\text{-}9)$$

We have already proved (14.8-8) in example 14.3-3, and (14.8-9) is a consequence of theorem 14.4-2. Using theorem 14.4-8, we can combine the analytic result from theorem 14.8-1 with the stochastic fact (14.8-9) to obtain the following fundamental asymptotic expansion of the maximum likelihood estimate.

THEOREM 14.8-3   *Under conditions 1–6 and assuming that* $\pi = \mathbf{f}(\varphi)$, *the MLE* $\hat{\theta}$ *satisfies the expansion*

$$\hat{\theta} = \varphi + (\hat{\mathbf{p}} - \pi)\mathbf{D}_\pi^{-1/2}\mathbf{A}(\mathbf{A}'\mathbf{A})^{-1} + o_p(N^{-1/2}), \qquad (14.8\text{-}10)$$

where $\mathbf{A}$ *is defined as in theorem* 14.8-1.

From theorem 14.8-3, we see that possible nonuniqueness of MLEs is a second-order effect, since they must all differ by at most an $o_p(N^{-1/2})$ term. An important consequence of theorem 14.8-3 is the asymptotic distribution of $\hat{\theta}$ under the hypothesis that the model is correct. This is given in the next theorem.

THEOREM 14.8-4   *Under conditions 1–6 and assuming that* $\pi = f(\varphi)$, *the asymptotic distribution of* $\hat{\theta}$ *is given by*

$$\mathscr{L}[\sqrt{N}(\hat{\theta} - \varphi)] \to \mathscr{N}(0, (\mathbf{A}'\mathbf{A})^{-1}), \qquad (14.8\text{-}11)$$

where $\mathbf{A}$ *is defined as in theorem* 14.8-1.

Proof   From the expansion of (14.8-10), we have

$$\sqrt{N}(\hat{\theta} - \varphi) = \mathbf{y}_N\mathbf{A}(\mathbf{A}'\mathbf{A})^{-1} + o_p(1), \qquad (14.8\text{-}12)$$

$$\mathbf{y}_N = \sqrt{N}(\hat{\mathbf{p}} - \pi)\mathbf{D}_\pi^{-1/2}.$$

But

$$\mathscr{L}[\mathbf{y}_N] \to \mathscr{N}(0, \mathbf{D}_\pi^{-1/2}(\mathbf{D}_\pi - \boldsymbol{\pi}'\boldsymbol{\pi})\mathbf{D}_\pi^{-1/2}),$$

$$\mathbf{D}_\pi^{-1/2}(\mathbf{D}_\pi - \boldsymbol{\pi}'\boldsymbol{\pi})\mathbf{D}_\pi^{-1/2} = \mathbf{I} - \sqrt{\boldsymbol{\pi}}'\sqrt{\boldsymbol{\pi}},$$

where $\sqrt{\boldsymbol{\pi}} = (\sqrt{\pi_1}, \ldots, \sqrt{\pi_T})$. Hence

$$\mathscr{L}[\mathbf{y}_N \mathbf{A}(\mathbf{A}'\mathbf{A})^{-1}] \to \mathscr{N}(0, \boldsymbol{\Sigma}),$$

where

$$\boldsymbol{\Sigma} = (\mathbf{A}'\mathbf{A})^{-1}\mathbf{A}'(\mathbf{I} - \sqrt{\boldsymbol{\pi}}\sqrt{\boldsymbol{\pi}'})\mathbf{A}(\mathbf{A}'\mathbf{A})^{-1}$$

$$= (\mathbf{A}'\mathbf{A})^{-1} - (\mathbf{A}'\mathbf{A})^{-1}\mathbf{A}'\sqrt{\boldsymbol{\pi}}\sqrt{\boldsymbol{\pi}'}\mathbf{A}(\mathbf{A}'\mathbf{A})^{-1}.$$

But we can easily show that $\mathbf{A}'\sqrt{\boldsymbol{\pi}} = 0$. Thus

$$\mathscr{L}[\mathbf{y}_N \mathbf{A}(\mathbf{A}'\mathbf{A})^{-1}] \to \mathscr{N}(0, (\mathbf{A}'\mathbf{A})^{-1}). \tag{14.8-13}$$

The conclusion of the theorem follows from (14.8-13) and theorem 14.4-3. ∎

We can use theorem 14.8-4 to provide an asymptotic variance estimate for the MLE $\hat{\boldsymbol{\theta}}$. The (i, j) element of the matrix $\mathbf{A}'\mathbf{A}$ can be written as

$$= \sum_{k=1}^{T} \frac{\partial \log(f_k(\boldsymbol{\varphi}))}{\partial \theta_i} \frac{\partial \log(f_k(\boldsymbol{\varphi}))}{\partial \theta_j} f_k(\boldsymbol{\varphi}), \tag{14.8-14}$$

so that $\mathbf{A}'\mathbf{A}$ is indeed the Fisher information matrix for the multinomial model $\mathbf{f}(\boldsymbol{\theta})$. An estimate of the covariance matrix of $\hat{\boldsymbol{\theta}}$ is then given by

$$\frac{1}{N}(\mathbf{A}(\hat{\boldsymbol{\theta}})'\mathbf{A}(\hat{\boldsymbol{\theta}}))^{-1}. \tag{14.8-15}$$

The estimated variances of the $\hat{\theta}_i$ can be used to set approximate confidence intervals for the individual φ_i, while the estimated covariances of the $\hat{\theta}_i$ can be used to set approximate confidence intervals for linear combinations of the φ_i.

Theorem 14.8-4 can also be used to obtain the asymptotic distribution of the fitted values $\mathbf{f}(\hat{\boldsymbol{\theta}})$ under the assumption that the model is correct. Using assumption 3 and theorem 14.8-4, we obtain the following expansion of $\mathbf{f}(\hat{\boldsymbol{\theta}})$ about $\mathbf{f}(\boldsymbol{\varphi})$:

$$\mathbf{f}(\hat{\boldsymbol{\theta}}) = \mathbf{f}(\boldsymbol{\varphi}) + (\hat{\boldsymbol{\theta}} - \boldsymbol{\varphi})\left(\frac{\partial \mathbf{f}}{\partial \boldsymbol{\theta}}\right)' + o_p(N^{-1/2}), \tag{14.8-16}$$

where $(\partial \mathbf{f}/\partial \boldsymbol{\theta})$ is the Jacobian matrix whose (i, j) element is

$$\frac{\partial f_i(\boldsymbol{\varphi})}{\partial \theta_j}. \tag{14.8-17}$$

Hence we have

$$\sqrt{N}(\mathbf{f}(\hat{\boldsymbol{\theta}}) - \mathbf{f}(\boldsymbol{\varphi})) = \sqrt{N}(\hat{\boldsymbol{\theta}} - \boldsymbol{\varphi})\left(\frac{\partial \mathbf{f}}{\partial \boldsymbol{\theta}}\right)' + o_p(1), \tag{14.8-18}$$

from which it follows that

$$\mathscr{L}[\sqrt{N}(\mathbf{f}(\hat{\boldsymbol{\theta}}) - \boldsymbol{\pi})] \to \mathscr{N}\left(0, \left(\frac{\partial \mathbf{f}}{\partial \boldsymbol{\theta}}\right)(\mathbf{A}'\mathbf{A})^{-1}\left(\frac{\partial \mathbf{f}}{\partial \boldsymbol{\theta}}\right)'\right). \tag{14.8-19}$$

The asymptotic covariance matrix can be used to find estimated asymptotic variances for the fitted values $\pi_i(\hat{\boldsymbol{\theta}})$ in just the same way $(\mathbf{A}'\mathbf{A})^{-1}$ is used to find estimated asymptotic variances for the $\hat{\theta}_i$.

14.8.3 *The MLE when the model is incorrect*

When M is not correct, it is still generally true that the MLE $\hat{\boldsymbol{\theta}}$ converges to some point $\boldsymbol{\varphi}$ in Θ. In this case, $\mathbf{f}(\boldsymbol{\varphi})$ is the closest point to $\boldsymbol{\pi}$ in M in the sense of the $G^2(\boldsymbol{\pi}, \mathbf{f}(\boldsymbol{\varphi}))$ distance measure. In order for this convergence to hold, we must assume that conditions (5) and (6) hold and, in addition, that for every $\varepsilon > 0$ there exists a $\delta > 0$ such that if $\|\boldsymbol{\theta} - \boldsymbol{\varphi}\| \geqq \varepsilon$ then $G^2(\boldsymbol{\pi}, \mathbf{f}(\boldsymbol{\theta})) - G^2(\boldsymbol{\pi}, \mathbf{f}(\boldsymbol{\varphi})) \geqq \delta$. This condition ensures that the closest point to $\boldsymbol{\pi}$ in M is strongly identified.

14.9 Asymptotic Distribution of Multinomial Goodness-of-Fit Tests

14.9.1 *Asymptotic equivalence of three goodness-of-fit statistics*

In Section 14.7, we discussed three different distance measures used elsewhere in this book to assess the goodness of fit of models. If $\mathbf{x}$ and $\mathbf{y}$ are arbitrary points in $\mathcal{S}_T$, then these three distance measures are:

$$G^2(\mathbf{x}, \mathbf{y}) = 2 \sum_i x_i \log\left(\frac{x_i}{y_i}\right), \tag{14.9-1}$$

$$X^2(\mathbf{x}, \mathbf{y}) = \sum_i \frac{(x_i - y_i)^2}{y_i}, \tag{14.9-2}$$

$$K_{FT}^2(\mathbf{x}, \mathbf{y}) = 4 \sum_i (\sqrt{x_i} - \sqrt{y_i})^2. \tag{14.9-3}$$

Continuing with the notation of Section 14.7, the corresponding goodness-of-fit statistics are

$$G^2 = NG^2(\hat{\mathbf{p}}, \hat{\boldsymbol{\pi}}), \tag{14.9-4}$$

$$X^2 = NX^2(\hat{\mathbf{p}}, \hat{\boldsymbol{\pi}}), \tag{14.9-5}$$

$$FT^2 = NK_{FT}^2(\hat{\mathbf{p}}, \hat{\boldsymbol{\pi}}). \tag{14.9-6}$$

When $\hat{\boldsymbol{\pi}}$ is the fitted value found by maximum likelihood, then G^2 is the log-likelihood ratio statistic. When $\hat{\boldsymbol{\pi}}$ is the fitted value found either by maximum likelihood or by minimum chi square, X^2 is the Pearson chi square statistic. We sometimes call FT^2 the Freeman–Tukey goodness-of-fit statistic when $\hat{\boldsymbol{\pi}}$ is the fitted value found by maximum likelihood. We observe that the definition of FT^2 in (14.9-6) differs from that in (14.7-20) by an order N^{-1} term which is irrelevant to the asymptotic discussion (but which can be important in small samples).

The main result of this section is that when $\mathbf{x}$ and $\mathbf{y}$ are close together, the three distance functions (14.9-1), (14.9-2), and (14.9-3) are nearly identical. This result has the consequence that when the model is correct and $\hat{\boldsymbol{\pi}}$ is estimated in any reasonable way (not merely by maximum likelihood), the three goodness-of-fit statistics (14.9-4), (14.9-5), and (14.9-6) all have the same limiting distribution. We note in passing that when the model is not correct, the three goodness-of-fit statistics do not have the same limiting behavior and may yield very different results.

We use the Chernoff–Pratt machinery developed in Section 14.4, and thus we separate the analytic and stochastic parts of the problem. The next lemma gives the necessary analytic result.

LEMMA 14.9-1   Let $G^2(\mathbf{x}, \mathbf{y})$, $X^2(\mathbf{x}, \mathbf{y})$, and $K_{FT}^2(\mathbf{x}, \mathbf{y})$ be as defined in (14.9-1), (14.9-2), and (14.9-3) and let $\{\mathbf{x}^{(n)}\}$ and $\{\mathbf{y}^{(n)}\}$ be two sequences of vectors in $\mathscr{S}_T$ such that $\mathbf{x}^{(n)} = \boldsymbol{\pi} + O(n^{-1/2})$ and $\mathbf{y}^{(n)} = \boldsymbol{\pi} + O(n^{-1/2})$ for $\boldsymbol{\pi} \in \mathscr{S}_T$ with $\pi_i > 0$. Then as $n \to \infty$, we have

$$nK_{FT}^2(\mathbf{x}^{(n)}, \mathbf{y}^{(n)}) = nX^2(\mathbf{x}^{(n)}, \mathbf{y}^{(n)}) + o(1), \tag{14.9-7}$$

$$nG^2(\mathbf{x}^{(n)}, \mathbf{y}^{(n)}) = nX^2(\mathbf{x}^{(n)}, \mathbf{y}^{(n)}) + o(1). \tag{14.9-8}$$

Proof   For (14.9-7), we observe that

$$K_{FT}^2(\mathbf{x}, \mathbf{y}) = \sum_i \frac{(x_i - y_i)^2}{y_i} \left(\frac{2\sqrt{y_i}}{\sqrt{x_i} + \sqrt{y_i}} \right)^2, \tag{14.9-9}$$

and that

$$\left(\frac{2\sqrt{y_i}}{\sqrt{x_i} + \sqrt{y_i}} \right)^2 = \frac{4}{\left(1 + \sqrt{1 + \dfrac{x_i - y_i}{y_i}} \right)^2}. \tag{14.9-10}$$

From the assumptions on $\mathbf{x}^{(n)}$ and $\mathbf{y}^{(n)}$, we have

$$\frac{x_i^{(n)} - y_i^{(n)}}{y_i^{(n)}} = \frac{o(1)}{\pi_i + o(1)} = o(1). \tag{14.9-11}$$

Hence, substituting $\mathbf{x}^{(n)}$ for $\mathbf{x}$ and $\mathbf{y}^{(n)}$ for $\mathbf{y}$ in (14.9-9), we have

$$K_{FT}^2(\mathbf{x}^{(n)}, \mathbf{y}^{(n)}) = \sum_i \frac{(x_i^{(n)} - y_i^{(n)})^2}{y_i^{(n)}} \left(\frac{4}{(1 + \sqrt{1 + o(1)})^2} \right)$$

$$= X^2(\mathbf{x}^{(n)}, \mathbf{y}^{(n)}) + o(1)X^2(\mathbf{x}^{(n)}, \mathbf{y}^{(n)}). \tag{14.9-12}$$

But it is easy to see that $X^2(\mathbf{x}^{(n)}, \mathbf{y}^{(n)}) = O(n^{-1})$, so that

$$nK_{FT}^2(\mathbf{x}^{(n)}, \mathbf{y}^{(n)}) = nX^2(\mathbf{x}^{(n)}, \mathbf{y}^{(n)}) + no(1)O(n^{-1}),$$

which proves (14.9-7).

For (14.9-8), we expand $G(\mathbf{x}, \mathbf{y})$ in a Taylor series about the point $(\boldsymbol{\pi}, \boldsymbol{\pi})$ out to the second order. This yields

$$G^2(\mathbf{x}, \mathbf{y}) = G^2(\boldsymbol{\pi}, \boldsymbol{\pi}) + 2\sum_i (x_i - \pi_i) + 2\sum_i (y_i - \pi_i)(-1)$$

$$+ \frac{1}{2} \cdot 2\sum_i \frac{(x_i - \pi_i)^2}{\pi_i} + \frac{1}{2} \cdot 2\sum_i \frac{(y_i - \pi_i)^2}{\pi_i}$$

$$- 2\sum_i \frac{(x_i - \pi_i)(y_i - \pi_i)}{\pi_i} + o(\|\mathbf{x} - \boldsymbol{\pi}\|^2 + \|\mathbf{y} - \boldsymbol{\pi}\|^2),$$

which simplifies to

$$G^2(\mathbf{x}, \mathbf{y}) = \sum_i \frac{(x_i - y_i)^2}{\pi_i} + o(\|\mathbf{x} - \boldsymbol{\pi}\|^2 + \|\mathbf{y} - \boldsymbol{\pi}\|^2). \tag{14.9-13}$$

From (14.9-13), an argument similar to that used in proving (14.9-7) yields the desired result and is left as an exercise. ∎

THEOREM 14.9-2   *Let $\hat{\pi}$ be any estimate of π $(\pi_i > 0)$ such that $\hat{p}$ and $\hat{\pi}$ have a joint limiting normal distribution, i.e.,*

$$\mathscr{L}[\sqrt{N}((\hat{p}, \hat{\pi}) - (\pi, \pi))] \to \mathscr{N}(0, \Sigma)$$

for some covariance matrix Σ. Then $NG^2(\hat{p}, \hat{\pi})$, $NX^2(\hat{p}, \hat{\pi})$, and $NK^2_{FT}(\hat{p}, \hat{\pi})$ all have the same limiting distribution.

Proof   In setting up the formal Chernoff–Pratt machinery for this problem, the only technical subtlety is that for some applications of this result we need to allow for the possibility that $\hat{\pi}$ is not necessarily a function of $\hat{p}$ but that both $\hat{p}$ and $\hat{\pi}$ are functions of still another random vector, which we denote by ω_N, with sample space Ω_N and a probability distribution given by P_N. When $\hat{\pi}$ is a function of $\hat{p}$, we can take Ω_N as $\mathscr{S}_T$ and ω_N as $\hat{p}$. To show that $NG^2(\hat{p}, \hat{\pi})$ and $NX^2(\hat{p}, \hat{\pi})$ have the same limiting distribution, we must show first that

$$NX^2(\hat{p}, \hat{\pi}) - NG^2(\hat{p}, \hat{\pi}) = o_p(1) \tag{14.9-14}$$

and then establish that $NX^2(\hat{p}, \hat{\pi})$ does have a limiting distribution. To show (14.9-14) we define $H_N^{(i)}$ on Ω_N, $i = 1, 2, 3$, by

$$H_N^{(1)}(\omega_N) = \hat{p},$$

$$H_N^{(2)}(\omega_N) = \hat{\pi},$$

$$H_N^{(3)}(\omega_N) = NX^2(\hat{p}, \hat{\pi}) - NG^2(\hat{p}, \hat{\pi}).$$

Now if $\mathbf{w}_N$ is a nonstochastic sequence such that $\mathbf{w}_N \in \Omega_N$ and

$$\mathbf{x}^{(N)} = H_N^{(1)}(\mathbf{w}_N),$$

$$\mathbf{y}^{(N)} = H_N^{(2)}(\mathbf{w}_N),$$

then from lemma 14.9-1 we see that if

$$H_N^{(i)}(\mathbf{w}_N) - \pi = O(N^{-1/2}) \quad i = 1, 2,$$

then

$$H_N^{(3)}(\mathbf{w}_N) = o(1).$$

But by hypothesis $\sqrt{N}((\hat{p}, \hat{\pi}) - (\pi, \pi))$ converges in distribution, so by theorem 14.4-2 we have

$$H_N^{(i)}(\omega_N) - \pi = O_p(N^{-1/2}) \quad i = 1, 2,$$

and hence from theorem 14.4-8 we conclude that

$$H_N^{(3)}(\omega_N) = o_p(1),$$

which proves (14.9-14).

To show that $NX^2(\hat{p}, \hat{\pi})$ does have a limiting distribution, we merely evaluate it as follows. First observe that

$$NX^2(\hat{p}, \hat{\pi}) = \mathbf{Y}_N(\mathbf{D}_\pi^{-1} + o_p(1))\mathbf{Y}_N', \tag{14.9-15}$$

where

$$\mathbf{Y}_N = \sqrt{N}(\hat{p} - \hat{\pi}). \tag{14.9-16}$$

But from the hypothesis it follows that

$$\mathscr{L}[\mathbf{Y}_N] \to \mathscr{L}[(\mathbf{Y})],$$

where $\mathbf{Y}$ has the $\mathcal{N}(0, \mathbf{\Sigma}_{11} - \mathbf{\Sigma}_{12} - \mathbf{\Sigma}_{21} + \mathbf{\Sigma}_{22})$ distribution with $\mathbf{\Sigma}$ partitioned in the obvious way. Thus $\mathbf{Y}_N \mathbf{Y}_N' = O_p(1)$, so that (14.9-15) can be written as

$$NX^2(\hat{\mathbf{p}}, \hat{\boldsymbol{\pi}}) = \mathbf{U}_N \mathbf{U}_N' + o_p(1), \qquad (14.9\text{-}17)$$

where $\mathbf{U}_N = \sqrt{N} \mathbf{D}_\pi^{-1/2}(\hat{\mathbf{p}} - \hat{\boldsymbol{\pi}}) = \mathbf{D}_\pi^{-1/2} \mathbf{Y}_N$. But $\mathscr{L}[\mathbf{U}_N] \to \mathscr{L}[\mathbf{U}]$, where $\mathbf{U}$ has the $\mathcal{N}(0, \mathbf{D}_\pi^{-1/2}(\mathbf{\Sigma}_{11} - \mathbf{\Sigma}_{12} - \mathbf{\Sigma}_{21} + \mathbf{\Sigma}_{22})\mathbf{D}_\pi^{-1/2})$ distribution. Hence the limiting distribution of $NX^2(\hat{\mathbf{p}}, \hat{\boldsymbol{\pi}})$ is the same as the distribution of $\mathbf{UU}'$. This proves that $NX^2(\hat{\mathbf{p}}, \hat{\boldsymbol{\pi}})$ and $NG^2(\hat{\mathbf{p}}, \hat{\boldsymbol{\pi}})$ have the same limiting distribution. The rest of the theorem is proved in the same way. ∎

From the last theorem we obtain a very general result on the distribution of chi square statistics under weak assumptions about the way $\hat{\boldsymbol{\pi}}$ is estimated. We state this result as a corollary.

COROLLARY 14.9-3   *Under the assumptions of theorem* 14.9-2, $\mathscr{L}[NX^2(\hat{\mathbf{p}}, \hat{\boldsymbol{\pi}})] \to \sum_{i=1}^{T-1} \lambda_i Z_i^2$, *where the* Z_i^2 *are independent chi square variables with one degree of freedom and the* λ_i *are the nonzero eigenvalues of* $\mathbf{D}_\pi^{-1/2}(\mathbf{\Sigma}_{11} - \mathbf{\Sigma}_{12} - \mathbf{\Sigma}_{21} + \mathbf{\Sigma}_{22})\mathbf{D}_\pi^{-1/2}$.

This corollary follows from the proofs of theorem 14.9-2 and theorem 14.3-7. It implies that the chi square distribution plays an important role in chi square statistics, but that unless the joint asymptotic distribution of $\hat{\mathbf{p}}$ and $\hat{\boldsymbol{\pi}}$ has particular properties, there is no reason to expect the usual simple result that we pay one degree of freedom for every estimated parameter. How much is paid depends on how the parameter is estimated.

14.9.2   *The limiting chi square distribution of* X^2

In this section we give a proof of the celebrated asymptotic chi square distribution of the Pearson chi square statistic X^2 under the assumptions that the model is correct, i.e., $\boldsymbol{\pi} \in M$, and that $\hat{\boldsymbol{\pi}}$ is estimated by maximum likelihood. From theorem 14.9-2 we see that we only need to investigate X^2 in this situation, since the limiting distributions of G^2 and FT^2 are the same as that of X^2 (because we assume $\boldsymbol{\pi} \in M$). We denote the chi square distribution with d degrees of freedom by

$$\chi_d^2. \qquad (14.9\text{-}18)$$

The basic result is that X^2 has an asymptotic χ_{T-s-1}^2 distribution. In example 14.3-5 we already examined the case $s = 0$. When there are no parameters to be estimated (i.e., M is a single fixed point in $\mathscr{S}_T$), that example shows that X^2 has an asymptotic χ_{T-1}^2 distribution.

THEOREM 14.9-4   *Assume that regularity conditions* 1–6 *of Section* 14.8.1 *hold and that* $\boldsymbol{\pi} \in M$. *If* $\hat{\boldsymbol{\pi}} = \mathbf{f}(\hat{\boldsymbol{\theta}})$, *where* $\hat{\boldsymbol{\theta}}$ *is estimated by maximum likelihood, and if* $X^2 = NX^2(\hat{\mathbf{p}}, \hat{\boldsymbol{\pi}})$ *as in* (14.9-2), *then*

$$\mathscr{L}[X^2] \to \chi_{T-s-1}^2. \qquad (14.9\text{-}19)$$

Proof   We use matrix notation, where $\boldsymbol{\theta}$, $\mathbf{p}$, $\boldsymbol{\pi}$, etc. are all row vectors. To apply corollary 14.9-3 we need to find the joint asymptotic distribution of $\hat{\boldsymbol{\pi}}$ and $\hat{\mathbf{p}}$. From theorem 14.8-3, we have

$$\hat{\boldsymbol{\theta}} = \boldsymbol{\varphi} + (\hat{\mathbf{p}} - \boldsymbol{\pi})\mathbf{D}_{\pi}^{-1/2}\mathbf{A}(\mathbf{A}'\mathbf{A})^{-1} + o_p(N^{-1/2}), \qquad (14.9\text{-}20)$$

and from regularity condition 3, $\mathbf{f}$ has the following expansion as $\boldsymbol{\theta} \to \boldsymbol{\varphi}$:

$$\mathbf{f}(\boldsymbol{\theta}) - \mathbf{f}(\boldsymbol{\varphi}) = (\boldsymbol{\theta} - \boldsymbol{\varphi})\left(\frac{\partial \mathbf{f}}{\partial \boldsymbol{\theta}}\right)' + o\left(\|\boldsymbol{\theta} - \boldsymbol{\varphi}\|\right).$$

From theorem 14.4-8, it then follows that

$$\hat{\boldsymbol{\pi}} - \boldsymbol{\pi} = (\hat{\mathbf{p}} - \boldsymbol{\pi})\mathbf{D}_{\pi}^{-1/2}\mathbf{A}(\mathbf{A}'\mathbf{A})^{-1}\left(\frac{\partial \mathbf{f}}{\partial \boldsymbol{\theta}}\right)' + o_p(N^{-1/2}), \qquad (14.9\text{-}21)$$

or

$$\hat{\boldsymbol{\pi}} - \boldsymbol{\pi} = (\hat{\mathbf{p}} - \boldsymbol{\pi})\mathbf{L} + o_p(N^{-1/2}), \qquad (14.9\text{-}22)$$

where

$$\mathbf{L} = \mathbf{D}_{\pi}^{-1/2}\mathbf{A}(\mathbf{A}'\mathbf{A})^{-1}\mathbf{A}'\mathbf{D}_{\pi}^{1/2} \qquad (14.9\text{-}23)$$

and

$$\mathbf{A} = \mathbf{D}_{\pi}^{-1/2}\left(\frac{\partial \mathbf{f}}{\partial \boldsymbol{\theta}}\right).$$

Thus we have shown that

$$(\hat{\mathbf{p}} - \boldsymbol{\pi}, \hat{\boldsymbol{\pi}} - \boldsymbol{\pi}) = (\hat{\mathbf{p}} - \boldsymbol{\pi})(\mathbf{I}, \mathbf{L}) + o_p(N^{-1/2}), \qquad (14.9\text{-}24)$$

and hence we have

$$\mathscr{L}[\sqrt{N}((\hat{\mathbf{p}}, \hat{\boldsymbol{\pi}}) - (\boldsymbol{\pi}, \boldsymbol{\pi}))] \to \mathscr{N}(0, (\mathbf{I}, \mathbf{L})'(\mathbf{D}_{\pi} - \boldsymbol{\pi}'\boldsymbol{\pi})(\mathbf{I}, \mathbf{L})). \qquad (14.9\text{-}25)$$

so that the joint asymptotic covariance matrix of $\hat{\mathbf{p}}$ and $\hat{\boldsymbol{\pi}}$ can be expressed as

$$\begin{pmatrix} \mathbf{D}_{\pi} - \boldsymbol{\pi}'\boldsymbol{\pi} & (\mathbf{D}_{\pi} - \boldsymbol{\pi}'\boldsymbol{\pi})\mathbf{L} \\ \mathbf{L}'(\mathbf{D}_{\pi} - \boldsymbol{\pi}'\boldsymbol{\pi}) & \mathbf{L}'(\mathbf{D}_{\pi} - \boldsymbol{\pi}'\boldsymbol{\pi})\mathbf{L} \end{pmatrix}. \qquad (14.9\text{-}26)$$

But from corollary 14.9-3, the relevant matrix whose eigenvalues we must find is

$$\mathbf{B} = \mathbf{D}_{\pi}^{-1/2}(\mathbf{D}_{\pi} - \boldsymbol{\pi}'\boldsymbol{\pi} - (\mathbf{D}_{\pi} - \boldsymbol{\pi}'\boldsymbol{\pi})\mathbf{L} - \mathbf{L}'(\mathbf{D}_{\pi} - \boldsymbol{\pi}'\boldsymbol{\pi}) + \mathbf{L}'(\mathbf{D}_{\pi} - \boldsymbol{\pi}'\boldsymbol{\pi})\mathbf{L})\mathbf{D}_{\pi}^{-1/2}$$

$$= \mathbf{J} - \mathbf{J}\mathbf{K} - \mathbf{K}'\mathbf{J} + \mathbf{K}'\mathbf{J}\mathbf{K}, \qquad (14.9\text{-}27)$$

where $\mathbf{J} = \mathbf{I} - \sqrt{\boldsymbol{\pi}}'\sqrt{\boldsymbol{\pi}}$ with $\sqrt{\boldsymbol{\pi}} = (\sqrt{\pi_1}, \ldots, \sqrt{\pi_T})$ and $\mathbf{K} = \mathbf{D}_{\pi}^{1/2}\mathbf{L}\mathbf{D}_{\pi}^{-1/2}$. But

$$\mathbf{K} = \mathbf{A}(\mathbf{A}'\mathbf{A})^{-1}\mathbf{A}' \qquad (14.9\text{-}28)$$

and it is easy to show that $\sqrt{\boldsymbol{\pi}}\mathbf{A} = 0$, so that $\mathbf{B}$ simplifies to

$$\mathbf{B} = \mathbf{I} - \sqrt{\boldsymbol{\pi}}'\sqrt{\boldsymbol{\pi}} - \mathbf{A}(\mathbf{A}'\mathbf{A})^{-1}\mathbf{A}'. \qquad (14.9\text{-}29)$$

To compute the eigenvalues of $\mathbf{B}$, we use the same trick that worked in example 14.3-5. First we note that $\mathbf{B}$ is symmetric ($\mathbf{B}' = \mathbf{B}$) and idempotent ($\mathbf{B}^2 = \mathbf{B}$) and then apply theorem 14.3-8 to deduce that $\mathbf{B}$ has only 1 and 0 for eigenvalues. We take the trace of $\mathbf{B}$ to find the number of unit eigenvalues:

$$\text{tr}(\mathbf{B}) = \text{tr}(\mathbf{I}) - \text{tr}(\sqrt{\boldsymbol{\pi}}'\sqrt{\boldsymbol{\pi}}) - \text{tr}(\mathbf{A}(\mathbf{A}'\mathbf{A})^{-1}\mathbf{A}')$$

$$= T - \text{tr}(\sqrt{\boldsymbol{\pi}}\sqrt{\boldsymbol{\pi}}') - \text{tr}((\mathbf{A}'\mathbf{A})^{-1}\mathbf{A}'\mathbf{A})$$

$$= T - 1 - s.$$

Thus

$$\mathcal{L}[X^2] \to \chi^2_{T-1-s}. \quad \blacksquare$$

We observe that the only property of $\hat{\boldsymbol{\theta}}$ used in this proof is the expansion (14.9-20). This same expansion is satisfied by a variety of estimators other than the MLE (see Holland [1967] for a proof that under assumptions 1–6 of Section 14.8.1, the minimum chi square estimate satisfies (14.9-20)). Estimators satisfying (14.9-20) are called "best asymptotically normal" (BAN) estimates, and the proof of theorem 14.9-4 shows that using BAN estimates in $NX^2(\hat{\mathbf{p}}, \hat{\boldsymbol{\pi}})$ leads to the same asymptotic chi square distribution for X^2 under the assumption $\boldsymbol{\pi} \in M$.

14.9.3   *The limiting noncentral chi square distribution of X^2*

In example 14.3-6, we examined the limiting distribution of X^2 when $M = \{\boldsymbol{\pi}\}$ is a single point in $\mathscr{S}_T$ (i.e., $s = 0$) and the true value of $\mathbf{p}$ is of the form

$$\mathbf{p}_N = \boldsymbol{\pi} + \boldsymbol{\mu} N^{-1/2},$$

where $\boldsymbol{\mu}$ is a vector such that $\sum_i \mu_i = 0$. In this case, X^2 has a limiting noncentral chi square distribution with $T - 1$ degrees of freedom and noncentrality parameter

$$\psi^2 = \boldsymbol{\mu} \mathbf{D}_\pi^{-1} \boldsymbol{\mu}' = NX^2(\mathbf{p}_N, \boldsymbol{\pi}). \tag{14.9-30}$$

In this formulation, we can study the behavior of X^2 when the model M is incorrect but not very incorrect. The noncentrality parameter ψ^2 measures the discrepancy between the true value of $\mathbf{p}$ and the model M in this simple case.

Now suppose M satisfies conditions 1–6 of Section 14.8.1 for some $0 < s < T - 1$, and suppose the true value of $\mathbf{p}$ is of the form

$$\mathbf{p}_N = \mathbf{f}(\boldsymbol{\varphi}) + \boldsymbol{\mu} N^{-1/2}. \tag{14.9-31}$$

In this situation, the maximum likelihood estimate and the minimum chi square estimate both converge to $\boldsymbol{\varphi}$, and if either is used, then $NX^2(\hat{\mathbf{p}}, \hat{\boldsymbol{\pi}})$ has a limiting noncentral chi square distribution with $T - 1 - s$ degrees of freedom and a noncentrality parameter given by

$$\psi^2 = \boldsymbol{\mu} \mathbf{D}_{\pi(\varphi)}^{-1} \boldsymbol{\mu}' = NX^2(\mathbf{p}_N, \mathbf{f}(\boldsymbol{\varphi})). \tag{14.9-32}$$

For further discussion of this problem and a proof, see Mitra [1958].

14.9.4   *The asymptotic variance of distance measures*

So far we have studied the limiting distribution of X^2 when $\boldsymbol{\pi} \in M$ in Section 14.9.2, and when $\boldsymbol{\pi} \notin M$ but $\boldsymbol{\pi}$ is "near" M in Section 14.9.3. We now examine the case when $\boldsymbol{\pi} \notin M$ and is not near M. This situation arises occasionally, especially when the model is known to be wrong and the actual distance from $\boldsymbol{\pi}$ to M is of interest. We formulate our discussion more generally here than in the two previous sections and consider a general distance function $K(\mathbf{x}, \mathbf{y})$.

We assume that whatever method of estimation is used, $\hat{\boldsymbol{\pi}}$ and $\hat{\mathbf{p}}$ have a limiting joint normal distribution given by

$$\mathcal{L}[\sqrt{N}((\hat{\mathbf{p}}, \hat{\boldsymbol{\pi}}) - (\boldsymbol{\pi}, \mathbf{f}(\boldsymbol{\varphi})))] \to \mathcal{N}(0, \Sigma), \tag{14.9-33}$$

where $\mathbf{f}(\boldsymbol{\varphi})$ is the point in M to which $\hat{\boldsymbol{\pi}}$ converges. We note that since $\boldsymbol{\pi} \notin M$, we must have $\mathbf{f}(\boldsymbol{\varphi}) \neq \boldsymbol{\pi}$. If we now examine $K(\hat{\mathbf{p}}, \hat{\boldsymbol{\pi}})$, we see that it converges to

$K(\pi, \mathbf{f}(\varphi)) > 0$ and that it has a limiting normal distribution which we can find using the δ method. The matrix Σ in (14.9-33) is of the form

$$\Sigma = \begin{pmatrix} \mathbf{D}_\pi - \pi'\pi & \Sigma_{12} \\ \Sigma_{21} & \Sigma_{22} \end{pmatrix}. \tag{14.9-34}$$

We let $(\partial K / \partial \mathbf{p})$ and $(\partial K / \partial \pi)$ be the $1 \times T$ vectors given by:

$$\begin{aligned} \left(\frac{\partial K}{\partial \mathbf{p}}\right)_j &= \frac{\partial K}{\partial x_j}\bigg|_{\mathbf{x}=\pi, \mathbf{y}=\mathbf{f}(\varphi)}, \\ \left(\frac{\partial K}{\partial \pi}\right)_j &= \frac{\partial K}{\partial y_j}\bigg|_{\mathbf{x}=\pi, \mathbf{y}=\mathbf{f}(\varphi)}. \end{aligned} \tag{14.9-35}$$

Then, applying the δ method of theorem 14.6-2, we have

$$\mathscr{L}\left[\sqrt{N}(K(\hat{\mathbf{p}}, \hat{\pi}) - K(\pi, \mathbf{f}(\varphi)))\right] \to \mathscr{N}(0, \sigma^2(\pi, \varphi)),$$

where

$$\sigma^2(\pi, \varphi) = \left(\frac{\partial K}{\partial \mathbf{p}}\right)(\mathbf{D}_\pi - \pi'\pi)\left(\frac{\partial K}{\partial \mathbf{p}}\right)' + 2\left(\frac{\partial K}{\partial \mathbf{p}}\right)\Sigma_{12}\left(\frac{\partial K}{\partial \pi}\right)'$$
$$+ \left(\frac{\partial K}{\partial \pi}\right)\Sigma_{22}\left(\frac{\partial K}{\partial \pi}\right)'. \tag{14.9-36}$$

Implicit in this analysis is the assumption that $K(\mathbf{x}, \mathbf{y})$ has a first-order Taylor expansion at $(\pi, \mathbf{f}(\varphi))$.

While (14.9-36) is fairly compact, it hides a multitude of sins in the matrices Σ_{12} and Σ_{22}. Its use in practice can lead to fairly complicated expressions. For example, Kondo [1929] gives the following expression for $\sigma^2(\pi, \varphi)$ in the special case where (i) $K(\mathbf{x}, \mathbf{y}) = X^2(\mathbf{x}, \mathbf{y})$; (ii) M is the manifold of independence for a two-way $I \times J$ table; (iii) $\hat{\pi}$ is found by maximum likelihood (i.e., $\hat{\pi}_{ij} = \hat{p}_{i+}\hat{p}_{+j}$):

$$\sigma^2(\pi) = 4\sum_{i,j}\frac{\pi_{ij}^3}{\pi_{i+}^2\pi_{+j}^2} - 3\sum_i\frac{\left(\sum_k\dfrac{\pi_{ik}^2}{\pi_{i+}\pi_{+k}}\right)^2}{\pi_{i+}} - 3\sum_j\frac{\left(\sum_l\dfrac{\pi_{lj}^2}{\pi_{l+}\pi_{+j}}\right)^2}{\pi_{+j}}$$
$$+ 2\sum_{i,j}\left(\frac{\pi_{ij}}{\pi_{i+}\pi_{+j}}\right)\left(\sum_k\frac{\pi_{ik}^2}{\pi_{i+}\pi_{+k}}\right)\left(\sum_l\frac{\pi_{lj}^2}{\pi_{l+}\pi_{+j}}\right). \tag{14.9-37}$$

We do not pursue these calculations because they are of limited value in most applications. One note of caution about using (14.9-36) and expressions like (14.9-37): it can happen that $\sigma^2(\pi, \varphi) = 0$, for instance, when $\pi = \mathbf{f}(\varphi)$ (i.e., the model is correct). This means that $K(\hat{\mathbf{p}}, \hat{\pi})$ must be multiplied by something larger than $\sqrt{N}$ in order for it to have a limiting distribution. For example, if we set $\pi_{ij} = \pi_{i+}\pi_{+j}$ in (14.9-37), we get $\sigma^2(\pi) = 0$, and we know that under the assumption of independence, the usual chi square statistic $NX^2(\hat{\mathbf{p}}, \hat{\pi})$ has a $\chi^2_{(I-1)(J-1)}$ distribution. So we should not be surprised to find that $\sqrt{N}X^2(\hat{\mathbf{p}}, \hat{\pi})$ converges to zero, as $\sigma^2(\pi) = 0$ indicates.

14.9.5   *The limiting distribution of X^2 in some complex cases*

In this section we examine what happens to the X^2 statistic when estimators are used which do not satisfy the expansion given in (14.9-30). Thus we consider estimators of φ essentially different from the estimator found by maximizing the likelihood function of the multinomial vector **X**. In all cases we assume that the model is correct, i.e., $\pi \in M$, so that from theorem 14.9-2 we know that the limiting distributions of $NX^2(\hat{\mathbf{p}}, \hat{\pi})$, $NG^2(\hat{\mathbf{p}}, \hat{\pi})$, and $NK_{FT}^2(\hat{\mathbf{p}}, \hat{\pi})$ are all the same. Hence we need only consider X^2 to obtain the limiting distribution for the other two good-ness-of-fit statistics.

When we look for ways of estimating φ that lead to estimators essentially different in their asymptotic properties from the maximum likelihood estimator based on **X**, we find that there are many. Here are some fairly common ones:

(i) $\hat{\pi}$ is found by the method of moments, where the statistics used are not sufficient statistics;

(ii) $\hat{\pi}$ is based on a function of **X** rather than **X** itself (e.g., some of the cells might be combined because the resulting likelihood function is easier to deal with even though some of the original information in **X** is lost);

(iii) $\hat{\pi}$ is based on a random vector of which **X** is a function (e.g., **X** is obtained by grouping the values of a random sample into T categories but $\hat{\theta}$ is estimated from the original sample);

(iv) $\hat{\theta}$ is based on a sample of data that is independent of **X**.

Most reasonable estimators $\hat{\theta}$ have an expansion about φ of the form

$$\hat{\theta} = \varphi + \mathbf{U}_N + o_p(N^{-1/2}), \qquad (14.9\text{-}38)$$

where

$$\mathscr{L}[\sqrt{N}\,\mathbf{U}_N] \to \mathscr{N}(0, \boldsymbol{\Sigma})$$

for some covariance matrix $\boldsymbol{\Sigma}$. However, only those estimates for which $\mathbf{U}_N$ agrees with the first-order term of (14.9-20) produce a goodness-of-fit statistic X^2 which has an asymptotic χ_{T-1-s}^2 distribution. This is because when $\hat{\theta}$ satisfies (14.9-20), it approximately minimizes $X^2(\hat{\mathbf{p}}, \mathbf{f}(\theta))$, and the error in this approximate minimization is small enough to be asymptotically negligible. But when $\hat{\theta}$ does not satisfy (14.9-20) and instead has some other asymptotic expansion about φ, then $NX^2(\hat{\mathbf{p}}, \mathbf{f}(\hat{\theta}))$ is not approximately minimized, even asymptotically. In this case, X^2 has a limiting distribution which is stochastically larger than χ_{T-1-s}^2, and therefore asymptotic significance levels found from the χ_{T-1-s}^2 distribution are too small.

Estimation based on a function of **X**

Methods (i) and (ii) of estimating φ can both be viewed as examples of the same thing. For instance, in example 14.7-2 the method of moments estimator $\hat{\theta} = -\log \hat{p}_1$ corresponds to the maximum likelihood estimator based on the likelihood function of the function of **X** given by

$$g(\mathbf{X}) = (X_1, X_2 + X_3). \qquad (14.9\text{-}39)$$

In either case (i) or (ii), $\hat{\boldsymbol{\theta}}$ is, in general, a function of $\hat{\mathbf{p}}$. Assuming that it is a sufficiently smooth function of $\hat{\mathbf{p}}$, we can assume without loss of generality that $\hat{\boldsymbol{\theta}}$ satisfies the following expansion about $\boldsymbol{\varphi}$:

$$\hat{\boldsymbol{\theta}} = \boldsymbol{\varphi} + (\hat{\mathbf{p}} - \boldsymbol{\pi})\mathbf{D}_\pi^{-1/2}\mathbf{C} + o_p(N^{-1/2}), \qquad (14.9\text{-}40)$$

for some $T \times s$ matrix $\mathbf{C}$. In particular cases $\mathbf{C}$ must be calculated explicitly, but we do not calculate it here except to note that for BAN estimators $\mathbf{C} = \mathbf{A}(\mathbf{A}'\mathbf{A})^{-1}$, i.e., see (14.9-20).

The following theorem, due to Watson [1959], gives some information about the distribution of $NX^2(\hat{\mathbf{p}}, \mathbf{f}(\hat{\boldsymbol{\theta}}))$ when $\hat{\boldsymbol{\theta}}$ satisfies (14.9-40).

THEOREM 14.9-5   *If* $\hat{\boldsymbol{\theta}}$ *satisfies* (14.9-40) *and* $\mathbf{f}$ *satisfies condition* (3) *of Section* 14.8.1, *then*

$$\mathcal{L}[NX^2(\hat{\mathbf{p}}, \mathbf{f}(\hat{\boldsymbol{\theta}}))] \to \chi^2_{T-2s-1} + \sum_{i=1}^{2s} \lambda_i Z_i^2,$$

where the Z_i^2 *are independent* χ_1^2 *variates which are also independent of* χ^2_{T-2s-1} *and the* λ_i *are the eigenvalues of* $\mathbf{I} - \mathbf{Q}$, *with*

$$\mathbf{Q} = \begin{pmatrix} \mathbf{A}'\mathbf{C} - \tfrac{1}{2}\mathbf{A}'\mathbf{A}\mathbf{D} & \mathbf{A}'\mathbf{A} \\ \mathbf{D} - \tfrac{1}{2}\mathbf{D}\mathbf{A}'\mathbf{C} - \tfrac{1}{2}\mathbf{C}'\mathbf{A}\mathbf{D} + \tfrac{1}{4}\mathbf{D}\mathbf{A}'\mathbf{A}\mathbf{D} & \mathbf{C}'\mathbf{A} - \tfrac{1}{2}\mathbf{D}\mathbf{A}'\mathbf{A} \end{pmatrix}, \qquad (14.9\text{-}41)$$

and

$$\mathbf{D} = \mathbf{C}'(\mathbf{I} - \sqrt{\boldsymbol{\pi}}'\sqrt{\boldsymbol{\pi}})\mathbf{C}. \qquad (14.9\text{-}42)$$

For a proof of theorem 14.9-5, see Watson [1959]. The proof is based on corollary 14.9-3 and the fact that from (14.9-40) and regularity condition 3 we can show that

$$\hat{\boldsymbol{\pi}} - \boldsymbol{\pi} = (\hat{\mathbf{p}} - \boldsymbol{\pi})\mathbf{L} + o_p(N^{-1/2}),$$

where $\mathbf{L} = \mathbf{D}_\pi^{-1/2}\mathbf{C}\mathbf{A}'\mathbf{D}_\pi^{1/2}$, and hence

$$\mathcal{L}[\sqrt{N}((\hat{\mathbf{p}}, \hat{\boldsymbol{\pi}}) - (\boldsymbol{\pi}, \boldsymbol{\pi}))] \to \mathcal{N}\left(0, \begin{pmatrix} \mathbf{D}_\pi - \boldsymbol{\pi}'\boldsymbol{\pi} & (\mathbf{D}_\pi - \boldsymbol{\pi}'\boldsymbol{\pi})\mathbf{L} \\ \mathbf{L}'(\mathbf{D}_\pi - \boldsymbol{\pi}'\boldsymbol{\pi}) & \mathbf{L}'(\mathbf{D}_\pi - \boldsymbol{\pi}'\boldsymbol{\pi})\mathbf{L} \end{pmatrix}\right).$$

Watson also shows that

$$\sum_{i=1}^{2s} \lambda_i \geqq s,$$

and hence

$$E\left(\chi^2_{T-2s-1} + \sum_{i=1}^{2s} \lambda_i Z_i^2\right) \geqq T - s - 1 = E(\chi^2_{T-s-1}).$$

This result is in agreement with the observation that if $\mathbf{C} \neq \mathbf{A}(\mathbf{A}'\mathbf{A})^{-1}$, then $\hat{\boldsymbol{\theta}}$ does not minimize $X^2(\hat{\mathbf{p}}, \mathbf{f}(\boldsymbol{\theta}))$, even asymptotically.

Example 14.9-1   Truncated Poisson variates (continued)
We continue examples 14.7-1, 14.7-2, and 14.7-3. Suppose θ is estimated by the simple method-of-moments estimator given by (14.7-14). Then $\hat{\theta} = -\log \hat{p}_1$, and

$$\hat{\boldsymbol{\pi}} = (\hat{p}_1, -\hat{p}_1 \log \hat{p}_1, 1 - \hat{p}_1(1 - \log \hat{p}_1)). \qquad (14.9\text{-}43)$$

Table 14.9-1

θ	$Q(\theta)$
0.6	1.30
0.8	1.41
1.0	1.51
1.2	1.62
1.4	1.73
2.0	2.09
4.0	3.51
6.0	5.24
10.0	9.10

Now if we form a goodness-of-fit statistic by using $X^2(\mathbf{x}, \mathbf{y})$ with this choice of $\hat{\pi}$, we obtain

$$NX^2(\hat{\mathbf{p}}, \hat{\boldsymbol{\pi}}) = N\left[\frac{(\hat{p}_1 - \hat{p}_1)^2}{\hat{p}_1} + \frac{(\hat{p}_2 + \hat{p}_1 \log \hat{p}_1)^2}{-\hat{p}_1 \log \hat{p}_1} + \frac{(\hat{p}_3 - 1 + \hat{p}_1(1 - \log \hat{p}_1))^2}{1 - \hat{p}_1(1 - \log \hat{p}_1)}\right]$$

$$= N\left[\frac{(\hat{p}_2 + \hat{p}_1 \log \hat{p}_1)^2}{-\hat{p}_1 \log \hat{p}_1} + \frac{(\hat{p}_2 + \hat{p}_1 \log \hat{p}_1)^2}{1 - \hat{p}_1(1 - \log \hat{p}_1)}\right]$$

$$= N(\hat{p}_2 + \hat{p}_1 \log \hat{p}_1)^2 \left[\frac{1 - \hat{p}_1}{-\hat{p}_1 \log \hat{p}_1(1 - \hat{p}_1(1 - \log \hat{p}_1))}\right].$$

We can show that $\hat{\theta}$ satisfies (14.9-40) for a particular 3×1 matrix $\mathbf{C}$ (see exercise 7 in Section 14.9.8). Hence from theorem 14.9-5 we have

$$\mathscr{L}[NX^2(\hat{\mathbf{p}}, \hat{\boldsymbol{\pi}})] \to \chi^2_{3-2-1} + \lambda_1(\theta)Z_1^2 + \lambda_2(\theta)Z_2^2,$$

so that $NX^2(\hat{\mathbf{p}}, \hat{\boldsymbol{\pi}})$ has the limiting distribution of $\lambda_1(\theta)Z_1^2 + \lambda_2(\theta)Z_2^2$ for two values $\lambda_1(\theta), \lambda_2(\theta)$, where Z_1^2 and Z_2^2 are independent χ_1^2 variates. We can use theorem 14.9-5 to compute the $\lambda_1(\theta)$ (see exercise 8 in Section 14.9.8), or we may proceed directly as follows.

As $N \to \infty$, we have

$$\frac{1 - \hat{p}_1}{-\hat{p}_1 \log \hat{p}_1(1 - \hat{p}_1(1 - \log \hat{p}_1))} \xrightarrow{p} \frac{1 - p_1}{-p_1 \log p_1(1 - p_1(1 - \log p_1))}. \quad (14.9\text{-}44)$$

Under the truncated Poisson model, the limit in (14.9-44) equals

$$\frac{1 - e^{-\theta}}{\theta e^{-\theta}(1 - (1 - \theta)e^{-\theta})}. \quad (14.9\text{-}45)$$

But if we let $g(\hat{\mathbf{p}}) = \hat{p}_2 + \hat{p}_1 \log \hat{p}_1$, then by the δ method (see exercise 6 in Section 14.9.8), $\sqrt{N}g(\hat{\mathbf{p}})$ is asymptotically normal with mean zero and variance given by

$$\sigma^2(\theta) = p_1(1 + \log p_1)^2 + p_2 - (p_1(1 + \log p_1) + p_2)^2$$

$$= e^{-\theta}[(1 - \theta)^2 + \theta - e^{-\theta}].$$

Hence

$$\mathscr{L}[NX^2(\hat{\mathbf{p}}, \hat{\boldsymbol{\pi}})] \to \mathscr{L}[\chi_1^2 Q(\theta)],$$

where

$$Q(\theta) = \sigma^2(\theta)\left(\frac{1 - e^{-\theta}}{\theta\, e^{-\theta}(1 - (1 + \theta)\, e^{-\theta})}\right)$$

$$= \left(\frac{1 - e^{-\theta}}{1 - (1 + \theta)\, e^{-\theta}}\right)\left(\frac{(1 - \theta)^2 + \theta - e^{-\theta}}{\theta}\right).$$

Hence we see that $\lambda_1(\theta) \equiv 0$ and $\lambda_2(\theta) = Q(\theta)$. We can show that $Q(0) = 1$ and $Q(\infty) = \infty$. Table 14.9-1 gives the values of $Q(\theta)$ for selected values of θ. ∎∎

Estimation based on ungrouped data

The most common example of method (iii) of estimating φ occurs when we group the observed values of a random sample $\mathbf{Y} = (Y_1, \ldots, Y_N)$ into T intervals (or classes) and let $\mathbf{X} = (X_1, \ldots, X_T)$ be the vector of class frequencies. If the Y_i have a common density function $g(y; \boldsymbol{\theta})$ and the T classes are given by a disjoint and exhaustive partition $S_1, \ldots, S_T$ of the sample space of the Y_i, then $\mathbf{f}(\boldsymbol{\theta})$ is given by

$$f_i(\boldsymbol{\theta}) = \int_{S_i} g(y; \boldsymbol{\theta})\, dy \quad i = 1, \ldots, T. \tag{14.9-46}$$

In this case, $\hat{\mathbf{p}} = N^{-1}\mathbf{X}$ is a function of the more basic random vector $\mathbf{Y}$.

Instead of estimating φ using only the multinomial information in $\mathbf{X}$, it is often more natural, easier, and certainly more efficient to maximize the original likelihood function, i.e.,

$$\prod_{i=1}^{N} g(Y_i; \boldsymbol{\theta}). \tag{14.9-47}$$

Let the resulting estimator be $\boldsymbol{\theta}^*$, and set $\boldsymbol{\pi}^* = \mathbf{f}(\boldsymbol{\theta}^*)$. Now $\boldsymbol{\pi}^*$ is a function of the basic variable $\mathbf{Y}$, and generally, $\boldsymbol{\theta}^*$ does not satisfy an expansion like (14.9-40), where the leading stochastic term involves only $\hat{\mathbf{p}}$.

Assuming sufficient regularity conditions on $g(y; \boldsymbol{\theta})$ and $\mathbf{f}(\boldsymbol{\theta})$, Chernoff and Lehmann [1954] prove the following theorem.

THEOREM 14.9-6

$$\mathscr{L}[NX^2(\hat{\mathbf{p}}, \boldsymbol{\pi}^*)] \to \chi_{T-s-1}^2 + \sum_{i=1}^{s} \lambda_i Z_i^2, \tag{14.9-48}$$

where the Z_i^2 are independent χ_1^2 variates which are independent of χ_{T-s-1}^2 and the λ_i satisfy $0 \leq \lambda_i \leq 1$.

The λ_i in (14.9-48) are the roots of a complicated equation. The condition $0 \leq \lambda_i \leq 1$ implies that, in this case, the limiting distribution of X^2 is bounded between χ_{T-s-1}^2 and χ_{T-1}^2. When T is large relative to s, these bounds mean that we may sometimes be able to ignore the λ_i in (14.9-48).

Example 14.9-2   Truncated Poisson variates (Continued)

We continue examples 14.7-1, 14.7-2, 14.7-3, and 14.9-1. If we do not estimate θ from the truncated data, but instead use the maximum likelihood estimate from the original sample (i.e., $\theta^* = \bar{Y}$), then theorem 14.9-6 applies. The limiting distribution of $NX^2(\hat{\mathbf{p}}, \mathbf{f}(\theta^*))$ in this case is

$$\chi_{3-1-1}^2 + \lambda(\theta)Z_2^2 \sim Z_1^2 + \lambda(\theta)Z_2^2,$$

where Z_1^2 and Z_2^2 are independent χ_1^2 variates. Chernoff and Lehmann compute these values for $\lambda(\theta)$:

$$\lambda(1) = 0.12, \qquad \lambda(2) = 0.35. \quad \blacksquare\blacksquare$$

Estimation from an independent sample

Occasionally, we need to know the limiting distribution of X^2 when φ is estimated using information from a second multinomial that is independent of $\mathbf{X}$. To discuss this problem, we set up the following notation. $\mathbf{X}$ is, as usual, an $\mathscr{M}_T(N, \mathbf{f}(\boldsymbol{\theta}))$ variable, and we let $\mathbf{X}^*$ be independent of X and have an $\mathscr{M}_T(N^*, \mathbf{f}(\boldsymbol{\theta}))$ distribution. Thus $\mathbf{X}$ and $\mathbf{X}^*$ have the same dimension and the same model $\mathbf{f}(\boldsymbol{\theta})$, but N and N^* may be different. We let $\tau = N/N^*$, and we always assume in our limiting processes that N and N^* grow together in the same ratio τ. We let $\hat{\mathbf{p}} = N^{-1}\mathbf{X}$ and $\mathbf{p}^* = N^{-1}\mathbf{X}^*$, and let $\hat{\boldsymbol{\theta}}^*$ denote the maximum likelihood estimate of φ based on $\mathbf{X}^*$ alone. Finally, we let $\hat{\boldsymbol{\theta}}^{**}$ be the maximum likelihood estimate of φ based on the pooled sample $\mathbf{X} + \mathbf{X}^*$.

THEOREM 14.9-7   *If regularity conditions 1–6 are satisfied, then*

(i) $\mathscr{L}[NX^2(\hat{\mathbf{p}}, \mathbf{f}(\boldsymbol{\theta}^*))] \to \chi_{T-s-1}^2 + (1 + \tau)\chi_s^2,$

(ii) $\mathscr{L}[NX^2(\hat{\mathbf{p}}, \mathbf{f}(\boldsymbol{\theta}^{**}))] \to \chi_{T-s-1}^2 + \left(1 + \dfrac{\tau}{1 + \tau}\right)\chi_s^2,$

where χ_{T-s-1}^2 and χ_s^2 are independent chi square variates.

Part (i) is due to Chase [1972], and part (ii) is due to Murthy and Gafarian [1970]. These authors consider other complications in addition to the two-sample feature of this problem. While part (ii) is like theorem 14.9-4 in that the actual limiting distribution of X^2 is bounded between the χ_{T-s-1}^2 and χ_{T-1}^2 distributions, in part (i) the situation is quite different, and the limiting distribution of X^2 is actually stochastically larger than that of the χ_{T-1}^2 distribution. We note that when $\tau = 0$ (i.e., $N^* \gg N$), θ^* is essentially equal to φ, so we are back to the case $s = 0$.

14.9.6   *Nested models*

When we use log-linear models, we often have many potential models to consider and need to compute G^2 (or some other goodness-of-fit statistic) for several models. We may then ask what relationships exist among these statistics for several models. In general, there is no simple relationship between the values of G^2 for two different models M_1 and M_2 in $\mathscr{S}_T$. However, there is one important circumstance in which a useful relationship does hold between the two values of G^2, and we devote this section to a discussion of this case.

Two models M_1 and M_2 are *nested*, or, more precisely, M_2 is nested in M_1, if M_2 is completely contained within M_1 when they are viewed as subsets of $\mathscr{S}_T$. The usual way this situation arises is when the parameter vector of M_1 is partitioned into two sets of components, say $(\boldsymbol{\theta}, \boldsymbol{\psi})$, and M_2 is obtained by setting the value of $\boldsymbol{\psi}$ equal to a fixed vector (which without loss of generality we may assume to be zero). Thus the parameter vector of M_2 is $(\boldsymbol{\theta}, 0)$. We denote the maximum likelihood estimate of $(\boldsymbol{\theta}, \boldsymbol{\psi})$ for M_1 by $(\hat{\boldsymbol{\theta}}, \hat{\boldsymbol{\psi}})$ and the corresponding value of $\mathbf{f}(\hat{\boldsymbol{\theta}}, \hat{\boldsymbol{\varphi}})$ by $\hat{\boldsymbol{\pi}}$. For M_2, we denote the maximum likelihood estimate of $\boldsymbol{\theta}$, by $\hat{\hat{\boldsymbol{\theta}}}$ and the corresponding value of $\mathbf{f}(\hat{\hat{\mathbf{\theta}}}, 0)$ by $\hat{\hat{\mathbf{\pi}}}$. Then we set

$$G^2(M_1) = NG^2(\hat{\mathbf{p}}, \hat{\mathbf{\pi}}), \qquad G^2(M_2) = NG^2(\hat{\mathbf{p}}, \hat{\hat{\mathbf{\pi}}}),$$

so that $G^2(M_i)$ is the G^2 goodness-of-fit statistic for model M_i, $i = 1, 2$.

Since M_2 is contained in M_1, we must have

$$G^2(M_2) \geqq G^2(M_1),$$

because $G^2(M_1)$ is minimized over a bigger parameter set. Hence we consider the difference between these two G^2 values.

THEOREM 14.9-8   *If regularity conditions 1–6 are satisfied and the true value of* $\pi = \mathbf{f}(\mathbf{\varphi}, 0)$, *then*

$$\mathscr{L}\,[G^2(M_2) - G^2(M_1)] \to \chi^2_{s_1 - s_2},$$

where s_1 is the dimension of $(\mathbf{\theta}, \mathbf{\psi})$ and s_2 is the dimension of $\mathbf{\theta}$.

Proof   It is easy to verify that

$$G^2(M_2) - G^2(M_1) = 2N \sum_i \hat{p}_i \log\!\left(\frac{\hat{\pi}_i}{\hat{\hat{\pi}}_i}\right). \tag{14.9-49}$$

One attack is to recognize that the right-hand side of (14.9-49) is of the form $-2 \log \lambda$, where λ is the likelihood ratio criterion for testing $\pi \in M_2$ versus $\pi \in M_1$, and to appeal to the general theory of the likelihood ratio test to prove the result (see for example, Rao [1965], p. 351). Since we have not developed this approach, we use a more direct approach. First, we write

$$2N \sum_i \hat{p}_i \log\!\left(\frac{\hat{\pi}_i}{\hat{\hat{\pi}}_i}\right) = NG^2(\hat{\mathbf{\pi}}, \hat{\hat{\mathbf{\pi}}}) + R, \tag{14.9-50}$$

where

$$R = 2N \sum_i \left(\frac{\hat{p}_i - \hat{\pi}_i}{\hat{\pi}_i}\right) \hat{\pi}_i \log(\hat{\pi}_i/\hat{\hat{\pi}}_i). \tag{14.9-51}$$

We note that if regularity conditions 1–6 hold for the M_1 and $\pi \in M_2 \subseteqq M_1$, then

$$\frac{\hat{p}_i - \hat{\pi}_i}{\hat{\pi}_i} = o_p(1), \tag{14.9-52}$$

so that R in (14.9-51) has the form

$$R = o_p(1)NG^2(\hat{\mathbf{\pi}}, \hat{\hat{\mathbf{\pi}}}). \tag{14.9-53}$$

We will show that $NG^2(\hat{\mathbf{\pi}}, \hat{\hat{\mathbf{\pi}}})$ has an asymptotic distribution, so we may conclude that

$$R = o_p(1)O_p(1) = o_p(1),$$

and hence we can ignore R. (In the special case of a log-linear model $R \equiv 0$, see Chapter 4, exercise 2 in Section 4.2.3.)

Before proceeding, we remark that, using an argument like that of theorem 14.9-2, we can show that the limiting distribution of $NG^2(\hat{\mathbf{\pi}}, \hat{\hat{\mathbf{\pi}}})$ is the same as that of $NX^2(\hat{\mathbf{\pi}}, \hat{\hat{\mathbf{\pi}}})$, and hence we will only look at the latter quantity.

From expansion (14.9-20), we have

$$(\hat{\boldsymbol{\theta}}, \hat{\boldsymbol{\psi}}) = (\boldsymbol{\varphi}, 0) + (\hat{\mathbf{p}} - \boldsymbol{\pi})\mathbf{D}_\pi^{-1/2}\mathbf{A}(\mathbf{A}'\mathbf{A})^{-1} + o_p(N^{-1/2}), \qquad (14.9\text{-}54)$$

where

$$\mathbf{A} = \mathbf{D}_\pi^{-1/2}\left(\left(\frac{\partial \mathbf{f}}{\partial \boldsymbol{\theta}}\right), \left(\frac{\partial \mathbf{f}}{\partial \boldsymbol{\psi}}\right)\right)\Bigg|_{\boldsymbol{\theta}=\boldsymbol{\varphi}, \boldsymbol{\psi}=0},$$

and

$$\hat{\hat{\boldsymbol{\theta}}} = \boldsymbol{\varphi} + (\hat{\mathbf{p}} - \boldsymbol{\pi})\mathbf{D}_\pi^{-1/2}\mathbf{B}(\mathbf{B}'\mathbf{B})^{-1} + o_p(N^{-1/2}), \qquad (14.9\text{-}55)$$

where

$$\mathbf{B} = \mathbf{D}_\pi^{-1/2}\left(\frac{\partial \mathbf{f}}{\partial \boldsymbol{\theta}}\right)\Bigg|_{\boldsymbol{\theta}=\boldsymbol{\varphi}, \boldsymbol{\psi}=0}.$$

Furthermore, from condition 3 we have, by the usual argument,

$$\hat{\boldsymbol{\pi}} - \boldsymbol{\pi} = ((\hat{\boldsymbol{\theta}}, \hat{\boldsymbol{\psi}}) - (\boldsymbol{\varphi}, 0))\left(\left(\frac{\partial \mathbf{f}}{\partial \boldsymbol{\theta}}\right), \left(\frac{\partial \mathbf{f}}{\partial \boldsymbol{\psi}}\right)\right)' + o_p(N^{-1/2})$$

$$= (\hat{\mathbf{p}} - \boldsymbol{\pi})\mathbf{D}_\pi^{-1/2}\mathbf{A}(\mathbf{A}'\mathbf{A})^{-1}\mathbf{A}'\mathbf{D}_\pi^{1/2} + o_p(N^{-1/2}).$$

Similarly, we have

$$\hat{\hat{\boldsymbol{\pi}}} - \boldsymbol{\pi} = (\hat{\hat{\boldsymbol{\theta}}} - \boldsymbol{\varphi})\left(\frac{\partial \mathbf{f}}{\partial \boldsymbol{\theta}}\right)' + o_p(N^{-1/2})$$

$$= (\hat{\mathbf{p}} - \boldsymbol{\pi})\mathbf{D}_\pi^{-1/2}\mathbf{B}(\mathbf{B}'\mathbf{B})^{-1}\mathbf{B}'\mathbf{D}_\pi^{1/2} + o_p(N^{-1/2}).$$

Hence we have

$$(\hat{\boldsymbol{\pi}} - \hat{\hat{\boldsymbol{\pi}}})\mathbf{D}_\pi^{-1/2} = (\hat{\boldsymbol{\pi}} - \boldsymbol{\pi})\mathbf{D}_\pi^{-1/2} - (\hat{\hat{\boldsymbol{\pi}}} - \boldsymbol{\pi})\mathbf{D}_\pi^{-1/2} + o_p(N^{-1/2})$$

$$= (\hat{\mathbf{p}} - \boldsymbol{\pi})\mathbf{D}_\pi^{-1/2}[\mathbf{A}(\mathbf{A}'\mathbf{A})^{-1}\mathbf{A}' - \mathbf{B}(\mathbf{B}'\mathbf{B})^{-1}\mathbf{B}'] + o_p(N^{-1/2})$$

$$= (\hat{\mathbf{p}} - \boldsymbol{\pi})\mathbf{D}_\pi^{-1/2}[\mathbf{Q}_A - \mathbf{Q}_B] + o_p(N^{-1/2}),$$

where $\mathbf{Q}_A = \mathbf{A}(\mathbf{A}'\mathbf{A})^{-1}\mathbf{A}'$ and $\mathbf{Q}_B = \mathbf{B}(\mathbf{B}'\mathbf{B})^{-1}\mathbf{B}'$. So we know that

$$\mathscr{L}[\sqrt{N}(\hat{\boldsymbol{\pi}} - \hat{\hat{\boldsymbol{\pi}}})\mathbf{D}_\pi^{-1/2}] \to \mathscr{N}(0, \boldsymbol{\Sigma}),$$

where

$$\boldsymbol{\Sigma} = (\mathbf{Q}_A - \mathbf{Q}_B)(\mathbf{I} - \sqrt{\boldsymbol{\pi}}'\sqrt{\boldsymbol{\pi}})(\mathbf{Q}_A - \mathbf{Q}_B)$$

$$= (\mathbf{Q}_A - \mathbf{Q}_B)(\mathbf{Q}_A - \mathbf{Q}_B), \qquad (14.9\text{-}56)$$

since $\sqrt{\boldsymbol{\pi}}\mathbf{A} = \sqrt{\boldsymbol{\pi}}\mathbf{B} = 0$ as usual. To simplify (14.9-56) we make use of the fact that $\mathbf{Q}_A$ and $\mathbf{Q}_B$ are orthogonal projection operators and the columns of $\mathbf{B}$ are a subset of the columns of $\mathbf{A}$. From this it follows that

$$\mathbf{Q}_A\mathbf{Q}_B = \mathbf{Q}_B\mathbf{Q}_A = \mathbf{Q}_B, \qquad (14.9\text{-}57)$$

and hence we see that $\mathbf{Q}_A - \mathbf{Q}_B$ is symmetric and idempotent. From theorems 14.3-7 and 14.3-9, we then deduce that the eigenvalues of $\mathbf{Q}_A - \mathbf{Q}_B$ are all 0 except for $s_1 - s_2$ unit values, and this fact proves the theorem. ∎

Various applications of this result are given in Chapter 4.

14.9.7 *Approximate percentage points for the χ^2 distribution*

The chi square distribution occurs throughout the analysis of discrete multivariate data, and we must be able to find the significance levels and percentage points for this distribution over a wide range of degrees of freedom. For small numbers of degrees of freedom (≤ 30), tables are readily available. For large numbers of degrees of freedom we can employ various approximations. In this section, we discuss four important normal approximations to the chi square distribution. We let χ_v^2 denote a chi square variate with v degrees of freedom. In all the approximations we let $v \to \infty$.

The easiest and most familiar approximation is based on the representation of a chi square variate as the sum of the squares of independent $\mathcal{N}(0, 1)$ variables. We recall that $E(\chi_v^2) = v$ and $\mathrm{Var}(\chi_v^2) = 2v$, so from the central limit theorem, we have

$$\mathscr{L}\left[\frac{\chi_v^2 - v}{\sqrt{2v}}\right] \to \mathcal{N}(0, 1) \qquad (14.9\text{-}58)$$

as $v \to \infty$. If we let Z_α denote the upper α level probability point for the $\mathcal{N}(0, 1)$ distribution, then (14.9-58) leads to the following approximation:

$$P\{\chi_v^2 > v + Z_\alpha\sqrt{2v}\} \approx \alpha. \qquad (14.9\text{-}59)$$

We call this "approximation 1."

A second approximation, due to Fisher, is based on a variance-stabilizing transformation for the chi square distribution. Employing a version of the δ method similar to the one used in example 14.6-3, we can show that

$$\mathscr{L}[\sqrt{2\chi_v^2} - \sqrt{2v - 1}] \to \mathcal{N}(0, 1) \qquad (14.9\text{-}60)$$

as $v \to \infty$. From (14.9-60) we are led to the following approximation:

$$P\{\chi_v^2 > \tfrac{1}{2}(\sqrt{2v - 1} + Z_\alpha)^2\} \approx \alpha. \qquad (14.9\text{-}61)$$

We call this "approximation 2."

The third approximation we discuss is due to Wilson and Hilferty [1931], and it is the most complicated of the ones we consider. It makes use of the following fact:

$$\mathscr{L}\left[\frac{\sqrt[3]{\dfrac{\chi_v^2}{v}} - \left(1 - \dfrac{2}{9v}\right)}{\sqrt{\dfrac{2}{9v}}}\right] \to \mathcal{N}(0, 1) \qquad (14.9\text{-}62)$$

as $v \to \infty$. This fact leads to the following approximation:

$$P\left\{\chi_v^2 > v\left(1 - \frac{2}{9v} + Z_\alpha\sqrt{\frac{2}{9v}}\right)^3\right\} \approx \alpha, \qquad (14.9\text{-}63)$$

which we call "approximation 3."

In order to give a qualitative comparison of these three approximations, we denote the right-hand sides of the inequalities in (14.9-59), (14.9-61), and (14.9-63) by $p_1(v)$, $p_2(v)$, $p_3(v)$, respectively. The $p_i(v)$ are approximations to the upper point for the χ_v^2 distribution. To compare them we expand each out to order $v^{-1/2}$.

This yields

$$p_1(v) = v + Z_\alpha\sqrt{2v},$$

$$p_2(v) = v + Z_\alpha\sqrt{2v} + \frac{1}{2}(Z_\alpha^2 - 1) - \frac{1}{2}Z_\alpha\frac{1}{\sqrt{2v}} + O(v^{-3/2}),$$

$$p_3(v) = v + Z_\alpha\sqrt{2v} + \frac{2}{3}(Z_\alpha^2 - 1) - \frac{4}{27}Z_\alpha(6 - Z_\alpha^2)\frac{1}{\sqrt{2v}} + O(v^{-1}).$$

We see that $p_2(v)$ and $p_3(v)$ can be viewed as corrections to $p_1(v)$ and that all of these approximations agree to order $v^{1/2}$. However, $p_3(v)$ is not a refinement of $p_2(v)$, as we might have conjectured, since $p_2(v)$ and $p_3(v)$ differ on the choice of $O(1)$ term. When $Z_\alpha > 1$, the $O(1)$ term of $p_3(v)$ is always larger than the corresponding term of $p_2(v)$. This suggests that $p_3(v)$ often yields a smaller probability value than $p_2(v)$ does.

The fourth approximation seeks a simple correction to approximation 1 and is based on the form of $p_2(v)$ and $p_3(v)$. If we ignore terms of order $v^{-1/2}$ and lower, then $p_2(v)$ and $p_3(v)$ are of the form:

$$v + Z_\alpha\sqrt{2v} + c(Z_\alpha^2 - 1), \tag{14.9-64}$$

where c is a free constant which we choose to fit the actual distribution best. The Fisher value $c = 1/2$ does not correct enough, while the Wilson–Hilferty value $c = 2/3$ corrects too much. A simple compromise is to split the difference and take $c = 7/12$. This gives us "approximation 4," as follows:

$$P\{\chi_v^2 > p_4(v)\} \approx \alpha, \tag{14.9-65}$$

where

$$p_4(v) = v + Z_\alpha\sqrt{2v} + \frac{7}{12}(Z_\alpha^2 - 1). \tag{14.9-66}$$

Table 14.9-2 Comparative Significance Levels for Approximations 1, 2, 3, and 4

v	Approx.	0.10	α 0.05	0.01
1	1	0.094	0.0682	0.0383
	2	0.107	0.0615	0.0187
	3	0.104	0.0529	0.0103
	4	0.074	0.0376	0.0088
10	1	0.108	0.0668	0.0257
	2	0.102	0.0546	0.0134
	3	0.101	0.0502	0.0099
	4	0.097	0.0493	0.0108
20	1	0.107	0.0636	0.0217
	2	0.102	0.0536	0.0126
	3	0.100	0.0501	0.0099
	4	0.099	0.0502	0.0108
30	1	0.106	0.0617	0.0197
	2	0.102	0.0530	0.0122
	3	0.100	0.0501	0.0099
	4	0.099	0.0504	0.0108

To give a more quantitative idea of the accuracy of these approximations, in table 14.9-2 we give the actual value of

$$P\{\chi_v^2 > p_i(v)\} \quad i = 1, 2, 3, 4 \tag{14.9-67}$$

for selected values of v and α. We see that, indeed, over this range, p_3 yields smaller (and more accurate) probability values than p_2. Approximations 2 and 3 work remarkably well even in the worst case ($v = 1$), while approximations 1 and 4 fare less well. For most practical work, all of these approximations are sufficiently accurate, except that for very small significance levels and low numbers of degrees of freedom (≤ 10), approximation 2 is quite accurate, while approximation 3 is the most accurate of all and gives very good values even for $v = 1$.

As table 14.9-2 indicates, approximation 4 goes the wrong way for $v = 1$, but by $v = 10$ it is a substantial improvement over approximation 1, which it was designed to correct.

14.9.8 Exercises

1. Fill in the details of the proof of lemma 14.9-1, formula (14.9-8).
2. Show that $\sigma^2(\pi) = 0$ if $\pi_{i+}\pi_{+j} = \pi_{ij}$ in (14.9-37).
3. For a two-way multinomially distributed table $\{X_{ij}\}$, find the limiting distribution of $\{\hat{p}_{i+}\hat{p}_{+j}\}$ for $i = 1, \ldots, I; j = 1, \ldots, J$.
4. Continuing exercise 3, find the joint asymptotic distribution of $\{\hat{p}_{i+}\hat{p}_{+j}\}$ and $\{\hat{p}_{ij}\}$.
5. Use exercises 3 and 4 to derive the Kondo formula (14.9-37) via expression (14.9-36).
6. In example 14.9-1, use the δ method to find the limiting distribution of $\hat{p}_2 + \hat{p}_1 \log \hat{p}_1$.
7. In example 14.9-1, find the 3×1 matrix $\mathbf{C}$ such that $\hat{\theta}$ satisfies (14.9-40).
8. In example 14.9-1, use theorem 14.9-5 to compute the $\lambda_i(\theta)$, and show that the answer agrees with $Q(\theta)$.
9. If χ_v^2 is a chi square variate with v degrees of freedom, show that $\mathscr{L}[\sqrt{2\chi_v^2} - \sqrt{2v - 1}] \to \mathscr{N}(0, 1)$ as $v \to \infty$.
10. Continuing exercise 9, show that

$$\mathscr{L}\left[\frac{\sqrt[3]{\dfrac{\chi_v^2}{v}} - \left(1 - \dfrac{2}{9v}\right)}{\sqrt{\dfrac{2}{9v}}} \right] \to \mathscr{N}(0, 1).$$

11. Let $\mathbf{X} = \{X_{ijk}\}$ be a $2 \times 2 \times K$ table with the multinomial distribution $\mathscr{M}(N, \mathbf{p})$ where

$$\mathbf{p} = \{p_{ijk}\}.$$

Let $u_{12} = u_{123} = 0$ denote the conditional independence model, where 1 and 2 are independent given 3 (see Chapter 3 for this notation). Thus

$$p_{ijk} = \frac{p_{i+k}p_{+jk}}{p_{++k}}.$$

The fitted values under $u_{12} = u_{123} = 0$ are given by

$$\hat{m}_{ijk} = N\frac{\hat{p}_{i+k}\hat{p}_{+jk}}{\hat{p}_{++k}},$$

where $\hat{\mathbf{p}} = \mathbf{X}/N$.

Consider the goodness-of-fit statistic

$$D^2 = \sum_{i,j}\frac{(X_{ij+} - \hat{m}_{ij+})^2}{\hat{m}_{ij+}}.$$

Show that, under the model $u_{12} = u_{123} = 0$,

$$\mathscr{L}[D^2] \to Q(\mathbf{p})\chi_1^2,$$

where

$$Q(\mathbf{p}) = \sigma^2(\mathbf{p})\sum_{i,j}(p_{ij+})^{-1}$$

and

$$\sigma^2(\mathbf{p}) = \sum_k\left(\frac{p_{1+k}p_{2+k}p_{+1k}p_{+2k}}{p_{++k}^3}\right).$$

(Hint: Use the δ method.)

References

ABELSON, R. P. AND TUKEY, J. W. [1963]. Efficient utilization of non-numerical information in quantitative analysis: general theory and the case of simple order. *Ann. Math. Statist.* 34, 1,347–1,369.

AITCHISON, J. AND SILVEY, S. D. [1957]. The generalization of probit analysis to the case of multiple responses. *Biometrika* 44, 131–140.

ALTHAM, P. M. E. [1969]. Exact Bayesian analysis of a 2 × 2 contingency table, and Fisher's "exact" significance test. *J. Roy. Statist. Soc. Ser. B* 31, 261–269.

ALTHAM, P. M. E. [1970a]. The measurement of association of rows and columns for an $r \times s$ contingency table. *J. Roy. Statist. Soc. Ser. B* 32, 63–73.

ALTHAM, P. M. E. [1970b]. The measurement of association in a contingency table: three extensions of the cross-ratios and metric methods. *J. Roy. Statist. Soc. Ser. B* 32, 395–407.

ALTHAM, P. M. E. [1971]. Exact Bayesian analysis of an intraclass 2 × 2 table. *Biometrika* 58, 679–680.

ANDERSON, T. W. [1954]. Probability models for analyzing time changes in attitudes. In *Mathematical Thinking in the Social Sciences*, edited by P. F. Lazarsfeld, pp. 17–66. Glencoe, Ill., The Free Press.

ANDERSON, T. W. AND GOODMAN, L. A. [1957]. Statistical inference about Markov chains. *Ann. Math. Statist.* 28, 89–110.

ANDREWARTHA, H. G. [1961]. *Introduction to the Study of Animal Populations.* Chicago, Univ. of Chicago Press.

ANSCOMBE, F. J. [1950]. Sampling theory of the negative binomial distribution. *Biometrika* 37, 358–382.

ARMITAGE, P., BLENDIS, L. M., AND SMYLLIE, H. C. [1966]. The measurement of observer disagreement in the recording of signs. *J. Roy. Statist. Soc. Ser. A* 129, 98–109.

BARTLETT, M. S. [1935]. Contingency table interactions. *J. Roy. Statist. Soc. Suppl.* 2, 248–252.

BARTLETT, M. S. [1947]. The use of transformations. *Biometrics* 3, 39–52.

BARTLETT, M. S. [1951]. The frequency goodness-of-fit test for probability chains. *Proc. Cambridge Phil. Soc.* 47, 86–95.

BECK, A. [1972]. The ecology of free-roving dogs in Baltimore City. Ph.D. dissertation, School of Hygiene and Public Health, Johns Hopkins University.

BENNETT, B. M. [1972]. Measures for clinicians' disagreements over signs. *Biometrics* 28, 607–612.

BENSON, L. AND OSLICK, A. [1969]. The uses and abuses of statistical methods in studies of legislative behavior: the 1836 Congressional Gag Rule decision as test case. Paper presented to the Conference on Applications of Quantitative Methods to Political, Social, and Economic History, University of Chicago.

BERKSON, J. [1944]. Application of the logistic function to bio-assay. *J. Amer. Statist. Assoc.* 39, 357–365.

BERKSON, J. [1953]. A statistically precise and relatively simple method of estimating the bio-assay with quantal response, based on the logistic function. *J. Amer. Statist. Assoc.* 48, 565–599.

BERKSON, J. [1955]. Maximum likelihood and minimum χ^2 estimates of the logistic function. *J. Amer. Statist. Assoc.* 50, 130–162.

BERKSON, J. [1968]. Application of minimum logit χ^2 estimate to a problem of Grizzle with a notation on the problem of "no interaction." *Biometrics* 24, 75–95.

BERKSON, J. [1972]. Minimum discrimination information, the "no interaction" problem and the logistic function. *Biometrics* 28, 443–468.

BHAPKAR, V. P. [1961]. Some tests for categorical data. *Ann. Math. Statist.* 29, 302–306.

BHAPKAR, V. P. [1966]. A note on the equivalence of two criteria for hypotheses in categorical data. *J. Amer. Statist. Assoc.* 61, 228–235.

BHAPKAR, V. P. [1970]. Categorical data analogs of some multivariate tests. In *Essays in Probability and Statistics*, edited by R. C. Bose et al., pp. 85–110. Chapel Hill, Univ. of North Carolina Press.

BHAPKAR, V. P. AND KOCH, G. [1968]. On the hypotheses of "no interaction" in contingency tables. *Biometrics* 24, 567–594.

BHAT, B. R. AND KULKARNI, S. R. [1966]. Lamp tests of linear and loglinear hypotheses in multinomial experiments. *J. Amer. Statist. Assoc.* 61, 236–245.

BIRCH, M. W. [1963]. Maximum likelihood in three-way contingency tables. *J. Roy. Statist. Soc. Ser. B25*, 220–233.

BIRCH, M. W. [1964a]. A new proof of the Pearson-Fisher theorem. *Ann. Math. Statist.* 35, 718–824.

BIRCH, M. W. [1964b]. The detection of partial association, I: the 2×2 case. *J. Roy. Statist. Soc. Ser. B* 26, 313–324.

BIRCH, M. W. [1965]. The detection of partial association, II: the general case. *J. Roy. Statist. Soc. Ser. B* 27, 111–124.

BISHOP, Y. M. M. [1967]. Multidimensional contingency tables: cell estimates. Ph.D. dissertation, Department of Statistics, Harvard University. Available from University Microfilm Service.

BISHOP, Y. M. M. [1969]. Full contingency tables, logits. and split contingency tables. *Biometrics* 25, 119–128.

BISHOP, Y. M. M. [1971]. Effects of collapsing multidimensional contingency tables. *Biometrics* 27, 545–562.

BISHOP. Y. M. M. AND FIENBERG, S. E. [1969]. Incomplete two-dimensional contingency tables. *Biometrics* 25, 119–128.

BISHOP, Y. M. M. AND MOSTELLER, F. [1969]. Smoothed contingency-table analysis. Chapter IV-3 in *The National Halothane Study*, edited by J. P. Bunker et al., pp. 237–286. Report of the subcommittee on the National Halothane Study of the Committee on anesthesia, Division of Medical Sciences, National Academy of Sciences–National Research Council. National Institutes of Health, National Institute of General Medical Sciences, Bethesda, Md. Washington, D.C., U.S. Government Printing Office.

BLACKWELL, D. AND GIRSHICK, M. A. [1954]. *Theory of Games and Statistical Decisions*, New York, John Wiley.

BLOCH, D. A. AND WATSON, G. S. [1967]. A Bayesian study of the multinomial distribution. *Ann. Math. Statist.* 38, 1,423–1,435.

BOCK, R. D. [1970]. Estimating multinomial response relations. In *Essays in Probability and Statistics*, edited by R. C. Bose et al., pp. 453–479. Chapel Hill, Univ. of North Carolina Press.

BOCK, R. D. [1972]. Estimating item parameters and latent ability when responses are scored in two or more nominal categories. *Psychometrika* 37, 29–51.

BORTKIEWICZ, L. V. [1898]. *Das Gesetz der Kleinen Zahlen*. Leipzig, Teubner.

BOSE, R. C. [1949]. Least squares aspects of the analysis of variance. Inst. Stat. Mimeo. Series 9, University of North Carolina, Chapel Hill.

BOWKER, A. H. [1948]. A test for symmetry in contingency tables. *J. Amer. Statist. Assoc.* 43, 572–574.

BRAUNGART, R. G. [1971]. Family status, socialization, and student politics: a multivariate analysis. *Amer. J. Sociol.* 77, 108–130.

BROWN, D. T. [1959]. A note on approximations to discrete probability distributions. *Information and Control* 2, 386–392.

BUSH, R. AND MOSTELLER, F. [1955]. *Stochastic Models for Learning*. New York, John Wiley.

CASSEDY, J. H. [1969]. *Demography in Early America*. Cambridge, Mass., Harvard Univ. Press.

CAUSSINUS, H. [1966]. Contribution à l'analyse statistique des tableaux de corrélation. *Ann. Fac. Sci. Univ. Toulouse* 29, 77–182.

CHAPMAN, D. C. [1951]. Some properties of the hypergeometric distribution with applications to zoological sample censuses. *Univ. Cal. Publ. Stat.* 1, 131–160.

CHASE, G. R. [1972]. On the chi-square test when the parameters are estimated independently of the sample. *J. Amer. Statist. Assoc.* 67, 609–611.

CHATFIELD, C. [1973]. Statistical inference regarding Markov chain models. *Appl. Statist.* 22, 7–21.

CHATFIELD, C. AND LEMON, R. E. [1970]. Analysing sequences of behavioral events. *J. Theoret. Biol.* 29, 427–445.

CHEN, W. Y., CRITTENDEN, L. B., MANTEL, N., AND CAMERON, W. R. [1961]. Site distribution of cancer deaths in husband-wife and sibling pairs. *J. Natl. Cancer Inst.* 27, 875–892.

CHERNOFF, H. AND LEHMANN, E. L. [1954]. The use of maximum likelihood estimates in χ^2 tests for goodness of fit. *Ann. Math. Statist.* 25, 579–586.

COCHRAN, W. [1950]. The comparison of percentages in matched samples. *Biometrika* 37, 256–266.

COCHRAN, W. [1952]. The χ^2 test of goodness-of-fit. *Ann. Math. Statist.* 23, 315–345.

COCHRAN, W. [1954]. Some methods for strengthening the common χ^2 tests. *Biometrics* 10, 417–451.

COHEN, J. [1960]. A coefficient of agreement for nominal scales. *Educational and Psych. Meas.* 20, 37–46.

COHEN, J. [1968]. Weighted kappa: nominal scale agreement with provision for scaled disagreement or partial credit. *Psych. Bull.* 70, 213–220.

COHEN, J. E. [1973]. Childhood mortality, family size, and birth order in pre-industrial Europe. Unpublished manuscript.

COHEN, J. E. and MacWILLIAMS, H. K. [1974]. The control of foot formation in transplantation experiments with *Hydra viridis*. Unpublished manuscript.

COLEMAN, J. S. [1966]. Measuring concordance in attitudes. Mimeograph, Department of Social Relations, Johns Hopkins University.

CONOVER, W. J. [1968]. Uses and abuses of the continuity correction. *Biometrics* 24, 1,028.

CORMACK, R. M. [1968]. The statistics of capture–recapture methods. *Oceanogr. Mar. Biol. Ann. Rev.* 6, 455–501.

CORNFIELD, J. [1956]. A statistical problem arising from retrospective studies. In *Proceedings of the Third Berkeley Symposium on Mathematical Statistics and Probability*, edited by J. Neyman, Vol. 4, pp. 135–148. Berkeley, Univ. of California Press.

CORNFIELD, J. [1962]. Joint dependence of risk of coronary heart disease on serum cholesterol and systolic blood pressure: a discriminant function analysis. *Federation Proc.* 21, 58–61.

CORNFIELD, J. [1970]. Bayesian estimation for higher order cross-classifications. *Milbank Mem. Fund Quart.* 48, 57–70.

COX, D. R. [1958]. The regression analysis of binary sequences. *J. Roy. Statist. Soc. Ser. B* 20, 215–242.

COX, D. R. [1970]. *The Analysis of Binary Data*. London, Methuen.

COX, D. R. [1972]. The analysis of multivariate binary data. *Appl. Statist.* 21, 113–120.

COX, D. R. and LAUH, E. [1967]. A note on the graphical analysis of multidimensional contingency tables. *Technometrics* 9, 481–488.

CRADDOCK, J. M. AND FLOOD, C. R. [1970]. The distribution of χ^2 statistics in small contingency tables. *Appl. Statist.* 19, 173–181.

CRAIG, W. [1943]. The song of the wood pewee. *Bull. N.Y. State Museum* 334, 1–186.

CRAMÉR, H. [1946]. *Mathematical Methods of Statistics*. Princeton, N.J., Princeton Univ. Press.

DARROCH, J. N. [1958]. The multiple-recapture census, I: estimation of a closed population. *Biometrika* 45, 343–359.

DARROCH, J. N. [1962]. Interaction in multi-factor contingency tables. *J. Roy. Statist. Soc. Ser. B* 24, 251–263.

DARROCH, J. N. AND RATCLIFF, D. [1972]. Generalized iterative scaling for loglinear models. *Ann. Math. Statist.* 43, 1,470–1,480.

DAS, T. [1945]. *The Purums: An Old Kuki Tribe of Manipur*. Calcutta, University of Calcutta.

DAS GUPTA, P. [1964]. On the estimation of the total number of events and the probabilities of detecting an event from information supplied by several agencies. *Calcutta Statist. Assoc. Bull.* 13, 89–100.

DAVIS, J. A. [1967]. A partial coefficient for Goodman and Kruskal's gamma. *J. Amer. Statist. Assoc.* 62, 189–193.

DAVIS, J. A. [1971]. *Elementary Survey Analysis*. Englewood Cliffs, N.J., Prentice-Hall.

DAVIS, P. J. AND RABINOWITZ, P. [1967]. *Numerical Integration*. Waltham, Mass., Blaisdell.

DAWBER, T. R., KANNEL, W. B., AND LYELL, L. P. [1963]. An approach to longitudinal studies in a community: the Framingham study. *Ann. N.Y. Acad. Sci.* 107, 539–556.

DEMING, W. E. AND KEYFITZ, N. [1967]. Theory of surveys to estimate total population. *Proc. World Pop. Conf. 1965* 3, 141–144.

DEMING, W. E. AND STEPHAN, F. F. [1940]. On a least squares adjustment of a sampled frequency table when the expected marginal totals are known. *Ann. Math. Statist.* 11, 427–444.

DICKEY, J. M. [1968a]. Smoothed estimates for multinomial cell probabilities. *Ann. Math. Statist.* 39, 561–566.

DICKEY, J. M. [1968b]. Estimation of disease probabilities conditioned on symptom variables. *Math. Biosci.* 3, 249–269.

DICKEY, J. M. [1969]. Smoothing by cheating. *Ann. Math. Statist.* 40, 1,477–1,482.

ECCLESTON, J. AND HEDAYAT, A. [1972]. Connectedness in experimental design—local and global. (abstract). *Biometrics* 28, 1,167.

ECKLER, A. R. [1973]. The similarity of two poems. In *Statistics by Example: Finding Models*, edited by F. Mosteller et al., pp. 75–87. Reading, Mass., Addison-Wesley.

EFRON, B. AND MORRIS, C. [1971]. Limiting the risk of Bayes and empirical Bayes estimators, I: the Bayes case. *J. Amer. Statist. Assoc.* 66, 807–815.

EFRON, B. AND MORRIS, C. [1972]. Limiting the risk of Bayes and empirical Bayes estimators, II: the empirical Bayes case. *J. Amer. Statist. Assoc.* 67, 130–139.

EINHORN, H. [1972]. Alchemy in the behavioral sciences. *Public Opinion Quart.* 36, 367–378.

ELSTON, R. C. AND BUSH, N. [1964]. The hypotheses that can be tested when there are interactions in an analysis of variance model. *Biometrics* 20, 681–698.

FELLER, W. [1966]. *An Introduction to Probability Theory and Its Applications.* Vol. 2, New York, John Wiley.

FELLER, W. [1968]. *An Introduction to Probability Theory and Its Applications.* Vol. 1, third edition, New York, John Wiley.

FIENBERG, S. E. [1968]. The geometry of an $r \times c$ contingency table. *Ann. Math. Statist.* 39, 1,186–1,190.

FIENBERG, S. E. [1969]. Preliminary graphical analysis and quasi-independence for two-way contingency tables. *Appl. Statist.* 18, 153–168.

FIENBERG, S. E. [1970a]. An iterative procedure for estimation in contingency tables. *Ann. Math. Statist.* 41, 907–917. Corrig. [1971] 42, 1,778.

FIENBERG, S. E. [1970b]. The analysis of multidimensional contingency tables. *Ecology* 51, 419–433.

FIENBERG, S. E. [1970c]. Quasi-independence and maximum likelihood estimation in incomplete contingency tables. *J. Amer. Statist. Assoc.* 65, 1,610–1,616.

FIENBERG, S. E. [1971]. A statistical technique for historians: standardizing tables of counts. *J. Interdisciplinary Hist.* 1, 305–315.

FIENBERG, S. E. [1972a]. The analysis of incomplete multi-way contingency tables. *Biometrics* 28, 177–202. Corrig. [1973] 29, 829.

FIENBERG, S. E. [1972b]. The multiple recapture census for closed populations and incomplete 2^k contingency tables. *Biometrika* 59, 591–603.

FIENBERG, S. E. AND GILBERT, J. P. [1970]. The geometry of a 2×2 contingency table. *J. Amer. Statist. Assoc.* 65, 694–701.

FIENBERG, S. E. AND HOLLAND, P. W. [1970]. Methods for eliminating zero counts in contingency tables. In *Random Counts on Models and Structures*, edited by G. P. Patil, pp. 233–260. University Park, Pennsylvania State Univ. Press.

FIENBERG, S. E. AND HOLLAND, P. W. [1972]. On the choice of flattening constants for estimating multinomial probabilities. *J. Multivariate Anal.* 2, 127–134.

FIENBERG, S. E. AND HOLLAND, P. W. [1973]. Simultaneous estimation of multinomial cell probabilities. *J. Amer. Statist. Assoc.* 68, 683–691.

FINNEY, D. J. [1971]. *Probit Analysis.* Third edition, Cambridge, Cambridge Univ. Press.

FISHER, R. A. [1922]. On the interpretation of chi-square from contingency tables, and the calculation of P. *J. Roy. Statist. Soc.* 85, 87–94.

FISHER, R. A. [1935]. The logic of inductive inference (with discussion). *J. Roy. Statist. Soc.* 98, 39–54.

FISHER, R. A. [1936]. Has Mendel's work been rediscovered? *Ann. Sci.* 1, 115–137.

FISHER, R. A. [1941]. The negative binomial distribution. *Ann. Eugenics* 11, 182–187.

FISHER, R. A. [1953]. *The Design of Experiments.* Edinburgh, Oliver and Boyd.

FLEISS, J. L. [1971]. Measuring nominal scale agreement among many raters. *Psych. Bull.* 76, 378–382.

FLEISS, J. L. [1973]. *Statistical Methods for Rates and Proportions.* New York, John Wiley.

FLEISS, J. L., COHEN, J., AND EVERETT, B. S. [1969]. Large sample standard errors of kappa and weighted kappa. *Psych. Bull.* 72, 323–327.

FOA, U. [1971]. Interpersonal and economic resources. *Science* 171, 345–351.

FRANK, M. and PFAFFMAN, C. [1969]. Taste nerve fibers: a random distribution of sensitivities to four tastes. *Science* 164, 1,183–1,185.

FREEMAN, M. F. AND TUKEY, J. W. [1950]. Transformations related to the angular and the square root. *Ann. Math. Statist.* 21, 607–611.

FRIEDLANDER, D. [1961]. A technique for estimating a contingency table given the marginal totals and some supplementary data. *J. Roy. Statist. Soc. Ser. A* 124, 412–420.

FRYER, J. G. [1971]. On the homogeneity of the marginal distributions of a multidimensional contingency table. *J. Roy. Statist. Soc. Ser. A* 134, 368–371.

GART, J. J. [1962]. On the combination of relative risks. *Biometrics* 18, 601–610.

GART, J. J. [1971]. The comparison of proportions: a review of significance tests, confidence intervals, and adjustments for stratification. *Rev. Int. Statist. Inst.* 39, 148–169.

GEIGER, H. AND WERNER, A. [1924]. Die Zahl der ion radium ausgesandsen α-Teilchen. *Z. Physik* 21, 187–203.

GINI, C. [1912]. Variabilità e mutabilità, contributo allo studio delle distribuzioni; relazione statische. In *Studi Economico—Giuridici della R. Università di Cagliari.*

GLASS, D. V. (ed.) [1954]. *Social Mobility in Britain.* Glencoe, Ill., The Free Press.

GOLDMAN, A. I. [1971]. The comparison of multidimensional rate tables: a simulation study. Ph.D. dissertation, Department of Statistics, Harvard University.

GOOD, I. J. [1953]. On the population frequencies of species and the estimation of population parameters. *Biometrika* 40, 237–264.

GOOD, I. J. [1956]. On the estimation of small frequencies in contingency tables. *J. Roy. Statist. Soc. Ser. B* 18, 113–124.

GOOD, I. J. [1963]. Maximum entropy for hypotheses formulation especially for multidimensional contingency tables. *Ann. Math. Statist.* 34, 911–934.

GOOD, I. J. [1965]. *The Estimation of Probabilities.* Cambridge, Mass., The MIT Press.

GOOD, I. J. [1967]. A Bayesian significance test for multinomial distributions (with discussion). *J. Roy. Statist. Soc. Ser. B* 29, 399–431.

GOODMAN, L. A. [1953]. A further note on Miller's "Finite Markov processes in psychology." *Psychometrika* 18, 245–248.

GOODMAN, L. A. [1961]. Statistical methods for the mover-stayer model. *J. Amer. Statist. Assoc.* 56, 841–868.

GOODMAN, L. A. [1962]. Statistical methods for analyzing processes of change. *Amer. J. Sociol.* 68, 57–78.

GOODMAN, L. A. [1963a]. Statistical methods for the preliminary analysis of transaction flows. *Econometrica* 31, 197–208.

GOODMAN, L. A. [1963b]. On Plackett's test for contingency table interactions. *J. Roy. Statist. Soc. Ser. B* 25, 179–188.

GOODMAN, L. A. [1964a]. A short computer program for the analysis of transaction flows. *Behavioral Sci.* 9, 176–186.

GOODMAN, L. A. [1964b]. Simple methods of analyzing three-factor interaction in contingency tables. *J. Amer. Statist. Assoc.* 59, 319–352.

GOODMAN, L. A. [1965]. On the statistical analysis of mobility tables. *Amer. J. Sociol.* 70, 564–585.

GOODMAN, L. A. [1968]. The analysis of cross-classified data: independence, quasi-independence, and interaction in contingency tables with or without missing cells. *J. Amer. Statist. Assoc.* 63, 1,091–1,131.

GOODMAN, L. A. [1969a]. On the measurement of social mobility: an index of status persistence. *Amer. Social. Rev.* 34, 832–850.

GOODMAN, L. A. [1969b]. On partitioning chi-square and detecting partial association in three-way contingency tables. *J. Roy. Statist. Soc. Ser. B* 31, 486–498.

GOODMAN, L. A. [1970]. The multivariate analysis of qualitative data: interactions among multiple classifications. *J. Amer. Statist. Assoc.* 65, 226–256.

GOODMAN, L. A. [1971a]. The partitioning of chi-square, the analysis of marginal contingency tables, and the estimation of expected frequencies in multidimensional contingency tables. *J. Amer. Statist. Assoc.* 66, 339–344.

GOODMAN, L. A. [1971b]. The analysis of multidimensional contingency tables: stepwise procedures and direct estimation methods for building models for multiple classifications. *Technometrics* 13, 33–61.

GOODMAN L. A. [1972a]. A modified multiple regression approach to the analysis of dichotomous variables. *Amer. Sociol. Rev.* 37, 28–46.

GOODMAN, L. A. [1972b]. Some multiplicative models for the analysis of cross-classified data. In *Proceedings of the Sixth Berkeley Symposium on Mathematical Statistics and Probability*, edited by L. Le Cam et al., Vol. 1, pp. 649–696. Berkeley, Univ. of California Press.

GOODMAN, L. A. AND KRUSKAL, W. H. [1954]. Measures of association for cross-classifications. *J. Amer. Statist. Assoc.* 49, 732–764.

GOODMAN, L. A. AND KRUSKAL, W. H. [1959]. Measures of association for cross-classifications, JI: further discussion and references. *J. Amer. Statist. Assoc.* 54, 123–163.

GOODMAN, L. A. AND KRUSKAL, W. H. [1963]. Measures of association for cross-classifications, III: approximate sampling theory. *J. Amer. Statist. Assoc.* 58, 310–364.

GOODMAN, L. A. AND KRUSKAL, W. H. [1972]. Measures of association for cross-classifications, IV: simplification of asymptotic variances. *J. Amer. Statist. Assoc.* 67, 415–421.

GRIZZLE, J. E. [1967]. Continuity correction in the χ^2 test for 2×2 tables. *Amer. Statist.* 21 (No. 4), 28–32.

GRIZZLE, J. E., STARMER, C. F., AND KOCH, G. G. [1969]. Analysis of categorical data by linear models. *Biometrics* 25, 489–504.

GRIZZLE, J. E. AND WILLIAMS, O. D. [1972]. Loglinear models and tests of independence for contingency tables. *Biometrics* 28, 137–156.

GROSS, P. [1971]. A study of supervisor reliability. Mimeograph, Laboratory of Human Development, Harvard Graduate School of Education, Cambridge, Mass.

GUMBEL, E. J. [1968]. Ladislaus Von Bortkiewicz. In *International Encyclopedia of the Social Sciences*, Vol. 2, pp. 128–131. New York, Macmillan.

HABERMAN, S. J. [1972]. Loglinear fit for contingency tables (Algorithm AS 51). *Appl. Statist.* 21, 218–225.

HABERMAN, S. J. [1973a]. Loglinear models for frequency data: sufficient statistics and likelihood equations. *Ann. Statist.* 1, 617–632.

HABERMAN, S. J. [1973b]. Printing multidimensional tables (Algorithm AS 57). *Appl. Statist.* 22, 118–126.

HABERMAN, S. J. [1974]. *The Analysis of Frequency Data*. Chicago, Univ. of Chicago Press.

HAIGHT, F. A. [1967]. *Handbook of the Poisson Distribution*. New York, John Wiley.

HALDANE, J. B. S. [1937]. The exact value of the moments of the distribution of χ^2, used as a test of goodness-of-fit, when expectations are small. *Biometrika* 29, 133–143.

HALKKA, O. [1964]. Geographical, spatial, and temporal variability in the balanced polymorphism of *Philaenus spumarius*. *Heredity* 19, 383–401.

HANSEN, E. W. [1966]. The development of initial and infant behavior in the rhesus monkey. *Behaviour* 27, 107–149.

HARRIS, J. A. [1910]. On the selective elimination occurring during the development of the fruits of *Staphylea. Biometrika* 7, 452–504.

HARVARD UNIVERSITY COMPUTATION LABORATORY [1955]. *Tables of the Cumulative Binomial Probability Distribution*. Cambridge, Mass., Harvard Univ. Press.

HITCHCOCK, S. E. [1966]. Tests of hypotheses about the parameters of the logistic function. *Biometrika* 53, 535–544.

HOEL, P. G. [1954]. A test for Markov chains. *Biometrika* 41, 430–433.

HOGG, R. V. AND CRAIG, A. T. [1968]. *Introduction to Mathematical Statistics*. Second edition, pp. 223–236. New York, Macmillan.

HOLLAND, P. W. [1967]. A variation on the minimum chi-square test. *J. Math. Psych.* 4, 377–413.

HOLLAND, P. W. [1973]. Covariance stabilizing transformations. *Ann. Statist.* 1, 84–92.

HOLLAND, P. W. AND SUTHERLAND, M. R. [1972a]. The risk of the Fienberg-Holland estimator. Memo NS-160, Department of Statistics, Harvard University.

HOLLAND, P. W. AND Sutherland, M. R. [1972b]. Minimal multinomial risk. Memo NS-163, Department of Statistics, Harvard University.

IRELAND, C. T., KU, H. H., AND KULLBACK, S. [1969]. Symmetry and marginal homogeneity of an $r \times r$ contingency table. *J. Amer. Statist. Assoc.* 64, 1,323–1,341.

IRELAND, C. T. AND KULLBACK, S. [1968a]. Minimum discrimination information estimation. *Biometrics* 24, 707–713.

IRELAND, C. T. AND KULLBACK, S. [1968b]. Contingency tables with given marginals. *Biometrika* 55, 179–188.

JAMES, W. AND STEIN, C. [1961]. Estimation with quadratic loss. In *Proceedings of the Fourth Berkeley Symposium on Mathematical Statistics and Probability*, edited by J. Neyman, Vol. 1, pp. 361–379. Berkeley, Univ. of California Press.

JOHNSON, B. M. [1971]. On the admissible estimators for certain fixed sample binomial problems. *Ann. Math. Statist.* 42, 1,579–1,587.

JOHNSON, N. J. AND KOTZ, S. [1969]. *Discrete Distributions*. Boston, Houghton Mifflin.

JOHNSON, W. D. AND KOCH, G. G. [1971]. A note on the weighted least squares analysis of the Ries-Smith contingency table data. *Technometrics* 13, 438–447.

KALTON, G. [1968]. Standardization: a technique to control for extraneous variables. *Appl. Statist.* 17, 118–136.

KASTENBAUM, M. A. [1958]. Estimation of relative frequencies of four sperm types in *Drosophila melanogaster*. *Biometrics* 14, 223–228.

KATZ, L. AND PROCTOR, C. H. [1959]. The concept of configuration of interpersonal relations in a group as a time-dependent stochastic process. *Psychometrika* 24, 317–327.

KEMENY, J. G. [1953]. The use of simplicity in induction. *Phil. Rev.* 62, 391–408.

KITAGAWA, E. M. [1964]. Standardized comparisons in population research. *Demography* 1, 296–315.

KLEINMAN, J. C. [1973]. Proportions with extraneous variance: single and independent samples. *J. Amer. Statist. Assoc.* 68, 46–55.

KOCH, G. G. AND REINFURT, D. W. [1971]. The analysis of categorical data from mixed models. *Biometrics* 27, 157–173.

KONDO, T. [1929]. On the standard error of the mean square contingency. *Biometrika* 21, 377–428.

KORFF, F. A., TABACK, M. A. M., AND BEARD, J. H. [1952]. A coordinated investigation of a food poisoning outbreak. *Public Health Reports* 67, No. 6, Sept., 909–913.

KRUSKAL, J. B. [1964a]. Multidimensional scaling by optimizing goodness-of-fit to a nonmetric hypothesis. *Psychometrika* 29, 1–28.

KRUSKAL, J. B. [1964b]. Nonmetric multidimensional scaling: a numerical method. *Psychometrika* 29, 115–130.

KRUSKAL, W. H. [1958]. Ordinal measures of association. *J. Amer. Statist. Assoc.* 53, 814–861.

KU, H. H. AND KULLBACK, S. [1968]. Interaction in multidimensional contingency tables: an information theoretic approach. *J. Res. Nat. Bur. Standards Sect. B* 72, 159–199.

KU, H. H., VARNER, R. N., AND KULLBACK, S. [1971]. Analysis of multidimensional contingency tables. *J. Amer. Statist. Assoc.* 66, 55–64.

KUČERA, H. AND FRANCIS, W. N. [1967]. *Computational Analysis of Present-Day American English*. Providence, R.I., Brown Univ. Press.

KULLBACK, S. [1959]. *Information Theory and Statistics*, New York, John Wiley. Reprinted [1968]. New York, Dover Publications.

KULLBACK, S. [1971]. Marginal homogeneity of multidimensional contingency tables. *Ann. Math. Statist.* 42, 594–606.

KULLBACK, S., KUPPERMAN, M., AND KU, H. H. [1962a]. An application of information theory to the analysis of contingency tables, with a table of $2N \ln N$, $N = 1\ (1)10{,}000$. *J. Res. Nat. Bur. Standards Sect. B* 66, 217–243.

KULLBACK, S., KUPPERMAN, M., and KU, H. H. [1962b]. Tests for contingency tables in Markov chains. *Technometrics* 4, 573–608.

LANCASTER, H. O. [1951]. Complex contingency tables treated by partition of χ^2. *J. Roy. Statist. Soc. Ser. B* 13, 242–249.

LANCASTER, H. O. [1969]. *The Chi-Squared Distribution*. New York, John Wiley.

LANCASTER, H. O. [1971]. The multiplicative definition of interaction. *Austral. J. Statist.* 13, 36–44.

LARNTZ, K. [1974]. Reanalysis of Vidmar's data on the effects of decision alternatives on verdicts of simulated jurors. *J. Personality and Social Psych.* 24, in press.

LAW ENFORCEMENT ASSISTANCE ADMINISTRATION [1972]. San Jose methods test of known crime victims. Statist. Tech. Report No. 1, U.S. Govt. Printing Office.

LAZARSFELD, P. F. [1969]. Problems in the analysis of mutual interaction between two variables. Bureau of Applied Social Research, Columbia University.

LAZARSFELD, P. F., BERELSON, B., AND GAUDET, H. [1948]. *The People's Choice*. Second edition, New York, Columbia Univ. Press.

LAZARSFELD, P. F. AND HENRY, N. W. [1968]. *Latent Structure Analysis*. Boston, Houghton Mifflin.

LEE, T. C., JUDGE, G. G., AND ZELLNER, A. [1970]. *Estimating the Parameters of the Markov Probability Model from Aggregate Time Series Data*. Amsterdam, North Holland.

LEHMANN, E. L. [1959]. *Testing Statistical Hypotheses*. New York, John Wiley.

LEONARD, T. [1972]. Bayesian methods for binomial data. *Biometrika* 59, 581–589.

LEVINE, J. H. [1967]. A measure of association for intergenerational status mobility. Unpublished manuscript, Harvard University.

LI, C. C. AND MANTEL, N. [1968]. A simple method of estimating the segregation ratio under complete ascertainment. *Amer. J. Hum. Genet.* 20, 61–81.

LIEBERMAN, G. J. AND OWEN, D. B. [1961]. *Tables of the Hypergeometric Probability Distribution*. Stanford, Calif., Stanford Univ. Press.

LIGHT, R. J. [1969]. Analysis of variance for categorical data, with applications to agreement and association. Ph.D. dissertation, Department of Statistics, Harvard University.

LIGHT, R. J. [1971]. Measures of response agreement for qualitative data: some generalizations and alternatives. *Psych. Bull.* 76, 365–377.

LIGHT, R. J. AND MARGOLIN, B. H. [1971]. An analysis of variance for categorical data. *J. Amer. Statist. Assoc.* 66, 534–544.

LIN, Y. S. [1974]. Statistical measurement of agreement. Ph.D. dissertation, School of Statistics, University of Minnesota.

LINCOLN, F. C. [1930]. Calculating waterfowl abundance on the basis of banding returns. *Cir. U.S. Dept. Agric.* 118, 1–4.

LINDLEY, D. V. [1962]. Discussion of a paper by C. Stein. *J. Roy. Statist. Soc. Ser. B* 24, 285–287.

LINDLEY, D. V. [1964]. The Bayesian analysis of contingency tables. *Ann. Math. Statist.* 35, 1,622–1,643.

LINDLEY, D. V. [1971]. The estimation of many parameters. In *Foundations of Statistical Inference*, edited by V. P. Godambe and D. A. Sprott, pp. 435–455. Toronto, Holt, Rinehart, and Winston.

LINDLEY, D. V. AND SMITH, A. F. M. [1972]. Bayesian estimates for the linear model (with discussion). *J. Roy. Statist. Soc. Ser. B* 34, 1–42.

LOEVE, M. [1963]. *Probability Theory.* Princeton, N.J., Van Nostrand.

MADANSKY, A. [1959]. Least squares estimation in finite Markov processes. *Psychometrika* 24, 137–144.

MADANSKY, A. [1963]. Tests of homogeneity for correlated samples. *J. Amer. Statist. Assoc.* 58, 97–119.

MANTEL, N. [1951]. Evaluation of a class of diagnostic tests. *Biometrics* 7, 240–246.

MANTEL, N. [1970]. Incomplete contingency tables. *Biometrics* 26, 291–304.

MANTEL, N. AND GREENHOUSE, S. W. [1968]. What is the continuity correction? *Amer. Statist.* 22 (No. 5), 27–30.

MANTEL, N. AND HAENSZEL, W. [1959]. Statistical aspects of the analysis of data from retrospective studies of disease. *J. Natl. Cancer Inst.* 22, 719–748.

MANTEL, N. AND HALPERIN, M. [1963]. Analysis of birth-rank data. *Biometrics* 19, 324–340.

MARGOLIN, B. H. AND LIGHT, R. J. [1974]. An analysis of variance for categorical data, II: small sample comparisons with chi-square and other competitors. *J. Amer. Statist. Assoc.* 69, in press.

MARX, T. J. [1973]. Statistical measurement of agreement for data in the nominal scale. Ed. D. dissertation, Harvard Graduate School of Education, Cambridge, Mass.

MARX, T. J. AND LIGHT, R. J. [1973]. A many observer agreement measure for qualitative response data. Mimeograph, Laboratory of Human Development, Harvard Graduate School of Education, Cambridge, Mass.

MAXWELL, A. E. [1961]. *Analysing Qualitative Data.* London, Methuen.

MENDEL, G. [1967]. *Experiments in Plant Hybridisation.* Cambridge, Mass., Harvard Univ. Press.

MENZERATH, P. [1950]. Typology of languages. *J. Acous. Soc. Amer.* 22, 698–701.

MILLER, G. A. [1952]. Finite Markov processes in psychology. *Psychometrika* 18, 149–169.

MITRA, S. K. [1958]. On the limiting power function of the frequency chi-square test. *Ann. Math. Statist.* 29, 1,221–1,233.

MOLINA, E. C. [1942]. *Poisson's Exponential Binomial Limit.* Princeton, N.J., Van Nostrand.

MORRIS, C. [1966]. Admissible Bayes procedures and classes of epsilon Bayes procedures for testing hypotheses in a multinomial distribution. Technical Report ONR-55, Department of Statistics, Stanford University.

MORRIS, C. [1969]. Central limit theorems for multinomial sums. Rand Corporation Technical Report RM-6026-PR, Santa Monica, Calif.

MORRISON, A. S., BLACK, M. M., LOWE, C. R., MACMAHON, B., and YUASA, S. [1973]. Some international differences in histology and survival in breast cancer. *Int. J. Cancer* 11, 261–267.

MOSTELLER, F. [1968]. Association and estimation in contingency tables. *J. Amer. Statist. Assoc.* 63, 1–28.

MOSTELLER, F. [1970]. Collegiate football scores, U.S.A. *J. Amer. Statist. Assoc.* 65, 35–48.

MOSTELLER, F., AND ROURKE, R. E. K. [1973]. *Sturdy Statistics.* Reading, Mass., Addison-Wesley.

MOSTELLER, F., ROURKE, R. E. K., AND THOMAS, G. B., JR. [1970]. *Probability with Statistical Applications.* Second edition, Reading, Mass., Addison-Wesley.

MOSTELLER, F. AND TUKEY, J. W. [1968]. Data analysis, including statistics. In *Revised Handbook of Social Psychology*, edited by G. Lindzey and E. Aronson, Chapter 10. Reading, Mass., Addison-Wesley.

MOSTELLER, F. AND WALLACE, D. L. [1964]. *Inference and Disputed Authorship: The Federalist.* Reading, Mass., Addison-Wesley.

MOSTELLER, F. AND YOUTZ, C. [1961]. Tables of the Freeman-Tukey transformations for the binomial and Poisson distributions. *Biometrika* 48, 433–440.

MULLER, T. P. AND MAYHALL, J. T. [1971]. Analysis of contingency table data on *torus mandibularis* using a loglinear model. *Amer. J. Phys. Anthrop.* 34, 149–154.

MURTHY, V. K. AND GAFARIAN, A. V. [1970]. Limiting distributions of some variations of the chi-square statistics. *Ann. Math. Statist.* 41, 188–194.

NATIONAL BUREAU OF STANDARDS [1950]. *Tables of the Binomial Probability Distribution.* Washington, U.S. Govt. Printing Office.

NERLOVE, M. AND PRESS, S. J. [1973]. Univariate and multivariate log-linear and logistic models. Rand Corporation Technical Report R-1306-EDA/NIH, Santa Monica, Calif.

NEWMAN, E. B. [1951a]. Computational methods useful in analyzing series of binary data. *Amer. J. Psych.* 64, 252–262.

NEWMAN, E. B. [1951b]. The pattern of vowels and consonants in various languages. *Amer. J. Psych.* 64, 369–379.

NEYMAN, J. [1949]. Contributions to the theory of the χ^2 test. *Proceedings of the First Berkeley Symposium on Mathematical Statistics and Probability*, edited by J. Neyman, pp. 230–273. Berkeley, Univ. of California Press.

NEYMAN, J. [1965]. Certain chance mechanisms involving discrete distributions. In *Classical and Contagious Discrete Distributions*, edited by G. P. Patil, pp. 4–14. Oxford, Pergamon.

NOVICK, M. R. [1970]. Bayesian considerations in educational information systems. ACT Research Report no. 38, Iowa City, American College Testing Program.

NOVICK, M. R., JACKSON, P. H., THAYER, D. T., AND COLE, N. S. [1971]. Applications of Bayesian methods to the prediction of educational performance. ACT Research Report no. 42, Iowa City, American College Testing Program.

NOVICK, M. R., LEWIS, C., AND JACKSON, P. H. [1973]. Estimation of proportions in m groups. *Psychometrika* 38, 19–46.

NOVITSKI, E. AND SANDLER, I. [1957]. Are all products of spermatogenesis regularly functional? *Proc. Nat. Acad. Sci.* 43, 318–324.

ODOROFF, C. L. [1970]. A comparison of minimum logit chi-square estimation and maximum likelihood estimation in $2 \times 2 \times 2$ and $3 \times 2 \times 2$ contingency tables: tests for interaction. *J. Amer. Statist. Assoc.* 65, 1,617–1,631.

OLSON, W. C. [1929]. *The Measurement of Nervous Habits in Normal Children.* Minneapolis, Univ. of Minneapolis Press.

OWEN, D. B. [1963]. *Handbook of Statistical Tables.* Reading, Mass., Addison-Wesley.

PATIL, G. P. AND JOSHII, S. W. [1968]. *A Dictionary and Bibliography of Discrete Distributions.* London, Oliver and Boyd.

PEARSON, E. S. AND HARTLEY, H. O. [1958]. *Biometrika Tables for Statisticians.* Second edition, London, Cambridge Univ. Press.

PEARSON, K. [1904a]. On the laws of inheritance in man, II: on the inheritance of the mental and moral characters in man, and its comparison with the inheritance of physical characters. *Biometrika* 3, 131–190.

PEARSON, K. [1904b]. On the theory of contingency and its relation to association and normal correlation. *Draper's Co. Res. Mem. Biometric Ser. 1.* Reprinted [1948] in *Karl Pearson's Early Papers*, Cambridge, Cambridge University Press.

PEIZER, D. B. AND PRATT, J. [1968]. A normal approximation for binomial, F, beta, and other common related tail probabilities, I. *J. Amer. Statist. Assoc.* 63, 1,416–1,456.

PETERSEN, C. G. J. [1896]. The yearly immigration of young plaice into the Limfjord from the German Sea. *Rep. Dan. Biol. Stn. (1895)* 6, 5–84.

PLACKETT, R. L. [1962]. A note on interactions in contingency tables. *J. Roy. Statist. Soc. Ser. B* 24, 162–166.

PLACKETT, R. L. [1964]. The continuity correction in 2×2 tables. *Biometrika* 5, 327–337.

PLACKETT, R. L. [1965]. A class of bivariate distributions. *J. Amer. Statist. Assoc.* 60, 516–522.

PLACKETT, R. L. [1971]. The analysis of contingency tables. Unpublished manuscript.

PLOOG, D. W. [1967]. The behavior of squirrel monkeys (*Saimiri sciureus*) as revealed by sociometry, bioacoustics, and brain stimulation. In *Social Communication Among Primates*, edited by S. Altmann, pp. 149–184. Chicago, Univ. of Chicago Press.

PRATT, J. [1959]. On a general concept of "in probability." *Ann. Math. Statist.* 20, 549–558.

RAO, C. R. [1961]. Asymptotic efficiency and limiting information. In *Proceedings of the Fourth Berkeley Symposium on Mathematical Statistics and Probability*, edited by J. Neyman, Vol. 1, pp. 531–545. Berkeley, Univ. of California Press.

RAO, C. R. [1962]. Efficient estimates and optimum inference procedures in large samples (with discussion). *J. Roy. Statist. Soc. Ser. B* 24, 46–72.

RAO, C. R. [1965]. *Linear Statistical Inference and Its Applications.* New York, John Wiley.

RASCH, G. [1960]. *Probabilistic Models for Some Intelligence and Attainment Tests.* Copenhagen, Nielson and Lydiche.

RIES, P. N. AND SMITH, H. [1963]. The use of chi-square for preference testing in multidimensional problems. *Chem. Eng. Progress* 59, 39–43.

ROBBINS, H. [1955]. An empirical Bayes approach to statistics. In *Proceedings of the Third Berkeley Symposium on Mathematical Statistics and Probability*, edited by J. Neyman, Vol. 1, pp. 157–164. Berkeley, Univ. of California Press.

ROMIG, H. G. [1947]. *50–100 Binomial Tables.* New York, John Wiley.

ROSCOE, J. T. AND BYARS, J. A. [1971]. Sample size restraints commonly imposed on the use of the chi-square statistics. *J. Amer. Statist. Assoc.* 66, 755–759.

ROSEN, B. [1969]. Exchange theory and the evaluation of helping behavior. Ph.D. dissertation, Department of Psychology, Wayne State University.

ROY, S. N. AND KASTENBAUM, M. A. [1956]. On the hypothesis of no "interaction" in a multi-way contingency table. *Ann. Math. Statist.* 27, 749–757.

RUDOLF, G. J. [1967]. A quasi-multinomial type of contingency table. *S. Afr. J. Statist.* 1, 59–65.

SAMEJIMA, F. [1969]. Estimation of latent ability using a response pattern of graded scores. *Psychometrika Mono. Suppl.*, no. 17.

SAMEJIMA, F. [1972]. A general model for free-response data. *Psychometrika Mono. Suppl.*, no. 18.

SANATHANAN, L. [1972a]. Estimating the size of a multinomial population. *Ann. Math. Statist.* 43, 142–152.

SANATHANAN, L. [1972b]. Models and estimation methods in visual scanning experiments. *Technometrics* 14, 813–829.

SANATHANAN, L. [1973]. A comparison of some models in visual scanning experiments. *Technometrics* 15, 67–78.

SAVAGE, I. R. [1973]. Incomplete contingency tables: condition for the existence of unique MLE. In *Mathematics and Statistics. Essays in Honor of Harold Bergström*, edited by P. Jagars and L. Råde, pp. 87–99. Göteborg, Sweden, Chalmers Institute of Technology.

SAVAGE, I. R. AND DEUTSCH, K. W. [1960]. A statistical model: the gross analysis of transaction flows. *Econometrika* 28, 551–572.

SCHOENER, T. W. [1970]. Nonsynchronous spatial overlap of lizards in patchy habitats. *Ecology* 51, 408–418.

SCOTT, W. A. [1955]. Reliability of content analysis; the case of nominal scale coding. *Public Opinion Quart.* 19, 321–324.

SEKAR, C. C. AND DEMING, W. E. [1949]. On a method of estimating birth and death rates and the extent of registration. *J. Amer. Statist. Assoc.* 44, 101–115.

SELESNICK, H. L. [1970]. The diffusion of crisis information: a computer simulation of Soviet mass media exposure during the Cuban missile crisis and the aftermath of President Kennedy's assassination. Ph.D. dissertation, Department of Political Science, Massachusetts Institute of Technology.

SHEPARD, R. N. [1962a]. The analysis of proximities: multidimensional scaling with an unknown distance function, I. *Psychometrika* 27, 125–140.

SHEPARD, R. N. [1962b]. The analysis of proximities: multidimensional scaling with an unknown distance function, II. *Psychometrika* 27, 219–246.

SKARIN, A. T., PINKUS, G. S., MYEROWITZ, R. L., BISHOP, Y. M. M., AND MOLONEY, W. C. [1973]. Combination chemotherapy of advanced lymphocytic lymphoma: importance of histologic classification in evaluating response. *Cancer* 34, 1023–1029.

SMITH, D. H. [1973]. Volunteer activity in eight Massachusetts towns. Center for a Voluntary Society, Washington, D.C.

SONQUIST, J. A. [1969]. Finding variables that work. *Public Opinion Quart.* 33, 83–95.

SONQUIST, J. A. AND MORGAN, J. N. [1964]. The detection of interaction effects; a report on a computer program for the selection of optimal combinations of explanatory variables. Institute for Social Research, University of Michigan.

SPAGNUOLO, M., PASTERNACK, B., AND TARANTA, A. [1971]. Risk of rheumatic fever recurrences after streptococcal infections, prospective study of clinical and social factors. *New Eng. J. Med.* 285, 641–647.

STEIN, C. [1962]. Confidence sets for the mean of a multivariate normal distribution (with discussion). *J. Roy. Statist. Soc. Ser. B* 24, 265–296.

STEYN, H. S. [1955]. On discrete multivariate probability functions of hypergeometric type. *Indag. Math.* 17, 588–595.

STEYN, H. S. [1959]. On χ^2 tests for contingency tables of negative multinomial types. *Statistica Neerlandica* 13, 433–444.

STOUFFER, S. A., SUCHMAN, E. A. DEVINNEY, L. C., STAR, S. A., AND WILLIAMS, R. M., JR. [1949]. *The American Soldier.* Vol. 1, Princeton, N.J., Princeton Univ. Press.

STUART, A. [1953]. The estimation and comparison of strengths of association in contingency tables. *Biometrika* 40, 105–110.

STUART, A. [1955]. A test for homogeneity of the marginal distributions in a two-way classification. *Biometrika* 42, 412–416.

SUTHERLAND, M. [1974]. Estimation in large sparse multinomials. Ph.D. dissertation, Department of Statistics, Harvard University.

SUTHERLAND, M., HOLLAND, P., AND FIENBERG, S. E. [1974]. Combining Bayes and frequency approaches to estimate a multinomial parameter. In *Studies in Bayesian Econometrics and Statistics,* edited by S. E. Fienberg and A. Zellner. pp. 585–617. Amsterdam, North Holland.

SVALASTOGA, K. [1959]. *Prestige, Class, and Mobility.* London, Heinemann.

TALLIS, G. M. [1962]. The maximum likelihood estimation of correlation from contingency tables. *Biometrics* 18, 342–353.

THEIL, H. [1970]. On estimation of relationships involving qualitative variables. *Amer. J. Sociol.* 76, 103–154.

TRUETT, J., CORNFIELD, J., AND KANNEL, W. [1967]. A multivariate analysis of the risk of coronary heart disease in Framingham. *J. Chron. Dis.* 20, 511–524.

TRYBULA, S. [1958]. Some problems of simultaneous minimax estimation. *Ann. Math. Statist.* 29, 245–253.

TUKEY, J. W. [1971]. *Exploratory Data Analysis.* Vol. 3. Reading, Mass., Addison-Wesley.

U.S. OFFICE OF THE CHIEF OF ORDNANCE [1952]. *Tables of the Cumulative Binomial Probabilities.* Washington, U.S. Govt. Printing Office.

VAN EEDEN, C. [1965]. Conditional limit distributions for the entries in a $2 \times k$ contingency table. In *Classical and Contagious Discrete Distributions,* edited by G. P. Patil, pp. 123–126. Oxford, Pergamon.

VECHIO, T. J. [1966]. Predictive value of a single diagnostic test in unselected populations. *New Eng. J. Med.* 274, 1,171–1,173.

VIDMAR, N. [1972]. Effects of decision alternatives on the verdicts and social preceptions of simulated jurors. *J. Personality and Social Psych.* 22, 211–218.

WAITE, H. [1915]. Association of finger prints. *Biometrika* 10, 421–478.

WALKER, S. H. AND DUNCAN, D. B. [1967]. Estimation of the probability of an event as a function of several independent variables. *Biometrika* 54, 167–179.

WATSON, G. S. [1959]. Some recent results in chi-square goodness-of-fit tests. *Biometrics* 15, 440–468.

WATSON, G. S. [1965]. Some Bayesian methods related to χ^2. *Bull. Internatl. Statist. Inst. (Proc. 35th Session)* 61, 64–76.

WEEKS, D. L. AND WILLIAMS, D. R. [1964]. A note on the determination of connectedness in an N-way cross-classification. *Technometrics* 6, 319–324. Errata [1965], *Technometrics* 7, 281.

WHITE, H. C. [1963]. *An Anatomy of Kinship,* p. 138. Englewood Cliffs, N.J., Prentice-Hall.

WILKS, S. S. [1962]. *Mathematical Statistics.* New York, John Wiley.

WILLICK, D. H. [1970]. Foreign affairs and party choice. *Amer. J. Sociol.* 75, 530–549.

WILSON, E. B. AND HILFERTY, M. M. [1931]. The distribution of chi-square. *Proc. Nat. Acad. Sci.* 17, 694.

WINSOR, C. P. [1947]. Quotations: Das Gesetz der Kleinen Zahlen. *Human Biol.* 19, 154–161.

WITTES, J. T. [1970]. Estimation of population size: the Bernoulli census. Ph.D. dissertation, Department of Statistics, Harvard University.

WITTES, J. T. [1974]. Applications of a multinomial capture-recapture model to epidemiological data. *J. Amer. Statist. Assoc.* 69, 93–97.

WITTES, J. T., COLTON, T., AND SIDEL, V. W. [1974]. Capture-recapture methods for assessing the completeness of case ascertainment when using multiple information sources. *J. Chron. Dis.* 27, 25–36.

WORCESTER, J. [1971]. The relative odds in the 2^3 contingency table. *Amer. J. Epidemiology* 93, 145–149.

YARNOLD, J. K. [1970]. The minimum expectation of χ^2 goodness-of-fit tests and the accuracy of approximations for the null distribution. *J. Amer. Statist. Assoc.* 65, 864–886.

YATES, F. [1934]. Contingency tables involving small numbers and the χ^2 test. *J. Roy. Statist. Soc. Suppl.* 1, 217–235.

YEE, A. H. AND GAGE, N. L. [1968]. Techniques for estimating the source and direction of causal influence in panel data. *Psych. Bull.* 70, 115–126.

YERUSHALMY, J. [1951]. A mortality index for use in place of the age-adjusted death rate. *Amer. J. Public Health* 41, 907–922.

YULE, G. U. [1900]. On the association of attributes in statistics. *Phil. Trans. Ser. A* 194, 257–319.

YULE, G. U. [1912]. On the methods of measuring association between two attributes. *J. Roy. Statist. Soc.* 75, 579–642.

ZACKS, S. [1971]. *The Theory of Statistical Inference*, pp. 60–68. New York, John Wiley.

ZANGWILL, W. I. [1969]. *Nonlinear Programming: A Unified Approach*. Englewood Cliffs, N.J., Prentice-Hall.

ZELEN, M. [1971]. The analysis of several 2×2 contingency tables. *Biometrika* 58, 128–138.

ZELLNER, A. AND LEE, T. H. [1965]. Joint estimation of relationships involving discrete random variables. *Econometrica* 33, 382–394.

ZELLNER, A. AND VANDAELE, W. [1974]. Bayes-Stein estimators for k-means, regression and simultaneous equation models. In *Studies in Bayesain Econometrics and Statistics*, edited by S. E. Fienberg and A. Zellner, pp. 627–653. Amsterdam, North Holland.

Index to Data Sets

Names of data sets are boldface, followed by a source reference and a short structural description. Then the names of each example, problem, or section where the data set is used are listed, with a short description of this use in parentheses, if necessary.

Author Index

Subject Index